REAL TIME DIGITAL SIGNAL PROCESSING APPLICATIONS WITH MOTOROLA'S DSP56000 FAMILY

MOHAMED EL-SHARKAWY, Ph.D.
BUCKNELL UNIVERSITY

With Appendices Provided by the Applications Engineering Staff of Motorola's DSP Operation

PRENTICE HALL, Englewood Cliffs, New Jersey 07632

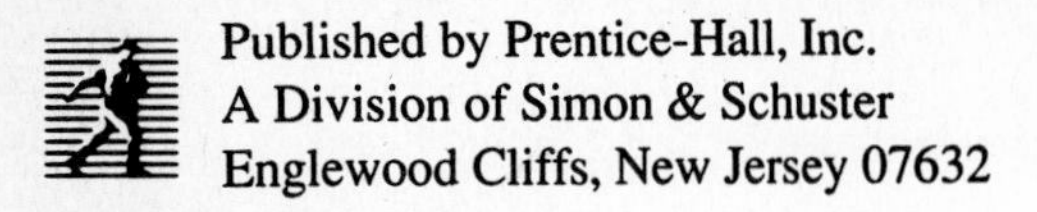
Published by Prentice-Hall, Inc.
A Division of Simon & Schuster
Englewood Cliffs, New Jersey 07632

The Publisher offers discounts on this book when ordered in bulk quantities. For more information write:

Special Sales/College Marketing
Prentice-Hall, Inc.
College/Technical and Reference Division
Englewood Cliffs, New Jersey 07632

Printed in the United States of America

10 9 8 7 6 5 4 3

ISBN 0-13-767138-5

Prentice-Hall International (UK) Limited, *London*
Prentice-Hall of Australia Pty. Limited, *Sydney*
Prentice-Hall Canada Inc., *Toronto*
Prentice-Hall Hispanoamericana, S.A., *Mexico*
Prentice-Hall India Private Limited, *New Delhi*
Prentice-Hall of Japan, Inc., *Tokyo*
Simon & Schuster Asia Pte. Ltd., *Singapore*
Editora Prentice-Hall do Brasil, Ltda., *Rio de Janeiro*

CONTENTS

PREFACE

OBJECTIVE

The objective of this book is to explain and demonstrate the operation of Motorola's DSP56000 Family of digital signal processors and peripherals and its DSP56000 Application Development Tools. The DSP56000 Family and its development tools are used to implement solutions to common real-time digital signal processing problems.

This book is intended for use by both undergraduate and graduate engineering students as well as industry professionals. In the academic environment, it can be used in a one-semester three-credit-hour digital signal processing projects course or in a laboratory course accompanying a first course in digital signal processing. In the academic or industrial environments, it provides a hands on means through which instructors, researchers, students and professionals can quickly acquaint themselves with the functions of the DSP56000 Family and its development tools.

ORGANIZATION

This book consists of four major parts. The first part includes Chapters 1 and 2 wherein the operation of the DSP56001 digital signal processor, the DSP56000ADS Application Development System (ADS), and the DSP56000CLAS Assembler/Linker/Simulator Software programs are explained and demonstrated. The second part includes Chapters 3 and 4 wherein the DSP56001 processor's architecture, addressing modes and instruction set are explained and demonstrated. The third part includes Chapters 5, 6, 7, and 8 wherein common real-time digital signal processing routines are explained, implemented, and demonstrated. The fourth part includes Chapters 9, 10, 11, 12, and 13 wherein applications specific topics are explained, implemented and demonstrated.

ACKNOWLEDGMENTS

This book is the result of the generous support of Motorola's Digital Signal Processor Operation and its DSP University Support Program. In particular, I am thankful to Bryant Wilder, Garth D. Hillman, and

Robert M. Bergeler of Motorola's DSP Operation for providing Bucknell University with multiple sets of DSP56000 Applications Development Tools for facilitization of a complete digital signal processing laboratory based on the DSP56000 Family. Additionally, I thank Robert M. Bergeler for his significant and careful efforts in editing and revising the draft of this book and thereby helping make a clearer presentation of my ideas.

I would like to thank Thomas Rich, Edward Mastascusa and Maurice Aburdene of Bucknell University for adjusting my schedule so I could finish this book on time. I gratefully acknowledge my students at Bucknell University for their commentary and assistance throughout my writing of this book. I especially wish to acknowledge my students Jinfeng Wang, John Yavis, Zhilie Li, and Bob Hentosh for completing some of the projects described in this book.

I also thank Jerry Purcell, President and Technical Director of Momentum Data Systems Inc., for permission to describe and use a demonstration version of Momentum Data Systems' *Filter Design and Analysis System* (FDAS) program in this book.

OPTIONAL SOFTWARE PROGRAMS

Two highly recommended software programs are optionally available for use with this book. The *DSP56000/1 Demonstration Software* program includes all of the DSP56000 Family programs presented in this book. It also includes easy to follow screen menus to present the topics and exercises of this book. A demonstration version of the *Filter Design and Analysis System* (FDAS) software program by Momentum Data Systems Inc. is used in this book to design finite and infinite impulse response filters.

PC PLATFORM REQUIREMENTS

The minimum hardware configuration to run the DSP56000 Application Development Tools and the *DSP56000/1 Demonstration Software* program is an IBM PC, XT or AT* with:

a) 384K bytes of RAM,
b) two 5.25" diskette drives,
c) one I/O expansion slot, and
d) PC-DOS* / MS-DOS** V2.1 or later.

The preferred hardware configuration to run the demonstration version of the *Filter Design And Analysis System* (FDAS) program is an IBM PC, XT or AT* with:

a) 640K bytes of RAM,
b) a 5.25" diskette drive,
c) a hard disk,
d) one I/O expansion slot,
e) PC-DOS* / MS-DOS** V2.1 or later, and
f) a math coprocessor chip.

* IBM PC,XT,AT and PC-DOS are trademarks of International Business Machines.

** MS-DOS is a trademark of Microsoft Corp.

Note that without 640K bytes of RAM and a hard disk, the demonstration version of the *Filter Design and Analysis System* (FDAS) Software program cannot be run; without a math coprocessor chip the FDAS software can be run, but at a significantly slower execution speed.

PROCURING THE OPTIONAL SOFTWARE PROGRAMS

The *DSP56000/1 Demonstration Software* program and the demonstration version of the *Filter Design And Analysis System* software program are sold together at a package price of $40.00. The programs are provided on six 5.25" diskettes. University instructors and students as well as industry professionals can procure the programs under the part number DSP56000/1 Textbook Software by completing the enclosed order form in the back of the book and mailing it along with a check, money order, or purchase order to the address on the form.

Each university instructor need only purchase a single DSP56000/1 Textbook Software package. Each university instructor is licensed to have a copy of the DSP56000/1 Textbook Software installed on each platform in his or her DSP laboratory. However, each instructor is asked to remove all copies whenever the associated DSP course(s) or research project(s) become inactive. For each DSP56000/1 Textbook Software package purchased, university students and industry professionals are licensed to have only one copy installed on a single host platform at any given time.

1

INTRODUCTION TO THE DSP56000 AND DSP56001 DIGITAL SIGNAL PROCESSORS AND THE DSP56000ADS APPLICATION DEVELOPMENT SYSTEM

The DSP56000 and DSP56001 are general purpose, single chip Digital Signal Processors (DSPs). The DSP56000ADS Application Development System (ADS) is a hardware development tool which uses the DSP56001 processor for designing real time DSP systems.

In this chapter, a general description of the DSP56000 and DSP56001 processors is presented followed by a general description of the ADS hardware. Step by step instructions covering the setup of the ADS, the installation of the ADS56000 User Interface Software program, and the installation of the DSP56000/1 Demonstration Software program are provided (the DSP56000/1 Demonstration Software program alternatively presents selected topics of this book via easy to follow screen menus). ADS User Interface Commands are then described and used to assemble, execute, and test a DSP56000/1 object code program.

1.1 DSP56000 AND DSP56001 PROCESSORS GENERAL DESCRIPTION

The DSP56000 and DSP56001 processors are high speed, general purpose DSPs which are implemented with high density, low power, 5-volt HCMOS technology. Both the DSP56000 and DSP56001 processors feature a dual Harvard architecture which includes an on-chip program memory and two independent on-chip data memories, all three of which are externally expandable to 64 Kwords (192 Kwords total) and accessible with zero wait states. The DSP56000 processor has 3.75 Kwords of securable, on-chip program ROM while the DSP56001 processor has 512 words of on-chip program RAM supported by an on-chip Bootstrap Loader. Both the DSP56000 and DSP56001 processors offer abundant features which combine to provide the processing strength necessary for efficient realization of high performance DSP or microcontrol systems. These features include:

1.1.1 High Performance Architecture

The DSP56000 and DSP56001 processors each have three multimode on-chip peripherals which enhance performance, and contribute to a compact system layout and reduced overall cost. The DSP56000 and DSP56001 processors use a non-obtrusive three-stage instruction fetch/decode/execute pipeline combined with three independent execution units connected by seven independent buses to three independent on-chip memory blocks to provide the parallelism needed for high performance digital signal processing. This parallelism allows, for example, a four coefficient Infinite Impulse Response (IIR) filter section to be executed in only four instruction cycles, the theoretical minimum for a single multiplier architecture.

The major architectural components of both the DSP56000 and DSP56001 processors are shown in Figure 1.1-A and Figure 1.1-B respectively, and include:

a. *Three independent Execution Units:*

   1) the Data ALU,
   2) the Address Generation Unit, and
   3) the Program Controller.

b. *Six on-chip Memories:*

   DSP56000: one 3.75 Kword Program ROM
   one 256 word X-Data RAM
   one 256 word Y-Data RAM
   two 256 word user defined ROMs
   one 32 word ROM (not used)

   DSP56001: one 512 word Program RAM
   one 256 word X-Data RAM
   one 256 word Y-Data RAM
   two 256 word programmed ROMs
   one 32 word Bootstrap ROM

c. *Four independent 24-bit Data Buses*:

   1) the XD Data Bus,
   2) the YD Data Bus,
   3) the PD Program Data Bus, and
   4) the GD Global Data Bus.

d. *Three independent 16-bit Address Buses*:

   1) the XA Address Bus,
   2) the YA Address Bus, and
   3) the PA Program Address Bus.

e. *One memory expansion Port A with*:

   - 24 package pins assigned for 24-bit data words;
   - 16 package pins assigned for 16-bit, memory block linear addressing to 64 Kwords;
   - 3 package pins assigned for segmented selection of a 64 Kword Program memory block, a 64 Kword X-Data memory block, and or a 64 Kword Y-Data memory block; and
   - 4 package pins assigned for handshaking functions.

f. *Three multimode on-chip Peripherals*:

   1) one Serial Communication Interface or general purpose I/O Port C,
   2) one Synchronous Serial Interface or general purpose I/O Port C, and
   3) one parallel Host Interface or general purpose I/O Port B.

g. One Clock Generator.

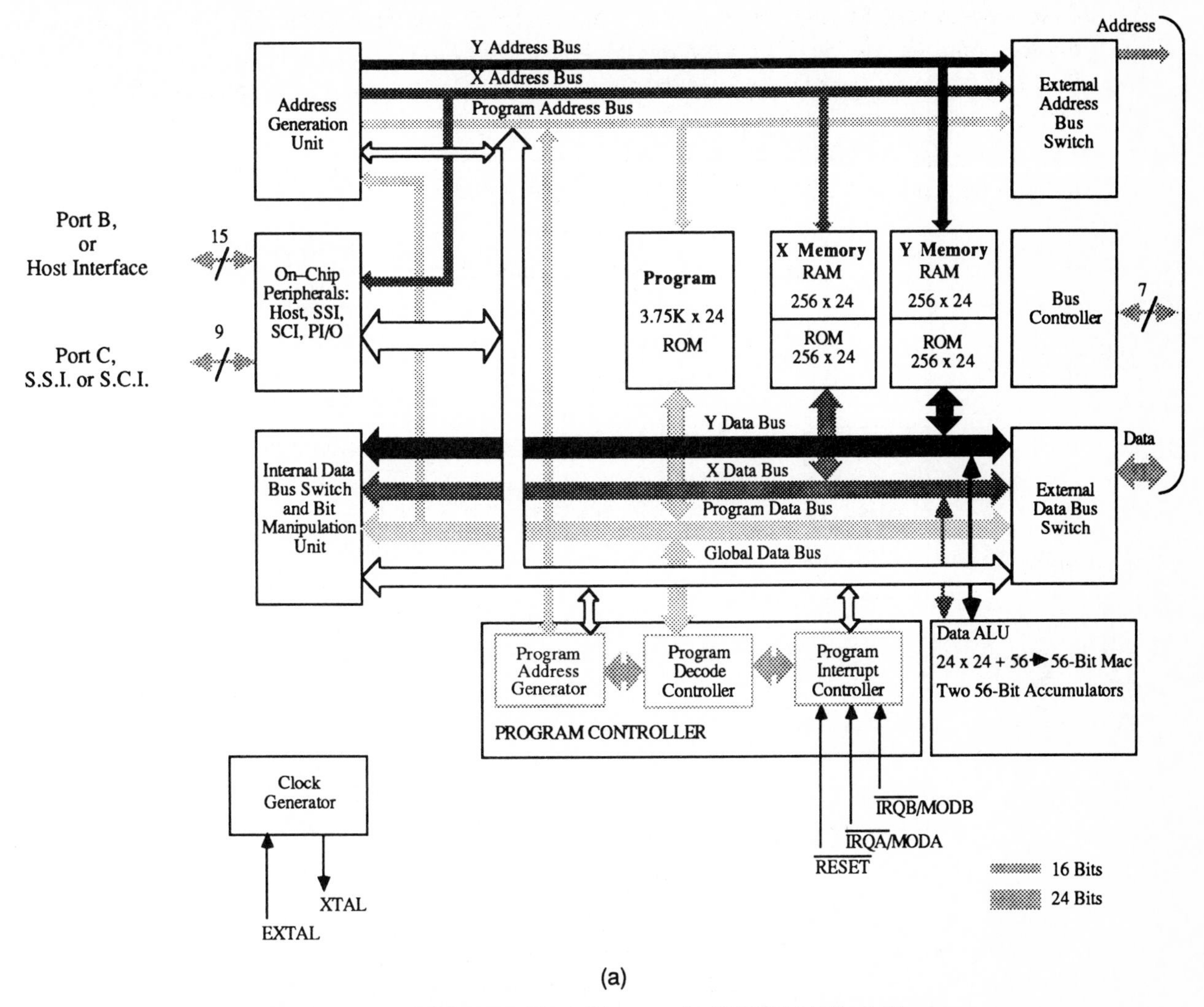

(a)

Figure 1-1 (a) DSP56000 Block Diagram (b) DSP56001 Block Diagram

1.1.2 High Performance Speed

At a clock rate of 20.48 MHz, the DSP56000 and DSP56001 processors can each execute 10.24 million instructions per second (MIPS) including up to 30.72 million concurrent arithmetic and dual data move operations per second (MOPS). At a clock rate of 27 MHz, this performance becomes 13.5 MIPS and 40.5 MOPS respectively. For example, at a 20.48 MHz clock rate the DSP56000/1 can execute a 1024 point complex Fast Fourier Transform (FFT) with bit reversed data addressing in 3.39 milliseconds using 24-bit fixed point arithmetic; at 27 MHz this same FFT can be executed in 2.57 milliseconds.

1.1.3 High Performance Precision

Single precision 24-bit data movement in the DSP56000 and DSP560001 processors takes place over 24-bit buses; double precision 48-bit data movement takes place over concatenated 24-bit buses. The 24-bit data buses provide 144 dB of data dynamic range; the 48-bit concatenated data buses provide 288 dB of data dynamic range. This data dynamic range is 50% greater than that provided by 16-bit DSPs. Data dynamic range is calculated as:

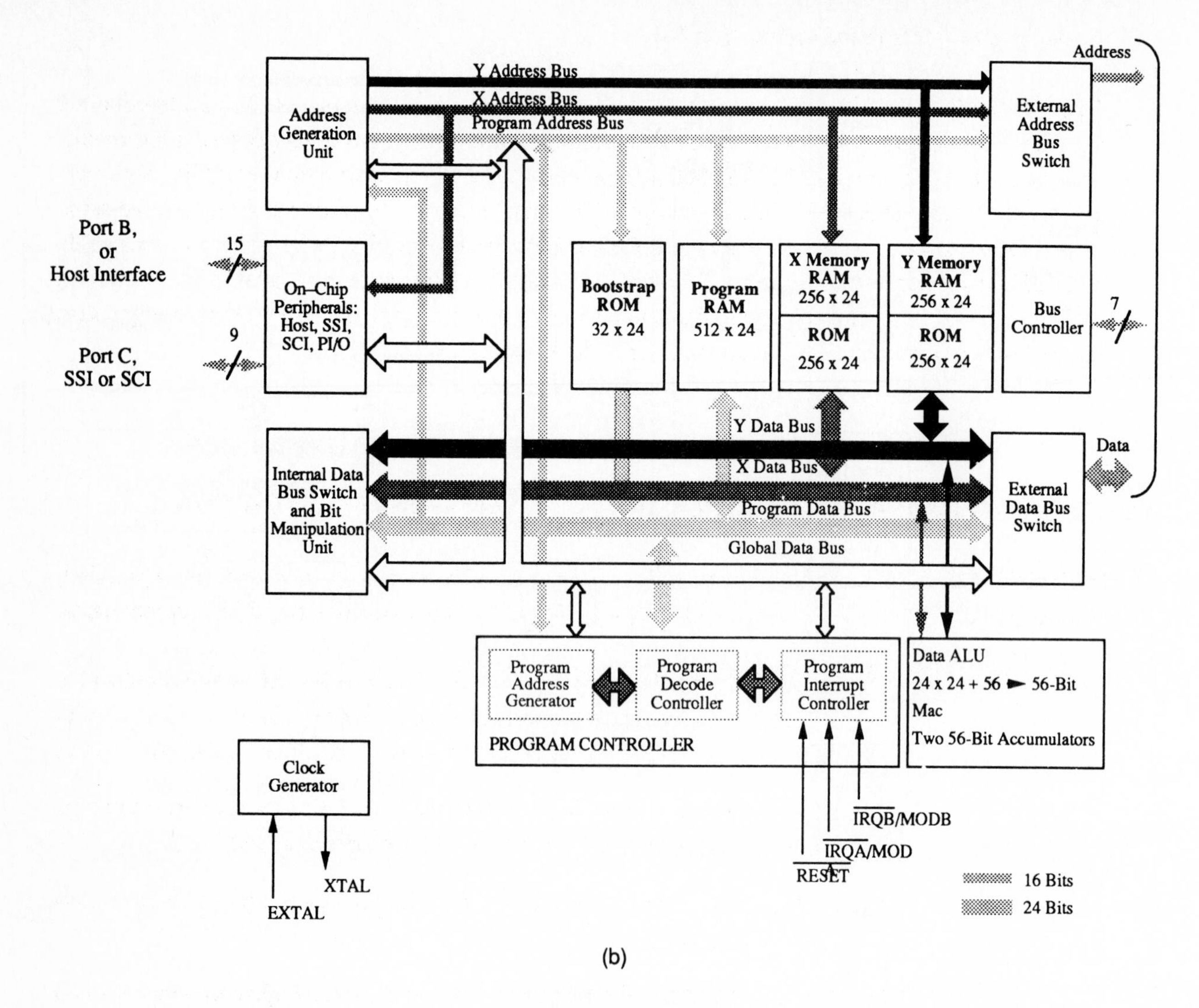

(b)

Figure 1-1 *(cont.)*

$$\text{Data dynamic range in dB} = 20\log_{10}(2^{\text{number of bits}})$$

$$144\text{dB} = 20\log_{10}(2^{24})$$

$$288\text{dB} = 20\log_{10}(2^{48})$$

$$96\text{dB} = 20\log_{10}(2^{16})$$

The Data Arithmetic and Logic Unit (Data ALU) of both the DSP56000 and DSP56001 processors uses a 24-bit by 24-bit hardware multiplier (which generates a 48-bit product) in conjunction with one of two 56-bit accumulators. This arrangement allows, for example, up to 256 Multiply-Accumulate instructions to be executed without any loss of precision due to truncation or rounding. This capability is directly applicable, for example, to high precision and efficient implementation of Finite Impulse Response (FIR) filters. The two 56-bit accumulators each provide 336 dB of data dynamic range.

1.1.4 High Performance Instruction Set

The DSP56000 and DSP56001 processors execute a common instruction set. The instruction set includes 62 microprocessor-like mnemonics which make the transition from programming microprocessors to programming the DSP56000/1 processors easy. The instruction set offers high performance features which take full advantage of the parallelism of the architecture and, at the same time, yields compact code. For example a Repeat Next Instruction (REP) instruction is repeatable n-times. A single REP (n) instruction followed by a single Multiply-Accumulate (MACR) instruction (with two concurrent data move operations) allows an n-tap FIR filter algorithm to be compactly coded with only two instructions and executed in only 2(n+1) clock cycles.

1.2 "DSP56000ADS APPLICATION DEVELOPMENT SYSTEM" GENERAL DESCRIPTION

The DSP56000ADS Application Development System (ADS) is a hardware tool for designing, debugging, and evaluating DSP56000-based systems. As shown in Figure 1.2, the ADS consists of three components:

1. an Application Development Module (ADM) board which contains a DSP56001 processor, off-chip expansion memory, interface and control circuitry, and several connectors for hook-up to application specific boards;

2. a Host Interface board which fits into the backplane of a user provided host platform, and connects to the ADM via a ribbon cable; and

3. a ADS56000 User Interface Software program which runs on the host platform, and controls the ADM.

   The IBM PC™ is used in this book as the host platform between the user and the ADM. Therefore, a Host Interface board compatible with the IBM PC's backplane and ADS56000 User Interface Software program compatible with MS-DOS is required and used. Versions of the Host Interface board and ADS56000 User Interface Software program which are compatible with the Macintosh™ II PC or SUN-3™ Workstation are also available from Motorola.

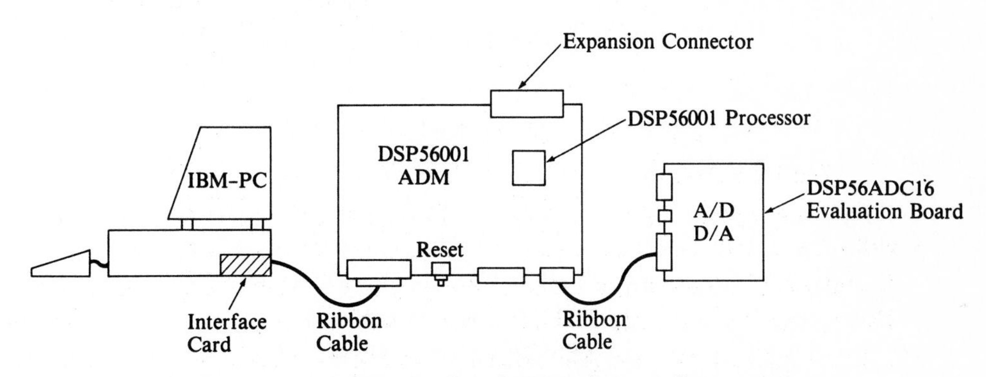

Figure 1-2 Application Development System Components

1.3 ADS SET-UP

1. TURN OFF THE POWER TO YOUR PC AND UNPLUG IT FROM ITS POWER SOURCE.
2. TURN OFF THE POWER TO YOUR PC'S PERIPHERALS (MONITOR, PRINTER, ETC.) AND UNPLUG THEM FROM THEIR POWER SOURCES.
3. CHECK JUMPERS JG1 AND JG2 ON THE "HOST INTERFACE" BOARD.

   The PC's I/O Address Map provides for the addressing of peripherals which are undefined with respect to standard PC configurations. The Host Interface board is factory assigned to decode PC I/O Address Map locations 100 to 200 hexadecimal. If those address locations are already used by some other peripheral board, you can change the address decoder of the Host Interface board to start decoding at either address location 200 or 300 hexadecimal. To reassign the address decoder to start decoding at address location 200 hexadecimal, simply remove the clip shorting pins 1 and 2 of JG1 by pulling it up. Then place it over pins 3 and 4 of JG1 and push it down until firmly seated. To reassign the address decoder to start decoding at address location 300 hexadecimal, remove the clip shorting pins 1 and 2 of JG1 and save it for future use. For any of the above address decoder assignments, JG2 should not have a shorting clip installed; if it does, remove it and save it for future use.
4. INSTALL THE "HOST INTERFACE" BOARD INTO THE PC'S BACKPLANE.

   Remove the cover from your PC and press the Host Interface board's edge connector (the gold striped end) into an expansion slot in the PC's backplane. Assure that the board is well seated. Secure the board to the PC's chassis by using a bracket retaining screw. If you need help, refer to your PC's Operation Manual for instructions on how to remove the cover, choose an expansion slot, or install a board.
5. CONNECT THE RIBBON CABLE BETWEEN THE "HOST INTERFACE" BOARD AND THE ADM.
6. PLUG YOUR PC AND ITS PERIPHERALS INTO THEIR RESPECTIVE POWER SOURCES.
7. TURN ON THE POWER TO YOUR PC AND ITS PERIPHERALS.

1.4 INSTALLING THE "ADS56000 USER INTERFACE SOFTWARE" PROGRAM

The ADS56000 User Interface Software program allows you to select and control the ADS's resources from your keyboard, view menu options on the screen of your monitor, and display ADS generated results on the screen of your monitor.

1.4.1 PC with Two Diskette Drives and No Hard Disk Drive

1. PLACE THE "DOS" DISKETTE INTO DRIVE "A" AND BOOT THE PC.
2. PLACE THE "DSP56000ADS USER INTERFACE PROGRAM" DISKETTE CONTAINING THE "ADS56000 USER INTERFACE SOFTWARE" PROGRAM INTO DRIVE "B".

3. TYPE **b:ads56000<return>** TO ENTER THE "ADS56000 USER INTERFACE SOFTWARE" PROGRAM.

   The following menu should be on the bottom of your screen:

```
ADM0>
asm break change copy device disassemble <space>=more
```

4. TYPE **pc<return>** TO DISPLAY THE PC'S I/O MAP BEGINNING ADDRESS LOCATION CURRENTLY ASSIGNED TO THE "HOST INTERFACE" BOARD.

   The display on your screen should be:

```
PC
PC Interface Card uses PC I/O address 100
```

5. PROCEED TO STEP 6 OR 7 AS APPLICABLE.
6. IF IN STEP 3 OF SECTION 1.3 YOU CHANGED THE ADDRESS DECODER ON THE "HOST INTERFACE" BOARD TO START ADDRESS DECODING AT ADDRESS LOCATION $200 OR $300, THEN TYPE **pc io $200<return>** OR **pc io $300<return>** AS APPLICABLE TO NOTIFY THE "ADS56000 USER INTERFACE SOFTWARE" PROGRAM OF THIS CHANGE. PROCEED TO STEP 8.
7. IF IN STEP 3 OF SECTION 1.3 YOU DID NOT CHANGE THE ADDRESS DECODER TO START ADDRESS DECODING AT ADDRESS LOCATION $200 OR $300, PROCEED TO STEP 8.
8. TYPE **quit<return>** TO EXIT THE "ADS56000 USER INTERFACE SOFTWARE" PROGRAM.

1.4.2 PC with a Hard Disk Drive and one or more Diskette Drives

For simplicity, it is assumed that Drive "C" is assigned to the hard disk and that DOS resides in directory c:\dos.

1. BOOT THE PC.
2. PLACE THE "DSP56000ADS USER INTERFACE PROGRAM" DISKETTE CONTAINING THE "ADS56000 USER INTERFACE SOFTWARE" PROGRAM INTO DRIVE "A".
3. TYPE **md c:\dsptools<return>** TO CREATE A **dsptools** DIRECTORY ON DRIVE "C".

   Optionally, you can give the **dsptools** directory any name you wish and locate it anywhere you wish on Drive "C".
4. TYPE **md c:\dsptools\ads56000<return>** TO CREATE A **dsptools\ads56000** DIRECTORY ON DRIVE "C".

5. TYPE **copy a: c:\dsptools\ads56000<return>** TO COPY THE CONTENTS OF THE "DSP56000ADS USER INTERFACE PROGRAM" DISKETTE TO THE SPECIFIED DIRECTORY ON DRIVE "C".
6. TYPE **cd c:\dsptools\ads56000<return>** TO ENTER THIS DIRECTORY.
7. TYPE **ads56000<return>** TO ENTER THE "ADS56000 USER INTERFACE SOFTWARE" PROGRAM.

   The following menu should appear at the bottom of your screen:

```
ADM0>
asm break change copy device disassemble <space>=more
```

8. PROCEED TO STEP 9 OR 10 AS APPLICABLE.
9. IF IN STEP 3 OF SECTION 1.3 YOU CHANGED THE ADDRESS DECODER ON THE "HOST INTERFACE" BOARD TO START ADDRESS DECODING AT ADDRESS LOCATION $200 OR $300, THEN TYPE **pc io $200<return>** OR **pc io $300<return>** AS APPLICABLE TO NOTIFY THE "ADS56000 USER INTERFACE SOFTWARE" PROGRAM OF THIS CHANGE. PROCEED TO STEP 11.
10. IF IN STEP 3 OF SECTION 1.3 YOU DID NOT CHANGE THE ADDRESS DECODER TO START ADDRESS DECODING AT ADDRESS LOCATION $200 OR $300, PROCEED TO STEP 11.
11. TYPE **quit<return>** TO EXIT THE "ADS56000 USER INTERFACE SOFTWARE" PROGRAM.

1.5 INSTALLING THE "DSP56000/1 DEMONSTRATION SOFTWARE" PROGRAM

The DSP56000/1 Demonstration Software program is an easy to follow, screen menu driven means through which selected topics of this book are presented. For example, the user can either directly type the commands presented in Sections 1.7 and 1.8 of this chapter or, alternatively, run the DSP56000/1 Demonstration Software program file named Chapter1, make selections from the various screen menus, and follow the screen prompts to step through the commands presented in Sections 1.7 and 1.8.

1.5.1 PC with two Diskette Drives and no Hard Disk

1. PLACE THE "DOS" DISKETTE INTO DRIVE "A" AND BOOT THE PC.
2. PROCEED DIRECTLY TO STEPS 4 AND 7 IF STEPS 3, 4, 5 AND 6 HAVE BEEN PREVIOUSLY ACCOMPLISHED; IF NOT, PROCEED TO STEP 3.
3. REMOVE THE "DOS" DISKETTE FROM DRIVE "A" AND PLACE IT INTO DRIVE "B".
4. PLACE THE "DSP56000/1 DEMONSTRATION SOFTWARE #N" DISKETTE INTO DRIVE "A".
5. TYPE **copy b:command.com a:<return>** TO COPY THE SPECIFIED "DOS" UTILITIES FILE ONTO THE "DSP56000/1 DEMONSTRATION SOFTWARE #N" DISKETTE.
6. REPEAT STEPS 4 AND 5 FOR ALL OF THE "DSP56000/1 DEMONSTRATION SOFTWARE" DISKETTES.

7. TYPE EITHER **a:chapter1<return>** OR **a:chapter2<return>**, etc., TO ENTER THE DESIRED "CHAPTERx" FILE.
8. TO EXIT THE "DSP56000/1 DEMONSTRATION SOFTWARE" PROGRAM, SELECT THE "EXIT" COMMANDS FROM THE MENUS.

1.5.2 PC with a Hard Disk Drive and one or more Diskette Drives

1. BOOT THE PC.
2. PLACE THE "DSP56000/1 DEMONSTRATION SOFTWARE #N" DISKETTE INTO DRIVE "A".
3. TYPE EITHER **a:chapter1<return>** OR **a:chapter2<return>**, etc.,TO ENTER THE DESIRED "CHAPTERx" FILE.
4. TO EXIT THE "DSP56000/1 DEMONSTRATION SOFTWARE" PROGRAM, SELECT THE "EXIT" COMMANDS FROM THE MENUS.

1.6 SUMMARY OF "ADS56000 USER INTERFACE COMMANDS"

The ADS56000 User Interface Commands are permuted by many optional parameters specific to each command to yield a large and powerful command set. This book does not attempt to use all of the numerous combinations of commands and specific optional parameters. Rather, only those commands and command parameters necessary to demonstrate the concepts of this book are covered. For detailed information on the ADS56000 commands and their optional parameters, refer to Motorola's DSP56000ADS Application Development System User's Manual. The ADS56000 User Interface Commands (without optional parameters) are summarized as follows:

Command	*Description*
ASM	Single line assembler
BREAK	Set breakpoint
CHANGE	Modify registers/memory values
COPY	Copy memory block to another place
DEVICE	Select default ADM
DISPLAY	Display registers/memory values
DISASSEMBLE	Disassemble memory
EVALUATE	Calculator
FORCE	Force hardware reset or break
GO	Execute ADM program real-time
HELP	On-line help text
INPUT	Open file for ADM program input
LOAD	Download object file into ADM memory
LOG	Save commands/entire session to file
OUTPUT	Open file for ADM program output
PATH	Set directory path for ADM(s)
PC	Set PC interface address/interrupt
QUIT	Exit user interface program
RADIX	Set default radix for command entry
SAVE	Save ADM state/memory to file
STEP	Multiple instruction step of ADM
SYSTEM	Execute operating system command(s)
TRACE	Single instruction step of ADM
WAIT	Wait before continuing

1.7 BASICS OF SOME "ADS56000 USER INTERFACE COMMANDS"

i. If you want to use the **DSP56000 Demonstration Software** program to cover the topics of Section 1.7, load the software by following Steps 1, 4 and 7 of Section 1.5.1 or Steps 1-3 of Section 1.5.2. Then select from the demonstration software's menus to cover the topics of Section 1.7.

ii. If you do *not* want to use the DSP56000 Demonstration Software program, load the **ADS56000 User Interface Software** program by following Steps 1-3 of Section 1.4.1 or Steps 1 , 6 and 7 of Section 1.4.2.

1.7.1 "Help" Command:

The **help** command allows the user to review all of the ADS56000 User Interface Commands or get information on a particular command.

1. TYPE **help<return>** TO DISPLAY A LISTING OF ALL OF THE "ADS56000 USER INTERFACE COMMANDS".

   The display on your screen should be:

```
-- DSP56001 APPLICATION DEVELOPMENT SYSTEM COMMANDS --

ASM [Dx] [(beginning at) addr] [I(interactive)]
         [assembler_mnemonic]
BREAK [D0..7] [#bn] [addr|OFF] T[expr]    [H(halt)
                     /In(inc. CNTn)/N(note)/S(show)]
CHANGE [D0..7] [reg[_block]/addr[_block]     (to)
               [expression]
COPY (from) [Dx] addr[_block] (to)[D0..7] addr
DISPLAY [D0..7] [ON/OFF/ALL/IO][reg[_block/_group]
                 /addr[_block]]...
DISPLAY W(active ADMs)/V(version)
DEVICE D0..7 [ON/OFF]
DISASSEMBLE [D0..7] [addr[_block]]
EVALUATE [Dx] [B(binary)/D(decimal) /F(fractional)
              /H(hexadecimal)] expression
FORCE [D0..7] R(hardware reset)|B(Return to monitor)
GO [D0..7] [(from)addr/R(reset)] [(to break #)#bn]
           [(occurrence):count]
HELP [Dx] [command|register]
INPUT [D0..7] [#number] OFF/TERM/filename
              [-rd|-rf|-rh]
LOAD [D0..7|S(state)] (from) filename
LOG [D0..7] [OFF] [C(commands)/S(session) [filename]]
OUTPUT [D0..7] [#number] OFF/TERM/filename
               [-rd|-rf|-rh]
PATH [D0..7] [pathname]
PC (address)[IO [$100|$200|$300]]
QUIT
RADIX [D0..7] [B(bin)/D(dec)/H(hex)/F(frac)]
              [reg[_block]/addr[_block]
SAVE [D0..7] S(state)/addr_block... filename
STEP [D0..7] [count]
SYSTEM [system_command [parameter_list]]
TRACE [D0..7] [count]
WAIT [count(seconds)]
 Macro filename
;comment string entry
Note: Trace/Step/Go commands cause ADM(s) Service
      Requests
```

2. USE THE "UP ARROW" AND "DOWN ARROW" KEYS TO SURVEY THE "ADS56000 USER INTERFACE COMMANDS".
3. TYPE **help display<return>** TO DISPLAY A LISTING OF ALL OF THE **display** COMMAND COMBINATIONS (EXAMPLES ARE INCLUDED IN THE LISTING).

   The display on your screen should be:

```
--------DISPLAY: Display Register or Memory ---------

DISPLAY [D0..7] [ON/OFF/ALL/IO] [reg[_block/_group]
                /addr[_block]]...
DISPLAY W(active ADMs)/V(version)

display
Display all currently enabled registers and memory.

display w
Display the active ADMs.

display v
Display ADS56000 rev. number, date of release, and
default ADM monitor rev.

display d2 v
Display ADS56000 rev. number, data of release, and
ADM#2 monitor rev.

 -------- Register and Memory Display Modes -------

 ON  Always display the following registers and
     memory locations.
 OFF Never display the following registers and
     memory locations.

 ----------- Examples -----------

 display on
 Enable all registers and stacked levels for display

 display on p:0..20 x:30..40 y:$100
 Display enable p memory address block 0 to 20, x
 memory address block 30 to 40 and y memory address
 hexadecimal 100.

 display off
 Disable all display of registers, memory and stacked
 levels.

 display on dsp host sci
 Enable display of the DSP56000 programming model
 registers, the HOST peripheral registers, and the
 SCI peripheral registers.

 ----------- Register Group Names -----------

   ALL    all registers.
   DSP    all of the DSP56000 programming model
          registers.
   IO     all peripheral register names.
   HOST   all registers of the Host peripheral
          interface.
   PORTB  all registers of the Port B interface.
   PORTC  all registers of the Port C interface.
```

```
  SCI   all registers of the SCI peripheral
        interface.
  SSI   all registers of the SSI peripheral
        interface.
  STACK all current stacked levels.

display p:0..300
Immediate display of p memory addresses 0 through
300.

display on host
Immediate display of all host port registers plus
currently enabled registers and memory locations.

display stack
Immediate display of all stacked levels up to the
current stack pointer value.  Example: sp=5 shows
stacked values from 1 to 5.
```

4. USE THE "UP ARROW" AND "DOWN ARROW" KEYS TO SURVEY THE **display** COMMANDS AND THE EXAMPLES.
5. TYPE **help asm<return>** TO DISPLAY A LISTING OF ALL OF THE **asm** COMMAND COMBINATIONS (EXAMPLES ARE INCLUDED IN THE LISTING).

   The display on your screen should be:

```
---------- ASM: Single Line Assembler ----------

ASM [Dx] [(beginning at) addr] [I(interactive)]
         [assembler_mnemonic]

asm p:$50
Start interactive assembler at program memory address
50 (hex) of default ADM.

asm x:0 move r0,a1
Assemble single instruction at x memory address 0 of
default ADM.

asm
Start assembler at current program counter value of
default ADM.

asm d3 p:100
Start assembler at program memory address 100
(decimal) of ADM 3.

asm i
Start assembler at current program counter value of
default ADM.
After instruction is assembled and placed in memory,
trace the instruction.

asm p:$50 i move #1,x0
Assemble a move #1,x0 instruction, place the opcode at
program address 50 hex, and immediately execute the
instruction.
```

6. SURVEY THE **asm** COMMANDS AND THE EXAMPLES.
7. USE THE **help** COMMAND TO SURVEY THE **go, break,** AND **trace** COMMANDS, e.g., **help go<return>** , **help** break<return>, **help trace<return>**.

1.7.2 "Radix" Command:

A hexadecimal constant can be specified by preceding the constant with a dollar ($) symbol. Likewise, a decimal constant can be specified with a preceding grave accent (') symbol; a binary constant can be specified with a percent (%) symbol. When the ADS56000 User Interface Software program is entered, the default radix is decimal. That is, decimal constants can be directly entered without the need for a preceding radix specification symbol. The **radix** command allows you to change the default radix to that of your convenience and then enter constants without the need for a preceding radix specification symbol. For example, if you need to enter a large number of hexadecimal constants, the **radix** command can be used to change the default radix to hexadecimal. However, after doing so you need to precede decimal constants with a grave accent (') radix specification symbol and precede binary constants with a percent (%) radix specification symbol.

1. TYPE **radix<return>** TO DISPLAY THE DEFAULT RADIX.
   The display on your screen should say:

```
The current default radix is decimal.
```

2. TYPE **radix h<return>** TO CHANGE THE DEFAULT RADIX TO HEXADECIMAL.
3. TYPE **radix<return>** TO DISPLAY THE NEW DEFAULT RADIX.
   The display on your screen should say:

```
The current default radix is hexadecimal.
```

4. TYPE **radix b<return>** TO CHANGE THE DEFAULT RADIX TO BINARY.
5. TYPE **radix<return>** TO DISPLAY THE NEW DEFAULT RADIX.
   The display on your screen should say:

```
The current default radix is binary.
```

6. TYPE **radix d<return>** TO CHANGE THE DEFAULT RADIX BACK TO DECIMAL.
7. TYPE **radix<return>** TO DISPLAY THE NEW DEFAULT RADIX.
   The display on your screen should say:

```
The current default radix is decimal.
```

1.7.3 "Display" Command:

The **display** command can be used to define (and redefine) which DSP56000/1 and ADS registers and or memory locations are displayed. You can define any combination of registers, register groups, and or memory locations for display. You are limited only by the size of the screen's display area. Motorola defined register groups can be selected for display by using the following DSP56000/1 register group names with the **display** command:

all	all registers
dsp	all programmer's model registers (default)
io	all registers of the on-chip peripherals
host	all registers of the on-chip Host peripheral
portb	all registers of general purpose Port B
portc	all registers of general purpose Port C
sci	all registers of the on-chip SCI peripheral
ssi	all registers of the on-chip SSI peripheral

For example only, execution of the **display on host ssi** command will enable the DSP56000/1's Host and SSI on-chip peripheral registers for display (DO NOT EXECUTE THIS COMMAND AT THIS TIME).

1. TYPE **display<return>** TO DISPLAY THE VALUES ASSOCIATED WITH ALL OF THE DSP56000/1 AND ADS REGISTERS AND/OR MEMORY LOCATIONS WHICH ARE ENABLED FOR DISPLAY.
   The notations used for the registers currently displayed on your screen (default display of the DSP56000/1 programmer's model registers) are:
   i. *DSP56000/1 Data ALU Registers:*
      x,x1,x0,y,y1,y0 for input registers
      a,a2,a1,a0,b,b2,b1,b0 for accumulator registers
   ii. *DSP56000/1 Address Generation Unit Registers:*
      r0-r7 for address registers
      n0-n7 for offset registers
      m0-m7 for modifier registers
   iii. *DSP56000/1 Program Controller Registers:*
      pc for Program Counter register
      sr for Status Register
      omr for Operating Mode Register
      la for Loop Address register
      lc for Loop Counter register
      sp for Stack Pointer register
      ssh for System Stack High register
      ssl for System Stack Low register
   iv. ADS56000 User Interface Program:

cntl-cnt4 for software Counter registers

The values in the display enabled DSP56000/1 and ADS registers and/or memory locations might not be the same each time you execute the **display** command or the DSP56000/1 Demonstration Software. If the values of any display enabled registers and/or memory locations have changed since the last time the **display** command was executed, the most recent values are displayed in bold type. The display on your screen should be similar to:

```
x=$000000000000      y=$000000000000
a=$00000000000000  b=$00000000000000
x1=$000000    x0=$000000            r7=$FFFF    n7=$FFFF
                                                m7=$FFFF
y1=$000000    y0=$000000            r6=$FFFF    n6=$FFFF
                                                m6=$FFFF
a2=$00 a1=$000000  a0=$000000       r5=$FFFF    n5=$FFFF
                                                m5=$FFFF
b2=$00 b1=$000000  b0=$000000       r4=$FFFF    n4=$FFFF
                                                m4=$FFFF
                                    r3=$FFFF    n3=$FFFF
                                                m3=$FFFF
pc=$0040   sr= $0300   omr=$02      r2=$FFFF    n2=$FFFF
                                                m2=$FFFF
la=$0000   lc= $FDFD                r1=$FFFF    n1=$FFFF
                                                m1=$FFFF
ssh=$FFFF ssl=$FFFF   sp=$0000      r0=$FFFF    n0=$FFFF
                                                m0=$FFFF
cnt1=000000 cnt2=000000  cnt3=000000   cnt4=000000
ADM#0 P:$0040 FDFFFF       = MACR -Y1,X1,B X:(R7)+,B Y:(R3)+,Y1
```

Remember, the **help** command can be used to determine the meaning of the register abbreviations used in the display associated with the **display** command.

TYPE **help lc<return>** TO DETERMINE THE MEANING OF THE ABBREVIATION lc.

The display on your screen should say:

```
Program Controller Loop Count Register
```

TYPE **help la<return>** TO DETERMINE THE MEANING OF THE ABBREVIATION la.

The display on your screen should say:

```
Program Controller Loop Address Register
```

2. TYPE **display p:0..8<return>** TO DISPLAY THE VALUES OF P-MEMORY LOCATIONS 0-8.

   The display on your screen should be similar to:

```
P:$0000 $FFFFFF $FDFFDF $FFFFFF $FAE3F7
P:$0004 $0BF080 $00E1D6 $0BF080 $00E1D6
P:$0008 $777EFF
ADM#0 P:$0040 FDFFFF     = MACR -Y1,X1,B X:(R7)+,B Y:(R3)+,Y1
```

3. TYPE **display p:0..8 x:30..40 y:$100<return>** TO DISPLAY THE VALUES OF P-MEMORY LOCATIONS 0-8, X-MEMORY LOCATIONS 30-40, AND Y-MEMORY LOCATION $100.

   The display on your screen should be similar to:

```
P:$0000 $FFFFFF $FDFFDF $FFFFFF $FAE3F7
P:$0004 $0BF080 $00E1D6 $0BF080 $00E1D6
P:$0008 $777EFF
X:$001E $000000 $000000
X:$0020 $000000 $001001 $000100 $240808
X:$0024 $000000 $000000 $400000 $000800
X:$0028 $000000
Y:$0100 $FA3EDE
ADM#0 P:$0040 FDFFFF     = MACR -Y1,X1,B X:(R7)+,B Y:(R3)+,Y1
```

4. TYPE **display on p:0..8 x:30..40 y:$100<return>** TO ADD THE SPECIFIED MEMORY LOCATIONS TO THE ENABLED DISPLAY.

5. TYPE **display<return>** TO SHOW THE VALUES OF ALL OF THE DISPLAY ENABLED DSP56000/1 AND ADS REGISTERS AND/OR MEMORY LOCATIONS.

   The display on your screen should be similar to:

```
x=$000000000000      y=$000000000000
a=$00000000000000    b=$00000000000000
x1=$000000   x0=$000000           r7=$FFFF    n7=$FFFF
                                              m7=$FFFF
y1=$000000   y0=$000000           r6=$FFFF    n6=$FFFF
                                              m6=$FFFF
a2=$00 a1=$000000   a0=$000000    r5=$FFFF    n5=$FFFF
                                              m5=$FFFF
b2=$00 b1=$000000   b0=$000000    r4=$FFFF    n4=$FFFF
                                              m4=$FFFF
                                  r3=$FFFF    n3=$FFFF
                                              m3=$FFFF
pc=$0040   sr= $0300   omr=$02    r2=$FFFF    n2=$FFFF
                                              m2=$FFFF
la=$0000   lc= $FDFD              r1=$FFFF    n1=$FFFF
                                              m1=$FFFF
ssh=$FFFF ssl=$FFFF   sp=$0000    r0=$FFFF    n0=$FFFF
                                              m0=$FFFF
```

```
cnt1=000000 cnt2=000000  cnt3=000000   cnt4=000000
P:$0000 $FFFFFF $FDFFDF $FFFFFF $FAE3F7
P:$0004 $0BF080 $00E1D6 $0BF080 $00E1D6
P:$0008 $777EFF
X:$001E $000000 $000000
X:$0020 $000000 $001001 $000100 $240808
X:$0024 $000000 $000000 $400000 $000800
X:$0028 $000000
Y:$0100 $FA3EDE
ADM#0 P:$0040 FDFFFF   = MACR -Y1,X1,B X:(R7)+,B Y:(R3)+,Y1
```

6. TYPE **display pc sr<return>** TO DISPLAY THE VALUES OF THE DSP56000/1'S PROGRAM COUNTER (PC) AND STATUS (SR) REGISTERS.

   The display on your screen should be similar to:

```
pc= $0040    sr=  $0300
ADM#0 P:$0040 FDFFFF   = MACR -Y1,X1,B X:(R7)+,B Y:(R3)+,Y1
```

7. TYPE **display off<return>** TO DISABLE THE ENTIRE DISPLAY.

   When you execute this command, you cancel any display structure which you might have previously established (enabled), e.g., the display structure established in Step 4 will be cancelled.

8. TYPE **display on<return>** TO ENABLE THE DEFAULT DISPLAY, e.g., THE DSP56000/1'S PROGRAMMER'S MODEL REGISTERS.

1.7.4 "Change" Command:

The **change** command can be used to examine or modify the values of registers and/or memory locations.

1. TYPE **change x:$55<return>** TO DISPLAY THE VALUE OF X-MEMORY LOCATION $55, AND GET A PROMPT FOR A NEW VALUE.
2. TYPE **$8<return>** TO CHANGE THE VALUE OF X-MEMORY LOCATION $55 TO $8.

   The display on your screen should say:

```
change X:$0055 $000008 ;Dec:8  Fract: 0.0000010
```

3. TYPE **$12<return>** TO CHANGE THE VALUE OF THE NEXT X-MEMORY LOCATION ($56) TO $12.

   The display on your screen should say:

```
change X:$0056 $000012 ;Dec:18 Fract: 0.0000021
```

4. TYPE **<escape>** TO EXIT THE **change** COMMAND.
5. TYPE **display x:$55..$56<return>** TO DISPLAY THE VALUES OF X-MEMORY LOCATIONS $55 AND $56.

   The display on your screen should be similar to:

```
X:$0055 $000008 $000012
ADM#0 P:$0040 FDFFFF    = MACR -Y1,X1,B X:(R7)+,B Y:(R3)+,Y1
```

6. TYPE **change pc $10<return>** TO CHANGE THE VALUE OF THE DSP56000/1'S PROGRAM COUNTER (PC) REGISTER TO $10.
7. TYPE **display pc<return>** TO DISPLAY THE NEW VALUE ($10) OF THE PC.

   The display on your screen should be similar to:

```
pc= $0010
ADM#0 P:$0010 EFFFFE         = DC $EFFFFE
```

8. TYPE **display p:$14<return>** TO DISPLAY THE VALUE OF P-MEMORY LOCATION $14.

   The display on your screen should be similar to:

```
P:$0014 $FFFFFF
ADM#0 P:$0010 EFFFFE         = DC $EFFFFE
```

9. TYPE **change p:$14 $123<return>** TO CHANGE THE VALUE OF P-MEMORY LOCATION $14 TO $123.
10. TYPE **display p:$14<return>** TO DISPLAY THE NEW VALUE ($123) OF P-MEMORY LOCATION $14.

    The display on your screen should be similar to:

```
P:$0014 $000123
ADM#0 P:$0010 EFFFFE         = DC $EFFFFE
```

11. TYPE **change r0..r7 0 x:$08 $134 pc 200<return>** TO CHANGE THE VALUES OF THE DSP56000/1'S ADDRESS REGISTERS R0-R7 TO 0, CHANGE THE VALUE OF X-MEMORY LOCATION $08 TO $134, AND CHANGE THE VALUE OF THE DSP56000/1'S PROGRAM COUNTER (PC) TO 200.
12. TYPE **display on x:$08<return>** TO ADD THE VALUE OF X-MEMORY LOCATION $08 TO THE ENABLED DISPLAY.
13. TYPE **display<return>** TO SHOW THE RESULTS OF EXECUTING THE ABOVE **change** COMMAND.

    The display on your screen should be similar to:

```
x=$000000000000        y=$000000000000
a=$00000000000000      b=$00000000000000
x1=$000000  x0=$000000              r7=$0000   n7=$FFFF
                                               m7=$FFFF
y1=$000000  y0=$000000              r6=$0000   n6=$FFFF
                                               m6=$FFFF
a2=$00 a1=$000000  a0=$000000       r5=$0000   n5=$FFFF
                                               m5=$FFFF
b2=$00 b1=$000000  b0=$000000       r4=$0000   n4=$FFFF
                                               m4=$FFFF
                                    r3=$0000   n3=$FFFF
                                               m3=$FFFF
pc=$00C8     sr=$0300    omr=$02    r2=$0000   n2=$FFFF
                                               m2=$FFFF
la=$0000     lc=$FDFD               r1=$0000   n1=$FFFF
                                               m1=$FFFF
ssh=$FFFF    ssl=$FFFF   sp=$0000   r0=$0000   n0=$FFFF
                                               m0=$FFFF
cnt1=000000  cnt2=000000  cnt3=000000   cnt4=000000
X:$0008 $000134
ADM#0 P:$00C8 FFFEFF          = DC $FFFEFF
```

1.7.5 "Copy" Command:

The **copy** command can be used to copy the values of one memory block to another.

1. TYPE **change p:0..10 $ff<return>** TO CHANGE THE VALUES OF P-MEMORY LOCATIONS 0-10 ($0-$a) TO $ff.
2. TYPE **display p:0..10 p:100..110<return>** TO DISPLAY THE VALUES OF P-MEMORY LOCATIONS 0-10 ($0-$a) AND 100-110 ($64-$6e).

    The display on your screen should be similar to:

```
P:$0000 $0000FF $0000FF $0000FF $0000FF
P:$0004 $0000FF $0000FF $0000FF $0000FF
P:$0008 $0000FF $0000FF $0000FF
P:$0064 $FF5FFF $FF7FFF $FFFFFF $7FFFFF
P:$0068 $FFF77F $7FFFFD $FBFBEF $5FFDCF
P:$006C $FFFDFF $FFFFFF $FFFFFF
ADM#0 P:$00C8 FFFEFF          = DC $FFFEFF
```

3. TYPE **copy p:0..10 p:100<return>** TO COPY THE VALUES OF P-MEMORY LOCATIONS 0-10 ($0-$a) TO THE ELEVEN CONSECUTIVE P-MEMORY LOCATIONS BEGINNING AT 100 ($64).

4. TYPE **display p:0..10 p:100..110<return>** TO SHOW THAT THE VALUES ($ff) OF P-MEMORY LOCATIONS 0-10 ($0-$a) HAVE BEEN COPIED TO P-MEMORY LOCATIONS 100-110 ($64-$6e).

   The display on your screen should be similar to:

```
P:$0000 $0000FF $0000FF $0000FF $0000FF
P:$0004 $0000FF $0000FF $0000FF $0000FF
P:$0008 $0000FF $0000FF $0000FF
P:$0064 $0000FF $0000FF $0000FF $0000FF
P:$0068 $0000FF $0000FF $0000FF $0000FF
P:$006C $0000FF $0000FF $0000FF
ADM#0 P:$00C8 FFFEFF        = DC $FFFEFF
```

5. TYPE **change x:0..10 $aa<return>** TO CHANGE THE VALUES OF X-MEMORY LOCATIONS 0-10 ($0-$a) TO $aa.

6. TYPE **display x:0..10 p:0..10<return>** TO DISPLAY THE VALUES OF X-MEMORY LOCATIONS 0-10 ($0-$a) AND P-MEMORY LOCATIONS 0-10 ($0-$a).

   The display on your screen should be similar to:

```
P:$0000 $0000FF $0000FF $0000FF $0000FF
P:$0004 $0000FF $0000FF $0000FF $0000FF
P:$0008 $0000FF $0000FF $0000FF
X:$0000 $0000AA $0000AA $0000AA $0000AA
X:$0004 $0000AA $0000AA $0000AA $0000AA
X:$0008 $0000AA $0000AA $0000AA
ADM#0 P:$C0C8 FFFEFF        = DC $FFFEFF
```

7. TYPE **copy x:0#11 p:0<return>** TO COPY THE VALUES OF THE ELEVEN CONSECUTIVE X-MEMORY LOCATIONS BEGINNING AT 0 TO THE ELEVEN CONSECUTIVE P-MEMORY LOCATIONS BEGINNING AT 0.

8. TYPE **display x:0..10 p:0..10<return>** TO SHOW THAT THE VALUES ($aa) IN X-MEMORY LOCATIONS 0-10 ($0-$a) HAVE BEEN COPIED TO P-MEMORY LOCATIONS 0-10 ($0-$a).

   The display on your screen should be similar to:

```
P:$0000 $0000AA $0000AA $0000AA $0000AA
P:$0004 $0000AA $0000AA $0000AA $0000AA
P:$0008 $0000AA $0000AA $0000AA
X:$0000 $0000AA $0000AA $0000AA $0000AA
X:$0004 $0000AA $0000AA $0000AA $0000AA
X:$0008 $0000AA $0000AA $0000AA
ADM#0 P:$00C8 FFFEFF         = DC $FFFEFF
```

9. TYPE **quit<return>** TO EXIT THE "ADS56000 USER INTERFACE SOFTWARE" PROGRAM.

1.8 USING THE ADS TO PROGRAM THE DSP56000/1

i. If you want to use the **DSP56000 Demonstration Software** program to cover the topics of Section 1.8, load the software by following Steps 1, 4 and 7 of Section 1.5.1 or Steps 1-3 of Section 1.5.2. Then select from the demonstration software's menus to cover the topics of Section 1.8.

ii. If you do *not* want to use the DSP56000 Demonstration Software program, load the **ADS56000 User Interface Software** program by following Steps 1-3 of Section 1.4.1 or Steps 1, 6 and 7 of Section 1.4.2.

1.8.1 Enter, Assemble, and Disassemble a Canned DSP56000/1 Program

The ADS56000 User Interface Software program has a single line, interactive assembler which accepts DSP56000/1 assembly language mnemonics. It can be used to enter or edit memory resident DSP56000/1 object code programs.

1. TYPE **asm p:$217<return>** TO BEGIN THE ASSEMBLY PROCESS AT P-MEMORY LOCATION $217.

2. TYPE:
```
move x:0300,b<return>
move b,y:(r5)<return>
clr b x:(r0)+,x0 y:(r5)+,y0<return>
mac -x0,y0,b x:(r0)+,x0 y:(r5)+,y0<return>
mac x0,y0,b y:(r5),y0<return>
mac -x0,y0,b x:(r0)+,x0 y:(r4)+,y0<return>
macr x0,y0,b<return>
move b,y:(r4)-<return>
move b,x:301<return>
<escape>
```

You have just entered your first DSP56001 program into the P-Memory. For now, don't concern yourself with the program itself, the DSP56000/1's instruction set will be covered in later chapters.

3. TYPE **disassemble p:$217..$220<return>** TO DISASSEMBLE P-MEMORY LOCATIONS $217-$220 AND DISPLAY THE RESULTING DSP56000/1 ASSEMBLY LANGUAGE MNEMONICS.

Note that this is the same program which you entered in Step 2. The display on your screen should be:

```
ADM#0 P:$0217 57F000 00012C = MOVE X:>$12C,B
ADM#0 P:$0219 5F6500        = MOVE B,Y:(R5)
ADM#0 P:$021A F0B81B        = CLR B X:(R0)+,X0 Y:(R5)+,Y0
ADM#0 P:$021B F0B8DE        = MAC -Y0,X0,B X:(R0)+,X0 Y:(R5)+,Y0
ADM#0 P:$021C 4EE5DA        = MAC Y0,X0,B Y:(R5),Y0
ADM#0 P:$021D F098DE        = MAC -Y0,X0,B X:(R0)+,X0 Y:(R4)+,Y0
ADM#0 P:$021E 2000DB        = MACR Y0,X0,B
ADM#0 P:$021F 5F5400        = MOVE B,Y:(R4)-
ADM#0 P:$0220 57F000 00012D = MOVE X:>$12D,B
```

Note that the disassembler has no way to determine the radix which was used by the programmer during the program entry (assembly) process. Consequently, the disassembler displays disassembled constants as hexadecimal values.

1.8.2 Saving and Loading or Reloading a Canned DSP56000/1 Program

The DSP56000/1 program which you entered in Section 1.8.1 can be named **progl** and saved on your diskette or hard disk using your PC's floppy or hard disk drive, respectively. Later, **progl** can be reloaded to the ADS.

1. TYPE **path a:\<return>** TO SET UP A PATH TO THE DIRECTORY OF DISKETTE DRIVE "A".

   Although Drive "A" is used here, you can use any drive which is available in your PC.

2. TYPE **save p:$217..$220 progl.lod<return>** TO SAVE **progl** ON THE DISKETTE IN DRIVE "A".

   P-Memory block $217-$220 (**progl**) is saved in a Object Module Format (OMF) indicated by the **.lod** extension to the file name **progl.** If **progl** is used as the file name without the **.lod** extension, the **save** command will automatically append the **.lod** extension to the **progl** file name. The OMF is used so **progl.lod** can be reloaded to the ADS.

   If **progl.lod** has been previously saved at the selected destination, the ADS56000 User Interface Software program will prompt you to enter **0, 1,** or **2** to select **Append**, **Replace**, or **Abort** respectively. Once **progl.lod** is saved, a wordprocessor program can be used to call, edit, and resave it.

3. TYPE **load progl.lod<return>** TO RELOAD **progl.lod** INTO THE ADS.

4. TYPE **disassemble p:$217..$220<return>** TO DISASSEMBLE AND DISPLAY P-MEMORY LOCATIONS $217-$220.

   The display on your screen should be:

```
ADM#0 P:$0217 57F000 00012C = MOVE X:>$12C,B
ADM#0 P:$0219 5F6500        = MOVE B,Y:(R5)
ADM#0 P:$021A F0B81B        = CLR B X:(R0)+,X0 Y:(R5)+,Y0
ADM#0 P:$021B F0B8DE        = MAC -Y0,X0,B X:(R0)+,X0 Y:(R5)+,Y0
ADM#0 P:$021C 4EE5DA        = MAC Y0,X0,B Y:(R5),Y0
ADM#0 P:$021D F098DE        = MAC -Y0,X0,B X:(R0)+,X0 Y:(R4)+,Y0
ADM#0 P:$021E 2000DB        = MACR Y0,X0,B
ADM#0 P:$021F 5F5400        = MOVE B,Y:(R4)-
ADM#0 P:$0220 57F000 00012D = MOVE X:>$12D,B
```

1.8.3 Testing and Executing a Canned DSP56000/1 Program

To test an executing DSP56000/1 program, the contents of user specified registers and/or memory locations can be displayed through use of the **break** and **trace** commands.

1. TYPE **break #1 p:$219 s<return>** TO SHOW THE VALUES OF THE DISPLAY ENABLED REGISTERS AND/OR MEMORY LOCATIONS WHICH OCCUR WHEN THE DSP56000/1 PROGRAM INSTRUCTION AT P-MEMORY LOCATION $219 IS REACHED DURING PROGRAM EXECUTION.

   The displayed register and/or memory location values are those that occur at the completion of the program instruction preceding the one at P-Memory location $219. When using the **break** command *without* the Halt (h) parameter as above, program execution continues past the breakpoint ($219).

2. TYPE **break #2 p:$21f s h<return>** TO SHOW THE VALUES OF THE DISPLAY ENABLED REGISTERS AND/OR MEMORY LOCATIONS WHICH OCCUR WHEN THE DSP56000/1 PROGRAM INSTRUCTION AT P-MEMORY LOCATION $21F IS REACHED DURING PROGRAM EXECUTION.

   The displayed registers and/or memory location values are those that occur at the completion of the program instruction preceding the one at P-Memory location $21f. When using the **break** command *with* the Halt (h) parameter as above, program execution halts when the breakpoint is reached.

3. TYPE **break<return>** TO DISPLAY THE BREAKPOINTS WHICH ARE ENABLED.

   The display on your screen should be:

```
ADM 0  Break #1 p:$219 s
ADM 0  Break #2 p:$21f s h
```

4. TYPE **go $217<return>** TO START PROGRAM EXECUTION AT P-MEMORY LOCATION $217.

The display enabled registers and/or memory locations are displayed when the breakpoint locations $219 and $21f are reached during program execution. The display on your screen should be similar to that shown below. Remember, the UP ARROW and DOWN ARROW keys can be used to peruse through the display.

```
ADM#0 break #1
x=$000000000000           y=$000000000000
a=$00000000000000         b=$FF8BFBAF000000
x1=$000000        x0=$000000          r7=$0000    n7=$FFFF
                                                  m7=$FFFF
y1=$000000        y0=$000000          r6=$0000    n6=$FFFF
                                                  m6=$FFFF
a2=$00  a1=$000000    a0=$000000      r5=$0000    n5=$FFFF
                                                  m5=$FFFF
b2=$FF  b1=$8BFBAF    b0=$000000      r4=$0000    n4=$FFFF
                                                  m4=$FFFF
                                      r3=$0000    n3=$FFFF
                                                  m3=$FFFF
pc=$0219      sr=$0300    omr=$02     r2=$0000    n2=$FFFF
                                                  m2=$FFFF
la=$0000      lc=$FDFD                r1=$0000    n1=$FFFF
                                                  m1=$FFFF
ssh=$FFFF     ssl=$FFFF   sp=$0000    r0=$0000    n0=$FFFF
                                                  m0=$FFFF
cnt1=000000   cnt2=000000  cnt3=000000   cnt4=000000
ADM#0 P:$0219 5F6500          = MOVE B,Y:(R5)
wait
ADM#0 break #2
x=$000000000000          y=$000000000000
a=$00000000000000        b=$FF8BFBAF000000
x1=$000000       x0=$000000            r7=$0000   n7=$FFFF
                                                  m7=$FFFF
y1=$000000       y0=$000000            r6=$0000   n6=$FFFF
                                                  m6=$FFFF
a2=$00  a1=$000000    a0=$000000       r5=$0000   n5=$FFFF
                                                  m5=$FFFF
b2=$FF  b1=$8BFBAF    b0=$000000       r4=$0000   n4=$FFFF
                                                  m4=$FFFF
                                       r3=$0000   n3=$FFFF
                                                  m3=$FFFF
pc=$0219      sr=$0300   omr=$02       r2=$0000   n2=$FFFF
                                                  m2=$FFFF
la=$0000      lc=$FDFD                 r1=$0000   n1=$FFFF
                                                  m1=$FFFF
ssh=$FFFF     ssl=$FFFF sp=$0000       r0=$0000   n0=$FFFF
                                                  m0=$FFFF
cnt1=000000    cnt2=000000   cnt3=000000   cnt4=000000
ADM#0 P:$021F 5F5400          = MOVE B,Y:(R4)-
```

5. TYPE **break #2 off<return>** TO CANCEL BREAKPOINT #2.
6. TYPE **go $217 #1<return>** TO START PROGRAM EXECUTION AT P-MEMORY LOCATION $217, DISPLAY THE ENABLED REGISTERS AND/OR MEMORY LOCATIONS WHEN BREAKPOINT #1 IS REACHED, AND STOP PROGRAM EXECUTION WHEN BREAKPOINT #1 IS REACHED.

   The display on your screen should be similar to:

```
ADM#0 break #1
x=$0000000000AA      y=$0000008BFBAF
a=$00000000000000    b=$FF8BFBAF000000
x1=$000000        x0=$0000AA              r7=$0000  n7=$FFFF
                                                    m7=$FFFF
y1=$000000        y0=$8BFBAF              r6=$0000  n6=$FFFF
                                                    m6=$FFFF
a2=$00     a1=$000000    a0=$000000       r5=$0002  n5=$FFFF
                                                    m5=$FFFF
b2=$FF     b1=$8BFBAF    b0=$000000       r4=$0000  n4=$FFFF
                                                    m4=$FFFF
                                          r3=$0000  n3=$FFFF
                                                    m3=$FFFF
pc=$0219   sr=$0220         omr=$02       r2=$0000  n2=$FFFF
                                                    m2=$FFFF
la=$0000   lc=$FDFD                       r1=$0000  n1=$FFFF
                                                    m1=$FFFF
ssh=$FFFF ssl=$FFFF      sp=$0000         r0=$0003  n0=$FFFF
                                                    m0=$FFFF
cnt1=000000   cnt2=000000   cnt3=000000   cnt4=000000
ADM#0 P:$0219 000006         = SWI
```

7. TYPE **break #1 off<return>** TO CANCEL BREAKPOINT #1.
8. TYPE **change pc $217<return>** TO PLACE THE VALUE $217 INTO THE DSP56000/1'S PROGRAM COUNTER (PC).

   The pc now points to P-Memory location $217.
9. TYPE **trace<return>** TO EXECUTE ONE INSTRUCTION AND SHOW THE VALUES OF THE DISPLAY ENABLED REGISTERS AND/OR MEMORY LOCATIONS.

   The display on your screen should be similar to:

```
ADM#0 TRACE COUNT=0
x=$0000000000AA         y=$0000008BFBAF
a=$00000000000000       b=$FF8BFBAF000000
x1=$000000           x0=$0000AA         r7=$0000   n7=$FFFF
                                                   m7=$FFFF
y1=$000000           y0=$8BFBAF         r6=$0000   n6=$FFFF
                                                   m6=$FFFF
a2=$00    a1=$000000    a0=$000000      r5=$0002   n5=$FFFF
                                                   m5=$FFFF
b2=$FF    b1=$8BFBAF    b0=$000000      r4=$0000   n4=$FFFF
                                                   m4=$FFFF
                                        r3=$0000   n3=$FFFF
                                                   m3=$FFFF
pc=$0219    sr=$0220    omr=$02         r2=$0000   n2=$FFFF
                                                   m2=$FFFF
la=$0000    lc=$FDFD                    r1=$0000   n1=$FFFF
                                                   m1=$FFFF
ssh=$FFFF   ssl=$FFFF   sp=$0000        r0=$0003   n0=$FFFF
                                                   m0=$FFFF
cnt1=000000   cnt2=000000   cnt3=000000   cnt4=000000
ADM#0 P:$0219 000006         = SWI
```

10. TYPE **trace 3<return>** TO EXECUTE THE NEXT THREE INSTRUCTIONS AND DISPLAY THE VALUES OF THE ENABLED REGISTER AND/OR MEMORY LOCATIONS WHICH OCCUR FROM EXECUTION OF EACH INSTRUCTION.

    The display on your screen should be similar to:

ADM#0 TRACE COUNT=2
x=$0000000000AA y=$0000008BFBAF
a=$00000000000000 b=$FF8BFBAF000000
x1=$000000 x0=$0000AA r7=$0000 n7=$FFFF
m7=$FFFF
y1=$000000 y0=$8BFBAF r6=$0000 n6=$FFFF
m6=$FFFF
a2=$00 a1=$000000 a0=$000000 r5=$0002 n5=$FFFF
m5=$FFFF
b2=$FF b1=$8BFBAF b0=$000000 r4=$0000 n4=$FFFF
m4=$FFFF
r3=$0000 n3=$FFFF
m3=$FFFF
pc=**$021A** sr=$0220 omr=$02 r2=$0000 n2=$FFFF
m2=$FFFF
la=$0000 lc=$FDFD r1=$0000 n1=$FFFF
m1=$FFFF
ssh=$FFFF ssl=$FFFF sp=$0000 r0=$0003 n0=$FFFF
m0=$FFFF
cnt1=000000 cnt2=000000 cnt3=000000 cnt4=000000
ADM#0 P:$021A F0B81B = CLR B X:(R0)+,X0 Y:(R5)+,Y0
wait
ADM#0 TRACE COUNT=1
x=**$000000000000** y=**$000000000000**
a=$00000000000000 b=**$00000000000000**
x1=$000000 x0=**$000000** r7=$0000 n7=$FFFF
m7=$FFFF
y1=$000000 y0=**$000000** r6=$0000 n6=$FFFF
m6=$FFFF
a2=$00 a1=$000000 a0=$000000 r5=**$0003** n5=$FFFF
m5=$FFFF
b2=**$00** b1=**$000000** b0=$000000 r4=$0000 n4=$FFFF
m4=$FFFF
r3=$0000 n3=$FFFF
m3=$FFFF
pc=**$021B** sr=**$0214** omr=$02 r2=$0000 n2=$FFFF
m2=$FFFF
la=$0000 lc=$FDFD r1=$0000 n1=$FFFF
m1=$FFFF
ssh=$FFFF ssl=$FFFF sp=$0000 r0=**$0004** n0=$FFFF
m0=$FFFF
cnt1=000000 cnt2=000000 cnt3=000000 cnt4=000000
ADM#0 P:$021B F0B8DE = MAC -Y0,X0,B X:(R0)+,X0
Y:(R5)+,Y0
wait
ADM#0 TRACE COUNT=0
x=**$0000000000AA** y=**$000000000000**
a=$00000000000000 b=**$00000000000000**
x1=$000000 x0=**$0000AA** r7=$0000 n7=$FFFF
m7=$FFFF
y1=$000000 y0=**$000000** r6=$0000 n6=$FFFF
m6=$FFFF
a2=$00 a1=$000000 a0=$000000 r5=**$0004** n5=$FFFF
m5=$FFFF
b2=**$00** b1=**$000000** b0=$000000 r4=$0000 n4=$FFFF
m4=$FFFF
r3=$0000 n3=$FFFF
m3=$FFFF

```
pc=$021C     sr=$0214    omr=$02      r2=$0000     n2=$FFFF
                                                   m2=$FFFF
la=$0000     lc=$FDFD                 r1=$0000     n1=$FFFF
                                                   m1=$FFFF
ssh=$FFFF    ssl=$FFFF   sp=$0000    r0=$0005      n0=$FFFF
                                                   m0=$FFFF
cnt1=000000    cnt2=000000  cnt3=000000    cnt4=000000
ADM#0 P:$021C 4EE5DA        = MAC Y0,X0,B Y:(R5),Y0
```

11. TYPE **quit<return>** TO EXIT THE "ADS56000 USER INTERFACE SOFTWARE" PROGRAM.

2

INTRODUCTION TO THE "DSP56000CLAS ASSEMBLER/SIMULATOR SOFTWARE" PACKAGE

The DSP56000CLAS Assembler/Simulator Software package is a DSP56000/1 program (algorithm) development tool consisting of:

1. the ASM56000 Relocatable Macro Cross Assembler,
2. the LNK56000 Linker,
3. the LIB56000 Librarian, and
4. the SIM56000 Multi-DSP56000/1 Clock-By-Clock Simulator.

The DSP56000CLAS Software package is provided by Motorola in host-specific form to run on either the IBM PC, Macintosh II PC, Sun-3 Workstation, or VAX™/VMS computer system.

In this chapter, the DSP56000CLAS Software package and the IBM PC Line Editor (EDLIN) are used to develop canned DSP56000/1 programs and macros. The program development process begins by using EDLIN to enter the DSP56000/1 assembly language source statements of the canned DSP56000/1 programs and macros to produce an assembler source statement file for each program or macro. Next, the ASM56000 Relocatable Macro Cross Assembler is used to translate the file of assembler source statements for each program or macro into a relocatable file of DSP56000/1 object code statements. The LNK56000 Linker is then used to process the relocatable object code file generated by the assembler for each program or macro to produce an absolute load file which can be either loaded directly into the ADS for execution and testing or input to the SIM56000 Simulator program for simulated execution and testing. Figure 2.1 shows the DSP56000/1 program development process. Alternatively, the DSP56000/1 Demonstration Software program can be used to present selected topics of this chapter.

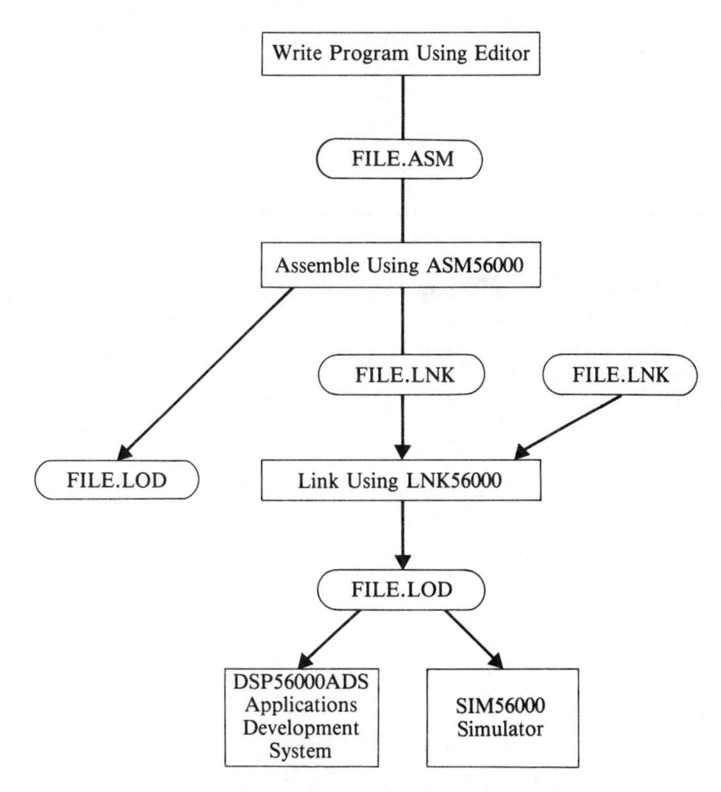

Figure 2-1 DSP56000/1 Program Development Process

2.1 INSTALLING THE "ASM56000 ASSEMBLER SOFTWARE" PROGRAM

2.1.1 PC with two Diskette Drives and no Hard Disk Drive

1. PLACE THE "DOS" DISKETTE INTO DRIVE "A" AND BOOT THE PC.
2. PLACE THE "DSP56000 MACRO CROSS ASSEMBLER" DISKETTE CONTAINING THE "ADS56000 ASSEMBLER SOFTWARE" PROGRAM INTO DRIVE "B".
3. TYPE **b:asm56000<return>**.

   Do not concern yourself with the details displayed on your screen. This step is performed only to show you that you have successfully invoked the ASM56000 Assembler Software program. The display on your screen should be similar to:

```
Motorola DSP56000 Macro Cross Assembler Version 3.0
(C) Copyright Motorola, Inc. 1987, 1988, 1989.
All rights reserved.
Usage: ASM56000 [-A] [-B[<lodfil>]] [-D<symbol> <string>]
[-F<argfil>] [-I<ipath>] [-L[<lstfil>]] [-M<mpath>]
[-O<opt>[,<oppt>...]] [-V] <srcfil>...
where:
```

```
lodfil - optional load file name
symbol - user-defined symbol
string - string associated with symbol
ipath  - include file directory path
lstfil - optional list file name
mpath  - macro library directory path
opt    - assembler option
srcfil - assembler source file(s)
```

2.1.2 PC with a Hard Disk Drive and one or more Diskette Drives

For simplicity, it is assumed that Drive "C" is assigned to the hard disk and that DOS resides in the directory c:\dos.

1. BOOT THE PC.
2. PLACE THE "DSP56000 MACRO CROSS ASSEMBLER" DISKETTE CONTAINING THE "ASM56000 ASSEMBLER SOFTWARE" PROGRAM INTO DRIVE "A".
3. TYPE **md c:\dsptools<return>** TO CREATE A dsptools DIRECTORY ON DRIVE "C".

   Optionally, you can give the dsptools directory any name you wish and locate it anywhere you wish on Drive "C".
4. TYPE **md c:\dsptools\asm56000<return>** TO CREATE A dsptools\asm56000 DIRECTORY ON DRIVE "C".
5. TYPE **copy a: c:\dsptools\asm56000<return>** TO COPY THE CONTENTS OF THE "DSP56000 MACRO CROSS ASSEMBLER" DISKETTE TO THE SPECIFIED DIRECTORY ON DRIVE "C".
6. TYPE **cd c:\dsptools\asm56000<return>** TO ENTER THIS DIRECTORY.
7. TYPE **asm56000<return>**.

   Do not concern yourself with the details displayed on your screen. This step is performed only to show you that you have successfully invoked the ASM56000 Assembler Software program. The display on your screen should be similar to that shown in Step 3 of Section 2.1.1.

2.2 INSTALLING THE "LNK56000 LINKER SOFTWARE" AND "LIB56000 LIBRARIAN SOFTWARE" PROGRAMS

2.2.1 PC with two Diskette Drives and no Hard Disk Drive

1. PLACE THE "DOS" DISKETTE INTO DRIVE "A" AND BOOT THE PC.
2. PLACE THE "DSP56000 LINKER/LIBRARIAN" DISKETTE CONTAINING THE "LNK56000 LINKER SOFTWARE" AND "LIB56000 LIBRARIAN SOFTWARE" PROGRAMS INTO DRIVE "B".
3. TYPE **b:lnk56000<return>**.

   Do not concern yourself with the details displayed on your screen. This step is performed only to show you that you have successfully accessed the LNK56000 Linker Software program. The display on your screen should be similar to:

```
Motorola DSP56000 Cross Linker Version 3.0
(C) Copyright Motorola, Inc. 1987, 1988, 1989.
All rights reserved.
Usage: LNK56000 [-B[<lodfil>]] [-D] [-F<argfil>]
[-L<library>] [-M[<map>]] [-O<mem>[<ctr>][<map>]:<origin>]
[-R[<memfil>]] [-V] [<lnkfil>...
where:

   lodfil  - optional load file name
   argfil  - command line argument file name
   library - library file name
   mapfil  - optional load map file name
   mem     - memory space (X, Y, L, P)
   ctr     - location counter (L, H)
   map     - memory mapping (I, E, B)
   origin  - memory space origin
   memfil  - optional memory map file name
   lnkfil  - link input file name
```

2.2.2 PC with a Hard Disk Drive and one or more Diskette Drives

For simplicity, it is assumed that Drive "C" is assigned to the hard disk and that DOS resides in the directory c:\dos.

1. BOOT THE PC.
2. PLACE THE "DSP56000 LINKER/LIBRARIAN" DISKETTE CONTAINING THE "LNK56000 LINKER SOFTWARE" AND "LIB56000 LIBRARIAN SOFTWARE" PROGRAMS INTO DRIVE "A".
3. TYPE **md c:\dsptools\lnk56000<return>** TO CREATE A dsptools\lnk56000 DIRECTORY ON DRIVE "C". The dsptools DIRECTORY WAS CREATED IN SECTION 2.1.2.
4. TYPE **copy a: c:\dsptools\lnk56000<return>** TO COPY THE CONTENTS OF THE "DSP56000 LINKER/LIBRARIAN" DISKETTE TO THE SPECIFIED DIRECTORY ON DRIVE "C".
5. TYPE **cd c:\dsptools\lnk56000<return>** TO ENTER THIS DIRECTORY.
6. TYPE **lnk56000<return>**.

   Do not concern yourself with the details displayed on your screen. This step is performed only to show you that you have successfully invoked the LNK56000 Linker Software program. The display on your screen should be similar that shown in Step 3 of Section 2.2.1.

2.3 INSTALLING THE "SIM56000 SIMULATOR SOFTWARE" PROGRAM

2.3.1 PC with two Diskette Drives and no Hard Disk Drive

1. PLACE THE "DOS" DISKETTE INTO DRIVE "A" AND BOOT THE PC.
2. PLACE THE "DSP56000 SIMULATOR" DISKETTE (1 OF 3) CONTAINING THE "SIM56000 SIMULATOR SOFTWARE" PROGRAM INTO DRIVE "B".
3. TYPE **b:sim56000<return>**.

The display on your screen should be as shown below. Note that the menus for the SIM56000 Simulator Software and ADS56000 User Interface Software (see Chapter 1) programs are very similar. As a result, you can easily move from one to the other in using these DSP56000/1 DSP development tools.

```
0>
asm break change copy disassemble display <space>=more
```

4. TYPE **quit<return>** TO EXIT THE "SIM56000 SIMULATOR SOFTWARE" PROGRAM.

2.3.2 PC with a Hard Disk Drive and one or more Diskette Drives

For simplicity, it is assumed that Drive "C" is assigned to the hard disk and that DOS resides in the directory c:\dos.

1. BOOT THE PC.
2. PLACE THE "DSP56000 SIMULATOR" DISKETTE (1 OF 3) CONTAINING THE "SIM56000 SOFTWARE" PROGRAM INTO DRIVE "A".
3. TYPE **md c:\dsptools\sim56000<return>** TO CREATE A dsptools\sim56000 DIRECTORY ON DRIVE "C". The dsptools DIRECTORY WAS CREATED IN SECTION 2.1.2.
4. TYPE **copy a: c:\dsptools\sim56000<return>** TO COPY THE CONTENTS OF THE "DSP56000 SIMULATOR" DISKETTE TO THE SPECIFIED DIRECTORY ON DRIVE "C".
5. TYPE **cd c:\dsptools\sim56000<return>** TO ENTER THIS DIRECTORY.
6. TYPE **sim56000<return>** TO ENTER THE "SIM56000 SIMULATOR SOFTWARE" PROGRAM.

   The display on your screen should be as shown below. Note that the menus for the SIM56000 Simulator Software and ADS56000 User Interface Software (see Chapter 1) programs are very similar. As a result, you can easily move from one to the other in using these DSP56000/1 DSP development tools.

```
0>
asm break change copy disassemble display <space>=more
```

7. TYPE **quit<return>** TO EXIT THE "SIM56000 SOFTWARE" PROGRAM.

2.4 PROGRAM SOURCE STATEMENT FORMAT

The ASM56000 Assembler Software program allows program source statements to use up to six fields. The fields are separated (delimited) by one or more spaces or tabs. The six fields are:

1. *Label Field*

   The label field appears as the first field of a source statement, and can take one of the following forms:

   a. A space or tab used as the first character of the Label Field indicates that the field is empty, and the source statement has no label.

      Example: **<TAB>**CLR<TAB>B<RETURN>

   b. An alphabetic character used as the first character of the Label Field indicates that the source statement contains a symbol called a label.

      Example: **LOOP**<TAB>MOVE<TAB>B,L:-(R0)<RETURN>

   c. An underscore (_) used as the first character of the Label Field indicates that the label is "local".

      Example: **_ENDP**<RETURN>

2. *Operation Field*

   The Operation Field appears after the Label Field, and must be preceded by at least one space or tab. Entries in the Operation Field may be one of three types:

   a. *Opcode*: DSP56000/1 instruction mnemonic which is recognized by the ASM56000 Assembler.

      Example: ENTRY<TAB>**ADD** B,A<RETURN>

   b. *Directive*: Special opcodes which are recognized by the ASM56000 Assembler and are used to control the assembly process itself.

      Example: CNST<TAB>**EQU**<TAB>$5<RETURN>

   c. *Macro Call*: Invokes a previously defined macro to be inserted in place of the macro call.

3. *Operand Field*:

   The interpretation of the Operand Field is dependent on the contents of the Operation Field. The Operand Field, if present, must follow the Operation Field, and must be preceded by at least one space or tab. The Operand Field may contain a symbol, an expression, or a combination of symbols and expressions separated by commas.

   Example: ENTRY<TAB>ADD **B,A**<RETURN>

4. *Data Transfer Fields #1 and #2*:

   Most DSP56000/1 Data ALU opcodes can be appended to allow one or two data transfer operations to occur concurrently with the execution of the opcode itself. If used, Data Transfer Field #1 specifies one data transfer operation; if used, Data Transfer Field #2 specifies a second data transfer operation. A data transfer operation is specified by two addressing mode operands separated by a comma, with no embedded blanks. A Data Transfer Field must be separated (delimited) from the preceding field by at least one space or tab. The following example shows specification of one such data transfer:

Example: LOOP<TAB>MOVE<TAB>**Y:-(R0),Y0**<RETURN>

The following example shows specification of two such data transfers using Data Transfer Fields #1 and #2:

Example:
<TAB>RND<TAB>A<TAB>**X:(R0)+,X0**<TAB>**Y:(R4)+,Y0**<RETURN>

5. *Comment Field*:

   Comments are not considered significant to the ASM56000 Assembler Software program, but can be included in the source statement for documentation purposes. A comment field is composed of any characters that are preceded by a semicolon (;). The semicolon can be the first character of the source statement. In this case the Comment Field becomes the entire source statement. The Comment Field can also be preceded by a space and inserted after the last of the source statement fields which are significant to the assembler program.

   Examples:
   ;PROGRAM TO CLEAR THE RAM IN X and Y MEMORY <RETURN>

   <TAB>CLR<TAB>B<TAB>**;USED TO CLEAR LONG MEMORY**
   <RETURN>

2.5 ENTERING AND EDITING A CANNED DSP56000/1 PROGRAM

DSP56000/1 programs can be entered and edited by using any available word processor program. Here, the DOS line editor **edlin** is used to enter, alter, display, list, load and save a canned DSP56000/1 program.

1. BOOT THE PC.

   If you are using a PC with two diskette drives and no hard disk drive, proceed to Step 2.

   If you are using a PC with a hard disk drive and one or more diskette drives, skip Step 2 and proceed to Step 3.

2. PLACE THE "DOS DISKETTE" INTO DRIVE "B" AND A FORMATTED DISKETTE INTO DRIVE "A".

   TYPE **b:edlin a:\prog2.asm<return>** TO PLACE THE FILENAME **prog2.asm** ONTO THE DISKETTE IN DRIVE "A".

   The message and prompt shown below should be on your screen. Proceed to Step 4.

```
New file
*_
```

3. PLACE A FORMATTED DISKETTE INTO DRIVE "A" AND TYPE **c:\dos\edlin a:\prog2.asm<return>** TO PLACE THE FILENAME **prog2.asm** ONTO THE DISKETTE IN DRIVE "A".

   The message and prompt shown below should be on your screen. Proceed to Step 4.

```
New file
*_
```

4. TYPE **I<return>** TO ENTER THE "INSERT MODE" OF THE "EDLIN" PROGRAM.

5. TYPE THE FOLLOWING SOURCE STATEMENTS FOR "PROG2.ASM".

   ;PROGRAM TO CLEAR THE RAM IN X and Y MEMORY <RETURN>
   <TAB>MOVE<TAB>#$100,R0<TAB>;POINTER TO LONG MEMORY <RETURN>
   <TAB>CLR<TAB>B<TAB>;USED TO CLEAR LONG MEMORY <RETURN>
   LOOP<TAB>MOVE<TAB>B,L:-(R0)<TAB>;DECREMENT THEN CLEAR ONE LOCATION<RETURN>
   <TAB>MOVE<TAB>R0,A<TAB>;IS POINTER ZERO?<RETURN>
   <TAB>TST<TAB>A<TAB>;TEST POINTER<RETURN>
   <TAB>JNE<TAB>LOOP<TAB>;LOOP IF NOT DONE<RETURN>
   ^Z<RETURN>

   You have just typed your second DSP56000/1 program. The **prog2.asm** clears the DSP56000/1's on chip X-Data and Y-Data memories. For now, don't concern yourself with the program itself, the DSP56000/1's instruction set will be covered in later chapters.

6. Type **L<return>** TO LIST "PROG2.ASM" ON YOUR SCREEN.

   The **prog2.asm** listing on your screen should be as shown below.

   VERIFY THAT YOU HAVE ENTERED "PROG2.ASM" CORRECTLY. IF YOU FIND AN ERROR, YOU CAN "DELETE" OR "INSERT" A SOURCE STATEMENT LINE BY TYPING THE LINE NUMBER FOLLOWED BY A "D" (DELETE) OR "I" (INSERT) COMMAND.

```
1:*;PROGRAM TO CLEAR THE  RAM IN X and Y MEMORY
2:*<TAB>MOVE<TAB>#$100,R0<TAB>;POINTER TO LONG MEMORY
3:*<TAB>CLR<TAB>B<TAB>;USED TO CLEAR LONG MEMORY
4:*LOOP<TAB>MOVE<TAB>B,L:-(R0)<TAB>;DECREMENT THEN
   CLEAR ONE LOCATION
5:*<TAB>MOVE<TAB>R0,A<TAB>;IS POINTER ZERO?
6:*<TAB>TST<TAB>A<TAB>;TEST POINTER
7:*<TAB>JNE<TAB>LOOP<TAB>;LOOP IF NOT DONE
8:*^Z<RETURN>
```

7. TYPE **E<return>** TO SAVE "PROG2.ASM" AT THE DESTINATION NAMED IN STEP 2 OR 3.

   A copy of **prog2.asm** can be found on the DSP56000/1 Demonstration Software #1 diskette. For more information on the **edlin** editor refer to your DOS reference manual.

2.6 ASSEMBLING AND LINKING A CANNED DSP56000/1 PROGRAM

1. If you want to use the DSP56000/1 Demonstration Software program to cover the topics of Section 2.6, load the software by following Steps 1 and 6 of Chapter 1 Section 1.5.1 or Steps 1-3 of Chapter 1 Section 1.5.2. Then select from the demonstration software's menus to cover the topics of Section 2.6.
2. If you do *not* want to use the DSP56000/1 Demonstration Software program, proceed to Section 2.6.1.

2.6.1 Using the "ASM56000 Assembler Software" Program:

The ASM56000 Assembler Software program processes DSP56000/1 assembly language source statements into DSP56000/1 object code, and produces a source statement listing.

If you are using a PC with two diskette drives and no hard disk drive, perform Steps 1-4 below. If you are using a PC with one or more diskette drives and a hard disk drive, perform Steps 5-8 below.

1. PLACE THE DISKETTE CONTAINING **prog2.asm** INTO DRIVE "A".
2. PLACE THE "DSP56000 MACRO ASSEMBLER" DISKETTE CONTAINING THE "ASM56000 ASSEMBLER SOFTWARE" PROGRAM INTO DRIVE "B".
3. TYPE **b:asm56000<return>** TO SHOW THE "ASM56000 ASSEMBLER SOFTWARE" PROGRAM'S COMMAND OPTIONS.

   The display on your screen should be similar to that shown below. Using the **-B[<lodfil>]** command option without a **-A** command option instructs the ASM56000 Assembler Software program to operate in its relative mode and create a relocatable link (.lnk) file. Using the **-A -B[<lodfil>]** command options together instructs the ASM56000 Assembler to operate in its absolute mode and create an absolute load (.lod) file.

```
Motorola DSP56000 Macro Cross Assembler Version 3.0
(C) Copyright Motorola, Inc. 1987, 1988, 1989.
All rights reserved.
Usage: ASM56000 [-A] [-B[<lodfil>]] [-D<symbol> <string>]
[-F<argfil>] [-I<ipath>] [-L[<lstfil>]] [-M<mpath>]
[-O<opt>[,<oppt>...]] [-V] <srcfil>...
where:

   lodfil - optional load file name
   symbol - user-defined symbol
   string - string associated with symbol
   ipath  - include file directory path
   lstfil - optional list file name
   mpath  - macro library directory path
   opt    - assembler option
   srcfil - assembler source file(s)
```

4. TYPE **b:asm56000 -ba:prog2.lnk a:prog2.asm<return>** TO ASSEMBLE THE FILE a:prog2.asm IN THE RELATIVE MODE TO PRODUCE THE ASSEMBLED RELOCATABLE FILE a:prog2.lnk.
5. PLACE THE DISKETTE CONTAINING **prog2.asm** INTO DRIVE "A".
6. TYPE **cd c:\dsptools\asm56000<return>** TO ENTER THE SPECIFIED DIRECTORY.
7. TYPE **asm56000<return>** TO SHOW THE "ASM56000 ASSEMBLER SOFTWARE" PROGRAM'S COMMAND OPTIONS.

   The display on your screen should be similar to that shown in Step 3. Using the **-B[<lodfil>]** command option without a **-A** command option instructs the ASM56000 Assembler Software program to operate in its relative mode and create a relocatable link (.lnk) file. Using the **-A -B[<lodfil>]** command options together instructs the ASM56000 Assembler Software program to operate in its absolute mode and create an absolute load (.lod) file.
8. TYPE **asm56000 -ba:prog2.lnk a:prog2.asm<return>** TO ASSEMBLE THE FILE a:prog2.asm IN THE RELATIVE MODE TO PRODUCE THE ASSEMBLED RELOCATABLE FILE a:prog2.lnk.

2.6.2 Using the "LNK56000 Linker Software" Program:

The LNK56000 Linker Software program processes relocatable link (.lnk) files, produced by the ASM56000 Assembler Software program, to generate an absolute load (.lod) file which can be directly loaded for execution onto the DSP56000ADS Application Development System or into the SIM56000 Simulator Software program environment.

If you are using a PC with two diskette drives and no hard disk drive, perform Steps 1-4 below. If you are using a PC with one or more diskette drives and a hard disk drive, perform Steps 5-8 below.

1. PLACE THE DISKETTE CONTAINING **prog2.lnk** INTO DRIVE "A".
2. PLACE THE "DSP56000 LINKER/LIBRARIAN" DISKETTE CONTAINING THE "LNK56000 LINKER SOFTWARE" PROGRAM INTO DRIVE "B".
3. TYPE **b:lnk56000<return>** TO SHOW THE "LNK56000 LINKER SOFTWARE" PROGRAM'S COMMAND OPTIONS.

   The display on your screen should be similar to:

```
Motorola DSP56000 Cross Linker Version 3.0
(C) Copyright Motorola, Inc. 1987, 1988, 1989.
All rights reserved.
Usage: LNK56000 [-B[<lodfil>]] [-D] [-F<argfil>]
[-L<library>] [-M[<map>]]
[-O<mem>[<ctr>][<map>]:<origin>]
[-R[<memfil>]] [-V] [<lnkfil>...
where:

   lodfil  - optional load file name
   argfil  - command line argument file name
   library - library file name
   mapfil  - optional load map file name
   mem     - memory space (X, Y, L, P)
   ctr     - location counter (L, H)
   map     - memory mapping (I, E, B)
   origin  - memory space origin
   memfil  - optional memory map file name
   lnkfil  - link input file name
```

4. TYPE **b:lnk56000 -ba:prog2.lod a:prog2.lnk<return>** TO LINK THE SINGLE FILE **a:prog2.lnk** AND PRODUCE THE ABSOLUTE LOAD FILE **a:prog2.lod.**

   Usually the LNK56000 Linker Software program is invoked to link several **.lnk** files and produce a single **.lod** file, e.g., **lnk56000 -ba:prgm.lod a:prgm1.lnk a:prgm2.lod**. However, as shown above, the LNK56000 Linker Software program can also be invoked to link a single **.lnk** file and produce a single **.lod** file.
5. PLACE THE DISKETTE CONTAINING **prog2.lnk** INTO DRIVE "A".
6. TYPE **cd c:\dsptools\lnk56000<return>** TO ENTER THE SPECIFIED DIRECTORY.
7. TYPE **lnk56000<return>** TO SHOW THE "LNK56000 LINKER SOFTWARE" PROGRAM'S COMMAND OPTIONS.

   The display on your screen should be similar to that shown in Step 3.
8. TYPE **lnk56000 -ba:prog2.lod a:prog2.lnk<return>** TO LINK THE SINGLE FILE **a:prog2.lnk** AND PRODUCE THE ABSOLUTE LOAD FILE **a:prog2.lod.**

   Usually the LNK56000 Linker Software program is invoked to link several **.lnk** files and produce a single **.lod** file, e.g., **lnk56000 -Ba:prgm.lod a:prgm1.lnk a:prgm2.lod**. However, as shown above, the LNK56000 Linker Software program can also be invoked to link a single **.lnk** file and produce a single **.lod** file.

2.7 PROGRAM DEVELOPMENT AND EXECUTION USING THE "SIM56000 SIMULATOR SOFTWARE" PROGRAM

The SIM56000 Simulator Software program is a software tool for developing and executing DSP56000/1 programs and algorithms. On a clock by clock basis, the SIM56000 Simulator Software program simulates the external memory blocks of the DSP56000/1 processors and all of the functions of the DSP56000/1 processors themselves including the operation of the DSP56000/1 processors': on-chip peripherals, Data Execution Unit, Address Generation Unit, Program Control Unit (including exception processing), internal memory blocks, and instruction fetch/decode/execute pipeline.

1. If you want to use the DSP56000/1 Demonstration Software program to cover the topics of Sections 2.7, load the software by following Steps 1 and 6 of Chapter 1 Section 1.5.1 or Steps 1-3 of Chapter 1 Section 1.5.2. Then select from the demonstration software's menus to cover the topics of Section 2.7.
2. If you do *not* want to use the DSP56000/1 Demonstration Software program, proceed to Section 2.7.1.

2.7.1 Summary of "SIM56000 Simulator Software Commands":

The SIM56000 Simulator Software Commands are permuted by many optional parameters specific to each command to yield a large and powerful command set. This book does not attempt to use all of the numerous combinations of commands and specific optional parameters. Rather, only those commands and command parameters necessary to demonstrate the concepts of this book are covered. For detailed information on the SIM56000 commands and their optional parameters, refer to the latest version of Motorola's *DSP56000 Digital Signal Processor Simulator Reference Manual.*

Note the similarity between the SIM56000 Simulator Software Commands (without optional parameters) shown below and the ADS56000 User Interface Commands shown in Section 1.6 of Chapter 1. Also note the similarity between the SIM56000 "Reset" command and the ADS56000 "Force" command.

Command	*Description*
ASM	Single line assembler
BREAK	Set, modify, or clear breakpoint
CHANGE	Modify register/memory values
COPY	Copy one memory block to another
DISASSEMBLE	Disassemble memory
DISPLAY	Display registers/memory values
EVALUATE	Calculator
FORCE	Force hardware reset or break
GO	Execute ADM program real-time
HELP	On-line help text
HISTORY	Disassemble and display previous 20 executed instructions
INPUT	Open file for ADM program input
LOAD	Download object file into ADM memory
LOG	Save commands/entire session to file
OUTPUT	Open file for ADM program output
PATH	Set directory path for ADM(s)
QUIT	Exit user interface program
RADIX	Set default radix for command entry
RESET	Reset registers or entire simulator
SAVE	Save ADM state/memory to file
STEP	Multiple instruction step of ADM
SYSTEM	Execute operating system command(s)
TRACE	Single instruction step of ADM
WAIT	Wait before continuing

2.7.2 "SIM56000 Simulator Software" Program Basics:

1. ENTER THE "SIM56000 SIMULATOR SOFTWARE" PROGRAM BY FOLLOWING STEPS 1-3 OF SECTION 2.3.1 OR STEPS 1, 5, AND 6 OF SECTION 2.3.2.
2. TYPE **change r0 $10<return>** TO CHANGE THE VALUE OF THE DSP56000/1 PROCESSOR'S ADDRESS REGISTER r0 TO $10.
3. TYPE **change p:$10 $2<return>** TO CHANGE THE VALUE OF P-MEMORY LOCATION $10 TO $2.
4. TYPE **evaluate r0+p:$10<return>** TO CALCULATE THE SUM OF THE VALUES IN THE DSP56000/1 PROCESSOR'S ADDRESS REGISTER r0 AND P-MEMORY LOCATION $10.

   The display on your screen should be:

```
Hex:000012    Dec:000018    Fract: 0.0000021
Bin:000000000000000000010010
```

5. TYPE **evaluate b $32<return>** TO CONVERT THE HEXADECIMAL VALUE 32 TO A BINARY VALUE.

   The display on your screen should be:

```
Bin:000000000000000000110010
```

6. TYPE **evaluate h %101010&p:r0<return>** TO CALCULATE IN HEXADECIMAL THE BITWISE LOGICAL "AND" OF THE BINARY VALUE 101010 AND THE P-MEMORY LOCATION POINTED TO BY THE DSP56000/1 PROCESSOR'S ADDRESS REGISTER r0.

   The display on your screen should say:

```
Hex:000002
```

7. TYPE **radix h<return>** TO CHANGE THE DEFAULT RADIX TO HEXADECIMAL.

   A hexadecimal constant can be specified by preceding the constant with a dollar ($) symbol. Likewise, a decimal constant can be specified with a preceding grave accent (') symbol; a binary constant can be specified with a percent (%) symbol. When the SIM56000 Simulator Software program is entered, the default radix is decimal. That is, decimal constants can be directly entered without the need for a preceding radix specification symbol. The **radix** command allows you to change the default radix to that of your convenience and then enter constants without the need for a preceding radix specification symbol. For example, if you need to enter a large number of hexadecimal constants, the **radix** command can be used to change the default radix to hexadecimal. However, after doing so you need to precede decimal constants with a grave accent (‘) radix specification symbol and precede binary constants with a percent (%) radix specification symbol.

8. TYPE **radix d<return>** TO CHANGE THE DEFAULT RADIX TO DECIMAL.
9. TYPE **change x:0..10 $10<return>** TO CHANGE THE VALUES OF X-MEMORY LOCATIONS 0-10 ($0-$a) TO $10.
10. TYPE **display x:0..10<return>** TO SHOW THE VALUES ($10) OF X-MEMORY LOCATIONS 0-10 ($0-$a).

    The display on your screen should say:

```
X:$0000     $000010     $000010     $000010     $000010
X:$0004     $000010     $000010     $000010     $000010
X:$0008     $000010     $000010     $000010
P:$E000 000000          = NOP
```

11. TYPE **radix f x:0..10<return>** TO CHANGE THE DEFAULT RADIX OF THE VALUES IN X-MEMORY LOCATIONS 0-10 ($0-$a) FROM DECIMAL TO "FRACTIONAL" ARITHMETIC.
12. TYPE **display x:0..10<return>** TO SHOW THE "FRACTIONAL" ARITHMETIC VALUES OF X-MEMORY LOCATIONS 0-10 ($0-$a).

    The display on your screen should be:

```
X:$0000  0.0000019  0.0000019  0.0000019  0.0000019
X:$0004  0.0000019  0.0000019  0.0000019  0.0000019
X:$0008  0.0000019  0.0000019  0.0000019
P:$E000 000000          = NOP
```

13. Type **display c<return>** TO SHOW THAT THE "SIM56000 SIMULATOR SOFTWARE" PROGRAM IS OPERATING IN THE DSP56001 CONFIGURATION.

    The display on your screen should say:

```
Device Index:0, Device Type:56001
```

14. TYPE **wait<return>** TO CAUSE A PAUSE CONDITION.

    When the **wait** command is encountered by the SIM56000 Simulator Software program, simulated processing of commands is halted until any key on the keyboard is typed.

15. TYPE **quit<return>** TO EXIT THE "SIM56000 SIMULATOR SOFTWARE" PROGRAM.

2.7.3 Using the "SIM56000 Simulator Software" Program to Load, Execute, and Test a Canned DSP56000/1 Program:

1. ENTER THE "SIM56000 SIMULATOR SOFTWARE" PROGRAM BY FOLLOWING STEPS 1-3 OF SECTION 2.3.1 OR STEPS 1, 5, AND 6 OF SECTION 2.3.2.
2. PLACE THE DISKETTE CONTAINING **prog2.lod** INTO DRIVE "A".
3. TYPE **load a:prog2.lod<return>** TO COPY **prog2.lod** FROM THE DISKETTE IN DRIVE "A" INTO THE "SIM56000 SIMULATOR SOFTWARE'S" PROGRAM ENVIRONMENT.

   Remember, execution of **prog2.lod** clears the DSP56000/1 processor's on-chip X-Memory and Y-Memory.
4. TYPE **disassemble p:0..10<return>** TO DISASSEMBLE AND DISPLAY P-MEMORY LOCATIONS 0-10 ($0-$a).

   The display on your screen should be:

```
P:$0000 60F400 000100 = MOVE #>$100,R0
P:$0002 20001B        = CLR B
P:$0003 497800        = MOVE B,L:-(R0)
P:$0004 220E00        = MOVE R0,A
P:$0005 200003        = TST A
P:$0006 0AF0A2 000003 = JNE >$3
P:$0008 000000        = NOP
P:$0009 000000        = NOP
P:$000A 000000        = NOP
```

5. TYPE **break #1 pc>=$8<return>** TO HALT THE SIMULATION OF **prog2.lod** EXECUTION WHEN THE DSP56001 PROCESSOR'S PROGRAM COUNTER (PC) HAS A VALUE OF GREATER THAN OR EQUAL TO $8. AT THAT TIME, SHOW THE VALUES OF THE DISPLAY ENABLED REGISTERS AND MEMORY LOCATIONS.
6. TYPE **change x:$e0..$ff $10 y:$e0..$ff $10<return>** TO CHANGE THE VALUES OF X-MEMORY LOCATIONS $e0-$ff AND Y-MEMORY LOCATIONS $e0-$ff to $10.
7. TYPE **display off<return>** TO TURN OFF AND RESET THE DISPLAY.
8. TYPE **display on x:$e0..$ff y:$e0..$ff<return>** TO ENABLE THE DISPLAY TO SHOW THE VALUES IN X-MEMORY LOCATIONS $0e-$ff AND Y-MEMORY LOCATIONS $e0-$ff.
9. TYPE **display<return>** TO SHOW THE VALUES OF THE DISPLAY ENABLED MEMORY LOCATIONS.

   The display on your screen should be:

```
X:$00E0     $000010     $000010     $000010     $000010
X:$00E4     $000010     $000010     $000010     $000010
X:$00E8     $000010     $000010     $000010     $000010
X:$00EC     $000010     $000010     $000010     $000010
X:$00F0     $000010     $000010     $000010     $000010
X:$00F4     $000010     $000010     $000010     $000010
X:$00F8     $000010     $000010     $000010     $000010
X:$00FC     $000010     $000010     $000010     $000010
Y:$00E0     $000010     $000010     $000010     $000010
Y:$00E4     $000010     $000010     $000010     $000010
Y:$00E8     $000010     $000010     $000010     $000010
Y:$00EC     $000010     $000010     $000010     $000010
Y:$00F0     $000010     $000010     $000010     $000010
Y:$00F4     $000010     $000010     $000010     $000010
Y:$00F8     $000010     $000010     $000010     $000010
Y:$00FC     $000010     $000010     $000010     $000010
P:$0000 60F400 000100 = MOVE #>$100,R0
```

10. TYPE **go #1<return>** TO SIMULATE EXECUTION OF **prog2.lod** AND STOP SIMULATION WHEN THE CONDITIONS OF BREAKPOINT #1 ARE MET.

    The display on your screen should be as shown below. Note that the values of X-Memory and Y-Memory locations $ff have changed from $10 to $0.

```
Break #1 pc>=$8 h ;dev:0 pc:0008 cyc:20
X:$00E0     $000010     $000010     $000010     $000010
X:$00E4     $000010     $000010     $000010     $000010
X:$00E8     $000010     $000010     $000010     $000010
X:$00EC     $000010     $000010     $000010     $000010
X:$00F0     $000010     $000010     $000010     $000010
X:$00F4     $000010     $000010     $000010     $000010
X:$00F8     $000010     $000010     $000010     $000010
X:$00FC     $000010     $000010     $000010     $000000
Y:$00E0     $000010     $000010     $000010     $000010
Y:$00E4     $000010     $000010     $000010     $000010
Y:$00E8     $000010     $000010     $000010     $000010
Y:$00EC     $000010     $000010     $000010     $000010
Y:$00F0     $000010     $000010     $000010     $000010
Y:$00F4     $000010     $000010     $000010     $000010
Y:$00F8     $000010     $000010     $000010     $000010
Y:$00FC     $000010     $000010     $000010     $000000
P:$0003 497800          = MOVE B,L:-(R0)
```

11. TYPE **go #1 :10<return>** TO CONTINUE SIMULATION OF **prog2.lod** EXECUTION FROM BREAKPOINT #1 AND HALT ON THE TENTH OCCURRENCE OF BREAKPOINT #1.

    The display on your screen should be as that shown below. Note that the values of X-Memory and Y-Memory locations $fb-$fe have changed from $10 to $0.

```
Break #1 pc>=$8 h ;dev:0 pc:0008 cyc:76
X:$00E0     $000010    $000010    $000010    $000010
X:$00E4     $000010    $000010    $000010    $000010
X:$00E8     $000010    $000010    $000010    $000010
X:$00EC     $000010    $000010    $000010    $000010
X:$00F0     $000010    $000010    $000010    $000010
X:$00F4     $000010    $000010    $000010    $000010
X:$00F8     $000010    $000010    $000010    $000000
X:$00FC     $000000    $000000    $000000    $000000
Y:$00E0     $000010    $000010    $000010    $000010
Y:$00E4     $000010    $000010    $000010    $000010
Y:$00E8     $000010    $000010    $000010    $000010
Y:$00EC     $000010    $000010    $000010    $000010
Y:$00F0     $000010    $000010    $000010    $000010
Y:$00F4     $000010    $000010    $000010    $000010
Y:$00F8     $000010    $000010    $000010    $000000
Y:$00FC     $000000    $000000    $000000    $000000
P:$0003 497800         = MOVE B,L:-(R0)
```

12. TYPE **step<return>** TO SIMULATE THE EXECUTION OF THE NEXT PROGRAM INSTRUCTION AND SHOW THE DISPLAY ENABLED MEMORY LOCATIONS.

    The DSP56001's Program Counter (PC) increments from $3 to $4 and X-Memory and Y-Memory locations $fa have changed from $10 to $0. The display on your screen should be:

```
X:$00E0     $000010    $000010    $000010    $000010
X:$00E4     $000010    $000010    $000010    $000010
X:$00E8     $000010    $000010    $000010    $000010
X:$00EC     $000010    $000010    $000010    $000010
X:$00F0     $000010    $000010    $000010    $000010
X:$00F4     $000010    $000010    $000010    $000010
X:$00F8     $000010    $000010    $000000    $000000
X:$00FC     $000000    $000000    $000000    $000000
Y:$00E0     $000010    $000010    $000010    $000010
Y:$00E4     $000010    $000010    $000010    $000010
Y:$00E8     $000010    $000010    $000010    $000010
Y:$00EC     $000010    $000010    $000010    $000010
Y:$00F0     $000010    $000010    $000010    $000010
Y:$00F4     $000010    $000010    $000010    $000010
Y:$00F8     $000010    $000010    $000000    $000000
Y:$00FC     $000000    $000000    $000000    $000000
P:$0004 220E00         = MOVE R0,A
```

13. TYPE **step 50<return>** TO SIMULATE EXECUTION OF THE NEXT 50 PROGRAM INSTRUCTIONS AND SHOW THE VALUES OF THE DISPLAY ENABLED X-MEMORY AND Y-MEMORY LOCATIONS AFTER THE FIFTIETH INSTRUCTION HAS BEEN EXECUTED.

    The display on your screen should be as that shown below. Note that the values of X-Memory and Y-Memory locations $ee-$f9 have changed from $10 to $0.

```
X:$00E0     $000010     $000010     $000010     $000010
X:$00E4     $000010     $000010     $000010     $000010
X:$00E8     $000010     $000010     $000010     $000010
X:$00EC     $000010     $000010     $000000     $000000
X:$00F0     $000000     $000000     $000000     $000000
X:$00F4     $000000     $000000     $000000     $000000
X:$00F8     $000000     $000000     $000000     $000000
X:$00FC     $000000     $000000     $000000     $000000
Y:$00E0     $000010     $000010     $000010     $000010
Y:$00E4     $000010     $000010     $000010     $000010
Y:$00E8     $000010     $000010     $000010     $000010
Y:$00EC     $000010     $000010     $000000     $000000
Y:$00F0     $000000     $000000     $000000     $000000
Y:$00F4     $000000     $000000     $000000     $000000
Y:$00F8     $000000     $000000     $000000     $000000
Y:$00FC     $000000     $000000     $000000     $000000
P:$0006 0AF0A2 000003 = JNE >$3
```

14. TYPE **display cyc<return>** TO SHOW THE NUMBER OF DSP56001 CLOCK CYCLES WHICH HAVE BEEN SIMULATED.

    The display on your screen should say:

```
cyc=000256
P:$0006 0AF0A2 000003 = JNE >$3
```

15. TYPE **display on cyc<return>** TO ENABLE THE DISPLAY TO SHOW THE NUMBER OF SIMULATED DSP56001 CLOCK CYCLES IN ADDITION TO THE VALUES OF THE DISPLAY ENABLED MEMORY LOCATIONS.
16. TYPE **step 52 cy<return>** TO SIMULATE THE EXECUTION OF THE NEXT 52 DSP56001 CLOCK CYCLES AND SHOW THE VALUES OF THE DISPLAY ENABLED ITEMS AFTER THE FIFTYSECOND CLOCK CYCLE.

    The display on your screen should be as that shown below. Note that the values of X-Memory and Y-Memory locations $ea-$ed have changed from $10 to $0.

```
cyc=000308
X:$00E0     $000010     $000010     $000010     $000010
X:$00E4     $000010     $000010     $000010     $000010
X:$00E8     $000010     $000010     $000000     $000000
X:$00EC     $000000     $000000     $000000     $000000
X:$00F0     $000000     $000000     $000000     $000000
X:$00F4     $000000     $000000     $000000     $000000
X:$00F8     $000000     $000000     $000000     $000000
X:$00FC     $000000     $000000     $000000     $000000
```

```
Y:$00E0    $000010    $000010    $000010    $000010
Y:$00E4    $000010    $000010    $000010    $000010
Y:$00E8    $000010    $000010    $000000    $000000
Y:$00EC    $000000    $000000    $000000    $000000
Y:$00F0    $000000    $000000    $000000    $000000
Y:$00F4    $000000    $000000    $000000    $000000
Y:$00F8    $000000    $000000    $000000    $000000
Y:$00FC    $000000    $000000    $000000    $000000
P:$0004 220E00        = MOVE R0,A
```

17. TYPE **display cyc<return>** TO SHOW THE NUMBER OF DSP56001 CLOCK CYCLES WHICH HAVE BEEN SIMULATED.

    The display on your screen should say:

```
cyc=000308
P:$0004 220E00        = MOVE R0,A
```

18. TYPE **trace 14 cy<return>** TO SIMULATE THE EXECUTION OF THE NEXT 14 DSP56001 CLOCK CYCLES AND SHOW THE DISPLAY ENABLED ITEMS AFTER THE FOURTEENTH CLOCK CYCLE.

    The display on your screen should be as that shown below. Note that the trace count is decremented from 13 to 0 and the values of X-Memory and Y-Memory locations $e9 have changed from $10 to $0.

```
Trace cycle count=0
cyc=000322
X:$00E0    $000010    $000010    $000010    $000010
X:$00E4    $000010    $000010    $000010    $000010
X:$00E8    $000010    $000000    $000000    $000000
X:$00EC    $000000    $000000    $000000    $000000
X:$00F0    $000000    $000000    $000000    $000000
X:$00F4    $000000    $000000    $000000    $000000
X:$00F8    $000000    $000000    $000000    $000000
X:$00FC    $000000    $000000    $000000    $000000
Y:$00E0    $000010    $000010    $000010    $000010
Y:$00E4    $000010    $000010    $000010    $000010
Y:$00E8    $000010    $000000    $000000    $000000
Y:$00EC    $000000    $000000    $000000    $000000
Y:$00F0    $000000    $000000    $000000    $000000
Y:$00F4    $000000    $000000    $000000    $000000
Y:$00F8    $000000    $000000    $000000    $000000
Y:$00FC    $000000    $000000    $000000    $000000
P:$0004 220E00        = MOVE R0,A
```

19. TYPE **quit<return>** TO EXIT THE "SIM56000 SIMULATOR SOFTWARE" PROGRAM.

2.8 DEFINING AND USING MACROS

Programs often repeat a group(s) of instructions. A macro provides a shorthand means through which a group of DSP56000/1 instructions can be specified by a name. Therefore when typing a program, a repeated group of instructions can be compactly specified by its macro name rather than having to retype the instructions themselves.

A macro is defined by: a header, a body, and a terminator. The header assigns a name to the macro and defines dummy arguments, i.e., symbolic names which will be replaced with actual arguments when the macro is called. The body contains a group of DSP56000/1 source instructions. The terminator is the End Macro (ENDM) directive.

A macro is specified within a program by a macro call statement. The macro call statement has three fields: a label field, an operation field, and an operand field. The label field, if used, corresponds to the value of the software location counter at the beginning of the macro expansion, i.e., substitution of the related instruction group in place of the macro call statement. The operation field contains the name of the macro. The operand field contains actual arguments which will be substituted in place of the dummy arguments used in the header part of the macro's definition.

1. If you want to use the DSP56000/1 Demonstration Software program to cover the topics of Sections 2.8, load the software by following Steps 1 and 6 of Chapter 1 Section 1.5.1 or Steps 1-3 of Chapter 1 Section 1.5.2. Then select from the demonstration software's menus to cover the topics of Section 2.8.
2. If you do *not* want to use the DSP56000/1 Demonstration Software program, proceed to Section 2.8.1.

2.8.1 Programming a Canned DSP56000/1 Macro

DSP56000/1 programs can be entered and edited by using any available word processor program. Here, the DOS line editor **edlin** is used to enter, display, and save a canned DSP56000/1 macro.

1. BOOT THE PC.

   If you are using a PC with two diskette drives and no hard disk drive, proceed to Step 2.

   If you are using a PC with a hard disk drive and one or more diskette drives, skip Step 2 and proceed to Step 3.
2. PLACE THE "DOS DISKETTE" INTO DRIVE "B" AND A FORMATTED DISKETTE INTO DRIVE "A".

   TYPE **b:edlin a:\ma1.asm<return>** TO PLACE THE FILENAME **ma1.asm** ONTO THE DISKETTE IN DRIVE "A".

   The message and prompt shown below should be on your screen. Proceed to Step 4.

```
New file
*_
```

3. PLACE A FORMATTED DISKETTE INTO DRIVE "A" AND TYPE **c:\dos\edlin a:\ma1.asm<return>** TO PLACE THE FILENAME **ma1.asm** ONTO THE DISKETTE IN DRIVE "A".

The message and prompt shown below should be on your screen. Proceed to Step 4.

```
New file
*_
```

4. TYPE **I<return>** TO ENTER THE "INSERT MODE" OF THE "EDLIN" PROGRAM.
5. TYPE **MA1<tab>MACRO<tab>VALUE<return>** TO DEFINE THE MACRO'S HEADER.

   This header assigns the name **MA1** and the dummy argument **VALUE** to the macro.
6. TYPE:

   ;MACRO TO CLEAR THE RAM IN X AND Y MEMORY<return>
   <tab>MOVE<tab>#VALUE,R0<tab>; POINTER TO LONG MEMORY<return>
   <tab>CLR<tab>B<tab>;USED TO CLEAR LONG MEMORY<return>
   LOOP<tab>MOVE<tab>B,L:-(R0)<tab>;DECREMENT THEN CLEAR ONE LOCATION<return>
   <tab>MOVE<tab>R0,A<tab>;DOES POINTER EQUAL ZERO?<return>
   <tab>TST<tab>A<tab>;TEST POINTER<return>
   <tab>JNE<tab>LOOP<tab>;LOOP IF NOT DONE<return>

   You have just typed the body of the macro **MA1. MA1** clears the DSP56000/1 processor's on-chip X-Memory and Y-Memory. For now, don't concern yourself with the instructions themselves. The DSP56000/1 processor's instruction set will be covered in later chapters.
7. TYPE **<tab>ENDM<return>^Z<return>** TO ENTER THE MACRO'S TERMINATOR AND EXIT THE INSERT MODE OF "EDLIN".
8. TYPE **E<return>** TO SAVE **ma1.asm** AT THE DESTINATION NAMED IN STEP 2 OR 3.

   A copy of **ma1.asm** can be found on the **DSP56000/1 Demonstration Software #1** diskette.
9. DEPENDING ON YOUR SYSTEM'S CONFIGURATION, TYPE EITHER **b:edlin a:prog3.asm<return>** OR **c:\dos\edlin a:prog3.asm<return>** TO PLACE THE FILENAME **prog3.asm** ONTO THE DISKETTE IN DRIVE "A".

   The following message and prompt should be on your screen:

```
New file
*_
```

10. TYPE **<tab>INCLUDE<tab>'MA1'<return>** TO "INCLUDE" THE **ma1** MACRO INTO **prog3.asm.**

    Assembler directives are instructions to the ASM56000 Assembler Software program itself. When writing programs, the ASM56000 Assembler Software directives can be used for storage allocation, symbol and data definition, assembly control, output listing control, object file control, macros and conditional assembly, and structured programming. One such assembler directive is named **INCLUDE** and is used as shown above. The subset of ASM56000 assembler directives which is used in this book is:

Directive	*Description*
DC	Define constant
DS	Define Storage
DSM	Define modulo storage
DSR	Define reverse carry storage
DUP	Duplicate a sequence of source lines
END	End of source program
EQU	Equate symbol to a value
INCLUDE	Include secondary file
MACRO	Macro Definition
ORG	Initialize memory space and location counters
SET	Set symbol to a value

EXAMPLES

1. The assembler directive **ORG P:$100** sets the runtime memory space to the P-Memory starting at location $100.

2. The assembler directive **COUNT EQU $12** assigns the value $12 to the symbol COUNT.

3. The assembler directives **ORG X:$100** and **TABLE DC 0.1, 0.2,0.3** store the decimal values 0.1, 0.2, and 0.3 in consecutive X-Memory locations beginning at location $100.

4. The assembler directives **ORG X:$100** and **STATES DS 6** reserve for STATES variables six consecutive X-Memory locations beginning at location $100.

11. TYPE **VALUE<tab>EQU<tab>$100<return>** TO ASSIGN THE VALUE $100 TO THE SYMBOL "VALUE".

    Note that the EQU assembler directive is used to assign the value $100 to the symbol VALUE.

12. TYPE **<tab>MA1<tab>VALUE<return>** TO ENTER THE MACRO CALL STATEMENT WHICH INVOKES THE **ma1** MACRO.

13. TYPE **<tab>END<return>^Z** TO INDICATE THE LOGICAL END OF **prog3.asm**.

    Note the use of the assembler END directive.

14. TYPE **E<return>** TO SAVE **prog3.asm** AT THE DESTINATION NAMED IN STEP 9.

A copy of **prog3.asm** can be found on the **DSP56000/1 Demonstration Software #1** diskette.

2.8.2 Assembling a Canned DSP56000/1 Macro

1. BOOT THE PC.

   If you are using a PC with two diskette drives and no hard disk drive, proceed to Step 2.

   If you are using a PC with a hard disk drive and one or more diskette drives, skip Steps 2 and 3, and proceed to Step 4.

2. PLACE THE "DSP56000 MACRO CROSS ASSEMBLER" DISKETTE CONTAINING THE "ADS56000 ASSEMBLER SOFTWARE" PROGRAM INTO DRIVE "B".

   Proceed to Step 3.

3. TYPE **b:asm56000 -ba:prog3.lnk -ia:prog3.asm<return>**
   TO ASSEMBLE **prog3.asm** AND OUTPUT THE RELOCATABLE LINK FILE **prog3.lnk** TO DRIVE "A".

   The **-I** option above directs the ASM56000 Assembler Software program to search the **a:** directory for the **include** file **ma1.asm.**

   Proceed to Section 2.8.3.

4. TYPE **cd c:\dsptools\asm56000<return>** TO ENTER THIS DIRECTORY.

5. TYPE **c:asm56000 -ba:prog3.lnk -ia:prog3.asm<return>**
   TO ASSEMBLE **prog3.asm** AND OUTPUT THE RELOCATABLE LINK FILE **prog3.lnk** TO DRIVE "A".

   The **-I** option above directs the ASM56000 Assembler Software program to search the **a:** directory for the **include** file **ma1.asm.**

   Proceed to Section 2.8.3.

2.8.3 Linking a Canned DSP56000/1 Macro

1. BOOT THE PC.

   If you are using a PC with two diskette drives and no hard disk drive, proceed to Step 2.

   If you are using a PC with a hard disk drive and one or more diskette drives, skip Steps 2 and 3, and proceed to Step 4.

2. PLACE THE "DSP56000 MACRO CROSS ASSEMBLER" DISKETTE CONTAINING THE "LNK56000 LINKER/LIBRARIAN SOFTWARE" PROGRAM INTO DRIVE "B".

   Proceed to Step 3.

3. TYPE **b:lnk56000 -ba:prog3.lod a:prog3.lnk<return>**
   TO LINK **prog3.lnk** AND OUTPUT THE ABSOLUTE LOAD FILE **prog3.lod** TO DRIVE "A".

   Proceed to Section 2.8.4.

4. TYPE **cd c:\dsptools\lnk56000<return>** TO ENTER THIS DIRECTORY.

5. TYPE **c:lnk56000 -ba:prog3.lod a:prog3.lnk<return>**
   TO LINK **prog3.lnk** AND OUTPUT THE ABSOLUTE LOAD FILE **prog3.lod** TO DRIVE "A".

   Proceed to Section 2.8.4.

2.8.4 Loading a Canned DSP56000/1 Macro

1. BOOT THE PC.

   If you are using a PC with two diskette drives and no hard disk drive, proceed to Step 2.

   If you are using a PC with a hard disk drive and one or more diskette drives, skip Steps 2 and 3, and proceed to Step 4.
2. PLACE THE "DSP56000 SIMULATOR" DISKETTE (1 OF 3) CONTAINING THE "SIM56000 SIMULATOR SOFTWARE" PROGRAM INTO DRIVE "B".
3. TYPE **b:sim56000<return>**.

   The display on your screen should be as shown below. Proceed to Step 6.

```
0>
asm break change copy disassemble display <space>=more
```

4. TYPE **cd c:\dsptools\sim56000<return>** TO ENTER THIS DIRECTORY.
5. TYPE **sim56000<return>** TO ENTER THE "SIM56000 SIMULATOR SOFTWARE" PROGRAM.

   The display on your screen should be as shown in Step 3. Proceed to Step 6.
6. PLACE THE DISKETTE CONTAINING **prog3.lod** INTO DRIVE "A".
7. TYPE **load a:prog3.lod<return>** TO LOAD **prog3.lod** FROM THE DISKETTE IN DRIVE "A" INTO THE "SIM56000 SIMULATOR SOFTWARE" PROGRAM ENVIRONMENT.
8. TYPE **disassemble p:0..10<return>** TO DISASSEMBLE AND DISPLAY THE CONTENTS OF P-MEMORY LOCATIONS 0-10 ($0-$a).

   The display on your screen should be as that shown below. Note its similarity to the display in Step 4 of Section 2.7.2.

```
P:$0000 60F400 000100 = MOVE #>$100,R0
P:$0002 20001B        = CLR B
P:$0003 497800        = MOVE B,L:-(R0)
P:$0004 220E00        = MOVE R0,A
P:$0005 200003        = TST A
P:$0006 0AF0A2 000003 = JNE >3
P:$0008 000000        = NOP
P:$0009 000000        = NOP
P:$000A 000000        = NOP
```

9. TYPE **quit<return>** TO EXIT THE "SIM56000 SIMULATOR SOFTWARE" PROGRAM.

For more information on writing DSP56000/1 assembly language programs, refer to Motorola's *DSP56000 Macro Assembler Reference Manual* and *DSP56000 Linker/Librarian Reference Manual.* For more information on the SIM56000 Simulator Software program, refer to Motorola's *DSP56000/1 Digital Signal Processor Simulator Reference Manual.*

3

DSP56001 ARCHITECTURE AND ADDRESSING MODES

In this chapter, the DSP56001 processor's architecture, addressing modes, and some of its machine instructions are introduced. The chapter begins with a review of the DSP56001 processor's architecture. Its three concurrent processing execution units, the Data Arithmetic/Logic Unit, the Program Controller, and the Address Generation Unit, are then described. The DSP56001 processor's addressing modes, Register Direct, Special, and Register Indirect, are also discussed.

The *DSP56000/1 Demonstration Software* program can be used to cover the topics of Sections 3.3 and 3.6. A detailed description of the DSP56001 processor's instruction set will be covered in Chapter 4.

3.1 DSP56001 ARCHITECTURE REVIEW:

The major components of the DSP56001 processor's architecture are shown in Figure 1.1.b of Chapter 1 and are listed below.

1. ***Buses:***

   The internal multiple-bus architecture of the DSP56001 processor consists of four bidirectional 24-bit data buses and three unidirectional 16-bit address buses.

   Data Buses: The data buses include: the XD Data Bus, the YD Data Bus, the PD Program Data Bus, and the GD Global Data Bus. The XD Data Bus and YD Data Bus transfer data between the Data ALU and the X Data Memory and Y Data Memory, respectively. The XD Data Bus and YD Data Bus can also be concatenated, by certain instructions, to function as one 48-bit bus. The PD Program Data Bus transfers pre-fetched instruction words. The GD Global Data Bus handles "other" data transfers such as I/O transfers to and from peripheral devices.

Address Buses: The address buses include: the XA Address Bus, the YA Address Bus, and the PA Program Address Bus. The XA Address Bus and YA Address Bus provide address data which points to specific locations in the X Data Memory and Y Data Memory, respectively. The PA Program Address Bus provides address data which points to a specific location in the P Program Memory.

2. ***Internal Memories:***

   The DSP56001 processor has six on-chip memories: the X Data RAM, the X Data ROM, the Y Data RAM, the Y Data ROM, the P Program Memory RAM, and the Bootstrap ROM. The X Data RAM and Y Data RAM are 24-bit internal memories which occupy the lowest 256 locations in the X-Memory address space and the Y-Memory address space, respectively. The X Data ROM and Y Data ROM are 24-bit internal memories which, when enabled by the Operating Mode Register, occupy the next lowest 256 locations in the X-Memory address space and Y-Memory address space, respectively. The P Program RAM is a 24-bit internal memory which occupies the lowest 512 locations in the P-Memory address space. The P Program RAM contains machine instructions, constants, and data tables which are fixed at assembly time. Unused locations in the P Program RAM, X Data RAM, and Y Data RAM can be used for general purpose storage. For example, unused locations in the X Data Memory and Y Data Memory could be used to store program overlays for the P Program Memory. The Bootstrap ROM is a 32 location by 24-bit factory programmed ROM which is used only in the bootstrap mode. The bootstrap mode provides a convenient, low cost, automatic means through which, after power-up or reset, a program of DSP56001 machine instructions can be transferred from an external memory to the internal P Program Memory.

3. ***Execution Units:***

   The core of the DSP56001 processor consists of three execution units which operate concurrently: the Data Arithmetic/Logic Unit, the Program Controller, and the Address Generation Unit. The execution units are described in Sections 3.3, 3.4 and 3.5, respectively.

4. ***On-chip Peripherals:***

   The DSP56001 processor's on-chip peripherals provide a microcomputer-style I/O capability. The on-chip peripherals include: a parallel Host MPU/DMA port, an asynchronous Serial Communication Interface (SCI) port, a Synchronous Serial Interface (SSI) port, and a Programmable I/O port. The 8-bit parallel Host MPU/DMA port provides a means through which a host processor or DMA controller can access the DSP56001 processor. The SCI port generally provides a standard serial I/O interface, using asynchronous character protocols, for communication with MCUs, other DSP chips, and RS232C-type lines. The SSI port generally provides a high speed synchronous serial data interface for communication with other DSPs or other synchronous serial devices. The Programmable I/O port provides a general purpose interface.

5. ***Memory Expansion Port :***

   The Memory Expansion port is composed of: a 16-bit address bus, a 24-bit bidirectional data bus, a set of memory bank selection signals, and a set of memory handshaking signals. It is used to access external X Data Memory, Y Data Memory, Program Memory, and I/O devices.

3.2 HOW TO PROCEED:

If you want to use the *DSP56000/1 Demonstration Software* program to cover the topics of Sections 3.3 and 3.6, load the software by following steps 1 and 6 of Chapter 1 Section 1.5.1 or steps 1-3 of Chapter 1 Section 1.5.2. Then select from the demonstration software's menus to cover the topics of Sections 3.3 and 3.6.

If you do *not* want to use the *DSP56000/1 Demonstration Software* program, load the SIM56000 Simulator Software by following steps 1, 2, and 3 of Chapter 2 Section 2.3.1 or steps 1, 5, and 6 of Chapter 2 Section 2.3.2. Proceed to Section 3.3.

3.3 DATA ALU EXECUTION UNIT:

The Data ALU execution unit, shown in Figure 3.1, performs arithmetic and logical operations on data operands. The major components of the Data ALU include: ten data registers, a 24x24 multiplier and arithmetic/logic unit, a bit manipulation unit, an accumulator shifter, and two shifter/limiters.

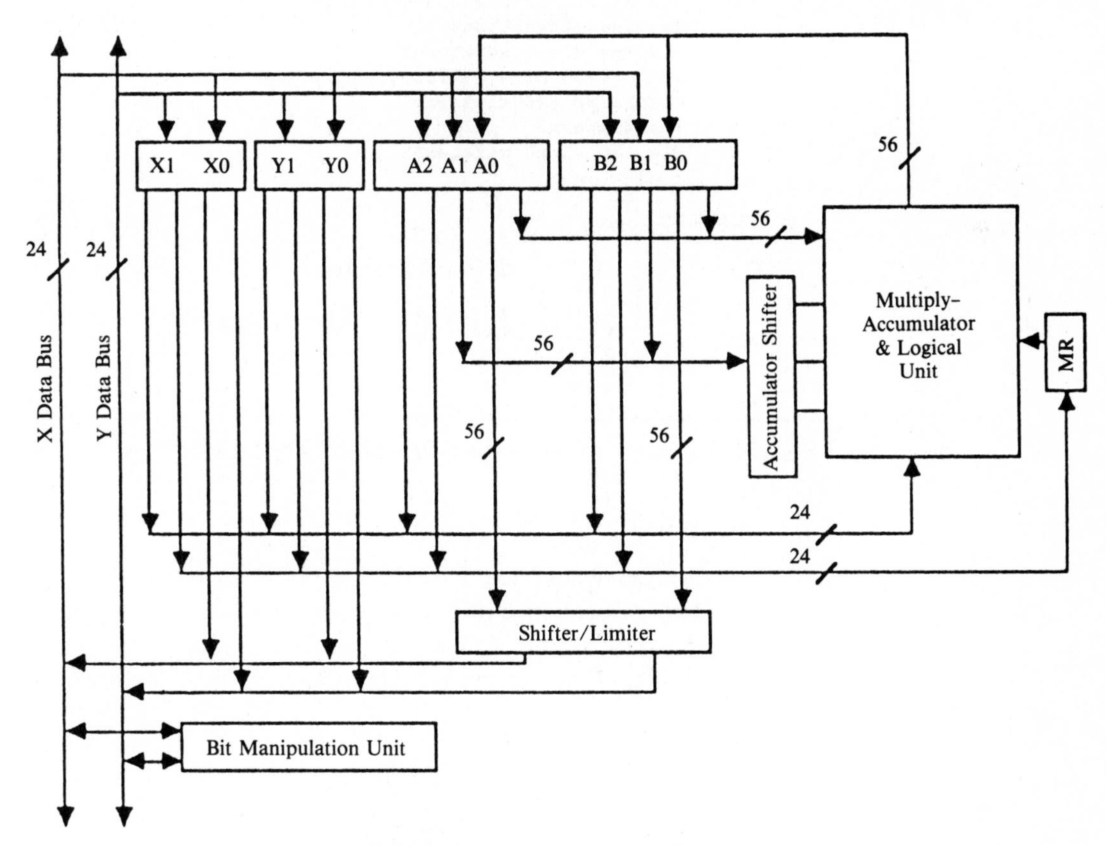

Figure 3-1 Data ALU Block Diagram

3.3.1 Data Registers:

The Data ALU has four input data registers and six data accumulator registers as shown in Figure 3.2.

1) Data Input Registers: X1,X0; Y1,Y0

   The Data Input Registers can be treated as four independent 24-bit registers, X1, X0, Y1, and Y0, or as two 48-bit registers, X and Y. The 48-bit X and Y registers are developed by the concatenation of X1:X0 and Y1:Y0, respectively.

EXAMPLE 3.1: Moving Data to a 24-bit Data Input Register.

PROBLEM:

Execute the instruction **MOVE #$111100,X1** to move the value $111100 to the X1 Data Input Register. Assume that the contents of the X1, X0, and X data registers before execution of the instruction are:

x = $333333333333

*x*1 = $333333   *x*0 = $333333

PROCEDURE:

```
TYPE: change x $333333333333<return>
      display off<return>
      display x1 x0 x<return>
      asm p:$200 move #$111100,x1<return>
      change pc $200<return>
      step<return>
      display x1 x0 x<return>
```

RESULTS:

x = $111100333333

*x*1 = $111100   *x*0 = $333333

EXAMPLE 3.2: Moving Data to a 48-bit Data Input Register.

PROBLEM:

Execute the instruction **MOVE L:$100,X** to move the contents of the L-Memory location $100 (concatenation of X-Memory and Y-Memory locations $100) to the X Data Input Register. Assume that the contents of the X Data Input Register and L-Memory location $100 before execution of the instruction are:

l:$100 $333333444444

x = $111111222222

Data ALU Input Registers

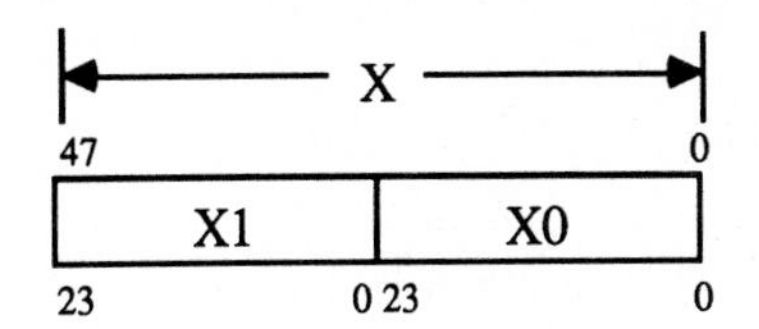

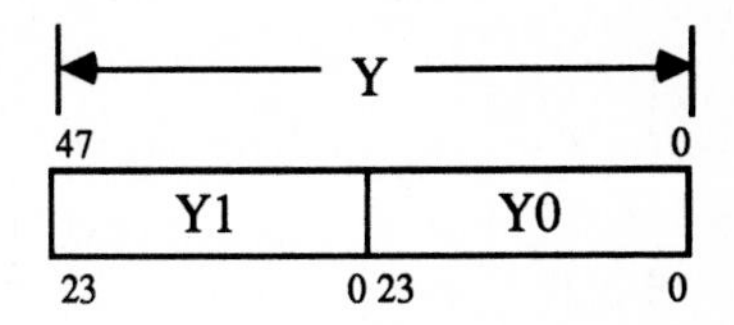

Data ALU Accumulators

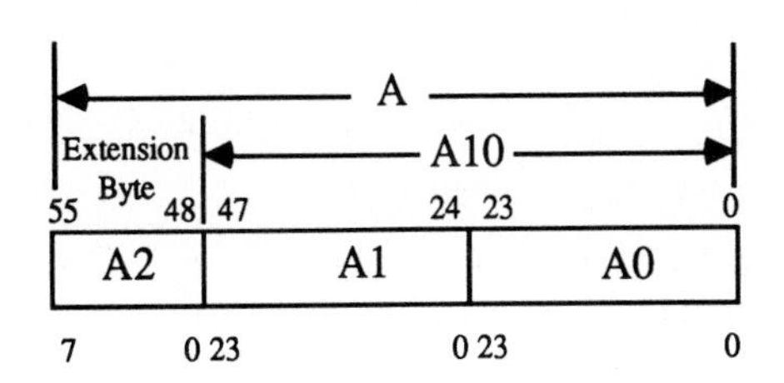

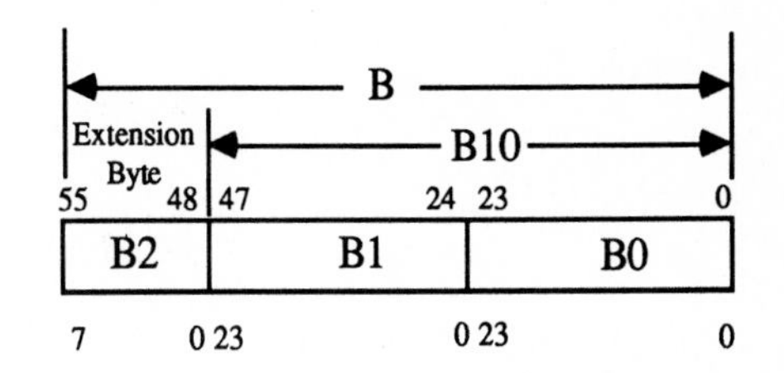

Figure 3-2 Data Registers

PROCEDURE:

```
TYPE: change l:$100 $333333444444 x $111111222222<return>
      display l:$100 x<return>
      asm p:$202 move l:$100,x<return>
      change pc $202<return>
      step<return>
      display l:$100 x<return>
```

RESULTS:

l:$100 $333333444444

x = $333333444444

---

2) Data Accumulator Registers: A2,A1,A0; B2,B1,B0

The six data accumulator registers, A2, A1, A0, B2, B1, and B0, form two general purpose 56-bit accumulators, A and B. The A-Accumulator is the concatenation of registers A2:A1:A0; the B-Accumulator is the concatenation of registers B2:B1:B0. Registers A1, A0, B1, and B0 are each 24-bits wide; registers A2 and B2 are each 8-bits wide. The two accumulators can be viewed as being 48-bits wide with 8-bit extensions. As a result, A2 and B2 are called extension registers.

EXAMPLE 3.3: Moving a 24-Bit Operand to a 56-bit Accumulator.

PROBLEM:

Execute the two instructions **MOVE X0,A** and **MOVE X1,B** to demonstrate that automatic sign-bit extension occurs when moving a 24-bit data operand to the 56-bit A-Accumulator and when moving a 24-bit data operand to the 56-bit B-Accumulator. Assume that the contents of the A-Accumulator and B-Accumulator, and the X0 and X1 Data Input Registers, before execution of the two instructions are:

$$x0 = \$888888$$

$$x1 = \$111111$$

$$a = \$00222222222222$$

$$a2 = \$00 \quad a1 = \$222222 \quad a0 = \$222222$$

$$b = \$00222222222222$$

$$b2 = \$00 \quad b1 = \$222222 \quad b0 = \$222222$$

PROCEDURE:

```
TYPE: change x0 $888888 x1 $111111<return>
      change a $00222222222222<return>
      change b $00222222222222<return>
      display a a2 a1 a0 b b2 b1 b0 x0 x1<return>
      asm p:$204 move x0,a<return>
      asm p:$206 move x1,b<return>
      change pc $204<return>
      step 4<return>
      display a a2 a1 a0 b b2 b1 b0 x0 x1<return>
```

RESULTS:

$$x0 = \$888888$$

$$x1 = \$111111$$

$$a = \$ff888888000000$$

$$a2 = \$ff \quad a1 = \$888888 \quad a0 = \$000000$$

$$b = \$00111111000000$$

$$b2 = \$00 \quad b1 = \$111111 \quad b0 = \$000000$$

Note that the A2 and B2 Accumulator Registers are sign-extended from A1 and B1, respectively, and that the A0 and B0 Accumulator Registers are automatically zeroed.

EXAMPLE 3.4: Moving a 16-bit Operand to a 56-bit Accumulator.

PROBLEM:

Execute the instruction **MOVE R0,A** to move the contents of the 16-bit R0 address register to the 56-bit A-Accumulator. Assume that the contents of Address Register R0 and the A-Accumulator before execution of the instruction are:

$$r0 = \$8888 \qquad a = \$00111111111111$$

PROCEDURE:

```
TYPE:  change r0 $8888 a $00111111111111<return>
       display r0 a<return>
       asm p:$208 move r0,a<return>
       change pc $208<return>
       step<return>
       display r0 a<return>
```

RESULTS:

$$r0 = \$8888 \qquad a = \$00008888000000$$

Note that A-Accumulator Register A0 is zeroed and that sign-extension did not occur.

EXAMPLE 3.5: Moving a 56-bit A-Accumulator Operand to both a 16-Bit Destination and a 24-Bit Destination.

PROBLEM:

Execute the two instructions **MOVE A,R2** and **MOVE A,X1** to move the contents of the 56-bit A-Accumulator to the 16-bit R2 Address Register and the 24-bit X1 Data Input Register, respectively. Assume that the contents of the A-Accumulator, the R2 Address Register, and the X1 Data Input Register before execution of the two instructions are:

$$a = \$00123456789abc$$

$$r2 = \$0000$$

$$x1 = \$000000$$

PROCEDURE:

```
TYPE:  change a $00123456789abc r2 $0000<return>
       change x1 $000000<return>
       display a r2 x1<return>
       asm p:$209 move a,r2<return>
       asm p:$20a move a,x1<return>
       change pc $209<return>
       step 2<return>
       display a r2 x1<return>
```

RESULTS:

$$a = \$00123456789abc$$

$$r2 = \$3456$$

$$x1 = \$123456$$

EXAMPLE 3.6: Moving a Data Operand to an Accumulator without Automatic Sign-Extension.

PROBLEM:

Execute the instruction **MOVE #$111111,A1** to show that by specifying the Destination Accumulator Register, A1, A0, B1, or B0, a 24-bit data operand can be moved to an accumulator *without* automatic sign-extension and zero filling. Assume that the contents of the A, A2, A1 and A0 Accumulator Registers before execution of the instruction are:

a = \$ff888888222222

$a2$ = \$ff $a1$ = \$888888 $a0$ = \$222222

PROCEDURE:

```
TYPE: change a $ff888888222222<return>
      display a a2 a1 a0<return>
      asm p:$20b move #$111111,a1<return>
      change pc $20b<return>
      step<return>
      display a a2 a1 a0<return>
```

RESULTS:

a = \$ff111111222222

$a2$ = \$ff $a1$ = \$111111 $a0$ = \$222222

3.3.2 Multiplier and Arithmetic/Logic Unit:

The Multiplier and Arithmetic/Logic Unit performs all of the DSP56001 processor's data operand calculations such as multiplication, addition, subtraction, AND, OR, XOR, and NOT. It accepts up to three input data operands and produces one 56-bit result, where the result is always stored in either the A-Accumulator or the B-Accumulator.

EXAMPLE 3.7: Addition

PROBLEM:

Execute the instruction **ADD A,B** to add the A-Accumulator to the B-Accumulator and store the result in the B-Accumulator. Assume that the contents of the A-Accumulator and the B-Accumulator before execution of the instruction are:

a = \$00111111111111 b = \$00222222222222

PROCEDURE:

```
TYPE: change a $00111111111111 b $00222222222222<return>
      display a b<return>
      asm p:$210 add a,b<return>
      change pc $210<return>
      step<return>
      display a b<return>
```

RESULTS:

$$a = \$00111111111111 \qquad b = \$00333333333333$$

EXAMPLE 3.8: Subtraction

PROBLEM:

Execute the instruction **SUB X,A** to subtract the 48-bit operand of the X Data Input Register from the A-Accumulator and store the result in the A-Accumulator. Assume that the contents of the X Data Input Register and the A-Accumulator before execution of the instruction are:

$$x = \$111111111111 \qquad a = \$00333333333333$$

PROCEDURE:

```
TYPE: change x $111111111111 a $00333333333333<return>
      display x a<return>
      asm p:$211 sub x,a<return>
      change pc $211<return>
      step<return>
      display x a<return>
```

RESULTS:

$$x = \$111111111111 \qquad a = \$00222222222222$$

EXAMPLE 3.9 Logical AND

PROBLEM:

Execute the instruction **AND X1,A** to logically AND the 24-bit operand of the X1 Data Input Register with the A-Accumulator and store the result in the A-Accumulator. Assume that the contents of the X1 Data Input Register and the A-Accumulator before execution of the instruction are:

$$x1 = \$f0f0f0 \qquad a = \$ff888888888888$$

PROCEDURE:

```
TYPE: change x1 $f0f0f0  a $ff888888888888<return>
      display x1 a<return>
      asm p:$212 and x1,a<return>
      change pc $212<return>
      step<return>
      display x1 a<return>
```

RESULTS:

$x1 = \$f0f0f0 \quad a = \$ff808080888888$

3.3.2.1 Two's-Complement Fractional Data Representation:

The Multiplier and Arithmetic/Logic Unit uses the two's-complement fractional data representation commonly used in DSP algorithms, where a fraction is any number whose magnitude is greater than or equal to zero and less than one. Figure 3.3 shows the bit-weighting of fractional data operands.

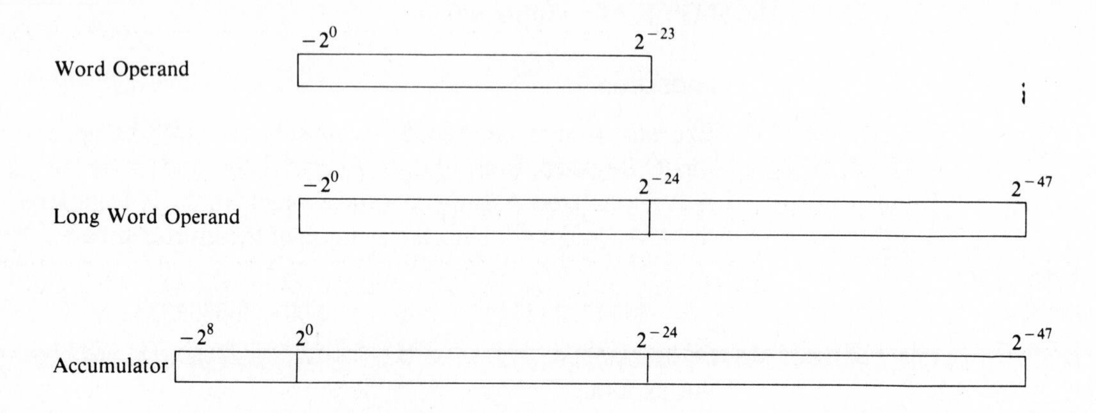

Figure 3-3 Bit Weighting of Operands

EXAMPLE 3.10: Positive Two's-Complement Fractional Data.

PROBLEM:

Execute the instruction **MOVE #0.875,Y0.**

PROCEDURE:

```
TYPE: asm p:$214 move #0.875,y0<return>
      change pc $214<return>
      step<return>
      display y0<return>
```

RESULTS:

$$y0 = \$700000 = \%011100000000000000000000$$

$$= 0*2^{-0} + 1*2^{-1} + 1*2^{-2} + 1*2^{-3} = \text{‘}0.875$$

EXAMPLE 3.11: Negative Two's-Complement Fractional Data.

PROBLEM:

Execute the instruction **MOVE #-0.875,X0.**

PROCEDURE:

```
TYPE: asm p:$216 move #-0.875,x0<return>
      change pc $216<return>
      step<return>
      display x0<return>
```

RESULTS:

$$x0 = \$900000 = \%100100000000000000000000$$

$$= \text{`}-0.875$$

To convert the above two's-complement binary number 100100...00 to decimal, each bit of the number must be complemented to form the one's-complement number 011011...11. Then a one is added to the one's-complement number to form the binary number 011100..00, which is equal to the decimal number 0.875. The sign of the decimal number 0.875 is negative because the sign-bit of its equivalent two's-complement binary number is a one.

3.3.2.2 Rounding:

The Multiplier and Arithmetic/Logic Unit can convergently round the least significant portion of an accumulator, A0 or B0, into the most significant portion, A1 or B1.

EXAMPLE 3.12: Value of A-Accumulator Register A0 is Greater Than 0.5.

PROBLEM:

Execute the instruction **RND A** to show that a one is added to A-Accumulator Register A1 when the value of A-Accumulator Register A0 is greater than 0.5 ($800000). Assume that the contents of the A1 and A0 Accumulator Registers, and the SR Status Register before execution of the instruction are:

$$sr = \$0300$$

$$a1 = \$111111 \qquad a0 = \$808080$$

PROCEDURE:

```
TYPE: change a1 $111111    a0 $808080<return>
      change sr $300
      display a1 a0 sr<return>
      asm p:$220 rnd a<return>
      change pc $220<return>
      step<return>
      display a1 a0 sr<return>
```

RESULTS:

$$sr = \$0310$$

$$a1 = \$111112 \qquad a0 = \$000000$$

EXAMPLE 3.13: Value of B-Accumulator Register B0 is Less Than 0.5.

PROBLEM:

Execute the instruction **RND B** to show that the B-Accumulator Register B1 is not changed when the value of the B-Accumulator Register B0 is less than 0.5 ($800000). Assume that the contents of the B1 and B0 accumulator registers before execution of the instruction are:

$$b1 = \$111111 \qquad b0 = \$777777$$

PROCEDURE:

```
TYPE: change b1 $111111  b0 $777777<return>
      display b1 b0<return>
      asm p:$221 rnd b<return>
      change pc $221<return>
      step<return>
      display b1 b0<return>
```

RESULTS:

$$b1 = \$111111 \qquad b0 = \$000000$$

EXAMPLE 3.14: The Values of A- and B-Accumulator Registers A0 and B0 are both 0.5.

PROBLEM:

Execute the two instructions **RND A** and **RND B** to show that:

1. a one is added to the A-Accumulator Register A1 when the least significant bit of A1 is equal to one and the value of A0 is equal to 0.5 ($800000), and
2. B-Accumulator Register B1 is unchanged when the least significant bit of B1 is equal to zero and the value of B0 is equal to 0.5 ($800000).

Assume that the contents of the A1, A0, B1, and B0 Accumulator Registers before execution of the two instructions are:

$$a1 = \$111111 \qquad a0 = \$800000$$

$$b1 = \$101010 \qquad b0 = \$800000$$

PROCEDURE:

```
TYPE: change a1 $111111  a0 $800000<return>
      change b1 $101010  b0 $800000<return>
      display a1 a0 b1 b0<return>
      asm p:$222 rnd a<return>
      asm p:$223 rnd b<return>
      change pc $222<return>
      step 2<return>
      display a1 a0 b1 b0<return>
```

RESULTS:

$$a1 = \$111112 \qquad a0 = \$000000$$

$$b1 = \$101010 \qquad b0 = \$000000$$

EXAMPLE 3.15: Multiplication With and Without Rounding.

PROBLEM:

Execute the instructions **MPYR X1,X0,A** and **MPY X1,X0,B** to:

1. multiply the value of the X1 Input Data Register by the value of the X0 Input Data Register and store the convergently rounded result in the A-Accumulator, and
2. multiply the value of the X1 Input Data Register by the value of the X0 Input Data Registers and store the non-rounded result in the B-Accumulator.

Assume that the contents of the X1 and X0 Data Registers and the A and B Accumulator before execution of the two instructions are:

$x1$ = \$400000 $x0$ = \$000007

a = \$00000000000000

b = \$00000000000000

PROCEDURE:

```
TYPE:  change a 0 b 0<return>
       change x1 $400000 x0 $000007<return>
       display x1 x0 a b<return>
       asm p:$224 mpyr x1,x0,a<return>
       asm p:$225 mpy x1,x0,b<return>
       change pc $224<return>
       step 2<return>
       display x1 x0 a b<return>
```

RESULTS:

$x1$ = \$400000 $x0$ = \$000007

a = \$00000004000000

b = \$00000003800000

Note that the contents of the A-Accumulator are convergently rounded and that the contents of the B-Accumulator are not.

3.3.3 Accumulator Shifter:

The Accumulator Shifter accepts a 56-bit input and produces a 56-bit output. The Accumulator Shifter either shifts the input data operand one bit position to the left, one bit position to the right, or passes it unshifted.

EXAMPLE 3.16: Left Shifting.

PROBLEM:

Execute the instruction **ASL A** to arithmetically shift the A-Accumulator one bit position to the left. Assume that the contents of the A-Accumulator and the Status Register (SR) before execution of the instruction are:

$$a = \$ff888888333333 = \text{`}-0.933333373069765$$

$$sr = \$0300$$

PROCEDURE:

```
TYPE: change a $ff888888333333 sr $0300<return>
      display a sr<return>
      radix f a<return>
      display a<return>
      radix h a<return>
      asm p:$230 asl a<return>
      change pc $230<return>
      step<return>
      display a sr<return>
      radix f a<return>
      display a<return>
      radix h a<return>
```

RESULTS:

$$a = \$ff111110666666 = \text{`}-1.866666746139529$$

$$sr = \$0339$$

Note that the C-Bit of the Status Register (least significant bit) receives the most significant bit shifted out of the A-Accumulator and that a zero is shifted into the least significant bit of the A-Accumulator. Also note that the contents of the A-Accumulator are scaled up by a factor of two.

EXAMPLE 3.17: Right Shifting.

PROBLEM:

Execute the instruction **ASR B** to arithmetically shift the B-Accumulator one bit position to the right. Assume that the contents of the B-Accumulator and the Status Register (SR) before execution of the instruction are:

$$b = \$ff888888333333 = \text{`}-0.933333373069765$$

$$sr = \$0300$$

PROCEDURE:

```
TYPE: change b $ff888888333333   sr $0300<return>
      display b sr<return>
      radix f b<return>
      display b<return>
      radix h b<return>
      asm p:$231 asr b<return>
      change pc $231<return>
      step<return>
      display b sr<return>
      radix f b<return>
      display b<return>
      radix h b<return>
```

RESULTS:

b = \$ffc44444199999 = '−0.466666686534886

sr = \$0319

Note that the C-Bit of the Status Register (least significant bit) receives the least significant bit shifted out of the B-Accumulator and that the most significant bit of the B-Accumulator is held constant. Also note that the contents of the B-Accumulator are scaled down by a factor of two.

3.3.4 Data Shifter/Limiters:

Two Data Shifters/Limiters provide special post processing on data which is moved from the accumulators to the XD or YD data buses. One 56-bit Shifter/Limiter is associated with the A-Accumulator; the other is associated with the B-Accumulator. Each shifter/limiter consists of a shifter followed by a limiter.

3.3.4.1 Data Shifters:

Each of two Data Shifters are capable of shifting a data operand one bit position to the left, one bit position to the right, or passing the data unshifted. One Data Shifter is associated with the A-Accumulator; the other is associated with the B-Accumulator. The output of each Data Shifter passes through an associated Data Limiter, where the Data Limiter has a 24-bit output and an overflow output indicator. The Data Shifters are controlled by the S1 and S0 Scaling Bits in the Status Register (SR), where S1 and S0 are the eleventh and tenth bits respectively. S1 and S0 permit dynamic scaling of fixed point data without the need for the user to modify the program's instructions. The scaling modes are shown in Table 3.1.

TABLE 3.1. Scaling Modes.

S1	*S0*	*Scaling Mode*
0	0	No Scaling
0	1	Scale Down (1 bit arithmetic right shift)
1	0	Scale Up (1 bit arithmetic left shift)

EXAMPLE 3.18: No Scaling.

PROBLEM:

Execute the instructions **MPY X1,Y1,A** and **MOVE a,x0** to show that the Data Shifter is controlled by the Scaling Bits (S1 and S0) in the Status Register. Assume that the contents of the X0, X1 and Y1 Data Registers, the A-Accumulator, and the Scaling Bits before execution of the instructions are:

$$x1 = \$400000$$

$$y1 = \$400000$$

$$a = \$00000000000000$$

$$x0 = \$000000$$

$$S1 = 0, S0 = 0\ (noscaling)$$

PROCEDURE:

```
TYPE: change x1 $400000 y1 $400000<return>
      change sr 0 a 0 x0 0<return>
      display x1 y1 sr a x0<return>
      asm p:$235 mpy x1,y1,a<return>
      asm p:$236 move a,x0<return>
      change pc $235<return>
      step 2<return>
      display x1 y1 sr a x0<return>
```

RESULTS:

$$x1 = \$400000$$

$$y1 = \$400000$$

$$sr = \$0010$$

$$a = \$00200000000000$$

$$x0 = \$200000$$

EXAMPLE 3.19: Down Scaling

PROBLEM:

Execute the instructions **MPY X1,Y1,A** and **MOVE a,x0** to show that the Data Shifter is controlled by the Scaling Bits S1 and S0 in the Status Register. Assume that the contents of the X0, X1 and Y1 Data Registers, the A-Accumulator, and the Scaling Bits before execution of the instructions are:

$$x1 = \$400000$$

$$y1 = \$400000$$

$$a = \$00000000000000$$

$$x0 = \$000000$$

$$S1 = 0, S0 = 1 \text{ (down scaling)}$$

PROCEDURE:

```
TYPE: change a 0 pc $235 x0 0<return>
      change sr $0400<return>
      display x0 x1 y1 sr a
      asm p:$235 mpy x1,y1,a
      asm p:$236 move a,x0
      step 2<return>
      display x1 y1 sr a x0<return>
```

RESULTS:

$$x1 = \$400000$$

$$y1 = \$400000$$

$$sr = \$0410$$

$$a = \$00200000000000$$

$$x0 = \$100000$$

Note that the contents of the X0 Data Input Register are down scaled by a factor of two.

EXAMPLE 3.20: Up Scaling

PROBLEM:

Execute the instructions **MPY X1,Y1,A** and **MOVE a,x0** to show that the Data Shifter is controlled by the Scaling Bits S1 and S0 in the Status Register. Assume that the contents of the X0, X1 and Y1 Data Registers, the A-Accumulator, and the Scaling Bits before execution of the instructions are:

$$x1 = \$400000$$

$$y1 = \$400000$$

$$a = \$00000000000000$$

$$x0 = \$000000$$

$$S1 = 1, S0 = 0 \text{ (up scaling)}$$

PROCEDURE:

```
TYPE: change a 0   pc $235 x0<return>
      change sr $0800<return>
      display x0 x1 y1 sr a
      asm p:235 mpy x1,y1,a
      asm p:$236 move a,x0
      step 2<return>
      display x1 y1 sr a x0<return>
```

RESULTS:

$$x1 = \$400000$$

$$y1 = \$400000$$

$$sr = \$0800$$

$$a = \$00200000000000$$

$$x0 = \$400000$$

Note that the contents of the X0 Data Input Register are up scaled by a factor of two.

3.3.4.2 Limiters:

Each of two Data Limiters are capable of automatically performing, as necessary, saturation arithmetic on data operands which are moved from the accumulators to the XD and YD data buses. If the contents of the selected source accumulator can be represented in the destination operand size without overflow, the data limiter is disabled and the operand is not modified. If the contents of the selected source accumulator cannot be represented without overflow in the destination operand size, the data limiter will substitute a "limited" data value having maximum magnitude and the same sign as the source accumulator. The limiting is performed on the output of the associated Data Shifter. The value of the source accumulator is not changed. One Data Limiter is associated with the A-Accumulator; the other Data Limiter is associated with the B-Accumulator.

With two Data Shifter/Limiters, two 24-bit operands can be limited independently in the same instruction cycle. Also, the two Data Shifters/Limiters can be concatenated to form one 48-bit shifter/limiter for long-word operands. Limited data values are shown in Table 3.2.

TABLE 3.2. Limited Data Values.

			Limited Value (Hex)		
Destination Memory Reference	*source operand*	*Acc sign*	*XD Bus*	*YD Bus*	*type of access*
X	X:A X:B	+ -	7FFFFF 800000	---- ----	one 24 bit word
Y	Y:A Y:B	+ -	---- ----	7FFFFF 800000	one 24 bit word
X and Y	X:A Y:A X:A Y:B X:B Y:A X:B Y:B L:AB L:BA	+ -	7FFFFF 800000	7FFFFF 800000	two 24 bit words
L (X:Y)	L:A L:B	+ -	7FFFFF 800000	FFFFFF 000000	one 48 bit long word

EXAMPLE 3.21: Transferring Data with Limiting.

PROBLEM:

Execute the two instructions **MOVE A,X0** and **MOVE B,Y1** to show that limiting occurs automatically when transferring data from the 56-bit A- and B-Accumulators to the 24-bit X0 and Y1 Data Input Registers, respectively. Assume that the contents of the A- and B-Accumulators, and the X0 and Y1 Data Input Registers, before execution of the two instructions are:

$$a = \$0100000f800000$$

$$a2 = \$01 \quad a1 = \$00000f \quad a0 = \$800000$$

$$x0 = \$000000$$

$$b = \$8000000f800004$$

$$b2 = \$80 \quad b1 = \$00000f \quad b0 = \$800004$$

$$y1 = \$000000$$

PROCEDURE:

```
TYPE: change a $0100000f800000 x0 $000000<return>
      change b $8000000f800004 y1 $000000<return>
      display a b x0 y1<return>
      asm p:$237 move a,x0<return>
      asm p:$238 move b,y1<return>
      change pc $237<return>
      step 2<return>
      display a b x0 y1<return>
```

RESULTS:

$$a = \$0100000f800000$$

$$a2 = \$01 \quad a1 = \$00000f \quad a2 = \$800000$$

$$x0 = \$7fffff$$

$$b = \$8000000f800004$$

$$b2 = \$80 \quad b1 = \$00000f \quad a2 = \$800004$$

$$y1 = \$800000$$

Note that the contents of the X0 and Y1 registers are limited to $7FFFFF and $800000, respectively.

EXAMPLE 3.22: Transferring Data without Limiting.

PROBLEM:

Execute the instruction **MOVE A1,X0** to show that limiting is not performed when the A1 (A2 or A0) Accumulator Register is specified as the source for a data move over the XD or YD data bus. Assume that the contents of the A-Accumulator and the X0 Data Input Register before execution of the instruction are:

$$a = \$00800000000000$$

$$a2 = \$00 \quad a1 = \$800000 \quad a0 = \$000000$$

$$x0 = \$000000$$

PROCEDURE:

```
TYPE: change a $00800000000000 x0 $000000<return>
      display a a2 a1 a0 x0<return>
      asm p:$239 move a1,x0<return>
      change pc $239<return>
      step<return>
      display a a2 a1 a0 x0<return>
```

RESULTS:

$$a = \$00800000000000$$

$$a2 = \$00 \quad a1 = \$800000 \quad a0 = \$000000$$

$$x0 = \$800000$$

Note that the error between the contents of the A-Accumulator and the X0 Data Input Register after execution of the MOVE instruction is relatively large [1-(-1) = 2 decimal]. To reduce this error, the contents of the full A-Accumulator, with limiting, should be transferred to the X0 Data Input Register. In that case the value of X0 would be $7FFFFF and the error between the contents of the A-Accumulator and X0 would be relatively small (1 - 0.9999999 = 0.0000001 decimal).

3.3.5 Bit Manipulation Unit:

The Bit Manipulation Unit performs bit manipulation operations on X-Memory or Y-Memory operands.

EXAMPLE 3.23: Bit Test and Clear.

PROBLEM:

Execute the instruction **BCLR #2,X:$100** to test and clear bit 2 of X-Memory location $100. Assume that the contents of X-Memory location $100 and the Status Register (SR) before execution of the instruction are:

x:$100 $00000*f* *sr* = $0300

PROCEDURE:

```
TYPE: change x:$100 $00000f  sr $0300<return>
      display x:$100 sr<return>
      asm p:$240 bclr #2,x:$100<return>
      change pc $240<return>
      step<return>
      display x:$100 sr<return>
```

RESULTS:

x:$100 $00000*b* *sr* = $0301

Note that the state of bit 2 of X-Memory location $100 is reflected in the C-Bit of the Status Register (least significant bit). Also note that bit 2 of X-Memory location $100 is cleared after it is tested.

EXAMPLE 3.24: Bit Test and Change.

PROBLEM:

Execute the instruction **BCHG #3,Y:$200** to test and change bit 3 of Y-Memory location $200. Assume that the contents of Y-Memory location $200 and the Status Register (SR) before execution of the instruction are:

y:$200 $00000*f* *sr* = $0300

PROCEDURE:

```
TYPE: Change y:$200 $00000f sr $0300<return>
      display y:$200 sr<return>
      asm p:$242 bchg #3,y:$200<return>
      change pc $242<return>
      step<return>
      display y:$200 sr<return>
```

RESULTS:

y:$200 $000007 *sr* = $0301

Note that the state of bit 3 of Y-Memory location $200 is reflected in the C-Bit of the Status Register (SR) and the state of bit 3 of Y-Memory location $200 is changed after it is tested.

3.4 PROGRAM CONTROLLER:

The Program Controller, shown in Figure 3.4, is an independent execution unit which provides standard program flow-control resources such as the Program Counter, Status Register, and System Stack. It also includes an Operating Mode Register, and a Loop Address Register and a Loop Counter dedicated to support of the DSP56001 processor's hardware DO loop instruction.

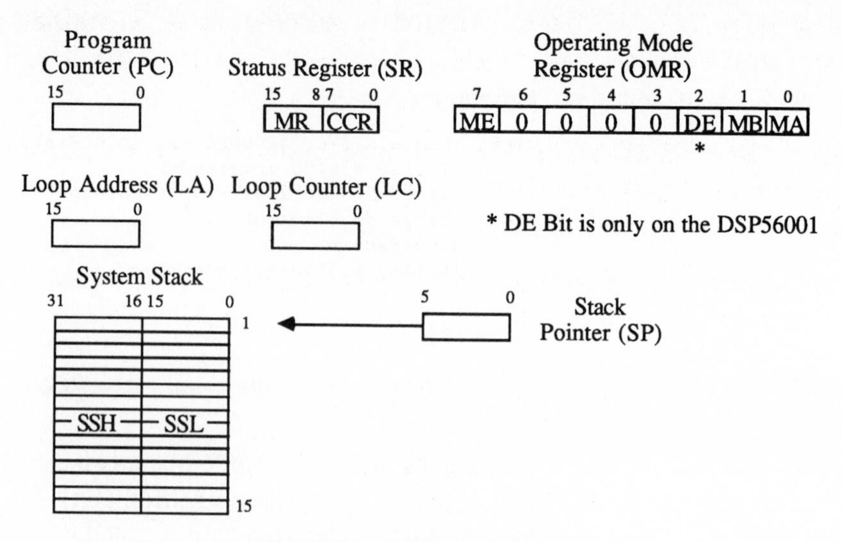

Figure 3-4 Program Controller

For more information on the Program Controller refer to the *DSP56000/DSP56001 Digital Signal Processor User's Manual.*

Program Counter Register (PC): The 16-bit Program Counter Register (PC) points to the Program Memory (P-Memory) location of the next instruction word, immediate data operand, or immediate address operand.

Status Register (SR): The 16-bit Status Register (SR), shown in Figure 3.5, consists of an 8-bit Mode Register (MR) and an 8-bit Condition Code Register (CCR). The MR is located in the high-order 8-bits of the Status Register; the CCR is located in the low-order 8-bits. The MR contains information about the system state of the DSP56001 processor; the CCR indicates the results of operations on data operands.

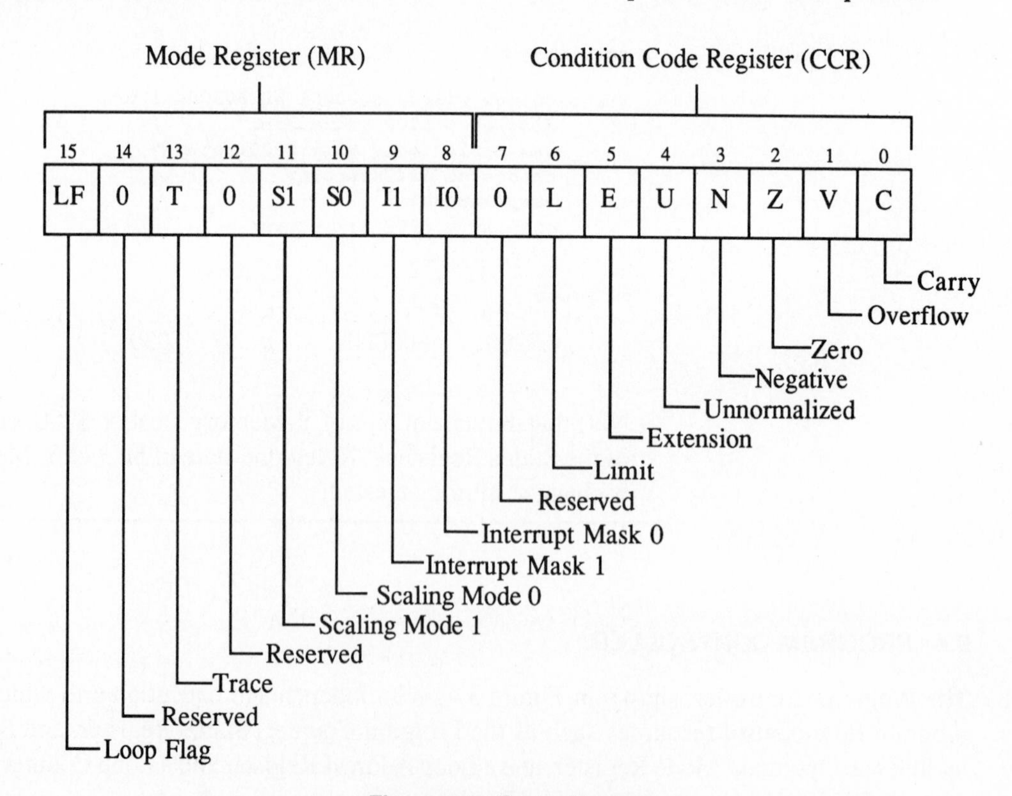

Figure 3-5 Status Register

System Stack (SS): The System Stack (SS) is a separate 32x15 internal memory which stores the Program Counter (PC) and Status Register (SR) for subroutine calls and long interrupts. The System Stack can also store the Loop Counter and Loop Address Register.

Stack Pointer (SP): The 6-bit Stack Pointer (SP), shown in Figure 3.6, points to the location of the top of the System Stack and indicates the System Stack status conditions such as underflow, empty, full, and overflow.

Loop Counter (LC): The 16-bit Loop Counter (LC) specifies the number of times a hardware DO loop instruction or hardware REPEAT instruction is to be repeated.

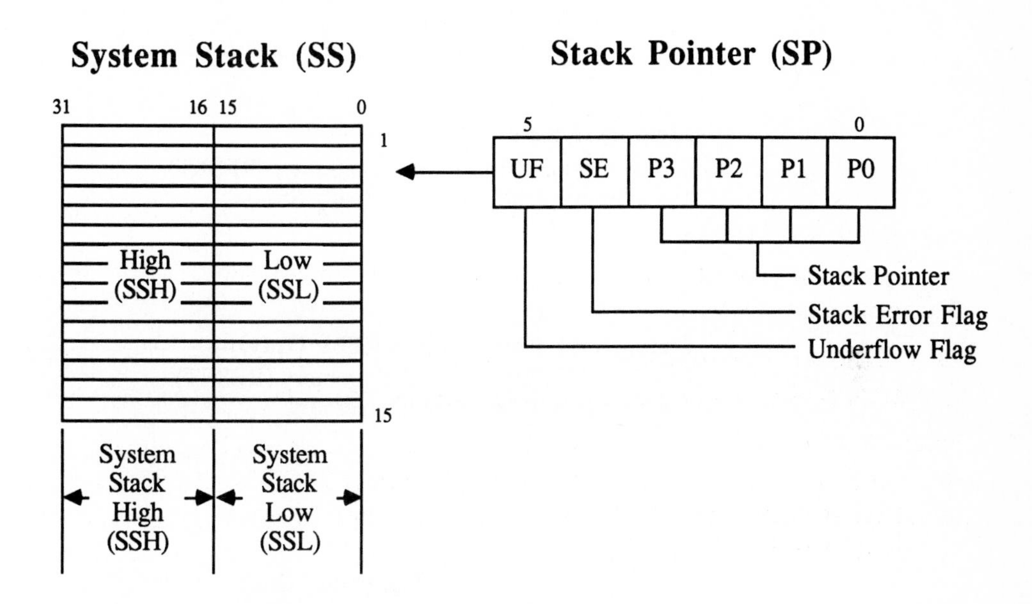

Uses:
- Subroutines
- Long Interrupts
- Hardware "DO LOOPS"
- Not recommended for data storage

UF	SE	P3	P2	P1	P0		
1	1	1	1	1	0	← Stack Underflow condition after double pull	Stack Error Exception
1	1	1	1	1	1	← Stack Underflow condition	
0	0	0	0	0	0	← Stack Empty (reset). Pull causes underflow.	
0	0	0	0	0	1	← Stack location 1	
.	.	.	.	.	.	.	
.	.	.	.	.	.	.	
.	.	.	.	.	.	.	
0	0	1	1	1	0	← Stack location 14.	
0	0	1	1	1	1	← Stack location 15. Push causes overflow	
0	1	0	0	0	0	← Stack Overflow condition	Stack Error Exception
0	1	0	0	0	1	← Stack Overflow condition after double push	

Figure 3-6 Stack Pointer

Loop Address Register (LA): The 16-bit Loop Address (LA) Register points to the location of the last instruction word in a hardware DO loop.

Operating Mode Register (OMR): The 8-bit Operating Mode Register (OMR) defines the current operating mode of the DSP56001 processor. It defines how the various memories are mapped and it defines the startup procedure. The OMR format is shown in Figure 3.7. Table 3.3 summarizes the DSP56001 processor's operating modes. Tables 3.4 and 3.5, and Figure 3.8, show the program and data memory spaces.

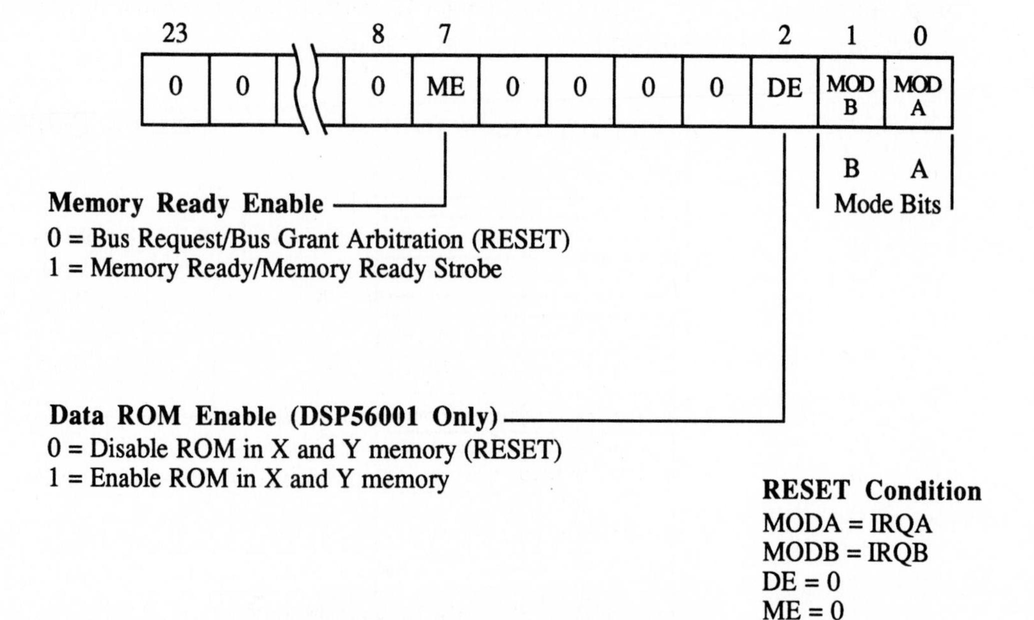

Figure 3-7 Operating Mode Register

TABLE 3.3. Operating Mode Summary.

Operating Mode	*M B*	*M A*	*Description*
0	0	0	PRAM enabled, Reset at $0000 (internal).
1	0	1	Special Bootstrap mode, after PRAM loading mode 2 is automatically selected.
2	1	0	PRAM enabled, Reset at $E000 (external).
3	1	1	PRAM disabled, Reset at $0000 (external).

TABLE 3.4. Program Memory Space.

Operating Mode	*M B*	*M A*	*Description*
0 and 2	X	0	internal RAM: $0000-$01FF external: $0200-$FFFF
3	1	1	external: $0000-$FFFF

TABLE 3.5. Data Memory Space.

DROM Enable DE	*Y Data Memory Space Map*	*X Data Memory Space Map*
0	internal RAM: $0000-$00FF external: $0100-$FFFF	internal RAM: $0000-$00FF external: $0100-$FFBF on-chip peripherals: $FFC0-$FFFF
1	internal RAM: $0000-$00FF internal ROM: $0100-$01FF external: $0200-$FFFF	internal RAM: $0000-$00FF internal ROM: $0100-$01FF external: $0200-$FFBF on-chip peripherals: $FFC0-$FFFF

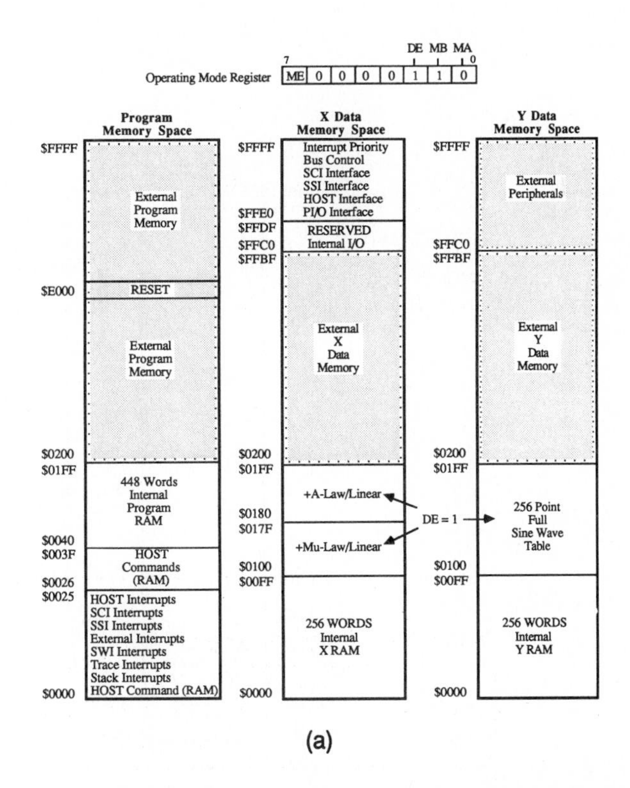

(a)

Figure 3-8 DSP56001 Memory Maps: (a) Mode 0 -- Normal Expanded and Internal Reset, (b) Mode 2 -- Normal Expanded and External Reset, (c) Mode 3 -- Development Mode External Program Memory

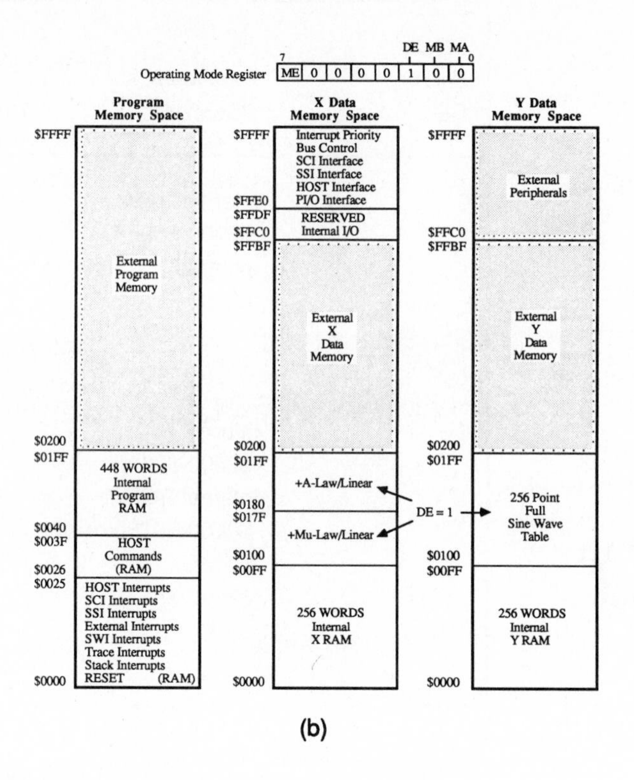

(b)

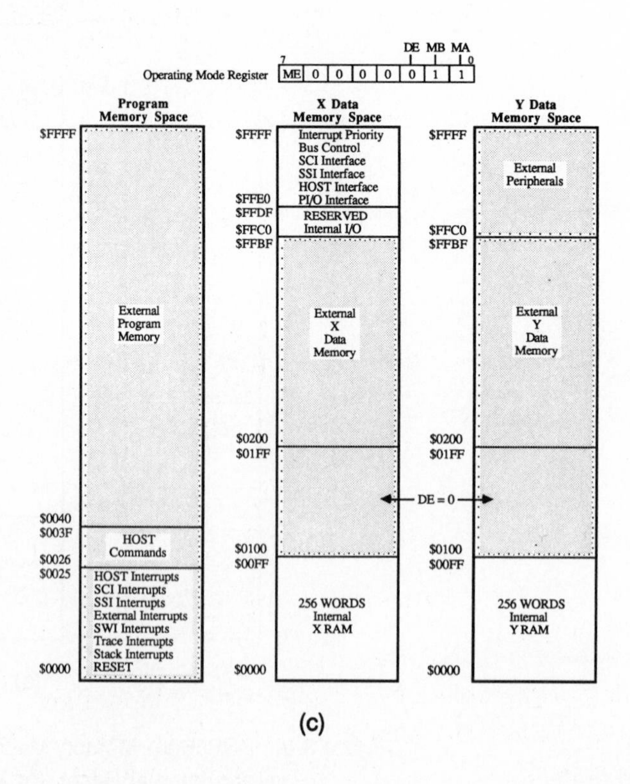

(c)

Figure 3-8 *(cont.)*

3.5 ADDRESS GENERATION UNIT:

The Address Generation Unit is an independent execution unit which generates addresses to locate data operands in X-Memory, Y-Memory, or P-Memory. It provides fourteen addressing modes and uses three types of address generation arithmetic. Its major components, shown in Figure 3.9, are: twenty-four 16-bit addressing registers, two 16-bit address arithmetic units, and three-address output-multiplexers.

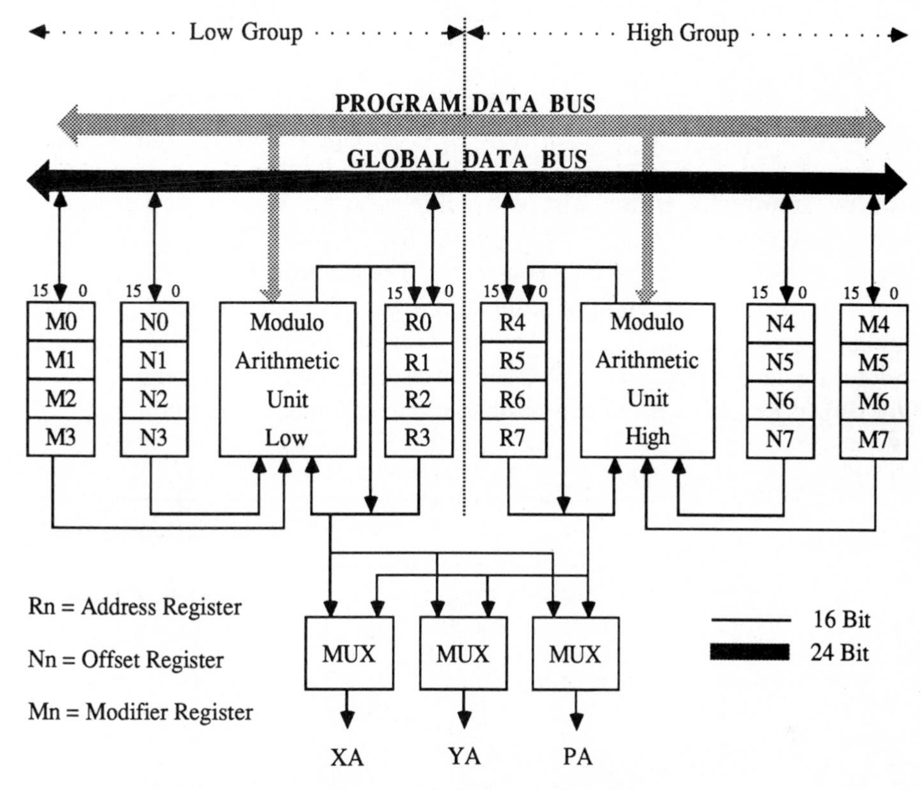

Figure 3-9 Address Arithmetic Units

Address Registers: The 24 addressing registers, shown in Figure 3.10, are organized into three sets of eight registers:

$$\begin{aligned} AddressRegisters \quad & Rn \quad n = 0,1,..,7 \\ OffsetRegisters \quad & Nn \quad n = 0,1,..,7 \\ ModifierRegisters \quad & Mn \quad n = 0,1,..,7 \end{aligned}$$

Each Address Register Rn has an associated Offset Register Nn and an associated Modifier Register Mn, the three having the same number n. The Address Registers Rn are used as address pointers to locate data operands in memory. The Offset Registers Nn are used to provide an offset value for offset updating of the Address Registers. The Modifier Registers Mn select the type of address arithmetic to be performed when an Address Register is updated.

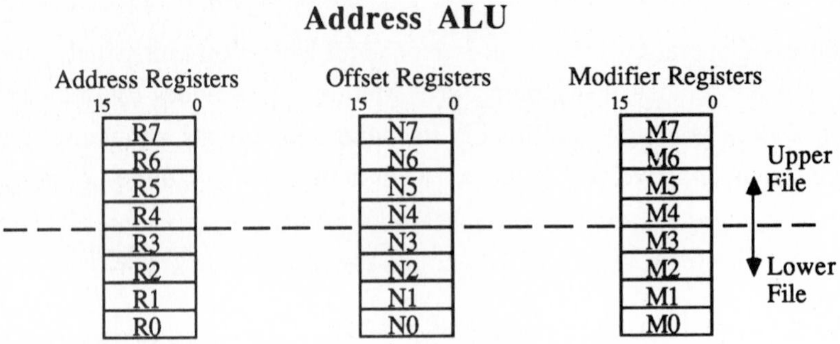

Address ALU

Address Registers 15 0: R7 R6 R5 R4 R3 R2 R1 R0

Offset Registers 15 0: N7 N6 N5 N4 N3 N2 N1 N0

Modifier Registers 15 0: M7 M6 M5 M4 M3 M2 M1 M0

Upper File

Lower File

Figure 3-10 Address Registers

Address Arithmetic Units: The two Address Arithmetic Units perform address arithmetic for the DSP56000/1 processor's addressing modes. The Address Arithmetic Units use three types of address arithmetic: linear, modulo, and reverse- carry. The Modifier Registers define the type of address arithmetic to be performed. Linear address arithmetic is used for standard MPU-type addressing. Modulo address arithmetic is useful to support circular buffers as described in Chapter 6. Reverse-carry arithmetic is useful for ordering data in Fast Fourier Transforms as described in Chapter 8. Table 3.6 shows how the contents of the Offset Register select the type of address arithmetic to be performed.

TABLE 3.6. Modifier Type Encoding Summary

Modifier	*Address Arithmetic Type*
0000000000000000	Reverse Carry (Bit Reversed Update)
0000000000000001	Modulo 2
0000000000000010	Modulo 3
0000000000000011	Modulo 4
.	. .
.	. .
.	. .
0111111111111110	Modulo 32767
0111111111111111	Modulo 32768
1000000000000000	Undefined
.	.
.	.
1111111111111110	.
1111111111111111	Linear (Modulo 65536)

Address Output Multiplexers: The three Address Output Multiplexers allow any Address Register Rn to be used as a pointer to any memory space.

3.6 ADDRESSING MODES:

The DSP56001 processor's instructions consist of one or two 24-bit words: the operation word and the extension word. The operation word contains an 8-bit opcode field and a 16-bit data bus movement field. The opcode field includes the instruction opcode with its source and destination register operands. The data bus movement field provides the direction of transfer and the effective addresses for data movement on the XD and YD data buses. The effective address specifies an addressing mode, and for some addressing modes the effective address will further specify an Address Register Rn.

The addressing modes specify whether the operands are in registers and/or memory locations, and provide the specific address of the operands. The addressing modes can be grouped into three categories: Register Direct, Special, and Register Indirect. The Register Indirect modes require additional address modifier information which is specified in the Modifier Registers. Both the Register Direct and Register Indirect addressing modes are associated with single word instructions. The Special addressing modes are associated with one or two word instructions.

3.6.1 Register Direct Addressing Modes:

The Register Direct addressing modes specify that the operand(s) is in one (or more) of the Data Input Registers, Address Registers, or Control Registers.

EXAMPLE 3.25: Transferring Register Data into a 24-Bit Accumulator Register.

PROBLEM:

Execute the instruction **MOVE X1,A0** to move the contents of the X1 Data Input Register to the A0 Accumulator Register using the Register Direct addressing mode. Assume that the contents of the X1 Data Input Register and the A0 Accumulator Register before execution of the instruction are:

$$x1 = \$111111 \qquad a0 = \$000000$$

PROCEDURE:

```
TYPE: change x1 $111111 a0 0<return>
      display x1 a0<return>
      asm p:$100 move x1,a0<return>
      change pc $100<return>
      step<return>
      display x1 a0<return>
```

RESULTS:

$$x1 = \$111111 \qquad a0 = \$111111$$

3.6.2 Special Addressing Modes:

The Special addressing modes specify the operand or the address of the operand in a field of the instruction or they implicitly reference an operand. The Special addressing modes are: Immediate Data, Immediate Short Data, Absolute Address, Absolute Short Address, I/O Short Address, Short Jump Address, and Implicit. The Absolute Short, I/O Short and Short Jump addresses, and the Immediate Short Data addressing modes are associated with single word instructions. The Absolute Address and Immediate Data addressing modes are associated with two word instructions.

Immediate Data Addressing Mode: The Immediate Data addressing mode points to a 24-bit operand located in the extension word of the instruction.

EXAMPLE 3.26: Transferring Immediate Data into a 24-Bit Accumulator Register.

PROBLEM:

Execute the instruction **MOVE #$818181,A0** to transfer the immediate value $818181 to Accumulator Register A0. Assume that the contents of the A-Accumulator before execution of the instruction are:

$$a = \$00000000000000$$

$$a2 = \$00 \quad a1 = \$000000 \quad a0 = \$000000$$

PROCEDURE:

```
TYPE: change a 0<return>
      display a a2 a1 a0<return>
      asm p:$101 move #$818181,a0<return>
      change pc $101<return>
      step<return>
      display a a2 a1 a0<return>
```

RESULTS:

$$a = \$00000000818181$$

$$a2 = \$00 \quad a1 = \$000000 \quad a0 = \$818181$$

EXAMPLE 3.27: Transferring Immediate Data into the 56-Bit Accumulators.

PROBLEM:

Execute the two instructions **MOVE #$818181,A** and **MOVE #$121212,B.** Assume that the contents of the A-Accumulator and B-Accumulator before execution of the two instructions are:

$$a = \$00000000000000$$

$$a2 = \$00 \quad a1 = \$000000 \quad a0 = \$000000$$

$$b = \$00000000000000$$

$$b2 = \$00 \quad b1 = \$000000 \quad b0 = \$000000$$

PROCEDURE:

```
TYPE: change a 0  b 0<return>
      display a a2 a1 a0 b b2 b1 b0<return>
      asm p:$103 move #$818181,a<return>
      asm p:$105 move #$121212,b<return>
      change pc $103<return>
      step 2<return>
      display a a2 a1 a0 b b2 b1 b0<return>
```

RESULTS:

$$a = \$ff818181000000$$

$$a2 = \$ff \quad a1 = \$818181 \quad a0 = \$000000$$

$$b = \$00121212000000$$

$$b2 = \$00 \quad b1 = \$121212 \quad b0 = \$000000$$

Note that Accumulator Registers A1 and B1 are loaded with the respective immediate data operands and that Accumulator Registers A2 and B2 are sign-extended from A1 and B1, respectively.

Immediate Short Addressing Mode: The Immediate Short addressing mode points to an 8-bit or a 12-bit immediate data operand located in the instruction operation word. The immediate data is interpreted as an unsigned integer if the destination register is one of the Registers A2,A1,A0, B2,B1,B0, R0-R7, N0-N7. The immediate data is transferred to the least significant bits of the destination with the most significant bits zeroed. The immediate data is interpreted as a signed fraction if the destination is one of the Registers X1,X0, Y1,Y0, or the A- and B-Accumulators.

EXAMPLE 3.28: Transferring Immediate Short Data into 24-Bit Registers.

PROBLEM:

Execute the two instructions **MOVE #$81,A1** and **MOVE #$81,X1**. Assume that the contents of the Data Input Registers X1,X0 and the A-Accumulator before execution of the two instructions are:

$$a = \$00000000000000$$

$$a2 = \$00 \quad a1 = \$000000 \quad a0 = \$000000$$

$$x1 = \$000000 \quad x0 = \$000000$$

PROCEDURE:

```
TYPE: change a 0 x 0<return>
      display a a2 a1 a0 x1 x0<return>
      asm p:$107 move #$81,a1<return>
      asm p:$108 move #$81,x1<return>
      change pc $107<return>
      step 2<return>
      display a a2 a1 a0 x1 x0<return>
```

RESULTS:

$$a = \$00000081000000$$

$$a2 = \$00 \quad a1 = \$000081 \quad a0 = \$000000$$

$$x1 = \$810000 \quad x0 = \$000000$$

Note that when Accumulator Register A1 is the destination, the immediate data $81 is interpreted as an unsigned integer and is transferred into the least significant bits of A1. When Data Input Register X1 is the destination, the immediate data $81 is interpreted as signed fraction and is transferred into the most significant bits of X1.

Example 3.29: Transferring Immediate Short Data into the Accumulators.

PROBLEM:

Execute the two instructions **MOVE #12,A** and **MOVE #81,B**. Assume that the contents of the A- and B-Accumulators before execution of the two instructions are:

a = $00000000000000

*a*2 = $00 *a*1 = $000000 *a*0 = $000000

b = $00000000000000

*b*2 = $00 *b*1 = $000000 *b*0 = $000000

PROCEDURE:

```
TYPE: change a 0 b 0<return>
      display a a2 a1 a0 b b2 b1 b0<return>
      asm p:$109 move #$12,a<return>
      asm p:$10a move #$81,b<return>
      change pc $109<return>
      step 2<return>
      display a a2 a1 a0 b b2 b1 b0<return>
```

RESULTS:

a = $00120000000000

*a*2 = $00 *a*1 = $120000 *a*0 = $000000

b = $*ff*810000000000

*b*2 = $*ff* *b*1 = $810000 *b*0 = $000000

Note that the immediate data operands $12 and $81 are interpreted as signed fractions and transferred into the most significant bits of the Accumulator Registers A1 and B1, respectively. Also note that Accumulator Registers A2 and B2 are sign-extended from A1 and B1, respectively.

Absolute Address Addressing Mode: The Absolute Address addressing mode uses the 16-bit address operand located in the instruction extension word as a pointer to the location of the data operand.

EXAMPLE 3.30: Transferring a Data Operand Pointed to by an Absolute Address Operand into the A-Accumulator.

PROBLEM:

Execute the instruction **MOVE X:$2000,A0** to transfer the contents of the X-Memory location pointed to by the instruction extension word to Accumulator Register A0. Assume that the contents of the A-Accumulator and X-Memory location $2000 before execution of the instruction are:

a = $00000000000000

*a*2 = $00 *a*1 = $000000 *a*0 = $000000

x:$2000 $121212

PROCEDURE:

```
TYPE: change x:$2000 $121212  a 0<return>
      display x:$2000 a a2 a1 a0<return>
      asm p:$10b move x:$2000,a0<return>
      change pc $10b<return>
      step<return>
      display x:$2000 a a2 a1 a0<return>
```

RESULTS:

a = $00000000121212

*a*2 = $00 *a*1 = $000000 *a*0 = $121212

x:$2000 $121212

Absolute Short Addressing Mode: The Absolute Short addressing mode uses an immediate 6-bit address operand, located in the instruction operation word, to form a 16-bit pointer to the data operand. The 6-bit immediate address operand is zero-extended to form the 16-bit pointer.

EXAMPLE 3.31: Transferring An Accumulator Register to a Memory Location Pointed to by an Absolute Short Address Operand.

PROBLEM:

Execute the instruction **MOVE A1,X:$2** to transfer the contents of Accumulator Register A1 to X-Memory location $0002 pointed to by the 6-bit (zero-extended to 16-bits) Absolute Short address $02. Assume that the contents of the A-Accumulator and X-Memory location $0002 before execution of the instruction are:

a = $00818181000000

*a*2 = $00 *a*1 = $818181 *a*0 = $000000

x:$0002 $000000

PROCEDURE:

```
TYPE: change x:$2 0  a 0  a1 $818181<return>
      display x:$2 a a2 a1 a0<return>
      asm p:$10d move a1,x:$2<return>
      change pc $10d<return>
      step<return>
      display x:$2 a a2 a1 a0<return>
```

RESULTS:

a = $00818181000000

$a2$ = $00 $a1$ = $818181 $a0$ = $000000

x:$0002 $818181

I/O Short Addressing Mode: The I/O Short addressing mode is similar to the Absolute Short addressing mode. The Absolute Short addressing mode uses an immediate 6-bit address operand, located in the instruction operation word, to form a 16-bit pointer to the data operand. The 6-bit immediate address operand is ones-extended to form a 16-bit pointer to the I/O sections ($FFC0-$FFFF) of X- or Y-Memory, respectively. It is used with the **bit manipulation** and **move peripheral data** instructions.

EXAMPLE 3.32: Transferring An Accumulator Register to an I/O Memory Location Pointed to by an I/O Short Address Operand.

PROBLEM:

Execute the instruction **MOVEP A1,X:$FFFE** to transfer the contents of Accumulator Register A1 to X-Memory I/O location $FFFE pointed to by the 6-bit (ones-extended to 16-bits) I/O Short address $3E. Assume that the contents of the A-Accumulator and X-Memory I/O location $FFFE before execution of the instruction are:

a = $00112233000000

$a2$ = $00 $a1$ = $112233 $a0$ = $000000

x:$*fffe* $000000

PROCEDURE:

```
TYPE: change x:$fffe 0  a 0  a1 $112233<return>
      display x:$fffe a a2 a1 a0<return>
      asm p:$10e movep a1,x:$fffe<return>
      change pc $10e<return>
      step<return>
      display x:$fffe a a2 a1 a0<return>
```

RESULTS:

a = $00112233000000

$a2$ = $00 $a1$ = $112233 $a0$ = $000000

x:$*fffe* $002233

Note that the least significant 16-bits of Accumulator Register A1 is transferred to the X-Memory I/O location $FFFE.

Short Jump Addressing Mode: The Short Jump addressing mode uses a 12-bit immediate jump operand, located in the instruction operation word, to form a 16-bit "jump to" operand. The 12-bit jump operand is zero-extended to 16-bits and replaces the contents of the Program Counter (PC).

EXAMPLE 3.33: Short Jumping to a P-Memory Location.

PROBLEM:

Execute the instruction **JMP $222** to jump to P-Memory location $0222. Assume that the contents of the Program Counter before execution of the instruction are:

$$pc = \$010f$$

PROCEDURE:

```
TYPE: change pc $10f<return>
      display pc<return>
      asm p:$10f jmp $222<return>
      change pc $10f<return>
      step<return>
      display pc<return>
```

RESULT:

$$pc = \$0223$$

Note that the above instruction **JMP $222** not only jumps to P-Memory location $0222, but the instruction contained in that location is fetched. Therefore the PC, after execution of the **JMP $222** instruction, is incremented by one (PC = $0222 + 1 = $0223) to point to the next instruction to be fetched from P-Memory.

Implicit Addressing Mode: The Implicit addressing mode is used by some instructions to implicitly reference the Program Controller Registers. The Program Controller Registers are implied in the mnemonic and opcode of the instructions.

3.6.3 Address Register Indirect Addressing Modes:

In the Address Register Indirect addressing modes, the instruction operation word specifies an Address Register Rn to point to an operand located in memory. The instruction operation word can also specify an address calculation to be performed either pre or post instruction execution, e.g., post-increment by one, post-decrement by an offset.

Each Address Register Rn is associated with an Offset Register Nn and a Modifier Register Mn. The Offset Register Nn contains an offset value which can be added to Rn to update its contents. The Modifier Register Mn specifies the type of address arithmetic to be performed when Address Register

Rn is updated. Each Modifier Register Mn is set to $FFFF upon processor reset to specify linear address arithmetic as the default address arithmetic.

No Update Address Register Indirect Addressing Mode: In this Address Register Indirect addressing mode, Address Register Rn points to the operand located in memory. The contents of Address Register Rn are not changed.

EXAMPLE 3.34: Transferring an Accumulator Register to a Memory Location Pointed to by an Address Register. The contents of the Address Register are Not Changed.

PROBLEM:

Execute the instruction **MOVE B1,Y:(R0)** to transfer the contents of Accumulator Register B1 to the Y-Memory location pointed to by Address Register R0. Assume that the contents of Accumulator Register B1, Address Register R0, Offset Register N0, Modifier Register M0, and Y-Memory location $1200 before execution of the instruction are:

$$b1 = \$010101$$

$$r0 = \$1200 \quad n0 = \$0000 \quad m0 = \$ffff$$

$$y{:}\$1200 \quad \$000000$$

PROCEDURE:

```
TYPE: change b1 $010101  y:$1200 0<return>
      change r0 $1200 n0 0  m0 $ffff<return>
      display b1 r0 n0 m0 y:$1200<return>
      asm p:$110 move b1,y:(r0)<return>
      change pc $110<return>
      step<return>
      display b1 r0 n0 m0 y:$1200<return>
```

RESULTS:

$$b1 = \$010101$$

$$r0 = \$1200 \quad n0 = \$0000 \quad m0 = \$ffff$$

$$y{:}\$1200 \quad \$010101$$

Post-Increment by One Address Register Indirect Addressing Mode: In this Address Register Indirect addressing mode, Address Register Rn points to the operand located in memory. After the operand address is used, the contents of Address Register Rn are incremented by one and the result is stored in Rn.

EXAMPLE 3.35: Transferring an Accumulator Register to a Memory Location Pointed to by an Address Register. The contents of the Address Register are Post-Incremented by One.

PROBLEM:

Execute the instruction **MOVE B0,Y:(R1)+** to transfer the contents of Accumulator Register B0 to the Y-Memory location pointed to by Address Register R1. After the transfer is complete, R1 is incremented by one. Assume that the contents of Accumulator Register B0, Address Register R1, Offset Register N1, Modifier Register M1, and Y-Memory location $1000 before execution of the instruction are:

*b*0 = \$010101

*r*1 = \$1000 *n*1 = \$0000 *m*1 = \$*ffff*

y:\$1000 \$000000

PROCEDURE:

```
TYPE: change b0 $010101 y:$1000 0<return>
      change r1 $1000 n1 0  m1 $ffff<return>
      display b0 r1 n1 m1 y:$1000<return>
      asm p:$111 move b0,y:(r1)+<return>
      change pc $111<return>
      step<return>
      display b0 r1 n1 m1 y:$1000<return>
```

RESULTS:

*b*0 = \$010101

*r*1 = \$1001 *n*1 = \$0000 *m*1 = \$*ffff*

y:\$1000 \$010101

Post-Decrement by One Address Register Indirect Addressing Mode: In this Address Register Indirect addressing mode, Address Register Rn points to the operand located in memory. After the operand address is used, the contents of Address Register Rn are decremented by one and the result is stored in Rn.

EXAMPLE 3.36: Transferring a Data Input Register to a Memory Location Pointed to by an Address Register. The contents of the Address Register are Post-Decremented by One.

PROBLEM:

Execute the instruction **MOVE Y0,X:(R2)-** to transfer the contents of Data Input Register Y0 to the X-Memory location pointed to by Address Register R2. After the transfer is complete, R2 is decremented by one. Assume that the contents of Data Input Register Y0, Address Register R2, Offset Register N2, Modifier Register M2, and X-Memory location $1001 before execution of the instruction are:

$y0 = \$010101$

$r2 = \$1001 \quad n2 = \$0000 \quad m2 = \$ffff$

x:\$1001 \$000000

PROCEDURE:

```
TYPE: change y0 $010101 x:$1001 0<return>
      change r2 $1001 n2 0  m2 $ffff<return>
      display  y0 r2 n2 m2 x:$1001<return>
      asm p:$112 move y0,x:(r2)-<return>
      change pc $112<return>
      step<return>
      display  y0 r2 n2 m2 x:$1001<return>
```

RESULTS:

$y0 = \$010101$

$r2 = \$1000 \quad n2 = \$0000 \quad m2 = \$ffff$

x:\$1001 \$010101

Post-Increment by Offset Nn Address Register Indirect Addressing Mode: In this Address Register Indirect addressing mode, Address Register Rn points to the operand located in memory. After the operand address is used, Rn is updated by adding the offset contained in Offset Register Nn to the contents of Rn. The contents of Offset Register Nn are unchanged.

EXAMPLE 3.37: Transferring a Data Input Register to a Memory Location Pointed to by an Address Register. The contents of the Address Register are Post-Updated by Adding the Offset Value in the Associated Offset Register.

PROBLEM:

Execute the instruction **MOVE X0,Y:(R3)+N3** to transfer the contents of Data Input Register X0 to the Y-Memory location pointed to by Address Register R3. After the transfer is complete, R3 is updated by adding the offset contained in Offset Register Nn to the contents of Rn. Assume that the contents of Data Input Register X0, Address Register R3, Offset Register N3, Modifier Register M3, and Y-Memory locations \$1000 and \$1003 before execution of the instruction are:

$x0 = \$010101$

$r3 = \$1000 \quad n3 = \$0003 \quad m3 = \$ffff$

y:\$1000 \$00000

y:\$1003 \$00000

PROCEDURE:

```
TYPE: change x0 $010101 <return>
      change y:$1000 0 y:$1003  0<return>
      change r3 $1000 n3 3  m3 $ffff<return>
      display  x0 r3 n3 m3 y:$1000 y:$1003<return>
      asm P:$113 move x0,y:(r3)+n3<return>
      change pc $113<return>
      step<return>
      display  x0 r3 n3 m3 y:$1000 y:$1003<return>
```

RESULTS:

$x0 = \$010101$

$r3 = \$1003 \quad n3 = \$0003 \quad m3 = \$ffff$

y:\$1000 \$010101

y:\$1003 \$000000

Post-Decrement by Offset Nn Address Register Indirect Addressing Mode: In this Address Register Indirect addressing mode, Address Register Rn points to the operand located in memory. After the operand address is used, Rn is updated by subtracting the offset contained in Offset Register Nn from the contents of Rn. The contents of Offset Register Nn are unchanged.

EXAMPLE 3.38: Transferring a Memory Location Pointed to by an Address Register to an Accumulator Register. The contents of the Address Register are Post-Updated by Subtracting the Offset Value in the Associated Offset Register.

PROBLEM:

Execute the instruction **MOVE Y:(R4)-N4,A0** to transfer the contents of the Y-Memory location pointed to by Address Register R4 to Accumulator Register A0. After the transfer is complete, R4 is updated by subtracting the offset contained in Offset Register Nn from the contents of Rn. Assume that the contents of Accumulator Register A0, Address Register R4, Offset Register N4, Modifier Register M4, and Y-Memory locations \$1000 and \$1004 before execution of the instruction are:

$a0 = \$000000$

$r4 = \$1004 \quad n4 = \$0004 \quad m4 = \$ffff$

y:\$1000 \$222222

y:\$1004 \$111111

PROCEDURE:

```
TYPE: change a0 0 <return>
      change y:$1004 $111111 y:$1000  $222222<return>
      change r4 $1004 n4 4  m4 $ffff<return>
      display  a0 r4 n4 m4 y:$1000 y:$1004<return>
      asm P:$114 move y:(r4)-n4,a0<return>
```

```
change pc $114<return>
step<return>
display  a0  r4 n4 m4 y:$1000 y:$1004<return>
```

RESULTS:

$a0 = \$111111$

$r4 = \$1000 \quad n4 = \$0004 \quad m4 = \$ffff$

y:$1000 $222222

y:$1004 $111111

Indexed by Offset Nn Address Register Indirect Addressing Mode: In this Address Register Indirect addressing mode, Address Register Rn is added to Offset Register Nn to form a pointer to an operand located in memory. The contents of Rn and Nn are unchanged.

EXAMPLE 3.39: Transferring a Data Input Register to a Memory Location Pointed to by the Sum of an Address Register and its Associated Offset Register. The contents of the Address Register and Offset Register are Not Changed.

PROBLEM:

Execute the instruction **MOVE X1,Y:(R5+N5)** to transfer the contents of Data Input Register X1 to the Y-Memory location pointed to by the summation of the contents of the Address Register R5 and the Offset Register N5. Assume that the contents of Data Input Register X1, Address Register R5, Offset Register N5, Modifier Register M5, and Y-Memory locations $1300 and $1306 before execution of the instruction are:

$x1 = \$010101$

$r5 = \$1300 \quad n5 = \$0006 \quad m5 = \$ffff$

y:$1300 $000000

y:$1306 $000000

PROCEDURE:

```
TYPE: change x1 $010101 y:$1300 0 y:$1306 0<return>
      change r5 $1300  n5 $0006  m5 $ffff<return>
      display x1 r5 n5 m5 y:$1300 y:$1306<return>
      asm p:$115 move x1,y:(r5+n5)<return>
      change pc $115<return>
      step<return>
      display x1 r5 n5 m5 y:$1300 y:$1306<return>
```

RESULTS:

$x1 \;=\; \$010101$

$r5 \;=\; \$1300 \quad n5 \;=\; \$0006 \quad m5 \;=\; \$ffff$

$y{:}\$1300 \quad \$000000$

$y{:}\$1306 \quad \$010101$

Pre-Decrement by One Address Register Indirect Addressing Mode: In this Address Register Indirect addressing mode, Address Register Rn points to the operand located in memory, but Rn is decremented by one before the operand is accessed, e.g., Rn = Rn - 1.

EXAMPLE 3.40: Transferring a Memory Location Pointed to by a Pre-Decremented Address Register to an Accumulator Register. The contents of the Address Register are Changed.

PROBLEM:

Execute the instruction **MOVE X:-(R6),A1** to pre-decrement Address Register R6 by one and transfer the X-Memory location pointed to by the decremented Address Register R6 to Accumulator Register A1. Assume that the contents of the A-Accumulator, Address Register R6, Offset Register N6, Modifier Register M6, and X-Memory locations $1000 and $1001 before execution of the instruction are:

$a \;=\; \$00000000000000$

$a2 \;=\; \$00 \quad a1 \;=\; \$000000 \quad a0 \;=\; \$000000$

$r6 \;=\; \$1001 \quad n6 \;=\; \$0000 \quad m6 \;=\; \$ffff$

$x{:}\$1000 \quad \$121212$

$x{:}\$1001 \quad \$343434$

PROCEDURE:

```
TYPE: change a 0 x:$1000 $121212 x:$1001 $343434<return>
      change r6 $1001 n6 0  m6 $ffff<return>
      display  a a2 a1 a0 r6 n6 m6 x:$1000..$1001<return>
      asm P:$116 move x:-(r6),a1<return>
      change pc $116<return>
      step<return>
      display  a a2 a1 a0 r6 n6 m6 x:$1000..$1001<return>
```

RESULTS:

$a \;=\; \$00121212000000$

$a2 \;=\; \$00 \quad a1 \;=\; \$121212 \quad a0 \;=\; \$000000$

$r6 \;=\; \$1000 \quad n6 \;=\; \$0000 \quad m6 \;=\; \$ffff$

$x{:}\$1000 \quad \$121212$

3.6.4 DSP56000/1 Addressing Mode Summary:

The DSP56000/1 processor's addressing modes are summarized in Table 3.7.

TABLE 3.7. Summary of DSP56000/1 Addressing Modes.

Addressing Mode	*Assembler Syntax*	*Example*
Register Direct		*move x1,a0*
Special		
Immediate Data	#xxxxxx	move #$818181,a0
Immediate Short Data	#xx	move #$81,a1
		move #$81,X1
		move #$12,a
		move #$81,b
Absolute Address	xxxx	move x:$2000,a0
Absolute Short Address	aa	move a1,x:$2
I/O Short Address	pp	movep a1,x:$fffe
Short Jump Address	xxx	jmp $222
Implicit		
Special Register Indirect		
No Update	(Rn)	move b1,y:(r0)
Postincrement by 1	(Rn)+	move b0,y:(r1)+
Postdecrement by 1	(Rn)-	move y0,x:(r2)-
Postincrement by Offset Nn	(Rn)+Nn	move x0,y:(r3)+n3
Postdecrement by Offset Nn	(Rn)-Nn	move y:(r4)-n4,a0
Indexed by Offset Nn	(Rn+Nn)	move x1,y:(r5+n5)
Predecrement by 1	-(Rn)	move x:-(r6),a1

4

DSP56001 INSTRUCTION SET

This chapter describes, in detail, the DSP56001 processor's instruction set. The DSP56001 processor's instruction set consists of 62 instructions. Each instruction is one or two words in length. Thirty of the instructions can specify one or two data transfers to be executed in parallel with execution of the instruction's opcode. These data transfers are called parallel-move operations. The instruction set is divided into the following groups: Move, Arithmetic, Logical, Bit Manipulation, Loop and Program Control. The instruction set features many digital signal processing oriented instructions such as CMPM, NORM, RND, MACR, SUBL, SUBR, ADDL, ADDR, DO, and REP.

The Chapter begins by describing the instruction format and the parallel-move operations. The instruction set groups are then presented. The DSP56000/1 Demonstration Software can be used to demonstrate the materials covered in Sections 4.3 and 4.4.

4.1 INSTRUCTION FORMAT:

Each DSP56001 instruction consists of one or two words. The first word specifies the instruction and its length. The second word can contain an absolute address or immediate data.

a) ***One Word Instruction:***

All the DSP56001 processor's addressing modes, except the immediate data and absolute address addressing modes, are available in one word instructions. The assembly language source code for a typical one word instruction can be organized into four fields: opcode field, operands field, X-Bus data transfer field and Y-Bus data transfer field.

Opcode: The opcode field typically indicates the data ALU operation to be performed; it can also specify a move, address generation, or program control operation.

Operands Field: The operands field specifies the operands to be used by the opcode.

Data Transfer Fields: The X-Bus and Y-Bus data transfer fields specify optional data transfers over the X- and Y-Buses, and the associated addressing modes.

EXAMPLE 4.1:

The following instructions are examples of one word instructions:

Opcode	*Operands*	*X-Bus Data Transfer*	*Y-Bus Data Transfer*
MOVE	X1,A		
MOVE		X0,X:(R2)-	
MOVE			X0,Y:(R3)+N3
RND	A	X:(R0)+,R0	Y:(R4)+,Y0
MAC	X0,Y0,B	X:(R0)+,R0	Y:(R4)+,Y0
ADD	X,B	X:(R6)+,X1	Y0,Y:(R7)-

---

b) ***Two Word Instruction:***

The absolute address and immediate data addressing modes are available in two word instructions to provide a full 16-bit absolute address or a full 24-bit immediate data value.

EXAMPLE 4.2:

The following instructions are examples of two word instructions:

$$\text{MOVE } \#\$123456, A0$$

$$\text{MOVE } X{:}\$2000, A0$$

The 24-bit immediate data operand ($123456) and the 16-bit address operand ($2000) are encoded in the second word of the instructions, respectively.

4.2 PARALLEL-MOVE OPERATIONS:

Thirty DSP56001 instructions can specify one or two parallel data move operations to be executed in one instruction cycle, in parallel, with the instruction's opcode. Data moves from register to register, register to memory, and memory to register are allowed. If the A-Accumulator or B-Accumulator is the source register, then shifting or limiting may occur with the data transfer. If the A-Accumulator or B-Accumulator is the destination register, then sign extension and least significant zero fill occurs automatically with the data transfer. When two data transfers are executed in one instruction, duplicate sources are permitted but duplicate destinations are not permitted.

EXAMPLE 4.3:

a) An instruction with the same source transferring to several destinations is permitted:

SUBL B,A B,X:(R0) B,Y:(R4)

b) An instruction with several sources transferring to the same destination is not permitted:

ADD B,A X1,B Y1,A (Not Permitted)

4.3 PARALLEL-MOVE TYPES:

The parallel-move types are:

1. Immediate Short Data Move
2. Register to Register Move
3. Address Register Update
4. X or Y Memory Move
5. X or Y Memory and Register Move
6. L Memory Move
7. XY Memory Move

If you want to use the DSP56000/1 Demonstration Software to cover the topics in the following sections, load the software by following steps 1 and 6 of Chapter 1 Section 1.5.1 or steps 1-3 of Chapter 1 Section 1.5.2. Then select from the demonstration software's menus to cover the topics of the following sections.

If you do not want to use the DSP56000/1 Demonstration Software program, load the SIM56000 Simulator Software by following steps 1, 2 and 3 of Chapter 2 Section 2.3.1 or steps 1, 5 and 6 of Chapter 2 Section 2.3.2. Proceed to Section 4.3.1.

4.3.1 Immediate Short Data Move:

This parallel-move operation transfers an 8-bit immediate short operand to a destination register. The 8-bit operand can be interpreted as an unsigned integer or as a signed fraction, depending on the destination register.

a) ***Unsigned Integer Immediate Short Operand:***

An immediate short operand is interpreted as an unsigned integer if the destination register is A2, A1, A0, B2, B1, B0, R0..R7, or N0..N7. The 8-bit data is stored in the least significant bits of the destination register with the high order bits of the destination register automatically reset to zero.

EXAMPLE 4.4:

PROBLEM:

Execute the instruction **ADD B,A** in parallel with the immediate short data move **#81,B0**. Assume that the contents of the A and B Accumulator Registers before execution of the instruction are:

$$a = \$00111111000000$$

$$a2 = \$00 \quad a1 = \$111111 \quad a0 = \$000000$$

$$b = \$00222222ff0000$$

$$b2 = \$00 \quad b1 = \$222222 \quad b0 = \$ff0000$$

PROCEDURE:

```
TYPE:   change a  $00111111000000<return>
        change b  $00222222ff0000<return>
        display a a2 a1 a0 b b2 b1 b0<return>
        asm p:$300 add b,a #$81,b0<return>
        change pc $300<return>
        step<return>
        display a a2 a1 a0 b b2 b1 b0<return>
```

RESULTS:

$$a = \$00333333ff0000$$

$$a2 = \$00 \quad a1 = \$333333 \quad a0 = \$ff0000$$

$$b = \$00222222000081$$

$$b2 = \$00 \quad b1 = \$222222 \quad b0 = = \$000081$$

Note that the immediate short operand $81 is interpreted as an unsigned integer and is moved into the lower portion of b0, with the high order bits of b0 reset to zero.

---

b) ***Signed Fraction Immediate Short Operand:***

An 8-bit immediate short operand is interpreted as a signed fraction if the destination register is X0, X1, Y0, Y1, A, or B. The 8-bit operand is stored in the most significant bits of the destination register with the low order bits of the destination register automatically reset to zero.

EXAMPLE 4.5:

PROBLEM:

Execute the instruction **ADD B,A** in parallel with the immediate short data move **#81,B**. Assume that the contents of the A and B Accumulator Registers before execution of the instruction are:

a = \$00111111000000

$a2$ = \$00 $a1$ = \$111111 $a0$ = \$000000

b = \$00222222222222

$b2$ = \$00 $b1$ = \$222222 $b0$ = \$222222

PROCEDURE:

```
TYPE:  change a $00111111000000<return>
       change b $00222222222222<return>
       display a a2 a1 a0 b b2 b1 b0<return>
       asm p:$301 add b,a #$81,b<return>
       change pc $301<return>
       step<return>
       display a a2 a1 a0 b b2 b1 b0<return>
```

RESULTS:

a = \$00333333222222

$a2$ = \$00 $a1$ = \$333333 $a0$ = \$222222

b = \$*ff*810000000000

$b2$ = \$*ff* $b1$ = \$810000 $b0$ = \$000000

Note that the immediate short operand \$81 is interpreted as a signed fraction and is moved into the higher order bits of b1. Register b2 is sign-extended from b1; the lower order bits of b1 and all of the bits of b0 are reset to zero.

4.3.2 Register to Register Data Move:

This parallel-move operation transfers the contents of a source register to a destination register.

EXAMPLE 4.6:

PROBLEM:

Execute the instruction **ADD B,A** in parallel with the register to register data move operation **X1,B**. Assume that the contents of the A and B Accumulator Registers, and the Input Register X1, before execution of the instruction are:

$a = \$00111111000000$

$a2 = \$00 \quad a1 = \$111111 \quad a0 = \$000000$

$b = \$00123456111111$

$b2 = \$00 \quad b1 = \$123456 \quad b0 = \$111111$

$X1 = \$900000$

PROCEDURE:

```
TYPE:   change a $00111111000000 b $00123456111111<return>
        change x1 $900000<return>
        display a a2 a1 a0 b b2 b1 b0 x1 <return>
        asm p:$302 add b,a  x1,b<return>
        change pc $302<return>
        step<return>
        display a a2 a1 a0 b b2 b1 b0 x1 <return>
```

RESULTS:

$a = \$00234567111111$

$a2 = \$00 \quad a1 = \$234567 \quad a0 = \$111111$

$b = \$ff9000000000000$

$b2 = \$ff \quad b1 = \$900000 \quad b0 = \$000000$

$x1 = \$900000$

Note that the destination Register b0 is automatically zeroed and the destination Register b2 is sign-extended from the destination Register b1.

EXAMPLE 4.7:

PROBLEM:

Execute the two instructions **ADD B,A** and **ADD X1,A** with the register to register data move operations **B,X1** and **B,R0**, respectively. Assume that the contents of the A and B Accumulators Registers, the X1 Data Register, and the R0 Address Register before execution of the two instructions are:

$a = \$00111111000000$

$a2 = \$00 \quad a1 = \$111111 \quad a0 = \$000000$

$b = \$00800001000000$

$b2 = \$00 \quad b1 = \$800001 \quad b0 = \$000000$

$x1 = \$000000$

$r0 = \$0000$

PROCEDURE:

```
TYPE:  change a $00111111000000 b $00800001000000<return>
       change x1 0 r0 0<return>
       display a a2 a1 a0 b b2 b1 b0 x1 r0 <return>
       asm p:$303 add b,a  b,x1<return>
       asm p:$304 add x1,a  b,r0<return>
       change pc $303<return>
       step 2<return>
       display a a2 a1 a0 b b2 b1 b0 x1 r0 <return>
```

RESULTS:

a = \$00911112000000

$a2$ = \$01 $a1$ = \$111111 $a0$ = \$000000

b = \$00800001000000

$b2$ = \$00 $b1$ = \$800001 $b0$ = \$000000

$x1$ = \$7*fffff*

$r0$ = \$*ffff*

Note that limiting is performed on the output of the B-Accumulator. The limited values of the destination Registers X1 and R0 are \$7FFFFF and \$FFFF, respectively.

4.3.3 Address Register Update:

This parallel-move operation updates the Address Register Rn according to the addressing mode.

EXAMPLE 4.8:

PROBLEM:

Execute the instruction **ADD B,A** in parallel with the address register update parallel move operation **(R1)+N1,** where the postincrement by offset Nn indirect addressing mode is used to update the R1 Address Register. Assume that the contents of the A and B Accumulator Registers, the R1 Address Register, and the N1 Offset Register before execution of the instruction are:

a = \$00111111000000

b = \$00222222000000

$r1$ = \$1000

$n1$ = \$0006

PROCEDURE:

```
TYPE: change a $00111111000000  b $00222222000000<return>
      change r1 $1000  n1 $0006<return>
      display a b r1 n1<return>
      asm p:$305 add b,a (r1)+n1<return>
      change pc $305<return>
      step<return>
      display a b r1 n1<return>
```

RESULTS:

$$a = \$00333333000000$$

$$b = \$00222222000000$$

$$r1 = \$1006$$

$$n1 = \$0006$$

4.3.4 X- or Y-Memory Data Move:

This parallel-move operation transfers a one word operand to/from X- or Y-Memory. The X- or Y-Memory to register, or register to X- or Y-Memory, direction can be specified.

EXAMPLE 4.9: Register to X-Memory.

PROBLEM:

Execute the instruction **ADD A,B** in parallel with the X-Memory data move operation **A,X:$1000**, where the absolute address addressing mode is used to specify the X-Memory address $1000. Assume that the contents of the A and B Accumulator Registers, and the X-Memory location $1000 before execution of the instruction are:

$$a = \$00123456000000$$

$$a2 = \$00 \quad a1 = \$123456 \quad a0 = \$000000$$

$$b = \$00111111000000$$

$$b2 = \$00 \quad b1 = \$111111 \quad b0 = \$000000$$

$$x{:}\$1000 \quad \$000000$$

PROCEDURE:

```
TYPE: change a $00123456000000  b $00111111000000<return>
      change x:$1000 0<return>
      display a a2 a1 a0 b b2 b1 b0 x:$1000<return>
      asm p:$306 ADD A,B  A,X:$1000<return>
      change pc $306<return>
      step<return>
      display a a2 a1 a0 b b2 b1 b0 x:$1000<return>
```

RESULTS:

$a = \$00123456000000$

$a2 = \$00 \quad a1 = \$123456 \quad a0 = \$000000$

$b = \$00234567000000$

$b2 = \$00 \quad b1 = \$234567 \quad b0 = \$000000$

$x{:}\$1000 \quad \$123456$

EXAMPLE 4.10: Y-Memory to Register.

PROBLEM:

Execute the instruction **ADD A,B** in parallel with the Y-Memory data move operation **Y:-(R3),A**, where the predecrement by one addressing mode is used to specify the Y-Memory address. Assume that the contents of the A and B Accumulator Registers, the R3 Address Register and the Y-Memory location $1000 before execution of the instruction are:

$a = \$00123456000000$

$a2 = \$00 \quad a1 = \$123456 \quad a0 = \$000000$

$b = \$00111111000000$

$b2 = \$00 \quad b1 = \$111111 \quad b0 = \$000000$

$y{:}\$1000 \quad \$999999$

$r3 = \$1001$

PROCEDURE:

```
TYPE:  change a $00123456000000 b $00111111000000<return>
       change r3 $1001 y:$1000 $999999<return>
       display a a2 a1 a0 b b2 b1 b0 y:$1000 r3<return>
       asm p:$307 ADD A,B  y:-(r3),a<return>
       change pc $307<return>
       step<return>
       display a a2 a1 a0 b b2 b1 b0 y:$1000 r3<return>
```

RESULTS:

$a = \$ff999999000000$

$a2 = \$ff \quad a1 = \$999999 \quad a0 = \$000000$

$b = \$00234567000000$

$b2 = \$00 \quad b1 = \$234567 \quad b0 = \$000000$

$y{:}\$1000 \quad \$999999$

$r3 = \$1000$

4.3.5 X- or Y-Memory and Register Data Move:

This parallel-move operation transfers a one word operand to/from X- or Y-Memory and a one word operand from register to register.

EXAMPLE 4.11:

PROBLEM:

Execute the instruction **ADD X,A** in parallel with the X-Memory data move operation **A,X:(R3+N3)** and the register data move operation **A,Y1**. Then, execute the signed multiply instruction **MPY Y1,X1,A**. Assume that the contents of the A-Accumulator, the X and Y1 Data Registers, the R3 Address Register, the N3 Offset Register, and X-Memory location $1006 before execution of the two instructions are:

$$x = \$111111111111$$

$$a = \$00123456000000$$

$$y1 = \$000000$$

$$r3 = \$1000 \qquad n3 = \$0006$$

$$x{:}\$1006 \qquad \$000000$$

PROCEDURE:

The register to register move operation **A,Y1** allows the contents of the A-Accumulator Register to be moved to the 24-bit Y1 Data Input Register. The contents of Y1 will be used as an input operand in the **MPY Y1,X1,A** instruction. The multiplier input operands are each 24-bits wide.

```
TYPE:   change x $111111111111 a $00123456000000<return>
        change r3 $1000 n3 $6 x:$1006 0 y1 0<return>
        display x a y1 r3 n3 x:$1006<return>
        asm p:$308 add x,a a,x:(r3+n3) a,y1<return>
        asm p:$309 mpy y1,x1,a<return>
        change pc $308<return>
        step 2<return>
        display x a y1 r3 n3 x:$1006<return>
```

RESULTS:

$$x = \$111111111111$$

$$a = \$00026D60CA5F6C$$

$$y1 = \$123456$$

$$r3 = \$1000 \qquad n3 = \$0006$$

$$x{:}\$1006 \qquad \$123456$$

4.3.6 Long Memory Data Move:

This parallel-move operation transfers one long word operand to/from L (X:Y) memory. Two Data ALU registers are concatenated to form the long word operand. This allows one double precision (high:low) data value or one complex (real:imaginary) data value to be transferred to/from one L-Memory location.

EXAMPLE 4.12:

PROBLEM:

Execute the instruction **ADD Y,B** in parallel with the long memory move operation **B10,L:$1000**, where the absolute address addressing mode is used to specify the L-Memory address $1000. Assume that the contents of the B-Accumulator Register, the Y Data Register, and the L-, X- and Y- memory locations $1000 before execution of the instruction are:

y = \$111111222222

b = \$00111111111111

$b2$ = \$00 $b1$ = \$111111 $b0$ = \$111111

l:\$1000   \$000000000000

x:\$1000   \$000000

y:\$1000   \$000000

PROCEDURE:

```
TYPE:  change y $111111222222  b $00111111111111<return>
       change l:$1000 $000000000000<return>
       display y b l:$1000 x:$1000  y:$1000<return>
       asm p:$30A add y,b b10,l:$1000<return>
       change pc $30A<return>
       step<return>
       display y b l:$1000 x:$1000 y:$1000<return>
```

RESULTS:

y = \$111111222222

b = \$00222222333333

l:\$1000   \$111111111111

x:\$1000   \$111111

y:\$1000   \$111111

Note that the 48-bit value in the b1 and b0 Accumulator Registers is stored as a double precision value in L-Memory location $1000.

4.3.7 XY Memory Data Move:

This parallel-move operation transfers two single word operands to/from X- and Y-Memory.

EXAMPLE 4.13:

PROBLEM:

Execute the instruction **ADD X,B** in parallel with the XY memory move operations **X:(R0)+,X1** and **Y0,Y:(R7)-**, where the postincrement by one and postdecrement by one addressing modes are used to specify the X- and Y-Memory locations, respectively. Assume that the contents of the B-Accumulator, X and Y0 Data Registers, the R0 and R7 Address Registers, the X-Memory location $1000, and the Y-Memory location $1500 before execution of the instruction are:

x = $111111222222

*y*0 = $333333

b = $00111111000000

*r*0 = $1000 *r*7 = $1500

x:$1000 $444444 *y*:$1500 $000000

PROCEDURE:

```
TYPE: change x $111111222222  b $00111111000000<return>
      change r0 $1000  r7 $1500  y0 $333333<return>
      change x:$1000 $444444  y:$1500 0<return>
      display x y0 b r0 r7 x:$1000 y:$1500<return>
      asm p:$30B add x,b x:(r0)+,x1 y0,y:(r7)-<return>
      change pc $30B<return>
      step<return>
      display x y0 b r0 r7 x:$1000 y:$1500<return>
```

RESULTS:

x = $444444222222

*y*0 = $333333

b = $00222222222222

*r*0 = $1001

*r*7 = $14*ff*

x:$1000 $444444 *y*:$1500 $333333

4.3.8 Parallel-Moves Summary:

Table 4.1 summarizes the parallel-move operations.

TABLE 4.1. Summary of Parallel-Move Operations.

Parallel Move Operations	*Opcode Operands*	*Parallel Move Examples*
Immediate Short Data	ADD B,A	#$81,B0
	ADD B,A	#$81,B
Register to Register	ADD B,A	X1,B
	ADD B,A	B,X1
	ADD X1,A	B,R0
Address Register Update	ADD B,A	(R1)+N1
X or Y Memory	ADD A,B	A,X:$1000
	ADD A,B	Y:(-R3),A
X or Y Memory plus register	ADD X,A	A,X:(R3+N3) A,Y1
L Memory	ADD Y,B	B10,L:$1000
XY Memory	ADD X,B	X:(R0)+,X1 Y0,Y:(R7)-

The addressing modes that may be used with the parallel-move operations are summarized in Table 4.2, where:

$$U = \text{Address Register Update}$$

$$X = X \text{ Memory Move}$$

$$Y = Y \text{ Memory Move}$$

$$X+R = X \text{ Memory and Register Move}$$

$$Y+R = Y \text{ Memory and Register Move}$$

$$L = L \text{ Memory Move}$$

$$XY = XY \text{ Memory Move}$$

4.4 DSP56001 INSTRUCTION SET:

The DSP56001 instruction set is divided into the following groups:

1. Move
2. Arithmetic
3. Logical
4. Bit Manipulation
5. Loop
6. Program control

TABLE 4.2. Addressing Modes for the U, X, Y, X+R, Y+R, L, and XY Parallel-Moves.

Addressing Mode	*Parallel Moves*				
	U	*X* *Y*	*X+R* *Y+R*	*L*	*XY*
Register Direct	No	No	No	No	No
Address Register Indirect					
No Update	No	Yes	Yes	Yes	Yes
Postincrement by 1	Yes	Yes	Yes	Yes	Yes
Postdecrement by 1	Yes	Yes	Yes	Yes	Yes
Postincrement by Offset Nn	Yes	Yes	Yes	Yes	Yes
Postdecrement by Offset Nn	Yes	Yes	Yes	Yes	No
Indexed by Offset Nn	No	Yes	Yes	Yes	No
Predecrement by 1	No	Yes	Yes	Yes	No
Special					
Immediate Data	No	Yes	Yes	No	No
Absolute Address	No	Yes	Yes	Yes	No
Immediate Short Data	No	No	No	No	No
Short Jump Address	No	No	No	No	No
Absolute Short Address	No	Yes	No	Yes	No
I/O Short Address	No	No	No	No	No
Implicit	No	No	No	No	No

4.4.1 Move Instructions:

The Move instructions Transfer data over the XD and YD Data Buses, the Global Data Bus and the Program Data Bus. The Move instructions can be considered to be a data ALU NOP (no operation) with the ability to perform parallel-move operations. The Move instructions affect only the limit bit L in the Condition Code Register CCR, where the limit bit is the sixth bit. The CCR occupies the low-order 8-bits of the Status Register (SR). The Move instructions are:

LUA	Load updated address
MOVE	Move data
MOVEC or Move	Move control register
MOVEM or Move	Move program memory
MOVEP or Move	Move peripheral data

EXAMPLE 4.14: Load Updated Address (LUA)

PROBLEM:

Execute the instruction **LUA (R0)+N0,N1** to update Address Register R0 and store the updated address in Offset Register N1. Assume that the contents of Address Register R0, and Offset Registers N0 and N1 before execution of the instruction are:

$$r0 = \$1000 \qquad n0 = \$0006 \qquad n1 = \$0000$$

PROCEDURE:

```
TYPE: change r0 $1000 n0 $6 n1 0<return>
      display r0 n0 n1<return>
      asm p:$400 lua (r0)+n0,n1<return>
      change pc $400<return>
      step<return>
      display r0 n0 n1<return>
```

RESULTS:

$$r0 = \$1000 \qquad n0 = \$0006 \qquad n1 = \$1006$$

EXAMPLE 4.15: Move Control Register (MOVEC)

PROBLEM:

Execute the instruction **MOVEC A,LC** to transfer the A-Accumulator to the Loop Control (LC) Register. Assume that the contents of the A-Accumulator, the Loop Control Register, and the Status Register (SR) before execution of the instruction are:

$$a = \$00800001000000$$

$$lc = \$0000$$

$$sr = \$0000$$

PROCEDURE:

```
TYPE:  change a $00800001000000  lc $0  sr $0<return>
       display a lc sr<return>
       asm p:$401 movec a,lc<return>
       change pc $401<return>
       step<return>
       display a lc sr<return>
```

RESULTS:

$$a = \$00800001000000$$

$$lc = \$ffff$$

$$sr = \$0040$$

Note that limiting is performed on the output of the A-Accumulator. The limited value transferred to the Loop Control Register is $FFFF. The limit bit in the Status Register is set, where the limit bit is bit 6.

EXAMPLE 4.16: Move Program Memory (MOVEM)

PROBLEM:

Execute the two instructions **MOVEM R3,P:(R2)-N2** and **MOVEM P:$0000,LC** to transfer the contents of Address Register R3 to the P-Memory location pointed to by Address Register R2; and transfer the contents of P-Memory location $0000 to the Loop Counter (LC) Register. Assume that the contents of Address Registers R2 and R3, Offset Register N2, P-Memory locations $0 and $500, and the Loop Counter Register before execution of the two instructions are:

*r*3 = $1000 *r*2 = $0500 *n*2 = $0002

p:$0000 $123456

p:$0500 $000000

lc = $0000

PROCEDURE:

```
TYPE: change r3 $1000  r2 $500  n2 $2<return>
      change p:$500 0  p:$0  $123456  lc 0<return>
      display r3 r2 n2 p:$500 p:$0 lc<return>
      asm p:$402 movem r3,p:(r2)-n2<return>
      asm p:$403 movem p:$0000,lc<return>
      change pc $402<return>
      step 2<return>
      display r3 r2 n2 p:$500 p:$0 lc<return>
```

RESULTS:

*r*3 = $1000 *r*2 = $04*fe* *n*2 = $0002

p:$0000 $123456

p:$0500 $001000

lc = $3456

EXAMPLE 4.17: Move Peripheral Data (MOVEP)

PROBLEM:

Execute the instruction **MOVEP X:$FFFF,A** to transfer the contents of X-Memory I/O peripheral location $FFFF to the A-Accumulator. Assume that the contents of X-Memory I/0 peripheral location $FFFF and the A-Accumulator before execution of the instruction are:

a = \$00000000000000

$a2$ = \$00 $a1$ = \$000000 $a2$ = \$000000

x:\$*ffff* \$001234

PROCEDURE:

```
TYPE:   change x:$ffff $001234 a 0<return>
        display x:$ffff a a2 a1 a0<return>
        asm p:$404 movep x:$ffff,a<return>
        change pc $404<return>
        step<return>
        display x:$ffff a a2 a1 a0<return>
```

RESULTS:

a = \$00001234000000

$a2$ = \$00 $a1$ = \$001234 $a2$ = \$000000

x:\$*ffff* \$001234

4.4.2 Arithmetic Instructions:

The Arithmetic instructions use the Data ALU to perform all arithmetic-type operations. Source operands for the Arithmetic instructions are contained in the Data ALU's Input Registers or Accumulators; destinations for the results generated by execution of the Arithmetic instructions are either the A-Accumulator or the B-Accumulator. The Arithmetic instructions allow up to two parallel-move operations to be executed concurrently with execution of the instruction's opcode. The parallel-move operations use the XD and YD Data Buses. As a result, the parallel-move operations can transfer new pre-fetched data from X- and or Y-Memory to the Data ALU for use in subsequent instructions; and can transfer the results generated by the execution of previous instructions from the Data ALU to X- and or Y-Memory.

The Arithmetic instructions execute in one instruction cycle and can affect all of the bits in the Condition Code Register. The Arithmetic instructions are:

ABS	Absolute value
ADC	Add with carry
ADD	Add
ADDL	Shift left then add
ADDR	Shift right then add
ASL	Arithmetic shift left
ASR	Arithmetic shift right
CLR	Clear
CMP	Compare
CMPM	Compare magnitude
DIV	divide iteration
MAC	Multiply/accumulate
MACR	Multiply/accumulate and round
MPY	Multiply

MPYR	Multiply and round
NEG	Negate
NORM	Normalize iteration
RND	Round
SBC	Subtract with carry
SUB	Subtract
SUBL	Shift left then subtract
SUBR	Shift right then subtract
Tcc	Transfer Conditionally
TFR	Transfer
TST	Test

In the following Arithmetic instruction examples, the default no-scaling mode is used, e.g. the scaling mode bits S1 and S0 are cleared, where S1 and S0 are the fourth and third bits, respectively, in the Mode Register (MR) and where the MR occupies the upper 8-bit portion of the Status Register (SR). To clear S0 and S1, the *DSP56000ADS User Interface Software* program's CHANGE command can be used to modify the contents of the MR register (Example 4.18) or the **AND immediate to control register** instruction **andi #$f3,mr** can be executed to logic AND the contents of the MR with the 8-bit immediate operand $f3, and store the result in the MR (Example 4.30).

a) ***Addition and Subtraction Arithmetic Instructions:***

The addition and subtraction Arithmetic instructions are: **add with carry (ADC), add (ADD), shift left then add (ADDL), shift right then add (ADDR), subtract with carry (SBC), subtract (SUB), shift left then subtract (SUBL),** and **shift right then subtract (SUBR).** The addition and subtraction arithmetic instructions are summarized in Table 4.3.

TABLE 4.3. Addition and Subtraction Arithmetic Instructions.

Mnemonic	*Operation*	*Assembler Syntax*	*Source S*	*Destination D*
ADC	S+D+C* → D	ADC S,D [PM]**	X,Y	A,B
ADD	S+D → D	ADD S,D [PM]	A	B
			B	A
			X,X1,X0	A,B
			Y,Y1,Y0	A,B
ADDL	S+2D → D	ADDL S,D [PM]	A	B
			B	A
ADDR	S+D/2 → D	ADDR S,D [PM]	A	B
			B	A
SBC	D-S-C → D	SBC S,D [PM]	X,Y	A,B
SUB	D-S → D	SUB S,D [PM]	A	B
			B	A
			X,X1,X0	A,B
			Y,Y1,Y0	A,B
SUBL	2D-S → D	SUBL S,D [PM]	A	B
			B	A
SUBR	D/2-S → D	SUBR S,D [PM]	A	B
			B	A

* C = Carry Bit
** PM = Parallel Move

EXAMPLE 4.18: Shift Left then Subtract

PROBLEM:

Execute the instruction **SUBL B,A** to arithmetically shift the A-Accumulator one bit position to the left with a zero shifted into the least significant bit, subtract the B-Accumulator from the shifted A-Accumulator, and store the result (2A - B) in the A-Accumulator. Assume that the contents of the A-Accumulator and B-Accumulator before execution of the instruction are:

a = $00222222222222

a2 = $00 *a1* = $222222 *a2* = $222222

b = $00111111111111

b2 = $00 *b1* = $111111 *b2* = $111111

PROCEDURE:

```
TYPE: change sr $0300<return>
      change b $00111111111111  a $00222222222222<return>
      display a a2 a1 a0 b b2 b1 b0<return>
      asm p:$500 subl b,a<return>
      change pc $500<return>
      step<return>
      display a a2 a1 a0 b b2 b1 b0<return>
```

RESULTS:

a = $00333333333333

a2 = $00 *a1* = $333333 *a2* = $333333

b = $00111111111111

b2 = $00 *b1* = $111111 *b2* = $111111

EXAMPLE 4.19: Add Long with Carry

PROBLEM:

Execute the instruction **ADC Y,A** to add the 48-bit Y Data Input Register plus the Carry Bit C to the A-Accumulator, and store the result in the A-Accumulator. The Carry Bit C is the least significant bit in the Status Register (SR). Assume that the contents of the 48-bit Y Data Input Register, the A-Accumulator, and the Status Register before execution of the instruction are:

y = $222222000000

a = $00555555000000

sr = $0001

PROCEDURE:

```
TYPE: change y $222222000000  a $00555555000000<return>
      change sr $0001<return>
      display y a sr<return>
      asm P:$501 adc y,a<return>
      change pc $501<return>
      step<return>
      display y a sr<return>
```

RESULTS:

$$y = \$222222000000$$

$$a = \$00777777000001$$

$$sr = \$0000$$

Note that a new Carry Bit, resulting from execution of the above instruction, is now in the SR. The carry bit is cleared after the execution.

EXAMPLE 4.20: Shift Right then Add

PROBLEM:

Execute the instruction **ADDR B,A** to arithmetically shift the A-Accumulator one bit position to the right with the most significant bit held constant, add the B-Accumulator to the shifted A-Accumulator, and store the result (A/2 + B) in the A-Accumulator. Assume that the contents of the A-Accumulator and B-Accumulator before execution of the instruction are:

$$a = \$00555555555555$$

$$a2 = \$00 \quad a1 = \$555555 \quad a0 = \$555555$$

$$b = \$00222222222222$$

$$b2 = \$00 \quad b1 = \$222222 \quad b0 = \$222222$$

PROCEDURE:

```
TYPE: change  a $00555555555555 b $00222222222222<return>
      display a a2 a1 a0 b b2 b1 b0<return>
      asm p:$502 addr b,a<return>
      change pc $502<return>
      step<return>
      display a a2 a1 a0 b b2 b1 b0<return>
```

RESULTS:

$$a = \$004cccccccccc$$

$$a2 = \$00 \quad a1 = \$4ccccc \quad a0 = \$cccccc$$

$$b = \$00222222222222$$

$$b2 = \$00 \quad b1 = \$222222 \quad b0 = \$222222$$

---

b) ***Multiplication Arithmetic Instructions:***

The multiplication Arithmetic instructions are: **multiply-accumulate (MAC), multiply-accumulate and round (MACR), multiply (MPY),** and **multiply and round (MPYR).** The multiplication Arithmetic instructions are summarized in Table 4.4. The destination (D) for the results generated by the execution of these instructions is either the A-Accumulator or the B-Accumulator.

TABLE 4.4. Multiplication Arithmetic Instructions.

Mnemonic	*Operation*	*Assembler Syntax*	*Sources*	
			S1	*S2*
MAC	D±(S1•mS2)→D	MAC ±S1,S2,D [PM]**	X0	X0
			Y0	Y0
			X1	X0
			Y1	Y0
			X0	Y1
			Y0	X0
			X1	Y0
			Y1	X1
MACR	D±(S1•S2)+r*→D	MACR ±S1,S2,D [PM]	X0	X0
			Y0	Y0
			X1	X0
			Y1	Y0
			X0	Y1
			Y0	X0
			X1	Y0
			Y1	X1
MPY	±(S1•S2)→D	MPY ±S1,S2,D [PM]	X0	X0
			Y0	Y0
			X1	X0
			Y1	Y0
			X0	Y1
			Y0	X0
			X1	Y0
			Y1	X1
MPYR	±(S1•S2)+r→D	MPYR ±S1,S2,D [PM]	X0	X0
			Y0	Y0
			X1	X0
			Y1	Y0
			X0	Y1
			Y0	X0
			X1	Y0
			Y1	X1

* r = Round
** PM = Parallel Move

EXAMPLE 4.21: Signed Multiply Without and With Rounding

PROBLEM:

Execute the two instructions **MPY X1,Y1,A** and **MPYR X1,Y1,B** to multiply without and with rounding, respectively, the two signed operands X1 and Y1, and store the products in the A-Accumulator and B-Accumulator, respectively. Assume that the contents of the X1 and Y1 Data Input Registers, the A-Accumulator, and the B-Accumulator before execution of the two instructions are:

$x1 = \$000007$

$y1 = \$400000$

$a = \$00000000000000$

$a2 = = \$00 \quad a1 = \$000000 \quad a0 = \$000000$

$b = \$00000000000000$

$b2 = \$00 \quad b1 = \$000000 \quad b0 = \$000000$

PROCEDURE:

```
TYPE: change x1 $000007  y1 $400000  a 0  b 0<return>
      display x1 y1 a a2 a1 a0 b b2 b1 b0<return>
      asm p:$503 mpy x1,y1,a<return>
      asm p:$504 mpyr x1,y1,b<return>
      change pc $503<return>
      step 2<return>
      display x1 y1 a a2 a1 a0 b b2 b1 b0<return>
```

RESULTS:

$x1 = \$000007$

$y1 = \$400000$

$a = \$00000003800000$

$a2 = \$00 \quad a1 = \$000003 \quad a0 = \$800000$

$b = \$00000004000000$

$b2 = \$00 \quad b1 = \$000004 \quad b0 = \$000000$

Note that the product in the B-Accumulator is a convergently rounded version of the product in the A-Accumulator.

Example 4.22: Signed Multiply-Accumulate with and without Rounding

PROBLEM:

Execute the two instructions **MAC -X1,Y1,A** and **MACR -X1,Y1,B** to multiply without and with rounding, respectively, the two signed operands X1 and Y1, and store the results in the A-Accumulator and the B-Accumulator, respectively. Assume that the contents of the X1 and Y1 Data Input Registers, the A-Accumulator, and the B-Accumulator before execution of the two instructions are:

$$x1 = \$400000$$

$$y1 = \$000007$$

$$a = \$00000005100000$$

$$b = \$00000005100000$$

PROCEDURE:

```
TYPE: change x1 $400000 y1 $000007<return>
      change a $00000005100000 b $00000005100000<return>
      display x1 y1 a b<return>
      asm p:$505 mac -x1,y1,a<return>
      asm p:$506 macr -x1,y1,b<return>
      change pc $505<return>
      step 2<return>
      display x1 y1 a b<return>
```

RESULTS:

$$x1 = \$400000$$

$$y1 = \$000007$$

$$a = \$00000001900000$$

$$b = \$00000002000000$$

Note that the result in the B-Accumulator ($00000002000000) is a convergently rounded version of the result in the A-Accumulator ($00000001900000).

c) ***Other Arithmetic Instructions:***

The **transfer data ALU register (TFR)**, **negate accumulator (NEG)**, **absolute value (ABS)**, **normalize accumulator iteration (NORM)**, and **clear accumulator (CLR)** instructions are summarized in Table 4.5.

TABLE 4.5. Transfer Data ALU Register, Negate Accumulator, Absolute Value, Normalize Accumulator Iteration, and Clear Accumulator Instructions.

Mnemonic	*Operation*	*Assembler Syntax*	*Source S*	*Destination D*
TFR	S → D	TFR S,D [PM]	A	B
			B	A
			X1,X0	A,B
			Y1,Y0	A,B
NEG	0-D → D	NEG D [PM]		A,B
ABS	|D| → D	D [PM]		A,B
NORM*	if $\bar{E}.U.\bar{Z}$=1,then ASL → D Rn-1 → Rn Else if E=1,then ASR → D RN+1 → Rn Else NOP	NORM Rn,D		A,B
CLEAR	0 → D	CLR D [PM]		A,B

* Rn = Address register R0-R7
E = Extension bit (fifth bit of the status register)
U = Unnormalized bit (fourth bit of the status register)
Z = Zero bit (second bit of the status register)

EXAMPLE 4.23: Transfer Data ALU Register

PROBLEM:

Execute the instruction **TFR A,B** to transfer data from the A-Accumulator to the B-Accumulator. Assume that the contents of the A-Accumulator and the B-Accumulator before execution of the instruction are:

$$a = \$00111111222222$$

$$b = \$00000000000000$$

PROCEDURE:

```
TYPE: change a $00111111222222  b 0<return>
      display a b<return>
      asm p:$507 tfr a,b<return>
      change pc $507<return>
      step<return>
      display a b<return>
```

RESULTS:

$$a = \$00111111222222$$

$$b = \$00111111222222$$

EXAMPLE 4.24: Negate accumulator

PROBLEM:

Execute the instruction **NEG A** to subtract the A-Accumulator from zero and store the result in the A-Accumulator. Assume that the contents of the A-Accumulator before execution of the instruction are:

$$a = \$00333333333333$$

PROCEDURE:

```
TYPE: change a $00333333333333<return>
      display a<return>
      asm p:$508 neg a<return>
      change pc $508<return>
      step<return>
      display a<return>
```

RESULT:

$$a = \$ffcccccccccccd$$

Note that the result can be checked by complementing each bit of the number $00333333333333 to form its one's complement $ffcccccccccccc, then adding one to the one's complement number to form the two's complement number $ffcccccccccccd.

EXAMPLE 4.25: Absolute value

PROBLEM:

Execute the instruction **ABS A** to take the absolute value of the A-Accumulator and store the result in the A-Accumulator. Assume that the contents of the A-Accumulator before execution of the instruction are:

$$a = \$ff900000000000$$

PROCEDURE:

```
TYPE: change a $ff900000000000<return>
      display a<return>
      asm p:$509 abs a<return>
      change pc $509<return>
      step<return>
      display a<return>
```

RESULT:

$$a = \$00700000000000$$

Note that the absolute value of the decimal number -0.875 ($ff900000000000) is the decimal number 0.875 ($00700000000000).

EXAMPLE 4.26: Normalize Accumulator Iteration

PROBLEM:

Execute the instruction **NORM R0,B** to perform one iteration of the normalization of the B-Accumulator.

i. *Accumulator Extension Bit is Set*:

Assume that the contents of the B-Accumulator, Address Register R0, and the Status Register (SR) before execution of the instruction are:

b = $22888888000000

$b2$ = $22 $b1$ = $888888 $b0$ = $000000

$r0$ = $*fffa*

sr = $0020

PROCEDURE:

If the Extension Bit (bit-5) of the Status Register (SR) is set before execution of the instruction, then the B-Accumulator is arithmetically shifted right one bit position and the Address Register R0 is incremented by one.

```
TYPE: change b $22888888000000  r0 $fffa sr $0020<return>
      display b b2 b1 b0 r0 sr<return>
      asm p:$50a norm r0,b<return>
      change pc $50a<return>
      step<return>
      display b b2 b1 b0 r0 sr<return>
```

RESULTS:

b = $11444444000000

$b2$ = $11 $b1$ = $444444 $b0$ = $000000

$r0$ = $*fffb*

sr = $0020

ii. *Accumulator Extension is not in Use*:

Assume that the contents of the B Accumulator Register, the R0 Address Register, and the Status Register before execution of the instruction are:

b = $00111111000000

$r0$ = $*ffff*

sr = $0010

PROCEDURE:

If the Extension Bit (bit-5) of the Status Register (SR) is cleared, and the Unnormalized Bit (bit-4) is set, before execution of the instruction; then the B-Accumulator is arithmetically shifted left one bit position and the Address Register R0 is decremented by one.

```
TYPE: change r0 $ffff b $00111111000000 sr $0010<return>
      display b r0 sr<return>
      asm p:$50b norm r0,b<return>
      change pc $50b<return>
      step<return>
      display b r0 sr<return>
```

RESULTS:

$$b = \$00222222000000$$

$$r0 = \$fffe$$

$$sr = \$0010$$

EXAMPLE 4.27: Clear Accumulator

PROBLEM:

Execute the instruction **CLR A** to clear the A-Accumulator. Assume that the contents of the A-Accumulator before execution of the instruction are:

$$a = \$11222222333333$$

PROCEDURE:

```
TYPE: change a $11222222333333<return>
      display a<return>
      asm p:$50c clr a<return>
      change pc $50c<return>
      step<return>
      display a<return>
```

RESULT:

$$a = \$00000000000000$$

4.4.3 Logic Instructions:

The Logic instructions use the Data ALU to perform all logic-type operations. Source operands for the Logic instructions, except ANDI and ORI, are contained in the Data ALU's Input Registers or Accumulators; destinations for the results generated by execution of the Logic instructions, except ANDI and ORI, are either the A-Accumulator or the B-Accumulator. The destination for the results generated by execution of the ANDI or ORI instructions is the Mode Register MR, Condition Code Register CCR, or Operating Mode Register OMR. The Logic instructions, except ANDI and ORI, allow up to two parallel-move operations to be executed concurrently with execution of the instruction's opcode. The parallel-move operations use the XD and YD Data Buses. As a result, the parallel-move operations can transfer new pre-fetched data from X- and or Y-Memory to the Data ALU for use in subsequent instructions; and can transfer the results generated by the execution of previous instructions from the Data ALU to X- and or Y-Memory.

The Logic instructions execute in one instruction cycle and can affect all of the bits in the Condition Code Register. The Logic instructions are:

AND	logical AND
ANDI	AND immediate control register
EOR	Logical exclusive OR
LSL	Logical shift left
LSR	Logical shift right
NOT	Complement
OR	Logical inclusive OR
ORI	OR immediate control register
ROL	Rotate left
ROR	Rotate right

a) ***Logic Instructions:***

The Logic instructions **AND (AND), logic inclusive OR (OR), logic exclusive OR (EOR)**, and **logic complement NOT (NOT)** are summarized in Table 4.6. These instructions are 24-bit operations which are performed on bits 24-47 of the A-Accumulator or B-Accumulator.

TABLE 4.6. Logic Instructions.

Mnemonic	*Operation*	*Assembler Syntax*	*Source S*	*Destination D*
AND	$S . D \rightarrow D$	AND S,D [PM]	X1,X0	A,B
			Y1,Y0	A,B
OR	$S + D \rightarrow D$	OR S,D [PM]	X1,X0	A,B
			Y1,Y0	A,B
EOR	$S + D \rightarrow D$	EOR S,D [PM]	X1,X0	A,B
			Y1,Y0	A,B
NOT	$\overline{D} \rightarrow D$	NOT D [PM]		A,B

b) ***Shift and Rotate Logic Instructions:***

The **logic-shift accumulator left (LSL), logic-shift accumulator right (LSR), rotate accumulator left (ROL)**, and **rotate accumulator right (ROR)** logic instructions are summarized in Table 4.7.

TABLE 4.7. Shift and Rotate Logic Instructions.

Mnemonic	Operation	Assembler Syntax	Destination D
LSL	Logical Shift Accumulator Left (C ← [47 ← 24] ← 0)	LSL D [PM]	A, B
ROL	Rotate Accumulator Left (C ← [47 ← 24] ← C)	ROL D [PM]	A, B
LSR	Logical Shift Accumulator Right (0 → [47 → 24] → C)	LSR D [PM]	A, B
ROR	Rotate Accumulator Right (C → [47 → 24] → C)	ROR D [PM]	A, B

EXAMPLE 4.28: Logic-Shift Accumulator Right

PROBLEM:

Execute the instruction **LSR A** to logic-shift bits 24-47 of the A-Accumulator one bit position to the right and store the result in the A-Accumulator. Assume that the contents of the A-Accumulator and the Status Register (SR) before execution of the instruction are:

$$a = \$00222222444444$$

$$sr = \$0001$$

PROCEDURE:

The **logic-shift accumulator right** instruction **LSR A** is used to logic-shift bits 24-47 of the A-Accumulator one bit position to the right and store the result in the A-Accumulator. The remaining bits of the A-Accumulator are not affected. The Carry Bit (bit 0) of the SR receives the least significant bit (bit 24) which is shifted out of the A-Accumulator. A zero is shifted into bit-47 of the A-Accumulator.

```
TYPE: change a $00222222444444<return>
      change sr $0001<return>
      display a sr<return>
      asm p:$600 lsr a<return>
      change pc $600<return>
      step<return>
      display a sr<return>
```

RESULTS:

$$a = \$00111111444444$$

$$sr = \$0000$$

Note that the Carry Bit (bit 0) of the SR receives the zero which is shifted out of bit-24 of the A-Accumulator.

EXAMPLE 4.29: Rotate Accumulator Left

PROBLEM:

Execute the instruction **ROL A** to rotate bits 24-47 of the A-Accumulator one bit position to the left and store the results in the A-Accumulator. Assume that the contents of the A-Accumulator and the Status Register (SR) before execution of the instruction are:

$$a = \$00222222000000$$

$$sr = \$0001$$

PROCEDURE:

The **rotate accumulator left** instruction **ROL A** is used to rotate bits 24-47 of the A-Accumulator one bit position to the left and store the results in the A-Accumulator. The remaining bits of the A-Accumulator are not affected. The Carry Bit (bit 0) of the SR receives the bit which is shifted out of the bit-47 position of the A-Accumulator. The previous value of the carry bit is shifted into bit-24 of the A-Accumulator.

```
TYPE: change a $00222222000000  sr $0001<return>
      display a sr<return>
      asm p:$601 rol a<return>
      change pc $601<return>
      step<return>
      display a sr<return>
```

RESULTS:

$$a = \$00444445000000$$

$$sr = \$0000$$

Note that the Carry Bit of the SR receives the "zero" which is shifted out of bit-47 of the A-Accumulator and the previous value of the Carry Bit is shifted into bit-24 of the A-Accumulator.

c) ***Immediate Logic Instructions:***

The **AND immediate control register (ANDI)** and the **OR immediate control register (ORI)** logical instructions are summarized in Table 4.8.

TABLE 4.8. Immediate Logic Instructions.

Mnemonic	*Operation*	*Assembler Syntax*	*Destination D*
ANDI	#xx.D → D	AND(I) #xx,D	MR,CCR,OMR
ORI	#xx+D → D	OR(I) #xx,D	MR,CCR,OMR

EXAMPLE 4.30: AND Immediate to Control Register

PROBLEM:

Execute the instruction **ANDI #$f3,MR** to logic-AND the contents of the Mode Register (MR) with the 8-bit immediate operand $f3 and store the results in the MR. Assume that the contents of the Status Register before execution of the instruction are:

$$sr = \$af7f$$

PROCEDURE:

```
TYPE: change sr $af7f<return>
      display sr<return>
      asm p:$602 andi #$f3,mr<return>
      change pc $602<return>
      step<return>
      display sr<return>
```

RESULT:

$$sr = \$a37f$$

4.4.4 Bit Manipulation Instructions:

There are two basics groups of bit manipulation instructions.

a) ***First Group:***

The first group of bit manipulation instructions is:

BCLR	Bit test and clear
BSET	Bit test and set
BCHG	Bit test and change
BTST	Bit test on memory

This group tests the state of any single bit in a memory location and then depending on the instruction's opcode sets, clears, inverts, or does not change, the memory bit. The Carry Bit of the Condition Code Register will reflect the result of the bit test. The first group of bit manipulation instructions is summarized in Table 4.9.

EXAMPLE 4.31: Bit Test on Memory

PROBLEM:

Execute the instruction **BTST #2,x:$200** to test bit-2 of X-Memory location $200. The contents of X-Memory location $200 and the Status Register before execution of the instruction are:

$$x{:}\$200 \quad \$000004$$

$$sr = \$0000$$

TABLE 4.9. First Group of Bit Manipulation Instructions.

Mnemonic	*Operation*	*Assembler Syntax*
BCLR*	$D(n) \rightarrow C$ $0 \rightarrow D(n)$	BCLR #n,X:<ea> BCLR #n,Y:<ea>
BSET*	$D(n) \rightarrow C$ $1 \rightarrow D(n)$	BSET #n,X:<ea> BSET #n,Y:<ea>
BCHG*	$D(n) \rightarrow C$ $\overline{D(n)} \rightarrow D(n)$	BCHG #n,X:<ea> BCHG #n,Y:<ea>
BTST*	$D(n) \rightarrow C$	BTST #n,X:<ea> BTST #n,Y:<ea>

* D(n) = n^{th} bit of the 24-bit destination operand field.
$0 \leq n \leq 23$.

PROCEDURE:

The **bit test on memory** instruction **BTST #2,x:$200** is used to test bit-2 of X-Memory location $200. After the bit is tested, its state is reflected in the Carry Bit of the CCR.

```
TYPE: change x:$200 $000004  sr $0000<return>
      display x:$200 sr<return>
      asm p:$603 btst #2,x:$200<return>
      change pc $603<return>
      step<return>
      display x:$200 sr<return>
```

RESULTS:

x:$200 $000004

sr = 0001

Note that the value of bit-2 of X-Memory location $200 is one. Therefore the Carry Bit of the CCR (bit 0 of SR) is set to a "one".

---

b) ***Second Group:***

The second group of the bit manipulation instructions is:

JCLR	Jump if bit clear
JSET	Jump if bit set
JSCLR	Jump to subroutine if bit clear
JSSET	Jump to subroutine if bit set

This group tests the state of any single bit in a memory location and jumps (or jumps to subroutine) if the bit is "set" or "clear". The second group of the bit manipulation instructions is summarized in Table 4.10.

TABLE 4.10. Second Group of Bit Manipulation Instructions.

Mnemonic	*Operation*	*Assembler Syntax*
JSET	If S(n)* = 1 Then xxxx → PC Else PC+1 → PC	JSET #n,X:<ea>,xxxx JSET #n,Y:<ea>,xxxx
JCLR	If S(n)* = 0 Then xxxx → PC Else PC+1 → PC	JCLR #n,X:<ea>,xxxx JCLR #n,Y:<ea>,xxxx
JSSET	If S(n)* = 1 Then SP+1 → SP PC → SSH SR → SSL xxxx → PC Else PC+1 → PC	JSSET #n,X:<ea>,xxxx JSSET #n,Y:<ea>,xxxx
JSCLR	If S(n)* = 0 Then SP+1 → SP PC → SSH SR → SSL xxxx → PC Else PC+1 → PC	JSCLR #n,X:<ea>,xxxx JSCLR #n,Y:<ea>,xxxx

* S(n) = n^{th} bit in the source operand.

EXAMPLE 4.32: Jump if Bit Set

PROBLEM:

Execute the instruction **JSET #2,X:(R0),$1000** to test bit-2 of the X-Memory location pointed to by Address Register R0. The contents of X-Memory location $200, the Program Counter (PC), and Address Register R0 before execution of the instruction are:

x:$200 $000004

pc = $604

*r*0 = $0200

PROCEDURE:

The **jump if bit set** instruction **JSET #2,X:(R0),$1000** is used to test bit-2 of the X-Memory location pointed to by the Address Register R0. If the bit is set, the instruction in P-Memory location $1000 is fetched. Otherwise, the Program Counter is simply incremented by one.

```
TYPE: change x:$200 $000004  r0 $200<return>
      asm p:$604 jset #2,x:(r0),$1000<return>
      change pc $604<return>
      display x:$200 pc r0<return>
      step<return>
      display x:$200 pc r0<return>
```

RESULTS:

x:$200 $000004

pc = $1001

$r0$ = $0200

Note that bit-2 of the X-Memory location $200 is "set". Therefore the instruction in P-Memory location $1000 is fetched and, afterward, the value of the PC is incremented by one from $1000 to $1001. If bit-2 was "clear", the PC would have been simply incremented by one from $604 to $605.

4.4.5 Loop Instructions:

The loop instructions are **DO** and **ENDDO**. The interruptible **DO** instruction initiates the beginning of a hardware **DO** loop. The **ENDDO** instruction can be used to terminate a hardware **DO** loop before that loop is completed. The contents of the Loop Counter (LC) specifies the number of times a hardware **DO** loop is to repeated. The contents of the Loop Address (LA) Register indicates the location of the last instruction word in a hardware **DO** loop. The LC and LA can be pushed to the System Stack, thereby allowing hardware **DO** loops to be interrupted or nested; they can be unstacked by a **RET** instruction or an **ENDDO** instruction.

EXAMPLE 4.33:

PROBLEM:

Execute the program **DOLOOP.LOD** to multiply the signed fraction data A(i) by the signed fraction data B(i), for i=1, 2, and 3. The contents of X-Memory locations $1000-$1002, Y-Memory locations $2000-$2002, and X-Memory locations $1500-$1502 before execution of the program are:

x:$1000 $400000 $400000 $400000

x:$1500 $000000 $000000 $000000

y:$2000 $300000 $300000 $300000

PROCEDURE:

The signed fraction data A(1),A(2),A(3) and B(1),B(2),B(3) are stored in the X-Memory locations $1000-$1002 and Y-Memory locations $2000-$2002, respectively. The multiplication results are stored in X-Memory locations $1500-$1502. The **DOLOOP.ASM** assembler program is shown in Figure 4.1.

```
        move #$1000,r0
        move #$2000,r4
        move #$1500,r1
        move              x:(r0),x0
        move                                y:(r4),y0
        do   #3,_end
        mpyr x0,y0,a      x:(r0)+,x0        y:(r4)+,y0
        move              a,x:(r1)+
_end
```

Figure 4-1 DOLOOP.ASM Assembler Program

1. POWER-ON AND BOOT THE PC.
2. INSERT the *DSP56000/1 Demonstration Software #2* diskette into Drive "A".
3. TYPE:

```
display off<return>
change x:$1000..$1002 $400000<return>
change x:$1500..$1502 0<return>
change y:$2000..$2002 $300000<return>
display on x:$1000..$1002 y:$2000..$2002<return>
display on x:$1500..$1502<return>
load a:doloop<return>
step 6<return>
display la lc<return>
step 6<return>
display la lc<return>
```

The contents of the Loop Counter (LC), the Loop Address (LA) Register, X-Memory locations $1000-$1002, Y-Memory locations $2000-$2002, and X-Memory locations $1500-$1502 at the start and at the end of the DO LOOP are:

START OF DO LOOP

la = $000*b*

lc = $0002

x:$1000	$400000	$400000	$400000
y:$2000	$300000	$300000	$300000
x:$1500	$000000	$000000	$000000

END OF DO LOOP

la = $0000

lc = $0000

x:$1000	$400000	$400000	$400000
y:$2000	$300000	$300000	$300000
x:$1500	$180000	$180000	$180000

Note that the result of multiplying 1/2 ($400000) by 3/8 ($300000) is 3/16 ($180000). The DO instruction loads the LA register with the location of the last instruction word in the program loop ($000B).

EXAMPLE 4.34:

PROBLEM:

Execute the program DOLOOP1.LOD to multiply the signed integer data A(i) by the signed integer data B(i), for i=1,2, and 3. The contents of X-Memory locations $1000-$1002, Y-Memory locations $2000-$2002, and L-Memory locations $1500-$1502 before execution of the program are:

x:$1000 $000002 $000002 $000002

y:$2000 $000138 $000138 $000138

l:$1500 $000000000000 $000000000000

l:$1502 $000000000000

PROCEDURE:

The signed integer data A(1), A(2), A(3) and B(1), B(2), B(3) are stored in X-Memory locations $1000-$1002 and Y-Memory locations $2000-$2002, respectively. When multiplying two signed integer numbers, the contents of the accumulator is right-shifted immediately after the multiplication to obtain the correct results. The right-shift operation shifts the least significant bit out of the accumulator and shifts a sign-extension into the most significant bit. The results of the multiplications are stored in L-Memory locations $1500-$1502. A list of the assembler program **DOLOOP1.ASM** is shown in Figure 4.2.

```
TYPE: display off
      change x:$1000..$1002 $000002<return>
      change 1:$1500..$1502 0<return>
      change y:$2000..$2002 $000138<return>
      display on x:$1000..$1002 y:$2000..$2002<return>
      display on 1:$1500..$1502<return>
      display<return>
      load a:doloop1<return>
      break #1 pc>=$e
      go #1<return>
```

RESULTS:

The contents of X-Memory locations $1000-$1002, Y-Memory locations $2000-$2002, and L-Memory locations $1500-$1502 after execution of the program are:

x:$1000 $000002 $000002 $000002

y:$2000 $000138 $000138 $000138

l:$1500 $000000000270 $000000000270

l:$1502 $000000000270

Note that the result of multiplying 2 ($000002) by 312 ($000138) is 624 ($000000000270).

```
     move #$1000,r0
     move #$2000,r4
     move #$1500,r1
     move                x:(r0),x0
     move                              y:(r4),y0
     do   #3,_end
     mpy  x0,y0,a        x:(r0)+,x0    y:(r4)+,y0
     asr  a
     move                a10,l:(r1)+
_end
```

Figure 4-2 DOLOOP1.ASM Assembler Program

4.4.6 Program Control Instructions:

The Program Control instructions include jumps, conditional jumps, and other instructions which affect the PC and System Stack. These instructions do not allow parallel-move operations. These instructions can affect the Condition Code Register. The Program Control instructions are:

Jcc	Jump conditionally
JMP	Jump
JScc	Jump to subroutine conditionally
JSR	Jump to subroutine
NOP	No operation
REP	Repeat instruction
RESET	Reset on-chip peripheral devices
RTI	Return from interrupt
RTS	Return from subroutine
STOP	Stop instruction processing
SWI	Software interrupt
WAIT	Wait for interrupt

EXAMPLE 4.35: Repeat Next Instruction

PROBLEM:

Execute the two instructions **REP #3** and **ASR B** to repetitively execute the **ASR B** instruction three times. The **repeat next (REP)** instruction is summarized in Table 4.11. The contents of the B-Accumulator and the Loop Counter (LC) before execution of the two instructions are:

$$b = \$00888888000000 \qquad lc = \$0300$$

TABLE 4.11. Repeat Next Instruction.

Mnemonic	Operation	Assembler Syntax
REP	LC → TEMP; #xxx → LC Repeat Next Instruction Until LC = 1 TEMP → LC	REP #xxx REP X:<ea> REP Y:<ea> REP S

PROCEDURE:

```
TYPE: change b $00888888000000 lc $0300<return>
      display b lc<return>
      asm p:$700<return>
      rep #3<return>
      asr b<return>
      <esc>
      change pc $700<return>
      step<return>
      display b lc<return>
      step<return>
      display b lc<return>
      step<return>
      display b lc<return>
      step<return>
      display b lc<return>
```

RESULTS:

The contents of the B-Accumulator and the Loop Counter before and after the execution of the program, and after the execution of each **REP** instruction are:

Before program execution	b	=	\$00888888000000	lc	=	\$0300
After 1st REP instruction	b	=	\$00888888000000	lc	=	\$0002
After 2nd REP instruction	b	=	\$00444444000000	lc	=	\$0001
After 3rd REP instruction	b	=	\$00222222000000	lc	=	\$0300
After program execution	b	=	\$00111111000000	lc	=	\$0300

EXAMPLE 4.36:

PROBLEM:

Execute the program **PR.LOD** to test bit-7 of X-Memory location $100, and jump if the Carry Bit of the CCR is cleared. The **PR.LOD** assembler program is shown in Figure 4.3.

```
     org  p:$200
wait btst #7,x:$100
     jcc  wait
     end
```

Figure 4-3 PR.ASM Assembler Program

PROCEDURE:

The **PR.LOD** program is used to test bit-7 of X-Memory location $100. The state of the bit is reflected in the Carry Bit of the CCR. If the Carry Bit is cleared, then program execution continues at P-Memory location $200; if the Carry Bit is set, the PC is incremented by one to point to P-Memory location $203.

a) TYPE:

```
display off<return>
change x:$100 $40<return>
load a:pr<return>
disassemble<return>
```

```
P:$0200 0B7027 000100 = BTST #$7,X:>$100
P:$0202 0E0200        = JCC <$200
P:$0203 000000        = NOP
P:$0204 000000        = NOP
```

```
display on sr<return>
step 2<return>
```

RESULTS:

$$sr = \$0310$$

$$P{:}\$0200 \quad 0B7027 \quad 000100 \quad = \quad BTST \quad \$7,X{:}>\$100$$

Note that the Carry Bit (bit 0 of SR) of the CCR is cleared. Therefore program execution continues at the P-Memory location $200.

b) TYPE:

```
change x:$100 $80<return>
change pc $200<return>
step 2<return>
```

RESULTS:

$$sr = \$0311$$

$$P{:}\$0203 \quad 000000 = NOP$$

Note that the Carry Bit (bit 0 of SR) of the CCR is set. Therefore program execution continues at P-Memory location $203.

4.4.7 DSP56000/1 Instruction Set Summary:

The DSP56001 instruction set is summarized in Table 4.12.

TABLE 4.12. Instruction Set Summary.

Move Instructions	
LUA	Load updated address
MOVE	Move data
MOVEC	Move control register
MOVEM	Move program memory
MOVEP	Move peripheral data
Arithmetic Instructions	
ABS	Absolute value
ADC	Add with carry
ADD	Add
ADDL	Shift left then add
ADDR	Shift right then add
ASL	Arithmetic shift left
ASR	Arithmetic shift right
CLR	Clear
CMP	Compare
CMPM	Compare magnitude
DIV	divide iteration
MAC	Multiply/accumulate
MACR	Multiply/accumulate and round
MPY	Multiply
MPYR	Multiply and round
NEG	Negate
NORM	Normalize iteration
RND	Round
SBC	Subtract with carry
SUB	Subtract
SUBL	Shift left then subtract
SUBR	Shift right then subtract
Tcc	Transfer Conditionally
TFR	Transfer
TST	Test

TABLE 4.12. (*cont.*)

Logical Instructions	
AND	logical AND
ANDI	AND immediate control Register
EOR	Logical exclusive OR
LSL	Logical shift left
LSR	Logical shift right
NOT	Complement
OR	Logical inclusive OR
ORI	OR immediate control register
ROL	Rotate left
ROR	Rotate right
Bit Manipulation Instructions	
BCLR	Bit test and clear
BSET	Bit test and set
BCHG	Bit test and change
BTST	Bit test on memory
JCLR	Jump if bit clear
JSET	Jump if bit set
JSCLR	Jump to subroutine if bit clear
JSSET	Jump to subroutine if bit set
Loop Instructions	
DO	Start hardware loop
ENDDO	Exit from hardware loop
Program Control Instructions	
Jcc	Jump conditionally
JMP	Jump
JScc	Jump to subroutine conditionally
JSR	Jump to subroutine
NOP	No operation
REP	Repeat instruction
RESET	Reset on-chip peripheral devices
RTI	Return from interrupt
RTS	Return from subroutine
STOP	Stop instruction processing
SWI	Software interrupt
WAIT	Wait for interrupt

Note, this chapter covers most of the DSP5600/1 processor's instructions. For more information on the DSP56000/1 processor's instruction set, refer to Appendix A of the *DSP56000/DSP56001 Digital Signal Processor User's Manual.*

5

INTRODUCTION TO DIGITAL SIGNAL PROCESSING SYSTEMS

In this chapter, a *simple* DSP system consisting of an analog-to-digital (A/D) converter, a DSP56001 processor, and a digital-to-analog (D/A) converter is implemented and exercised. A DSP56000ADS Application Development System (ADS) is used to provide and control the DSP56001 processor; a DSP56ADC16EVB Evaluation Board (EVB) is used to provide and control the A/D converter and the D/A converter. The DSP56001 processor's Synchronous Serial Interface (SSI) port is used to accommodate serial data transfers from the A/D converter to the DSP56001 processor and from the DSP56001 processor to the D/A converter.

First, the set-up of the DSP56ADC16EVB Evaluation Board is described followed by a discussion of the DSP56001 processor's instructions which are needed to select the DSP56001 processor's on-chip SSI port and set the SSI port's control registers to receive data from the A/D converter and transmit data to the D/A converter. The DSP system is then exercised to demonstrate the "aliasing effect" produced by sampling an input signal at frequencies below the Nyquist sampling rate.

5.1 DSP SYSTEM

In the DSP System defined above and shown in Figures 5.1 and 5.2, an analog input signal is digitized by the A/D converter on the EVB. The digitized signal is output from the A/D converter as a 16-bit serial data stream delimited by a Frame Sync Output (FSO) signal. Via the receive channel of the DSP56001 processor's SSI port, the digitized signal is received by the DSP56001 processor located on the Application Development Module (ADM) of the ADS. The DSP56001 processor routes the digitized signal from the receive channel of its SSI port to the transmit channel of its SSI port. The algorithm which the DSP56001 processor executes to route the digitized signal without modification is developed in Section 5.4 of this chapter. (Note: in later chapters, this algorithm will be expanded to modify the digitized signal.) Via the transmit channel of the DSP56001 processor's SSI port, the digitized signal is output from the DSP56001 processor as a 16-bit serial data stream and routed to the serial input of the D/A converter located on the EVB. The D/A converter constructs an analog output signal from the digitized input signal. A reconstruction filter smooths the analog signal output from the D/A converter.

Figure 5-1 DSP System

In order for the D/A converter to be able to properly reconstruct the analog signal which was originally input to the A/D converter, the Nyquist Sampling Theorem must first be satisfied. That is, the analog signal input to the A/D converter must be sampled by the A/D converter at a rate greater than twice its highest frequency component (f_h). Sampling at this frequency ($2f_h$) is known as sampling at the Nyquist rate.

In practice, no practical analog input signal can be strictly bandlimited to the point where one can absolutely assure that the true Nyquist sampling rate is selected. Those high frequency components of the analog input signal which are greater than one half the Nyquist sampling rate introduce an error component into the digitized signal which diminishes a D/A converter's ability to exactly reconstruct the original analog input signal. In general, the better the analog input signal is bandlimited, the lower the energy of those high frequency components which violate the Nyquist Sampling Theorem and the better the reconstruction of the original analog input signal.

Another error component is introduced by the D/A converter itself through its use of infinite duration sinc functions in its signal reconstruction process. Because of the accumulative effect of these various error components, exact reconstruction of the original analog input signal can only be approximated.

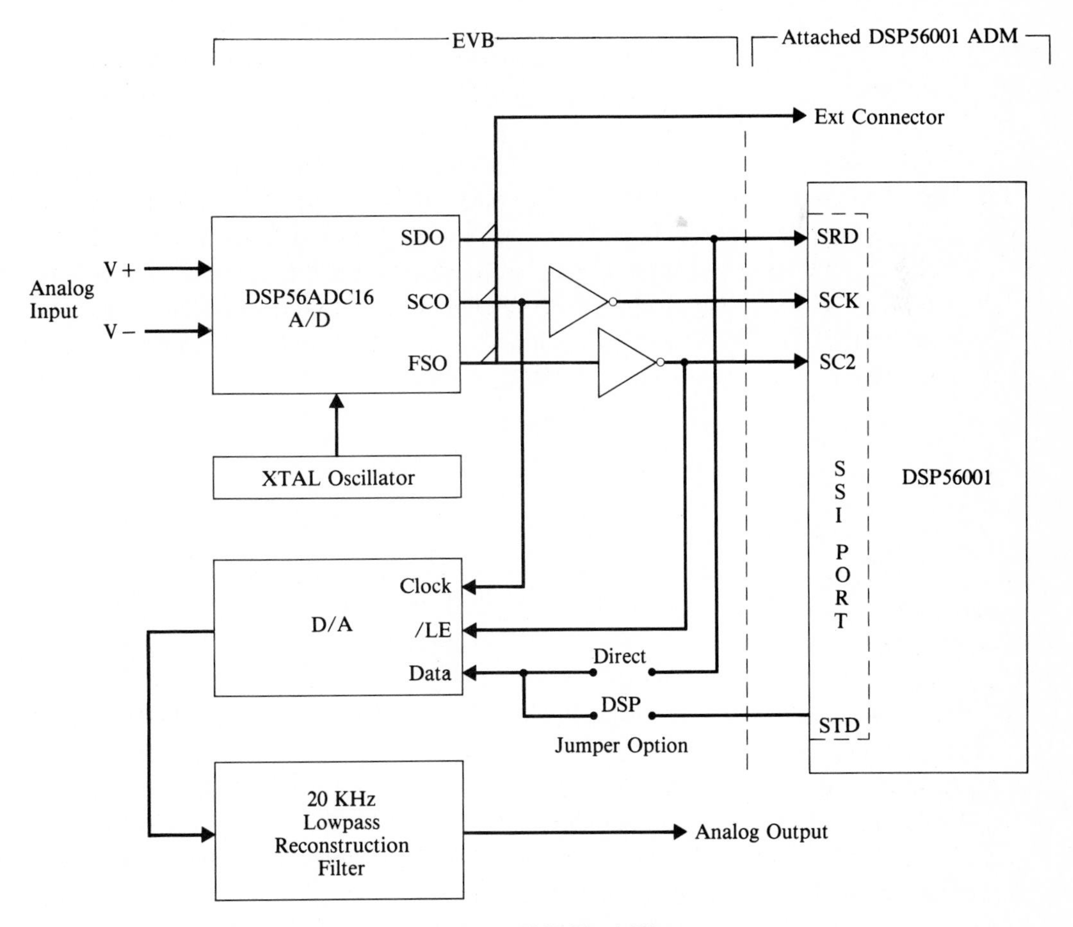

Figure 5-2 EVB Block Diagram

For the purposes of this book, if a practical analog input signal is sampled at the Nyquist rate, then a practical analog output signal can be reconstructed. If sampling of the input analog signal occurs below the Nyquist rate, then a practical analog output signal cannot be reconstructed. The effect of undersampling an analog input signal, i.e., sampling below the Nyquist rate, is known as "aliasing".

5.2 DSP56ADC16EVB EVALUATION BOARD (EVB)

5.2.1 EVB General Description

The EVB is an A/D and D/A conversion system that can be used in conjunction with the ADS. The major components of the EVB, shown in Figure 5.2, are:

1. one DSP56ADC16 16-bit, oversampling, Sigma- Delta A/D converter which can serially output 16-bit data samples at group rates up to 100 KHz per 16-bit sample,
2. one 16-bit D/A converter, and
3. one low pass reconstruction filter with a cut off frequency of 20 KHz.

5.2.2 Set-Up of the EVB for Operation with the ADS

1. As shown in Figure 5.3, connect one end of the ribbon cable supplied with the EVB to J2 on the EVB; connect the other end of the ribbon cable to J5 on the DSP56001 Application Development Module (ADM). If the cable does not insert easily, flip it over 180 degrees and try again.
2. The EVB can draw its power from the ADM. A power supply pigtail is supplied with the EVB for this purpose. As shown in Figure 5.3, connect the "forked" end of the pigtail's black wire to the GND screw terminal on the ADM's power block. Connect the "forked" end of the pigtail's red wire to the DIG +5V screw terminal on the ADM's power block. Connect the black banana plug on the other end of the pigtail's black wire to the J3 GND banana jack on the EVB. Connect the red banana plug on the other end of the pigtail's red wire to the J3 +5V banana jack on the EVB.
3. On the EVB, install jumpers JP2, JP4, JP6, JP8, JP9, JP11, and JP13, and remove jumpers JP1, JP3, JP5, JP7, JP10, JP12, and JP14. This jumper configuration allows operation of the EVB with:
   a. a 6.144 MHz clock supplied to the Clock Input (CLKIN) of the DSP56ADC16 A/D converter (the A/D converter serially outputs 16-bit samples at a rate which is equal to the clock frequency supplied to the Clock Input of the A/D converter divided by 128, e.g., a 6.144 MHz Clock Input frequency causes the A/D converter to serially output 16-bit samples at a group rate of 48 KHz per 16-bit sample),
   b. a 16-bit-wide Frame Sync Output (FSO) from the DSP56ADC16 A/D converter,
   c. a single ended analog input signal supplied via BNC2 to the Analog+Input (V_{in+}) of the DSP56ADC16 A/D converter,
   d. the DSP56001 processor's SSI port transmit data output routed via connector/pin J2-1 to the digital input of the D/A converter, and
   e. a low pass reconstruction filter at the analog output of the D/A converter.

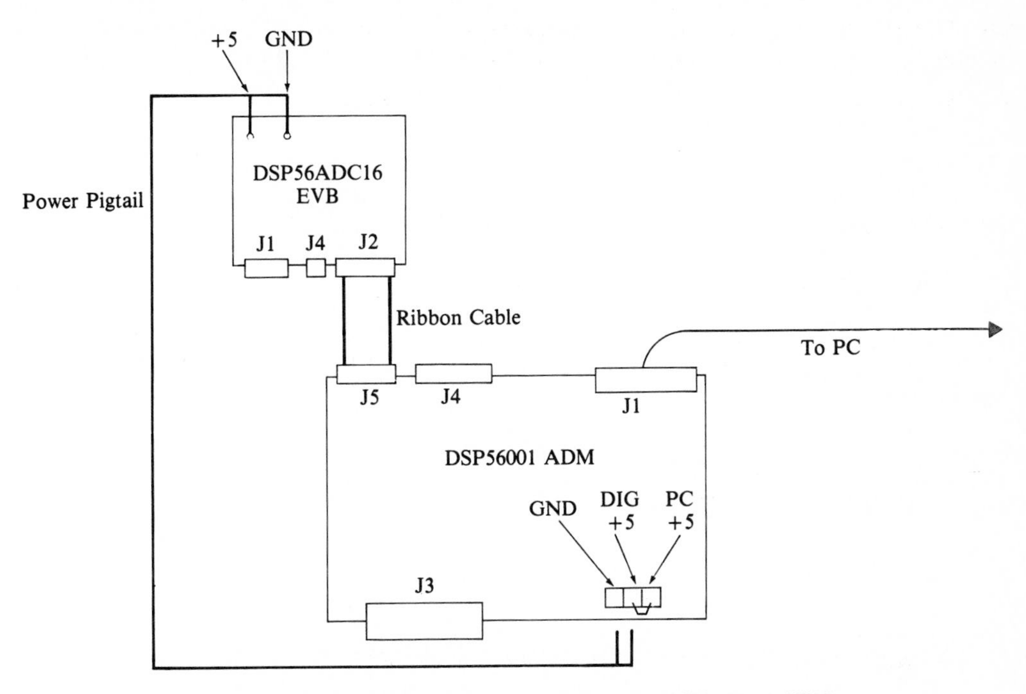

Figure 5-3 Connection of the EVB to the DSP56001 ADM

5.3 DSP56001 PROCESSOR SSI PORT PINS:

The SSI port has six dedicated pins as shown in Figure 5.4:

1. Serial Transmit Data (STD),
2. Serial Receive Data (SRD),
3. Serial Clock (SCK)
4. Serial Control (SC0),
5. Serial Control (SC1), and
6. Serial Control (SC2).

As shown in Figure 5.2, the SSI port's STD, SRD, SCK and SC2 pins are used to interface the DSP56001 processor on the ADM to the A/D converter and D/A converter on the EVB. The STD pin is used to output data from the SSI port's Serial Transmit Shift Register. This data is routed to the input of the D/A converter. The SRD pin is used to receive data into the SSI port's Receive Data Shift Register. This data comes from the A/D converter. The SCK pin is used by the SSI port to accept an external clock. The source of this external clock (inverted form) is the Serial Clock Output pin of the A/D converter. This external clock (non-inverted form) also clocks the D/A converter. The SC2 pin is used by the SSI port to accept an external frame sync. The source of this external frame sync (inverted form) is the Frame Sync Output pin of the A/D converter. This external frame sync (non-inverted form) also asserts the Latch Enable input pin of the D/A converter.

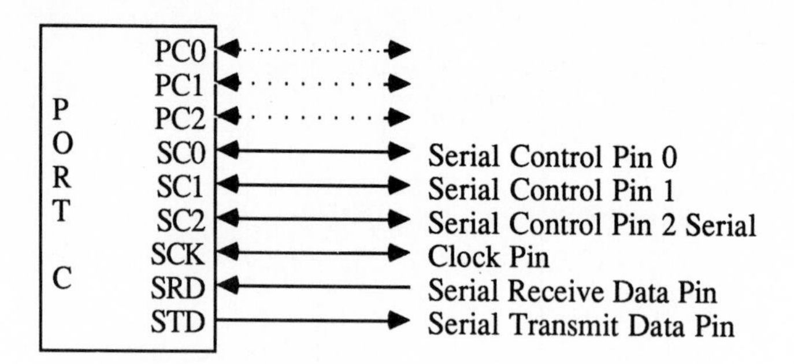

Figure 5-4 Synchronous Serial Interface Pins

5.4 SETTING UP THE DSP56001 PROCESSOR'S SSI PORT:

As shown in Figure 5.5, the DSP56001 processor has:

a. a 47-pin memory expansion Port A;

b. a 15-pin Port B which can be operated either as a general purpose I/O port or as a parallel Host MPU/DMA Interface on-chip peripheral port; and

c. a 9-pin Port C where each individual pin can be selected to operate as a general purpose I/O pin or as an on-chip peripheral function pin. The on-chip peripherals include a Serial Communication Interface (SCI) port and a Synchronous Serial Interface (SSI) port. Any mix of general purpose I/O pins or on-chip peripheral function pins can be selected.

Each of the Host, SCI, and SSI on-chip peripherals have their own control, status, data, etc., registers. The DSP56001 processor accesses each of these registers as if they were memory-mapped I/O. These on-chip memory-mapped registers are accessed at X-Memory locations $FFC0-$FFFF as shown in Figure 5.6.

5.4.1 SSI Port Selection

In the DSP System, the SSI port of the DSP56001 processor which is located on the ADM is used to receive serial data from the A/D converter on the EVB and transmit serial data to the D/A converter on the EVB. The DSP56001 processor's Port C Control Register (PCC) has nine bits which, in a one-on-one fashion, individually select each of the nine pins of Port C to operate either as a general purpose I/O pin or as a SCI and or SSI peripheral function pin. Port C pins can be configured as general purpose I/O pins by clearing the corresponding PCC register bits; Port C pins can be configured as SCI and or SSI function pins by setting the corresponding PCC register bits. The PCC Register can be accessed at X-Memory location $FFE1. The following DSP56001 processor instructions select Port C to operate as both a SCI port and a SSI port:

```
pcc     equ     $0001ff
        movep   #pcc,x:$FFE1   ;write PCC
```

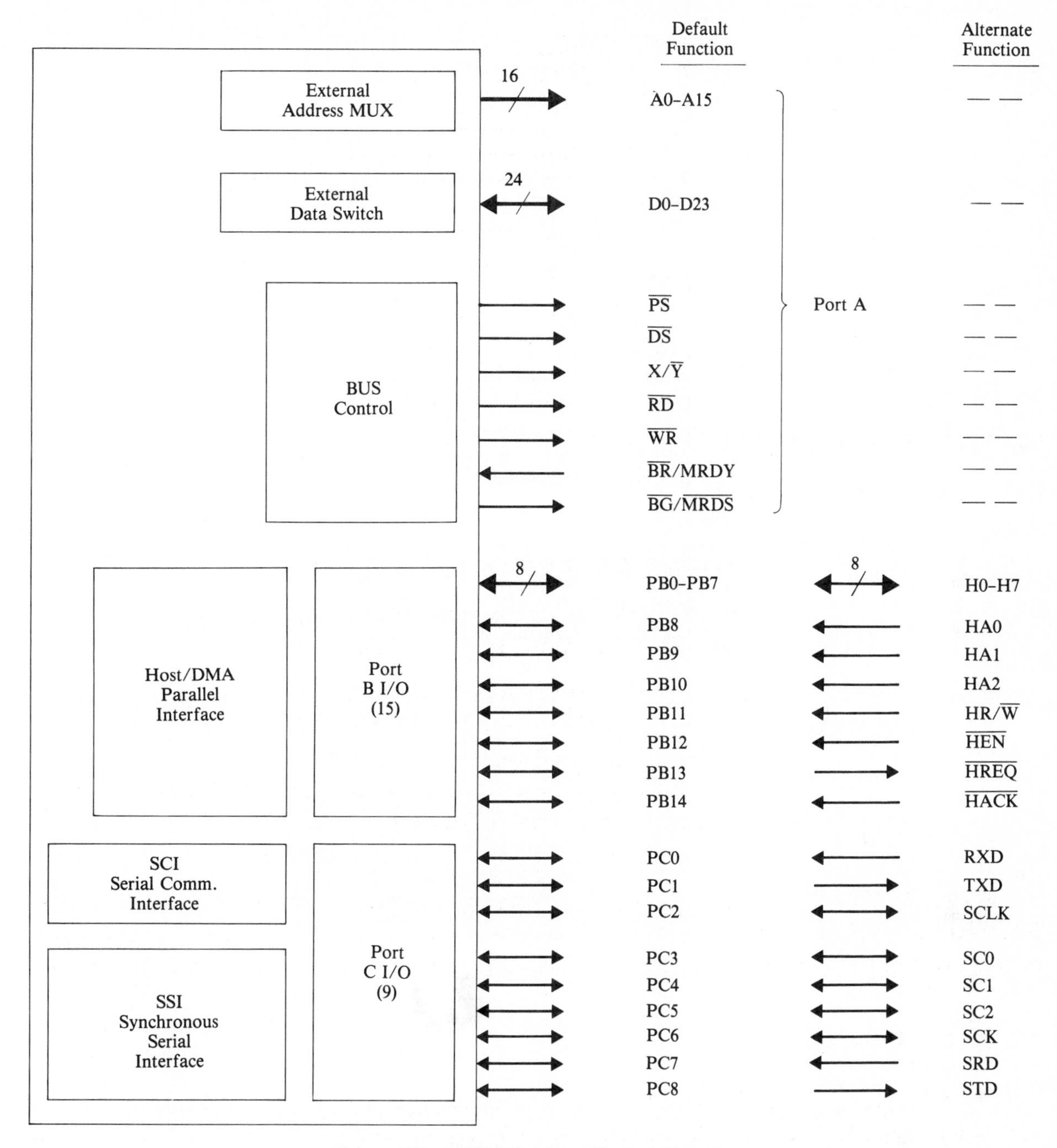

Figure 5-5 DSP56001 Input/Output Diagram

5.4.2 SSI Port Control Register A (CRA)

The operation of the SSI Port is directed by two 16-bit read/write Control Registers CRA and CRB. The SSI Registers are shown in Figure 5.7. CRA controls the SSI port's:

a. internal bit-clock generator rate,

b. internal frame divider rate, and

c. data word length.

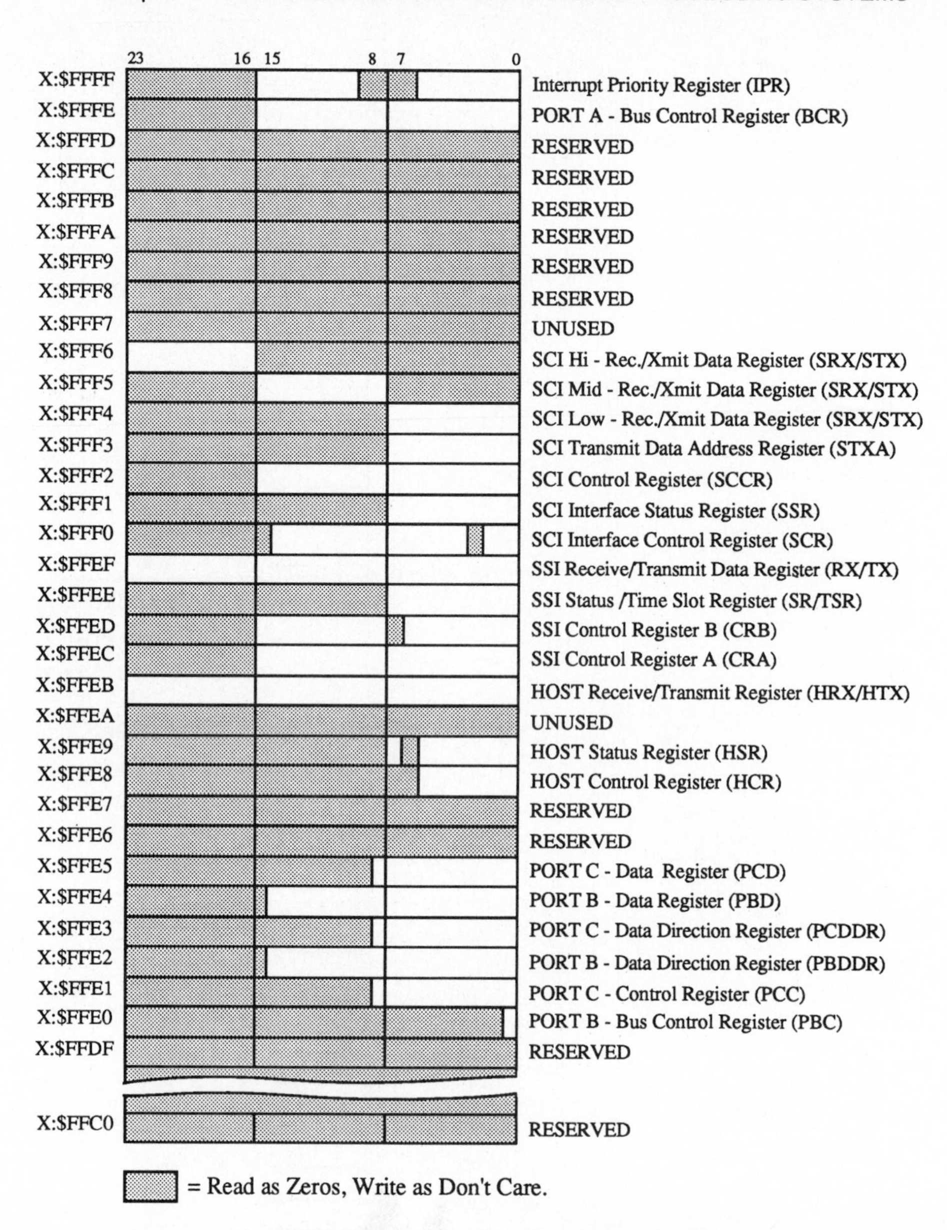

Figure 5-6 DSP56001 Peripheral Memory Map

The data word length can be 8, 12, 16, or 24 bits. CRA's Word Length control bits 13 (WL0) and 14 (WL1) select the word length according to Table 5.1.

TABLE 5.1. SSI Port Word Length.

WL1	*WL0*	*No. of Bits/Word*
0	0	8
0	1	12
1	0	16
1	1	24

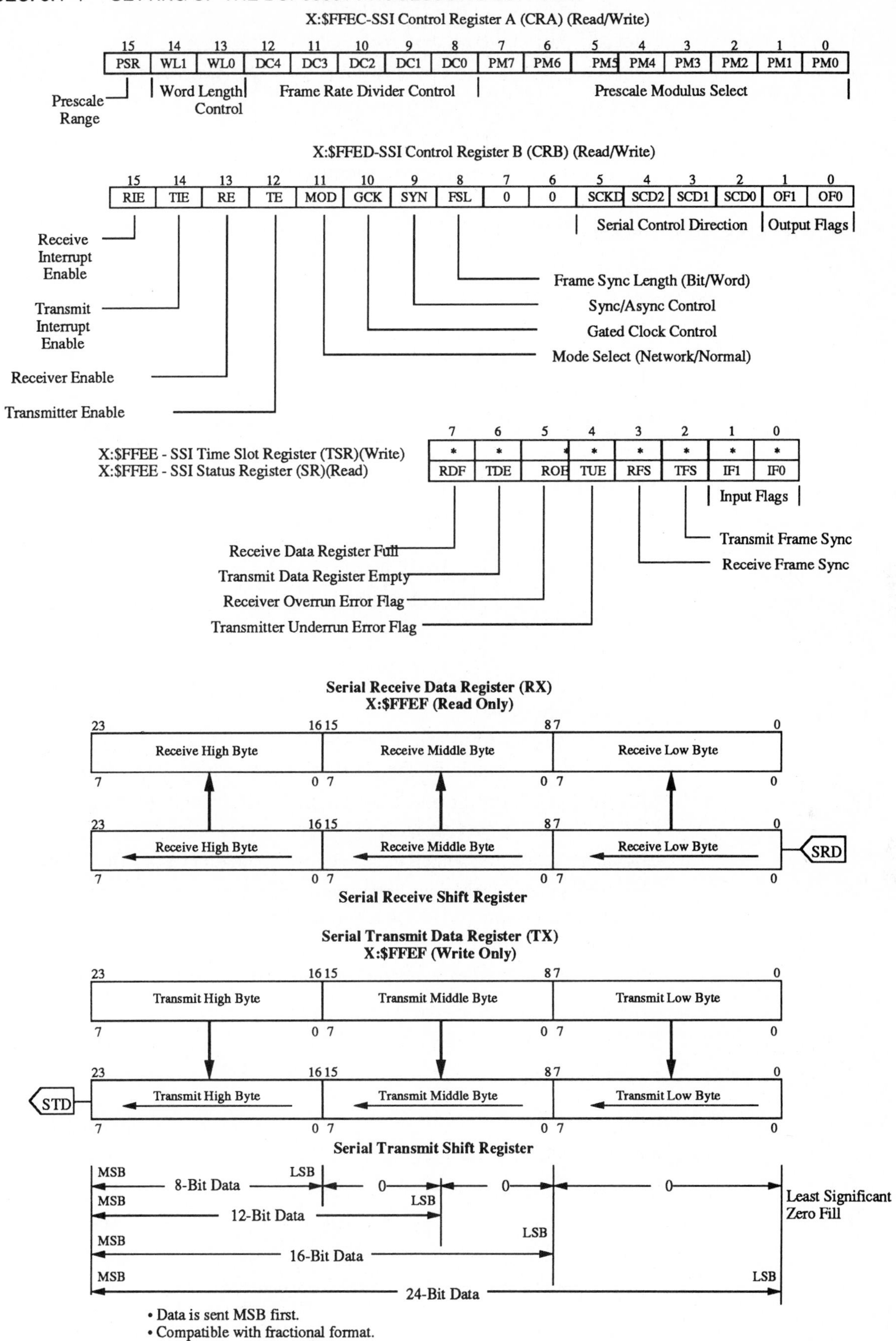

Figure 5-7 Synchronous Serial Interface Registers

In the DSP System, the SSI port's internal bit-clock and frame generators are not used. The DSP56001 instructions shown below set CRA to select a 16-bit word length. CRA occupies X-Memory location $FFEC.

```
cra     equ     $004000
        movep   #cra,x:$FFEC ;CRA pattern for word
                             ;length=16 bits
```

5.4.3 SSI Port Control Register B (CRB)

The operation of the SSI Port is directed by two 16-bit read/write Control Registers CRA and CRB. CRB controls the SSI port's:

a. multifunction pins SC2, SC1 and SC0;
b. serial output flag control bits;
c. input versus output direction of pins SCK, SC2, SC1, and SC0;
d. operating modes;
e. transmitter and receiver enables; and
f. transmitter and receiver interrupt enables.

In the DSP System shown in Figures 5.1 and 5.2, the SSI port's SC1 and SC0 pins are not used. CRB occupies X-Memory location $FFED. The DSP56001 instructions shown below set up CRB where:

a. CRB bit 4 is cleared to select the direction of the SC2 pin to be an input (external frame sync);
b. CRB bit 5 is cleared to select the direction of the SCK pin to be an input (external clock for both the Transmit and Receive Shift Registers);
c. CRB bit 6 is cleared to select the Transmit Shift Register to transmit data MSB first and the Receive Shift Register to receive data MSB first;
d. CRB bits 7 and 8 are cleared to select a word length (16-bit) frame sync;
e. CRB bit 9 is set to select both the transmitter and the receiver to operate in the synchronous mode and use a common clock (SCK pin) and a common frame sync (SC2 pin);
f. CRB bit 10 is cleared to select the "continuous" clock operating mode;
g. CRB bit 11 is cleared to select the "normal" operating mode where one data word (16-bits) is transmitted or received per frame;
h. CRB bits 12 and 13 are both set to enable the transmitter and receiver respectively;
i. CRB bits 14 and 15 are both cleared to disable the transmit and receive interrupts respectively; and
j. CRB bits 3-0 are not pertinent to this particular SSI port configuration, nevertheless they are cleared.

```
crb     equu    $003200
        movep   #crb,x:$FFED    ;CRB pattern for
                                ;continuous ck,
                                ;synch, normal mode
                                ;word long frame SYnc:
                                ;FSL=0;ext ck/fs
```

5.4.4 SSI Port Status Register

The SSI port's Status Register (SSISR) is an 8-bit read-only register used by the DSP56001 processor to interrogate the status and serial input flags of the SSI port. The Status Register occupies X-Memory location $FFEE.

5.4.5 SSI Port Receive Registers

The SSI port's Receive Shift Register is a 24-bit register that receives incoming data from the Serial Receive Data (SRD) pin. The Receive Data Register (RX) is a 24-bit register that accepts data in parallel from the Receive Shift Register when the Receive Shift Register becomes full. RX occupies X-Memory location $FFEF. The Receive Data Register Full Flag (RDF) of the SSI port's Status Register (bit 7) is set when the contents of the Receive Shift Register are transferred to RX. RDF is cleared when the DSP56001 processor reads the Receive Data Register or when the DSP56001 processor is reset.

In the DSP System, the Receive Shift Register accepts serial input data via the Serial Receive Data (SRD) pin when an external frame sync asserts Serial Control Pin 2 (SC2). The Receive Shift Register is externally clocked via the Serial Clock (SCK) pin.

5.4.6 SSI Port Transmit Registers

The SSI port's Transmit Shift Register is a 24-bit register which transmits (outputs) serial data via the Serial Transmit Data (STD) pin. The Transmit Data Register (TX) is a 24-bit register which supplies data in parallel to the Transmit Shift Register. Data to be transmitted is written into the TX Register and is automatically transferred to the Transmit Shift Register when a frame sync is asserted. TX occupies X-Memory location $FFEF. The Transmit Data Register Empty Flag (TDE) of the SSI port's Status Register (bit 6) is set when the contents of TX are transferred to the Transmit Shift Register. TDE is cleared when the DSP56001 processor writes new data to TX or when the DSP56001 processor is reset.

In the DSP System, the Transmit Shift Register transmits serial output data via the Serial Transmit Data (SRD) pin when an external frame sync asserts Serial Control Pin 2 (SC2). The Transmit Shift Register is externally clocked via the Serial Clock (SCK) pin.

5.4.7 Reading RX and Writing to TX

For the DSP System, the DSP56001 processor reads RX and transfers its contents without modification to TX. The DSP56001 instructions shown below:

1. polls the RDF in the SSI port's Status Register (bit 7),
2. reads RX when RDF becomes set (RX contains data supplied by the DSP56ADC16 A/D converter at a 48 KHz sampling rate),
3. moves the contents of RX to TX (TX contains data to be supplied to the D/A converter), and
4. loops back to Step 1.

```
poll    btst    #7,x:$FFEE  ;test for A/D data
        jcc     poll        ;loop until RDF bit = 1

        movep   x:$FFEF,a   ;move A/D data to "a"

        move    a,x:$FFEF   ;send A/D data to D/A
        jmp     poll        ;loop indefinitely
```

5.5 EXERCISING THE "DSP SYSTEM"

To exercise the DSP System, explained in Section 5.1, an analog input signal is provided from a function generator to the analog input of the EVB. The analog input signal is monitored by one channel of a dual trace oscilloscope. The analog output signal from the EVB is monitored by the other channel of the oscilloscope. The connection diagram is as shown in Figure 5.8.

The DSP56001 processor located on the ADM executes the absolute load file **evb.lod** to route the digitized form of the analog signal. The file **evb.lod** is the assembled and linked result of the assembler program shown in Figure 5.9 and developed in Section 5.4.

1. POWER DOWN THE PC AND FUNCTION GENERATOR.
2. CONNECT THE FUNCTION GENERATOR'S OUTPUT TO THE EVB'S "BNC2" INPUT (A/D INPUT).

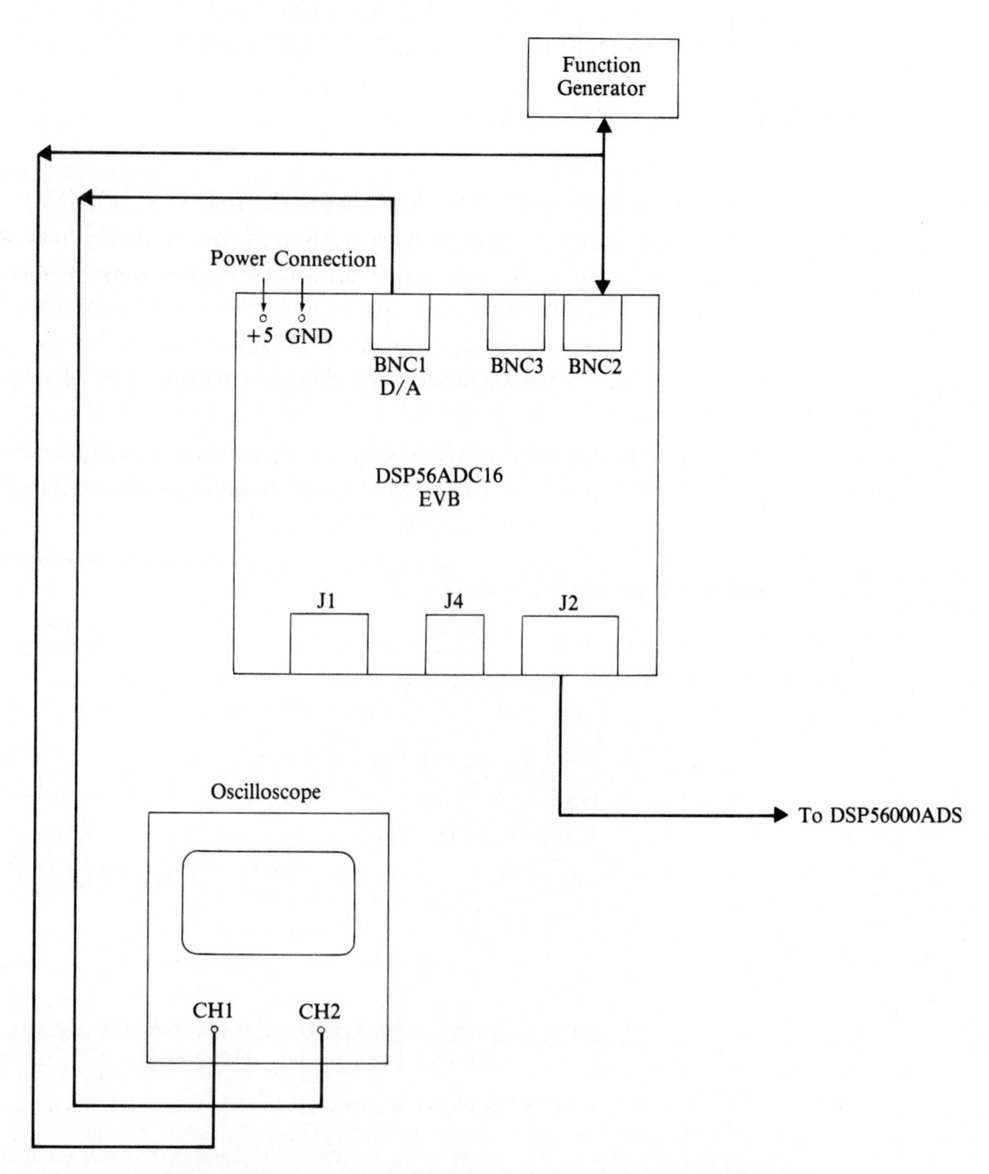

Figure 5-8 Connection Diagram for DSP System in Section 5.5

3. CONNECT THE FUNCTION GENERATOR'S OUTPUT TO THE OSCILLOSCOPE'S CHANNEL ONE INPUT.
4. CONNECT THE EVB'S "BNC1" OUTPUT (D/A OUTPUT) TO THE OSCILLOSCOPE'S CHANNEL TWO INPUT.
5. POWER ON THE FUNCTION GENERATOR.
6. SET THE FUNCTION GENERATOR TO OUTPUT A 2 VOLT (PEAK-TO-PEAK), 5 KHZ SINE WAVE.
7. POWER ON AND BOOT THE PC.
8. INSERT THE "DSP56000/1 DEMONSTRATION SOFTWARE #1" DISKETTE INTO DRIVE "A".
9. Depending on your PC's configuration, INVOKE THE "ADS56000 USER INTERFACE SOFTWARE" PROGRAM from Drive B or the hard disk.
10. TYPE **load a:evb.lod<return>** to load the assembled and linked version of the assembler program shown in Figure 5.9.
11. TYPE **go<return>** to begin execution of the program.
12. OBSERVE THE INPUT AND OUTPUT SIGNALS ON THE OSCILLOSCOPE. Note that the output signal is a good reconstruction of the input signal.
13. VARY THE FUNCTION GENERATOR'S OUTPUT FREQUENCY (f) FROM 0.5 TO 30 KHZ.
14. OBSERVE THE INPUT AND OUTPUT SIGNALS ON THE OSCILLOSCOPE. Note that the output signal is a good reconstruction of the input signal at frequencies from 0.5 KHz to just below 20 KHz. This should be the case since both of the following are (and must) be satisfied:
    a. the 48 KHz sampling rate selected for the A/D converter, and the D/A converter, satisfies the Nyquist Sampling Theorem for input frequencies up to 24 KHz, e.g., 2 x 24 KHz = 48 KHz; and
    b. input frequencies up to just below 20 KHz are less than the 20 KHz cutoff frequency of the low pass reconstruction filter at the output of the D/A converter.

    At input frequencies above 20 KHz, the 20 KHz cutoff frequency of the reconstruction filter itself defeats reconstruction of the input signal.
15. SET THE FUNCTION GENERATOR TO OUTPUT A 1 VOLT AMPLITUDE, 5 KHZ SQUARE WAVE.
16. OBSERVE THE INPUT AND OUTPUT SIGNALS ON THE OSCILLOSCOPE.
17. VARY THE FREQUENCY (f) OF THE SQUARE WAVE INPUT SIGNAL FROM 0.5 TO 20 KHZ.
18. OBSERVE THE INPUT AND OUTPUT SIGNALS ON THE OSCILLOSCOPE. Recall that a square wave signal can be decomposed into an infinite number of odd harmonic sine wave signals. It follows then that the output signal is reconstructed from odd harmonic sine wave signals having frequencies (nf) less than one half the 48 KHz sampling frequency *and* less than the 20 KHz cutoff frequency of the reconstruction filter. As the frequency of the input square wave signal approaches 20 KHz, fewer and fewer odd harmonic components are available to contribute to the reconstruction of the signal. As a result, the quality of the signal reconstruction suffers.
19. TYPE **force b<return>** to halt execution of the program.

20. TYPE **quit<return>** IF YOU WANT TO EXIT THE "ADS56000 USER INTERFACE SOFTWARE" PROGRAM.

```
; Program start address

    org     p:$40

; Set up ADS board in case of force break instead of force
; reset

         movep #0,x:$FFFE      ;set bcr to zero
         movec #0,sp           ;init stack pointer
         movec #0,sr           ;clear loop flag

; Set up the SSI for operation with the EVB
; The following code sets Port C to function as SCI/SSI

        move    #$0,a0           ;zero PCC to cycle it
        movep   a0,x:$FFE1

pcc     equ     $0001ff
        movep   #pcc,x:$FFE1     ;write PCC

; The following code sets the SSI port's CRA and CRB
; control registers for external continuous clock,
; synchronous, normal mode.


cra     equ     $004000
        movep   #cra,x:$FFEC  ;CRA pattern for word
                              ;length=16 bits

crb     equ     $003200
        movep   #crb,x:$FFED  ;CRB pattern for continuous
                              ;ck,synch,normal mode
                              ;word long frame sync:
                              ;FSL=0;ext ck/fs

;*********************************************************
; Actual read A/D and write D/A
;*********************************************************

; The following code polls the RDF flag in the SSI port's
; SR  and waits for RDF=1, then reads the RX register to
; retrieve data from the A/D converter.
; Sample rate is controlled by EVB.

poll       btst     #7,x:$FFEE
           jcc      poll         ;loop until RDF bit = 1

           movep    x:$FFEF,a    ;get A/D converter data

; Write DSP56ADC16 A/D converter data to the D/A converter

           move     a,x:$FFEF   ;write the D/A via SSI
                                ; TX reg.
           jmp      poll        ;loop indefinitely
           end
```

Figure 5-9 EVB.ASM Assembler Program for Reading A/D and Writing D/A.

5.6 EFFECT OF ALIASING ON AN ANALOG OUTPUT SIGNAL

The DSP System set-up of Section 5.5 can be used to show the effect of aliasing on the output signal, e.g., providing digitized samples to the D/A converter at less than the Nyquist sampling rate. The output sample rate can be reduced to 48/N KHz if the DSP56001 program writes one sample to the D/A converter for every N samples it reads from the A/D converter.

The DSP56001 instructions shown in Figure 5.10 are a modified version of the code used in Section 5.5. Note that the code segment

```
start

        do      #$2f,_end       ;48 iteration loop
poll1   btst    #7,x:$FFEE
        jcc     poll1           ;loop until RDF=1

        movep   x:$FFEF,a       ;get A/D data
_end

poll2   btst    #7,x:$FFEE
        jcc     poll2           ;loop until RDF=1
        movep   x:$FFEF,a       ;get A/D data
```

is included to reduce the output sample rate from 48 Khz to 1 KHz.

The DSP56001 processor located on the ADM executes the absolute load file **evbm.lod** to route the digitized form of the analog signal. The file **evbm.lod** is the assembled and linked result of the assembler program shown in Figure 5.10.

1. POWER DOWN THE PC AND FUNCTION GENERATOR.
2. CONNECT THE FUNCTION GENERATOR'S OUTPUT TO THE EVB'S "BNC2" INPUT (A/D INPUT).
3. CONNECT THE FUNCTION GENERATOR'S OUTPUT TO THE OSCILLOSCOPE'S CHANNEL ONE INPUT.
4. CONNECT THE EVB'S "BNC1" OUTPUT (D/A OUTPUT) TO THE OSCILLOSCOPE'S CHANNEL TWO INPUT.
5. POWER ON THE FUNCTION GENERATOR.
6. SET THE FUNCTION GENERATOR TO OUTPUT A 2 VOLT (PEAK-TO-PEAK), 5 KHZ SINE WAVE.
7. POWER ON AND BOOT THE PC.
8. INSERT THE "DSP56000/1 DEMONSTRATION SOFTWARE #1" DISKETTE INTO DRIVE "A".
9. Depending on your PC's configuration, INVOKE THE "ADS56000 USER INTERFACE SOFTWARE" PROGRAM from Drive B or the hard disk.
10. TYPE **load a:evbm.lod<return>** to load the assembled and linked version of the assembler program shown in Figure 5.10.
11. TYPE **go<return>** to begin execution of the program.
12. OBSERVE THE INPUT AND OUTPUT SIGNALS ON THE OSCILLOSCOPE. Note that since the 1 KHz output sample rate is less than the 10 KHz Nyquist sample rate (2 x 5 KHz = 10 KHz), aliasing occurs on the output signal and, as a result, the analog input signal cannot be reconstructed.
13. TYPE **force b<return>** to halt execution of the program.
14. TYPE **quit<return>** IF YOU WANT TO EXIT THE "ADS56000 USER INTERFACE SOFTWARE" PROGRAM.

```
; Program start address

     org     p:$40

; Set up ADS board in case of force break instead of force
; reset

            movep #0,x:$FFFE     ;set bcr to zero
            movec #0,sp          ;init stack pointer
            movec #0,sr          ;clear loop flag

; Set up the SSI for operation with the EVB
; The following code sets port C to function as SCI/SSI

        move     #$0,a0           ;zero PCC to cycle it
        movep    a0,x:$FFE1

pcc     equ      $0001ff
        movep    #pcc,x:$FFE1     ;write PCC

; The following code sets the SSI CRA and CRB control
; registers for external continuous clock, synchronous,
; normal mode.

cra     equ      $004000
        movep    #cra,x:$FFEC  ;CRA pattern for word
                               ;length=16 bits

crb     equ      $003200
        movep    #crb,x:$FFED  ;CRB pattern for continuous
                               ;ck,synch,normal mode
                               ;word long frame SYnc:
                               ;FSL=0;ext ck/fs

;**********************************************************
; Actual read A/D and write D/A
;**********************************************************

; The following code writes one sample to the D/A
; converter for every 48 samples it reads from the A/D
; converter.

start

            do       #$2f,end
poll1       btst     #7,x:$FFEE
            jcc      poll1        ;loop until RDF bit = 1

            movep    x:$FFEF,a    ;get A/D converter data
end

poll2       btst     #7,x:$FFEE
            jcc      poll2        ;loop until RDF bit = 1

            movep    x:$FFEF,a    ;get A/D converter data

; Write DSP56ADC16 A/D converter data to the D/A converter

            move     a,x:$FFEF   ;write the D/A via SSI
                                 ; TX reg.
            jmp      start       ;loop indefinitely

            end
```

Figure 5-10 EVBM.ASM Assembler Program for Reading A/D and Writing D/A.

5.7 HEARING THE EFFECTS OF ALIASING

If a keyboard, amplifier, and speaker are added to the DSP system, as shown if Figure 5.11, the effects of aliasing can be heard.

1. POWER DOWN THE PC.
2. CONNECT A KEYBOARD TO THE EVB'S "BNC2" INPUT.
3. CONNECT AN AMPLIFIER AND A SPEAKER TO THE EVB'S "BNC1" OUTPUT.
4. POWER ON AND BOOT THE PC.
5. INSERT THE "DSP56000/1 DEMONSTRATION SOFTWARE #1" DISKETTE INTO DRIVE "A".
6. Depending on your PC's configuration, INVOKE THE "ADS56000 USER INTERFACE SOFTWARE" PROGRAM from Drive B or the hard disk.
7. TYPE **load a:evb.lod<return>** to load the assembled and linked version of the assembler program (48 KHz sampling rate) shown in Figure 5.8.
8. TYPE **go<return>** to begin execution of the program.
9. PLAY SOME MUSICAL NOTES ON THE KEYBOARD AND LISTEN TO THE SOUNDS PRODUCED BY THE SPEAKER. Note that the musical quality of the sounds produced by the speaker at this time depend on the quality of the keyboard, amplifier, and speaker. By definition, the musical quality of these sounds is "good".

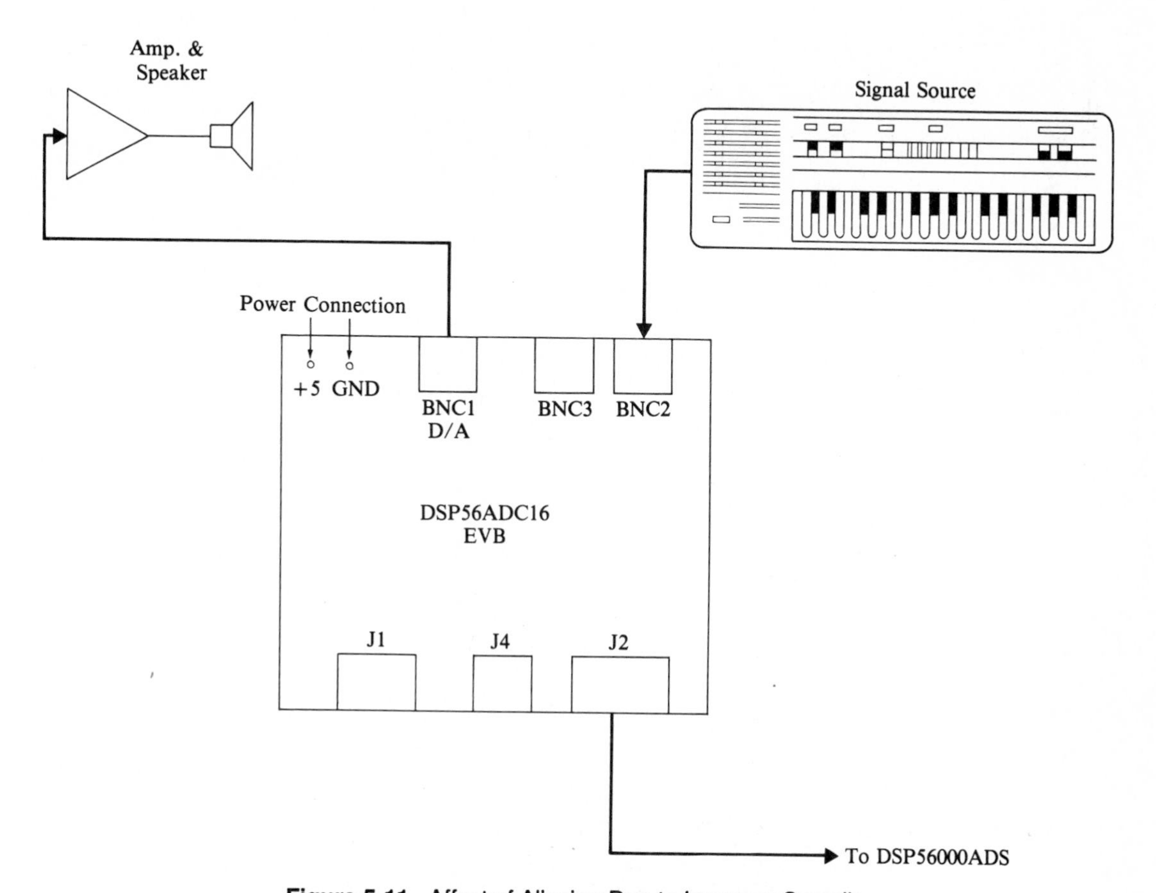

Figure 5-11 Affect of Aliasing Due to Improper Sampling

10 TYPE **force b<return>** to halt execution of the program.

11. TYPE **load a:evbm.lod<return>** to load the assembled and linked version of the assembler program (1 KHz sampling rate) shown in Figure 5.10.

12. PLAY SOME MUSICAL NOTES ON THE KEYBOARD AND LISTEN TO THE SOUNDS PRODUCED BY THE SPEAKER. Musical notes below 500 Hz, e.g., middle "A" is 440 Hz, can be reconstructed without aliasing. Musical notes above 500 Hz are aliased due to violation of the Nyquist sampling rate. By definition, aliased musical notes sound "bad" although some individuals might find that aliasing can produce "special effect" sounds that are not necessarily unpleasant.

13. TYPE **force b<return>** to halt execution of the program.

14. TYPE **quit<return>** IF YOU WANT TO EXIT THE "ADS56000 USER INTERFACE SOFTWARE" PROGRAM.

6

DESIGNING FIR FILTERS AND IMPLEMENTING THEM ON THE DSP56001 PROCESSOR

In this chapter, Finite Impulse Response (FIR) filters are designed, and implemented on the DSP56001 processor. The FIR filter has three distinct properties:

1. it is always stable,
2. it is always realizable, and
3. it can always be designed to have a linear phase response.

In this chapter, the transfer function and impulse response of a FIR filter are defined, followed by a description of how the ideal impulse response of an ideal FIR filter is obtained from its frequency response. Since the ideal impulse response sequence of a FIR filter is infinitely long, an appropriate "window" function is designed and used to create a practical, finite length impulse response sequence.

In the design process for a practical FIR filter, its desired frequency domain specifications are used with an appropriate window function to determine its impulse response h(i). The DSP56000/1 Demonstration Software program, and a special demonstration version of the Filter Design And Analysis System software program by Momentum Data Systems Inc., are used to demonstrate this design process. The DSP56001 instructions which implement the designed FIR filter on the DSP56001 processor are then described.

Low-pass, high-pass, band-pass, and band-stop FIR filters are then designed and implemented on the DSP system developed in Chapter 5 and again shown in Figure 6.1. The DSP56001 instructions which implement these four FIR filter types are included on the DSP56000/1 Demonstration Software #2 diskette.

Figure 6-1 DSP System

6.1 FIR FILTER FREQUENCY RESPONSE:

The relationship between a FIR filter's digital input sequence x(n) and digital output sequence y(n) can be written as:

$$y(n) = \sum_{i=0}^{N-1} b_i \; x(n-i) \tag{6.1}$$

where b_i are filter coefficients and N is the number of those coefficients. The filter coefficients b_i provide the characteristics of the filter through which the input sequence x(n) is passed to create the desired output sequence y(n). By both definition and design, the output sequence y(n) of a FIR filter (also known as a nonrecursive filter) is a function of its current and past inputs only.

The z-Transform transfer function H(z) corresponding to Equation 6.1 can be expressed as:

$$H(Z) = \frac{Y(z)}{X(z)} = \sum_{i=0}^{N-1} b_i \; z^{-i}$$

$$= \sum_{i=0}^{N-1} h(i) \; z^{-i} \tag{6.2}$$

where h(i), the impulse response of the FIR filter, is composed of b_i, the coefficients of the FIR filter.

To determine the frequency response of the FIR filter, the z-Transform variable (z) is evaluated on the unit circle as:

$$z = e^{j2\pi f} \tag{6.3}$$

where f = normalized frequency, i.e., the actual frequency divided by the sampling frequency.

By substituting Equation 6.3 into Equation 6.2, the frequency response of the FIR filter can be written as:

$$H(e^{j2\pi f}) = \sum_{i=0}^{N-1} h(i) \; e^{-j2\pi fi} \tag{6.4}$$

6.2 RELATING A PRACTICAL FIR FILTER TO AN IDEAL FIR FILTER:

In the design process for a practical FIR filter, its desired frequency domain specifications are used with an appropriate window function to determine its impulse response h(i). The coefficients h(i) (or b_i) of the transfer function H(z) in Equation 6.2 are designed so that the frequency response in Equation (6.4) meets certain frequency domain specifications. Once the impulse response is obtained, its b_i components are used to implement Equation 6.1 on the DSP56001 processor.

The first step in the design process is to obtain the frequency response of an ideal FIR filter. By definition, an ideal FIR filter has a magnitude of one in the passbands and zero elsewhere. The magnitude of the frequency response of an ideal low-pass, high-pass, band-pass, and band-stop filter are shown in Figure 6.2.

The frequency response of the ideal FIR filter can be written as:

$$H_1(e^{j2\pi f}) = \sum_{i=-\infty}^{\infty} h_1(i) \; e^{-j2\pi fi} \tag{6.5}$$

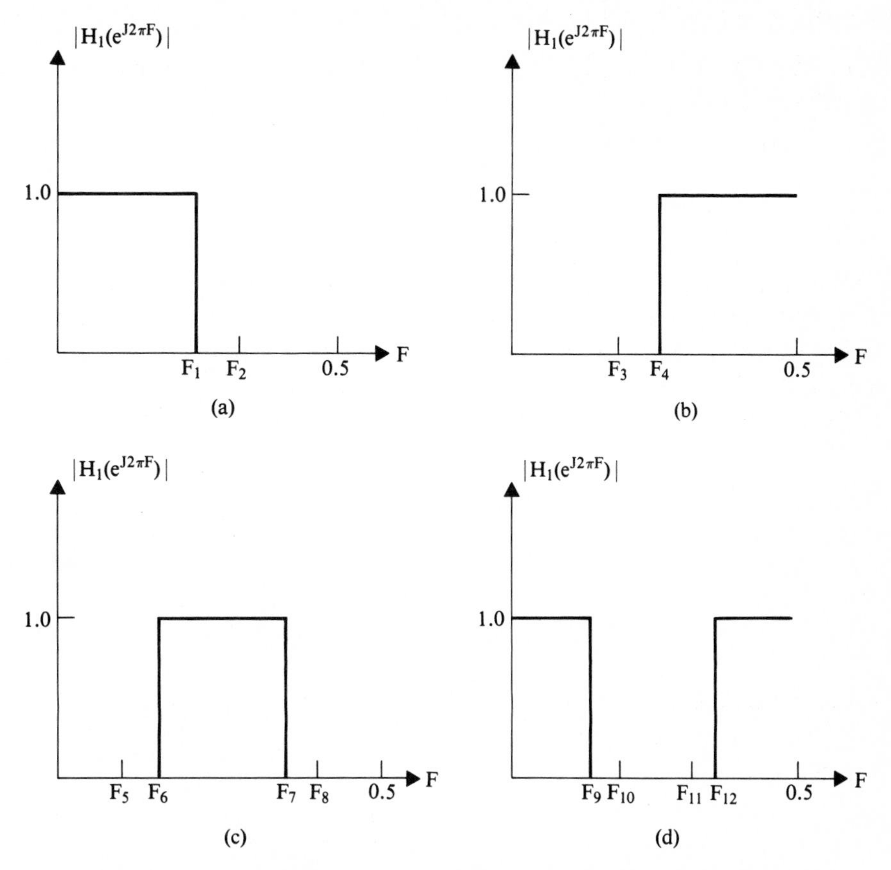

Figure 6-2 Ideal Filters. (a) Low-Pass (b) High-Pass (c) Band-Pass (d) Band-Stop

where $h_1(i)$, the ideal impulse response, can be calculated as:

$$h_1(i) \ = \ \int_{-0.5}^{0.5} H_1(e^{j2\pi f}) \ \ e^{j2\pi f i} df \tag{6.6}$$

and 0.5 is half the normalized sampling frequency.

A practical FIR filter can be implemented only if its impulse response is of finite length. Since the ideal impulse response $h_1(i)$ is an infinitely long sequence, it must be limited to a finite length. To do so, $h_1(i)$ is multiplied by an N-point window. The N-point impulse response $h_2(i)$ is then given by:

$$\begin{aligned} h_2(i) \ &= \ h_1(i)\, w(i) && -(N-1)/2 \ \le \ i \ \le \ (N-1)/2 \\ &= \ 0 && \text{otherwise} \end{aligned} \tag{6.7}$$

where N is an odd-valued integer. Some of the commonly used window types are shown in Figure 6.3.

Since a causal filter has zero impulse response for i less than zero, an FIR filter can be made to satisfy this condition by sufficiently delaying its impulse response (if necessary). That is, $h_2(i)$ can be delayed by (N-1)/2 so that the desired impulse response h(i) of a causal FIR filter can be obtained as:

$$h(i) \ = \ h_2[i - ((N-1)/2)], \text{ where } i \ = \ 0, 1, 2, \ldots, N-1 \tag{6.8}$$

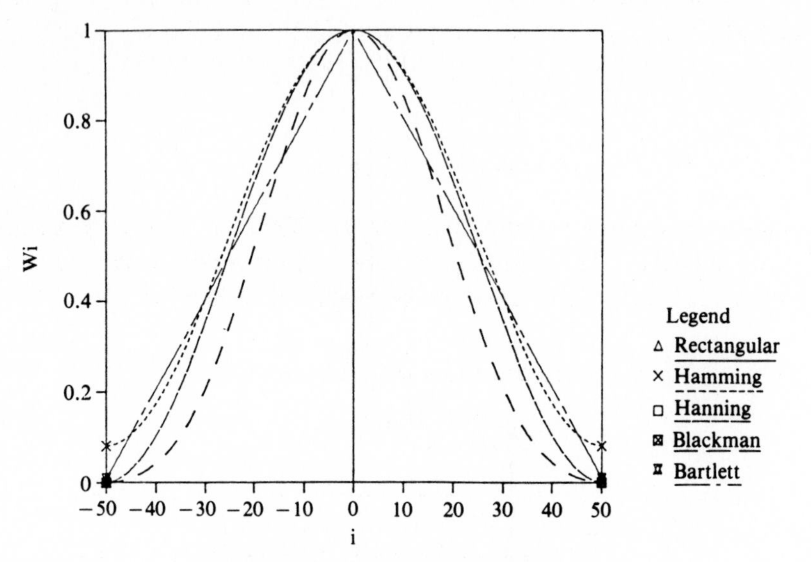

Figure 6-3 Commonly Used Windows

A causal FIR filter can be made to have a linear phase response by choosing h(i) to be equal to h(N-1-i) for i = 0,..,(N-1)/2. With this choice for h(i), Equation 6.4 can be rewritten as:

$$H(e^{j2\pi f}) = \sum_{i=0}^{(N-3)/2} h(i)\ e^{-j2\pi fi} + h(0.5(N-1))\ e^{-j2\pi f(N-1)/2} + \sum_{i=(N+1)/2}^{N-1} h(i)\ e^{-j2\pi fi} \tag{6.9}$$

By letting N-1-i=i, the last summation in Equation 6.9 can be expressed as:

$$\sum_{i=(N+1)/2}^{N-1} h(i)\ e^{-j2\pi fi} = \sum_{i=(N+1)/2}^{N-1} h(N-1-i)\ e^{-j2\pi fi} = \sum_{i=0}^{(N-3)/2} h(i)\ e^{-j2\pi f(N-1-i)} \tag{6.10}$$

From Equations 6.9 and 6.10:

$$\begin{aligned} H(e^{j2\pi f}) &= e^{-j\pi f(N-1)}\ [h(0.5(N-1)) + \sum_{i=0}^{(N-3)/2} h(i)(e^{j\pi f(N-1-2i)} + e^{-j\pi f(N-1-2i)})] \\ &= e^{-j\pi f(N-1)}\ [h(0.5(N-1)) + \sum_{i=0}^{(N-3)/2} 2h(i)\cos(\pi f(N-1-2i)) \end{aligned} \tag{6.11}$$

For real h(i), the term between the two brackets in Equation 6.11 is a real function of f. If this term is positive, then the phase of $H(e^{j2\pi f})$ is equal to $-\pi f(N-1)$. If this term is negative, then the phase of $H(e^{j2\pi f})$ is equal to $\pi-\pi f(N-1)$. In either case, the phase of $H(e^{j2\pi f})$ is a linear function of f.

6.3 SPECIFYING A PRACTICAL FIR FILTER:

The frequency domain specifications for practical FIR filters include the passband and stopband attenuations, and the passband and stopband cutoff frequencies. Figure 6.4 shows the frequency domain specifications for low-pass, high-pass, band-pass, and band-stop FIR filters, where:

α = Passband attenuation in dB.

β = Stopband attenuation in dB.

f_1 = Normalized passband cutoff frequency of the low-pass filter.

f_2 = Normalized stopband cutoff frequency of the low-pass filter.

f_3 = Normalized stopband cutoff frequency of the high-pass filter.

f_4 = Normalized passband cutoff frequency of the high-pass filter.

f_5 = Normalized lower stopband cutoff frequency of the band-pass filter.

f_6 = Normalized lower passband cutoff frequency of the band-pass filter.

f_7 = Normalized upper passband cutoff frequency of the band-pass filter.

f_8 = Normalized upper stopband cutoff frequency of the band-pass filter.

f_9 = Normalized lower passband cutoff frequency of the band-stop filter.

f_{10} = Normalized lower stopband cutoff frequency of the band-stop filter.

f_{11} = Normalized upper stopband cutoff frequency of the band-stop filter.

f_{12} = Normalized upper passband cutoff frequency of the band-stop filter.

6.4 USING THE WINDOW METHOD TO DESIGN A PRACTICAL FIR FILTER:

There are three well known methods for designing practical, linear phase response, FIR filters:

a. the frequency sampling method,

b. the "minimax" method, and

c. the window method.

The window method is used in this book. The procedure is:

1. COMPUTE THE IDEAL IMPULSE RESPONSE $h_I(i)$.

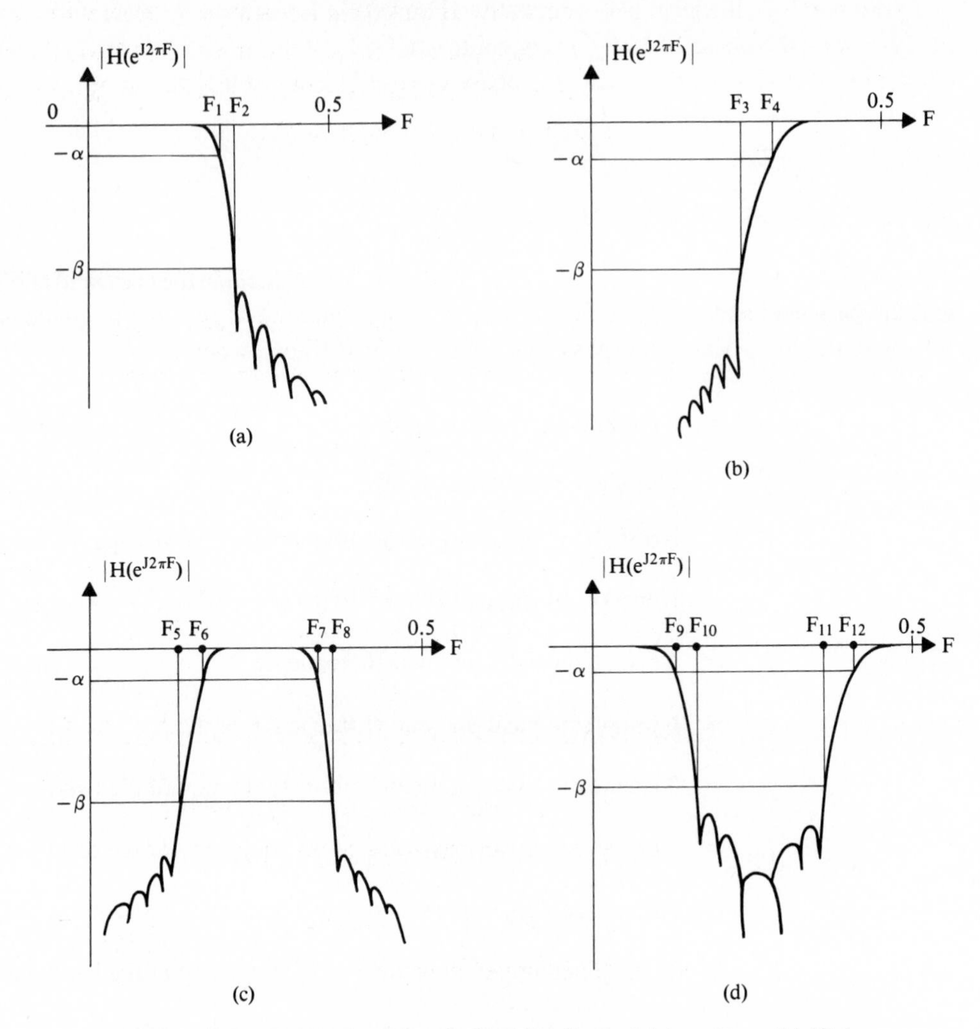

Figure 6-4 Frequency-Domain Specifications. (a) Low-Pass (b) High-Pass (c) Band-Pass (d) Band Stop

By using Equation 6.6, the ideal impulse response $h_1(i)$ of the low-pass, high-pass, band-pass and band-stop FIR filters can be calculated as:

$$Low-pass: \quad h_1(i) = \frac{\sin 2\pi f_1 i}{\pi i} \tag{6.12}$$

$$High-pass: \quad h_1(i) = -\frac{\sin 2\pi f_4 i}{\pi i} \tag{6.13}$$

$$Band-pass: \quad h_1(i) = \frac{\sin 2\pi f_7 i}{\pi i} - \frac{\sin 2\pi f_6 i}{\pi i} \tag{6.14}$$

$$Band-stop: \quad h_1(i) = \frac{\sin 2\pi f_9 i}{\pi i} - \frac{\sin 2p f_{12} i}{\pi i} \tag{6.15}$$

2. SELECT THE WINDOW TYPE.

To achieve a desired level of stopband attenuation, an appropriate window type must be selected from Table 6.1. For example, if the desired stopband attenuation is 50 dB, then, from Table 6.1, either the Hamming, Blackman or Kaiser windows are applicable.

TABLE 6.1. Window Type Selection Table.

Window	*Stopband Attenuation in dB*	*K*
Rectangular	21	2.00
Bartlett	25	4.00
Hanning	44	4.00
Hamming	53	4.00
Blackman	74	6.00
Kaiser (γ=2.12)	30	1.54
Kaiser (γ=4.54)	50	2.93
Kaiser (γ=7.76)	70	4.32
Kaiser (γ=8.96)	90	5.71

3. SELECT THE WINDOW LENGTH "N" WHICH CORRESPONDS WITH THE SELECTED WINDOW TYPE.

The following relationship can be used to approximate the necessary length N of the window:

$$N \geq K/\Delta f \tag{6.16}$$

where N is the least odd integer which satisfies Equation 6.16, K is a constant selected from Table 6.1, and

Δf = the transition band expressed in Hz.

$= f_2 - f_1$ for low-pass filter

$= f_4 - f_3$ for high-pass filter

$= \min[(f_6 - f_5), (f_8 - f_7)]$ for band pass filter

$= \min[(f_{10} - f_9), (f_{12} - f_{22})]$ for band-stop filter

4. FROM TABLE 6.2, SELECT THE WINDOW FUNCTION w(i) WHICH CORRESPONDS WITH THE SELECTED WINDOW TYPE.

TABLE 6.2. Window Functions.

Window Type	*Window Function* $w(i)$ $\|i\| \leq (N-1)/2$
Rectangular	1
Bartlett	$1 - 2(\|i\|/N)$
Hanning	$0.5 + 0.5\cos(2\pi i/N)$
Hamming	$0.54 + 0.46\cos(2\pi i/N)$
Blackman	$0.42 + 0.5\cos(2\pi i/N) + 0.08\cos(4\pi i/N)$
Kaiser*	$I_0[\gamma(1-(2i/(N-1))^2]^{0.5} / I_0(\gamma)$

*I_0 is the modified Bessel function of order zero. The FORTRAN subroutine "BESSEL" shown in Figure 6.5 can be used to calculate the modified Bessel function of order zero.

```
C*****************************************************
C*                                                   *
C*   SUBROUTINE BESSEL                               *
C*                                                   *
C*                                                   *
C*****************************************************
C
C        I0(X)  = XBESS
C
C
C
         SUBROUTINE BESSEL (X,XBESS)
         INTEGER K
         DOUBLE PRECISION IF
         S=1.0
         K=1
         IF=1
         E1=10000.0
         E2=E1
77       IF=IF*K
         E1=E2
         E2=((X/2.0)**K/IF)**2
         S=S+E2
         ERR1=E2
         ERR2=E2/E1
         IF (ERR1.LT.0.00001) GOTO 88
         IF (ERR2.LT.0.00001) GOTO 88
         IF (K.GT.100) GOTO 99
         K=K+1
         GOTO 77
88       XBESS=S
99       RETURN
         END
```

Figure 6-5 Figure 6.5 FORTRAN Subroutine "BESSEL" to Calculate the Modified Bessel Function of Order Zero

5. USING EQUATION 6.7, CALCULATE $h_2(i)$.

$h_2(i)$ is calculated by multiplying the appropriate $h_1(i)$ selected from Equations 6.12 through 6.15 by the appropriate $w(i)$ selected from Table 6.2.

6. USING EQUATION 6.8, CALCULATE h(i).

   Equation 6.4 can be used to compute the magnitude of the frequency response. The frequency response can be computed using the fast Fourier transform presented in Chapter 8. The designed FIR filter can then be tested to verify that its desired frequency domain specifications are met.

   Note that the value of N obtained from Equation 6.16 is higher than that which is necessary to meet the desired frequency domain specifications of the filter which is being designed. To make an adjustment, N can be decremented (incremented) by an even integer value to reduce (increase) the "order" of the FIR filter. This can be repeated until a filter order which best meets the desired frequency domain specifications is obtained.

6.5 USING A SOFTWARE PROGRAM TO DESIGN PRACTICAL FIR FILTERS:

The DSP56000/1 Demonstration Software program can be used to design practical, linear phase response FIR filters as follows:

1. BOOT THE PC.
2. INSERT THE "DSP56000/1 DEMONSTRATION SOFTWARE #2" PROGRAM DISKETTE INTO DRIVE "A".
3. TYPE **a:chapter6<return>** to load the DSP56000/1 Demonstration Software chapter 6 program.
4. SELECT FROM THE SCREEN MENUS to design the desired FIR filter.
5. DECREMENT (INCREMENT) THE FILTER ORDER UNTIL THE DESIRED FREQUENCY DOMAIN SPECIFICATIONS ARE OBTAINED.

NOTE:

h(i) for the designed filter are stored on the DSP56000/1 Demonstration Software #2 program diskette in a data file FIRWxx.DAT, where xx is replaced by LP (Low-Pass filter), HP (High-Pass filter), BP (Band-Stop filter), and BS (Band-Stop filter) as appropriate.

6.5.1 Low-Pass Filter Design Example:

User defined specifications:

Sampling frequency = 48000.000 Hertz

Stopband attenuation = 50.000 dB

Passband edge = 8000.000 Hertz

Stopband edge = 10000.000 Hertz

Window type = Hamming window

The designed passband attenuation in dB is .020
The designed stopband attenuation in dB is 52.745

Coefficients of the causal low-pass filter are:

h(1)	=	.11213E-02	= h(41)
h(2)	=	.13362E-02	= h(40)
h(3)	=	.95611E-10	= h(39)
h(4)	=	-.23447E-02	= h(38)
h(5)	=	-.31894E-02	= h(37)
h(6)	=	-.33537E-08	= h(36)
h(7)	=	.57094E-02	= h(35)
h(8)	=	.74605E-02	= h(34)
h(9)	=	.38783E-08	= h(33)
h(10)	=	-.12211E-01	= h(32)
h(11)	=	-.15372E-01	= h(31)
h(12)	=	-.52937E-09	= h(30)
h(13)	=	.23965E-01	= h(29)
h(14)	=	.29919E-01	= h(28)
h(15)	=	.75957E-08	= h(27)
h(16)	=	-.48045E-01	= h(26)
h(17)	=	-.63144E-01	= h(25)
h(18)	=	-.88328E-08	= h(24)
h(19)	=	.13488E+00	= h(23)
h(20)	=	.27418E+00	= h(22)
h(21)	=	.33333E+00	= h(21)

6.5.2 High-Pass Filter Design Example:

User defined specifications:

Sampling frequency	=	48000.000 Hertz
Stopband attenuation	=	50.000 dB
Passband edge	=	10000.000 Hertz
Stopband edge	=	8000.000 Hertz
Window type	=	Blackman window

The designed passband attenuation in dB is .026
The designed stopband attenuation in dB is 50.312
Coefficients of the causal high-pass filter are:

h(1)	=	-.19380E-05	= h(53)
h(2)	=	-.35399E-04	= h(52)
h(3)	=	.60145E-10	= h(51)
h(4)	=	.21958E-03	= h(50)
h(5)	=	.20330E-03	= h(49)
h(6)	=	-.46795E-03	= h(48)
h(7)	=	-.87685E-03	= h(47)
h(8)	=	.38364E-03	= h(46)
h(9)	=	.20988E-02	= h(45)

h(10)	=	.74902E-03	= h(44)
h(11)	=	-.33817E-02	= h(43)
h(12)	=	-.36583E-02	= h(42)
h(13)	=	.33748E-02	= h(41)
h(14)	=	.83951E-02	= h(40)
h(15)	=	-.30789E-08	= h(39)
h(16)	=	-.13498E-01	= h(38)
h(17)	=	-.87617E-02	= h(37)
h(18)	=	.15473E-01	= h(36)
h(19)	=	.23634E-01	= h(35)
h(20)	=	-.88315E-02	= h(34)
h(21)	=	-.43002E-01	= h(33)
h(22)	=	-.14249E-01	= h(32)
h(23)	=	.62815E-01	= h(31)
h(24)	=	.71223E-01	= h(30)
h(25)	=	-.77762E-01	= h(29)
h(26)	=	-.30570E+00	= h(28)
h(27)	=	.58333E+00	= h(27)

6.5.3 Band-Pass Filter Design Example:

User defined specifications:

Sampling frequency	=	48000.000 Hertz
Stopband attenuation	=	50.000 dB
Lower passband edge	=	6000.000 Hertz
Upper passband edge	=	12000.000 Hertz
Lower stopband edge	=	4000.000 Hertz
Upper stopband edge	=	14000.000 Hertz
Window type	=	Blackman window

The designed passband attenuation in dB is .021
The designed stopband attenuation in dB is 52.223
Coefficients of the causal band-pass filter are:

h(1)	=	.12388E-10	= h(57)
h(2)	=	-.50005E-04	= h(56)
h(3)	=	-.85938E-04	= h(55)
h(4)	=	.52566E-04	= h(54)
h(5)	=	.15195E-10	= h(53)
h(6)	=	-.15064E-03	= h(52)
h(7)	=	.78036E-03	= h(51)
h(8)	=	.19343E-02	= h(50)
h(9)	=	.18979E-09	= h(49)
h(10)	=	-.37179E-02	= h(48)

h(11)	=	-.29173E-02	= h(47)
h(12)	=	.11243E-02	= h(46)
h(13)	=	.17394E-08	= h(45)
h(14)	=	-.18634E-02	= h(44)
h(15)	=	.80464E-02	= h(43)
h(16)	=	.17209E-01	= h(42)
h(17)	=	.89654E-09	= h(41)
h(18)	=	-.26428E-01	= h(40)
h(19)	=	-.19047E-01	= h(39)
h(20)	=	.68508E-02	= h(38)
h(21)	=	.50265E-08	= h(37)
h(22)	=	-.10397E-01	= h(36)
h(23)	=	.44256E-01	= h(35)
h(24)	=	.95863E-01	= h(34)
h(25)	=	.17518E-08	= h(33)
h(26)	=	-.17317E+00	= h(32)
h(27)	=	-.15601E+00	= h(31)
h(28)	=	.92767E-01	= h(30)
h(29)	=	.25000E+00	= h(29)

6.5.4 Band-Stop Filter Design Example:

User defined specifications:

Sampling frequency	=	48000.000 Hertz
Stopband attenuation	=	50.000 dB
Lower passband edge	=	4000.000 Hertz
Upper passband edge	=	14000.000 Hertz
Lower stopband edge	=	6000.000 Hertz
Upper stopband edge	=	12000.000 Hertz
Window type	=	Hamming window

The designed passband attenuation in dB is .017
The designed stopband attenuation in dB is 54.043
Coefficients of the causal band-stop filter are:

h(1)	=	.52248E-03	= h(47)
h(2)	=	-.17633E-02	= h(46)
h(3)	=	-.27286E-02	= h(45)
h(4)	=	-.12283E-09	= h(44)
h(5)	=	-.64955E-03	= h(43)
h(6)	=	-.35163E-02	= h(42)
h(7)	=	.34528E-02	= h(41)
h(8)	=	.10078E-01	= h(40)
h(9)	=	.21528E-02	= h(39)

h(10)	=	.33598E-02	= h(38)
h(11)	=	.16636E-01	= h(37)
h(12)	=	.99570E-09	= h(36)
h(13)	=	-.24860E-01	= h(35)
h(14)	=	-.75339E-02	= h(34)
h(15)	=	-.73065E-02	= h(33)
h(16)	=	-.52450E-01	= h(32)
h(17)	=	-.28048E-01	= h(31)
h(18)	=	.45613E-01	= h(30)
h(19)	=	.13834E-01	= h(29)
h(20)	=	-.88802E-09	= h(28)
h(21)	=	.17452E+00	= h(27)
h(22)	=	.21386E+00	= h(26)
h(23)	=	-.14770E+00	= h(25)
h(24)	=	.58333E+00	= h(24)

6.6 IMPLEMENTING AN FIR FILTER:

Equation 6.1 is used to implement an FIR filter. In general, the following steps must be performed:

1. Save the input sample x(n) to be the first filter state.
2. Multiply the input sample x(n) by b_0 and accumulate the products of the filter states x(n-i) and the coefficients b_i to yield the output sample y(n).
3. Shift the filter states to obtain the next output sample y(n+1).

6.6.1 Implementing an FIR Filter on the DSP56001 Processor:

Steps 1-3 of Section 6.6 can be efficiently implemented on the DSP56001 processor by using its:

a. address pointers to mimic FIFO-like shifting of RAM data,
b. modulo addressing capability to provide wrap-around data buffers,
c. Multiply/Accumulate (MAC) instruction to both multiply two operands and add the product to a third operand in a single instruction cycle,
d. data move capability in parallel with the MAC instruction to keep the multiplier running at 100% capacity, and
e. Repeat Next Instruction (REP) instruction to provide compact filter code.

The DSP56001 processor's Address Generation Unit is composed of eight Address Registers (R_n), eight associated Offset Registers (N_n), and eight associated Modulo Registers (M_n). The DSP56001 processor's capability to perform modulo addressing allows an Address Register (R_n) value to be incremented (or decremented) and yet remain within an address range of size L, where L is defined by a lower and an upper address boundary.

For the FIR filter, L is equal to the number of coefficients (taps). The value L-1 is stored in the DSP56001 processor's Modifier Register (M_n). Because of the manner in which the DSP56001 processor's modulo addressing mechanism is implemented, the lower address boundary of L (base address) must have zeros in its K LSBs, where $2^{**}K \geq L$, e.g. , the base address value must be a multiple of 2**K. This requirement is applied to the DSP56001 processor's Address Register (R_n). The upper address boundary is the sum of the lower address boundary (base address) plus the modulo size minus one, e.g., the base address value plus L-1. Note, the upper address boundary is not stored in a register.

When modulo addressing is used, the Address Register (R_n) points to a modulo (circular) data buffer located in X-Memory and or Y-Memory. The address pointer (R_n) is not required to point at the lower address boundary; that is, it can point anywhere within the defined modulo address range L. If the address pointer increments past the upper address boundary (base address plus L-1 plus 1), it will wrap around to the base address.

6.6.2 DSP56001 Instructions for Implementing an FIR Filter:

The DSP56001 instructions shown in Figure 6.6 are used to implement the FIR filter algorithm where:

1. Modulo Register M0 is programmed to the value NTAPS-1 (modulo NTAPS). Address Register R0 is programmed to point to the state variable modulo buffer located in X-Memory. Modulo Register M4 is programmed to the value NTAPS-1 (modulo NTAPS). Address Register R4 is programmed to point to the coefficient buffer located in Y-Memory. Given that the FIR filter algorithm has been executing for some time and is ready to process the input sample x(n) in the Data ALU Input Register X0, the address in R4 is the base address (lower-boundary) of the coefficient buffer. The address in R0 is M, where M is greater than or equal to the lower-boundary X-Memory address and less than or equal to the upper-boundary X-Memory address. The X-Memory map for the filter states, the Y-Memory map for the coefficients, and the contents of the DSP56001 processor's A-Accumulator and Data ALU Input Registers X0 and Y0 are as shown in Figure 6.7.
2. The CLR instruction clears the A-Accumulator and simultaneously:
   a. moves the input sample x(n) from the Data ALU's Input Register X0 to the X-Memory location pointed to by Address Register R0, and
   b. moves the first coefficient from the Y-Memory location pointed to by Address Register R4 to the Data ALU's Input Register Y0.

```
;
;   FIR  Filter
;
;   x0 = input sample
;   a = output sample
;   ntaps = number of coefficient taps in the filter
;
        move    #states,r0  ;point to filter states
        move    #ntaps-1,m0 ;mod(ntaps)
        move    #coef,r4    ;point to filter coefficients
        move    #ntaps-1,m4 ;mod(ntaps)

        clr     a              x0,x:(r0)+      y:(r4)+,y0
        rep     #ntaps-1
        mac     x0,y0,a        x:(r0)+,x0      y:(r4)+,y0
        macr    x0,y0,a        (r0)-
```

Figure 6-6 DSP56001 Assembler Instructions to Implement the FIR Filter Algorithm

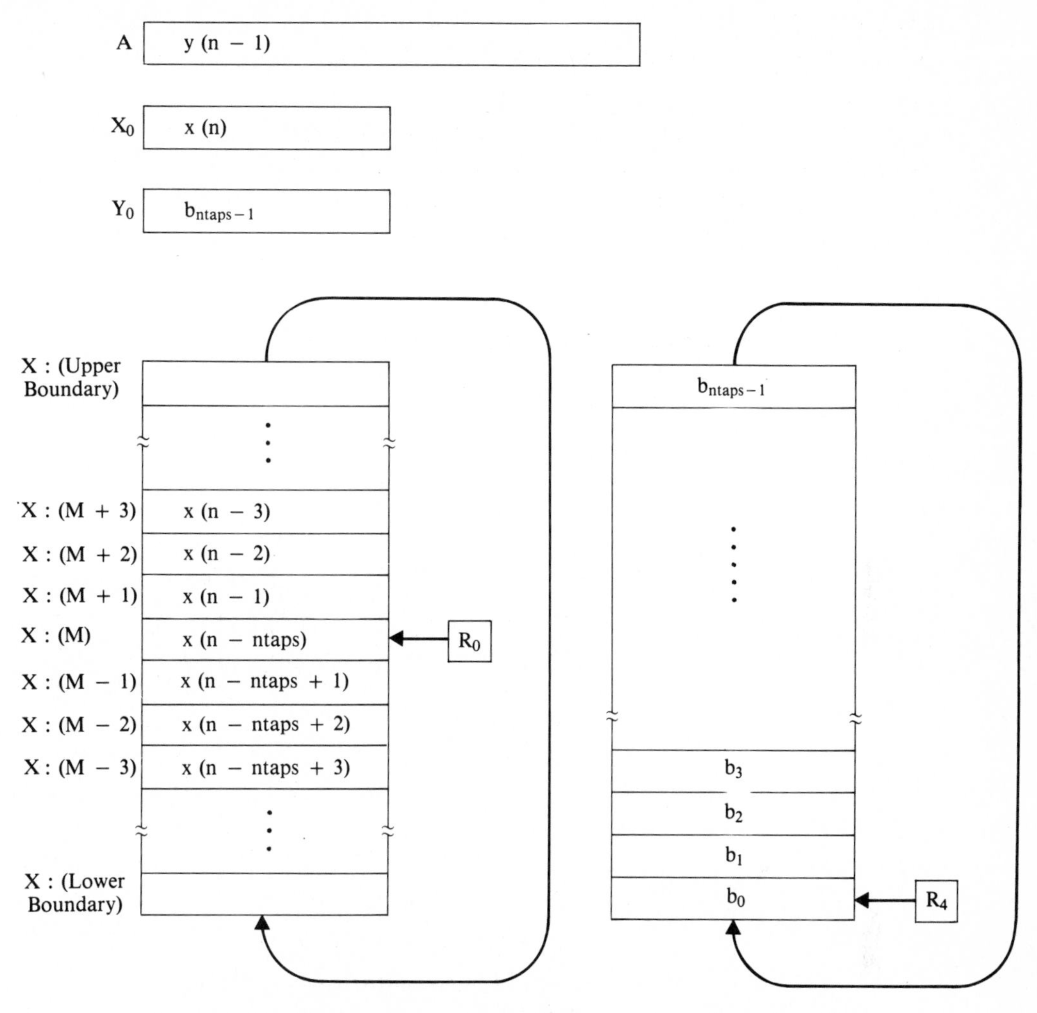

Figure 6-7 Memory Map and Data Registers at the Beginning of the n Iteration

Both Address Registers R0 and R4 are automatically incremented at the end of the CLR instruction (post incremented).

The X-Memory map for the filter states, the Y-Memory map for the coefficients, and the contents of the A-Accumulator and Data ALU Input Registers X0 and Y0 after execution of the CLR instruction are as shown in Figure 6.8.

3. The REP instruction regulates execution of NTAPS-1 iterations of the instruction:

```
mac      x0,y0,a          x:(r0)+,x0        y:(r4)+,y0
```

4. The MAC instruction multiplies the filter state variable in X0 by the coefficient in Y0, adds the product to the A-Accumulator, and simultaneously:
   a. moves the next state variable from the X-Memory location pointed to by the Address Register R0 to Input Register X0, and
   b. moves the next coefficient from the Y-Memory location pointed to by Address Register R4 to Input Register Y0.

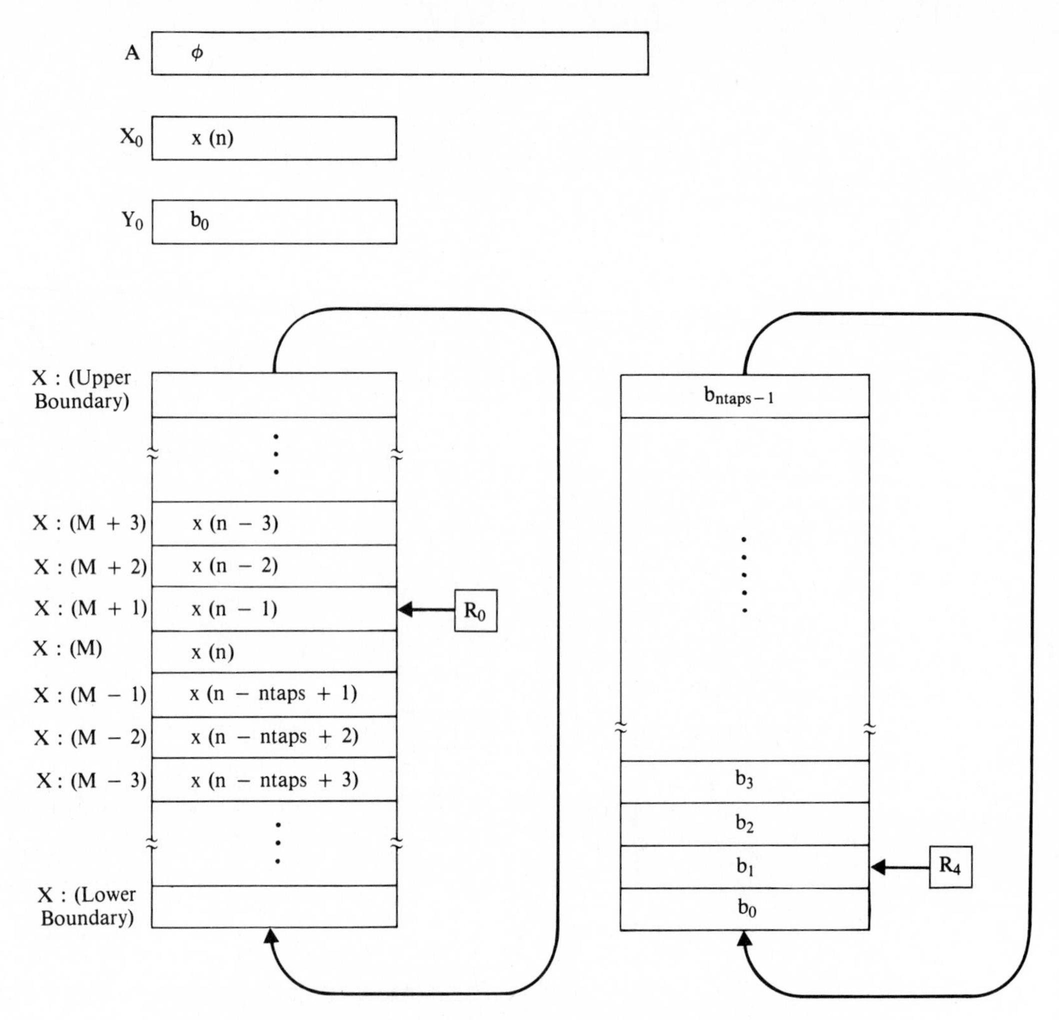

Figure 6-8 Memory Map and Data Registers after the CLR Instruction

Both Address Registers R0 and R4 are automatically incremented at the end of the MAC instruction (post incremented).

The X-Memory map for the filter states, the Y-Memory map for the coefficients, and the contents of the A-Accumulator and Data ALU Input Registers X0 and Y0 after execution of the first, second and last MAC instruction are as shown in Figures 6.9, 6.10 and 6.11 respectively.

5. The MACR instruction calculates the final tap of the filter algorithm, performs convergent rounding of the result, and simultaneously decrements the Address Register R0 (post decrementing). At the end of the MACR instruction, the A-Accumulator contains the filter output sample y(n).

   Note that during the execution of the filter algorithm, Address Register R4 is post incremented a total of NTAPS times; once in conjunction with the CLR instruction and NTAPS-1 times (due to the REP instruction) in conjunction with the MAC instruction. Since the modulus for R4 is NTAPS and R4 is incremented NTAPS times, the address value in R4 wraps around and points to the coefficient buffer's lower boundary location. Recall that the first coefficient b_0 resides in this buffer location.

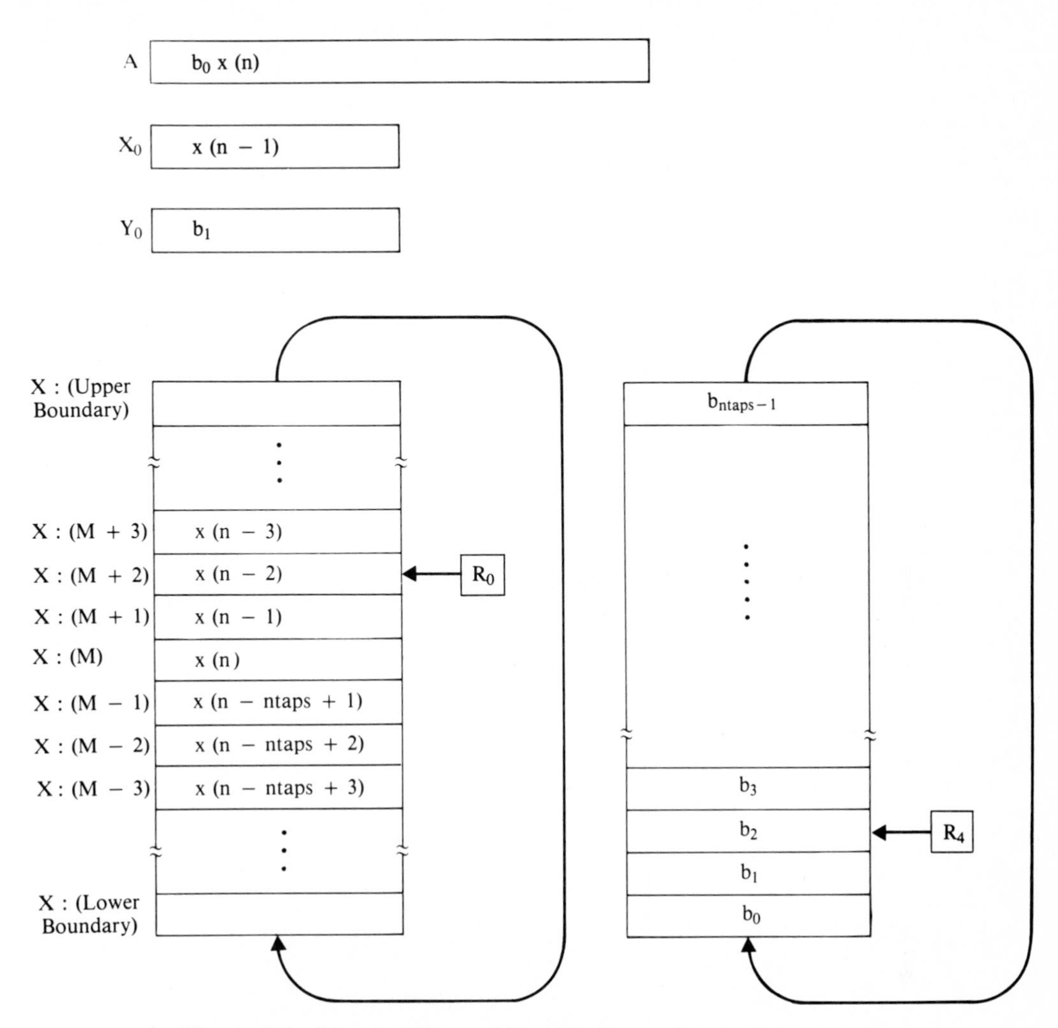

Figure 6-9 Memory Map and Data Registers after the Execution of the First MAC instruction

Note that during the execution of the filter algorithm, Address Register R0 is post incremented a total of NTAPS times; once in conjunction with the CLR instruction and NTAPS-1 times (due to the REP instruction) in conjunction with the MAC instruction. Also note that at the beginning of the algorithm, the input sample x(n) is moved from Data ALU Input Register X0 to the X-Memory location pointed to by R0. Since the modulus for R0 is NTAPS and R0 is incremented NTAPS times, the address value in R0 wraps around and points to the state variable buffer's X-Memory location M. In conjunction with the MACR instruction, R0 is post decremented by one. As a result, R0 then points to the state variable buffer's X-Memory location M-1. The next time the algorithm is executed, a new (next) input sample x(n+1) will then overwrite the value in X-Memory location M-1. Thus FIFO-like shifting of the filter state variables is accomplished by simply adjusting the R0 address pointer. The X-Memory map for the filter states, the Y-Memory map for the coefficients, and the contents of the A-Accumulator and Data ALU Input Registers X0 and Y0 after execution of the MACR instruction are as shown in Figure 6.12.

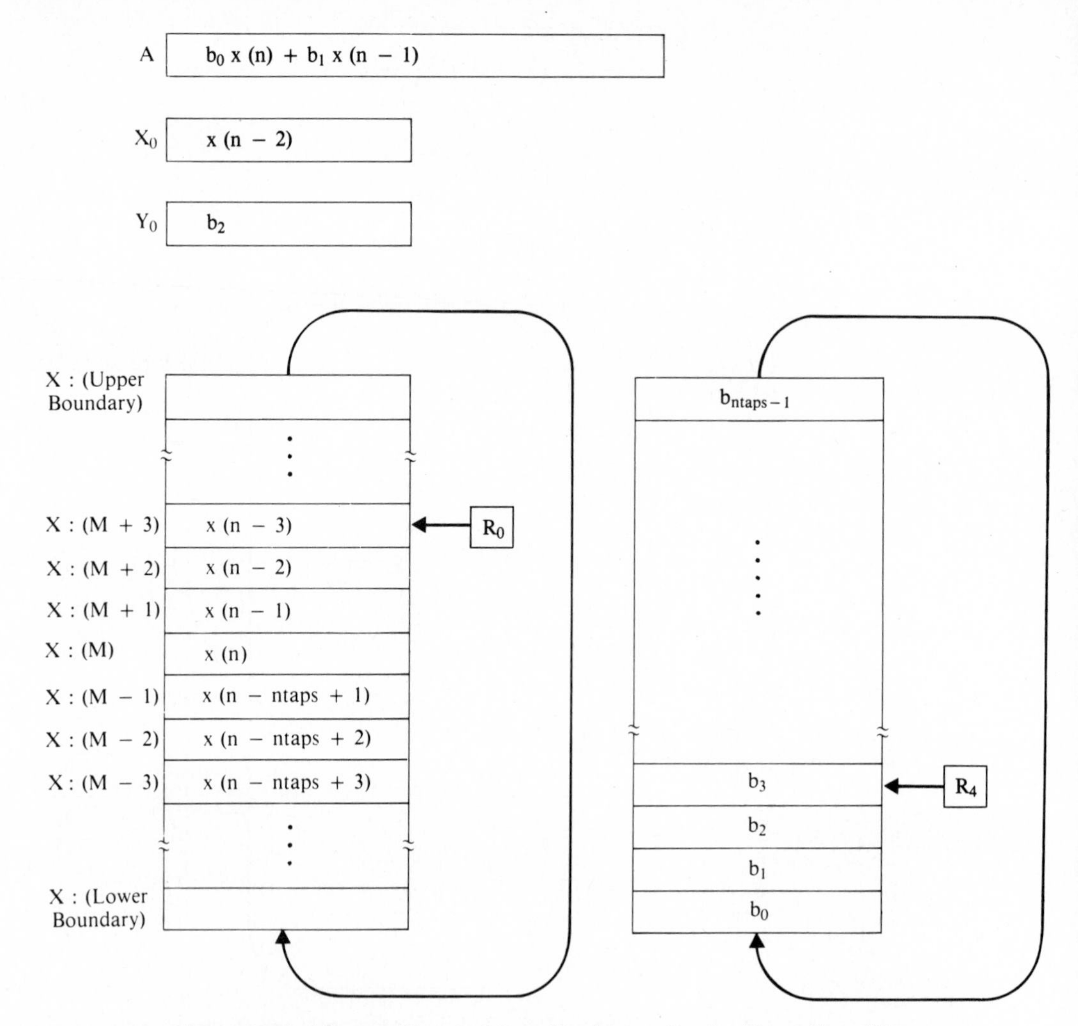

Figure 6-10 Memory Map and Data Registers after the Execution of the Second MAC Instruction

6.7 INTRODUCTION TO THE "FILTER DESIGN AND ANALYSIS SYSTEM" SOFTWARE PROGRAM:

The DSP56000/1 Demonstration Software program can be used to design a filter and output the designed filter coefficients. A demonstration version of the *Filter Design And Analysis System* (FDAS) software program by Momentum Data Systems Inc. is included with this book to not only design filters but also to graphically show the filters' response in both the frequency and time domains. The demonstration version of the FDAS software program is a complete working system except that its capability to output the designed filter coefficients is omitted. The FDAS program is entirely menu-driven for ease of use. There are no commands to memorize, and its unique screen-recycling feature will store previous data inputs thereby allowing filter specifications to be changed with a minimum number of keystrokes. For more information about the FDAS software program, please phone Momentum Data Systems Inc., Costa Mesa, California, at (714) 557-6884.

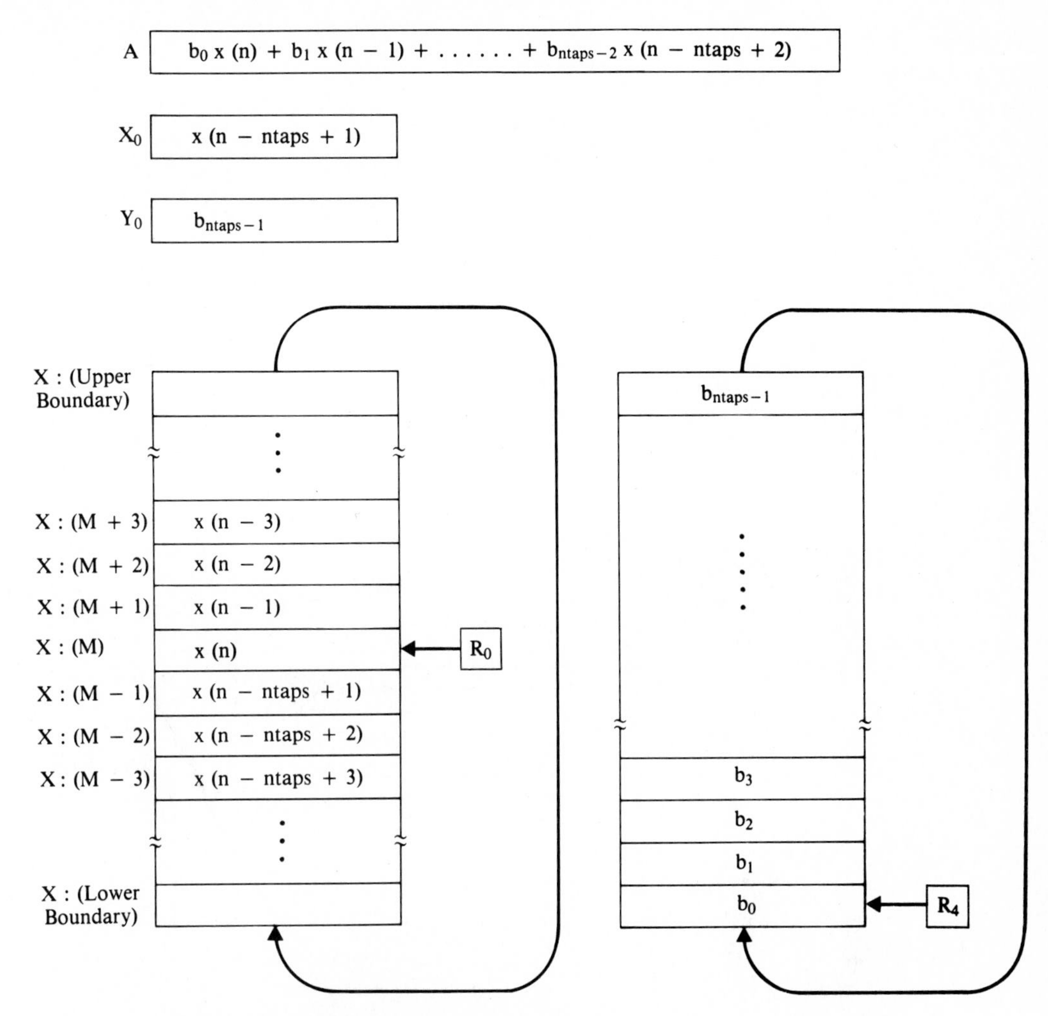

Figure 6-11 Memory Map and Data Registers after the Execution of the Last MAC Instruction

6.8 INSTALLING THE "FILTER DESIGN AND ANALYSIS SYSTEM"

Eight hundred (800) Kbytes of hard disk storage is required for installation of the *Filter Design And Analysis System* (FDAS) demonstration version software program; 640 Kbytes of RAM are required for its operation. The presence of a hardware arithmetic co-processor device within the PC is not a requirement, but it would greatly accelerate the mathematics calculations performed by the FDAS software program.

1. POWER-UP AND BOOT THE PC.
2. TYPE **md \dsptools\fdas<return>** to create an FDAS program directory.
3. TYPE **cd \dsptools\fdas<return>** to enter the FDAS program directory.
4. INSERT "FDAS DEMONSTRATION DISKETTE #1" INTO DISKETTE DRIVE "A".
5. TYPE **a:run_me c<return>** to begin installation of the FDAS program.
6. FOLLOW THE FDAS INSTALLATION INSTRUCTIONS WHICH APPEAR ON YOUR SCREEN.
7. CHOOSE THE "EXIT TO DOS" OPTION FROM THE MAIN MENU to exit the FDAS program.

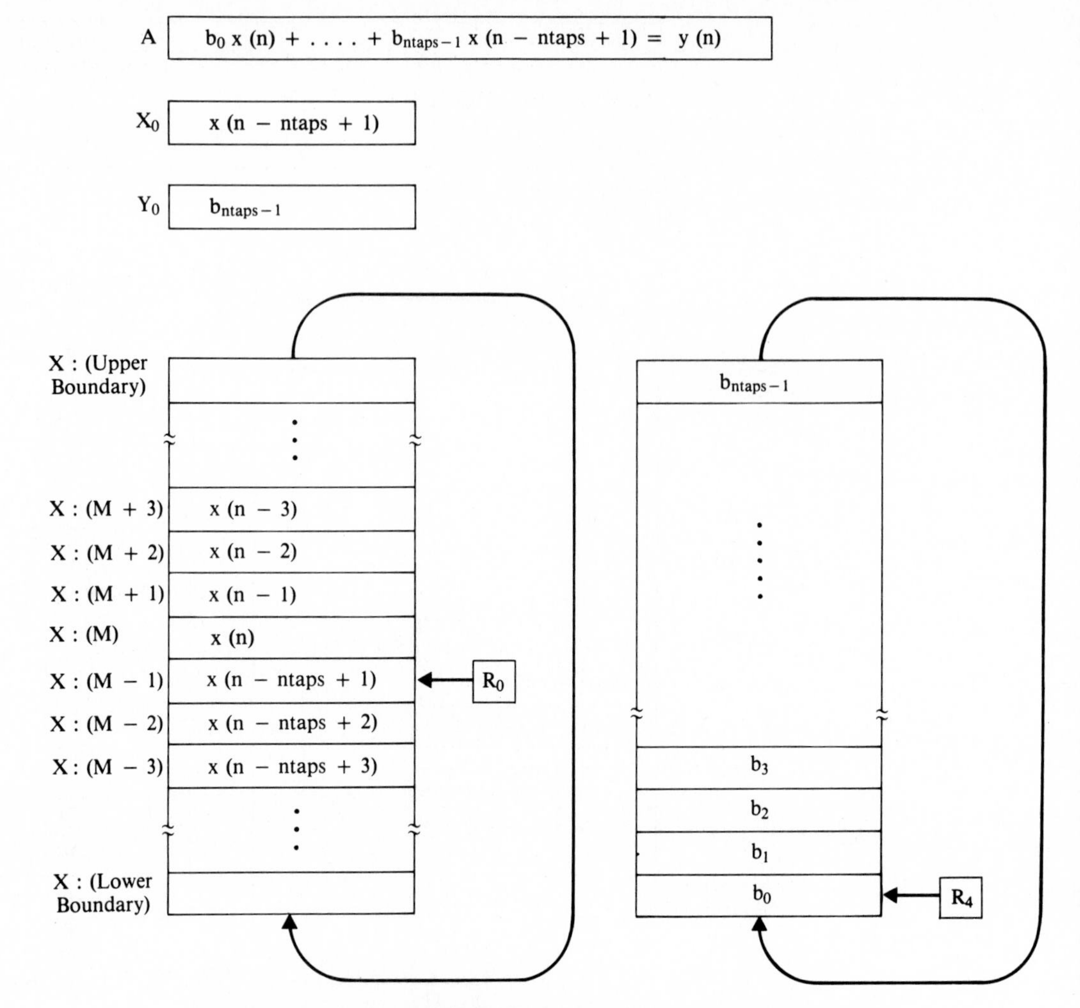

Figure 6-12 Memory Map and Data Registers after the Execution of the MACR Instruction

6.9 FIR FILTER DEMONSTRATION SYSTEM:

The FIR filtering process can be demonstrated by using the DSP system described in Chapter 5 and again shown in Figure 6.1. The DSP system consists of the DSP56000ADS application development system (ADS) and an A/D-D/A evaluation board (EVB). The analog input signal x(t) is first digitized using the A/D converter on the EVB. The DSP56001 processor then executes an FIR filter algorithm to process the digitized input signal x(n), thereby producing a filtered output signal y(n). The D/A converter on the EVB is then used to convert the digital output signal y(n) to the analog output signal y(t).

6.10 IMPLEMENTING FIR FILTERS:

The DSP56001 instructions shown in Figure 6.6 are combined with the DSP56001 instructions in Figure 5.8 to form the FIR.ASM assembler program macro shown in Figure 6.13. The FDAS program will be used in Sections 6.10.1 through 6.10.4 to design low-pass, high-pass, band-pass, and band-stop FIR filters respectively. The designed filter coefficients and the FIR.ASM program macro will be used to implement each FIR filter on the DSP system.

1. POWER-DOWN THE PC.
2. SET-UP THE DSP SYSTEM as described in Section 5.2.2 of Chapter 5.
3. POWER-UP AND BOOT THE PC.

```
fir  macro  states,coef,ntaps
;
;  states  = Location of filter states in X memory
;  coef    = Location of filter coefficients in Y memory
;  ntaps   = Number of taps
;
;*********************************************************
; Initialize routine
;*********************************************************
;
init2    reset
         movep  #0,x:$FFFE     ;set bcr to zero
         movep  #0,x:$FFEF     ;clear the transfer register
         movep  #0,x:$FFFF     ;stop all interrupts

;*********************************************************
; Set up the SSI for operation with the EVB.
; The following code sets port C to function as SSI
;*********************************************************

pcc   equu     $0001e0
      movep    #pcc,x:$FFE1  ;write PCC

;*********************************************************
; The following code sets the SSI CRA and CRB control
; registers for external continuous clock, synchronous,
; normal mode.
;*********************************************************

cra     equ     $004000
        movep   #cra,x:$FFEC  ;CRA pattern for word
                              ;length=16 bits

crb     equ     $003200
        movep   #crb,x:$FFED  ;CRB pattern for continuous
                              ;ck,synch,normal mode
                              ;word long frame sync

        move    #states,r0  ;point to filter states
        move    #ntaps-1,m0 ;mod(ntaps)
        move    #coef,r4    ;point to filter coefficients
        move    #ntaps-1,m4 ;mod(ntaps)

;*********************************************************
; Read A/D
;*********************************************************

; The following code polls the RDF flag in the SSI-SR and
; waits for RDF=1 and then reads the RX register to
; retrieve the data from the A/D converter.
; Sample rate is controlled by EVB.

wait2   nop
wait1   btst    #7,x:$ffee
        jcc     wait1        ;loop  until RDF bit = 1
        nop

        movep   x:$FFEF,x0   ;get A/D converter data

;*********************************************************
;  FIR Filter
```

```
;*******************************************************

        clr     a           x0,x:(r0)+      y:(r4)+,y0
        rep     #ntaps-1
        mac     x0,y0,a     x:(r0)+,x0      y:(r4)+,y0
        macr    x0,y0,a     (r0)-

;*******************************************************
; Write data to the D/A converter
;*******************************************************

      move    a,x:$FFEF     ;write D/A via SSI
      nop
      jmp     wait2         ;loop  indefinitely
      nop

      endm
```

Figure 6-13 FIR.ASM Assembler Program Macro

6.10.1 Implementing a Low-Pass FIR Filter:

1. TYPE **cd \dsptools\fdas<return>** to enter the FDAS program directory.
2. TYPE **demo<return>** to load and enter the FDAS program.
3. FOLLOW THE FDAS PROGRAM'S MENUS TO SELECT THE BELOW LISTED FILTER SPECIFICATIONS and design the following low-pass FIR filter.
4. EXIT THE FDAS PROGRAM.

FILTER SPECIFICATIONS:

Filter Type:	Low-Pass	
Filter Design Method:	FIR Design - Hamming Window	
Sampling Frequency:	48000	Hz
Passband Cutoff Frequency:	8000	Hz
Stopband Cutoff Frequency:	10000	Hz
Passband Attenuation:	.5	dB
Stopband Attenuation	50.0	dB

DESIGN RESULTS

NUMBER OF TAPS: 75

Ideal Impulse Coefficients:

H(1)	=	-.32922140E-02	= H(75)
H(2)	=	-.88419413E-02	= H(74)
H(3)	=	-.34803406E-02	= H(73)
H(4)	=	.66199729E-02	= H(72)
H(5)	=	.89115148E-02	= H(71)
H(6)	=	-.14624132E-16	= H(70)
H(7)	=	-.94864513E-02	= H(69)
H(8)	=	-.75026360E-02	= H(68)
H(9)	=	.42004110E-02	= H(67)
H(10)	=	.11368210E-01	= H(66)

```
H( 11)  =   .45115526E-02   = H( 65)
H( 12)  =  -.86568877E-02   = H( 64)
H( 13)  =  -.11763200E-01   = H( 63)
H( 14)  =   .14624132E-16   = H( 62)
H( 15)  =   .12786086E-01   = H( 61)
H( 16)  =   .10230867E-01   = H( 60)
H( 17)  =  -.58005676E-02   = H( 59)
H( 18)  =  -.15915494E-01   = H( 58)
H( 19)  =  -.64111537E-02   = H( 57)
H( 20)  =   .12504393E-01   = H( 56)
H( 21)  =   .17298823E-01   = H( 55)
H( 22)  =  -.14624132E-16   = H( 54)
H( 23)  =  -.19605333E-01   = H( 53)
H( 24)  =  -.16077077E-01   = H( 52)
H( 25)  =   .93701477E-02   = H( 51)
H( 26)  =   .26525824E-01   = H( 50)
H( 27)  =   .11073811E-01   = H( 49)
H( 28)  =  -.22507908E-01   = H( 48)
H( 29)  =  -.32675554E-01   = H( 47)
H( 30)  =   .14624132E-16   = H( 46)
H( 31)  =   .42011427E-01   = H( 45)
H( 32)  =   .37513180E-01   = H( 44)
H( 33)  =  -.24362384E-01   = H( 43)
H( 34)  =  -.79577472E-01   = H( 42)
H( 35)  =  -.40603973E-01   = H( 41)
H( 36)  =   .11253954E+00   = H( 40)
H( 37)  =   .29407999E+00   = H( 39)
H( 38)  =   .37500000E+00   = H( 38)
```

Window Coefficients:

```
W(  1)  =   .00000000E+00   = W( 75)
W(  2)  =   .18012557E-02   = W( 74)
W(  3)  =   .71920448E-02   = W( 73)
W(  4)  =   .16133527E-01   = W( 72)
W(  5)  =   .28561277E-01   = W( 71)
W(  6)  =   .44385755E-01   = W( 70)
W(  7)  =   .63492943E-01   = W( 69)
W(  8)  =   .85745175E-01   = W( 68)
W(  9)  =   .11098212E+00   = W( 67)
W( 10)  =   .13902195E+00   = W( 66)
W( 11)  =   .16966264E+00   = W( 65)
W( 12)  =   .20268341E+00   = W( 64)
W( 13)  =   .23784636E+00   = W( 63)
W( 14)  =   .27489813E+00   = W( 62)
W( 15)  =   .31357176E+00   = W( 61)
W( 16)  =   .35358861E+00   = W( 60)
W( 17)  =   .39466037E+00   = W( 59)
W( 18)  =   .43649109E+00   = W( 58)
W( 19)  =   .47877940E+00   = W( 57)
```

```
W( 20)   =   .52122060E+00   = W( 56)
W( 21)   =   .56350891E+00   = W( 55)
W( 22)   =   .60533963E+00   = W( 54)
W( 23)   =   .64641139E+00   = W( 53)
W( 24)   =   .68642824E+00   = W( 52)
W( 25)   =   .72510187E+00   = W( 51)
W( 26)   =   .76215364E+00   = W( 50)
W( 27)   =   .79731659E+00   = W( 49)
W( 28)   =   .83033736E+00   = W( 48)
W( 29)   =   .86097805E+00   = W( 47)
W( 30)   =   .88901788E+00   = W( 46)
W( 31)   =   .91425482E+00   = W( 45)
W( 32)   =   .93650706E+00   = W( 44)
W( 33)   =   .95561425E+00   = W( 43)
W( 34)   =   .97143872E+00   = W( 42)
W( 35)   =   .98386647E+00   = W( 41)
W( 36)   =   .99280796E+00   = W( 40)
W( 37)   =   .99819874E+00   = W( 39)
W( 38)   =   .10000000E+01   = W( 38)
```

Impulse Coefficients of Filter as Modified by the Window Coefficients:

```
H(  1)*W(  1)   =    .00000000E+00   = H( 75)*W( 75)
H(  2)*W(  2)   =   -.15854836E-04   = H( 74)*W( 74)
H(  3)*W(  3)   =   -.24914742E-04   = H( 73)*W( 73)
H(  4)*W(  4)   =    .10669231E-03   = H( 72)*W( 72)
H(  5)*W(  5)   =    .25451183E-03   = H( 71)*W( 71)
H(  6)*W(  6)   =    .00000000E+00   = H( 70)*W( 70)
H(  7)*W(  7)   =   -.60224533E-03   = H( 69)*W( 69)
H(  8)*W(  8)   =   -.64325333E-03   = H( 68)*W( 68)
H(  9)*W(  9)   =    .46610832E-03   = H( 67)*W( 67)
H( 10)*W( 10)   =    .15803576E-02   = H( 66)*W( 66)
H( 11)*W( 11)   =    .76532364E-03   = H( 65)*W( 65)
H( 12)*W( 12)   =   -.17545223E-02   = H( 64)*W( 64)
H( 13)*W( 13)   =   -.27977228E-02   = H( 63)*W( 63)
H( 14)*W( 14)   =    .00000000E+00   = H( 62)*W( 62)
H( 15)*W( 15)   =    .40092468E-02   = H( 61)*W( 61)
H( 16)*W( 16)   =    .36174059E-02   = H( 60)*W( 60)
H( 17)*W( 17)   =   -.22891760E-02   = H( 59)*W( 59)
H( 18)*W( 18)   =   -.69469213E-02   = H( 58)*W( 58)
H( 19)*W( 19)   =   -.30695200E-02   = H( 57)*W( 57)
H( 20)*W( 20)   =    .65175295E-02   = H( 56)*W( 56)
H( 21)*W( 21)   =    .97479820E-02   = H( 55)*W( 55)
H( 22)*W( 22)   =    .00000000E+00   = H( 54)*W( 54)
H( 23)*W( 23)   =   -.12673020E-01   = H( 53)*W( 53)
H( 24)*W( 24)   =   -.11035681E-01   = H( 52)*W( 52)
H( 25)*W( 25)   =    .67942142E-02   = H( 51)*W( 51)
H( 26)*W( 26)   =    .20216703E-01   = H( 50)*W( 50)
H( 27)*W( 27)   =    .88292360E-02   = H( 49)*W( 49)
H( 28)*W( 28)   =   -.18689156E-01   = H( 48)*W( 48)
```

```
H( 29)*W( 29)  =  -.28132915E-01  = H( 47)*W( 47)
H( 30)*W( 30)  =   .00000000E+00  = H( 46)*W( 46)
H( 31)*W( 31)  =   .38409114E-01  = H( 45)*W( 45)
H( 32)*W( 32)  =   .35131335E-01  = H( 44)*W( 44)
H( 33)*W( 33)  =  -.23280978E-01  = H( 43)*W( 43)
H( 34)*W( 34)  =  -.77304602E-01  = H( 42)*W( 42)
H( 35)*W( 35)  =  -.39948821E-01  = H( 41)*W( 41)
H( 36)*W( 36)  =   .11173010E+00  = H( 40)*W( 40)
H( 37)*W( 37)  =   .29355025E+00  = H( 39)*W( 39)
H( 38)*W( 38)  =   .37500000E+00  = H( 38)*W( 38)
```

Plots of the amplitude response and impulse response of the designed low-pass FIR filter are shown in Figure 6.14.

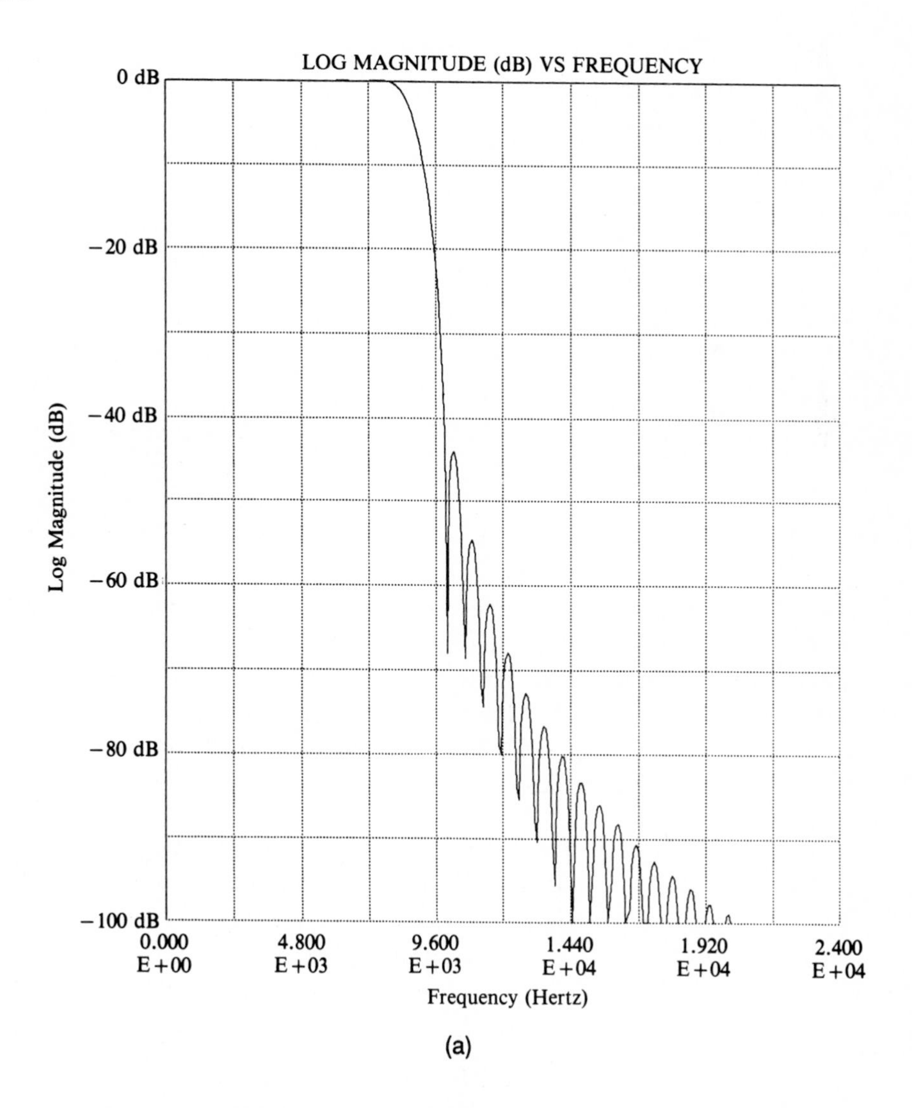

(a)

Figure 6-14 Plots of Amplitude Response and Impulse Response of Low-Pass Filter in Section 6.10.1. (a) Amplitude Response in dB (b) Impulse Response

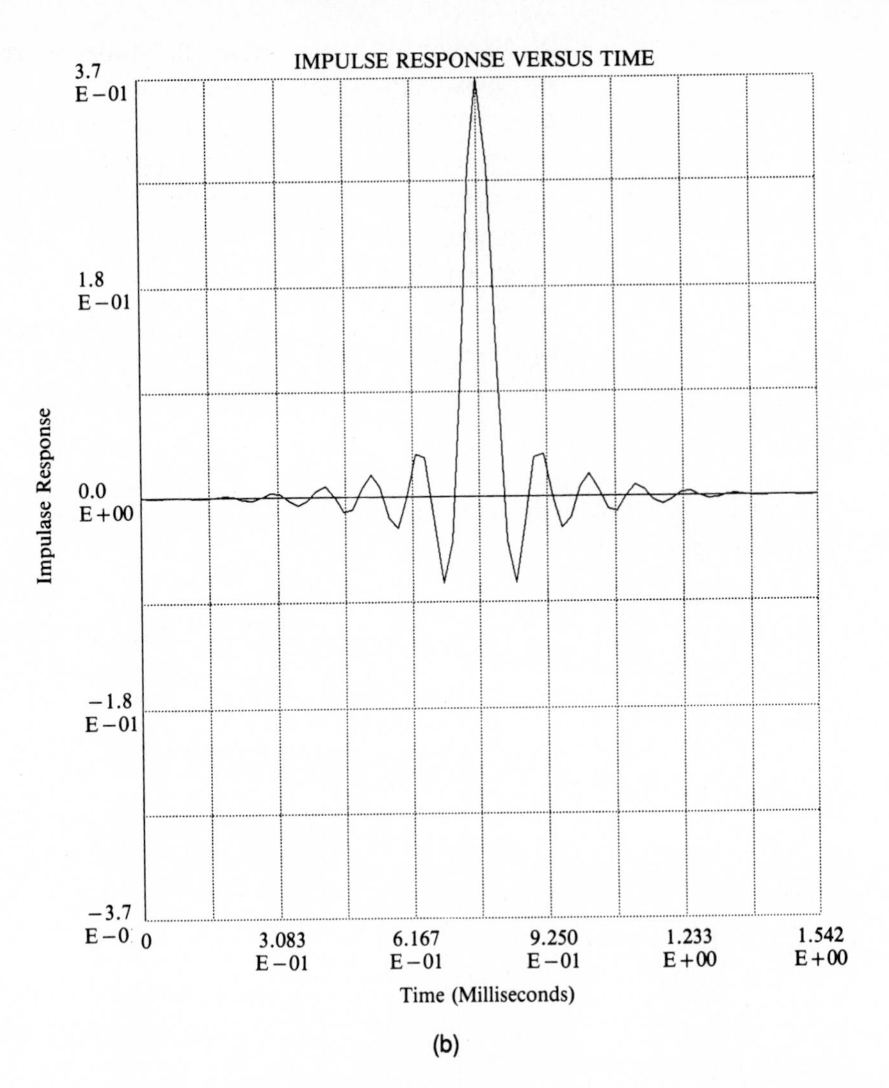

(b)

Figure 6-14 *(cont.)*

The "FIRLP.ASM" assembler program, shown in Figure 6.15, implements the designed low-pass FIR filter on the DSP56001 processor. Note that this program has an "include" statement which references the FIR.ASM assembler program macro shown in Figure 6.13. A copy of the FIRLP.ASM assembler program, FIR.ASM assembler program macro, and the "FIRLP.LOD" absolute load program can be found on the DSP56000/1 Demonstration Software #2 program diskette.

To demonstrate the designed low-pass FIR filter on the DSP system:

1. CONNECT A FUNCTION GENERATOR TO "BNC2" ON THE EVB AND TO CHANNEL ONE OF A DUAL TRACE OSCILLOSCOPE.
2. SET THE FUNCTION GENERATOR TO PRODUCE A 2-VOLT (PEAK TO PEAK) SINE WAVE.
3. CONNECT CHANNEL TWO OF THE OSCILLOSCOPE TO "BNC1" ON THE EVB.
4. LOAD THE "ADS56000 USER INTERFACE SOFTWARE" PROGRAM.
5. FROM WITHIN THE "ADS56000" PROGRAM, LOAD THE FILE **firlp.lod** FROM THE "DSP56001 DEMONSTRATION SOFTWARE #2" DISKETTE.

6. TYPE **go<return>** to execute the designed low-pass FIR filter program (algorithm) on the DSP56001 processor.
7. VARY THE SINUSOIDAL INPUT FREQUENCY FROM 0.5 TO 20 KHZ. View the analog input and output signals on the oscilloscope. Note that the FIR filter passes the frequencies from 0.5 to 8 KHz and stops the frequencies from 10 to 20 KHz as designed. Also note the increasing attenuation of the output signal as the input signal's frequency is increased from 8 to 10 KHz.
8. TYPE **force b<return>** to halt execution of the FIR filter algorithm and return control of the ADM to its monitor program for keyboard command entry.
9. TYPE **quit<return>** to exit the ADS56000 program.

```
;*****************************************************
;
; File: FIRLP.ASM
;
;*****************************************************
;
        include 'fir'
;
;****************************************************
; Coefficients of Low-Pass FIR
;****************************************************
;
;
ntaps   equ     75
;
        org     x:$0
states  dsm     ntaps          ;filter states
;
        org     y:$0
coef
        dc           0
        dc        -133
        dc        -209
        dc         895
        dc        2135
        dc           0
        dc       -5052
        dc       -5396
        dc        3910
        dc       13257
        dc        6420
        dc      -14718
        dc      -23469
        dc           0
        dc       33632
        dc       30345
        dc      -19203
        dc      -58275
        dc      -25749
        dc       54673
        dc       81772
        dc           0
        dc     -106309
        dc      -92574
        dc       56994
        dc      169590
        dc       74065
        dc     -156776
        dc     -235996
        dc           0
        dc      322199
```

```
        dc      294703
        dc     -195295
        dc     -648478
        dc     -335115
        dc      937260
        dc     2462478
        dc     3145728
        dc     2462478
        dc      937260
        dc     -335115
        dc     -648478
        dc     -195295
        dc      294703
        dc      322199
        dc           0
        dc     -235996
        dc     -156776
        dc       74065
        dc      169590
        dc       56994
        dc      -92574
        dc     -106309
        dc           0
        dc       81772
        dc       54673
        dc      -25749
        dc      -58275
        dc      -19203
        dc       30345
        dc       33632
        dc           0
        dc      -23469
        dc      -14718
        dc        6420
        dc       13257
        dc        3910
        dc       -5396
        dc       -5052
        dc           0
        dc        2135
        dc         895
        dc        -209
        dc        -133
        dc           0
;
; Program start address
;
        org     p:$100
;
        fir     states,coef,ntaps
;
        end
```

Figure 6-15 FIRLP.ASM Assembler Program

6.10.2 Implementing a High-Pass FIR Filter:

1. TYPE **cd \dsptools\fdas<return>** to enter the FDAS program directory.
2. TYPE **demo<return>** to load and enter the FDAS program.
3. FOLLOW THE FDAS PROGRAM'S MENUS TO SELECT THE BELOW LISTED FILTER SPECIFICATIONS and design the following high-pass FIR filter.
4. EXIT THE FDAS PROGRAM.

FILTER SPECIFICATIONS:

Filter Type:	High-Pass	
Filter Design Method:	FIR Design - Blackman Window	
Sampling Frequency:	48000	Hz
Passband Cutoff Frequency:	10000	Hz
Stopband Cutoff Frequency:	8000	Hz
Passband Attenuation:	.5	dB
Stopband Attenuation	50.0	dB

Design Results

Number of Taps: 131

Ideal Impulse Coefficients:

H(1)	=	-.45243075E-02	= H(131)
H(2)	=	.85303125E-16	= H(130)
H(3)	=	.46679363E-02	= H(129)
H(4)	=	.36303077E-02	= H(128)
H(5)	=	-.19969167E-02	= H(127)
H(6)	=	-.53051648E-02	= H(126)
H(7)	=	-.20646088E-02	= H(125)
H(8)	=	.38806738E-02	= H(124)
H(9)	=	.51592980E-02	= H(123)
H(10)	=	-.85303125E-16	= H(122)
H(11)	=	-.53469089E-02	= H(121)
H(12)	=	-.41681311E-02	= H(120)
H(13)	=	.22983381E-02	= H(119)
H(14)	=	.61213440E-02	= H(118)
H(15)	=	.23884690E-02	= H(117)
H(16)	=	-.45015816E-02	= H(116)
H(17)	=	-.60016324E-02	= H(115)
H(18)	=	.85303125E-16	= H(114)
H(19)	=	.62570210E-02	= H(113)
H(20)	=	.48930235E-02	= H(112)
H(21)	=	-.27069316E-02	= H(111)
H(22)	=	-.72343156E-02	= H(110)
H(23)	=	-.28328353E-02	= H(109)
H(24)	=	.53590257E-02	= H(108)
H(25)	=	.71726827E-02	= H(107)
H(26)	=	-.85303125E-16	= H(106)
H(27)	=	-.75405125E-02	= H(105)
H(28)	=	-.59231337E-02	= H(104)
H(29)	=	.32922140E-02	= H(103)
H(30)	=	.88419413E-02	= H(102)
H(31)	=	.34803406E-02	= H(101)
H(32)	=	-.66199729E-02	= H(100)
H(33)	=	-.89115148E-02	= H(99)
H(34)	=	.85303125E-16	= H(98)
H(35)	=	.94864513E-02	= H(97)

H(36)	=	.75026360E-02	= H(96)
H(37)	=	-.42004110E-02	= H(95)
H(38)	=	-.11368210E-01	= H(94)
H(39)	=	-.45115526E-02	= H(93)
H(40)	=	.86568877E-02	= H(92)
H(41)	=	.11763200E-01	= H(91)
H(42)	=	-.85303125E-16	= H(90)
H(43)	=	-.12786086E-01	= H(89)
H(44)	=	-.10230867E-01	= H(88)
H(45)	=	.58005676E-02	= H(87)
H(46)	=	.15915494E-01	= H(86)
H(47)	=	.64111537E-02	= H(85)
H(48)	=	-.12504393E-01	= H(84)
H(49)	=	-.17298823E-01	= H(83)
H(50)	=	.85303125E-16	= H(82)
H(51)	=	.19605333E-01	= H(81)
H(52)	=	.16077077E-01	= H(80)
H(53)	=	-.93701477E-02	= H(79)
H(54)	=	-.26525824E-01	= H(78)
H(55)	=	-.11073811E-01	= H(77)
H(56)	=	.22507908E-01	= H(76)
H(57)	=	.32675554E-01	= H(75)
H(58)	=	-.85303125E-16	= H(74)
H(59)	=	-.42011427E-01	= H(73)
H(60)	=	-.37513180E-01	= H(72)
H(61)	=	.24362384E-01	= H(71)
H(62)	=	.79577472E-01	= H(70)
H(63)	=	.40603973E-01	= H(69)
H(64)	=	-.11253954E+00	= H(68)
H(65)	=	-.29407999E+00	= H(67)
H(66)	=	.62500000E+00	= H(66)

Window Coefficients:

W(1)	=	-.14901161E-07	= W(131)
W(2)	=	.21040247E-03	= W(130)
W(3)	=	.84377786E-03	= W(129)
W(4)	=	.19064518E-02	= W(128)
W(5)	=	.34088941E-02	= W(127)
W(6)	=	.53655586E-02	= W(126)
W(7)	=	.77946819E-02	= W(125)
W(8)	=	.10718027E-01	= W(124)
W(9)	=	.14160578E-01	= W(123)
W(10)	=	.18150182E-01	= W(122)
W(11)	=	.22717153E-01	= W(121)
W(12)	=	.27893826E-01	= W(120)
W(13)	=	.33714084E-01	= W(119)
W(14)	=	.40212849E-01	= W(118)
W(15)	=	.47425545E-01	= W(117)
W(16)	=	.55387547E-01	= W(116)

```
W( 17)  =  .64133606E-01  = W( 115)
W( 18)  =  .73697274E-01  = W( 114)
W( 19)  =  .84110318E-01  = W( 113)
W( 20)  =  .95402143E-01  = W( 112)
W( 21)  =  .10759922E+00  = W( 111)
W( 22)  =  .12072454E+00  = W( 110)
W( 23)  =  .13479706E+00  = W( 109)
W( 24)  =  .14983121E+00  = W( 108)
W( 25)  =  .16583643E+00  = W( 107)
W( 26)  =  .18281668E+00  = W( 106)
W( 27)  =  .20077013E+00  = W( 105)
W( 28)  =  .21968873E+00  = W( 104)
W( 29)  =  .23955794E+00  = W( 103)
W( 30)  =  .26035652E+00  = W( 102)
W( 31)  =  .28205630E+00  = W( 101)
W( 32)  =  .30462208E+00  = W( 100)
W( 33)  =  .32801154E+00  = W(  99)
W( 34)  =  .35217528E+00  = W(  98)
W( 35)  =  .37705688E+00  = W(  97)
W( 36)  =  .40259298E+00  = W(  96)
W( 37)  =  .42871356E+00  = W(  95)
W( 38)  =  .45534214E+00  = W(  94)
W( 39)  =  .48239610E+00  = W(  93)
W( 40)  =  .50978713E+00  = W(  92)
W( 41)  =  .53742157E+00  = W(  91)
W( 42)  =  .56520101E+00  = W(  90)
W( 43)  =  .59302276E+00  = W(  89)
W( 44)  =  .62078047E+00  = W(  88)
W( 45)  =  .64836479E+00  = W(  87)
W( 46)  =  .67566397E+00  = W(  86)
W( 47)  =  .70256465E+00  = W(  85)
W( 48)  =  .72895253E+00  = W(  84)
W( 49)  =  .75471313E+00  = W(  83)
W( 50)  =  .77973257E+00  = W(  82)
W( 51)  =  .80389830E+00  = W(  81)
W( 52)  =  .82709990E+00  = W(  80)
W( 53)  =  .84922984E+00  = W(  79)
W( 54)  =  .87018422E+00  = W(  78)
W( 55)  =  .88986349E+00  = W(  77)
W( 56)  =  .90817318E+00  = W(  76)
W( 57)  =  .92502454E+00  = W(  75)
W( 58)  =  .94033523E+00  = W(  74)
W( 59)  =  .95402984E+00  = W(  73)
W( 60)  =  .96604051E+00  = W(  72)
W( 61)  =  .97630738E+00  = W(  71)
W( 62)  =  .98477902E+00  = W(  70)
W( 63)  =  .99141285E+00  = W(  69)
W( 64)  =  .99617541E+00  = W(  68)
W( 65)  =  .99904263E+00  = W(  67)
W( 66)  =  .99999999E+00  = W(  66)
```

Impulse Coefficients of Filter as Modified by the Window Coefficients:

H(1)*W(1) = .00000000E+00 = H(131)*W(131)
H(2)*W(2) = .00000000E+00 = H(130)*W(130)
H(3)*W(3) = .39339066E-05 = H(129)*W(129)
H(4)*W(4) = .69141388E-05 = H(128)*W(128)
H(5)*W(5) = -.67949295E-05 = H(127)*W(127)
H(6)*W(6) = -.28371811E-04 = H(126)*W(126)
H(7)*W(7) = -.15974045E-04 = H(125)*W(125)
H(8)*W(8) = .41484833E-04 = H(124)*W(124)
H(9)*W(9) = .72956085E-04 = H(123)*W(123)
H(10)*W(10) = .00000000E+00 = H(122)*W(122)
H(11)*W(11) = -.12135506E-03 = H(121)*W(121)
H(12)*W(12) = -.11622906E-03 = H(120)*W(120)
H(13)*W(13) = .77486038E-04 = H(119)*W(119)
H(14)*W(14) = .24604797E-03 = H(118)*W(118)
H(15)*W(15) = .11324883E-03 = H(117)*W(117)
H(16)*W(16) = -.24926662E-03 = H(116)*W(116)
H(17)*W(17) = -.38480759E-03 = H(115)*W(115)
H(18)*W(18) = .00000000E+00 = H(114)*W(114)
H(19)*W(19) = .52618980E-03 = H(113)*W(113)
H(20)*W(20) = .46670437E-03 = H(112)*W(112)
H(21)*W(21) = -.29122829E-03 = H(111)*W(111)
H(22)*W(22) = -.87332726E-03 = H(110)*W(110)
H(23)*W(23) = -.38182735E-03 = H(109)*W(109)
H(24)*W(24) = .80287457E-03 = H(108)*W(108)
H(25)*W(25) = .11894703E-02 = H(107)*W(107)
H(26)*W(26) = .00000000E+00 = H(106)*W(106)
H(27)*W(27) = -.15138388E-02 = H(105)*W(105)
H(28)*W(28) = -.13011694E-02 = H(104)*W(104)
H(29)*W(29) = .78856945E-03 = H(103)*W(103)
H(30)*W(30) = .23020506E-02 = H(102)*W(102)
H(31)*W(31) = .98156929E-03 = H(101)*W(101)
H(32)*W(32) = -.20165443E-02 = H(100)*W(100)
H(33)*W(33) = -.29230118E-02 = H(99)*W(99)
H(34)*W(34) = .00000000E+00 = H(98)*W(98)
H(35)*W(35) = .35768747E-02 = H(97)*W(97)
H(36)*W(36) = .30204058E-02 = H(96)*W(96)
H(37)*W(37) = -.18006563E-02 = H(95)*W(95)
H(38)*W(38) = -.51764250E-02 = H(94)*W(94)
H(39)*W(39) = -.21762848E-02 = H(93)*W(93)
H(40)*W(40) = .44131279E-02 = H(92)*W(92)
H(41)*W(41) = .63217878E-02 = H(91)*W(91)
H(42)*W(42) = .00000000E+00 = H(90)*W(90)
H(43)*W(43) = -.75824261E-02 = H(89)*W(89)
H(44)*W(44) = -.63511133E-02 = H(88)*W(88)
H(45)*W(45) = .37608147E-02 = H(87)*W(87)
H(46)*W(46) = .10753512E-01 = H(86)*W(86)
H(47)*W(47) = .45042038E-02 = H(85)*W(85)
H(48)*W(48) = -.91150999E-02 = H(84)*W(84)
H(49)*W(49) = -.13055563E-01 = H(83)*W(83)
H(50)*W(50) = .00000000E+00 = H(82)*W(82)
H(51)*W(51) = .15760660E-01 = H(81)*W(81)

```
H( 52)*W( 52)  =   .13297319E-01  = H( 80)*W( 80)
H( 53)*W( 53)  =  -.79573393E-02  = H( 79)*W( 79)
H( 54)*W( 54)  =  -.23082256E-01  = H( 78)*W( 78)
H( 55)*W( 55)  =  -.98540783E-02  = H( 77)*W( 77)
H( 56)*W( 56)  =   .20441055E-01  = H( 76)*W( 76)
H( 57)*W( 57)  =   .30225635E-01  = H( 75)*W( 75)
H( 58)*W( 58)  =   .00000000E+00  = H( 74)*W( 74)
H( 59)*W( 59)  =  -.40080070E-01  = H( 73)*W( 73)
H( 60)*W( 60)  =  -.36239147E-01  = H( 72)*W( 72)
H( 61)*W( 61)  =   .23785114E-01  = H( 71)*W( 71)
H( 62)*W( 62)  =   .78366160E-01  = H( 70)*W( 70)
H( 63)*W( 63)  =   .40255189E-01  = H( 69)*W( 69)
H( 64)*W( 64)  =  -.11210907E+00  = H( 68)*W( 68)
H( 65)*W( 65)  =  -.29379833E+00  = H( 67)*W( 67)
H( 66)*W( 66)  =   .62499988E+00  = H( 66)*W( 66)
```

Plots of the amplitude response and impulse response of the designed high-pass FIR filter are shown in Figure 6.16.

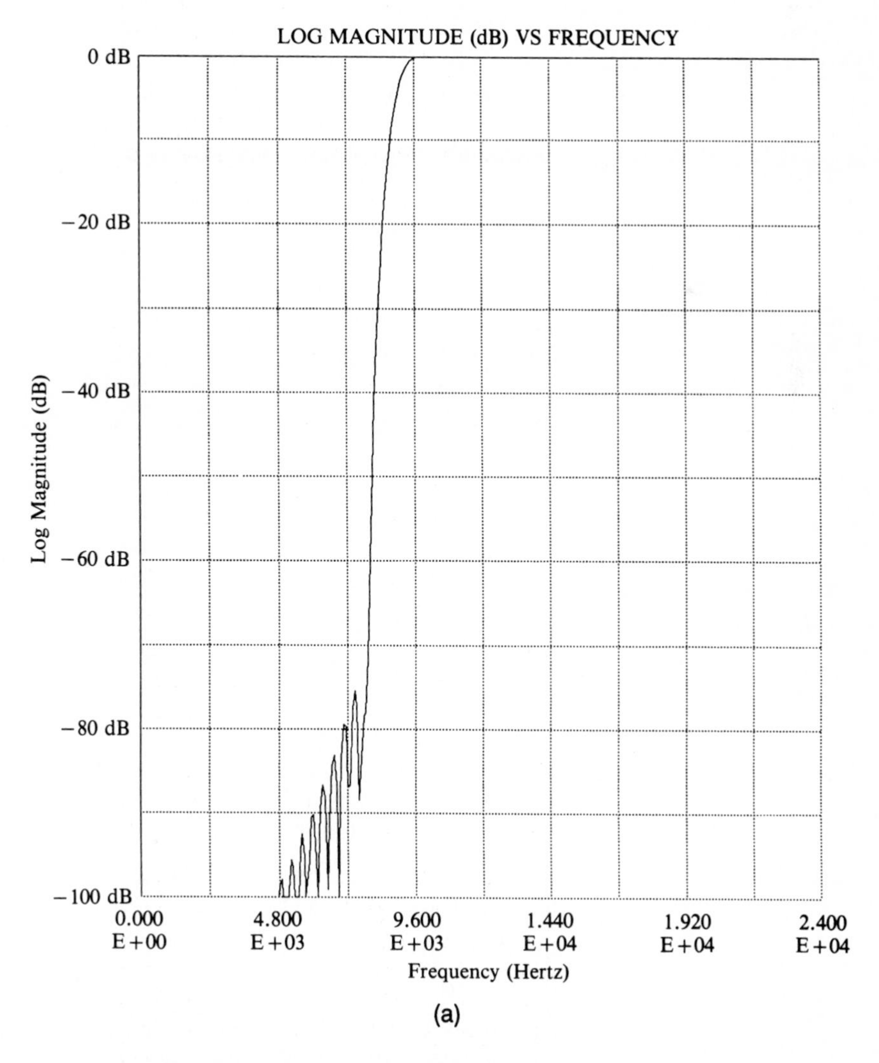

(a)

Figure 6-16 Plots of Amplitude Response and Impulse Response of High-Pass Filter in Section 6.10.2. (a) Amplitude Response in dB (b) Impulse Response

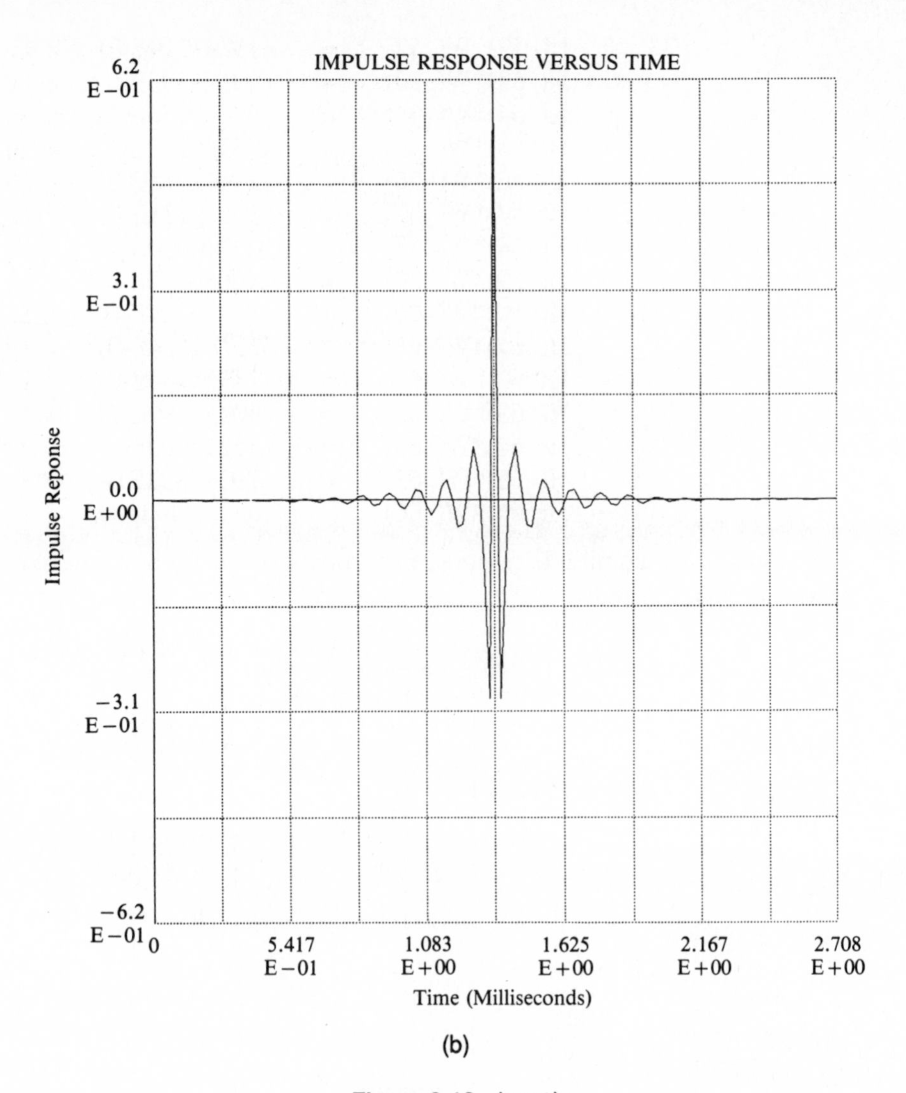

(b)

Figure 6-16 *(cont.)*

The FIRHP.ASM assembler program, shown in Figure 6.17, implements the designed high-pass FIR filter on the DSP56001 processor. Note that this program has an "include" statement which references the FIR.ASM assembler program macro shown in Figure 6.13. A copy of the FIRHP.ASM assembler program, FIR.ASM assembler program macro, and the FIRHP.LOD absolute load program can be found on the DSP56000/1 Demonstration Software #2 program diskette.

To demonstrate the designed high-pass FIR filter on the DSP system:

1. CONNECT A FUNCTION GENERATOR TO "BNC2" ON THE EVB AND TO CHANNEL ONE OF A DUAL TRACE OSCILLOSCOPE.
2. SET THE FUNCTION GENERATOR TO PRODUCE A 2-VOLT (PEAK TO PEAK) SINE WAVE.
3. CONNECT CHANNEL TWO OF THE OSCILLOSCOPE TO "BNC1" ON THE EVB.
4. LOAD THE "ADS56000 USER INTERFACE SOFTWARE" PROGRAM.
5. FROM WITHIN THE "ADS56000" PROGRAM, LOAD THE FILE **firhp.lod** FROM THE "DSP56001 DEMONSTRATION SOFTWARE #2" DISKETTE.

6. TYPE **go<return>** to execute the designed high-pass FIR filter program (algorithm) on the DSP56001 processor.
7. VARY THE SINUSOIDAL INPUT FREQUENCY FROM 0.5 TO 20 KHZ. View the analog input and output signals on the oscilloscope. Note that the FIR filter stops the frequencies from 0.5 to 8 KHz and passes the frequencies from 10 to 20 KHz as designed. Also note the decreasing attenuation of the output signal as the input signal's frequency is increased from 8 to 10 KHz.
8. TYPE **force b<return>** to halt execution of the FIR filter algorithm and return control of the ADM to its monitor program for keyboard command entry.
9. TYPE **quit<return>** to exit the ADS56000 program.

```
;*******************************************************
;
; File: FIRHP.ASM
;
;*******************************************************
;
        include 'fir'
;
;*******************************************************
; Coefficients of High Pass FIR
;*******************************************************
;
ntaps   equ     131
;
        org     x:$0
states  dsm     ntaps           ;filter states
;
        org     y:$0
coef
        dc        0
        dc        0
        dc       33
        dc       58
        dc      -57
        dc     -238
        dc     -134
        dc      348
        dc      612
        dc        0
        dc    -1018
        dc     -975
        dc      650
        dc     2064
        dc      950
        dc    -2091
        dc    -3228
        dc        0
        dc     4414
        dc     3915
        dc    -2443
        dc    -7326
        dc    -3203
        dc     6735
        dc     9978
        dc        0
        dc   -12699
        dc   -10915
        dc     6615
        dc    19311
        dc     8234
        dc   -16916
```

```
dc   -24520
dc        0
dc    30005
dc    25337
dc   -15105
dc   -43423
dc   -18256
dc    37020
dc    53031
dc        0
dc   -63606
dc   -53277
dc    31548
dc    90207
dc    37784
dc   -76463
dc  -109518
dc        0
dc   132210
dc   111546
dc   -66751
dc  -193628
dc   -82662
dc   171472
dc   253551
dc        0
dc  -336216
dc  -303996
dc   199524
dc   657383
dc   337685
dc  -940439
dc -2464559
dc  5242879
dc -2464559
dc  -940439
dc   337685
dc   657383
dc   199524
dc  -303996
dc  -336216
dc        0
dc   253551
dc   171472
dc   -82662
dc  -193628
dc   -66751
dc   111546
dc   132210
dc        0
dc  -109518
dc   -76463
dc    37784
dc    90207
dc    31548
dc   -53277
dc   -63606
dc        0
dc    53031
dc    37020
dc   -18256
dc   -43423
dc   -15105
dc    25337
dc    30005
dc        0
dc   -24520
dc   -16916
dc     8234
```

```
        dc      19311
        dc       6615
        dc     -10915
        dc     -12699
        dc          0
        dc       9978
        dc       6735
        dc      -3203
        dc      -7326
        dc      -2443
        dc       3915
        dc       4414
        dc          0
        dc      -3228
        dc      -2091
        dc        950
        dc       2064
        dc        650
        dc       -975
        dc      -1018
        dc          0
        dc        612
        dc        348
        dc       -134
        dc       -238
        dc        -57
        dc         58
        dc         33
        dc          0
        dc          0
;
; Program start address
;
        org     p:$100
;
        fir     states,coef,ntaps
;
        end
```

Figure 6-17 FIRHP.ASM Assembler Program

6.10.3 Implementing a Band-Pass FIR Filter:

1. TYPE **cd \dsptools\fdas<return>** to enter the FDAS program directory.
2. TYPE **demo<return>** to load and enter the FDAS program.
3. FOLLOW THE FDAS PROGRAM'S MENUS TO SELECT THE BELOW LISTED FILTER SPECIFICATIONS and design the following band-pass FIR filter.
4. EXIT THE FDAS PROGRAM.

Filter Specifications:

Filter Type:	Band-Pass
Filter Design Method:	FIR Design - Blackman Window
Sampling Frequency:	48000 Hz
Passband Cutoff Frequencies:	6000 and 12000 Hz
Stopband Cutoff Frequencies:	4000 and 14000 Hz
Passband Attenuation:	.5 dB
Stopband Attenuation	50.0 dB

Design Results
Number of Taps: 133
Ideal Impulse Coefficients:

H(1)	=	-.50839482E-16	= H(133)
H(2)	=	.18740295E-02	= H(132)
H(3)	=	.86145140E-02	= H(131)
H(4)	=	.38670451E-02	= H(130)
H(5)	=	-.62878774E-02	= H(129)
H(6)	=	-.48209834E-02	= H(128)
H(7)	=	.00000000E+00	= H(127)
H(8)	=	-.49844066E-02	= H(126)
H(9)	=	-.67215241E-02	= H(125)
H(10)	=	.42741024E-02	= H(124)
H(11)	=	.98451588E-02	= H(123)
H(12)	=	.22147622E-02	= H(122)
H(13)	=	.50838872E-16	= H(121)
H(14)	=	.55486790E-02	= H(120)
H(15)	=	-.55934569E-16	= H(119)
H(16)	=	-.11532549E-01	= H(118)
H(17)	=	-.77969680E-02	= H(117)
H(18)	=	.24859575E-02	= H(116)
H(19)	=	-.71898390E-16	= H(115)
H(20)	=	-.25917430E-02	= H(114)
H(21)	=	.84749652E-02	= H(113)
H(22)	=	.13070222E-01	= H(112)
H(23)	=	-.55934463E-16	= H(111)
H(24)	=	-.68390695E-02	= H(110)
H(25)	=	.50838735E-16	= H(109)
H(26)	=	-.29710224E-02	= H(108)
H(27)	=	-.13783222E-01	= H(107)
H(28)	=	-.62467651E-02	= H(106)
H(29)	=	.10259168E-01	= H(105)
H(30)	=	.79481078E-02	= H(104)
H(31)	=	.00000000E+00	= H(103)
H(32)	=	.84022854E-02	= H(102)
H(33)	=	.11466129E-01	= H(101)
H(34)	=	-.73825406E-02	= H(100)
H(35)	=	-.17229028E-01	= H(99)
H(36)	=	-.39294168E-02	= H(98)
H(37)	=	-.50838489E-16	= H(97)
H(38)	=	-.10140689E-01	= H(96)
H(39)	=	.55934379E-16	= H(95)
H(40)	=	.21783703E-01	= H(94)
H(41)	=	.14994169E-01	= H(93)
H(42)	=	-.48724768E-02	= H(92)
H(43)	=	.71898390E-16	= H(91)
H(44)	=	.52961704E-02	= H(90)
H(45)	=	-.17720382E-01	= H(89)
H(46)	=	-.28007618E-01	= H(88)

H(47)	=	.55934071E-16	= H(87)
H(48)	=	.15477894E-01	= H(86)
H(49)	=	-.50838872E-16	= H(85)
H(50)	=	.71654070E-02	= H(84)
H(51)	=	.34458056E-01	= H(83)
H(52)	=	.16241589E-01	= H(82)
H(53)	=	-.27846314E-01	= H(81)
H(54)	=	-.22621538E-01	= H(80)
H(55)	=	.00000000E+00	= H(79)
H(56)	=	-.26734544E-01	= H(78)
H(57)	=	-.38984840E-01	= H(77)
H(58)	=	.27069316E-01	= H(76)
H(59)	=	.68916112E-01	= H(75)
H(60)	=	.17401703E-01	= H(74)
H(61)	=	.50837913E-16	= H(73)
H(62)	=	.58815998E-01	= H(72)
H(63)	=	-.55938385E-16	= H(71)
H(64)	=	-.19605333E+00	= H(70)
H(65)	=	-.19492420E+00	= H(69)
H(66)	=	.12181192E+00	= H(68)
H(67)	=	.33333333E+00	= H(67)

Window Coefficients:

W(1)	=	-.14901161E-07	= W(133)
W(2)	=	.20406932E-03	= W(132)
W(3)	=	.81831963E-03	= W(131)
W(4)	=	.18487021E-02	= W(130)
W(5)	=	.33050712E-02	= W(129)
W(6)	=	.52010370E-02	= W(128)
W(7)	=	.75537812E-02	= W(127)
W(8)	=	.10383824E-01	= W(126)
W(9)	=	.13714742E-01	= W(125)
W(10)	=	.17572847E-01	= W(124)
W(11)	=	.21986814E-01	= W(123)
W(12)	=	.26987284E-01	= W(122)
W(13)	=	.32606421E-01	= W(121)
W(14)	=	.38877447E-01	= W(120)
W(15)	=	.45834154E-01	= W(119)
W(16)	=	.53510387E-01	= W(118)
W(17)	=	.61939521E-01	= W(117)
W(18)	=	.71153928E-01	= W(116)
W(19)	=	.81184433E-01	= W(115)
W(20)	=	.92059779E-01	= W(114)
W(21)	=	.10380610E+00	= W(113)
W(22)	=	.11644638E+00	= W(112)
W(23)	=	.12999999E+00	= W(111)
W(24)	=	.14448217E+00	= W(110)
W(25)	=	.15990362E+00	= W(109)
W(26)	=	.17627004E+00	= W(108)

```
W( 27)  =  .19358176E+00  = W( 107)
W( 28)  =  .21183343E+00  = W( 106)
W( 29)  =  .23101368E+00  = W( 105)
W( 30)  =  .25110493E+00  = W( 104)
W( 31)  =  .27208313E+00  = W( 103)
W( 32)  =  .29391767E+00  = W( 102)
W( 33)  =  .31657128E+00  = W( 101)
W( 34)  =  .33999999E+00  = W( 100)
W( 35)  =  .36415319E+00  = W(  99)
W( 36)  =  .38897371E+00  = W(  98)
W( 37)  =  .41439797E+00  = W(  97)
W( 38)  =  .44035618E+00  = W(  96)
W( 39)  =  .46677262E+00  = W(  95)
W( 40)  =  .49356598E+00  = W(  94)
W( 41)  =  .52064972E+00  = W(  93)
W( 42)  =  .54793249E+00  = W(  92)
W( 43)  =  .57531864E+00  = W(  91)
W( 44)  =  .60270870E+00  = W(  90)
W( 45)  =  .62999999E+00  = W(  89)
W( 46)  =  .65708720E+00  = W(  88)
W( 47)  =  .68386301E+00  = W(  87)
W( 48)  =  .71021877E+00  = W(  86)
W( 49)  =  .73604517E+00  = W(  85)
W( 50)  =  .76123294E+00  = W(  84)
W( 51)  =  .78567356E+00  = W(  83)
W( 52)  =  .80925996E+00  = W(  82)
W( 53)  =  .83188725E+00  = W(  81)
W( 54)  =  .85345340E+00  = W(  80)
W( 55)  =  .87385995E+00  = W(  79)
W( 56)  =  .89301269E+00  = W(  78)
W( 57)  =  .91082226E+00  = W(  77)
W( 58)  =  .92720484E+00  = W(  76)
W( 59)  =  .94208268E+00  = W(  75)
W( 60)  =  .95538464E+00  = W(  74)
W( 61)  =  .96704675E+00  = W(  73)
W( 62)  =  .97701261E+00  = W(  72)
W( 63)  =  .98523377E+00  = W(  71)
W( 64)  =  .99167014E+00  = W(  70)
W( 65)  =  .99629024E+00  = W(  69)
W( 66)  =  .99907141E+00  = W(  68)
W( 67)  =  .99999999E+00  = W(  67)
```

Impulse Coefficients of Filter as Modified by the Window Coefficients:

```
H(  1)*W(  1)  =   .00000000E+00  = H( 133)*W( 133)
H(  2)*W(  2)  =   .35762787E-06  = H( 132)*W( 132)
H(  3)*W(  3)  =   .70333481E-05  = H( 131)*W( 131)
H(  4)*W(  4)  =   .70333481E-05  = H( 130)*W( 130)
H(  5)*W(  5)  =  -.20742416E-04  = H( 129)*W( 129)
H(  6)*W(  6)  =  -.25033951E-04  = H( 128)*W( 128)
```

```
H(  7)*W(  7)   =   .00000000E+00   = H( 127)*W( 127)
H(  8)*W(  8)   =  -.51736832E-04   = H( 126)*W( 126)
H(  9)*W(  9)   =  -.92148781E-04   = H( 125)*W( 125)
H( 10)*W( 10)   =   .75101852E-04   = H( 124)*W( 124)
H( 11)*W( 11)   =   .21636486E-03   = H( 123)*W( 123)
H( 12)*W( 12)   =   .59723854E-04   = H( 122)*W( 122)
H( 13)*W( 13)   =   .00000000E+00   = H( 121)*W( 121)
H( 14)*W( 14)   =   .21564960E-03   = H( 120)*W( 120)
H( 15)*W( 15)   =   .00000000E+00   = H( 119)*W( 119)
H( 16)*W( 16)   =  -.61702728E-03   = H( 118)*W( 118)
H( 17)*W( 17)   =  -.48291683E-03   = H( 117)*W( 117)
H( 18)*W( 18)   =   .17678738E-03   = H( 116)*W( 116)
H( 19)*W( 19)   =   .00000000E+00   = H( 115)*W( 115)
H( 20)*W( 20)   =  -.23853779E-03   = H( 114)*W( 114)
H( 21)*W( 21)   =   .87964535E-03   = H( 113)*W( 113)
H( 22)*W( 22)   =   .15219450E-02   = H( 112)*W( 112)
H( 23)*W( 23)   =   .00000000E+00   = H( 111)*W( 111)
H( 24)*W( 24)   =  -.98800659E-03   = H( 110)*W( 110)
H( 25)*W( 25)   =   .00000000E+00   = H( 109)*W( 109)
H( 26)*W( 26)   =  -.52368641E-03   = H( 108)*W( 108)
H( 27)*W( 27)   =  -.26681423E-02   = H( 107)*W( 107)
H( 28)*W( 28)   =  -.13232231E-02   = H( 106)*W( 106)
H( 29)*W( 29)   =   .23699999E-02   = H( 105)*W( 105)
H( 30)*W( 30)   =   .19958019E-02   = H( 104)*W( 104)
H( 31)*W( 31)   =   .00000000E+00   = H( 103)*W( 103)
H( 32)*W( 32)   =   .24695396E-02   = H( 102)*W( 102)
H( 33)*W( 33)   =   .36298037E-02   = H( 101)*W( 101)
H( 34)*W( 34)   =  -.25099516E-02   = H( 100)*W( 100)
H( 35)*W( 35)   =  -.62739849E-02   = H(  99)*W(  99)
H( 36)*W( 36)   =  -.15283823E-02   = H(  98)*W(  98)
H( 37)*W( 37)   =   .00000000E+00   = H(  97)*W(  97)
H( 38)*W( 38)   =  -.44654608E-02   = H(  96)*W(  96)
H( 39)*W( 39)   =   .00000000E+00   = H(  95)*W(  95)
H( 40)*W( 40)   =   .10751605E-01   = H(  94)*W(  94)
H( 41)*W( 41)   =   .78066587E-02   = H(  93)*W(  93)
H( 42)*W( 42)   =  -.26696920E-02   = H(  92)*W(  92)
H( 43)*W( 43)   =   .00000000E+00   = H(  91)*W(  91)
H( 44)*W( 44)   =   .31919479E-02   = H(  90)*W(  90)
H( 45)*W( 45)   =  -.11163831E-01   = H(  89)*W(  89)
H( 46)*W( 46)   =  -.18403411E-01   = H(  88)*W(  88)
H( 47)*W( 47)   =   .00000000E+00   = H(  87)*W(  87)
H( 48)*W( 48)   =   .10992646E-01   = H(  86)*W(  86)
H( 49)*W( 49)   =   .00000000E+00   = H(  85)*W(  85)
H( 50)*W( 50)   =   .54545403E-02   = H(  84)*W(  84)
H( 51)*W( 51)   =   .27072668E-01   = H(  83)*W(  83)
H( 52)*W( 52)   =   .13143659E-01   = H(  82)*W(  82)
H( 53)*W( 53)   =  -.23164988E-01   = H(  81)*W(  81)
H( 54)*W( 54)   =  -.19306421E-01   = H(  80)*W(  80)
H( 55)*W( 55)   =   .00000000E+00   = H(  79)*W(  79)
H( 56)*W( 56)   =  -.23874283E-01   = H(  78)*W(  78)
```

```
H( 57)*W( 57)   =   -.35508156E-01   = H( 77)*W( 77)
H( 58)*W( 58)   =    .25098681E-01   = H( 76)*W( 76)
H( 59)*W( 59)   =    .64924598E-01   = H( 75)*W( 75)
H( 60)*W( 60)   =    .16625285E-01   = H( 74)*W( 74)
H( 61)*W( 61)   =    .00000000E+00   = H( 73)*W( 73)
H( 62)*W( 62)   =    .57463884E-01   = H( 72)*W( 72)
H( 63)*W( 63)   =    .00000000E+00   = H( 71)*W( 71)
H( 64)*W( 64)   =   -.19442022E+00   = H( 70)*W( 70)
H( 65)*W( 65)   =   -.19420099E+00   = H( 69)*W( 69)
H( 66)*W( 66)   =    .12169874E+00   = H( 68)*W( 68)
H( 67)*W( 67)   =    .33333325E+00   = H( 67)*W( 67)
```

Plots of the amplitude response and impulse response of the designed band-pass FIR filter are shown in Figure 6.18.

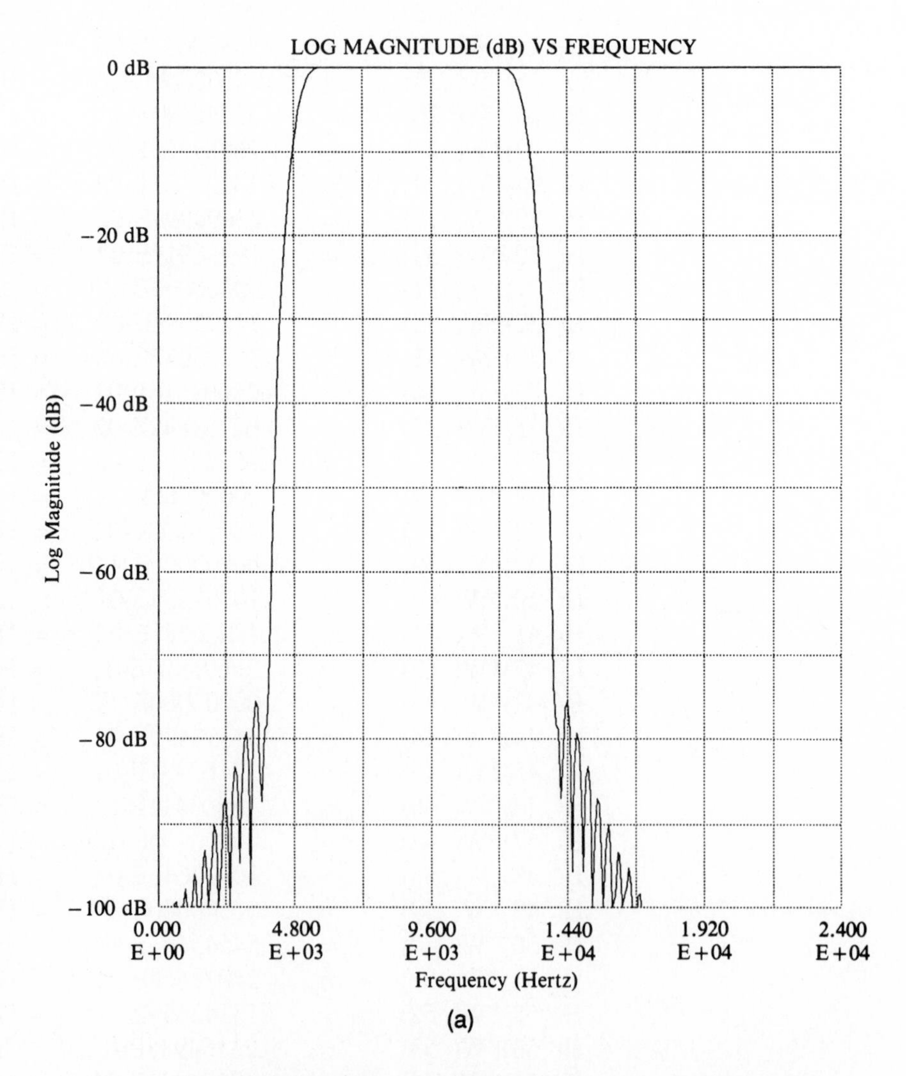

Figure 6-18 Plots of Amplitude Response and Impulse Response of Band-Pass Filter in Section 6.10.3. (a) Amplitude Response in dB (b) Impulse Response

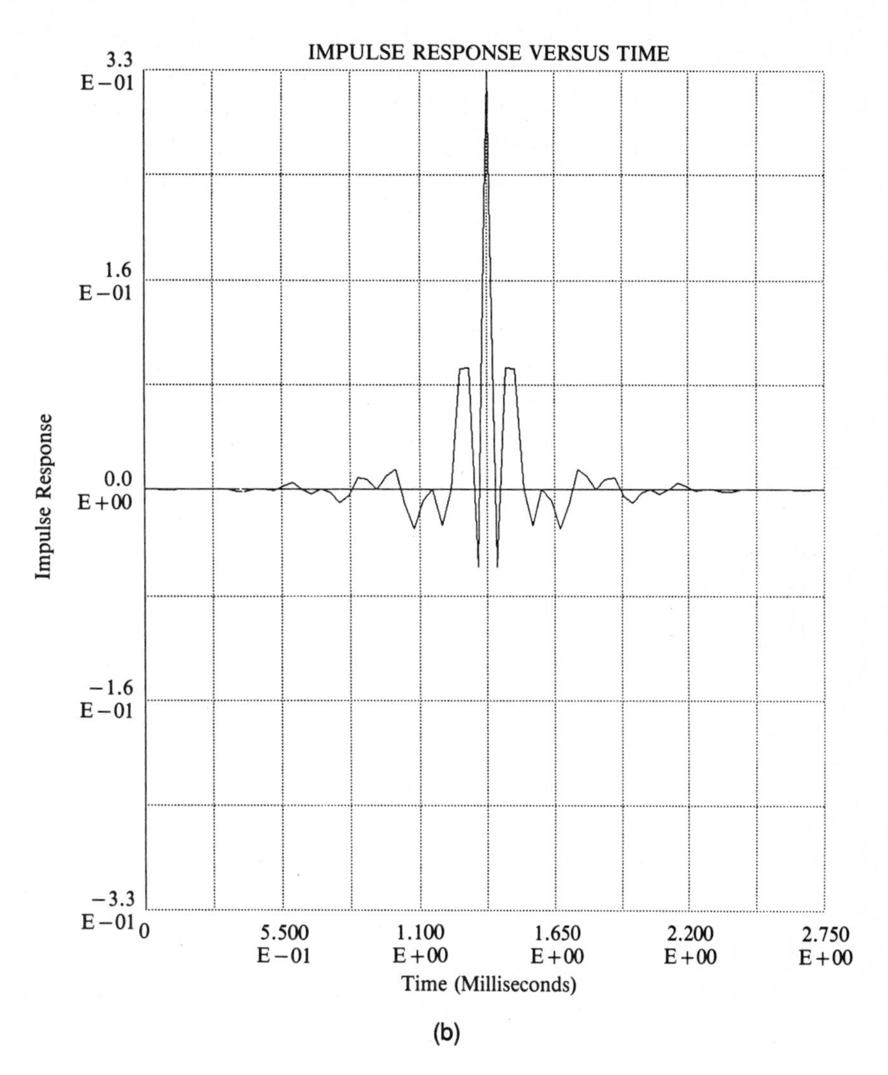

(b)

Figure 6-18 *(cont.)*

The FIRBP.ASM assembler program, shown in Figure 6.19, implements the designed band-pass FIR filter on the DSP56001 processor. Note that this program has an "include" statement which references the FIR.ASM assembler program macro shown in Figure 6.13. A copy of the FIRBP.ASM assembler program, FIR.ASM assembler program macro, and the FIRBP.LOD absolute load program can be found on the DSP56000/1 Demonstration Software #2 program diskette.

To demonstrate the designed band-pass FIR filter on the DSP system:

1. CONNECT A FUNCTION GENERATOR TO "BNC2" ON THE EVB AND TO CHANNEL ONE OF A DUAL TRACE OSCILLOSCOPE.
2. SET THE FUNCTION GENERATOR TO PRODUCE A 2-VOLT (PEAK TO PEAK) SINE WAVE.
3. CONNECT CHANNEL TWO OF THE OSCILLOSCOPE TO "BNC1" ON THE EVB.
4. LOAD THE "ADS56000 USER INTERFACE SOFTWARE" PROGRAM.
5. FROM WITHIN THE "ADS56000" PROGRAM, LOAD THE FILE **firbp.lod** FROM THE "DSP56001 DEMONSTRATION SOFTWARE #2" DISKETTE.

6. TYPE **go<return>** to execute the designed band-pass FIR filter program (algorithm) on the DSP56001 processor.
7. VARY THE SINUSOIDAL INPUT FREQUENCY FROM 0.5 TO 20 KHZ. View the analog input and output signals on the oscilloscope. Note that the FIR filter passes the frequencies from 6 to 12 KHz and stops the frequencies from 0.5 to 4 KHz and 14 to 20 KHz as designed. Also note the decreasing attenuation of the output signal as the input signal's frequency is increased from 4 to 6 KHz; note the increasing attenuation of the output signal as the input signal's frequency is increased from 12 to 14 KHz.
8. TYPE **force b<return>** to halt execution of the FIR filter algorithm and return control of the ADM to its monitor program for keyboard command entry.
9. TYPE **quit<return>** to exit the ADS56000 program.

```
;*******************************************************
;
; File: FIRBP.ASM
;
;*******************************************************
;
        include 'fir'
;
;*******************************************************
; Coefficients of Band Pass FIR
;*******************************************************
;
ntaps   equ     133
;
        org     x:$0
states  dsm     ntaps           ;filter states
;
        org     y:$0
coef
        dc          0
        dc          3
        dc         59
        dc         59
        dc       -174
        dc       -210
        dc          0
        dc       -434
        dc       -773
        dc        630
        dc       1815
        dc        501
        dc          0
        dc       1809
        dc          0
        dc      -5176
        dc      -4051
        dc       1483
        dc          0
        dc      -2001
        dc       7379
        dc      12767
        dc          0
        dc      -8288
        dc          0
        dc      -4393
        dc     -22382
        dc     -11100
        dc      19881
```

```
        dc      16742
        dc          0
        dc      20716
        dc      30449
        dc     -21055
        dc     -52630
        dc     -12821
        dc          0
        dc     -37459
        dc          0
        dc      90191
        dc      65487
        dc     -22395
        dc          0
        dc      26776
        dc     -93649
        dc    -154379
        dc          0
        dc      92213
        dc          0
        dc      45756
        dc     227102
        dc     110257
        dc    -194322
        dc    -161954
        dc          0
        dc    -200272
        dc    -297864
        dc     210543
        dc     544627
        dc     139463
        dc          0
        dc     482042
        dc          0
        dc   -1630915
        dc   -1629076
        dc    1020883
        dc    2796202
        dc    1020883
        dc   -1629076
        dc   -1630915
        dc          0
        dc     482042
        dc          0
        dc     139463
        dc     544627
        dc     210543
        dc    -297864
        dc    -200272
        dc          0
        dc    -161954
        dc    -194322
        dc     110257
        dc     227102
        dc      45756
        dc          0
        dc      92213
        dc          0
        dc    -154379
        dc     -93649
        dc      26776
        dc          0
        dc     -22395
        dc      65487
        dc      90191
        dc          0
        dc     -37459
        dc          0
        dc     -12821
```

```
        dc      -52630
        dc      -21055
        dc       30449
        dc       20716
        dc           0
        dc       16742
        dc       19881
        dc      -11100
        dc      -22382
        dc       -4393
        dc           0
        dc       -8288
        dc           0
        dc       12767
        dc        7379
        dc       -2001
        dc           0
        dc        1483
        dc       -4051
        dc       -5176
        dc           0
        dc        1809
        dc           0
        dc         501
        dc        1815
        dc         630
        dc        -773
        dc        -434
        dc           0
        dc        -210
        dc        -174
        dc          59
        dc          59
        dc           3
        dc           0
;
; Program start address
;
        org     p:$100
;
        fir     states,coef,ntaps
;
        end
```

Figure 6-19 FIRBP.ASM Assembler Program

6.10.4 Implementing a Band-Stop FIR Filter:

1. TYPE **cd \dsptools\fdas<return>** to enter the FDAS program directory.
2. TYPE **demo<return>** to load and enter the FDAS program.
3. FOLLOW THE FDAS PROGRAM'S MENUS TO SELECT THE BELOW LISTED FILTER SPECIFICATIONS and design the following band-stop FIR filter.
4. EXIT THE FDAS PROGRAM.

Filter Specifications:

Filter Type:	Band-Stop
Filter Design Method:	FIR Design - Hamming Window
Sampling Frequency:	48000 Hz
Passband Cutoff Frequency:	4000 and 14000 Hz

Stopband Cutoff Frequency:	6000 and 12000 Hz
Passband Attenuation:	.5 dB
Stopband Attenuation	50.0 dB

Design Results

Number of Taps: 75

Ideal Impulse Coefficients:

H(1)	=	-.79481078E-02	= H(75)
H(2)	=	.00000000E+00	= H(74)
H(3)	=	-.84022854E-02	= H(73)
H(4)	=	-.11466129E-01	= H(72)
H(5)	=	.73825406E-02	= H(71)
H(6)	=	.17229028E-01	= H(70)
H(7)	=	.39294168E-02	= H(69)
H(8)	=	-.24126224E-16	= H(68)
H(9)	=	.10140689E-01	= H(67)
H(10)	=	-.25329392E-16	= H(66)
H(11)	=	-.21783703E-01	= H(65)
H(12)	=	-.14994169E-01	= H(64)
H(13)	=	.48724768E-02	= H(63)
H(14)	=	.34120100E-16	= H(62)
H(15)	=	-.52961704E-02	= H(61)
H(16)	=	.17720382E-01	= H(60)
H(17)	=	.28007618E-01	= H(59)
H(18)	=	-.25329515E-16	= H(58)
H(19)	=	-.15477894E-01	= H(57)
H(20)	=	-.24125265E-16	= H(56)
H(21)	=	-.71654070E-02	= H(55)
H(22)	=	-.34458056E-01	= H(54)
H(23)	=	-.16241589E-01	= H(53)
H(24)	=	.27846314E-01	= H(52)
H(25)	=	.22621538E-01	= H(51)
H(26)	=	.00000000E+00	= H(50)
H(27)	=	.26734544E-01	= H(49)
H(28)	=	.38984840E-01	= H(48)
H(29)	=	-.27069316E-01	= H(47)
H(30)	=	-.68916112E-01	= H(46)
H(31)	=	-.17401703E-01	= H(45)
H(32)	=	.24123348E-16	= H(44)
H(33)	=	-.58815998E-01	= H(43)
H(34)	=	.25326927E-16	= H(42)
H(35)	=	.19605333E+00	= H(41)
H(36)	=	.19492420E+00	= H(40)
H(37)	=	-.12181192E+00	= H(39)
H(38)	=	.66666667E+00	= H(38)

Window Coefficients:

```
W( 1)  =  .00000000E+00  = W( 75)
W( 2)  =  .18012557E-02  = W( 74)
W( 3)  =  .71920448E-02  = W( 73)
W( 4)  =  .16133527E-01  = W( 72)
W( 5)  =  .28561277E-01  = W( 71)
W( 6)  =  .44385755E-01  = W( 70)
W( 7)  =  .63492943E-01  = W( 69)
W( 8)  =  .85745175E-01  = W( 68)
W( 9)  =  .11098212E+00  = W( 67)
W( 10) =  .13902195E+00  = W( 66)
W( 11) =  .16966264E+00  = W( 65)
W( 12) =  .20268341E+00  = W( 64)
W( 13) =  .23784636E+00  = W( 63)
W( 14) =  .27489813E+00  = W( 62)
W( 15) =  .31357176E+00  = W( 61)
W( 16) =  .35358861E+00  = W( 60)
W( 17) =  .39466037E+00  = W( 59)
W( 18) =  .43649109E+00  = W( 58)
W( 19) =  .47877940E+00  = W( 57)
W( 20) =  .52122060E+00  = W( 56)
W( 21) =  .56350891E+00  = W( 55)
W( 22) =  .60533963E+00  = W( 54)
W( 23) =  .64641139E+00  = W( 53)
W( 24) =  .68642824E+00  = W( 52)
W( 25) =  .72510187E+00  = W( 51)
W( 26) =  .76215364E+00  = W( 50)
W( 27) =  .79731659E+00  = W( 49)
W( 28) =  .83033736E+00  = W( 48)
W( 29) =  .86097805E+00  = W( 47)
W( 30) =  .88901788E+00  = W( 46)
W( 31) =  .91425482E+00  = W( 45)
W( 32) =  .93650706E+00  = W( 44)
W( 33) =  .95561425E+00  = W( 43)
W( 34) =  .97143872E+00  = W( 42)
W( 35) =  .98386647E+00  = W( 41)
W( 36) =  .99280796E+00  = W( 40)
W( 37) =  .99819874E+00  = W( 39)
W( 38) =  .10000000E+01  = W( 38)
```

Impulse Coefficients of Filter as Modified by the Window Coefficients:

```
H( 1)*W( 1)  =   .00000000E+00  = H( 75)*W( 75)
H( 2)*W( 2)  =   .00000000E+00  = H( 74)*W( 74)
H( 3)*W( 3)  =  -.60319901E-04  = H( 73)*W( 73)
H( 4)*W( 4)  =  -.18489361E-03  = H( 72)*W( 72)
H( 5)*W( 5)  =   .21076202E-03  = H( 71)*W( 71)
H( 6)*W( 6)  =   .76460838E-03  = H( 70)*W( 70)
H( 7)*W( 7)  =   .24938583E-03  = H( 69)*W( 69)
```

```
H(  8)*W(  8)   =    .00000000E+00   = H( 68)*W( 68)
H(  9)*W(  9)   =    .11253357E-02   = H( 67)*W( 67)
H( 10)*W( 10)   =    .00000000E+00   = H( 66)*W( 66)
H( 11)*W( 11)   =   -.36958456E-02   = H( 65)*W( 65)
H( 12)*W( 12)   =   -.30390024E-02   = H( 64)*W( 64)
H( 13)*W( 13)   =    .11588335E-02   = H( 63)*W( 63)
H( 14)*W( 14)   =    .00000000E+00   = H( 62)*W( 62)
H( 15)*W( 15)   =   -.16607046E-02   = H( 61)*W( 61)
H( 16)*W( 16)   =    .62656403E-02   = H( 60)*W( 60)
H( 17)*W( 17)   =    .11053443E-01   = H( 59)*W( 59)
H( 18)*W( 18)   =    .00000000E+00   = H( 58)*W( 58)
H( 19)*W( 19)   =   -.74104071E-02   = H( 57)*W( 57)
H( 20)*W( 20)   =    .00000000E+00   = H( 56)*W( 56)
H( 21)*W( 21)   =   -.40377378E-02   = H( 55)*W( 55)
H( 22)*W( 22)   =   -.20858765E-01   = H( 54)*W( 54)
H( 23)*W( 23)   =   -.10498643E-01   = H( 53)*W( 53)
H( 24)*W( 24)   =    .19114494E-01   = H( 52)*W( 52)
H( 25)*W( 25)   =    .16402841E-01   = H( 51)*W( 51)
H( 26)*W( 26)   =    .00000000E+00   = H( 50)*W( 50)
H( 27)*W( 27)   =    .21315813E-01   = H( 49)*W( 49)
H( 28)*W( 28)   =    .32370567E-01   = H( 48)*W( 48)
H( 29)*W( 29)   =   -.23306012E-01   = H( 47)*W( 47)
H( 30)*W( 30)   =   -.61267614E-01   = H( 46)*W( 46)
H( 31)*W( 31)   =   -.15909553E-01   = H( 45)*W( 45)
H( 32)*W( 32)   =    .00000000E+00   = H( 44)*W( 44)
H( 33)*W( 33)   =   -.56205392E-01   = H( 43)*W( 43)
H( 34)*W( 34)   =    .00000000E+00   = H( 42)*W( 42)
H( 35)*W( 35)   =    .19289029E+00   = H( 41)*W( 41)
H( 36)*W( 36)   =    .19352221E+00   = H( 40)*W( 40)
H( 37)*W( 37)   =   -.12159240E+00   = H( 39)*W( 39)
H( 38)*W( 38)   =    .66666663E+00   = H( 38)*W( 38)
```

Plots of the amplitude response and impulse response of the designed band-stop FIR filter are shown in Figure 6.20.

The FIRBS.ASM assembler program, shown in Figure 6.21, implements the designed band-stop FIR filter on the DSP56001 processor. Note that this program has an "include" statement which references the FIR.ASM assembler program macro shown in Figure 6.13. A copy of the FIRBS.ASM assembler program, FIR.ASM assembler program macro, and the FIRBS.LOD absolute load program can be found on the DSP56000/1 Demonstration Software #2 program diskette.

To demonstrate the designed band-stop FIR filter on the DSP system:

1. CONNECT A FUNCTION GENERATOR TO "BNC2" ON THE EVB AND TO CHANNEL ONE OF A DUAL TRACE OSCILLOSCOPE.
2. SET THE FUNCTION GENERATOR TO PRODUCE A 2-VOLT (PEAK TO PEAK) SINE WAVE.
3. CONNECT CHANNEL TWO OF THE OSCILLOSCOPE TO "BNC1" ON THE EVB.
4. LOAD THE "ADS56000 USER INTERFACE SOFTWARE" PROGRAM.

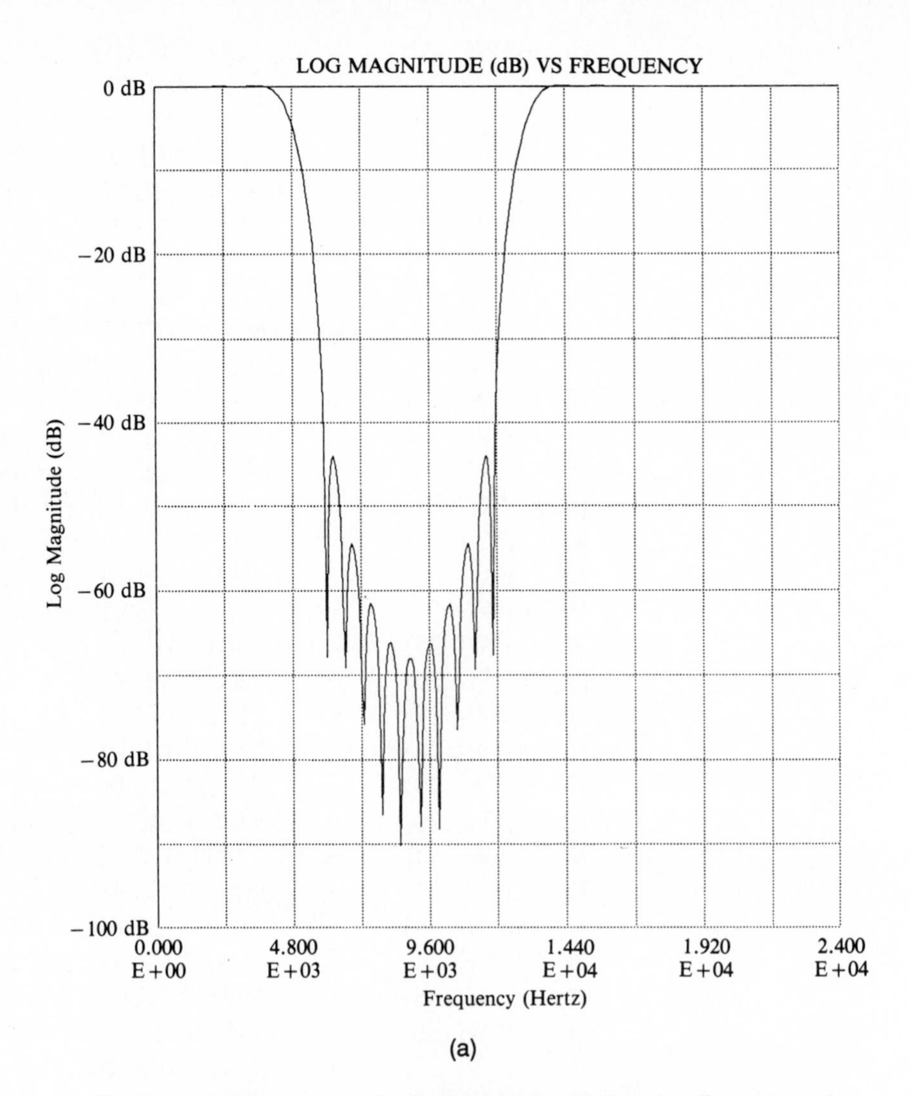

(a)

Figure 6-20 Plots of Amplitude Response and Impulse Response of Band-Stop Filter in Section 6.10.4. (a) Amplitude Response in dB (b) Impulse Response

5. FROM WITHIN THE "ADS56000" PROGRAM, LOAD THE FILE **firbs.lod** FROM THE "DSP56001 DEMONSTRATION SOFTWARE #2" DISKETTE.
6. TYPE **go<return>** to execute the designed band-stop FIR filter program (algorithm) on the DSP56001 processor.
7. VARY THE SINUSOIDAL INPUT FREQUENCY FROM 0.5 TO 20 KHZ. View the analog input and output signals on the oscilloscope. Note that the FIR filter stops the frequencies from 6 to 12 KHz and passes the frequencies from 0.5 to 4 KHz and 14 to 20 KHz as designed. Also note the increasing attenuation of the output signal as the input signal's frequency is increased from 4 to 6 KHz; note the decreasing attenuation of the output signal as the input signal's frequency is increased from 12 to 14 KHz.
8. TYPE **force b<return>** to halt execution of the FIR filter algorithm and return control of the ADM to its monitor program for keyboard command entry.
9. TYPE **quit<return>** to exit the ADS56000 program.

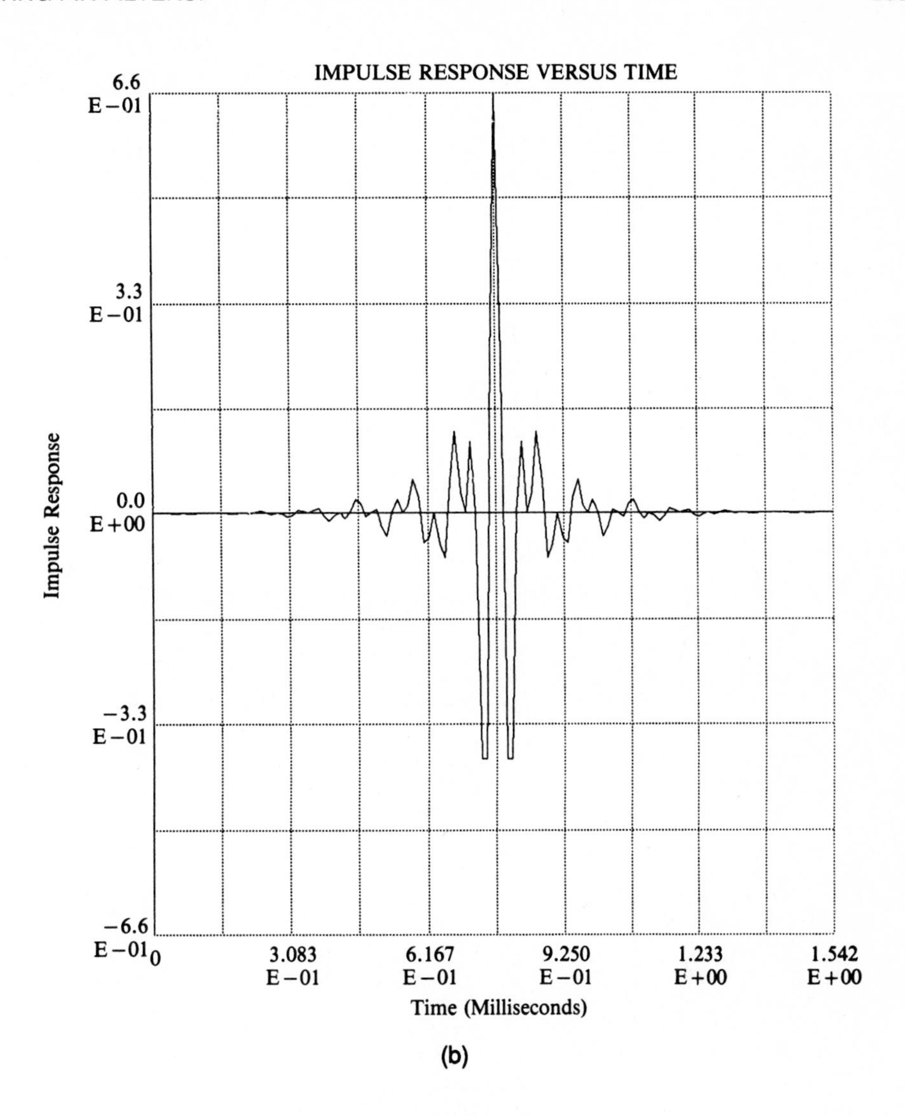

(b)

Figure 6-20 *(cont.)*

```
;*******************************************************
;
; File: FIRBS.ASM
;
;*******************************************************
;
        include 'fir'
;
;*******************************************************
; Coefficients of Band Stop FIR
;*******************************************************
;
ntaps   equ     75
;
        org     x:$0
states  dsm     ntaps       ;filter states
;
        org     y:$0
```

```
coef
        dc          0
        dc          0
        dc       -506
        dc      -1551
        dc       1768
        dc       6414
        dc       2092
        dc          0
        dc       9440
        dc          0
        dc     -31003
        dc     -25493
        dc       9721
        dc          0
        dc     -13931
        dc      52560
        dc      92723
        dc          0
        dc     -62163
        dc          0
        dc     -33871
        dc    -174976
        dc     -88069
        dc     160344
        dc     137597
        dc          0
        dc     178810
        dc     271544
        dc    -195505
        dc    -513950
        dc    -133459
        dc          0
        dc    -471485
        dc          0
        dc    1618081
        dc    1623382
        dc   -1019991
        dc    5592405
        dc   -1019991
        dc    1623382
        dc    1618081
        dc          0
        dc    -471485
        dc          0
        dc    -133459
        dc    -513950
        dc    -195505
        dc     271544
        dc     178810
        dc          0
        dc     137597
        dc     160344
        dc     -88069
        dc    -174976
        dc     -33871
        dc          0
        dc     -62163
        dc          0
        dc      92723
        dc      52560
        dc     -13931
        dc          0
        dc       9721
        dc     -25493
        dc     -31003
        dc          0
        dc       9440
        dc          0
```

```
        dc      2092
        dc      6414
        dc      1768
        dc     -1551
        dc      -506
        dc         0
        dc         0
;
; Program start address
;
        org     p:$100
;
        fir     states,coef,ntaps
;
        end
```

Figure 6-21 FIRBS.ASM Assembler Program

6.11 REFERENCES:

1. Alan V. Oppenheim and Ronald W. Shafer, Discrete-Time Signal Processing, Prentice-Hall, 1989.
2. N.K.Bose, Digital Filters Theory and Applications, North-Holland, 1985.
3. Johnny R. Johnson, Introduction to Digital Signal Processing, Prentice-Hall, 1989.
4. Samuel D. Stearns and Ruth A. David, Signal Processing Algorithms, Prentice-Hall, 1988.
5. Harry Y-F. Lam, Analog and Digital Filters, Prentice-Hall, 1979.
6. Chi-Tsong Chen, One-Dimensional Digital Signal Processing, Marcel Dekker, 1979.
7. Richard A. Roberts and Clifford T. Mullis, Digital Signal Processing, Addison Wesley, 1987.

7

DESIGNING IIR FILTERS AND IMPLEMENTING THEM ON THE DSP56001 PROCESSOR

In this chapter, Infinite Impulse Response (IIR) digital filters are designed and implemented on the DSP56001 processor. The design of IIR digital filters involves the determination of a transfer function in the z variable that meets certain frequency-domain specifications. The analog-filter-design techniques that have developed over the years are also based on frequency domain specifications. For this reason, the IIR digital filters are usually designed from analog filters by using various transformations. This approach is most useful for designing standard low-pass, high-pass, band-pass and band-stop filters, for which a considerable body of knowledge on analog filtering is available. In analog filtering, there are several analog-filter approximations that have been found to be satisfactory in designing analog filters that meet certain frequency-domain specifications. The most popular approximations are the Butterworth and the Chebyshev approximations.

The design of a digital filter begins by designing a suitable low-pass analog filter. Either of two different design procedures are usually used to obtain a digital filter from a low-pass analog filter. When employing the first design procedure, an analog frequency transformation is used to obtain another analog filter of the desired type. The digital filter can be determined from the desired type analog filter by using the Impulse Invariant Transformation or the Bilinear Transformation as explained below. When employing the second design procedure, the low-pass analog filter is converted into a low-pass digital filter by using the Impulse Invariant Transformation or the Bilinear Transformation. A digital frequency transformation is then used to obtain the digital filter. The stability of the filter then needs to be checked. The first design procedure is considered in this chapter.

The Impulse Invariant Transformation preserves the impulse response of the analog filter by specifying the digital filter's impulse response to be a sampled version of the continuous-time response. As a result, the frequency response of the digital filter is an aliased version of the frequency response of the analog filter. The Bilinear Transformation eliminates the aliasing problem by translating the analog transfer function to a digital transfer function with comparable frequency-response characteristics and by providing a one-to-one mapping of poles and zeros from the continuous time s-plane to the discrete time z-plane. Thus, the Bilinear Transformation is a better choice for designing digital IIR filters that can be implemented on the DSP56000/1 signal processor.

In the design process, the desired IIR filter's frequency domain specifications are first used to design normalized low-pass analog Butterworth and Chebyshev filters. Analog frequency transformations are then used to obtain the desired analog filters from the normalized low-pass analog Butterworth and Chebyshev filters. Digital filters are obtained from the analog filters by using the Bilinear Transformation.

By assuming that the filter order is an even integer, the last step in the design process is to decompose the transfer functions of the Butterworth and Chebyshev filters into products of second order sections. The second order sections are usually referred to as biquad sections. Figure 7.1 shows the procedure that is used in this chapter to design the digital filters. The coefficients of the biquad sections can be used to implement the digital filters on the DSP56001 processor.

In this chapter, a special demonstration version of the Filter Design and Analysis System (FDAS) software program by Momentum Data Systems, Inc. is used to demonstrate the design process. The DSP56001 instructions which implement a biquad section are described. Low-pass, high-pass, band-pass and band-stop IIR digital filters are then designed and implemented on the DSP System developed in Chapter 5 and again shown in Figure 7.2. The DSP56001 instructions which implement the four filter types are included on the DSP56000/1 Demonstration Software #3 diskette.

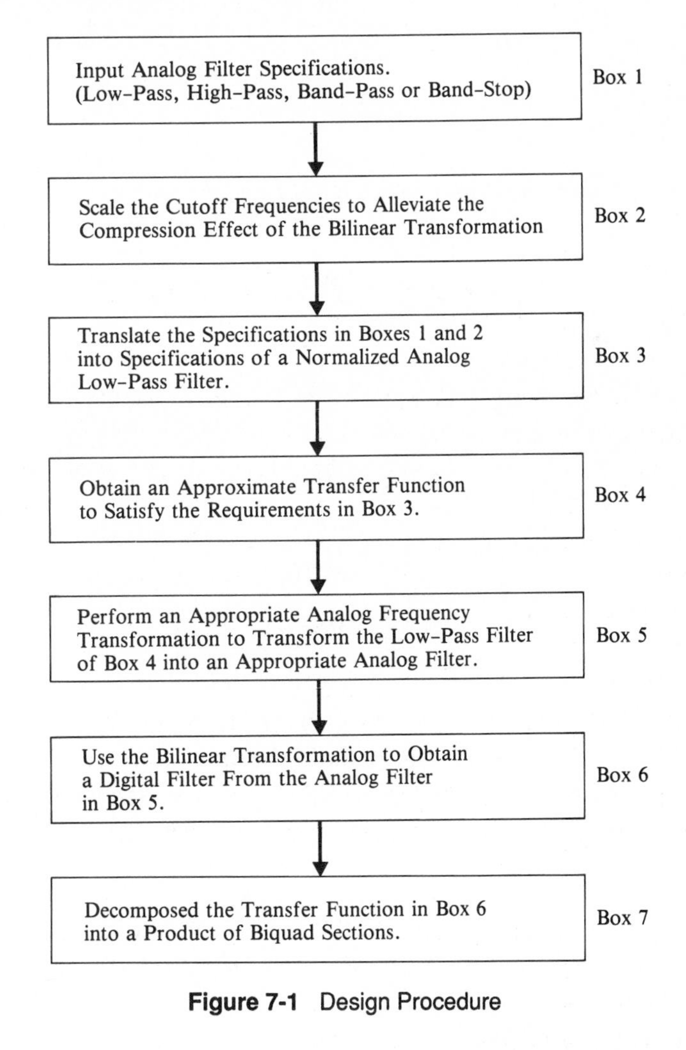

Figure 7-1 Design Procedure

Figure 7-2 DSP System

7.1 IIR FILTER TRANSFER FUNCTION:

The relationship between the IIR filter's input sequence x(n) and the output sequence y(n) can be written as the following difference equation:

$$y(n) = \sum_{i=0}^{N} b_i x(n-i) + \sum_{j=1}^{N} a_j y(n-j) \tag{7.1}$$

where b_i and a_j are the filter's coefficients, and at least one of the a_j coefficients is nonzero. The filter coefficients b_i and a_j provide the characteristics of the filter through which the input sequence is passed to create the desired output sequence y(n). By both definition and design, the output sequence y(n) of an IIR filter (also known as a recursive filter) is a function of the past outputs, and the current and past inputs.

The transfer function that corresponds to Difference Equation 7.1 can be expressed as:

$$H(z) = \frac{Y(z)}{X(z)} = \frac{b_0 + b_1 z^{-1} + \dots + b_N z^{-N}}{1 - a_1 z^{-1} - \dots - a_N z^{-N}} \tag{7.2}$$

In the remainder of this chapter, N is assumed to be an even integer. The transfer function in Equation 7.2 can then be decomposed into a product of biquad sections:

$$H(z) = k \prod_{i=1}^{M} H_i(z) \tag{7.3}$$

where

$$H_i(z) = \frac{Y_{i+1}(z)}{Y_i(z)} = \frac{1 + b_{1i} z^{-1} + b_{2i} z^{-2}}{1 - a_{1i} z^{-1} - a_{2i} z^{-2}} \tag{7.4}$$

$$Y_1(z) = K X(z) \tag{7.5}$$

$$Y_{M+1}(z) = Y(z) \tag{7.6}$$

and M is equal to 0.5 N, K is a positive gain, and the output of the ith biquad section $y_{i+1}(n)$ is the input of i+1th biquad section. Figure 7.3 shows the realization of the transfer function H(z) as a cascade of biquad sections. To guarantee the stability of the IIR digital filter, the magnitude of a_{1i} and a_{2i} must be less than 2.0 and 1.0 respectively. The biquad section $H_i(z)$ can be rewritten as:

$$H_i(z) = \frac{Y_{i+1}(z)}{Y_i(z)} = \frac{(1 + b_{1i} z^{-1} + b_{2i} z^{-2}) X_i(z)}{(1 - a_{1i} z^{-1} - a_{2i} z^{-2}) X_i(z)} \tag{7.7}$$

where $X_i(z)$ is chosen so that:

$$Y_i(z) = (1 - a_{1i} z^{-1} - a_{2i} z^{-2}) X_i(z) \tag{7.8}$$

$$Y_{i+1}(z) = (1 + b_{1i} z^{-1} + b_{2i} z^{-2}) X_i(z) \tag{7.9}$$

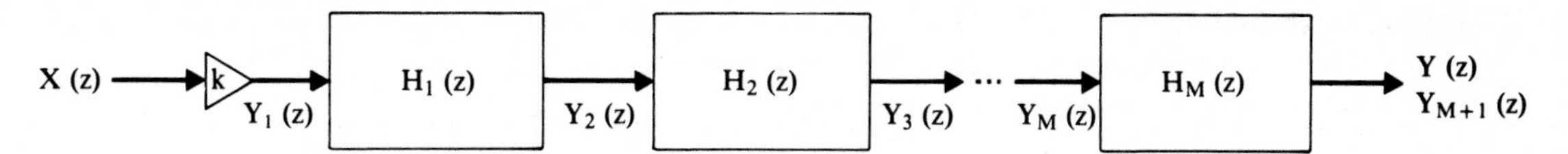

Figure 7-3 Realization of the Transfer Function as a Cascade of Biquad Sections

The realization of Equations 7.8 and 7.9 is shown in Figure 7.4. The realization of the transfer function of a sixth order digital IIR filter as a cascade of three biquad sections is shown in Figure 7.5.

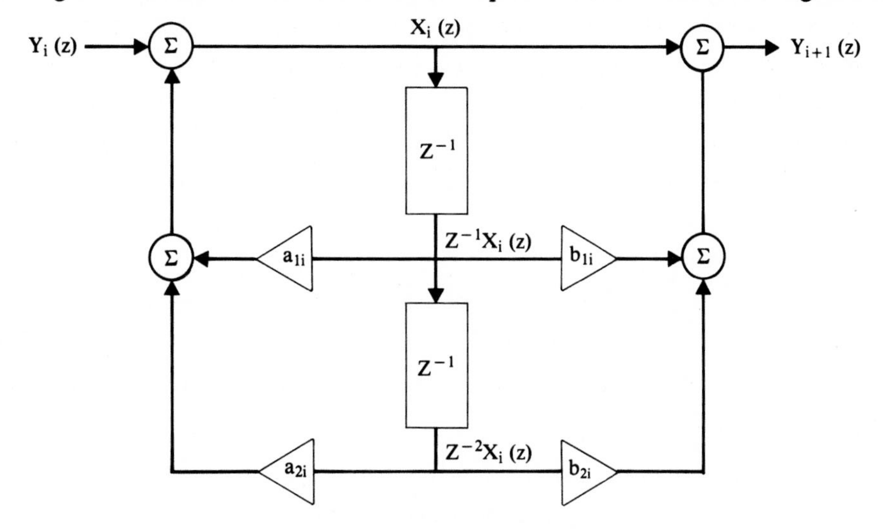

Figure 7-4 Biquad Section Structure

The main objective of this chapter is to design the transfer function H(z) of equations 7.3 and 7.4 to meet certain frequency-domain specifications, and then use the filter coefficients of equation 7.4 to implement the desired digital filter on the DSP56001 processor.

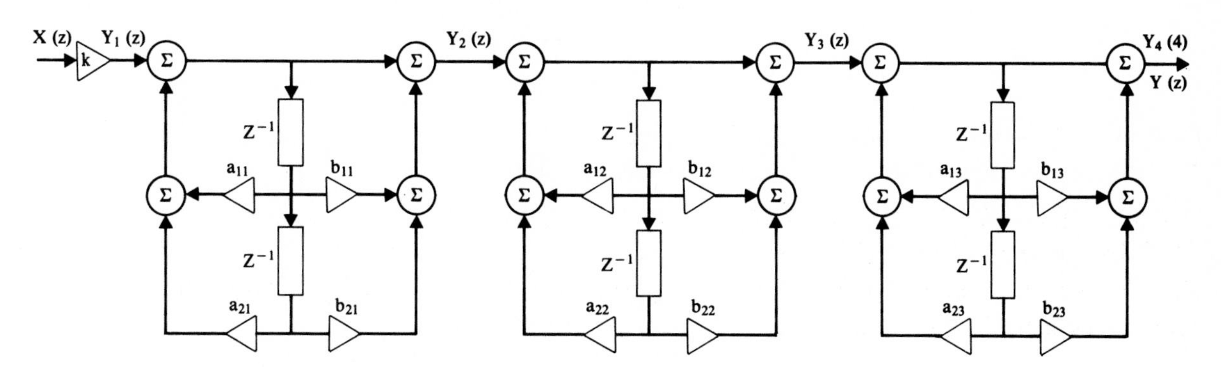

Figure 7-5 Realization of a Six-Order Digital Filter as a Cascade of Three Biquad Sections

7.2 REVIEW OF THE BILINEAR TRANSFORMATION:

The Bilinear Transformation is defined by the following substitution:

$$s = \frac{2}{T} \frac{z-1}{z+1} \tag{7.10}$$

where T is the sampling period. To consider the frequency response, the Laplace variable s is evaluated on the imaginary axis and the z transform variable is evaluated on the unit circle:

$$s = j\Omega \tag{7.11}$$

$$z = e^{jwT} \tag{7.12}$$

where Ω is the analog frequency and w is the digital frequency. The relationship between the analog and digital frequencies can be obtained using Equations 7.8, 7.9 and 7.10 as:

$$\begin{aligned}\Omega &= \frac{2}{j\,T}\,\frac{e^{jwT} - 1}{e^{jwT} + 1} \\ &= \frac{2}{j\,T}\,\frac{e^{jwT/2}\,(e^{jwT/2} - e^{-jwT/2})}{e^{jwT/2}(e^{jwT/2} + e^{-jwT/2})} \\ &= \frac{2}{j\,T}\,\frac{2\,j\,\sin wT/2}{2\cos wT/2} \\ &= \frac{2}{T}\,\tan wT/2\end{aligned} \tag{7.13}$$

The relationship between the analog frequency Ω and the digital frequency w is shown in Figure 7.6. It can be seen that the infinite imaginary axis in the s-plane ($-\infty \leq \Omega \leq +\infty$) is mapped onto the unit circle in the Z plane ($-\pi \leq wT \leq +\pi$). Furthermore, the frequencies Ω and w have a nonlinear relationship which has a compression effect on the frequency-response characteristics. The compression effect is known as frequency warping. This compression causes the transfer function at the high Ω frequencies to be highly distorted when it is translated to the w-domain. This compression can be alleviated by introducing a suitable prewarping to the Ω frequency scale to convert it to the Ω^* frequency scale, where

$$\Omega^* = \frac{2}{T}\tan \Omega T/2 \tag{7.14}$$

In the following sections, the cutoff frequencies of the analog filters are first converted to the Ω^* frequency scale and then are used to design the analog filters.

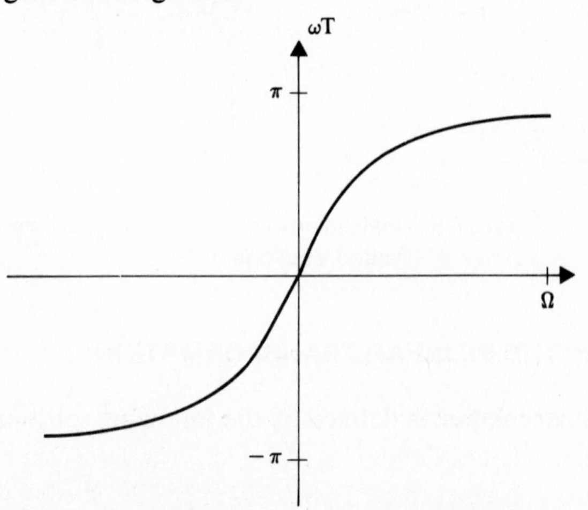

Figure 7-6 Relationship Between Analog and Digital Frequencies

7.3 IIR FILTER SPECIFICATIONS:

The frequency-domain specifications of standard IIR filters are the pass-band and stop-band attenuations, and the pass-band and stop-band cutoff frequencies. Figure 7.7. shows the frequency-domain specifications for the low-pass, high-pass, band-pass and band-stop filters, where:

α = Pass-band attenuation in dB.

β = Stop-band attenuation in dB.

Ω_c = 3 dB cutoff frequency.

Ω_1 = Pass-band cutoff frequency of low-pass filter.

Ω_2 = Stop-band cutoff frequency of low-pass filter.

Ω_3 = Stop-band cutoff frequency of high-pass filter.

Ω_4 = Pass-band cutoff frequency of high-pass filter.

Ω_5 = Stop-band lower cutoff frequency of band-pass filter.

Ω_6 = Pass-band lower cutoff frequency of band-pass filter.

Ω_7 = Pass-band upper cutoff frequency of band-pass filter.

Ω_8 = Stop-band upper cutoff frequency of band-pass filter.

Ω_9 = Pass-band lower cutoff frequency of band-stop filter.

Ω_{10} = Stop-band lower cutoff frequency of band-stop filter.

Ω_{11} = Stop-band upper cutoff frequency of band-stop filter.

Ω_{12} = Pass-band upper cutoff frequency of band-stop filter.

The design process begins by prewarping the cutoff frequencies of the desired analog filter to obtain the corresponding cutoff frequencies on the Ω^* scale, where the relationship between Ω^* and Ω is shown in Equation 7.14. The transfer function of the analog filter is then designed to meet the specifications on the Ω^* scale. The digital filter transfer function obtained from the analog transfer function, by the Bilinear Transformation, will satisfy the specifications on the Ω scale. This can be seen by obtaining w from Equations 7.13 and 7.14 as:

$$W = \frac{2}{T}\arctan\Omega^* T/2 = \Omega \tag{7.15}$$

7.4 NORMALIZED LOW-PASS FILTER DESIGN:

In this section, the cutoff frequencies of the analog filter on the Ω^* scale are converted to corresponding cutoff frequencies of a normalized analog low-pass filter $H_n(s)$ by using an analog frequency transformation selected from Table 7.1. $H_n(s)$ is an approximation of the normalized ideal low-pass filter that has a gain of one in the band of frequencies from zero to 1 rad/sec and a gain of zero for all frequencies above one rad/sec.

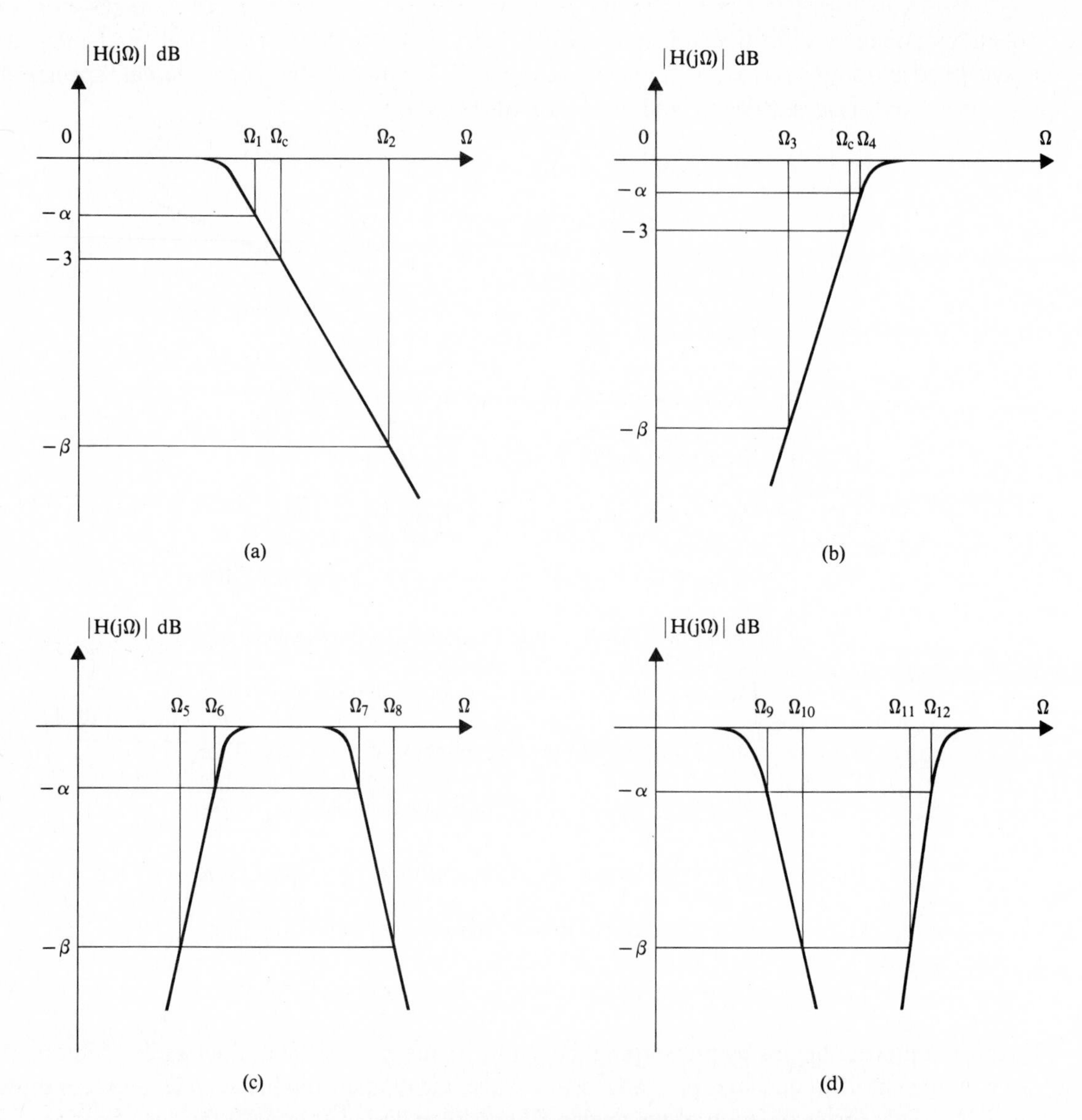

Figure 7-7 Typical Analog Filters. (a) Low-Pass Filter (b) High-Pass Filter (c) Band-Pass Filter (d) Band-Stop Filter

For given normalized low-pass analog filter specifications, there are numerous types of filters that can be designed. The most popular are the Butterworth and Chebyshev filters. For $w \geq 0$, the Butterworth low-pass filter is characterized by a monotonically decreasing magnitude function of w, while the Chebyshev low-pass filter is characterized by an equiripple magnitude function across the passband and a monotonically decreasing magnitude function in the stopband. The normalized low-pass Butterworth and Chebyshev filters are shown in Figure 7.8, where $\Omega_L{}^n$ and $\Omega_H{}^n$ are the normalized pass-band and stop-band cutoff frequencies, respectively.

The normalized pass-band cutoff frequency $\Omega_L{}^n$ is equal to one for the Chebyshev filter and is equal to the pass-band cutoff frequency divided by the 3 dB cutoff frequency for the Butterworth filter. If the actual 3 dB cutoff frequency is not given, then a guess must be made to calculate $\Omega_L{}^n$.

TABLE 7.1. Analog Frequency Transformations.

Desired Filter	Analog Frequency Transformation
Low-pass	$s' = s/\Omega^+$
High-pass	$s' = \Omega^+/s$
Band-pass	$s' = [(s^2+\Omega_7{}^*\Omega_6{}^*)/(s(\Omega_7{}^*-\Omega_6{}^*))]$
Band-stop	$s' = [s(\Omega_{11}{}^*-\Omega_{10}{}^*)\Omega_H{}^*/(s^2+\Omega_{11}{}^*\Omega_{10}{}^*)]$
where	
	$s = j\Omega^n$
	$s' = j\Omega^*$
	$\Omega^+ = \Omega_c{}^*$ for Butterworth filter
	$= \Omega_L{}^*$ for Chebyshev filter
	Ω^n = normalized analog low-pass frequency

The ratio of $\Omega_H{}^n$ to $\Omega_L{}^n$, referred to as R, can be calculated by using the analog frequency transformations and the low-pass, high-pass, band-pass or band-stop cutoff frequencies as:

a) *Low-pass to Normalized Low-pass*:

From Table 7.1, the low-pass to normalized low-pass transformation is given by:

$$s' = s/\Omega^+ \text{ or } \Omega^n = \Omega^*/\Omega^+ \qquad (7.16)$$

Hence, R can be calculated as:

$$R = \Omega_H^n/\Omega_L^n$$
$$= (\Omega_2^*/\Omega^+)/(\Omega_1^*/\Omega^+) = \Omega_2^*/\Omega_1^* \qquad (7.17)$$

b) *High-pass to Normalized Low-pass*:

From Table 7.1, the high-pass to normalized low-pass transformation is given by:

$$s' = \Omega^+/s \text{ or } \Omega^n = \Omega^+/\Omega^* \qquad (7.18)$$

Using Equation 7.18, R can be calculated as:

$$R = \Omega_H^n/\Omega_L^n$$
$$= (\Omega^+/\Omega_3^*)/(\Omega^+/\Omega_4^*) = \Omega_4^*/\Omega_3^* \qquad (7.19)$$

c) *Band-pass to Normalized Low-pass*:

From Table 7.1, the band-pass to normalized low-pass transformation is given by:

$$s' = \frac{s^2 + \Omega_7^*\Omega_6^*}{s\,(\Omega_7^* - \Omega_6^*)}$$

or

$$\Omega^n = \frac{\Omega^{*2} - \Omega_7^*\Omega_6^*}{\Omega^*(\Omega_7^* - \Omega_6^*)} \tag{7.20}$$

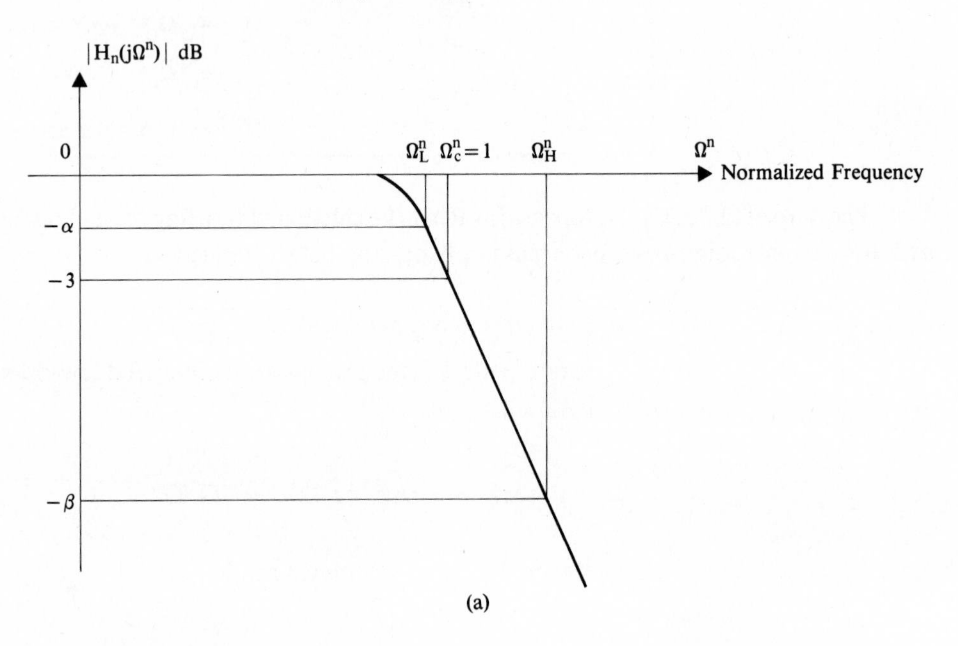

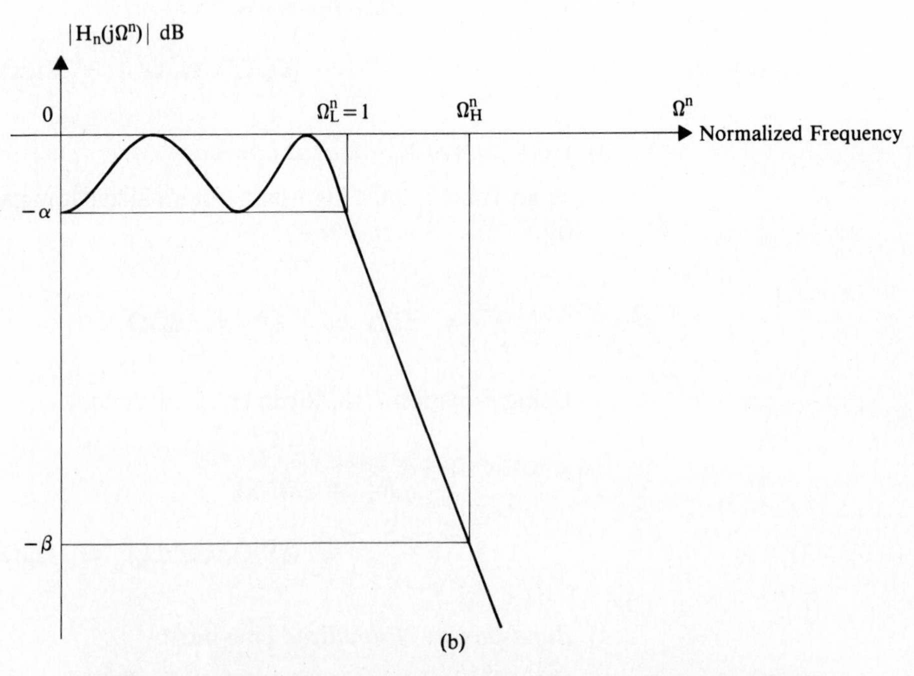

Figure 7-8 Normalized Analog Low-Pass Filter. (a) Butterworth Filter (b) Chebyshev Filter

The transformation will transform Ω_6^* and Ω_7^* into $\Omega_L^n = \pm 1$, and Ω_5^* and Ω_8^* into:

$$\Omega_{H1}^n = \frac{\Omega_5^{*2} - \Omega_7^*\Omega_6^*}{\Omega_5^*(\Omega_7^* - \Omega_6^*)} \tag{7.21}$$

$$\Omega_{H2}^n = \frac{\Omega_8^{*2} - \Omega_7^*\Omega_6^*}{\Omega_8^*(\Omega_7^* - \Omega_6^*)} \tag{7.22}$$

In order to have a stop-band attenuation of at least β dB, the stop-band cutoff frequency Ω_H^n must be chosen as the smaller of υθε absolute value οζ Ω_{H1}^n and Ω_{H2}^n, respectively. Hence, R can be calculated as:

$$R = \min\left[\left|\frac{\Omega_5^{*2} - \Omega_7^*\Omega_6^*}{\Omega_5^*(\Omega_7^* - \Omega_6^*)}\right|, \left|\frac{\Omega_8^{*2} - \Omega_7^*\Omega_6^*}{\Omega_8^*(\Omega_7^* - \Omega_6^*)}\right|\right] \tag{7.23}$$

d) *Band-stop to Normalized Low-pass*:

The band-stop to normalized low-pass transformation is given by:

$$s' = \frac{s\,(\Omega_{11}^* - \Omega_{10}^*)\;\Omega_H^*}{s^2 + \Omega_{11}^*\Omega_{10}^*}$$

or

$$\Omega^n = \frac{\Omega^*\;(\Omega_{11}^* - \Omega_{10}^*)\Omega_H^*}{\Omega_{11}^*\Omega_{10}^* - \Omega^{*2}} \tag{7.24}$$

The transformation will transform Ω_{11}^* and Ω_{10}^* into $\Omega_L^n = \pm\Omega_H^n$, and Ω_9^* and Ω_{12}^* into:

$$\Omega_{L1}^n = \frac{\Omega_9^*(\Omega_{11}^* - \Omega_{10}^*)\Omega_H^n}{\Omega_{11}^*\Omega_{10}^* - \Omega_9^{*2}} \tag{7.25}$$

$$\Omega_{L2}^n = \frac{\Omega_{12}^*(\Omega_{11}^* - \Omega_{10}^*)\Omega_H^n}{\Omega_{11}^*\Omega_{10}^* - \Omega_{12}^{*\,2}} \tag{7.26}$$

In order to have a pass-band attenuation of at least α dB, the pass-band cutoff frequency Ω_L^n must be chosen as the larger of the absolute value οζ Ω_{L1}^n and Ω_{L2}^n, respectively.

Now R can be calculated as:

$$R = \min\left[\left|\frac{\Omega_{11}^*\Omega_{10}^* - \Omega_9^{*2}}{\Omega_9^*(\Omega_{11}^* - \Omega_{10}^*)}\right|, \left|\frac{\Omega_{11}^*\Omega_{10}^* - \Omega_{12}^{*2}}{\Omega_{12}^*(\Omega_{11}^* - \Omega_{10}^*)}\right|\right] \tag{7.27}$$

Table 7.2 shows the values of R that are calculated by using the low-pass, high-pass, band-pass and band-stop cutoff frequencies. After calculating R, the normalized stop-band cutoff frequency $\Omega_H{}^n$ is obtained by multiplying $\Omega_L{}^n$ by R.

TABLE 7.2. Ratio of Stop-band to Pass-band cutoff Frequencies.

Desired Filter	R for Normalized Low-pass Filter $R = \Omega_H{}^n/\Omega_L{}^n$
Low-pass	$\Omega_2{}^*/\Omega_1{}^*$
High-pass	$\Omega_4{}^*/\Omega_3{}^*$
Band-pass	$\min\left[\left\lvert\frac{\Omega_5^{*2} - \Omega_6^{\times}\Omega_7^{*}}{\Omega_5^{*}\left(\Omega_7^{*} - \Omega_6^{*}\right)}\right\rvert , \left\lvert\frac{\Omega_8^{*2} - \Omega_6^{\times}\Omega_7^{*}}{\Omega_8^{*}\left(\Omega_7^{*} - \Omega_6^{*}\right)}\right\rvert\right]$
Band-stop	$\min\left[\left\lvert\frac{\Omega_{11}^{*}\Omega_{10}^{*} - \Omega^{*}\Omega_9^{*2}}{\Omega_9^{*}\left(\Omega_{11}^{*} - \Omega_{10}^{*}\right)}\right\rvert , \left\lvert\frac{\Omega_{11}^{*}\Omega_{10}^{*} - \Omega^{*}\Omega_{12}^{*2}}{\Omega_{12}^{*}\left(\Omega_{11}^{*} - \Omega_{10}^{*}\right)}\right\rvert\right]$

7.4.1 Butterworth Filter Design:

The amplitude response of an Nth-order normalized low-pass analog Butterworth filter can be written as:

$$|H_n(j\Omega^n)|^2 = \frac{1}{1 + (\Omega^n)^{2n}} \tag{7.28}$$

The first step in the design process is to find the order of the filter N which meets the normalized low-pass filter specifications ($\alpha, \beta, \Omega_L{}^n$, and $\Omega_H{}^n$). This can be accomplished by choosing N as the smallest even integer satisfying:

$$N \geq \max(N1, N2) \tag{7.29}$$

where N_1 and N_2 satisfy the following inequalities:

$$-10 \log \frac{1}{1 + (\Omega_L^n)_1^{2N}} < \alpha \tag{7.30}$$

$$-10 \log \frac{1}{1 + (\Omega_H^n)_2^{2N}} > \beta \tag{7.31}$$

The next step in the design process is to find the normalized transfer function $H_n(s)$. Substituting $s' = j\Omega^n$ into Equation 7.28 yields:

$$|H_n(s')|^2 = \frac{1}{1 + (-1)^N s'^{2N}} \tag{7.32}$$

The denominator has the following 2N roots:

$$s'_i \;=\; \exp\, j\pi(2i+N-1)/2N \qquad 1 \le i \le 2N \tag{7.33}$$

The 2N roots of are located on a unit circle in the s'-plane as shown in Figure 7.9. To form a stable Nth order normalized low-pass filter, the poles in the left half of the s'-plane are used. The related transfer function is:

$$H_n(s') \;=\; \frac{1}{(s'-s_1')(s'-s_2')\ldots(s'-s_N')} \tag{7.34}$$

For N even, all poles occur in complex conjugate pairs of the form s_i', s_{N+1-i}', where $1 \le +i \le N$. Using the pole definitions in Equation 7.33, the transfer function for a normalized filter can be rewritten as:

$$H_n(s') \;=\; \prod_{i=1}^{M=0.5N} H_i(s') \tag{7.35}$$

where

$$\begin{aligned} H_i(s') &= \frac{1}{(s'-s'_i)\;\;(s'-s'_{N+1-i})} \\ &= \frac{1}{s'^2 \;-\; 2s' \;\cos((2i+N-1)\pi/2N)+1} \end{aligned} \tag{7.36}$$

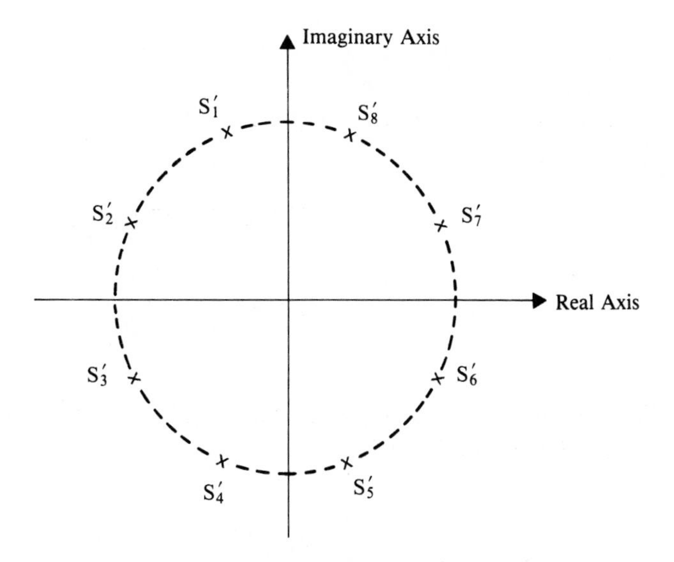

Figure 7-9 Butterworth Poles for N=4

7.4.2 Chebyshev Filter Design:

The magnitude response of an Nth-order Chebyshev filter can be written as:

$$|H_n(j\Omega^n)|^2 = \frac{1}{1 + \varepsilon^2 T_N^2(\Omega^n)} \tag{7.37}$$

where $T_N(\Omega^n)$ is the Nth-order Chebyshev polynomial:

$$\begin{aligned} T_N(\Omega^n) &= \cos(N\cos^{-1}\Omega^n) \qquad \Omega^n \le 1 \\ &= \cosh(N\cosh^{-1}\Omega^n) \qquad \Omega^n > 1 \end{aligned} \tag{7.38}$$

and ε is a parameter that controls the passband ripple. The passband amplitude oscillates between 1 and $1/(1+\varepsilon^2)^{0.5}$. The amplitude at $\Omega^n = 0$ is one if N is odd and $1/(1+\varepsilon^2)^{0.5}$ if N is even. The amplitude at the normalized cutoff frequency of 1 rad/sec is equal to $1/(1+\varepsilon^2)^{0.5}$ and, consequently, the passband attenuation α can be written as:

$$-10 \log \frac{1}{(1+\varepsilon^2)} = \alpha \tag{7.39}$$

Hence, the ripple parameter ε is determined as:

$$\varepsilon = (10_\alpha^{0.1} - 1)^{0.5} \tag{7.40}$$

Similarly, the stopband attenuation β can be written, by using Equation 7.37, as:

$$-10 \log \frac{1}{(1 + \varepsilon^2 T_N^2(\Omega_H^n))} = \beta \tag{7.41}$$

Using Equations 7.38 and 7.41, the Chebyshev polynomial $T_N(\Omega_H{}^n)$ at the normalized stopband cutoff frequency $\Omega_H{}^n$ can be written as:

$$\begin{aligned} T_N(\Omega_H^n) &= [(10^{0.1\beta} - 1)/\varepsilon]^{0.5} \\ &= \cosh(N\cosh^{-1}\Omega_H^n) \end{aligned} \tag{7.42}$$

From Equation 7.42, the order of the Chebyshev filter N is the smallest integer satisfying:

$$N \ge \frac{\cosh^{-1} [(10^{-0.1\beta} - 1)/\varepsilon]^{0.5}}{\cosh^{-1}\Omega_H^n} \tag{7.43}$$

As in the previously designed Butterworth filter, the normalized Chebyshev filter is specified in terms of the pole locations on the s'-plane. The poles of this filter lie on an ellipse in the s'-plane as shown in Figure 7.10. To form a stable normalized low-pass filter, the poles that lie in the stable region are selected. For even N, the stable poles s_i' are given by:

$$s'_i = \mu_i + j\eta_i \tag{7.44}$$

where

$$\mu_i = -\sinh\gamma \cos\frac{i\pi}{2N} \qquad i = \pm1, \pm3, \pm5, \ldots, \pm N-1 \tag{7.45}$$

$$= \cosh\gamma \sin\frac{i\pi}{2N} \qquad i = \pm1, \pm3, \pm5, \ldots, \pm N-1 \tag{7.46}$$

$$\gamma = \frac{1}{N} \sinh^{-1} \frac{1}{\varepsilon} \tag{7.47}$$

As in the previous section, the complex-conjugate roots can be combined to be the roots of a second-order polynomial. For N even, $H_n(s')$ has the form:

$$H_n(s') = \pi_i \frac{1}{s'^2 - 2\mu_i s' + (\mu_i^2 + \eta_i^2)} \tag{7.48}$$

where

$$i = 1, 3, 5, .., N-1$$

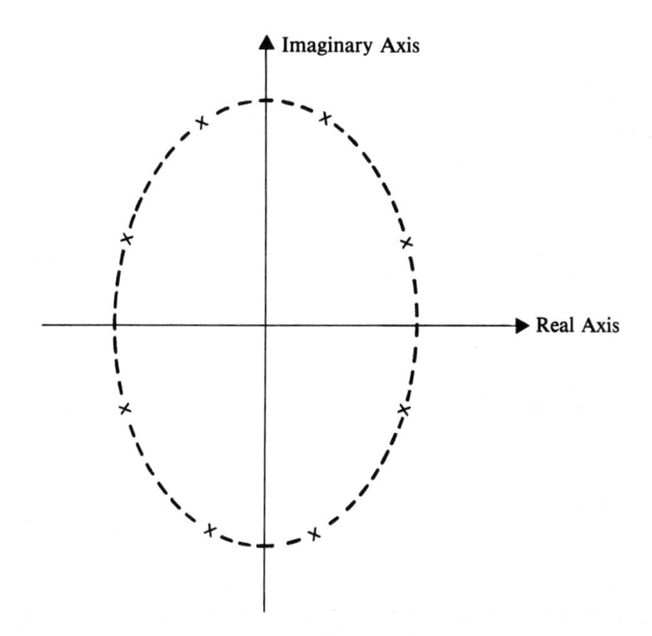

Figure 7-10 Chebyshev Poles for N=4

7.5 ANALOG FILTER DESIGN:

The analog frequency transformations are used to convert the normalized low-pass analog filter to the desired analog filter, where the transfer function of the analog low-pass, high-pass, band-pass and band-stop filter can be written as:

$$H_{LP}(s) = H_n(s') \downarrow_{s' = s/\Omega^+} \tag{7.49}$$

$$H_{HP}(s) = H_n(s') \downarrow_{s' = \Omega^+/s} \tag{7.50}$$

$$H_{BP}(s) = H_n(s') \downarrow_{s' = \frac{s^2+\Omega_7^*\Omega_6^*}{s\left(\Omega_7^*-\Omega_6^*\right)}} \tag{7.51}$$

$$H_{BS}(s) = H_n(s') \downarrow_{s' = \frac{s\left(\Omega_{11}^*-\Omega_{10}^*\right)\Omega_H^*}{s^2+\Omega_{22}^*\Omega_{10}^*}} \tag{7.52}$$

7.6 DIGITAL FILTER DESIGN:

The final step in the design process is to obtain the desired digital filter from the analog filter by using the Bilinear Transformation:

$$H(z) = H(s) \downarrow_{s = (2/T)(z - 1/z + 1)} \tag{7.53}$$

The transfer function H(z) can be decomposed into a product of biquad sections. The biquad sections are described by Equations 7.3 through 7.6. In the following chapter sections, the coefficients of the biquad sections b_{1i}, b_{2i}, a_{1i} and a_{2i} and the total gain K are used to implement the digital IIR filter on the DSP56001 processor.

7.7 IMPLEMENTING A BIQUAD SECTION ON THE DSP56001 PROCESSOR:

For the ith biquad section in Figure 7.4, the input sample $y_i(n)$ is in Register A and the output sample $y_{i+1}(n)$ will be in Register A. This allows $y_{i+1}(n)$ to be the input sample for the (i+1)th section.

From Equations 7.8 and 7.9, the difference equations for the ith biquad section can be written as:

$$x_i(n) = y_i(n) + a_{1i}x_i(n-1) + a_{2i}x_i(n-2) \tag{7.54}$$

$$y_{i+1}(n) = x_i(n) + b_{1i}x_i(n-1) + b_{2i}x_i(n-2) \tag{7.55}$$

Since the actual biquad coefficients vary from -2 to +2, the scaling mode is used and the coefficients a_{1i}, a_{2i}, b_{1i}, b_{2i} are the actual values for the filter divided by two.

The X-Memory map for the filter states and the Y-Memory map for the coefficients before the execution of the ith biquad section is shown in Figure 7.11. R0 is intialized to point to the filter states in X-Memory and R4 is initialized to point to the filter coefficients in Y-Memory. Note the sequence in which the filter states are stored. The first element of the sequence is the second filter state $x_i(n-2)$ and the second element is the first filter state $x_i(n-1)$. Similarly, the first element of the coefficients is a_{2i} then a_{1i}, followed by b_{2i} then b_{1i}. The initial values in R0 and R4 are assumed to be \$50 and \$100 respectively.

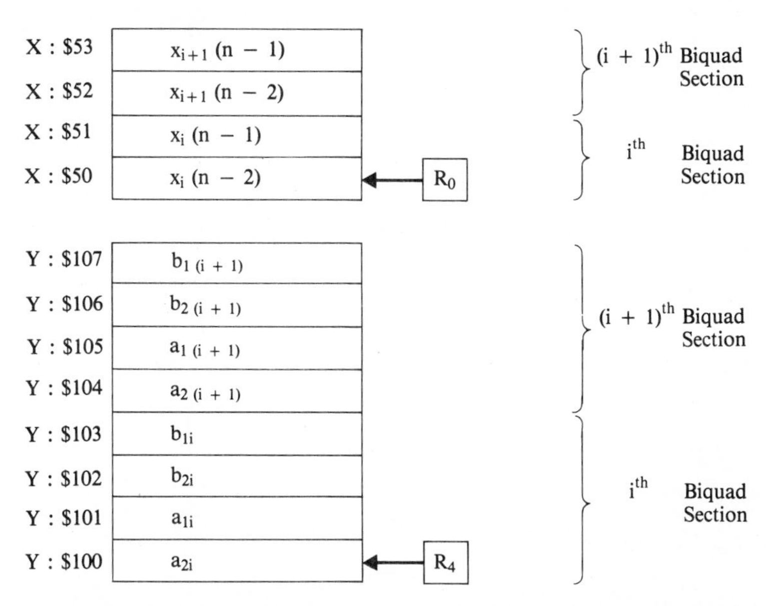

Figure 7-11 Memory Map Before the Execution of the ith Biquad Section

The following DSP56001 instructions are used to implement the ith biquad section:

```
ori     #$8,mr
move              x:(r0)+,x0      y:(r4)+,y0
mac     x0,y0,a   x:(r0)-,x1      y:(r4)+,y0
macr    x1,y0,a   x1,x:(r0)+      y:(r4)+,y0
mac     x0,y0,a   a,x:(r0)+       y:(r4)+,y0
mac     x1,y0,a   x:(r0)+,x0      y:(r4)+,y0
```

The ith biquad section operates as follows:

1. The **ORI** instruction selects the DSP56001 processor's up-scaling mode.
2. The **MOVE** instruction transfers $\mathbf{x_i(n-2)}$ from the X-Memory location \$50, pointed to by R0, to the Data ALU's Input Register X0 and transfers $\mathbf{a_{2i}}$ from the Y-Memory location \$100, pointed to by R4, to the Data ALU's Input Register Y0. Address Register R0 is post-incremented by one to point to $\mathbf{x_i(n-1)}$ in X-Memory location \$51; R4 is post-incremented by one to point to $\mathbf{a_{1i}}$ in Y-Memory location \$101. The X-Memory map for the filter states, the Y-Memory map for the coefficients, and the contents of the A-Accumulator and Data ALU Input Registers X0, X1, and Y0 after the execution of the **MOVE** instruction are as shown in Figure 7.12.
3. The first **MAC** instruction multiplies $\mathbf{x_i(n-2)}$ from X0 by $\mathbf{a_{2i}}$ from Y0, and adds the product to $\mathbf{y_i(n)}$ from the A-Accumulator, i.e., $\mathbf{[x_i(n-2)a_{2i}] + y_i(n) \rightarrow}$ **A-Accumulator**. In the parallel-move portion of the **MAC** instruction, $\mathbf{x_i(n-1)}$ is transferred from X-Memory location \$51, pointed to by R0, to X1, and the next coefficient $\mathbf{a_{1i}}$ is transferred from Y-Memory location \$101, pointed to by R4, to Y0. Also, R0 is post-decremented by one to point back to $\mathbf{x_i(n-2)}$

in X-Memory location \$50, and R4 is incremented by one to point to $\mathbf{b_{2i}}$ in Y-Memory location \$102. The X-Memory map for the filter states, the Y-Memory map for the coefficients, and the contents of the A-Accumulator and Data ALU Input Registers X0, X1, and Y0 after the execution of the first **MAC** instruction are as shown in Figure 7.13.

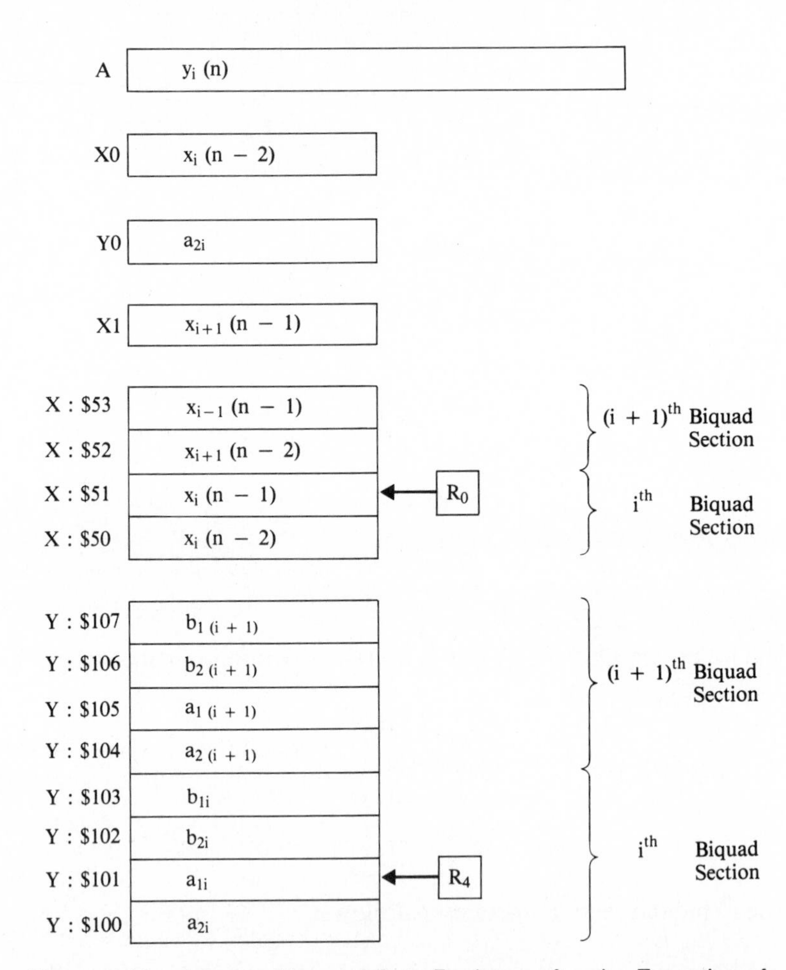

Figure 7-12 Memory Map and Data Registers after the Execution of the Move Instruction

4. The **MACR** instruction multiplies $\mathbf{x_i(n\text{-}1)}$ from X1 by $\mathbf{a_{1i}}$ from Y0, adds the product to $\mathbf{[y_i(n) + a_{2i}x_i(n\text{-}2)]}$ from the A-Accumulator, and convergently rounds the result, i.e., $\mathbf{x_i(n\text{-}1)a_{1i} + [y_i(n) + a_{2i}x_i(n\text{-}2)]}$ **-> A-Accumulator** with convergent rounding of the result. In the parallel-move portion of the **MACR** instruction, $\mathbf{x_i(n\text{-}1)}$ from X1 is transferred to X-Memory location \$50, pointed to by R0, and the next coefficient $\mathbf{b_{2i}}$ is transferred from Y-Memory location \$102, pointed to by R4, to Y0. Also, R0 is post-incremented by one to point to x_i(n-1) in X-Memory location \$51; R4 is post-incremented by one to point to $\mathbf{b_{1i}}$ in Y-Memory location \$103. Note that $\mathbf{x_i(n\text{-}1)}$ overwrites $\mathbf{x_i(n\text{-}2)}$ at X-Memory location \$50. The X-Memory map for the filter states, the Y-Memory map for the coefficients, and the contents of the A-Accumulator and Data ALU Input Registers X0, X1, and Y0 after the execution of the **MACR** instruction are as shown in Figure 7.14.

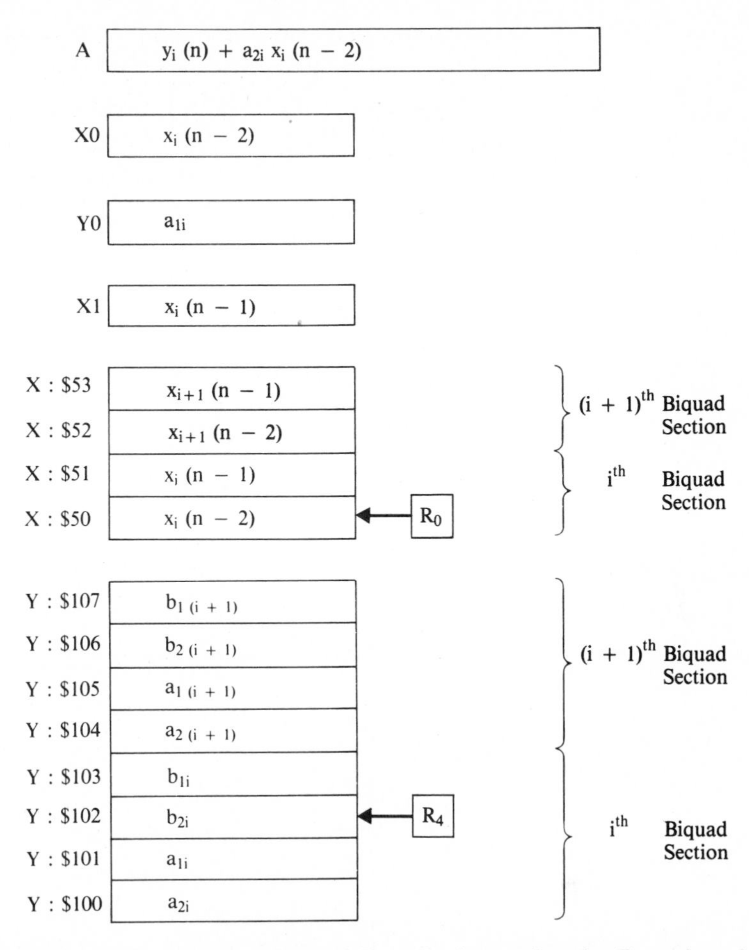

Figure 7-13 Memory Map and Data Registers after the Execution of the first MAC Instruction

5. The second **MAC** instruction multiplies $\mathbf{x_i(n\text{-}2)}$ from X0 by $\mathbf{b_{2i}}$ from Y0, and adds the product to the contents of the A-Accumulator. In the parallel-move portion of this **MAC** instruction, $\mathbf{x_i(n)}$ from the A-Accumulator is up-scaled by two and transferred to X-Memory location $51, pointed to by R0, and the next coefficient $\mathbf{b_{1i}}$ is transferred from Y-Memory location $103, pointed to by R4, to Y0. Also, R0 is post-incremented by one to point to $\mathbf{x_{i+1}(n\text{-}2)}$ of the $\mathbf{i{+}1^{th}}$ biquad filter section in X-Memory location $52; R4 is post-incremented by one to point to coefficient $\mathbf{a_{2(i+1)}}$ of the $\mathbf{i{+}1^{th}}$ biquad filter section in Y-Memory location $104. Note that $\mathbf{x_i(n)}$, up-scaled by two, overwrites $\mathbf{x_i(n\text{-}1)}$ at X-Memory location $51. The X-Memory map for the filter states, the Y-Memory map for the coefficients, and the contents of the A-Accumulator and Data ALU Input Registers X0, X1, and Y0 after the execution of the second **MAC** instruction are as shown in Figure 7.15.
6. The last **MAC** instruction multiplies $\mathbf{x_i(n\text{-}1)}$ from X1 by $\mathbf{b_{1i}}$ from Y0, and adds the product to the contents of the A-Accumulator. The result in the A-Accumulator, then, is the output of the biquad section $\mathbf{y_{i+1}(n)}$ and the input to the $\mathbf{i{+}1^{th}}$ biquad section. In the parallel-move portion of this **MAC** instruction, $\mathbf{x_{i+1}(n\text{-}2)}$ is transferred from X-Memory location $52, pointed to by R0, to X0 and the first coefficient $\mathbf{a_{2(i+1)}}$ of the $\mathbf{i{+}1^{th}}$ biquad section is transferred from Y-Memory location $104, pointed to by R4, to Y0. Also, R0

is post-incremented by one to point to $\mathbf{x_{i+1}(n-1)}$ in X-Memory location $53; R4 is post-incremented by one to point to $\mathbf{a_{1(i+1)}}$ in Y-Memory location $105. Note that the conditions at the beginning of the **i+1**th biquad section are exactly the same as the conditions that existed at the beginning of the **i**th biquad section. The X-Memory map for the filter states, the Y-Memory map for the coefficients, and the contents of the A-Accumulator and Data ALU Input Registers X0, X1, and Y0 after the execution of the last **MAC** instruction are as shown in Figure 7.16.

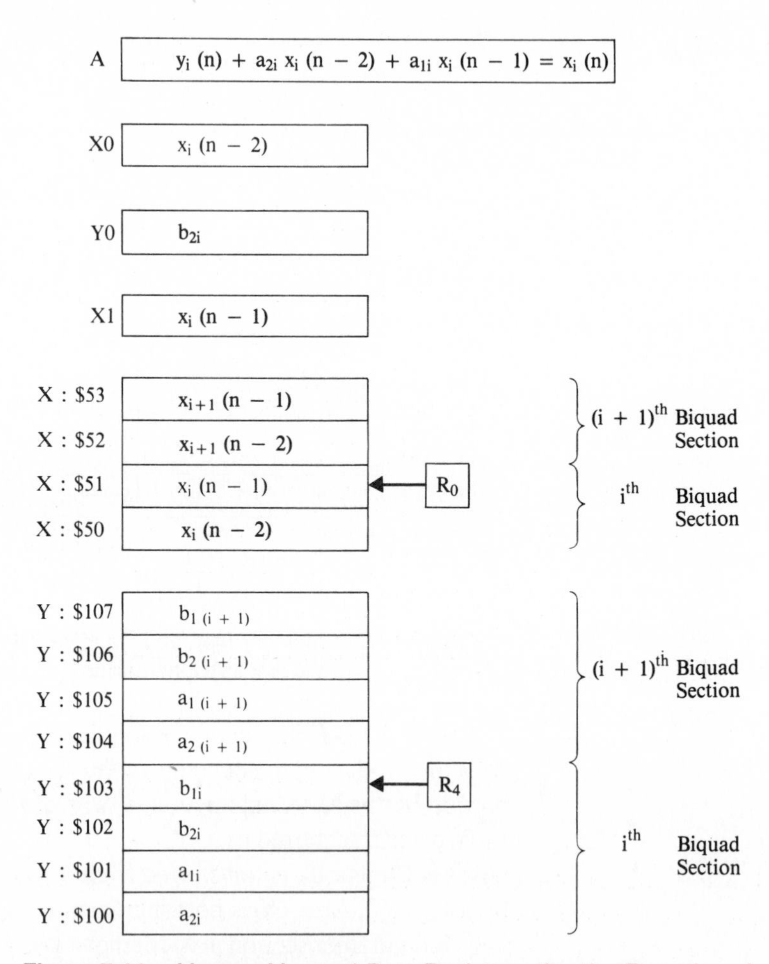

Figure 7-14 Memory Map and Data Registers after the Execution of the MACR Instruction

7.8 IMPLEMENTING AN IIR FILTER:

A desired IIR filter is implemented by cascading biquad sections as shown in Figure 7.5. The biquad sections themselves are implemented as described above.

The DSP56001 instructions which generate the input $\mathbf{y_1(n)}$ to the first biquad section are shown in Figure 7.17. Initially, the first element of the input sequence x(n) is in the A-Accumulator. The first **MOVE** instruction transfers the 8-bit immediate pointer value **gk** to R1. Since R1 has just been changed, the first **NOP** instruction is used to provide sufficient access time such that the X-Memory data pointed to by R1 is available for use during the second **MOVE** instruction. The second **MOVE** instruction transfers the **"total**

gain" value, from the X-Memory location pointed to by R1, to Y1. The third **MOVE** instruction transfers **x(n)** from the A-Accumulator to X1. In the **MPY** instruction, **x(n)** from X1 is multiplied by **"total gain"** from Y1 to place the product $\mathbf{y_1(n)}$ into the A-Accumulator.

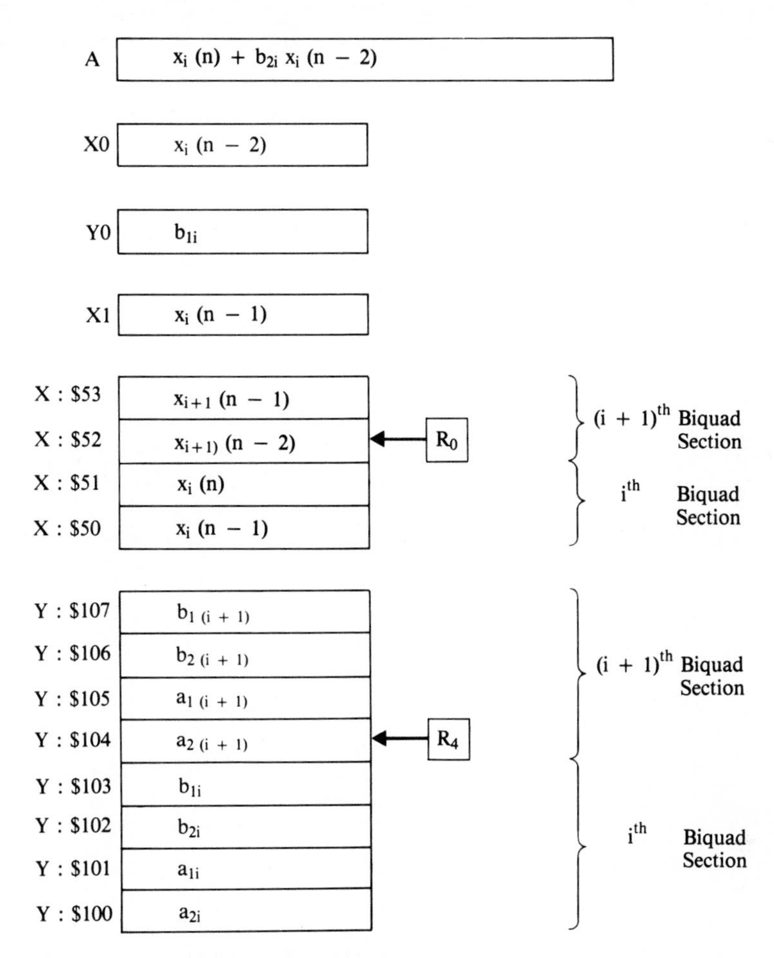

Figure 7-15 Memory Map and Data Registers after the Execution of the Second MAC Instruction

The DSP56001 instructions which implement **NSEC** cascaded biquad sections, where **NSEC** specifies the quantity of the biquad sections, are shown in Figure 7.18. Note, the core "biquad section" instructions were explained in Section 7.7. Here, the **DO** instruction sets up a hardware "do loop" to execute, NSEC times, the instruction group which implements a biquad section. The **RND** instruction convergently rounds the contents of the A-Accumulator to produce the final 24-bit output of the IIR filter. The **ANDI** instruction disables the up-scaling mode.

7.9 FILTER DESIGN AND ANALYSIS SYSTEM PACKAGE:

As previously stated in Section 6.7, a demonstration version of the *Filter Design And Analysis System* (FDAS) software program by Momentum Data Systems Inc. is included with this book to not only design filters but also to graphically show the filters' response in both the frequency and time domains. The demonstration version of the FDAS software program is a complete working system except that its capability to output the designed filter coefficients is omitted. The FDAS program is entirely menu-driven for ease of use. There are no commands to memorize, and its unique screen-recycling feature will store previous data inputs thereby allowing filter specifications to be changed with a minimum number of keystrokes. For more information about the FDAS software program, please phone Momentum Data Systems Inc., Costa Mesa, California, at (714) 557-6884. Installation of the FDAS program is covered in Section 6.8.

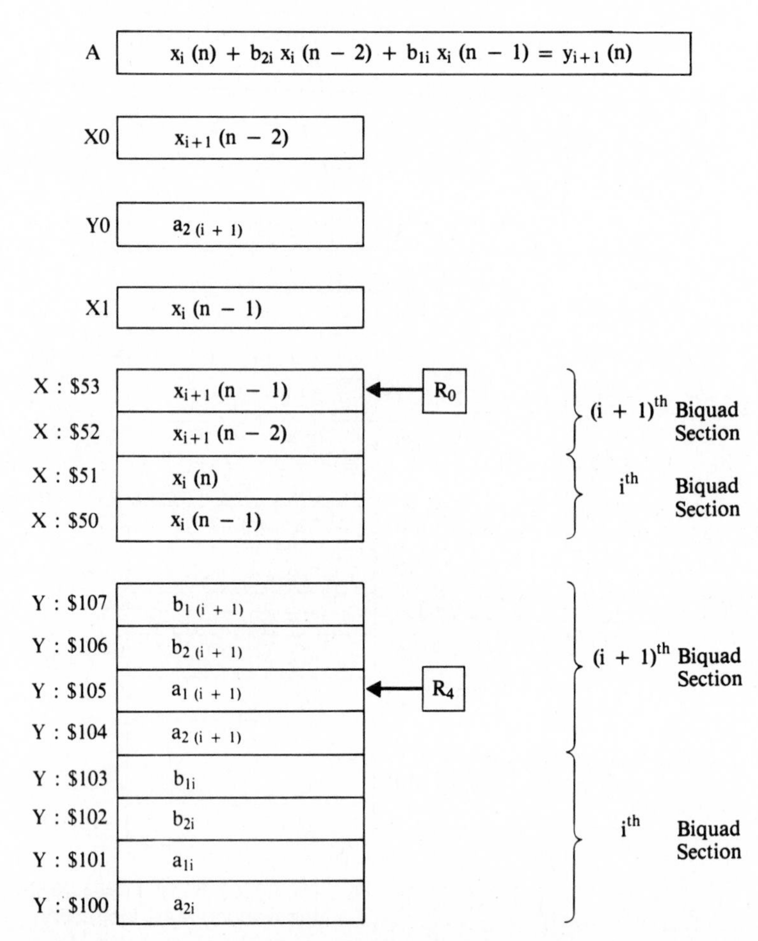

Figure 7-16 Memory Map and Data Registers after the Execution of the last MAC Instruction

```
move    #gk,r1
nop
move    x:(r1),y1
move    a,x1
mpy     x1,x0,a
```

Figure 7-17 DSP56001 Assembler Instructions to Generate the $y_1(n)$ Input to the First Biquad Section

```
        ori     #$8,mr
        nop
        move              x:(r0)+,x0       y:(r4)+,y0

        do      #NSEC,_iirend
        mac     x0,y0,a   x:(r0)-,x1       y:(r4)+,y0
        macr    x1,y0,a   x1,x:(r0)+       y:(r4)+,y0
        mac     x0,y0,a   a,x:(r0)+        y:(r4)+,y0
        mac     x1,y0,a   x:(r0)+,x0       y:(r4)+,y0
_iirend

        rnd     a
        andi    #$f7,mr
```

Figure 7-18 DSP56001 Assembler Instructions to Implement Cascaded Biquad Sections

7.10 IIR FILTER DEMONSTRATION SYSTEM:

The IIR filtering process can be demonstrated by using the DSP System described in Chapter 5 and again shown in Figure 7.2. The DSP System consists of the DSP56000ADS application development system (ADS) and an A/D-D/A evaluation board (EVB). The analog input signal x(t) is first digitized using the A/D converter on the EVB. The DSP56001 processor then executes an IIR filter algorithm to process the digitized input signal x(n), thereby producing a filtered output signal y(n). The D/A converter on the EVB is then used to convert the digital output signal y(n) to the analog output signal y(t).

7.11 DESIGNING AND IMPLEMENTING IIR FILTERS:

The DSP56001 instructions shown in Figures 7.17 and 7.18 are combined with the DSP56001 instructions shown in Figure 5.8 to form the IIR.ASM assembler program macro shown in Figure 7.19. The FDAS program by Momentum Data Systems Inc. will be used in Sections 7.11.1 through 7.11.4 to design a low-pass, high-pass, band-pass, and band-stop IIR filter, respectively. The designed filter coefficients and the IIR.ASM program macro will be used to implement each IIR filter on the DSP System.

1. POWER-DOWN THE PC.
2. SET-UP THE DSP SYSTEM as described in Section 5.2.2.
3. POWER-UP AND BOOT THE PC.

```
iir     macro    states,coef,gk,nsec

;     states =  Location of filter states in X memory
;     coef   =  Location of filter coefficients in Y memory
;     gk     =  Total gain K
;     nsec   =  Number of biquad sections

;
```

```
;************************************************************
; Initialize routine
;************************************************************
;
        movep   #0,x:$FFFF
;
init2   reset
;

; set up ads board in case of force break instead of force
; reset
;
        movep #0,x:$FFFE         ;set bcr to zero
        movec #0,sp              ;init stack pointer
        movec #0,sr              ;clear loop flag
;
;************************************************************
; Set up the SSI for operation with the EVB
; The following code sets port C to function as SCI/SSI
;************************************************************
;
pcc     equu    $0001ff
        movep   #pcc,x:$FFE1     ;write PCC
;
;************************************************************
; The following code sets the SSI CRA and CRB control
; registers for external continuous clock, synchronous,
; normal mode.
;************************************************************
;
cra     equ     $004000
        movep   #cra,x:$FFEC     ;CRA pattern for word
;                                ;length=16 bits
;
crb     equ     $003200
        movep   #crb,x:$FFED     ;CRB pattern for continuous
;                                ;ck,sych,normal mode
;                                ;word long frame SYNC:
;
;************************************************************
; Read A/D
;************************************************************
;
; The following code polls the RDF flag in the SSI-SR and
; waits for RDF=1 and then reads the RX register to retrieve
; the data from the A/D converter.
; Sample rate is controlled by the EVB.
;
wait2   btst    #7,x:$FFEE
        jcc     wait2            ;loop  until RDF bit = 1
;
        movep   x:$FFEF,x1       ;get A/D converter data
        move    #gk,r1
        nop
;
;*********************************************************
; Multiply the input by the overall gain
;********************************************************
;
        move    x:(r1),x0
        mpy     x1,x0,a
        nop
;
;************************************************************
; Initialize R0,R4,M0 and M4
;***********************************************************
;
        move    #$ffff,m0
        move    m0,m4
```

```
        move    #states,r0
        move    #coef,r4
;
;*********************************************************
;  IIR Filter
;*********************************************************
;
        ori     #$8,mr
        nop
        move            x:(r0)+,x0      y:(r4)+,y0
;
        do      #nsec,_iirend
        mac     x0,y0,a  x:(r0)-,x1     y:(r4)+,y0
        macr    x1,y0,a  x1,x:(r0)+     y:(r4)+,y0
        mac     x0,y0,a  a,x:(r0)+      y:(r4)+,y0
        mac     x1,y0,a  x:(r0)+,x0     y:(r4)+,y0
_iirend
;
        rnd     a
        andi    #$f7,mr
        rol     a
        nop
;
;*********************************************************
; Write D/A
;*********************************************************
        move    #$ffff00,y1
        and     y1,a
        movep   a,x:$FFEF       ;write D/A via SSI
        nop
        jmp     wait2           ;loop  indefinitely
        nop
;
        endm
```

Figure 7-19 IIR.ASM Assembler Macro to Implement an IIR Digital Filter

7.11.1 Designing and Implementing a Low-Pass IIR Filter:

a) *Designing the Desired Low-Pass IIR Filter:*

1. TYPE **cd \dsptools\fdas<return>** to enter the FDAS program directory.
2. TYPE **demo<return>** to load and enter the FDAS program.
3. FOLLOW THE FDAS PROGRAM'S MENUS TO SELECT THE BELOW LISTED IIR FILTER SPECIFICATIONS. Plots of the amplitude response, phase response, impulse response, poles and zeros of the designed low-pass IIR filter are shown in Figure 7.20.
4. EXIT THE FDAS PROGRAM.

Filter Specifications:

Filter Type:	Low-pass
Analog Filter Type:	Butterworth
Sampling Frequency:	48000 Hz
Passband Cutoff Frequency:	8000 Hz
Stopband Cutoff Frequency:	10000 Hz
Passband attenuation:	0.5 dB
Stopband Attenuation:	30.0 dB
Filter Design Method:	Bilinear Transformation

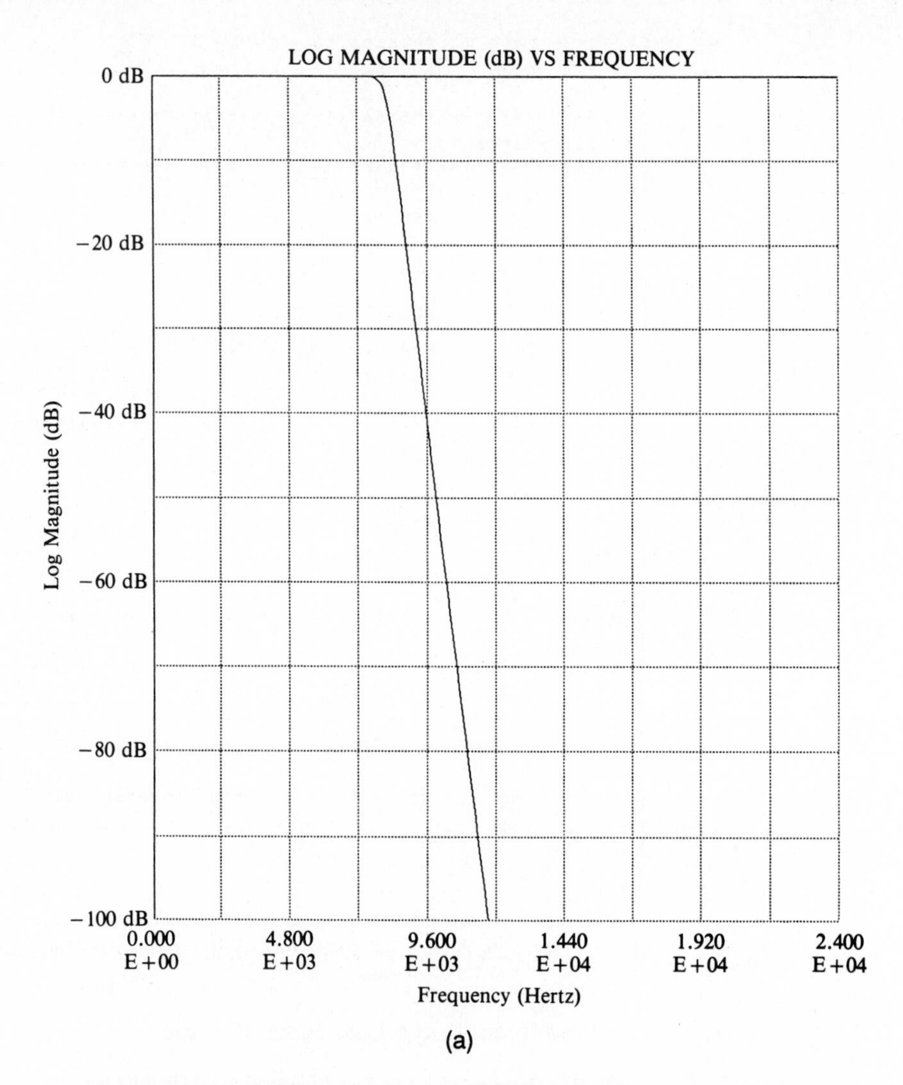

(a)

Figure 7-20 Plots of (a) Amplitude Response (b) Phase Response (c) Impulse Response (d) Poles and Zeros of Low-Pass Filter in Section 7.12.1

Design Results:

Filter Order: 16

Biquad Sections of the Normalized Analog Transfer Function:
(Numerator Coefficients Then Denominator Coefficients)

S**2 Term	S Term	Const Term
.000000E+00	.000000E+00	.100000E+01
.100000E+01	.199037E+01	.100000E+01
.000000E+00	.000000E+00	.100000E+01
.100000E+01	.191388E+01	.100000E+01
.000000E+00	.000000E+00	.100000E+01
.100000E+01	.176384E+01	.100000E+01

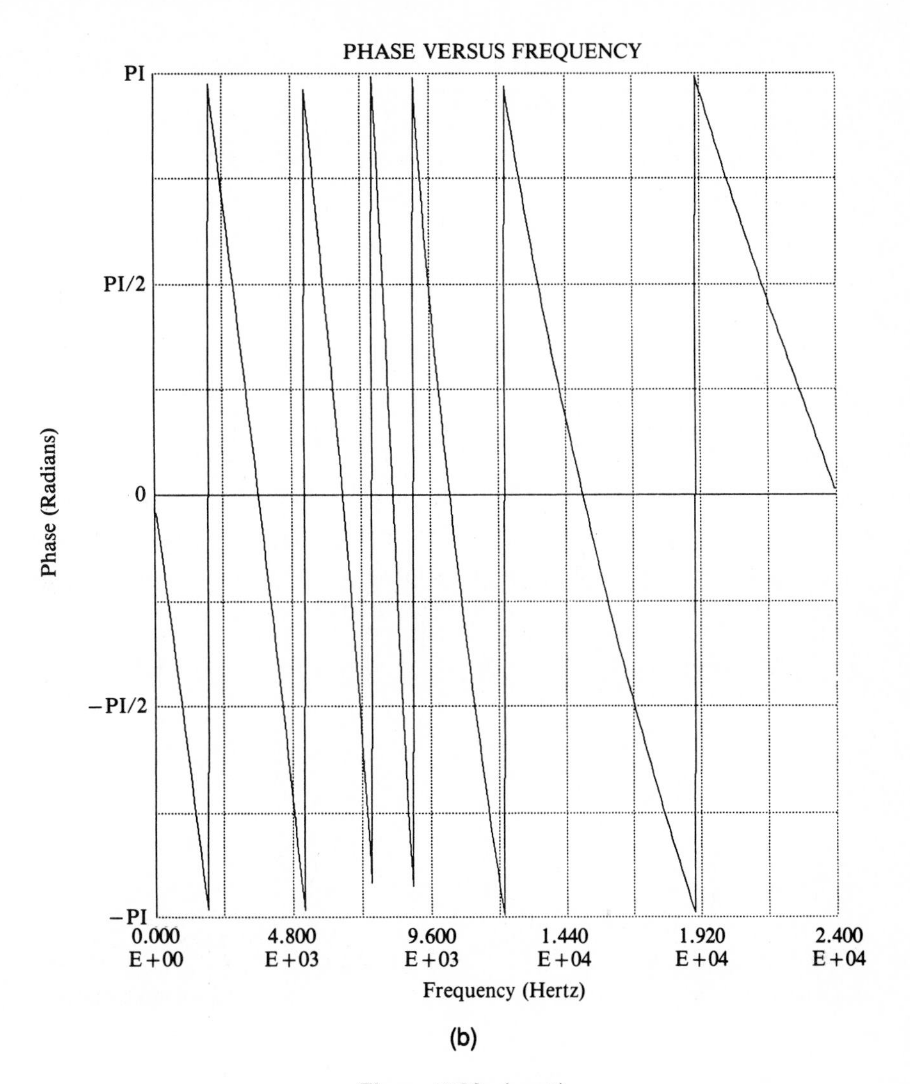

(b)

Figure 7-20 (*cont.*)

Biquad Sections of the Normalized Analog Transfer Function (cont.)		
S**2 Term	S Term	Const Term
.000000E+00	.000000E+00	.100000E+01
.100000E+01	.154602E+01	.100000E+01
.000000E+00	.000000E+00	.100000E+01
.100000E+01	.126879E+01	.100000E+01
.000000E+00	.000000E+00	.100000E+01
.100000E+01	.942793E+00	.100000E+01
.000000E+00	.000000E+00	.100000E+01
.100000E+01	.580569E+00	.100000E+01
.000000E+00	.000000E+00	.100000E+01
.100000E+01	.196034E+00	.100000E+01
Initial Gain 1.00000000		

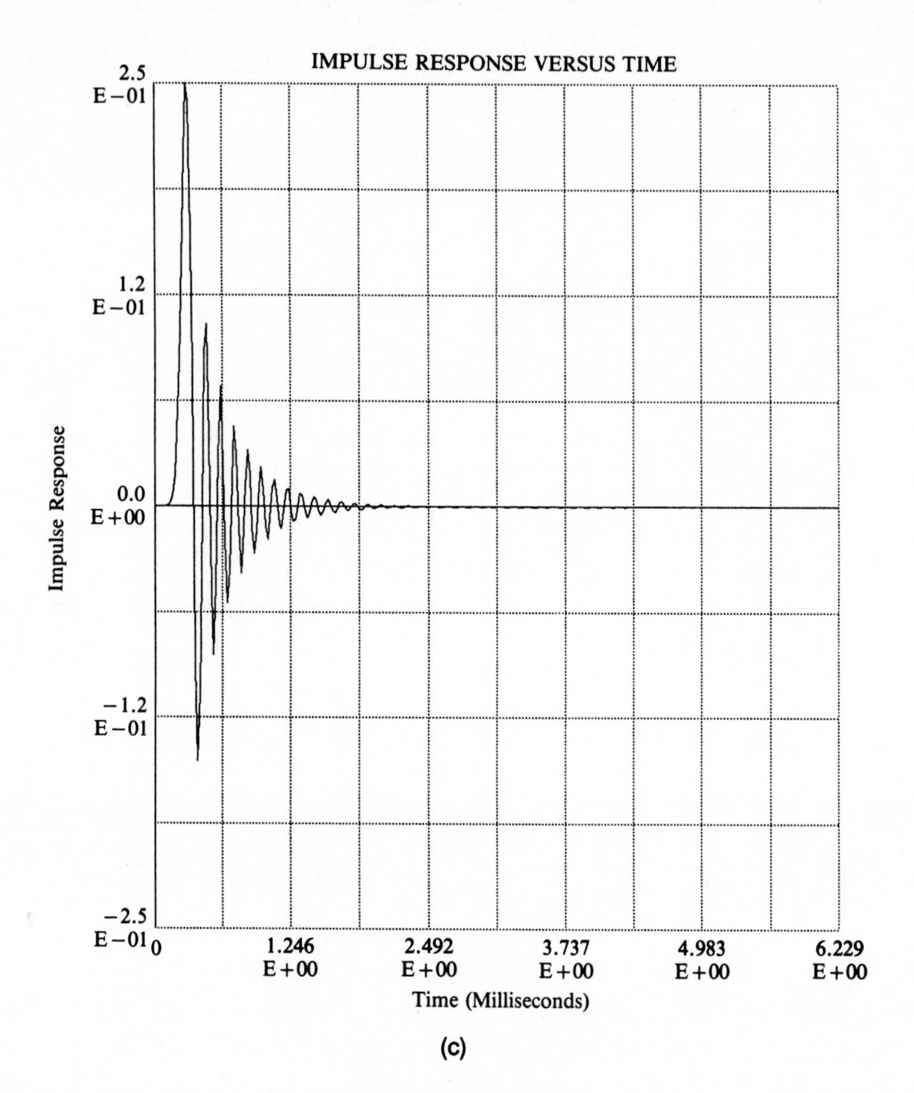

(c)

Figure 7-20 (*cont.*)

Biquad Sections of the Analog Transfer Function: (Numerator Coefficients Then Denominator Coefficients)		
*S**2 Term*	*S Term*	*Const Term*
.000000E+00	.000000E+00	.350364E+10
.100000E+01	.117813E+06	.350364E+10
.000000E+00	.000000E+00	.350364E+10
.100000E+01	.113286E+06	.350364E+10
.000000E+00	.000000E+00	.350364E+10
.100000E+01	.104405E+06	.350364E+10
.000000E+00	.000000E+00	.350364E+10
.100000E+01	.915114E+05	.350364E+10

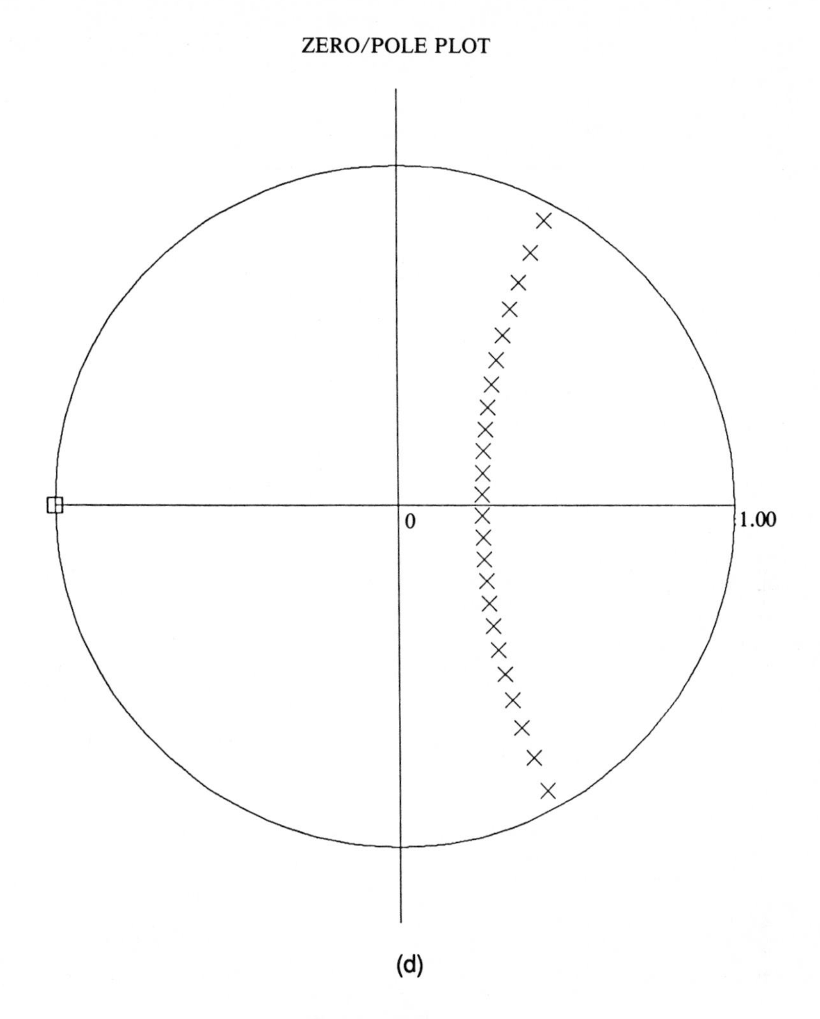

(d)

Figure 7-20 (*cont.*)

Biquad Sections of the Analog Transfer Function: (Cont.)		
*S**2 Term*	*S Term*	*Const Term*
.000000E+00	.000000E+00	.350364E+10
.100000E+01	.751015E+05	.350364E+10
.000000E+00	.000000E+00	.350364E+10
.100000E+01	.558054E+05	.350364E+10
.000000E+00	.000000E+00	.350364E+10
.100000E+01	.343648E+05	.350364E+10
.000000E+00	.000000E+00	.350364E+10
.100000E+01	.116036E+05	.350364E+10
Initial Gain 1.00000000		

Biquad Sections of the Digital Transfer Function: (Numerator Coefficients Then Denominator Coefficients)

*Z**2 Term*	*Z Term*	*Const Term*
1.0000000000	2.0000000000	1.0000000000
1.00000	-.4754416049	.0586601458
1.0000000000	2.0000000000	1.0000000000
1.00000	-.4841995835	.0781614780
1.0000000000	2.0000000000	1.0000000000
1.00000	-.5023513436	.1185798347
1.0000000000	2.0000000000	1.0000000000
1.00000	-.5312651992	.1829618961
1.0000000000	2.0000000000	1.0000000000
1.00000	-.5732600093	.2764712274
1.0000000000	2.0000000000	1.0000000000
1.00000	-.6320043206	.4072764516
1.0000000000	2.0000000000	1.0000000000
1.00000	-.7132129073	.5881026983
1.0000000000	2.0000000000	1.0000000000
1.00000	-.8258681893	.8389508724

Initial Gain .102573556E-05

Zeros of Digital Transfer Function:

Real Part	Imaginary Part
-.100000000E+01	.000000000E+00
-.100000000E+01	.000000000E+00
-.100000000E+01	.000000000E+00
-.100000000E+01	.000000000E+00
-.100000000E+01	.000000000E+00
-.100000000E+01	.000000000E+00
-.100000000E+01	.000000000E+00
-.100000000E+01	.000000000E+00
-.100000000E+01	.000000000E+00
-.100000000E+01	.000000000E+00
-.100000000E+01	.000000000E+00
-.100000000E+01	.000000000E+00
-.100000000E+01	.000000000E+00
-.100000000E+01	.000000000E+00
-.100000000E+01	.000000000E+00
-.100000000E+01	.000000000E+00

Poles of Digital Transfer Function:	
Real Part	Imaginary Part
.237720802E+00	.463569406E-01
.237720802E+00	-.463569406E-01
.242099792E+00	.139818342E+00
.242099792E+00	-.139818342E+00
.251175672E+00	.235564464E+00
.251175672E+00	-.235564464E+00
.265632600E+00	.335262909E+00
.265632600E+00	-.335262909E+00
.286630005E+00	.440811148E+00
.286630005E+00	-.440811148E+00
.316002160E+00	.554453863E+00
.316002160E+00	-.554453863E+00
.356606454E+00	.678921597E+00
.356606454E+00	-.678921597E+00
.412934095E+00	.817579541E+00
.412934095E+00	-.817579541E+00

Direct Form Digital Transfer Function:

i) *Numerator Coefficients - Highest Order First (in Z):*

.100000000E+01	.160000000E+02	.120000000E+03
.560000000E+03	.182000000E+04	.436800000E+04
.800800000E+04	.114400000E+05	.128700000E+05
.114400000E+05	.800800000E+04	.436800000E+04
.182000000E+04	.560000000E+03	.120000000E+03
.160000000E+02	.100000000E+01	

ii) *Denominator Coefficients - Highest Order First (in Z):*

.100000000E+01	-.473760316E+01	.123146660E+02
-.217678558E+02	.287481903E+02	-.296453528E+02
.244832550E+02	-.164083913E+02	.897489810E+01
-.400386702E+01	.144745184E+01	-.418464884E+00
.946350381E-01	-.161528643E-01	.195923087E-02
-.150660767E-03	.552632393E-05	

b) *Implementing the Designed Low-Pass IIR Filter:*

The "IIRLP.ASM" assembler program, shown in Figure 7.21, implements the designed low-pass IIR filter on the DSP56001 processor. Note that this program has an "include" statement which references the IIR.ASM assembler program macro shown in Figure 7.19. A copy of the IIRLP.ASM assembler program, IIR.ASM assembler program macro, and the "IIRLP.LOD" absolute load program can be found on the DSP56000/1 Demonstration Software #3 program diskette.

To demonstrate the designed low-pass IIR filter on the DSP System:

1. CONNECT A FUNCTION GENERATOR TO "BNC2" ON THE EVB AND TO CHANNEL ONE OF A DUAL TRACE OSCILLOSCOPE.
2. SET THE FUNCTION GENERATOR TO PRODUCE A 2-VOLT (PEAK-TO-PEAK) SINE WAVE.
3. CONNECT CHANNEL TWO OF THE OSCILLOSCOPE TO "BNC1" ON THE EVB.

4. LOAD THE "ADS56000 USER INTERFACE SOFTWARE" PROGRAM.
5. PLACE THE "DSP56001 DEMONSTRATION SOFTWARE #3" DISKETTE INTO DRIVE ""A".
6. TYPE **a:iirlp.lod<return>** to load the **iirlp.lod** program.
7. TYPE **go<return>** to execute the designed low-pass IIR filter program (algorithm) on the DSP56001 processor.
8. VARY THE SINUSOIDAL INPUT FREQUENCY FROM 0.5 TO 20 KHZ. View the analog input and output signals on the oscilloscope. Note that the IIR filter passes the frequencies from 0.5 to 8 KHz and stops the frequencies from 10 to 20 KHz as designed. Also note the increasing attenuation of the output signal as the input signal's frequency is increased from 8 to 10 KHz.
9. TYPE **force b<return>** to halt execution of the IIR filter algorithm and return control of the ADM to its monitor program for keyboard command entry.
10. TYPE **quit<return>** to exit the ADS56000 program.

```
;*****************************************************
;
; File: IIRLP.ASM
;
;*****************************************************
;
        include 'iir'
;
;*****************************************************
; Coefficients of Low Pass IIR Filter
;*****************************************************
;
nsec    equ     $8
;
        org     x:$0
states  ds      2*nsec
;
        org     x:$20
gk      dc      .10257356E-05
;
        org     y:$0
coef
        dc -.58660146E-01/2.0
        dc  .47544160/2.0
        dc  1.0000000/2.0
        dc  1.9999999/2.0
;
        dc -.78161478E-01/2.0
        dc  .48419958/2.0
        dc  1.0000000/2.0
        dc  1.9999999/2.0
;
        dc -.11857983/2.0
        dc  .50235134/2.0
        dc  1.0000000/2.0
        dc  1.9999999/2.0
;
        dc -.18296190/2.0
        dc  .53126520/2.0
        dc  1.0000000/2.0
```

```
        dc  1.9999999/2.0
;
        dc -.27647123/2.0
        dc  .57326001/2.0
        dc  1.0000000/2.0
        dc  1.9999999/2.0
;
        dc -.40727645/2.0
        dc  .63200432/2.0
        dc  1.0000000/2.0
        dc  1.9999999/2.0
;
        dc -.58810270/2.0
        dc  .71321291/2.0
        dc  1.0000000/2.0
        dc  1.9999999/2.0
;
        dc -.83895087/2.0
        dc  .82586819/2.0
        dc  1.0000000/2.0
        dc  1.9999999/2.0
;
; program start
;
        org     p:$200
;
        iir     states,coef,gk,nsec
;
        end
```

Figure 7-21 IIRLP.ASM Assembler Program to Implement a Designed Low-Pass IIR Filter

7.11.2 Designing and Implementing a High-Pass IIR Filter:

a) *High-Pass Filter Design:*

1. TYPE **cd \dsptools\fdas<return>** to enter the FDAS program directory.
2. TYPE **demo<return>** to load and enter the FDAS program.
3. FOLLOW THE FDAS PROGRAM'S MENUS TO SELECT THE BELOW LISTED IIR FILTER SPECIFICATIONS. Plots of the amplitude response, phase response, impulse response, poles and zeros of the high-pass IIR filter are shown in Figure 7.22.
4. EXIT THE FDAS PROGRAM.

Filter Specifications:

Filter Type :	High-pass
Analog Filter Type:	Chebyshev
Sampling Frequency:	48000 Hz
Passband Cutoff Frequency:	10000 Hz
Stopband Cutoff Frequency:	8000 Hz
Passband attenuation:	0.5 dB
Stopband Attenuation:	50.0 dB
Filter Design Method:	Bilinear Transformation

Design Results:

Filter Order: 10

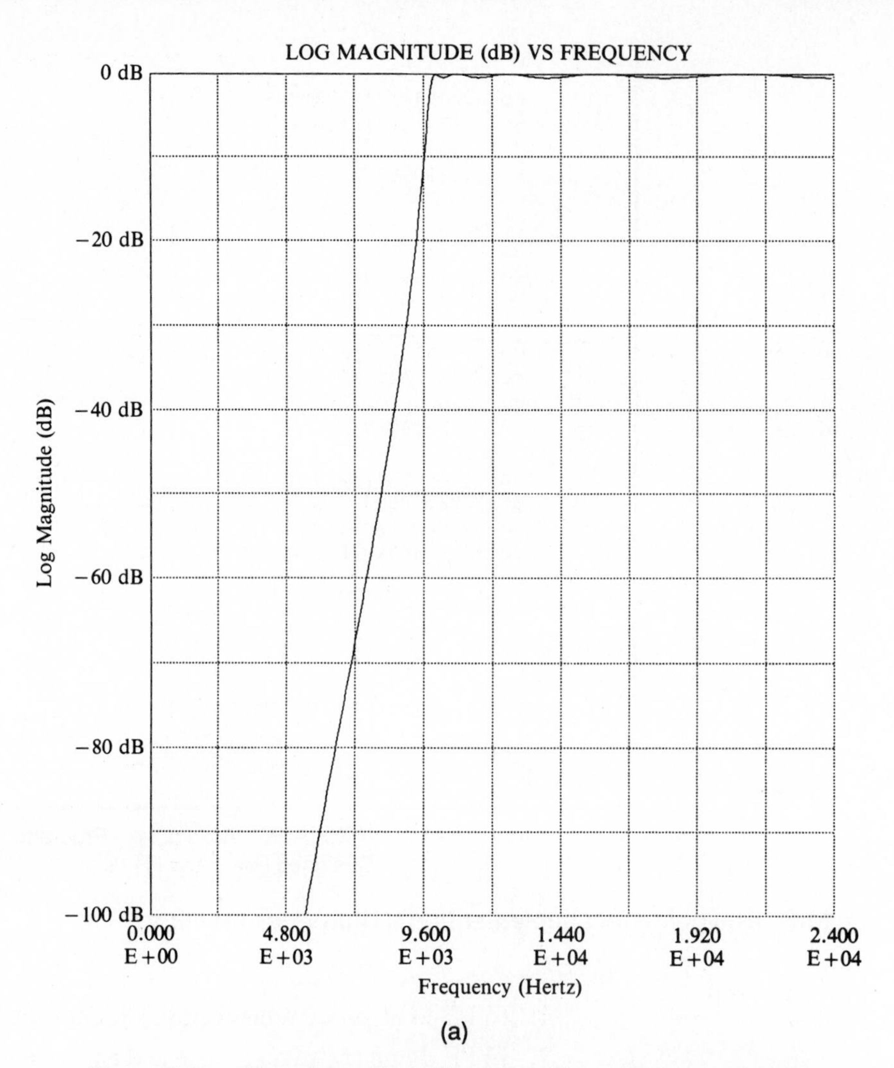

(a)

Figure 7-22 Plots of Amplitude Response, Phase Response, Impulse Response, Poles, and Zeros of High-Pass Filter in Section 7.12.2. (a) Amplitude Response in dB (b) Phase Response (c) Impulse Response (d) Poles and Zeros

Biquad Sections of the Normalized Analog Transfer Function: (Numerator Coefficients Then Denominator Coefficients)		
*S**2 Term*	*S Term*	*Const Term*
.000000E+00	.000000E+00	.100000E+01
.100000E+01	.557988E-01	.100734E+01
.000000E+00	.000000E+00	.100000E+01
.100000E+01	.161934E+00	.825700E+00
.000000E+00	.000000E+00	.100000E+01
.100000E+01	.252219E+00	.531807E+00
.000000E+00	.000000E+00	.100000E+01
.100000E+01	.317814E+00	.237915E+00
.000000E+00	.000000E+00	.100000E+01
.100000E+01	.352300E+00	.562789E-01
Initial Gain .559135769E-02		

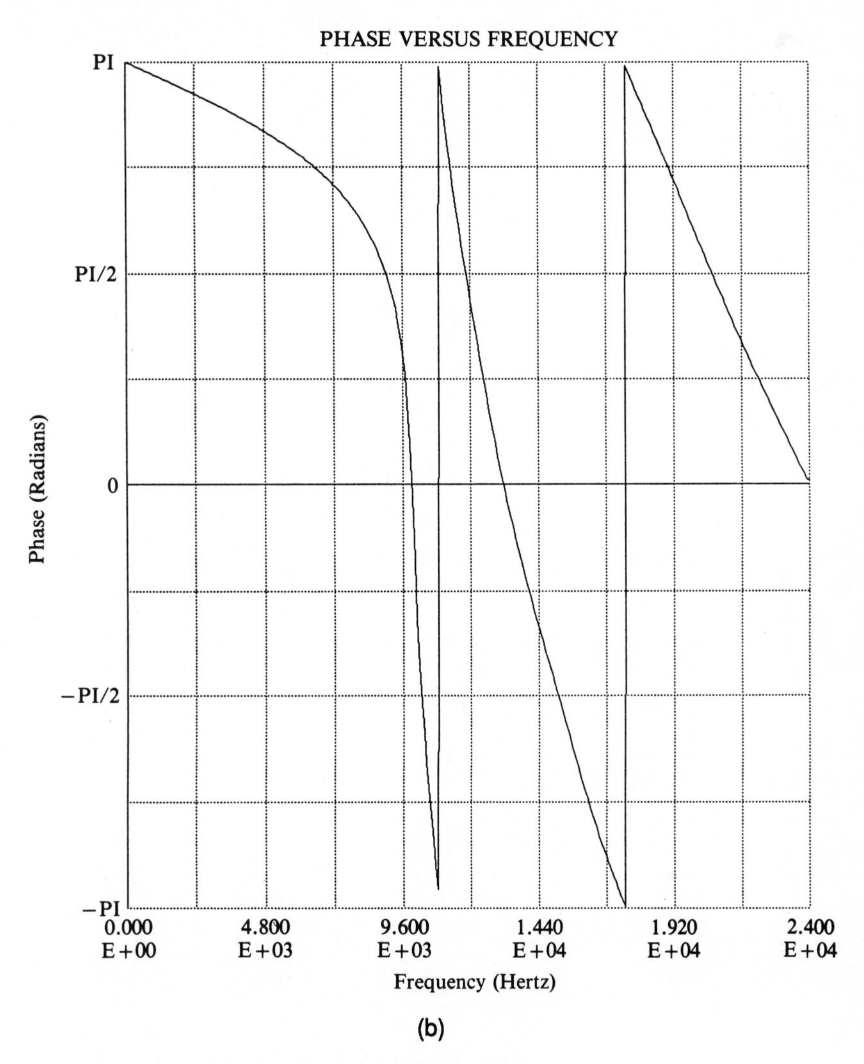

(b)

Figure 7-22 (*cont.*)

Biquad Sections of the Analog Transfer Function: (Numerator Coefficients Then Denominator Coefficients)

*S**2 Term*	*S Term*	*Const Term*
.100000E+01	.000000E+00	.000000E+00
.100000E+01	.408040E+04	.538678E+10
.100000E+01	.000000E+00	.000000E+00
.100000E+01	.144467E+05	.657175E+10
.100000E+01	.000000E+00	.000000E+00
.100000E+01	.349362E+05	.102035E+11
.100000E+01	.000000E+00	.000000E+00
.100000E+01	.984021E+05	.228077E+11
.100000E+01	.000000E+00	.000000E+00
.100000E+01	.461125E+06	.964179E+11

Initial Gain .944060875

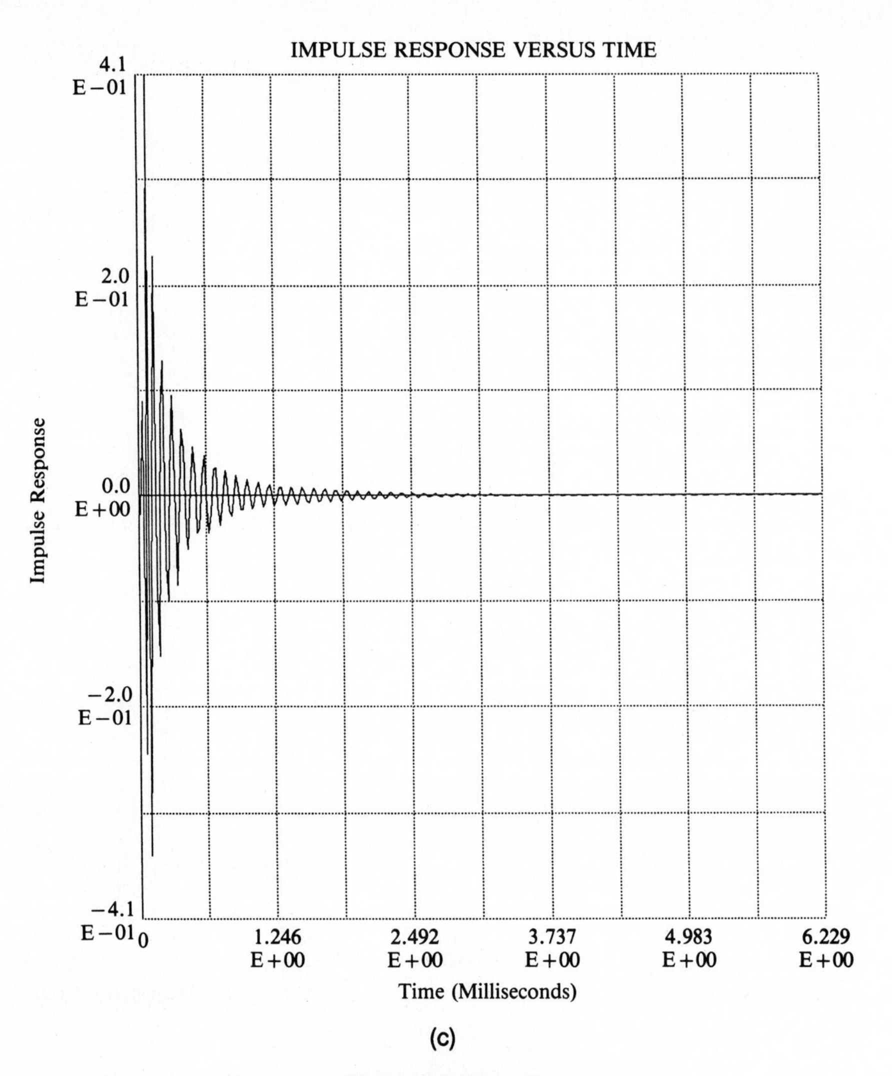

(c)

Figure 7-22 (*cont.*)

Biquad Sections of the Digital Transfer Function: (Numerator Coefficients Then Denominator Coefficients)

*Z**2 Term*	*Z Term*	*Const Term*
1.0000000000	-2.0000000000	1.0000000000
1.00000	1.1634529829	.4093739092
1.0000000000	-2.0000000000	1.0000000000
1.00000	.6554923654	.5444167256
1.0000000000	-2.0000000000	1.0000000000
1.00000	.0867241845	.7054567933
1.0000000000	-2.0000000000	1.0000000000
1.00000	-.3079246283	.8384962678
1.0000000000	-2.0000000000	1.0000000000
1.00000	-.5107498169	.9477517009

Initial Gain .172155447E-02

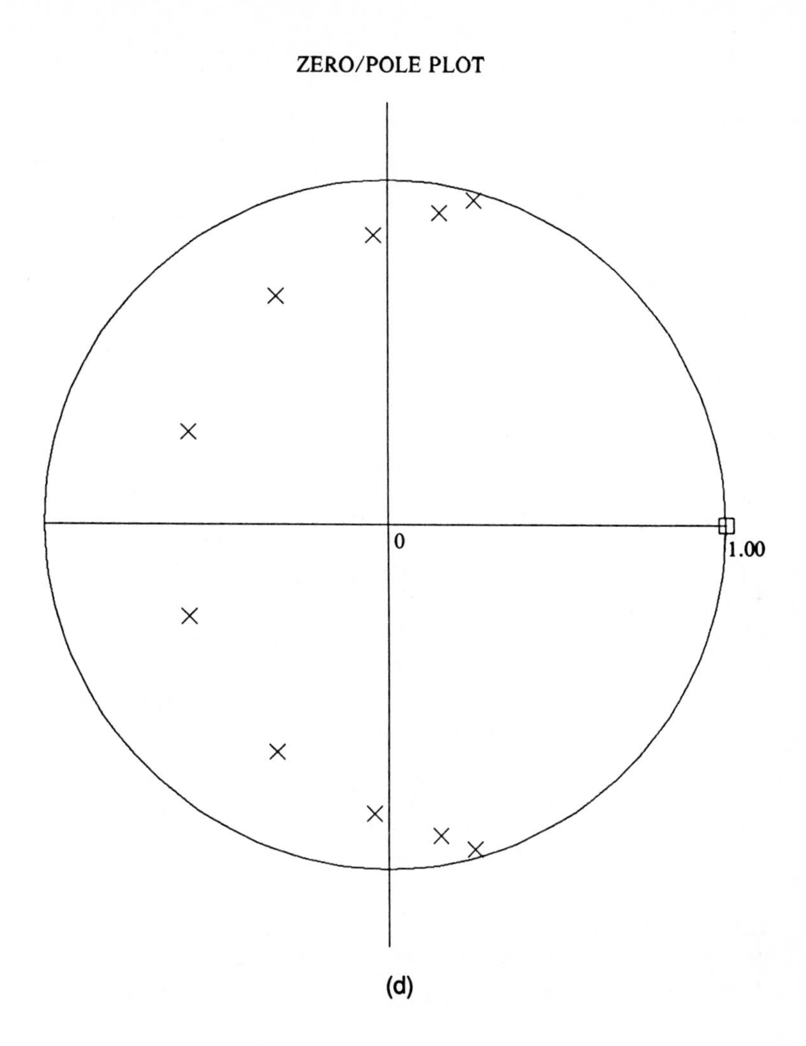

(d)

Figure 7-22 *(cont.)*

Zeros of Digital Transfer Function:	
Real Part	*Imaginary Part*
.100000000E+01	.000000000E+00
.100000000E+01	.000000000E+00
.100000000E+01	.000000000E+00
.100000000E+01	.000000000E+00
.100049163E+01	.000000000E+00
.999508610E+00	.000000000E+00
.100000000E+01	.000000000E+00
.100000000E+01	.000000000E+00
.100000000E+01	.000000000E+00
.100000000E+01	.000000000E+00

Poles of Digital Transfer Function:	
Real Part	*Imaginary Part*
-.581726432E+00	.266398421E+00
-.581726432E+00	-.266398421E+00
-.327746153E+00	.661059094E+00
-.327746153E+00	-.661059094E+00
-.433620811E-01	.838794649E+00
-.433620811E-01	-.838794649E+00
.153962314E+00	.902658193E+00
.153962314E+00	-.902658193E+00
.255374908E+00	.939433435E+00
.255374908E+00	-.939433435E+00

Direct Form Digital Transfer Function:

i) *Numerator Coefficients - Highest Order First (in Z):*

.175010905E-02	-.175010909E-01	.787549104E-01
-.210013097E+00	.367522923E+00	-.441027509E+00
.367522923E+00	-.210013097E+00	.787549104E-01
-.175010909E-01	.175010905E-02	

ii) *Denominator Coefficients - Highest Order First (in Z):*

.250000000E+00	.271748722E+00	.740756363E+00
.801380128E+00	.100932536E+01	.902568218E+00
.732131728E+00	.477772211E+00	.258770600E+00
.101918844E+00	.312361428E-01	

b) *Implementing the Designed High-Pass IIR Filter:*

The IIRHP.ASM assembler program, shown in Figure 7.23, implements the designed high-pass IIR filter on the DSP56001 processor. Note that this program has an "include" statement which references the IIR.ASM assembler program macro shown in Figure 7.19. A copy of the IIRHP.ASM assembler program, IIR.ASM assembler program macro, and the IIRHP.LOD absolute load program can be found on the DSP56000/1 Demonstration Software #3 diskette.

To demonstrate the designed high-pass IIR filter on the DSP System:

1. CONNECT A FUNCTION GENERATOR TO "BNC2" ON THE EVB AND TO CHANNEL ONE OF A DUAL TRACE OSCILLOSCOPE.
2. SET THE FUNCTION GENERATOR TO PRODUCE A 2-VOLT (PEAK-TO-PEAK) SINE WAVE.
3. CONNECT CHANNEL TWO OF THE OSCILLOSCOPE TO "BNC1" ON THE EVB.
4. LOAD THE "ADS56000 USER INTERFACE SOFTWARE" PROGRAM.
5. PLACE THE "DSP56001 DEMONSTRATION SOFTWARE #3" DISKETTE INTO DRIVE "A".
6. TYPE **a:iirhp.lod<return>** to load the **iirhp.lod** program.
7. TYPE **go<return>** to execute the designed high-pass IIR filter program (algorithm) on the DSP56001 processor.

8. VARY THE SINUSOIDAL INPUT FREQUENCY FROM 0.5 TO 20 KHZ. View the analog input and output signals on the oscilloscope. Note that the IIR filter stops the frequencies from 0.5 to 8 KHz and passes the frequencies from 10 to 20 KHz as designed. Also note the decreasing attenuation of the output signal as the input signal's frequency is increased from 8 to 10 KHz.
9. TYPE **force b<return>** to halt execution of the IIR filter algorithm and return control of the ADM to its monitor program for keyboard command entry.
10. TYPE **quit<return>** to exit the ADS56000 program.

```
;*****************************************************
;
; File: IIRHP.ASM
;
;*****************************************************
;
        include 'iir'
;
;*****************************************************
; Coefficients of High Pass IIR filter
;*****************************************************
;
nsec    equ     $5
;
        org     x:$0
states  ds      2*nsec
;
        org     x:$10
gk      dc      .17215545E-02
;
        org     y:$0
coef
        dc -.40937391/2.0
        dc -1.1634530/2.0
        dc  1.0000000/2.0
        dc -1.9999999/2.0
;
        dc -.54441673/2.0
        dc -.65549237/2.0
        dc  1.0000000/2.0
        dc -1.9999999/2.0
;
        dc -.70545679/2.0
        dc -.86724184E-01/2.0
        dc  1.0000000/2.0
        dc -1.9999999/2.0
;
        dc -.83849627/2.0
        dc  .30792463/2.0
        dc  1.0000000/2.0
        dc -1.9999999/2.0
;
        dc -.94775170/2.0
        dc  .51074982/2.0
        dc  1.0000000/2.0
        dc -1.9999999/2.0
;
; program start
;
        org     p:$200
;
```

```
        iir     states,coef,gk,nsec
;
        end
```

Figure 7-23 IIRHP.ASM Assembler Program to Implement a Designed High-Pass IIR Filter

7.11.3 Designing and Implementing a Band-Pass IIR Filter:

a) *Band-Pass Filter Design:*

1. TYPE **cd \dsptools\fdas<return>** to enter the FDAS program directory.
2. TYPE **demo<return>** to load and enter the FDAS program.
3. FOLLOW THE FDAS PROGRAM'S MENUS TO SELECT THE BELOW LISTED IIR FILTER SPECIFICATIONS. Plots of the amplitude response, phase response, impulse response, poles and zeros of the band-pass filter are shown in Figure 7.24.
4. EXIT THE FDAS PROGRAM.

Filter Specifications:

Filter Type:	Band-pass
Analog Filter Type:	Butterworth
Sampling Frequency:	48000 Hz
Passband Cutoff Frequencies:	6000 Hz 12000 Hz
Stopband Cutoff Frequencies:	4000 Hz 14000 Hz
Passband attenuation:	0.5 dB
Stopband Attenuation:	50.0 dB
Filter Design Method:	Bilinear Transformation

Design Results:
Filter Order: 28

Biquad Sections of the Normalized Analog Transfer Function: (Numerator Coefficients Then Denominator Coefficients)		
*S**2 Term*	*S Term*	*Const Term*
.000000E+00	.000000E+00	.100000E+01
.100000E+01	.198742E+01	.100000E+01
.000000E+00	.000000E+00	.100000E+01
.100000E+01	.188777E+01	.100000E+01
.000000E+00	.000000E+00	.100000E+01
.100000E+01	.169345E+01	.100000E+01
.000000E+00	.000000E+00	.100000E+01
.100000E+01	.141421E+01	.100000E+01

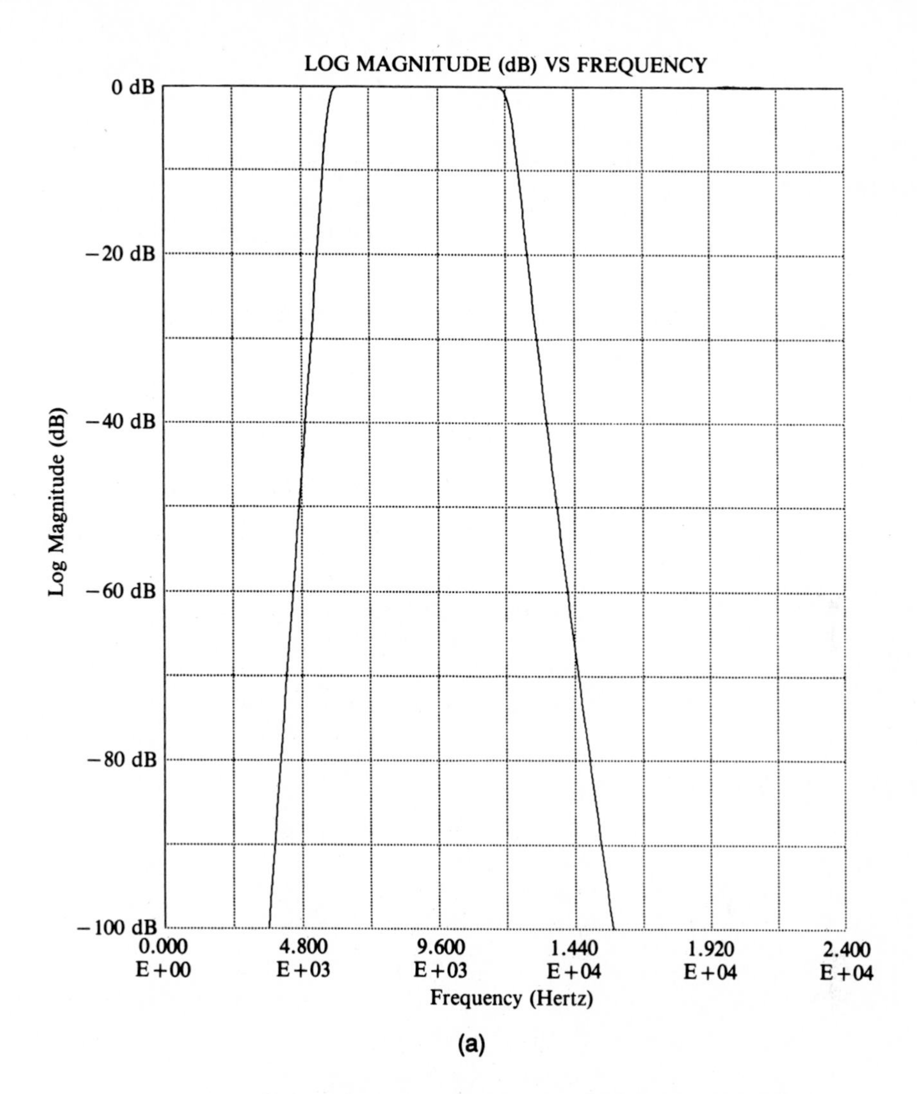

(a)

Figure 7-24 Plots of Amplitude Response, Phase Response, Impulse, Poles and Zeros of Band-Pass Filter in Section 7.12.3. (a) Amplitude Response in dB (b) Phase Response (c) Impulse Response (d) Poles and Zeros

Biquad Sections of the Normalized Analog Transfer Function: (Cont.)

*S**2 Term*	*S Term*	*Const Term*
.000000E+00	.000000E+00	.100000E+01
.100000E+01	.106406E+01	.100000E+01
.000000E+00	.000000E+00	.100000E+01
.100000E+01	.660558E+00	.100000E+01
.000000E+00	.000000E+00	.100000E+01
.100000E+01	.223929E+00	.100000E+01

Initial Gain 1.00000000

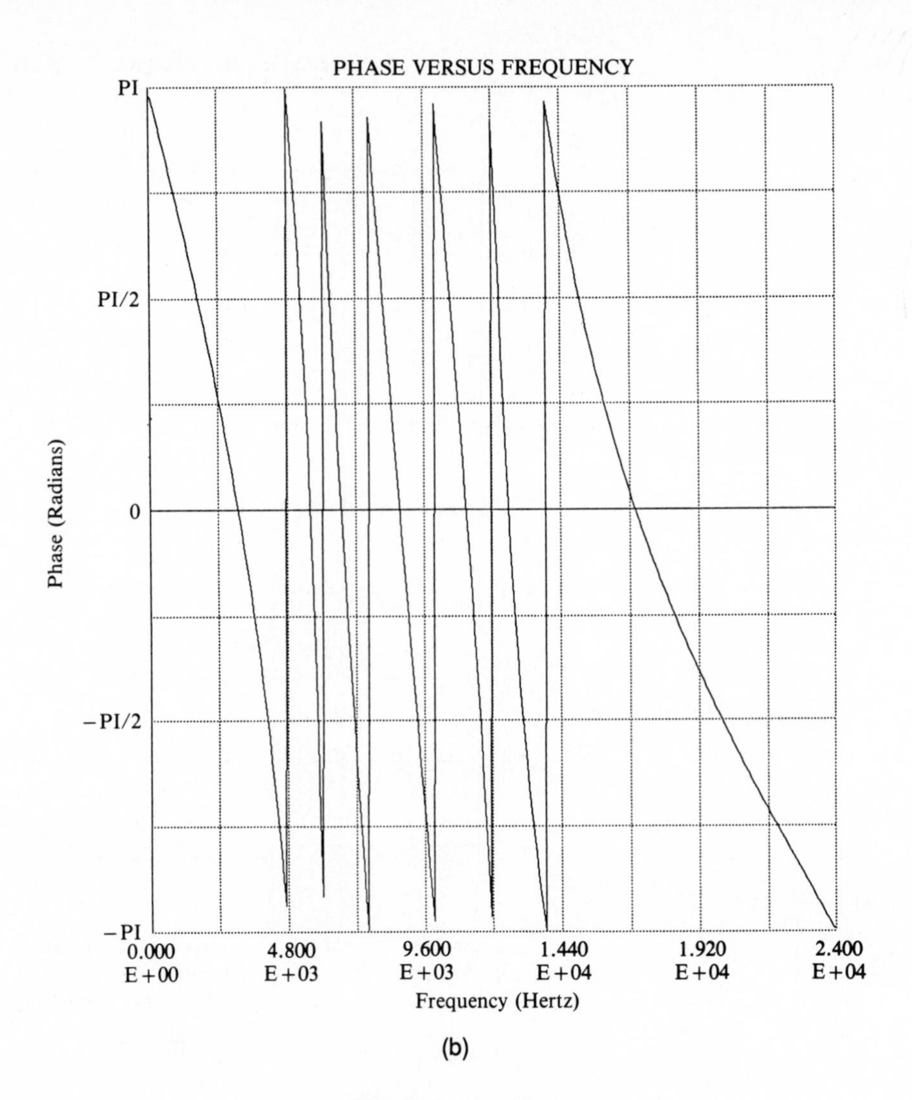

(b)

Figure 7-24 (*cont.*)

Biquad Sections of the Analog Transfer Function: (Numerator Coefficients Then Denominator Coefficients)		
*S**2 Term*	*S Term*	*Const Term*
.000000E+00	.606231E+05	.000000E+00
.100000E+01	.564618E+05	.336660E+10
.000000E+00	.606231E+05	.000000E+00
.100000E+01	.640221E+05	.432855E+10
.000000E+00	.606231E+05	.000000E+00
.100000E+01	.469741E+05	.265782E+10
.000000E+00	.606231E+05	.000000E+00
.100000E+01	.674682E+05	.548287E+10
.000000E+00	.606231E+05	.000000E+00
.100000E+01	.372745E+05	.217612E+10

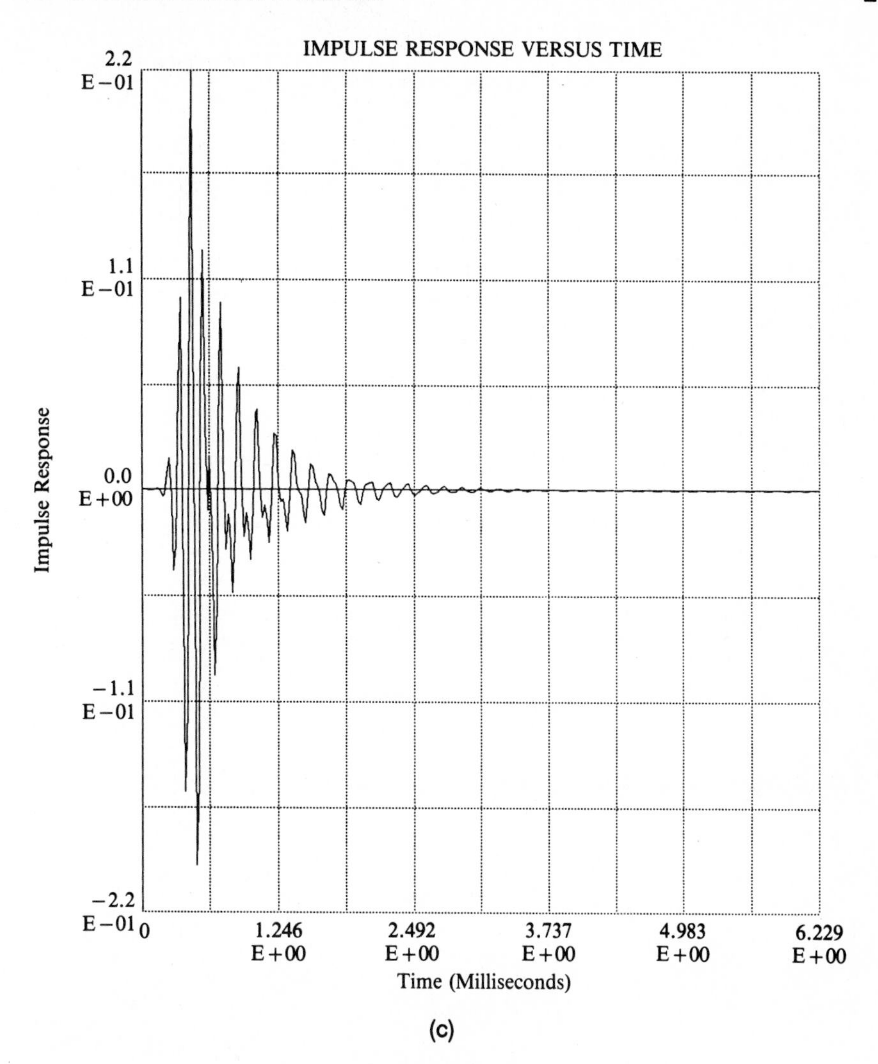

(c)

Figure 7-24 (*cont.*)

Biquad Sections of the Analog Transfer Function: (Cont.)		
*S**2 Term*	*S Term*	*Const Term*
.000000E+00	.606231E+05	.000000E+00
.100000E+01	.653876E+05	.669653E+10
.000000E+00	.606231E+05	.000000E+00
.100000E+01	.281013E+05	.186133E+10
.000000E+00	.606231E+05	.000000E+00
.100000E+01	.576327E+05	.782906E+10
.000000E+00	.606231E+05	.000000E+00
.100000E+01	.195649E+05	.166186E+10
.000000E+00	.606231E+05	.000000E+00
.100000E+01	.449419E+05	.876880E+10
.000000E+00	.606231E+05	.000000E+00
.100000E+01	.115335E+05	.154420E+10
.000000E+00	.606231E+05	.000000E+00
.100000E+01	.285116E+05	.943691E+10

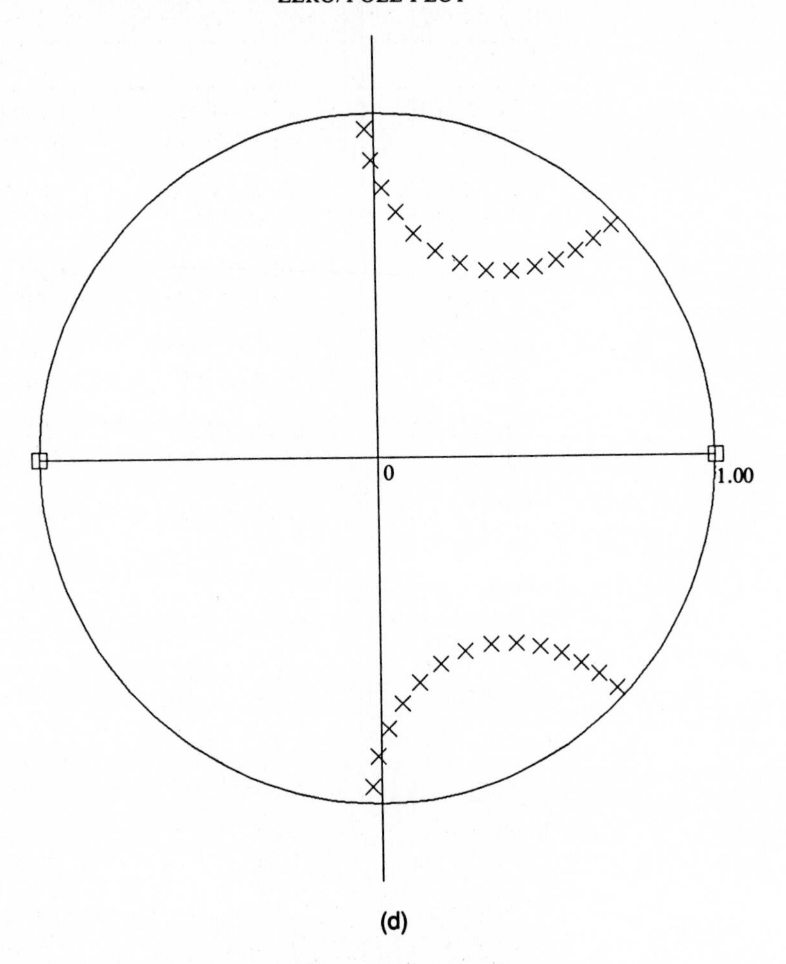

(d)

Figure 7-24 (*cont.*)

Biquad Sections of the Analog Transfer Function: (Cont.)		
*S**2 Term*	*S Term*	*Const Term*
.000000E+00	.606231E+05	.000000E+00
.100000E+01	.381029E+04	.148954E+10
.000000E+00	.606231E+05	.000000E+00
.100000E+01	.976498E+04	.978320E+10
	Initial Gain 1.00000000	

Biquad Sections of the Digital Transfer Function: (Numerator Coefficients Then Denominator Coefficients)		
*Z**2 Term*	*Z Term*	*Const Term*
1.0000000000	.0000000000	-1.0000000000
1.00000	-.4964235425	.3757326901
1.0000000000	.0000000000	-1.0000000000
1.00000	-.3525845110	.3882693052

Biquad Sections of the Digital Transfer Function: (Cont.)		
*Z**2 Term*	*Z Term*	*Const Term*
1.0000000000	.0000000000	-1.0000000000
1.00000	-.6498274207	.3978390992
1.0000000000	.0000000000	-1.0000000000
1.00000	-.2270842493	.4342242777
1.0000000000	.0000000000	-1.0000000000
1.00000	-.8005914688	.4495001435
1.0000000000	.0000000000	-1.0000000000
1.00000	-.1228588596	.5098953843
1.0000000000	.0000000000	-1.0000000000
1.00000	-.9405009747	.5219451189
1.0000000000	.0000000000	-1.0000000000
1.00000	-1.0678237677	.6083174348
1.0000000000	.0000000000	-1.0000000000
1.00000	-.0401085988	.6130425334
1.0000000000	.0000000000	-1.0000000000
1.00000	-1.1843982935	.7055158019
1.0000000000	.0000000000	-1.0000000000
1.00000	.0206551403	.7440753579
1.0000000000	.0000000000	-1.0000000000
1.00000	-1.2929184437	.8134030700
1.0000000000	.0000000000	-1.0000000000
1.00000	.0568998940	.9059581757
1.0000000000	.0000000000	-1.0000000000
1.00000	-1.3957597017	.9339216352
	Initial Gain .252520644E-06	

Zeros of Digital Transfer Function:	
Real Part	*Imaginary Part*
.100000000E+01	.000000000E+00
-.100000000E+01	.000000000E+00
.100000000E+01	.000000000E+00
-.100000000E+01	.000000000E+00
.100000000E+01	.000000000E+00
-.100000000E+01	.000000000E+00
.100000000E+01	.000000000E+00
-.100000000E+01	.000000000E+00
.100000000E+01	.000000000E+00
-.100000000E+01	.000000000E+00
.100000000E+01	.000000000E+00
-.100000000E+01	.000000000E+00
.100000000E+01	.000000000E+00
-.100000000E+01	.000000000E+00
.100000000E+01	.000000000E+00
-.100000000E+01	.000000000E+00

Zeros of Digital Transfer Function: (cont.)	
Real Part	*Imaginary Part*
.100000000E+01	.000000000E+00
-.100000000E+01	.000000000E+00
.100000000E+01	.000000000E+00
-.100000000E+01	.000000000E+00
.100000000E+01	.000000000E+00
-.100000000E+01	.000000000E+00
.100000000E+01	.000000000E+00
-.100000000E+01	.000000000E+00
.100000000E+01	.000000000E+00
-.100000000E+01	.000000000E+00
.100000000E+01	.000000000E+00
-.100000000E+01	.000000000E+00

Poles of Digital Transfer Function:	
Real Part	*Imaginary Part*
.248211741E+00	.560467298E+00
.248211741E+00	-.560467298E+00
.176292241E+00	.597654040E+00
.176292241E+00	-.597654040E+00
.324913681E+00	.540620171E+00
.324913681E+00	-.540620171E+00
.113542080E+00	.649101259E+00
.113542080E+00	-.649101259E+00
.400295734E+00	.537832138E+00
.400295734E+00	-.537832138E+00
.614293814E-01	.711422347E+00
.614293814E-01	-.711422347E+00
.470250487E+00	.548461118E+00
.470250487E+00	-.548461118E+00
.533911824E+00	.568555661E+00
.533911824E+00	-.568555661E+00
.200542808E-01	.782713421E+00
.200542808E-01	-.782713421E+00
.592199087E+00	.595664221E+00
.592199087E+00	-.595664221E+00
-.103275180E-01	.862536168E+00
-.103275180E-01	-.862536168E+00
.646459222E+00	.628882633E+00
.646459222E+00	-.628882633E+00
-.284498930E-01	.951393073E+00
-.284498930E-01	-.951393073E+00
.697879791E+00	.668494856E+00
.697879791E+00	-.668494856E+00

Direct Form Digital Transfer Function:

i) *Numerator Coefficients - Highest Order First (in Z):*

.276490596E-07	.000000000E+00	-.387086834E-06
.000000000E+00	.251606442E-05	.000000000E+00
-.100642577E-04	.000000000E+00	.276767086E-04
.000000000E+00	-.553534172E-04	.000000000E+00
.830301258E-04	.000000000E+00	-.948915724E-04
.000000000E+00	.830301258E-04	.000000000E+00
-.553534172E-04	.000000000E+00	.276767086E-04
.000000000E+00	-.100642577E-04	.000000000E+00
.251606442E-05	.000000000E+00	-.387086834E-06
.000000000E+00	.276490596E-07	

ii) *Denominator Coefficients - Highest Order First (in Z):*

.625000000E-01	-.530832767E+00	.251207986E+01
-.843389559E+01	.222337880E+02	-.484798859E+02
.902932943E+02	-.146681774E+03	.210903238E+03
-.271190807E+03	.314219049E+03	-.329838470E+03
.314876521E+03	-.274044842E+03	.217739867E+03
-.157987192E+03	.104612494E+03	-.631065002E+02
.345835497E+02	-.171460187E+02	.764673669E+01
-.304372247E+01	.106996685E+01	-.327313664E+00
.853465422E-01	-.183749442E-01	.310616169E-02
-.373261264E-03	.253873722E-04	

b) *Band-Pass Filter Implementation:*

The IIRBP.ASM assembler program, shown in Figure 7.25, implements the designed band-pass IIR filter on the DSP56001 processor. Note that this program has an "include" statement which references the IIR.ASM assembler program macro shown in Figure 7.19. A copy of the IIRBP.ASM assembler program, IIR.ASM assembler program macro, and the IIRBP.LOD absolute load program can be found on the DSP56000/1 Demonstration Software #3 program diskette.

To demonstrate the designed band-pass IIR filter on the DSP System:

1. CONNECT A FUNCTION GENERATOR TO "BNC2" ON THE EVB AND TO CHANNEL ONE OF A DUAL TRACE OSCILLOSCOPE.
2. SET THE FUNCTION GENERATOR TO PRODUCE A 2-VOLT (PEAK-TO-PEAK) SINE WAVE.
3. CONNECT CHANNEL TWO OF THE OSCILLOSCOPE TO "BNC1" ON THE EVB.
4. LOAD THE "ADS56000 USER INTERFACE SOFTWARE" PROGRAM.
5. PLACE THE "DSP56001 DEMONSTRATION SOFTWARE #3" DISKETTE INTO DRIVE ""A".
6. TYPE **a:iirbp.lod<return>** to load the **iirbp.lod** program.
7. TYPE **go<return>** to execute the designed band-pass IIR filter program (algorithm) on the DSP56001 processor.
8. VARY THE SINUSOIDAL INPUT FREQUENCY FROM 0.5 TO 20 KHZ. View the analog input and output signals on the oscilloscope. Note that the IIR filter passes the frequencies from 6 to 12 KHz and stops the frequencies from 0.5 to 4 KHz and 14 to 20 KHz as designed. Also note the decreasing attenuation of the output signal as the input signal's frequency is increased from 4 to 6 KHz; note the increasing attenuation of the output signal as the input signal's frequency is increased from 12 to 14 KHz.
9. TYPE **force b<return>** to halt execution of the IIR filter algorithm and return control of the ADM to its monitor program for keyboard command entry.
10. TYPE **quit<return>** to exit the ADS56000 program.

```
;*******************************************************
;
; File: IIRBP.ASM
;
;*******************************************************
;
        include 'iir'
;
;*******************************************************
; Coefficients of Band Pass IIR filter
;*******************************************************
;
nsec    equu    $e
;
        org     x:$0
states  ds      2*nsec
;
        org     x:$20
gk      dc      .25252064E-06
;
        org     y:$0
coef
         dc -.37573269/2.0
         dc  .49642354/2.0
         dc -1.0000000/2.0
         dc  .00000000/2.0
;
         dc -.38826931/2.0
         dc  .35258451/2.0
         dc -1.0000000/2.0
         dc  .00000000/2.0
;
         dc -.39783910/2.0
         dc  .64982742/2.0
         dc -1.0000000/2.0
         dc  .00000000/2.0
;
         dc -.43422428/2.0
         dc  .22708425/2.0
         dc -1.0000000/2.0
         dc  .00000000/2.0
;
         dc -.44950014/2.0
         dc  .80059147/2.0
         dc -1.0000000/2.0
         dc  .00000000/2.0
;
         dc -.50989538/2.0
         dc  .12285886/2.0
         dc -1.0000000/2.0
         dc  .00000000/2.0
;
         dc -.52194512/2.0
         dc  .94050097/2.0
         dc -1.0000000/2.0
         dc  .00000000/2.0
;
         dc -.60831743/2.0
         dc  1.0678238/2.0
         dc -1.0000000/2.0
         dc  .00000000/2.0
;
         dc -.61304253/2.0
         dc  .40108599E-1/2.0
         dc -1.0000000/2.0
         dc  .00000000/2.0
;
```

```
        dc -.70551580/2.0
        dc  1.1843983/2.0
        dc -1.0000000/2.0
        dc  .00000000/2.0
;
        dc -.74407536/2.0
        dc -.20655140E-1/2.0
        dc -1.0000000/2.0
        dc  .00000000/2.0
;
        dc -.81340307/2.0
        dc  1.2929184/2.0
        dc -1.0000000/2.0
        dc  .00000000/2.0
;
        dc -.90595818/2.0
        dc -.56899894E-1/2.0
        dc -1.0000000/2.0
        dc  .00000000/2.0
;
        dc -.93392164/2.0
        dc  1.3957597/2.0
        dc -1.0000000/2.0
        dc  .00000000/2.0
;
; program start
;
       org     p:$200
;
       iir     states,coef,gk,nsec
;
       end
```

Figure 7-25 IIRBP.ASM Assembler Program to Implement the Band-Pass IIR Filter

7.11.4 Designing and Implementing a Band-Stop IIR Filter:

a) *Band-Stop Filter Design:*

1. TYPE **cd \dsptools\fdas<return>** to enter the FDAS program directory.
2. TYPE **demo<return>** to load and enter the FDAS program.
3. FOLLOW THE FDAS PROGRAM'S MENUS TO SELECT THE BELOW LISTED IIR FILTER SPECIFICATIONS. Plots of the amplitude response, phase response, impulse response, poles and zeros of the band-stop filter are shown in Figure 7.26.
4. EXIT THE FDAS PROGRAM.

Filter Specifications:

Filter Type :	Band-stop			
Analog Filter Type:	Chebyshev			
Sampling Frequency:	48000	Hz		
Passband Cutoff Frequencies:	4000	Hz	14000	Hz
Stopband Cutoff Frequencies:	6000	Hz	12000	Hz
Passband attenuation:	0.5	dB		
Stopband Attenuation:	50.0	dB		
Filter Design Method:	Bilinear Transformation			

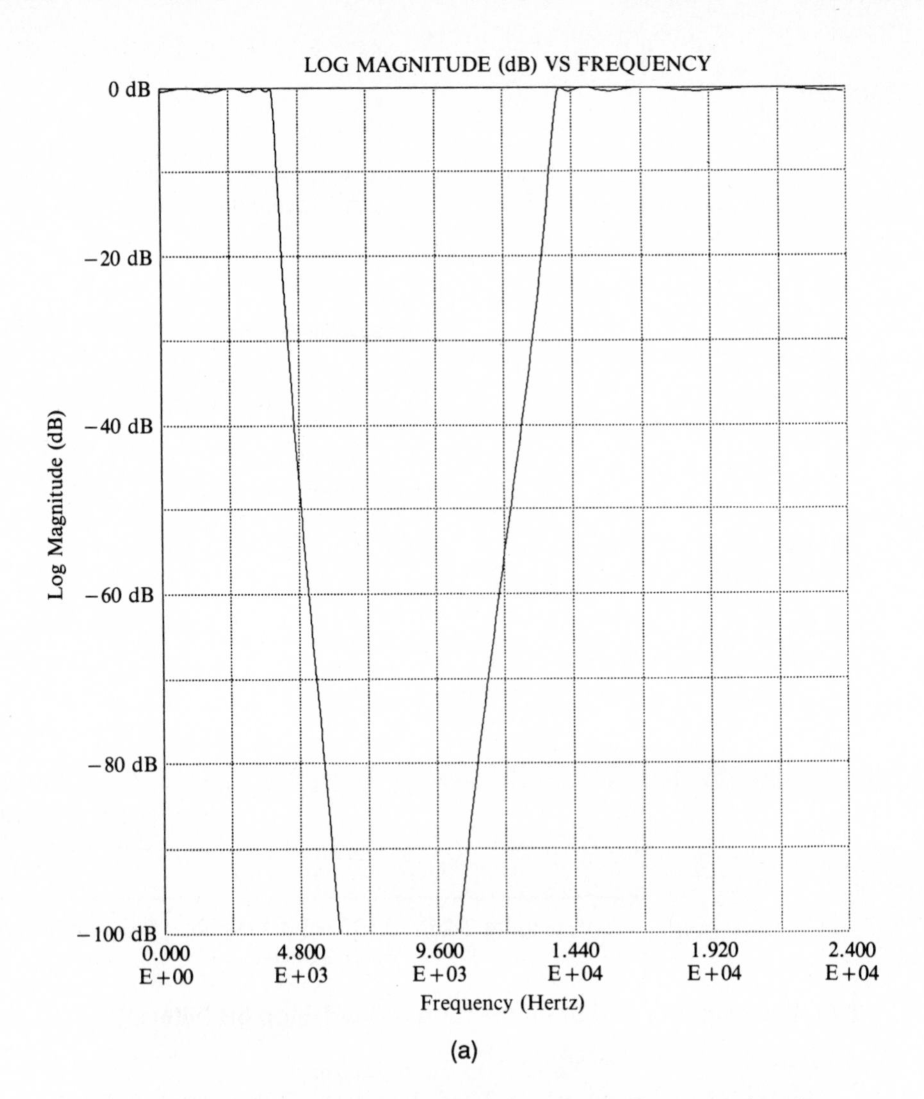

(a)

Figure 7-26 Plots of Amplitude Response, Phase Response, Impulse, Poles and Zeros of Band-Stop Filter in Section 7.12.4. (a) Amplitude Response in dB (b) Phase Response (c) Impulse Response (d) Poles and Zeros

Design Results:

Filter Order: 16

Biquad Sections of the Normalized Analog Transfer Function: (Numerator Coefficients Then Denominator Coefficients)		
*S**2 Term*	*S Term*	*Const Term*
.000000E+00	.000000E+00	.100000E+01
.100000E+01	.872402E-01	.101193E+01
.000000E+00	.000000E+00	.100000E+01
.100000E+01	.248439E+00	.741334E+00
.000000E+00	.000000E+00	.100000E+01
.100000E+01	.371815E+00	.358650E+00

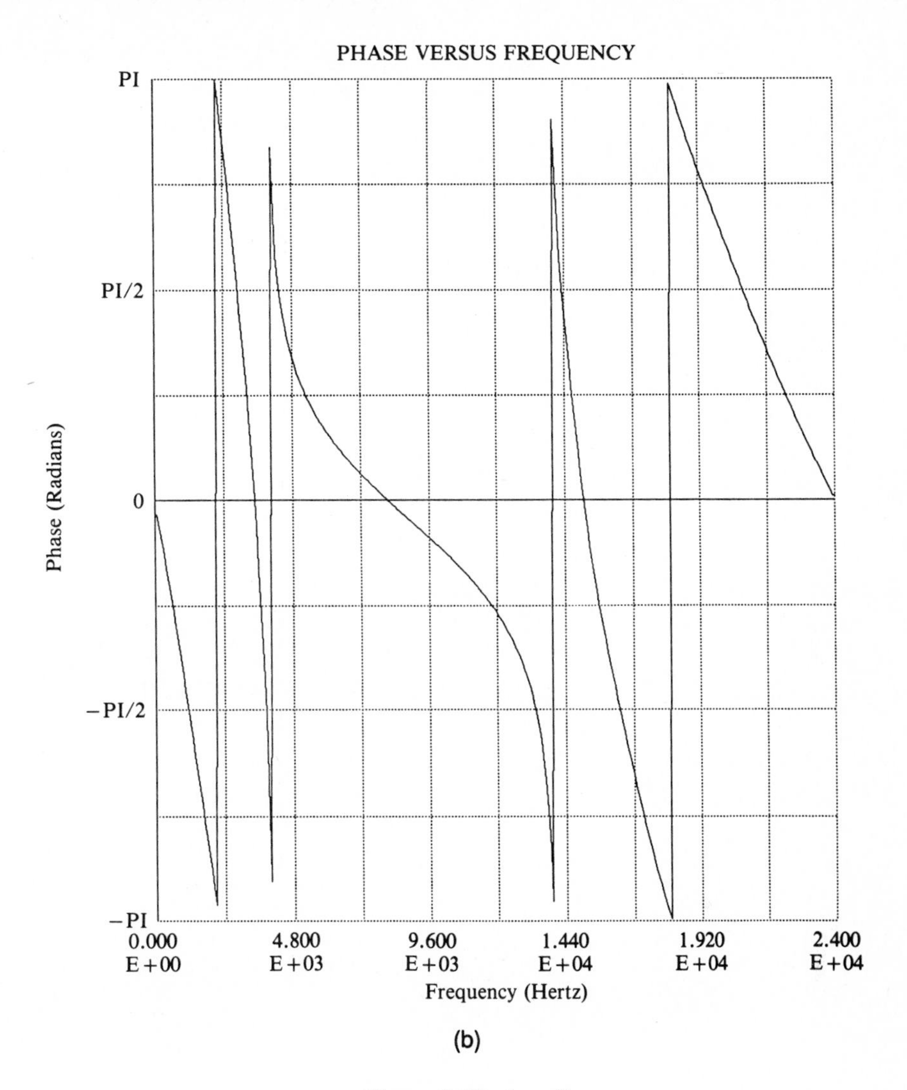

(b)

Figure 7-26 (*cont.*)

Biquad Sections of the Normalized Analog Transfer Function: (Cont.)		
*S**2 Term*	*S Term*	*Const Term*
.000000E+00	.000000E+00	.100000E+01
.100000E+01	.438586E+00	.880523E-01
Initial Gain .223654308E-01		

Biquad Sections of the Analog Transfer Function: (Numerator Coefficients Then Denominator Coefficients)		
*S**2 Term*	*S Term*	*Const Term*
.100000E+01	.000000E+00	.321821E+10
.100000E+01	.709668E+04	.155199E+11
.100000E+01	.000000E+00	.321821E+10
.100000E+01	.147157E+04	.667331E+09

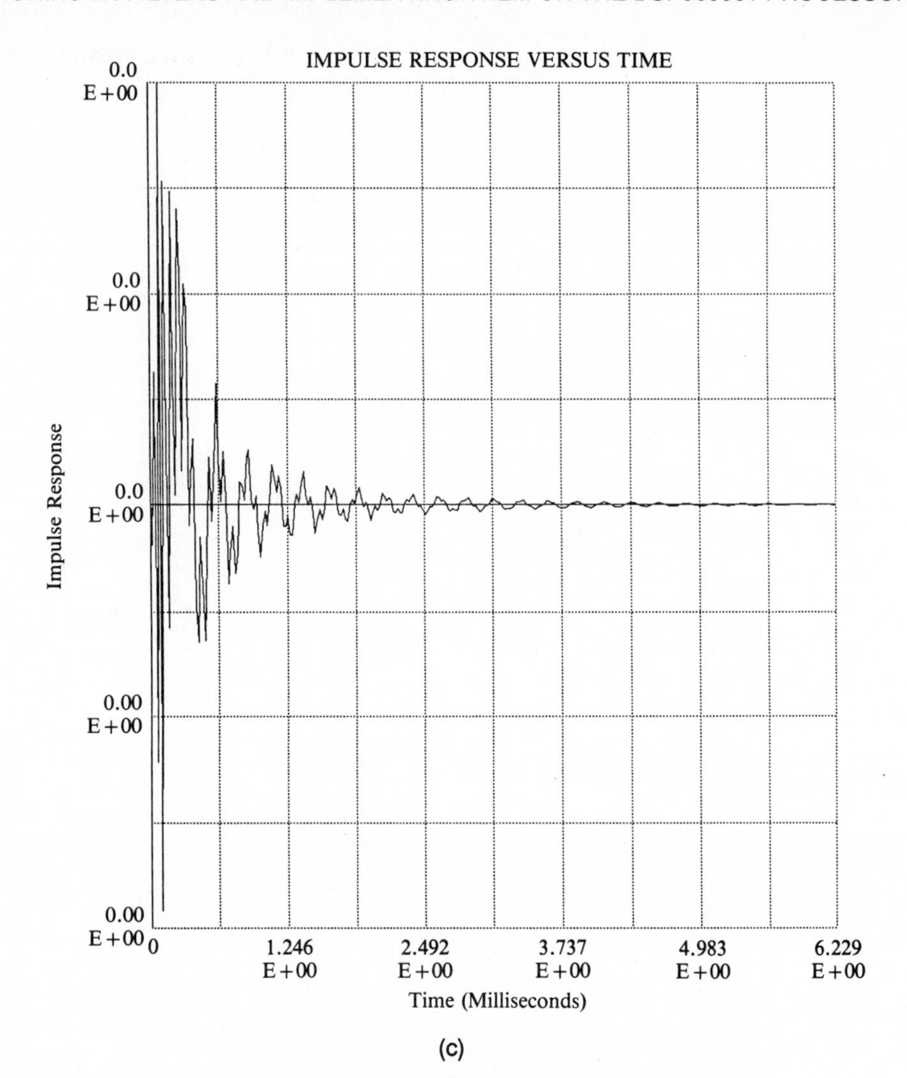

(c)

Figure 7-26 (*cont.*)

Biquad Sections of the Analog Transfer Function: (Cont.)		
*S**2 Term*	*S Term*	*Const Term*
.100000E+01	.000000E+00	.321821E+10
.100000E+01	.285000E+05	.190808E+11
.100000E+01	.000000E+00	.321821E+10
.100000E+01	.480686E+04	.542790E+09
.100000E+01	.000000E+00	.321821E+10
.100000E+01	.938280E+05	.327981E+11
.100000E+01	.000000E+00	.321821E+10
.100000E+01	.920659E+04	.315777E+09
.100000E+01	.000000E+00	.321821E+10
.100000E+01	.481197E+06	.111862E+12
.100000E+01	.000000E+00	.321821E+10
.100000E+01	.138438E+05	.925863E+08
	Initial Gain .944060875	

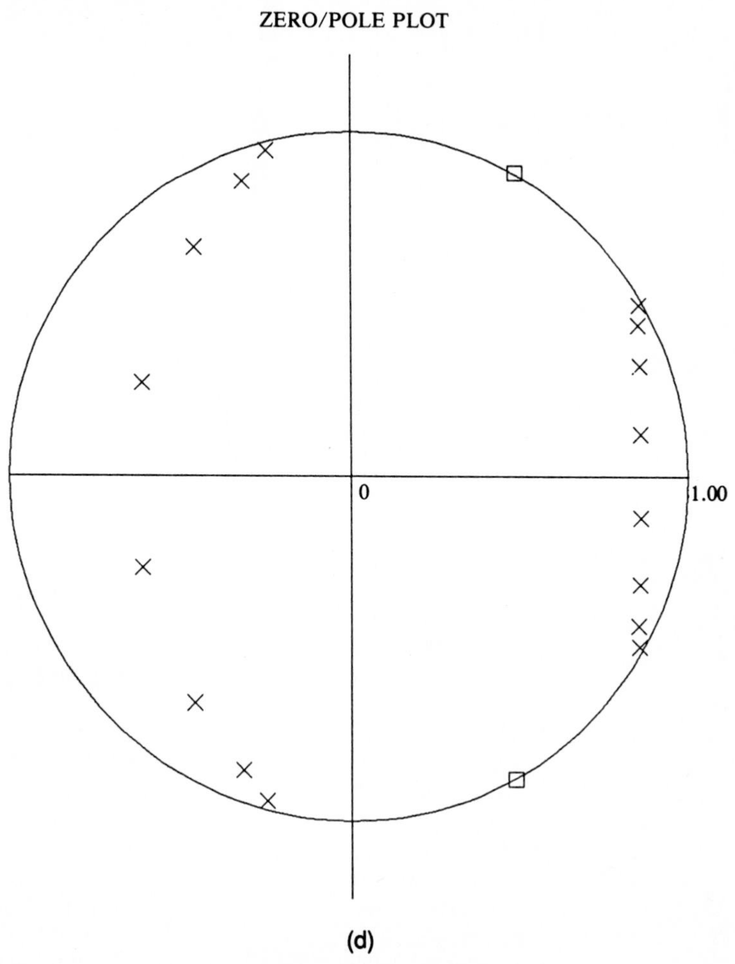

(d)

Figure 7-26 *(cont.)*

Biquad Sections of the Digital Transfer Function: (Numerator Coefficients Then Denominator Coefficients)		
*Z**2 Term*	*Z Term*	*Const Term*
1.0000000000	-.9647237659	1.0000000000
1.00000	1.2272869349	.4476695061
1.0000000000	-.9647237659	1.0000000000
1.00000	.9243958592	.6469141841
1.0000000000	-.9647237659	1.0000000000
1.00000	-1.7153155804	.7501303554
1.0000000000	-.9647237659	1.0000000000
1.00000	.6357675791	.8236707449
1.0000000000	-.9647237659	1.0000000000
1.00000	-1.7090160847	.8302868009

Biquad Sections of the Digital Transfer Function: (Cont.)		
*Z**2 Term*	*Z Term*	*Const Term*
1.0000000000	-.9647237659	1.0000000000
1.00000	-1.6972601414	.9096971750
1.0000000000	-.9647237659	1.0000000000
1.00000	.4960319400	.9463918805
1.0000000000	-.9647237659	1.0000000000
1.00000	-1.7055377960	.9718151093
	Initial Gain .705930730E-02	

Zeros of Digital Transfer Function:	
Real Part	Imaginary Part
.482361910E+00	.875972024E+00
.482361910E+00	-.875972024E+00
.482361943E+00	.875972006E+00
.482361943E+00	-.875972006E+00
.482361935E+00	.875972011E+00
.482361935E+00	-.875972011E+00
.482361845E+00	.875972060E+00
.482361845E+00	-.875972060E+00
.482362117E+00	.875971911E+00
.482362117E+00	-.875971911E+00
.482361829E+00	.875972069E+00
.482361829E+00	-.875972069E+00
.482361904E+00	.875972028E+00
.482361904E+00	-.875972028E+00
.482361947E+00	.875972004E+00
.482361947E+00	-.875972004E+00

Poles of Digital Transfer Function:	
Real Part	Imaginary Part
-.613643408E+00	.266666972E+00
-.613643408E+00	-.266666972E+00
-.462197900E+00	.658245476E+00
-.462197900E+00	-.658245476E+00
.857657790E+00	.120637024E+00
.857657790E+00	-.120637024E+00
-.317883730E+00	.850070915E+00
-.317883730E+00	-.850070915E+00
.854508042E+00	.316390181E+00
.854508042E+00	-.316390181E+00
.848630071E+00	.435343610E+00
.848630071E+00	-.435343610E+00
-.248015940E+00	.940680559E+00
-.248015940E+00	-.940680559E+00
.852768898E+00	.494570840E+00
.852768898E+00	-.494570840E+00

Direct Form Digital Transfer Function:

i) *Numerator Coefficients - Highest Order First (in Z):*

.250531358E-03	-.193354863E-02	.853294257E-02
-.261316097E-01	.613775331E-01	-.115312082E+00
.178377224E+00	-.230371775E+00	.250752174E+00
-.230371775E+00	.178377224E+00	-.115312082E+00
.613775331E-01	-.261316097E-01	.853294257E-02
-.193354863E-02	.250531358E-03	

ii) *Denominator Coefficients - Highest Order First (in Z):*

.781250000E-02	-.276847472E-01	.411992408E-01
-.446219697E-01	.571624223E-01	-.599295205E-01
.366658219E-01	-.242438138E-01	.257487886E-01
-.110718857E-01	-.995329283E-03	-.472807716E-02
.521939517E-02	.405981700E-03	.594297128E-03
-.242711375E-02	.971099719E-03	

b) *Band-Stop Filter Implementation:*

The IIRBS.ASM assembler program, shown in Figure 7.27, implements the designed band-stop IIR filter on the DSP56001 processor. Note that this program has an "include" statement which references the IIR.ASM assembler program macro shown in Figure 7.19. A copy of the IIRBS.ASM assembler program, IIR.ASM assembler program macro, and the IIRBS.LOD absolute load program can be found on the DSP56000/1 Demonstration Software #3 program diskette.

To demonstrate the designed band-stop IIR filter on the DSP System:

1. CONNECT A FUNCTION GENERATOR TO "BNC2" ON THE EVB AND TO CHANNEL ONE OF A DUAL TRACE OSCILLOSCOPE.
2. SET THE FUNCTION GENERATOR TO PRODUCE A 2-VOLT (PEAK-TO-PEAK) SINE WAVE.
3. CONNECT CHANNEL TWO OF THE OSCILLOSCOPE TO "BNC1" ON THE EVB.
4. LOAD THE "ADS56000 USER INTERFACE SOFTWARE" PROGRAM.
5. PLACE THE "DSP56001 DEMONSTRATION SOFTWARE #3" DISKETTE INTO DRIVE ""A".
6. TYPE **a:iirbs.lod<return>** to load the **iirbs.lod** program.
7. TYPE **go<return>** to execute the designed band-stop IIR filter program (algorithm) on the DSP56001 processor.
8. VARY THE SINUSOIDAL INPUT FREQUENCY FROM 0.5 TO 20 KHZ. View the analog input and output signals on the oscilloscope. Note that the IIR filter stops the frequencies from 6 to 12 KHz and passes the frequencies from 0.5 to 4 KHz and 14 to 20 KHz as designed. Also note the increasing attenuation of the output signal as the input signal's frequency is increased from 4 to 6 KHz; note the decreasing attenuation of the output signal as the input signal's frequency is increased from 12 to 14 KHz.
9. TYPE **force b<return>** to halt execution of the IIR filter algorithm and return control of the ADM to its monitor program for keyboard command entry.
10. TYPE **quit<return>** to exit the ADS56000 program.

```
;******************************************************
;
; File: IIRBS.ASM
;
;******************************************************
;
        include 'iir'
;
;******************************************************
; Coefficients of Band Stop IIR filter
;******************************************************
;
nsec    equu    $8
;
        org     x:$0
states  ds      2*nsec
;
        org     x:$20
gk      dc      .70593073E-02
;
        org     y:$0
coef
        dc  -.44766951/2.0
        dc  -1.2272869/2.0
        dc   1.0000000/2.0
        dc  -.96472377/2.0
;
        dc  -.64691418/2.0
        dc  -.92439586/2.0
        dc   1.0000000/2.0
        dc  -.96472377/2.0
;
        dc  -.75013036/2.0
        dc   1.7153156/2.0
        dc   1.0000000/2.0
        dc  -.96472377/2.0
;
        dc  -.82367074/2.0
        dc  -.63576758/2.0
        dc   1.0000000/2.0
        dc  -.96472377/2.0
;
        dc  -.83028680/2.0
        dc   1.7090161/2.0
        dc   1.0000000/2.0
        dc  -.96472377/2.0
;
        dc  -.90969718/2.0
        dc   1.6972601/2.0
        dc   1.0000000/2.0
        dc  -.96472377/2.0
;
        dc  -.94639188/2.0
        dc  -.49603194/2.0
        dc   1.0000000/2.0
        dc  -.96472377/2.0
;
        dc  -.97181511/2.0
        dc   1.7055378/2.0
        dc   1.0000000/2.0
        dc  -.96472377/2.0
;
; program start
```

```
;
        org     p:$200
;
        iir     states,coef,gk,nsec
;
        end
```

Figure 7-27 IIRBS.ASM Assembler Program to Implement the Band-Stop IIR Filter

7.12 QUANTIZATION EFFECTS:

The Filter Design and Analysis System (FDAS) Software program can also be used to compare the effects of quantization, e.g., the word length precision of the digital signal processor. Figure 7.28 shows the frequency response of a tenth order 50 Hz bandpass IIR filter implemented with a 24-bit DSP56001 processor and a 16-bit processor. It is clear that the filter implemented with the 24-bit DSP56001 processor has a better performance than the filter implemented with a 16-bit processor.

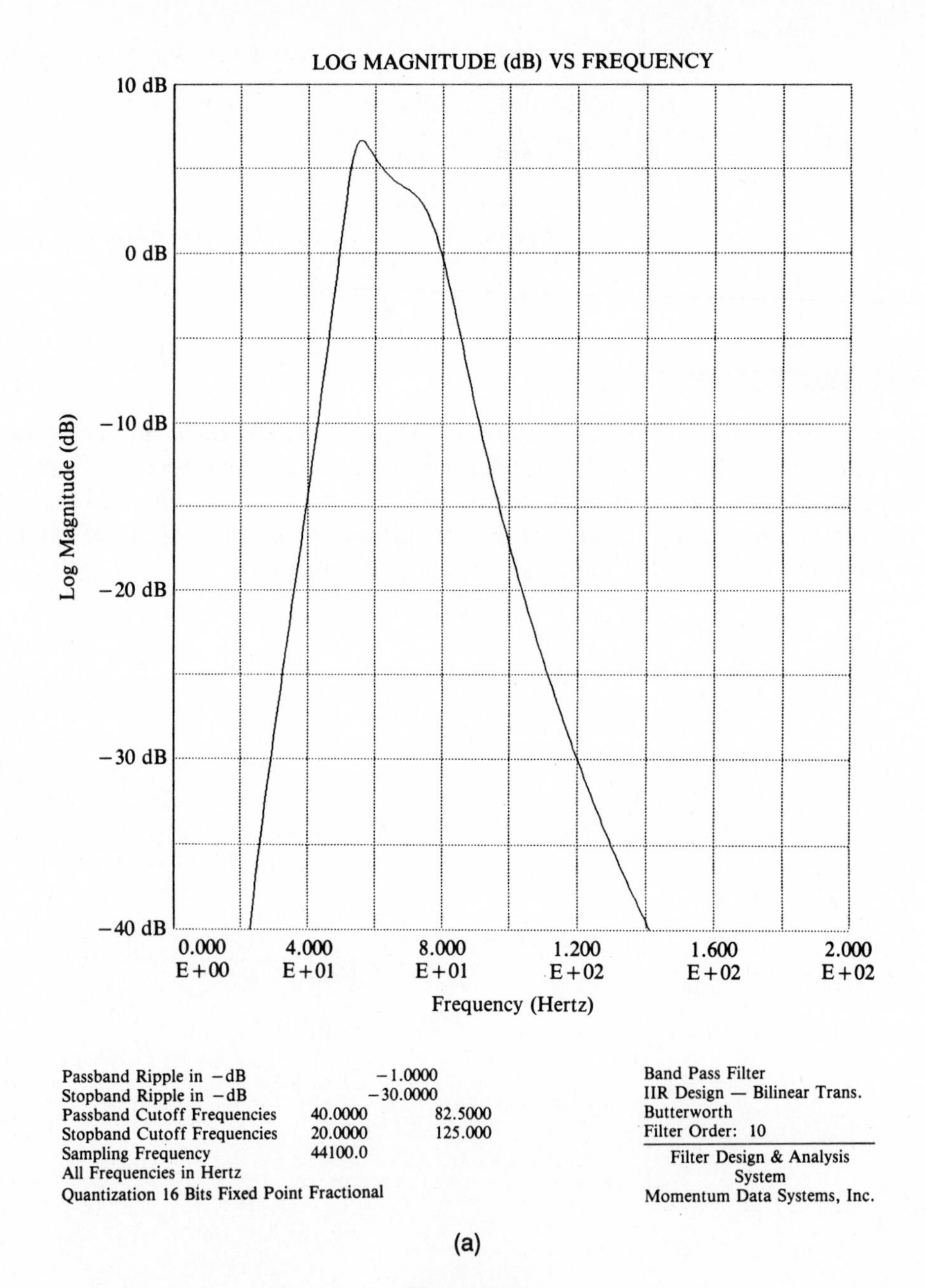

(a)

Figure 7-28

7.13 REFERENCES:

1. Alan V. Oppenheim and Ronald W. Schafer, Discrete-Time Signal Processing, Prentice-Hall, 1989.
2. N.K.Bose, Digital Filters Theory and Applications, North-Holland, 1985.
3. Johnny R. Johnson, Introduction to Digital Signal Processing, Prentice-Hall, 1989.
4. Samuel D. Stearns and Ruth A. David, Signal Processing Algorithms, Prentice-Hall, 1988.
5. Harry Y-F. Lam, Analog and Digital Filters, Prentice-Hall, 1979.

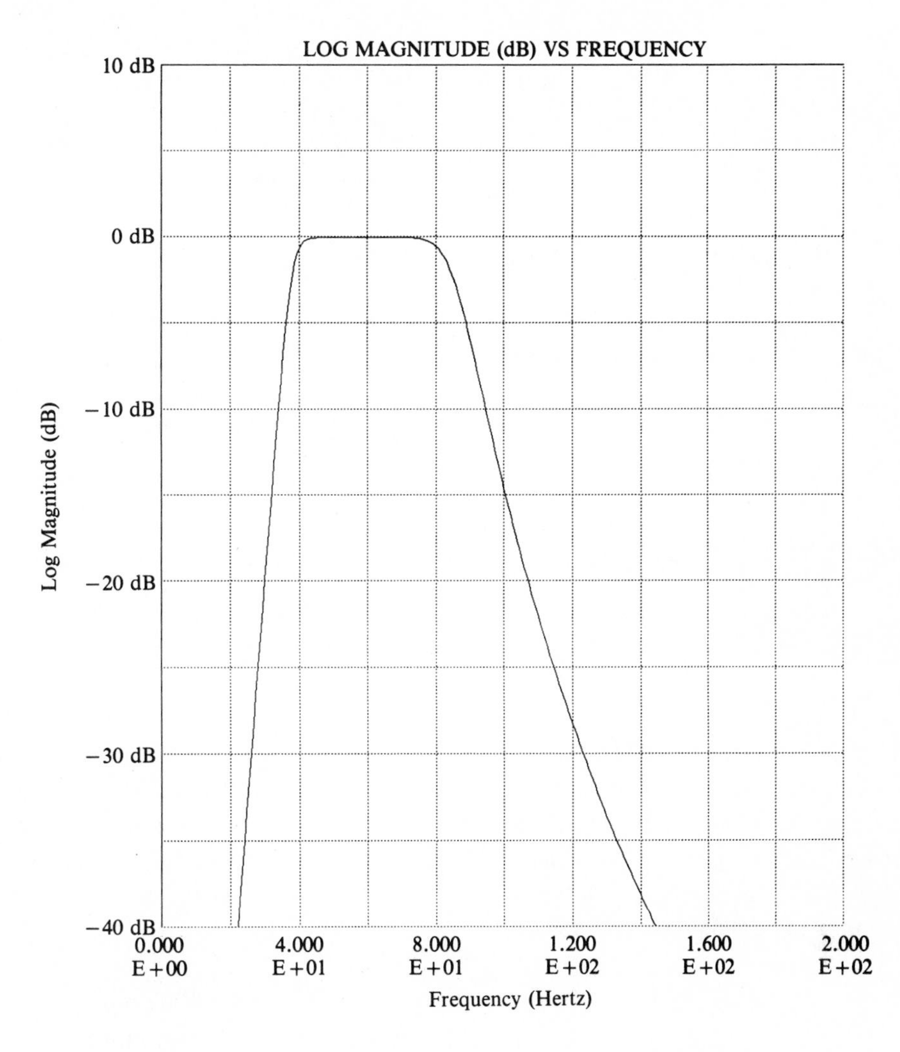

Passband Ripple in −dB		−1.0000
Stopband Ripple in −dB		−30.0000
Passband Cutoff Frequencies	40.0000	82.5000
Stopband Cutoff Frequencies	20.0000	125.000
Sampling Frequency	44100.0	
All Frequencies in Hertz		
Quantization 24 Bits Fixed Point Fractional		

Band Pass Filter
IIR Design — Bilinear Trans.
Butterworth
Filter Order: 10

Filter Design & Analysis System
Momentum Data Systems, Inc.

(b)

Figure 7-28 *(cont.)*

6. Chi-Tsong Chen, One-Dimensional Digital Signal Processing, Marcel Dekker, 1979.
7. John G. Proakis and Dimitris G. Manolakis, Introduction to Digital Signal Processing, Macmillan, 1988.
8. Richard A. Roberts and Clifford T. Mullis, Digital Signal Processing, Addison Wesley, 1987.

8

IMPLEMENTING THE FAST FOURIER TRANSFORM WITH THE DSP56001 PROCESSOR

The discrete Fourier transform (DFT) is widely used in digital signal processing applications to determine the frequency content of signals and to perform filtering of signals in the frequency domain. The fast Fourier transform (FFT) is an efficient, high speed algorithm for computing the discrete Fourier transform (DFT). The FFT achieves its computational high speed by decomposing the data sample set into smaller subsets, in an orderly manner, to eliminate redundant multiplication calculations. Two common methods for decomposing the data sample set are:

1. the decimation-in-time FFT algorithm, and
2. the decimation-in-frequency FFT algorithm.

The DSP56000/1 processors' instruction set lends itself particularly well for implementation of the decimation-in-time FFT algorithm. The decimation-in-frequency FFT algorithm can also be implemented on the DSP56000/1 processors, but the execution speed is somewhat slower than the decimation-in-time FFT algorithm.

This chapter begins with a review of the decimation-in-time FFT algorithm. The decimation-in-time FFT algorithm is then implemented on the DSP56001 processor. The DSP56001 assembler macros used to generate "twiddle" factors, store the input sequence in bit-reversed order, implement the FFT butterfly, calculate group and coefficient offsets, and perform complete N-point FFTs are presented. The DSP56001 FFT assembler macros can be found on the DSP56000/1 Demonstration Software #3 diskette. These macros can also be used in Chapter 13 to implement a spectrum analyzer algorithm on the DSP56001 processor.

8.1 FFT REVIEW:

The discrete Fourier transform (DFT) can be defined as:

$$X(k) = \sum_{k=0}^{N-1} x(n) \; W_N^{nk} \tag{8.1}$$

where

$$W_N = e^{-j2\pi/N} \tag{8.2}$$

and

$$W_N^{nk} = e^{-j2\pi nk/N} \tag{8.3}$$

For a real or complex discrete-time sequence x(n) of length N, the summation in Equation 8.1 gives a complex frequency X(k) of length N. W_N^{nk} is known as either the "twiddle" factor or the weighting factor. In the remainder of this chapter, N is considered to be an integral power of 2 , that is, $N = 2^m$, where m is a positive number.

The FFT can be computed by decomposing the data sample set into smaller subsets, where decomposition is accomplished by using the decimation-in-time FFT algorithm. The idea is to first separate the terms in the DFT summation into two groups: one group containing the terms for which n is even, and the other group containing the terms for which n is odd. Letting n = 2r for the even terms and n = 2r + 1 for the odd terms, the DFT can be written in the form:

$$\begin{aligned} X(k) &= \sum_{r=0}^{(N/2)-1} x(2r)\, W_N^{2rk} + \sum_{r=0}^{(N/2)-1} x(2r+1)\, W_N^{(2r+1)k} \\ &= \sum_{r=0}^{(N/2)-1} x(2r)\, W_N^{2rk} + W_N^{K} \sum_{r=0}^{(N/2)-1} x(2r+1)\, W_N^{2rk} \end{aligned} \tag{8.4}$$

Since N is an even integer and $W_N^{\,2} = W_{N/2} = e^{-j4\pi n/N}$, Equation 8.4 can be rewritten as:

$$X_{1,m}(k) = X_{1,m-1}(k) + W_N^k\, X_{2,m-1}(k) \tag{8.5}$$

where

$$X_{1,m}(k) = X(k) \tag{8.6}$$

$$X_{1,m-1}(k) = \sum_{r=0}^{(N/2)-1} X(2r)\, W_{N/2}^{rk} \tag{8.7}$$

$$X_{2,m-1}(k) = \sum_{r=0}^{(N/2)-1} x(2r+1)\, W_{N/2}^{rk} \tag{8.8}$$

and where $X_{1,m-1}(k)$ and $X_{2,m-1}(k)$ are each a N/2-point DFT.

Since an algorithm is available for computing a N/2-point DFT, $X_{1,m-1}(k)$ and $X_{2,m-1}(k)$ can be combined in accordance with Equation 8.5 and the signal-flow graph shown in Figure 8.1 to provide the N-point DFT of x(n).

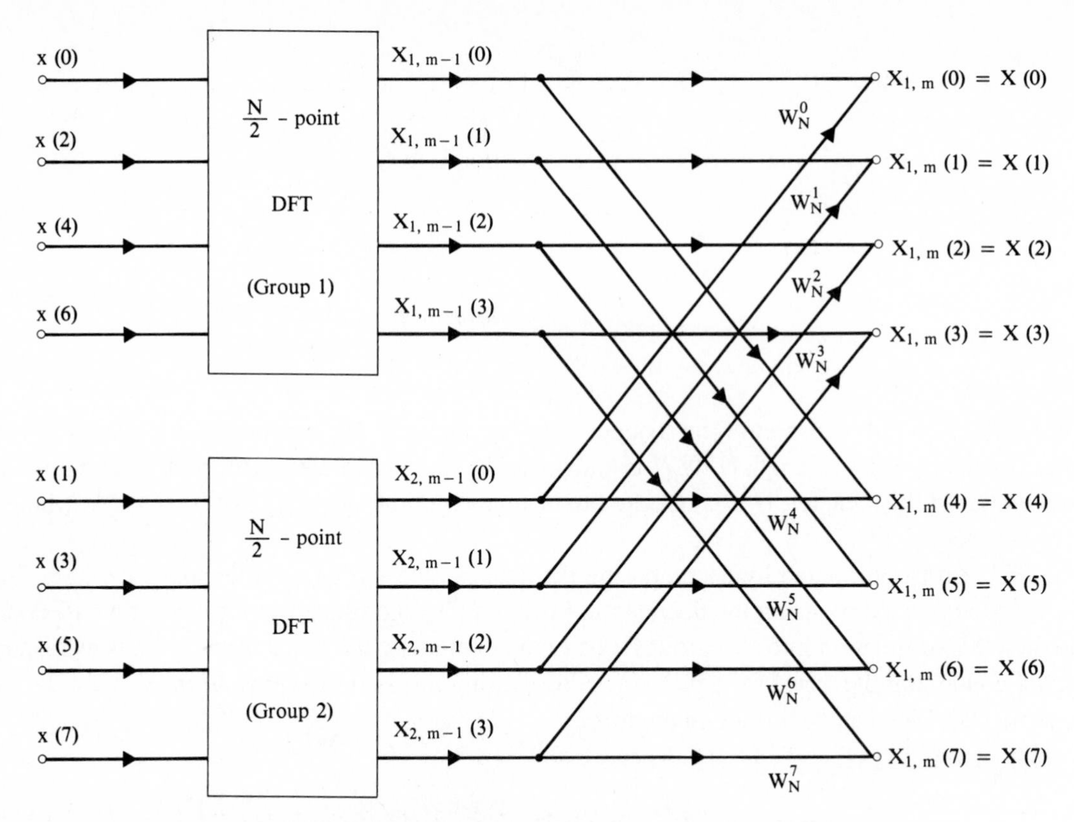

Figure 8-1 Decimation-in-Time Decomposition of N-Point DFT in Terms of Two $\frac{n}{2}$-point DFTs

In a similar manner, the two N/2-point DFTs can be written in the form:

$$X_{1,m-1}(k) = /x_{1,m-2}(k) + W_N^{2K} X_{2,m-2}(k) \tag{8.9}$$

$$X_{2,m-1}(k) = X_{3,m-2}(k) + W_N^{2K} X_{4,m-2}(k) \tag{8.10}$$

where $X_{1,m-2}(k)$, $X_{2,m-2}(k)$, $X_{3,m-2}(k)$ and $X_{4,m-2}(k)$ are each a N/4-point DFT.

Since an algorithm is available for computing a N/4-point DFT, $X_{1,m-2}(k)$, $X_{2,m-2}(k)$, $X_{3,m-2}(k)$, and $X_{4,m-2}(k)$ can be combined in accordance with Equations 8.9, 8.10, and 8.5, and the signal-flow graph shown in Figure 8.2, to provide an N-point DFT of x(n).

The decomposition process can be continued until a group of N/2 two-point (radix-2) DFTs result for the first stage (pass) of the N-point DFT computation, where the total number of stages (passes) m is equal to $\log_2(N)$, e.g., 2^m=N. The signal-flow graph for N=8 input samples, N/2=4 groups, and m=3 stages is shown in Figure 8.3, where each solid dot at the intersection of two or more lines indicates a memory location for storage of either an input data sample, the result of an intermediate computation, or a frequency domain output sample.

8.1.1 Butterfly Computation:

The common computation for each group of each stage of the DFT shown in Figure 8.3 is based on the radix-2 FFT Butterfly shown in Figure 8.4. Two complex values in one stage are derived from two complex values in the previous stage through the equations:

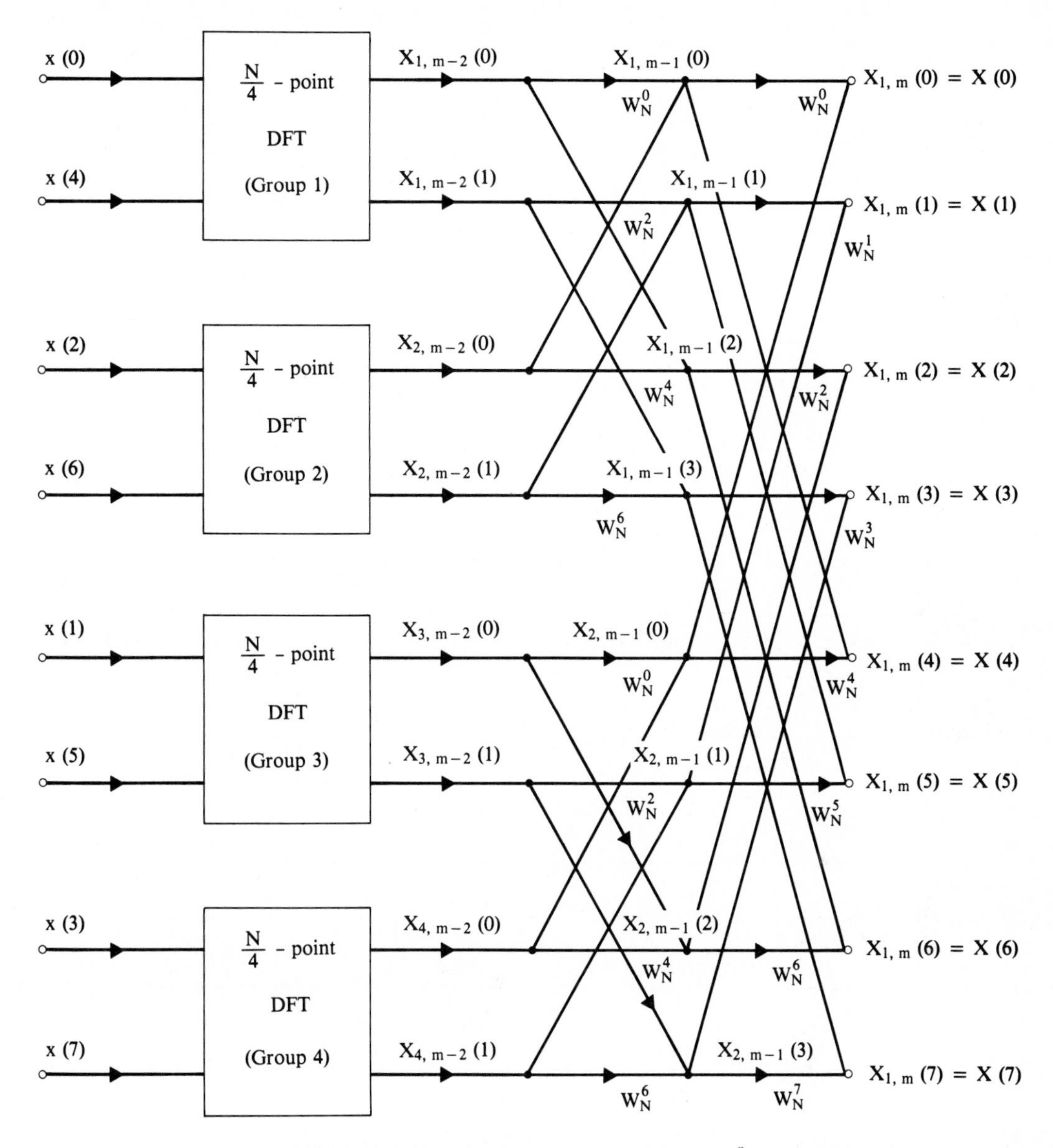

Figure 8-2 Decomposition of N-Point DFT in Terms of $\frac{n}{4}$-point DFTs

$$X_{i,j}[p] = X_{i,j-1}[p] + W_N^r X_{i,j-1}[q] \tag{8.11}$$

$$X_{i,j}[q] = X_{i,j-1}[p] + W_N^{r+N/2} X_{i,j-1}[q] \tag{8.12}$$

where i is the group number , j is the stage number, p and q are the RAM data locations for the butterfly.

The values of i,p,q and r vary from stage to stage as shown in Figure 8.3. Note that the values of $X_{i,j}[p]$ and $X_{i,j}[q]$ are stored in the same RAM locations as the values of $X_{i,j-1}[p]$ and $X_{i,j-1}[q]$, respectively. Therefore, this type of "in-place" FFT butterfly computation minimizes the total number of RAM locations necessary to compute the DFT.

The coefficients W_N of the butterfly shown in Figure 8.4 are always powers of r or r+N/2. Since $W_N^{(r+N/2)} = W_N^r W_N^{N/2} = e^{-j\pi} W_N^r = - W_N^r$, the butterfly shown in Figure 8.4 can be simplified to the butterfly shown in Figure 8.5, and Equations 8.11 and 8.12 can be rewritten as:

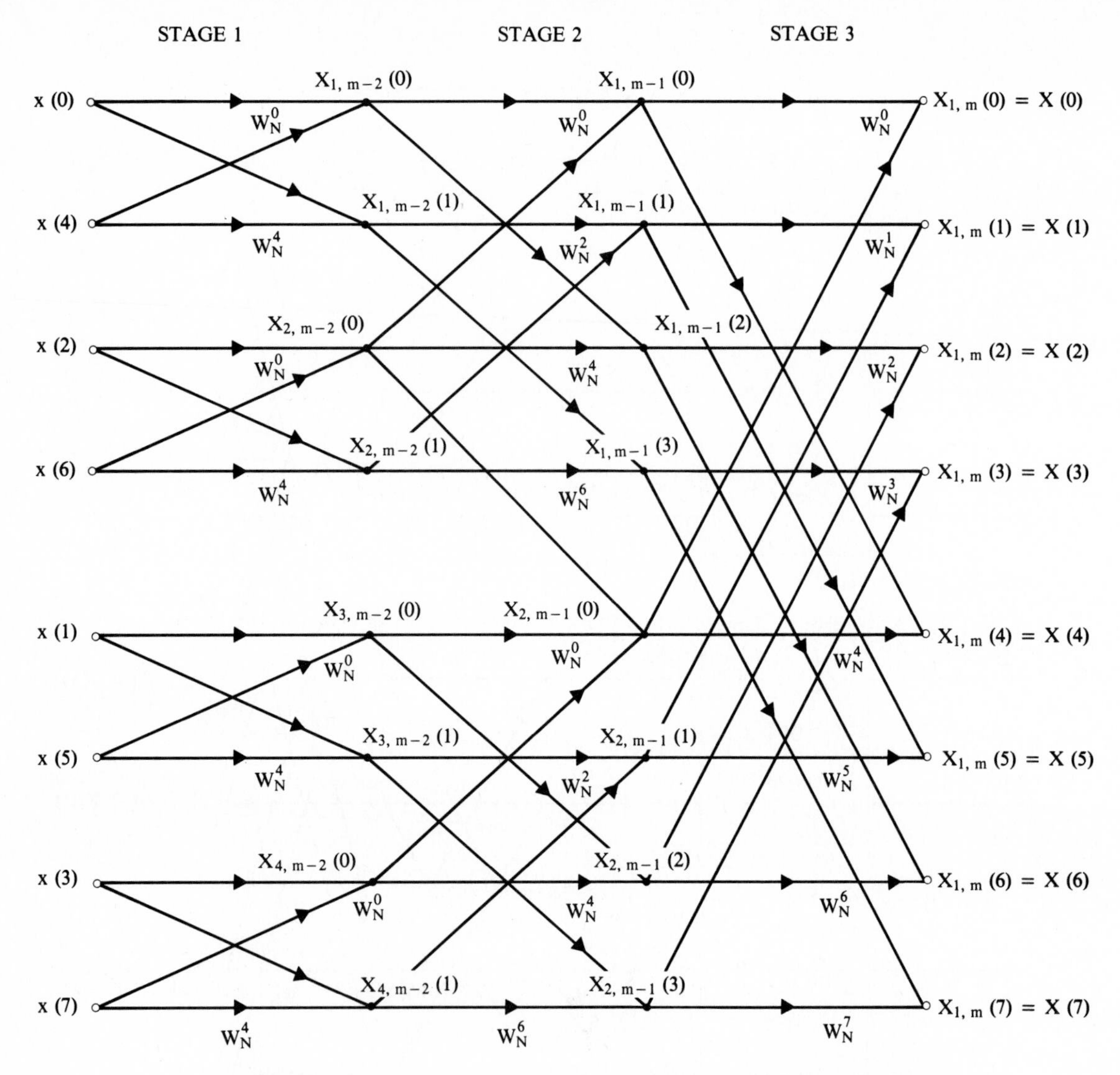

Figure 8-3 Decimation-in-Time Decomposition of N-Point DFT (N=8)

$$X_{i,j}[p] \quad = \quad X_{i,j-1}[p] + W_N^r X_{i,j-1}[q] \tag{8.13}$$

$$X_{i,j}[q] \quad = \quad X_{i,j-1}[p] - W_N^r X_{i,j-1}[q] \tag{8.14}$$

The stage number j varies from 1 to m. As shown in Figure 8.3, each stage j has 2^{m-j} groups and each group i within each stage j has 2^{j-1} butterflies. The distance between the RAM memory locations p and q within a butterfly is referred to as the group offset n_0, where n_0 is equal to 2^{j-1}. The difference between the values of the coefficient exponents r in group i of stage j is referred to as the coefficient offset n_2, where n_2 is equal to 2^{m-j}.

$$\text{Group offset } n_0 \quad = \quad 2^{j-1} \qquad j = 1, 2, .., m$$

$$\text{Coefficient offset } n_2 \quad = \quad i \quad = \quad 2^{m-j} \qquad j = 1, 2, .., m \tag{8.15}$$

In computing the DFT, j is incremented from 1 to m, i and n_2 are decremented from 2^{m-1} to 2^0, and n_0 is incremented from 2^0 to 2^{m-1}.

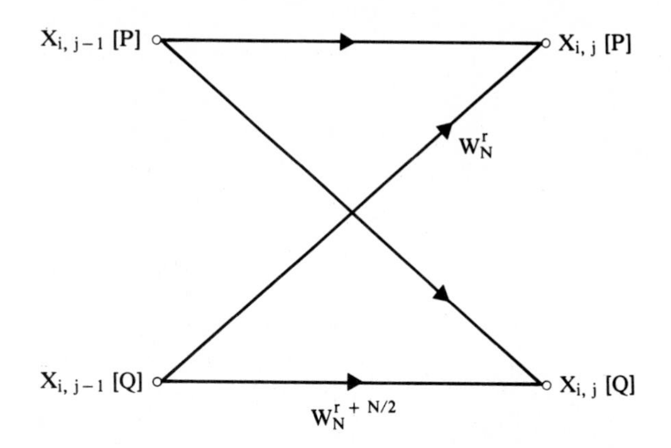

Figure 8-4 Butterfly Representing Equations (8.11) and (8.12)

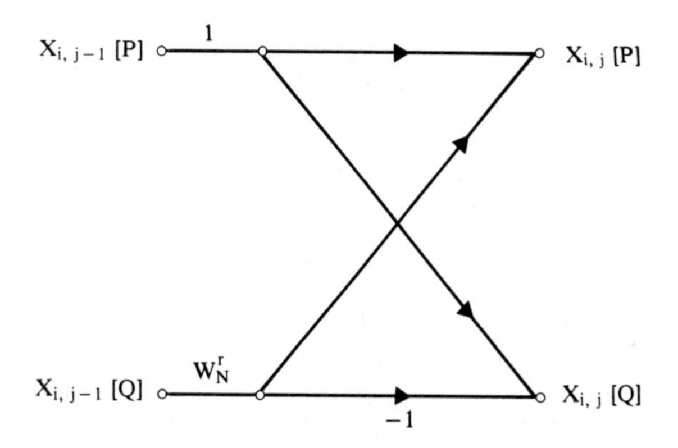

Figure 8-5 Simplified Butterfly Representing Equations (8.13) and (8.14)

8.1.2 "Bit-Reversed" Reordering of Input Data:

To perform the "in-place" computations described in Section 8.1.1, the elements of the input sequence x(n) must be reordered in a particular manner, and stored in consecutive RAM locations. To illustrate the reordering technique, consider each element of the input sequence x(n) to have an associated binary address, where the address of the first input element x(0) is called the base address L and where the m least significant bits of L are assumed to be zero. For an 8-point (eight element) input sequence

$$x(0), x(1), x(2), x(3), x(4), x(5), x(6), x(7)$$

the three least significant bits of the associated binary addresses are:

$$000, 001, 010, 011, 100, 101, 110, 111, \text{respectively.}$$

To reorder the input sequence, the m least significant bits of the address of each sequence element must be "bit-reversed" as shown in Table 8.1.

TABLE 8.1. Input Sequence in Bit-Reversed Order.

Input Sequence in Normal Order	*LSBs of Address*	*Bit-Reversed LSBs of Address*	*Desired Displacement from Base Address L*	*Input Sequence in Bit-Reversed Order*
x(0)	000	000	0	x(0)
x(1)	001	100	4	x(4)
x(2)	010	010	2	x(2)
x(3)	011	110	6	x(6)
x(4)	100	001	1	x(1)
x(5)	101	101	5	x(5)
x(6)	110	011	3	x(3)
x(7)	111	111	7	x(7)

8.2 RUNNING THE "DSP56000/1 DEMONSTRATION SOFTWARE":

To demonstrate the materials in the following sections, you can either:

1. type the commands that load and execute the FFT programs provided on the DSP56000/1 Demonstration Software #3 diskette, or
2. execute the file "Chapter8" provided on the DSP56000/1 Demonstration Software #3 diskette and follow the screen menus to load and execute the FFT programs also provided on the diskette.

To load and execute the file "Chapter8", insert the DSP56000/1 Demonstration Software #3 diskette into Drive "A" and type **a:chapter8<return>**.

8.3 FFT IMPLEMENTATION:

The following DSP56001 programs are used to implement the FFT on the DSP56001 processor:

1. A DSP56001 assembler macro to generate "twiddle" factors is developed in Section 8.3.1.
2. A DSP56001 assembler macro to perform bit-reversed reordering of the input sequence x(n) is developed in Section 8.3.2.
3. Three different procedures to implement an FFT butterfly on the DSP56001 processor are developed in Section 8.3.3.
4. Procedures to compute an FFT's group offset and compute an FFT's coefficient offset using the DSP56001 processor are developed in Section 8.3.4.
5. Three different DSP56001 assembler macros which each implement a complete N-point FFT are developed in Section 8.3.5. An 8-point FFT example is included.

8.3.1 Generating "Twiddle" Factors:

The "twiddle"" factor in Equation 8.2 can be rewritten as:

$$W_N^k = e^{-j2\pi k/N} = \cos(2\pi k/N) - j\sin(2\pi k/N)$$
$$\text{where } k = 0, 1, .., (N-1)/2. \quad (8.16)$$

The DSP56001 assembler macro **SINCOS.ASM** shown in Figure 8.6 generates lookup tables for the "twiddle" factors. The negative of the generated real-part values of Equation 8.16 (-cos(2πk/N)) are stored in X-Memory; the generated imaginary-part values (-sin(2πk/N)) are stored in Y-Memory. Each table contains one-half cycle of the -cos and sine waveform, respectively.

The sine and cosine transcendental functions built into the ASM56000 Assembler Software program are used by the assembler macro **SINCOS.ASM** to calculate the sine and cosine table values. Since both sine and cosine values can range from -1 to +1, and since the DSP56001 processor has a fractional-arithmetic range from -1 to $1\text{-}2^{-23}$ (short of +1), you may choose to avoid using cos(0) = +1 by using the negative of cos(0) instead.

However, if you do choose to use a cosine waveform instead of a negative cosine waveform, the following assembler warning message will appear on your screen when cos(0) is calculated:

WARNING - Floating point value outside fractional domain

Since the assembler will substitute the maximum positive fractional data value $7fffff for cos(0), and since negligible error is thereby introduced, this warning can be ignored.

```
;**********************************************************
;
; Sine-Cosine Table Generator for FFTs
;
;**********************************************************
;
;
sincos    macro      points,coef
;
;
;
;   points - number of points
;   coef   - base address of sine/cosine table
;          - negative cosine value in X memory
;          - negative sine value in Y memory
;
;
pi         equu       3.141592654
freq       equ        2.0*pi/@cvf(points)   ;freq=2pi/points

           org        x:coef          ; Calculation of
count      set        0               ; -cos(freq*count) for
           dup        points/2        ; count=0,.., (points/2)-1

           dc         -@cos(@cvf(count)*freq)
count      set        count+1
           endm
```

```
          org     y:coef          ; Calculation of
count     set     0               ; -sin(freq*count) for
          dup     points/2        ; count=0,..,(points/2)-1

          dc      -@sin(@cvf(count)*freq)
count     set     count+1
          endm

          endm                    ; end of sincos macro
```

Figure 8-6 SINCOS.ASM Assembler Macro for Generating Sine and Cosine Lookup Tables

---

EXAMPLE 8.1: Execute the Program "SINEX.LOD" to Generate Sine and Cosine Lookup Table Values for an 8-Point FFT.

In this example, the base addresses for the cosine and sine lookup tables in X-Memory and Y-Memory, respectively, are chosen to be equal to the FFT size N=8. Therefore, the table values of -cos($2\pi k/8$) for k=0,1,2,3 are in stored in X-Memory locations $8-$b; the table values of -sin ($2\pi k/8$) for k=0,1,2,3 are stored in Y-Memory locations $8-$b. The SINEX.ASM assembler program is shown in Figure 8.7.

1. LOAD THE "SIM56000 SIMULATOR SOFTWARE" PROGRAM.
2. INSERT THE "DSP56000/1 DEMONSTRATION SOFTWARE #3" DISKETTE INTO DRIVE "A".
3. TYPE THE SIM56000 COMMANDS:

```
display off<return>
load a:sinex<return>
step<return>
radix f x:$8..$b y:$8..$b<return>
display x:$8..$b<return>
display y:$8..$b<return>
```

4. TYPE **quit<return>** if you want to exit the SIM56000 program.

The values in the cosine and sine lookup tables after execution of the SINEX.LOD program are:

X:$0008 –1.0000000 –0.7071068 0.0000000 0.7071068

Y:$0008 0.0000000 –0.7071068 –1.0000000 –0.7071068

```
;
;   Twiddle factors for 8-point FFT
;
;

points        equ           $8
coef          equ           $8

;    points - number of points
;    coef   - base address of sine/cosine table
;    negative cosine value in X memory
;    negative sine value in Y memory
;
;

              include 'sincos'
              sincos            points,coef
              end
```

Figure 8-7 SINEX.ASM Assembler Program

8.3.2 Storing the Input Sequence X(n) in Bit-Reversed Order:

The DSP56001 processor's Address Generation Unit (AGU) has an addressing mode which automatically performs reverse-carry (bit-reversed) arithmetic. When using this addressing mode in conjunction with computing an FFT, the DSP56001 processor will automatically select the appropriate input (or output) sequence element, thereby enhancing the FFT's execution speed by making storing of the reordered input (or output) sequence unnecessary. For example, in Chapter 13 an FFT's input sequence is in normal order and its output sequence is in bit-reversed order. There, the Reverse Carry addressing capability is used to automatically address the stored output sequence elements in normal order for transfer to a D/A converter.

However, in order to fully illustrate the reverse-carry arithmetic concept as a subject in itself, the remainder of this section uses the DSP56001 processor's AGU to first reorder the input sequence x(n) and then store the reordered results. Reverse-carry arithmetic can be used to perform a bit-reversed modification of the m least significant bits of a memory address. This can be accomplished by adding an offset $N=2^{m-1}$ to the address value and then propagating any generated carry term in the reverse direction, e.g., the carry term propagates in the direction of the least significant bit.

The BITREV.ASM assembler macro shown in Figure 8.8 reorders an input sequence x(n) by performing bit-reversed address modifications as described above, and then stores the reordered results. The input sequence is assumed to be originally in normal order, with its real-part values stored in X-Memory locations L through L+N-1 and its imaginary-part values stored in Y-Memory locations L through L+N-1, where L is the base address of both. The m least significant bits of L must be zeros. After execution of the assembler macro, the reordered input sequence's real-part values are stored in X-Memory locations 0 through N-1 and its imaginary-part values are stored in Y-Memory locations 0 to N-1. Since the sine table is stored in X-Memory locations N through 3N/2-1 and the cosine table is stored in Y-Memory locations N through 3N/2-1, then L can be greater than or equal to 3N/2.

The BITREV.ASM assembler macro uses the DSP56001 processors ability to post-increment an address by an offset N with reverse-carry arithmetic, thereby performing a bit-reversed modification on the address. The base addresses L of the normally ordered elements of the input sequence are initially stored in AGU Registers R1 and R6, where R1 points to X-Memory and R6 points to Y-Memory; the base addresses of the reordered elements of the output sequence are initially stored in R0 and R5, where R0 points to X-Memory and R5 points to Y-Memory. R1 and R6 are each post-incremented by an offset N with reverse-carry arithmetic. Registers N1 and N6 are programmed with the offset value $N=2^{m-1}$; Registers M1 and M6 are programmed with the value $0000 to select the reverse-carry arithmetic mode. R0 and R5 are each post-incremented by one with linear arithmetic, where linear arithmetic is selected by programming M0 and M5 with the value $ffff. The assembler syntax itself defines the post-increment by one condition, consequently N0 and N5 need not be programmed. Table 8.2 shows the reverse-carry address modifications for input sequence elements 0.0,0.1,0.2,0.3,0.4,0.5,0.6,0.7, where N=8 and L=$18.

```
;
;*********************************************************
;  Storing  Input Data in Bit-Reversed Order
;
;*********************************************************

bitrev  macro   points,data,olddata

;
;     points  = Number of points
;     olddata = Input data in normal order
;     data    = Input data in bit-reversed order
;
;

          move      #olddata,r1      ; initialize the
          move      r1,r6            ; pointers
          move      #data,r0
          move      r0,r5

          move      #0,m1             ; m1 and m6 = bit
          move      m1,m6             ; reverse

          move      #-1,m0            ; m5 and m6 = linear
          move      m0,m5

          move      #points/2,n1
          move      n1,n6

          do        #points,_end_reverse
          move      x:(r1)+n1,a   y:(r6)+n6,b
          move      a,x:(r0)+     b,y:(r5)+

_end_reverse

          endm
```

Figure 8-8 BITREV.ASM Assembler Macro for Storing Input Sequence in Bit-Reversed Order

TABLE 8.2. Address Modifications for N=8 and L=$18.

Input Sequence in Normal Order	Address Modification	Input Sequence in Bit-Reversed Order
X:$18 0.0	00011000 = 18 + 1 --------	X:$0 0.0
X:$19 0.1	00011100 = 1C + 1 --------	X:$1 0.4
X:$1A 0.2	00011010 = 1A + 1 --------	X:$2 0.2
X:$1B 0.3	00011110 = 1E + 1 --------	X:$3 0.6
X:$1C 0.4	00011001 = 19 + 1 --------	X:$4 0.1
X:$1D 0.5	00011101 = 1D + 1 --------	X:$5 0.5
X:$1E 0.6	00011011 = 1B + 1 --------	X:$6 0.3
X:$1F 0.7	00011111 = 1F	x:$7 0.7

EXAMPLE 8.2: Execute the Program BITEX.LOD to Store an 8-Point Input Sequence in Bit-Reversed Order.

In this example, the original input sequence is in normal order with its real-part values stored in X-Memory locations $18-$1f and its imaginary-part values stored in Y-Memory locations $18-$1f. After execution of the program **BITEX.LOD**, input sequence is in bit-reversed order with its real-part values stored in X-Memory locations $0-$7 and its imaginary-part values stored in Y-Memory locations $0-$7. The BITEX.ASM assembler program is shown in Figure 8.9. Note that the it has an "include" statement for the **BITREV.ASM** assembler program shown in Figure 8.8.

1. LOAD THE "SIM56000 SIMULATOR SOFTWARE" PROGRAM.
2. INSERT THE "DSP56000/1 DEMONSTRATION SOFTWARE #3" DISKETTE INTO DRIVE "A".
3. TYPE THE SIM56000 COMMANDS:

```
display off<return>
load a:bitex<return>
break #1 pc>$10f<return>
go #1<return>
radix f x:$18..$1f<return>
display x:$18..$1f<return>
radix f y:$18..$1f<return>
display y:$18..$1f<return>
```

```
radix f x:$0..$7<return>
display x:$0..$7<return>
radix f y:$0..$7<return>
display y:$0..$7<return>
```

INPUT SEQUENCE IN NORMAL ORDER:

Real parts:

X:$0018 0.0000000 0.1000000 0.2000000 0.3000000

X:$001*C* 0.4000000 0.5000000 0.6000000 0.7000000

Imaginary Parts:

Y:$0018 0.0000000 0.1000000$0.2000000$0.3000000

Y:$001*C* 0.4000000 0.5000000 0.6000000 0.7000000

INPUT SEQUENCE IN BIT-REVERSED ORDER:

Real parts:

X:$0000 0.0000000 0.4000000 0.2000000 0.6000000

X:$0004 0.1000000 0.5000000 0.3000000 0.7000000

Imaginary parts:

Y:$0000 0.0000000 0.4000000 0.2000000 0.6000000

Y:$0004 0.1000000 0.5000000 0.3000000 0.7000000

4. TYPE **quit**<return> if you want to exit the SIM56000 program.

```
;
;*********************************************************
;
;  8-point Bit-Reversed  Example
;
;*********************************************************

         include 'bitrev'

;    Input Data
;
;    points  = Number of points
;    olddata = Input data in normal order
;    data    = Input data in bit-reversed order
;

data      equ      $0               ; Begin of data
points    equ      $8               ; FFT size
olddata   equ      $18
```

```
        org       x:$18                ; Begin of olddata
olddatar                               ; (real part)
        dc        0.0,0.1,0.2,0.3
        dc        0.4,0.5,0.6,0.7

        org       y:$18                ; Begin of olddata
olddataI                               ; (imaginary part)
        dc        0.0,0.1,0.2,0.3
        dc        0.4,0.5,0.6,0.7

; program start

        org       p:$100

        bitrev    points,data,olddata

        end
```

Figure 8-9 BITEX.ASM Assembler Program

8.3.3 Butterfly Computation:

Equations 8.13 and 8.14 are used to compute the butterfly shown in Figure 8.5. If the real and imaginary parts of $X_{i,j-1}[p]$, $X_{i,j-1}[q]$, $X_{i,j}[p]$, $X_{i,j}[q]$ and W_N^r are:

$$
\begin{aligned}
X_{i,j-1}[p] &= ar + jai \\
X_{i,j-1}[q] &= br + jbi \\
X_{i,j}[p] &= ar' + jai' \\
X_{i,j}[q] &= br' + jbi' \\
W_N^r &= wr + jwi
\end{aligned}
\tag{8.17}
$$

then the butterflies' outputs ar', ai', br', and bi' can be calculated from Equations 8.13, 8.14, and 8.17 as:

$$
\begin{aligned}
ar' &= ar + wr*br - wi*bi \\
ai' &= ai + wi*br + wr*bi \\
br' &= ar - wr*br + wi*bi = 2*ar - ar' \\
bi' &= ai - wi*br - wr*bi = 2*ai - ai'
\end{aligned}
\tag{8.18}
$$

where after computation of the butterfly, ar' and br' are stored in X-Memory and ai' and bi' are stored in Y-Memory.

Since the twiddle factor used in Equations 8.13 and 8.14 has the form $e^{-j2\pi r/N}$, which has unit magnitude, the magnitude of $X_{i,j}[p]$ and $X_{i,j}[q]$ can grow by a maximum factor of two per stage. For an N-point FFT, the magnitude can grow by a total factor of N. Consequently, the magnitude of $X_{i,j}[p]$, and $X_{i,j}[q]$, must be kept to less than one such that they can be stored without overflow or limiting. This can be accomplished by using one of the following three procedures:

1. Limit the magnitude of the input sequence elements to 1/N, where the magnitude is calculated as the square root of the sum of the squares of the real-part and imaginary-part values.
2. Automatically down-scale by a factor of 0.5 the outputs of all of the butterflies within the FFT. The magnitude of the FFT input data are less than one. A common scale-factor of N needs to be applied to the FFT results if the absolute (non-scaled) results are required. The DSP56001 processor can be selected to automatically down-scale data by a factor of 0.5. This is accomplished by programming bits 10 and 11 of its Status Register to 0 and 1, respectively.
3. Use a block floating point technique to accomplish conditional scaling. This technique allows complete FFT passes (stages) to be selectively scaled depending on the growth in magnitude which occurred in the previous pass. This technique can be implemented most efficiently by using the Scaling Bit S in the Status Register (bit 7) of full CMOS version DSP56001 processors. The Scaling Bit is set when the DSP56001 processor moves a result larger than 0.25 in absolute value to memory. The FFT algorithm tests the Scaling Bit upon completition of each FFT pass. If it is set, the DSP56001 processor's automatic scaling mode is turned on before the next pass. If the Scaling Bit is not set, no growth has occurred and the automatic scaling mode is turned off. If absolute (non-scaled) values are required, then a scaling factor needs to be calculated. The scaling factor is equal to 2^k, where k is the number of passes for which automatic down-scaling was turned on.

EXAMPLE 8.3: Limit to 1/N = 0.5 the Magnitude of the Input Values of a Single 2-Point Butterfly.

In this example, the **BUTER1.LOD** program is executed to compute the output values ar', ai', br', and bi' of a single 2-point butterfly. The real and imaginary parts of the input values are: ar = 0.4, ai = 0.25, br = 0.1, and bi = 0.15; the real and imaginary parts of the "twiddle" factor values are wr = -1 and wi = 0. The real-part input values ar and br are stored in X-Memory locations $0 and $1, respectively; the imaginary-part input values ai and bi are stored in Y-Memory locations $0 and $1, respectively. The real-part value of the "twiddle" factor is stored in X-Memory location $2; the imaginary-part value of the "twiddle" factor is stored in the Y-Memory location $2.

After the execution of the **BUTER1.LOD** program, the real-part output values ar' and br' are stored in X-Memory locations $0 and $1, respectively; the imaginary-part output values ai' and bi' are stored in Y-Memory locations $0 and $1. Note that since the output values overwrite the input values in both X-Memory and Y-Memory, the **BUTER1.LOD** program is an in-place butterfly algorithm. The BUTER1.ASM assembler program is shown in Figure 8.10.

1. LOAD THE "SIM56000 SIMULATOR SOFTWARE" PROGRAM.
2. INSERT THE "DSP56000/1 DEMONSTRATION SOFTWARE #3" DISKETTE INTO DRIVE "A".
3. TYPE THE SIM56000 COMMANDS:

```
change x:0 0.4 y:0 0.25<return>
change x:1 0.1 y:1 0.15<return>
change x:2 -'1.0 y:2 '0<return>
display off<return>
radix f x:0..1 y:0..1<return>
display x:0..1 y:0..1<return>
load a:buter1<return>
break #1 pc>$11d<return>
go<return>
display x:0..1 y:0..1<return>
```

4. TYPE **quit<return>** if you want to exit the SIM56000 program.

Butterfly inputs:

X:$0000 0.4000000 0.1000000

Y:$0000 0.2500000 0.1500000

Butterfly outputs:

X:$0000 0.5000000 0.3000000

Y:$0000 0.4000000 0.1000000

```
;*********************************************************
;   2-point FFT
;
;**********************************************************

passes    equ       $1
points    equ       $2
data      equ       $0
coef      equ       $2

          org       p:$100

          move      #1,n0             ; n0 = group offset
          move      #points/2,n2      ; n2 = group per pass

          move      #-1,m0            ; m0=m1=m2=m4=m5=m6
          move      m0,m1             ; = linear addressing
          move      m0,m2
          move      m0,m4
          move      m0,m5
          move      m0,m6


          move      n0,n1           ; n0 = n1 = n4 = n5
          move      n0,n4           ; group offset
          move      n0,n5
          move      n2,n6           ; n6 = coefficient offset

          move      #data,r0        ;r0 = ar,ai input pointer
          move      r0,r4           ;r4 = ar',ai'output pointer
          move      #coef,r6        ;r6 = wr,wi input pointer

          nop
          lua       (r0)+n0,r1      ;r1 = br,bi input pointer
          nop
          move      r1,r5           ;r5 = br',bi' output pointer
```

```
;***********************************************************
;
;   2-point butterfly
;
;***********************************************************

    move                            y:(r0),b   ;b  = ai

    move               x:(r1),x1   y:(r6),y0  ;x1 = br
                                              ;y0 = wi

    mac   x1,y0,b   x:(r6),x0   y:(r1),y1  ;b  = ai+br*wi
                                              ;x0 = -wr
                                              ;y1 = bi

    macr  -x0,y1,b                y:(r0),a   ;b  = ai+br*wi+bi*wr
                                              ;   = ai'
                                              ;a  = ai

    subl  b,a        x:(r0),b    b,y:(r4)   ;a  = 2ai-ai' = bi'
                                              ;b  = ar
                                              ;y(r4) = ai'

    mac   -x0,x1,b  x:(r0),a    a,y:(r5)   ;b  = ar+br*wr
                                              ;a  = ar
                                              ;y(r5) = bi'

    macr  -y0,y1,b                          ;b = ar+br*wr-bi*wi
                                              ;  = ar'

    subl  b,a        b,x:(r4)               ;a = 2ar-ar'= br'
                                              ;x:(r4) = ar'

    move             a,x:(r5)               ;x:(r5) = br'

    end
```

Figure 8-10 BUTER1.ASM Assembler Program

EXAMPLE 8.4: Automatically Down-Scale by a Factor of 0.5 the Outputs of a single 2-Point Butterfly.

In this example, the BUTER2.LOD program is executed to compute the output values ar', ai', br', and bi' of a 2-point butterfly, where the magnitudes of its input values are limited to be less than 1.0 and its output values are halved. The real and imaginary parts of the input values are: ar = 0.8, ai = 0.5, br = 0.2, and bi = 0.3; the real and imaginary parts of the "twiddle" factor values are wr = -1 and wi = 0. The real-part input values ar and br are stored in X-Memory locations $0 and $1, respectively; the imaginary-part input values ai and bi are stored in Y-Memory locations $0 and $1, respectively. The real-part value of the "twiddle" factor is stored in X-Memory location $2; the imaginary-part value of the "twiddle" factor is stored in the Y-Memory location $2.

After the execution of the **BUTER2.LOD** program, the real-part output values ar' and br' are stored in X-Memory locations $0 and $1, respectively; the imaginary-part output values ai' and bi' are stored in Y-Memory locations $0 and $1. Note that since the output values overwrite the input values in both X-Memory and Y-Memory, the **BUTER2.LOD** program is an in-place butterfly algorithm. The **BUTER2.ASM** program is shown in Figure 8.11. Note that use of the ORI instruction enables the down-scaling mode and use of the ANDI instruction disables the down-scaling mode.

1. LOAD THE "SIM56000 SIMULATOR SOFTWARE" PROGRAM.
2. INSERT THE "DSP56000/1 DEMONSTRATION SOFTWARE #3" DISKETTE INTO DRIVE "A".
3. TYPE THE SIM56000 COMMANDS:

```
change x:0 '0.8 y:0 '0.5<return>
change x:1 '0.2 y:1 '0.3<return>
change x:2 -'1.0 y:2 '0<return>
display off<return>
radix f x:0..1 y:0..1<return>
display x:0..1 y:0..1<return>
load a:buter2<return>
break #1 pc>$11f<return>
go<return>
display x:0..1 y:0..1<return>
```

4. TYPE **quit<return>** if you want to exit the SIM56000 program.

Butterfly inputs:

X:$0000 0.8000000 0.2000000

Y:$0000 0.5000000 0.3000000

Butterfly outputs:

X:$0000 0.5000000 0.3000000

Y:$0000 0.4000000 0.1000000

```
;**********************************************************
;    2-point FFT
;
;**********************************************************

passes     equ       $1
points     equ       $2
data       equ       $0
coef       equ       $2

           org       p:$100

           move      #1,n0             ; n0 = group offset
           move      #points/2,n2      ; n2 = group per pass
```

```
            move      #-1,m0          ; m0=m1=m2=m4=m5=m6
            move      m0,m1           ; = linear addressing
            move      m0,m2
            move      m0,m4
            move      m0,m5
            move      m0,m6


            move      n0,n1        ; n0 = n1 = n4 = n5
            move      n0,n4        ; group offset
            move      n0,n5
            move      n2,n6        ; n6 = coefficient offset

            move      #data,r0     ;r0 = ar,ai input pointer
            move      r0,r4        ;r4 = ar',ai'output pointer
            move      #coef,r6     ;r6 = wr,wi input pointer

            nop
            lua       (r0)+n0,r1   ;r1 = br,bi input pointer
            nop
            move      r1,r5        ;r5 = br',bi' input pointer

;************************************************************
;
;   2-point butterfly
;
;************************************************************

    ori   #$4,mr

    move                      y:(r0),b   ;b  = ai

    move            x:(r1),x1 y:(r6),y0  ;x1 = br
                                         ;y0 = wi

    mac   x1,y0,b   x:(r6),x0 y:(r1),y1  ;b  = ai+br*wi
                                         ;x0 = -wr
                                         ;y1 = bi

    macr  -x0,y1,b            y:(r0),a   ;b  = ai+br*wi+bi*wr
                                         ;   = ai'
                                         ;a  = ai

    subl  b,a       x:(r0),b  b,y:(r4)   ;a  = 2ai-ai' = bi'
                                         ;b  = ar
                                         ;y(r4) = ai'

    mac   -x0,x1,b  x:(r0),a  a,y:(r5)   ;b  = ar+br*wr
                                         ;a  = ar
                                         ;y(r5) = bi'

    macr  -y0,y1,b                       ;b = ar+br*wr-bi*wi
                                         ;  = ar'

    subl  b,a       b,x:(r4)             ;a = 2ar-ar'= br'
                                         ;x:(r4) = ar'

    move            a,x:(r5)             ;x:(r5) = br'

    andi  #$fb,mr

    end
```

Figure 8-11 Buter2.ASM Assembler Program

EXAMPLE 8.5: Use a Block Floating Point Technique to Perform Conditional Scaling of the Output Values of a Single 2-Point Butterfly

In this example, the **BUTER3.LOD** program is executed to compute the butterfly outputs ar', ai', br'and bi' of a 2-point butterfly, where the magnitudes of its input values are limited to 1/N = 0.5 and the block floating point technique is used to accomplish conditional scaling of its output values. The real and imaginary parts of the input values are: ar = 0.8, ai = 0.5, br = 0.2, and bi = 0.3; the real and imaginary parts of the "twiddle" factor values are wr = -1 and wi = 0. The real-part input values ar and br are stored in X-Memory locations $0 and $1, respectively; the imaginary-part input values ai and bi are stored in Y-Memory locations $0 and $1, respectively. The real-part value of the "twiddle" factor is stored in X-Memory location $2; the imaginary-part value of the "twiddle" factor is stored in the Y-Memory location $2.
The **BUTER3.LOD** program is executed two times: the first time with the Scaling Bit set to down-scale the butterfly output values by a factor of 0.5, and the second time with the Scaling Bit cleared to obtain the butterfly output values without scaling. After execution of the **BUTER3.LOD** program, the real-part output values ar' and br' are stored in X-Memory locations $0 and $1, respectively; the imaginary-part output values ai' and bi' are stored in Y-Memory locations $0 and $1. Note that since the output values overwrite the input values in both X-Memory and Y-Memory, the **BUTER3.LOD** program is an in-place butterfly algorithm. The **BUTER3.ASM** program is shown in Figure 8.12.

1. LOAD THE "SIM56000 SIMULATOR SOFTWARE" PROGRAM. Note the version of the SIM56000 program used must simulate the full CMOS version of the DSP56001 processor.
2. INSERT THE "DSP56000/1 DEMONSTRATION SOFTWARE #3" DISKETTE INTO DRIVE "A".
3. TYPE THE SIM56000 COMMANDS:
   a) *Scaling bit is set*:

```
change sr $80<return>
change x:0 '0.4 y:0 '0.25<return>
change x:1 '0.1 y:1 '0.15<return>
change x:2 -'1.0 y:2 '0<return>
display off<return>
radix f x:0..1 y:0..1<return>
display x:0..1 y:0..1<return>
load a:buter3<return>
break #1 pc>$127<return>
go<return>
display r7<return>
display x:0..1 y:0..1<return>
```

Butterfly inputs:

X:$0000 0.4000000 0.1000000

Y:$0000 0.2500000 0.1500000

Scaling factor: 2^k, where k is equal to 1.

r7 = $0001

Butterfly outputs:

X:$0000 0.2500000 0.1500000

Y:$0000 0.2000000 0.0500000

b) *Scaling bit is cleared*:

```
change sr $00<return>
change x:0 '0.4 y:0 '0.25<return>
change x:1 '0.1 y:1 '0.15<return>
change x:2 -'1.0 y:2 '0<return>
display off<return>
radix f x:0..1 y:0..1<return>
display x:0..1 y:0..1<return>
load a:buter3<return>
break #1 pc>$127<return>
go<return>
display r7<return>
display x:0..1 y:0..1<return>
```

Butterfly inputs:

X:$0000 0.4000000 0.1000000

Y:$0000 0.2500000 0.1500000

Scaling factor: 2^k, where k is equal to 0.

r7 = $0000

Butterfly outputs:

X:$0000 0.5000000 0.3000000

Y:$0000 0.4000000 0.1000000

4. TYPE **quit<return>** if you want to exit the SIM56000 program.

```
;***************************************************************
;    2-point FFT
;
;***************************************************************

passes     equ      $1
points     equ      $2
data       equ      $0
coef       equ      $2
accr       equ      $3

           org      p:$100

           move     #1,n0            ; n0 = group offset
           move     #points/2,n2     ; n2 = group per pass
```

```
        move      #-1,m0          ; m0=m1=m2=m4=m5=m6
        move      m0,m1           ; = linear addressing
        move      m0,m2
        move      m0,m4
        move      m0,m5
        move      m0,m6

        move      #accr,r3
        move      n0,n1         ; n0 = n1 = n4 = n5
        move      n0,n4         ; group offset
        move      n0,n5
        move      n2,n6         ; n6 = coefficient offset

        move      #data,r0      ;r0 = ar,ai input pointer
        move      r0,r4         ;r4 = ar',ai'output pointer
        move      #coef,r6      ;r6 = wr,wi input pointer

        nop
        lua       (r0)+n0,r1    ;r1 = br,bi input pointer
        nop
        move      r1,r5         ;r5 = br',bi' input pointer


        move      #$0,r7

;*********************************************************
;
;   2-point butterfly
;
;**********************************************************

   move  sr,x:(r3)
   jclr  #7,x:(r3),noscale
   ori   #$4,mr
   move              (r7)+
noscale
   move                        y:(r0),b  ;b  = ai

   andi  #$7f,ccr

   move            x:(r1),x1   y:(r6),y0 ;x1 = br
                                         ;y0 = wi
   mac   x1,y0,b   x:(r6),x0   y:(r1),y1 ;b  = ai+br*wi
                                         ;x0 = -wr
                                         ;y1 = bi
   macr -x0,y1,b               y:(r0),a  ;b  = ai+br*wi+bi*wr
                                         ;   = ai'
                                         ;a  = ai
   subl b,a        x:(r0),b    b,y:(r4)  ;a  = 2ai-ai' = bi'
                                         ;b  = ar
                                         ;y(r4) = ai'
   mac -x0,x1,b    x:(r0),a    a,y:(r5)  ;b  = ar+br*wr
                                         ;a  = ar
                                         ;y(r5) = bi'
   macr -y0,y1,b                         ;b = ar+br*wr-bi*wi
                                         ;  = ar'
   subl b,a        b,x:(r4)              ;a = 2ar-ar'= br'
                                         ;x:(r4) = ar'
   move            a,x:(r5)              ;x:(r5) = br'

   end
```

Figure 8-12 Buter3.ASM Assembler Program

8.3.4 Group And Coefficient Offsets:

The **PASSEX.ASM** assembler program shown in Figure 8.13 computes the group offset n_0 and the coefficient offset n_2, where n_0 and n_2 are defined by Equation 8.15.

EXAMPLE 8.6:

In this example, the **PASSEX.LOD** program is used to compute the group and coefficient offsets for the three stages of an 8-point FFT. The **PASSEX.ASM** assembler program is shown in Figure 8.13.

1. LOAD THE "SIM56000 SIMULATOR SOFTWARE" PROGRAM.
2. INSERT THE "DSP56000/1 DEMONSTRATION SOFTWARE #3" DISKETTE INTO DRIVE "A".
3. TYPE THE SIM56000 COMMANDS:

```
display off<return>
display n0 n2<return>
load a:passex<return>
step 2<return>
display n0 n2<return>
step 7<return>
display n0 n2<return>
step 4<return>
display n0 n2<return>
```

4. TYPE **quit<return>** if you want to exit the SIM56000 program.

The value of the group offset n_0 and the coefficient offset n_2 for each of the three stages of the FFT are:

First stage	$n_0 = 1$	$n_2 = 4$
Second Stage	$n_0 = 2$	$n_2 = 2$
Third Stage	$n_0 = 4$	$n_2 = 1$

```
;
;
;    passes = number of passes
;    points = FFT size
;

passes      equu        $3
points      equ         $8

            move        #1,n0               ; n0 = 1,2,4
            move        #points/2,n2        ; n2 = 4,2,1

            move        #-1,m0              ; m0=m1= linear
            move        m0,m2
```

```
            do      #passes,_end_pass

            move    n2,b1
            lsr     b           n0,a1
            lsl     a           b1,n2
            move    a1,n0

_end_pass

            end
```

Figure 8-13 PASSEX.ASM Assembler Program

8.3.5 FFT Macros:

The FFT macros **FFT1.ASM, FFT2.ASM** and **FFT3.ASM** can each perform complete FFTs on a complex input sequence. The three macros are shown in Figures 8.14, 8.15 and 8.16, respectively. The macros can be called with the following arguments:

1. number of FFT points,
2. location of the normally ordered or bit-reverse ordered input sequence, and
3. location of the sine-cosine table.

All DSP56001 register initialization is performed within the macros, and the output sequence is in normal order. In the **FFT1** macro, the elements of the input sequence are assumed to be limited to a magnitude of 1/N. In the **FFT2** macro, the outputs of all of the FFT butterflies are down-scaled by a factor of two. In the **FFT3** macro, the elements of the input sequence are limited to a magnitude of 0.5, the block floating point technique is used to accomplish conditional scaling, and a memory location is used to store the contents of the DSP56001 processor's Status Register.

```
;*********************************************************
;
;    Decimation in Time  FFT
;
;*********************************************************

fftl    macro   points,data,coef,olddata

;   passes  =   number of passes = log2 ( number of points)
;   points  =   number of points
;   coef    =   base address of sine/cosine table
;   olddata =   Starting address of input data before bit
;               reverse
;   data    =   Starting address of input data after bit
;               reverse

;
;
;**********************************************************
;
;  FFT Initialization
;
;**********************************************************

          move      #1,n0           ; n0 = group offset
          move      #points/2,n2    ; n2 = group per pass
```

```
          move      #-1,m0          ; m0=m1=m2=m4=m5=m6
          move      m0,m1           ; = linear addressing
          move      m0,m2
          move      m0,m4
          move      m0,m5
          move      m0,m6

;**********************************************************
;
;         FFT passes
;
;**********************************************************

          do        #@cvi(@log(points)/@log(2)+0.5),_end_pass

          move      #data,r0     ;r0 = ar,ai input pointer
          move      r0,r4        ;r4 = ar',ai'output pointer
          move      #coef,r6     ;r6 = wr,wi input pointer

          nop
          lua       (r0)+n0,r1   ;r1 = br,bi input pointer
          nop
          lua       (r1)-,r5     ; r5 = br',bi' input pointer


          move      n0,n1        ; n0 = n1 = n4 = n5
          move      n0,n4        ; group offset
          move      n0,n5
          move      n2,n6        ; n6 = coefficient offset

;**********************************************************
;
;     FFT Group
;
;**********************************************************

     do     n2,_end_grp

     move                 x:(r5),a        y:(r0),b

;**********************************************************
;
;     FFT Butterfly
;
;**********************************************************

     do     n0,_end_bfy

     move                x:(r1),x1       y:(r6),y0
     mac       x1,y0,b   x:(r6)+n6,x0    y:(r1)+,y1
     macr      -x0,y1,b  a,x:(r5)+       y:(r0),a
     subl      b,a       x:(r0),b        b,y:(r4)
     mac       -x0,x1,b  x:(r0)+,a       a,y:(r5)
     macr      -y0,y1,b  x:(r1),x1
     subl      b,a       b,x:(r4)+       y:(r0),b

_end_bfy

     move    #coef,r6
     move                 a,x:(r5)+n5      y:(r1)+n1,y1
     move                 x:(r0)+n0,x1     y:(r4)+n4,y1

_end_grp


     move     n2,b1
     lsr      b              n0,a1
     lsl      a              b1,n2
     move     a1,n0
```

```
_end_pass

     endm
```

Figure 8-14 FFT1.ASM Assembler Macro

```
;**********************************************************
;
;   Decimation in Time  FFT
;
;*********************************************************

fft2     macro     points,data,coef,olddata

;   passes  =  number of passes = log2 ( number of points)
;   points  =  number of points
;   coef    =  base address of sine/cosine table
;   olddata =  Starting address of input data before bit
;              reverse
;   data    =  Starting address of input data after bit
;              reverse

;
;
;**********************************************************
;
;  FFT Initialization
;
;**********************************************************

          move      #1,n0           ; n0 = group offset
          move      #points/2,n2    ; n2 = group per pass

          move      #-1,m0          ; m0=m1=m2=m4=m5=m6
          move      m0,m1           ; = linear addressing
          move      m0,m2
          move      m0,m4
          move      m0,m5
          move      m0,m6


;*********************************************************
;
;       FFT passes
;
;*********************************************************

          do     #@cvi(@log(points)/@log(2)+0.5),_end_pass

          move     #data,r0     ;r0 = ar,ai input pointer
          move     r0,r4        ;r4 = ar',ai'output pointer
          move     #coef,r6     ;r6 = wr,wi input pointer

          nop
          lua      (r0)+n0,r1   ;r1 = br,bi input pointer
          nop
          lua      (r1)-,r5     ; r5 = br',bi' input pointer


          move     n0,n1        ; n0 = n1 = n4 = n5
          move     n0,n4        ; group offset
          move     n0,n5
          move     n2,n6        ; n6 = coefficient offset
```

```
;*********************************************************
;
;    FFT Group
;
;*********************************************************

     ori       #$4,mr

     do       n2,_end_grp

     move                    x:(r5),a      y:(r0),b
     lsl     a

;*********************************************************
;
;    FFT Butterfly
;
;*********************************************************

     do       n0,_end_bfy

     move                   x:(r1),x1       y:(r6),y0
     mac      x1,y0,b       x:(r6)+n6,x0    y:(r1)+,y1
     macr     -x0,y1,b      a,x:(r5)+       y:(r0),a
     subl     b,a           x:(r0),b        b,y:(r4)
     mac      -x0,x1,b      x:(r0)+,a       a,y:(r5)
     macr     -y0,y1,b      x:(r1),x1
     subl     b,a           b,x:(r4)+       y:(r0),b

_end_bfy

     move     #coef,r6
     move                   a,x:(r5)+n5      y:(r1)+n1,y1
     move                   x:(r0)+n0,x1     y:(r4)+n4,y1

_end_grp

     andi      #$fb,mr

     move                    n2,b1
     lsr       b             n0,a1
     lsl       a             b1,n2
     move      a1,n0

_end_pass

     endm
```

Figure 8-15 FFT2.ASM Assembler Macro

```
;*********************************************************
;
;   Decimation in Time  FFT
;
;*********************************************************

fft3     macro      points,data,coef,olddata,accr

;   passes  =  number of passes = log2 ( number of points)
;   points  =  number of points
;   coef    =  base address of sine/cosine table
;   olddata =  Starting address of input data before bit
;              reverse
```

```
;   data     =  starting address of input data after bit
;               reverse
;   accr     =  address used to store the value of the ccr
;               register
;
;*********************************************************
;
;  FFT Initialization
;
;*********************************************************

          move      #0,r7
          move      #1,n0           ; n0 = group offset
          move      #points/2,n2    ; n2 = group per pass

          move      #-1,m0          ; m0=m1=m2=m4=m5=m6
          move      m0,m1           ; = linear addressing
          move      m0,m2
          move      m0,m4
          move      m0,m5
          move      m0,m6
          move      m0,m7

;*********************************************************
;
;       FFT passes
;
;*********************************************************

          do      #@cvi(@log(points)/@log(2)+0.5),_end_pass

          move      #data,r0     ;r0 = ar,ai input pointer
          move      r0,r4        ;r4 = ar',ai'output pointer
          move      #coef,r6     ;r6 = wr,wi input pointer

          nop
          lua       (r0)+n0,r1   ;r1 = br,bi input pointer
          nop
          lua       (r1)-,r5     ; r5 = br',bi' input pointer

          move      #accr,r3

          move      n0,n1        ; n0 = n1 = n4 = n5
          move      n0,n4        ; group offset
          move      n0,n5
          move      n2,n6        ; n6 = coefficient offset

;*********************************************************
;
;    FFT Group
;
;*********************************************************

      move  sr,x:(r3)
      jclr  #7,x:(r3),noscale
      ori   #$4,mr
      move              (r7)+
noscale

      do     n2,_end_grp
      move                  x:(r5),a       y:(r0),b

      jclr  #7,x:(r3),Continue
      lsl   a
Continue
      andi  #$7f,ccr
              ;*********************************************************
;
;    FFT Butterfly
```

```
;
;*********************************************************

        do      n0,_end_bfy

        move                x:(r1),x1       y:(r6),y0
        mac     x1,y0,b     x:(r6)+n6,x0    y:(r1)+,y1
        macr    -x0,y1,b    a,x:(r5)+       y:(r0),a
        subl    b,a         x:(r0),b        b,y:(r4)
        mac     -x0,x1,b    x:(r0)+,a       a,y:(r5)
        macr    -y0,y1,b    x:(r1),x1
        subl    b,a         b,x:(r4)+       y:(r0),b

_end_bfy

        move    #coef,r6
        move                a,x:(r5)+n5      y:(r1)+n1,y1
        move                x:(r0)+n0,x1     y:(r4)+n4,y1

_end_grp

        andi     #$fb,mr

        move     n2,b1
        lsr      b           n0,a1
        lsl      a           b1,n2
        move     a1,n0

_end_pass

        endm
```

Figure 8-16 FFT3.ASM Assembler Macro

EXAMPLE 8.7: Execute the FFTEX1.LOD, FFTEX2.LOD, and FFTEX3.LOD Programs, Where Each Program Uses a Different Approach to Perform an 8-point FFT on a Complex Input Sequence.

In the **FFTEX1.LOD**, **FFTEX2.LOD**, and **FFTEX3.LOD** programs, the real-part elements of the input sequence are stored in X-Memory locations \$18-\$1f and the imaginary-part elements are stored in Y-Memory locations \$18-\$1f. Each program calls the **SINCOS** macro to generate sine and cosine lookup tables, where the cosine table values are stored in X-Memory locations \$8-\$b and the sine table values are stored in Y-Memory locations \$8-\$b. Each program also calls the **BITREV** macro to store the input sequence in bit-reversed order, with the real-part bit-reverse ordered elements stored in X-Memory locations \$0-\$7 and the imaginary-part bit-reverse ordered elements stored in Y-Memory locations \$0-\$7. The FFT1, FFT2 and FFT3 macros are called by the FFTEX1.ASM, FFTEX2.ASM and FFTEX3.ASM assembler programs, respectively, to perform an 8-point FFT on the bit-reverse ordered input sequence, where the real-part elements of the normally ordered output sequence are stored in X-Memory locations \$0-\$7 and the imaginary-part elements are stored in Y-Memory locations \$0-\$7.

In the FFTEX1.LOD program, the elements of the input sequence are limited to a magnitude of 0.125. The FFTEX1.ASM assembler program is shown in Figure 8.17.

1. LOAD THE "SIM56000 SIMULATOR SOFTWARE" PROGRAM.
2. INSERT THE "DSP56000/1 DEMONSTRATION SOFTWARE #3" DISKETTE INTO DRIVE "A".
3. TYPE THE SIM56000 COMMANDS:

```
display off<return>
load a:fftex1<return>
break #1 pc>$138<return>
go #1<return>
radix f x:$18..$1f<return>
display x:$18..$1f<return>
radix f y:$18..$1f<return>
display y:$18..$1f<return>
radix f x:$0..$7<return>
display x:$0..$7<return>
radix f y:$0..$7<return>
display y:$0..$7<return>
```

4. TYPE **quit<return>** if you want to exit the SIM56000 program.

Real – part elements of the input sequence:

X:$0018 0.1000000 0.1000000 0.0500000 0.1000000

X:$001*C* 0.1000000 0.0500000 0.1000000 0.1000000

Imaginary – part elements of the input sequence:

Y:$0018 0.0000000 0.0000000 0.0000000 0.0000000

Y:$001*C* 0.0000000 0.0000000 0.0000000 0.0000000

Real – part elements of FFT output sequence:

X:$0000 0.7000000 0.0353554 0.0500001 –0.0353554

X:$0004 0.0000000 –0.0353554 0.0500001 0.0353554

Imaginary – part elements of the FFT output sequence:

Y:$0000 0.0000000 0.0146446 0.0500001 –0.0853555

Y:$0004 0.0000000 0.0853555 –0.0500001 –0.0146446

```
;***********************************************************
;    8-point FFT
;
;***********************************************************

        include 'sincos'
        include 'bitrev'
        include 'fftl'
```

```
reset     equ       0
start     equ       100
points    equ       8
data      equ       0
coef      equ       8
olddata   equ       $18

;   points  =  number of points
;   coef    =  base address of sine/cosine table
;   olddata =  Starting address of input data before bit
;              reverse
;   data    =  Starting address of input data after bit
;              reverse

          org       x:$18                ; Begin of old data
olddatar                                 ; (real part)
          dc        0.1,0.1,0.05,0.1
          dc        0.1,0.05,0.1,0.1

          org       y:$18                ; Begin of old data
olddataI                                 ; (imaginary part)
          dc        0,0,0,0
          dc        0,0,0,0

          sincos    points,coef

          org      p:reset
          jmp      start
          org      p:start

          bitrev   points,data,olddata
          nop

          fftl     points,data,coef,olddata
          end
```

Figure 8-17 FFTEX1.ASM Assembler Program

In the FFTEX2.LOD program, the outputs of all of the FFT butterflies are down-scaled by a factor of two. When absolute (non-scaled) results are required, a common scale-factor of 2^3 is used. The FFTEX2.ASM assembler program is shown in Figure 8.18.

1. LOAD THE "SIM56000 SIMULATOR SOFTWARE" PROGRAM.
2. INSERT THE "DSP56000/1 DEMONSTRATION SOFTWARE #3" DISKETTE INTO DRIVE "A".
3. TYPE THE SIM56000 COMMANDS:

```
display off<return>
load a:fftex2<return>
break #1 pc>$138<return>
go #1<return>
radix f x:$18..$1f<return>
display x:$18..$1f<return>
radix f y:$18..$1f<return>
display y:$18..$1f<return>
radix f x:$0..$7<return>
display x:$0..$7<return>
radix f y:$0..$7<return>
display y:$0..$7<return>
```

4. TYPE **quit<return>** if you want to exit the SIM56000 program.

Real – part elements of the input sequence:

X:\$0018 0.8000000 0.8000000 0.4000000 0.8000000

X:\$001C 0.8000000 0.4000000 0.8000000 0.8000000

Imaginary – part elements of the input sequence:

Y:\$0018 0.0000000 0.0000000 0.0000000 0.0000000

Y:\$001C 0.0000000 0.0000000 0.0000000 0.0000000

Real – part elements of the FFT output sequence:

X:\$0000 0.6999999 0.0353553 0.0500000 –0.0353553

X:\$0004 0.0000000 –0.0353553 0.0500001 0.0353553

Imaginary – part elements of the FFT output sequence:

Y:\$0000 0.0000000 0.0146446 0.0500000 –0.0853553

Y:\$0004 0.0000000 0.0853553 –0.0500000 –0.0146446

```
;*********************************************************
;   8-point FFT
;
;*********************************************************

          include 'sincos'
          include 'bitrev'
          include 'fft2'

reset     equu       0
start     equ        100
points    equ        8
data      equ        0
coef      equ        8
olddata   equ        $18

;   points  =  number of points
;   coef    =  base address of sine/cosine table
;   olddata =  Starting address of input data before bit
;              reverse
;   data    =  Starting address of input data after bit
;              reverse

          org        x:$18             ; Begin of old data
olddatar                               ; (real part)
          dc         0.8,0.8,0.4,0.8
          dc         0.8,0.4,0.8,0.8

          org        y:$18             ; Begin of old data
olddataI                               ; (imaginary part)
          dc         0,0,0,0
          dc         0,0,0,0

          sincos     points,coef

          org        p:reset
          jmp        start
          org        p:start

          bitrev     points,data,olddata
```

```
        nop

        fft2      points,data,coef,olddata
        end
```

Figure 8-18 FFTEX2.ASM Assembler Program

In the **FFTEX3.LOD** program, the block floating point technique is used to accomplish conditional scaling. In this case, the scale-factor is equal to 2^K, where k is the number of passes for which down-scaling is enabled. For N =3, K varies from 0 to 2. The **FFTEX3.ASM** assembler program is shown in Figure 8.19.

1. LOAD THE "SIM56000 SIMULATOR SOFTWARE" PROGRAM.
2. INSERT THE "DSP56000/1 DEMONSTRATION SOFTWARE #3" DISKETTE INTO DRIVE "A".
3. TYPE THE SIM56000 COMMANDS:

```
display off<return>
load a:fftex3<return>
break #1 pc>$218<return>
go #1<return>
radix f x:$18..$1f<return>
display x:$18..$1f<return>
radix f y:$18..$1f<return>
display y:$18..$1f<return>
display r7<return>
radix f x:$0..$7<return>
display x:$0..$7<return>
radix f y:$0..$7<return>
display y:$0..$7<return>
```

4. TYPE **quit<return>** if you want to exit the SIM56000 program.

Real − part elements of the input sequence:

X:$0018 0.2000000 0.2000000 0.1000000 0.2000000

X:$001*C* 0.2000000 0.1000000 0.2000000 0.2000000

Imaginary − part elements of the input sequence:

0.0000000 0.0000000 0.0000000 0.0000000

Y:$001*C* 0.0000000 0.0000000 0.0000000 0.0000000

Scaling factor: 2^k, where *k* is equal to 2.

r7 = $0002

Real − part elements of the FFT output sequence:

X:$0000 0.3500001 0.0176777 0.0250000 −0.0176777

X:$0004 0.0000000 −0.0176777 0.0250000 0.0176777

Imaginary – part elements of the FFT output sequence:

Y:\$0000 0.0000000 0.0073223 0.0250001 –0.0426776

Y:\$0004 0.0000000 0.0426776 –0.0250001 –0.0073223

```
;********************************************************
;
;   8-point FFT
;
;********************************************************

         include 'sincos'
         include 'bitrev'
         include 'fft3'

reset    equ      0
start    equ      100
points   equ      8
data     equ      0
coef     equ      8
olddata  equ      $18
accr     equ      $20

;   points  =  number of points
;   coef    =  base address of sine/cosine table
;   olddata =  Starting address of input data before bit
;              reverse
;   data    =  Starting address of input data after bit
;              reverse
;   accr     = address for the ccr register

         org      x:$18               ; Begin of old data
olddatar                              ; (real part)
         dc       0.2,0.2,0.1,0.2
         dc       0.2,0.1,0.2,0.2

         org      y:$18               ; Begin of old data
olddataI                              ; (imaginary part)
         dc       0,0,0,0
         dc       0,0,0,0

         sincos   points,coef

         org     p:reset
         jmp     start
         org     p:start

         bitrev  points,data,olddata
         nop

         fft3    points,data,coef,olddata,accr
         end
```

Figure 8-19 FFTEX3.ASM Assembler Program

8.4 REFERENCES:

1. Alan V. Oppenheim and Ronald W. Schafer, Discrete-Time Signal Processing, Prentice Hall, 1989.

2. Johnny R. Johnson, Introduction to Digital Signal Processing, Prentice Hall, 1989.
3. Richard A. Roberts and Clifford T. Mullis, Digital Signal Processing, Addison Wesley, 1987.
4. Chi-Tsong Chen, One-Dimensional Digital Signal Processing, Dekker, 1979.
5. Roman Kuc, Introduction to Digital Signal Processing, McGraw-Hill,1988.
6. John G. Proakis and Dimitris G. Manolakis, "Introduction to Digital Signal Processing, Mcmillan 1988.

9

DESIGNING ADAPTIVE FIR FILTERS AND IMPLEMENTING THEM ON THE DSP56001 PROCESSOR

In this chapter, finite impulse response adaptive filters are designed, and implemented on the DSP56001 signal processor. Adaptive filters have the ability to adjust their own parameters (coefficients) automatically. Hence, their design requires little or no a priori knowledge of the input signal or noise characteristics of the system. Adaptive filters have two inputs, x(n) and d(n), which are usually correlated in some manner. For the adaptive filter shown in Figure 9.1:

1. the filter's output y(n), which is computed with the current parameter estimates, is compared with the input signal d(n);
2. the resulting prediction error e(n) is feedback through a parameter adaptation algorithm which will produce a new estimate for the parameters and,
3. as the next input sample is received, a new prediction error will be generated, etc. The goal of the adaptive filter is the minimization of the prediction error.

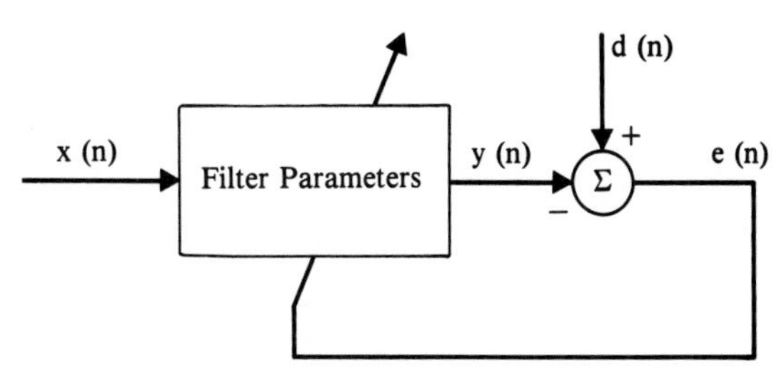

Figure 9-1 Adaptive Filter

Two aspects of adaptive filters are its internal structure and its adaptation algorithm. Its internal structure can be either that of a nonrecursive (FIR) filter or that of a recursive (IIR) filter. Its adaptation algorithm can be divided into two major classes: gradient algorithms and nongradient algorithms. In this chapter, a gradient algorithm is used to adjust the parameters of a FIR filter. In Chapter 10, a nongradient Supermartingale algorithm is used to adjust the parameters of a recursive filter.

Gradient algorithms are based on the method of steepest descent used for numerical solution of systems of linear and nonlinear equations. The Least-Mean-Square (LMS) Algorithm is probably the most widely applied gradient algorithm. The LMS algorithm adjusts the filter's parameters to minimize the mean square error between the filter's output y(n) and the desired response input d(n).

This chapter begins with a review of the LMS algorithm. DSP56001 instructions which implement the adaptive FIR filter on the DSP Demonstration System are then described. The DSP56001 instructions are included on the DSP56000/1 Demonstration Software #4 diskette.

9.1 LMS ALGORITHM REVIEW:

The signal x(n) is filtered to provide an estimate of the input signal d(n) according to the following equation:

$$y(n) = \sum_{i=0}^{N-1} b_i(n)\, x(n-i)$$

$$= B^T(n)\, X(n) \qquad (9.1)$$

where

$$B(n) = [\, b_0(n)\, b_1(n) \ldots b_{N-1}(n)\,]^T \qquad (9.2)$$

$$X(n) = [x(n)\, x(n-1) \ldots x(n-N+1)]^T \qquad (9.3)$$

The error in the estimate can be calculated as:

$$e(n) = d(n) - y(n)$$

$$= d(n) - B^T(n)\, X(n) \qquad (9.4)$$

The square of this error is equal to:

$$e^2(n) = d^2(n) - 2d(n)\, X^T(n)\, B(n) + B^T(n)\, X(n)\, X^T(n)\, B(n) \qquad (9.5)$$

The mean-square error is the expected value of $e^2(n)$. The mean-square error can be written as:

$$E[e^2(n)] = E[d2(n)] - 2\, R(x,d)\, B(n) + B^T(n)\, R(x,x)\, B(n) \qquad (9.6)$$

where the vector of cross correlations between X(n) and d(n) is defined as:

$$R(x,d) = E[d(n) X(n)^T] \qquad (9.7)$$

and where the correlation matrix of the input signals X(n) is defined as:

$$R(x,x) = E[X(n)\, X(n)^T] \qquad (9.8)$$

For stationary input signals, the mean-square error in Equation 9.6 is a second-order function of the parameters. The method of steepest descent seeks the minimum mean-square error by making each change in the parameter vector proportional to the gradient of the mean-square error with respect to the parameter vector. The method of steepest descent can then be described by the following equation:

$$B(n+1) = B(n) - 0.5\,K\,\nabla[E(e^2(n))] \tag{9.9}$$

where:

$B(n+1)$ = the updated parameter vector to be used during the next sample period.

$B(n)$ = the parameter vector during the current sample period.

K = The loop gain that controls the rate of convergence and the filter stability. K is a positive gain.

$\nabla[E(e^2(n))]$ = the gradient of the mean-square error with respect to the parameter vector $B(n)$.

Using Equation 9.6, the gradient of the mean-square error can be obtained as:

$$\nabla[E(e^2(n))] = -2R(x,d) + 2R(x,x)B(n) \tag{9.10}$$

By setting the gradient to zero to yield the least mean square error, the following solution can be obtained:

$$B_{LMS} = R^{-1}(x,x)\,R(x,d) \tag{9.11}$$

where Equation 9.11 is usually referred to as Wiener-Hopf equation.

The LMS algorithm is the most popular method used to find an approximate solution to Equation 9.11, in that it does not require explicit measurements of correlation functions, nor does it require matrix inversion. The LMS algorithm can be written as:

$$B(n+1) = B(n) - 0.5K\hat{\nabla}[E(e^2(n))] \tag{9.12}$$

where $\hat{\nabla}[E(e^2(n))]$ is an estimate of the gradient of the mean-square error with respect to the parameter vector B(n). This estimated gradient can be obtained by taking the gradient of a single time sample of the squared error.

$$\hat{\nabla}[E(e^2(n))] = \nabla[e^2(n)] = 2e(n)\nabla e(n) \tag{9.13}$$

From Equation 9.4,

$$\nabla e(n) = \nabla[d(n) - B^T(n)\,X(n)] = -X(n) \tag{9.14}$$

Thus

$$\hat{\nabla}[E(e^2(n))] = -2e(n)X(n) \tag{9.15}$$

Using Equations 9.12 and 9.15, the LMS algorithm can be written as:

$$B(n+1) = B(n) + Ke(n)X(n) \tag{9.16}$$

Equation 9.16 can be rewritten as:

$$b_i(n+1) = b_i(n) + Ke(n)x(n-i) \qquad i = 0, 1, .., N-1 \tag{9.17}$$

where:

$b_i(n+1)$ = the value of the updated "i^{th}" parameter to be used during the next sample period.

$b_i(n)$ = the value of the "i^{th}" parameter during the current sample period.

At the beginning of the next sample period, two new input samples are shifted into the system and the process repeats. After several iterations, the FIR parameters converge to values which consistently minimize the mean square error.

The loop gain K controls the rate of convergence and stability of the FIR filter. K has an optimum value depending on the FIR tap length used. The following values, found empirically, are recommended maximum values and we should start out with a value less than or equal to the K value shown for the corresponding FIR length chosen. The FIR tap length should be rounded to the nearest FIR tap length shown in Table 9.1.

TABLE 9.1. Maximum Recommended Loop Gains.

FIR Tap Length	*Maximum Recommended K*
≤ 32 (20 Hex)	0.75 (600000 Hex)
≤ 64 (40 Hex)	0.375 (300000 Hex)
≤ 128 (80 Hex)	0.125 (100000 Hex)
≤ 192 (C0 Hex)	0.078125 (A0000 Hex)
≤ 256 (FF Hex)	0.0703125 (90000 Hex)

In many digital signal processing applications such as adaptive line enhancement (ALE), the input signal x(n) is a delayed version of the input signal d(n). The ALE detects and tracks a moving spectral line in a broadband noise while enhancing the signal to noise ratios. In the following sections, it is assumed that d(n) is equal to x(n). It is also assumed that d(n) consists of a desired signal s(n) and a white noise signal w(n).

$$d(n) = s(n) + w(n) \tag{9.18}$$

9.2 IMPLEMENTING AN ADAPTIVE FIR FILTER:

Equations 9.1, 9.4 and 9.17 are used to implement an adaptive FIR filter. In general, the following steps must be performed to obtain the output sample y(n), the error sample e(n) and the updated parameters $b_i(n+1)$:

1. Save the input sample x(n) to be the first filter state.
2. Multiply the input sample x(n) by $b_0(n)$ and accumulate the products of the filter states x(n-i) and the coefficients $b_i(n)$ to yield the output sample y(n).
3. Subtract the output sample y(n) from the input sample x(n) to obtain the error sample e(n).
4. Update the filter parameters to obtain the next sample period parameters $b_i(n+1)$.
5. Shift the filter states to obtain the next output sample y(n+1).

9.2.1 Implementing an Adaptive FIR Filter on the DSP56001 Processor:

Similar to the FIR filter presented in Chapter 6, steps 1-5 can be efficiently implemented on the DSP56001 processor by using its:

a. address pointers to mimic FIFO-like shifting of RAM data,
b. modulo addressing capability to provide wrap-around data buffers,
c. Multiply/Accumulate (MAC) instruction to both multiply two operands and add the product to a third operand in a single instruction cycle,
d. data move capability in parallel with the MAC instruction to keep the multiplier running at 100% capacity, and
e. Repeat Next Instruction (REP) instruction to provide compact filter code.

The DSP56001 processor's capability to perform modulo addressing allows an Address Register (R_n) value to be incremented (or decremented) and yet remain within an address range of size L, where L is defined by a lower and an upper address boundary.

For the adaptive FIR filter, L is equal to the number of coefficients (taps). The value L-1 is stored in the DSP56001 processor's Modifier Register (M_n). Because of the manner in which the DSP56001 processor's modulo addressing mechanism is implemented, the lower address boundary of L (base address) must have zeros in its J LSBs, where 2**J $\geq$ L , e.g., the base address value must be a multiple of 2**J. This requirement is applied to the DSP56001 processor's Address Register (R_n). The upper address boundary is the sum of the lower address boundary (base address) plus the modulo size minus one, e.g., the base address value plus L-1. Note, the upper address boundary is calculated by the DSP56001 processor and is not stored in a register.

When modulo addressing is used, the Address Register (R_n) points to a modulo (circular) data buffer located in X-Memory and or Y-Memory. The address pointer (R_n) is not required to point at the lower address boundary; that is, it can point anywhere within the defined modulo address range L. If the address pointer increments past the upper address boundary (base address plus L-1 plus 1), it will wrap around to the base address.

9.2.2 DSP56001 Instructions for Implementing an Adaptive FIR Filter:

The DSP56001 instructions shown in Figure 9.2. are used to implement the adaptive FIR algorithm where:

```
;
;  Adaptive FIR  Filter

; x0 = Input sample
; a  = Output sample
; b  = Error sample * loop gain
; ntaps = number of parameter taps in the filter
; lg    = loop gain

;   Initialize routine

    move     #states,r0
    move     #para,r4
    move     #ntaps-1,m0
    move     m0,m4

;   The following code obtains the output sample y(n):

    clr      a            x0,x:(r0)+       y:(r4)+,y0
    rep      #ntaps-1
    mac      x0,y0,a      x:(r0)+,x0       y:(r4)+,y0
    macr     x0,y0,a      x:(r0),b

;   The following code obtains the product of the error
;   sample and the loop gain:

    sub      a,b

    move     #lg,y1
    move     b,x1
    mpy      x1,y1,b
    move     b,y1

;    The following code updates the parameters:

      do       #ntaps,_update
      move     y:(r4),a  x:(r0)+,x0
      mac      x0,y1,a
      move     a,y:(r4)+
_update

;  The following instruction updates the address register R0

      lua      (r0)-,r0
```

Figure 9-2. Assembler Code to Implement the Adaptive FIR Filter.

1. Modulo Register M0 is programmed to the value NTAPS-1 (modulo NTAPS). Address Register R0 is programmed to point to the state variable modulo buffer located in X-Memory. Modulo Register M4 is programmed to the value NTAPS-1 (modulo NTAPS). Address Register R4 is programmed to point to the coefficient buffer located in Y-Memory. Given that the FIR filter algorithm has been executing for some time and is ready to process the input sample x(n) in the Data ALU Input Register X0, the address in R4 is the base address (lower-boundary) of the coefficient buffer. The address in R0 is M, where M is greater than or equal to the lower-boundary X-Memory address and less than or equal to the upper-boundary X-Memory address. The X-Memory map for the filter states, the Y-Memory map for the coefficients, and the contents of the DSP56001 processor's A and B Accumulators and Data ALU Input Registers X0, X1, Y0 and Y1 are as shown in Figure 9.3.

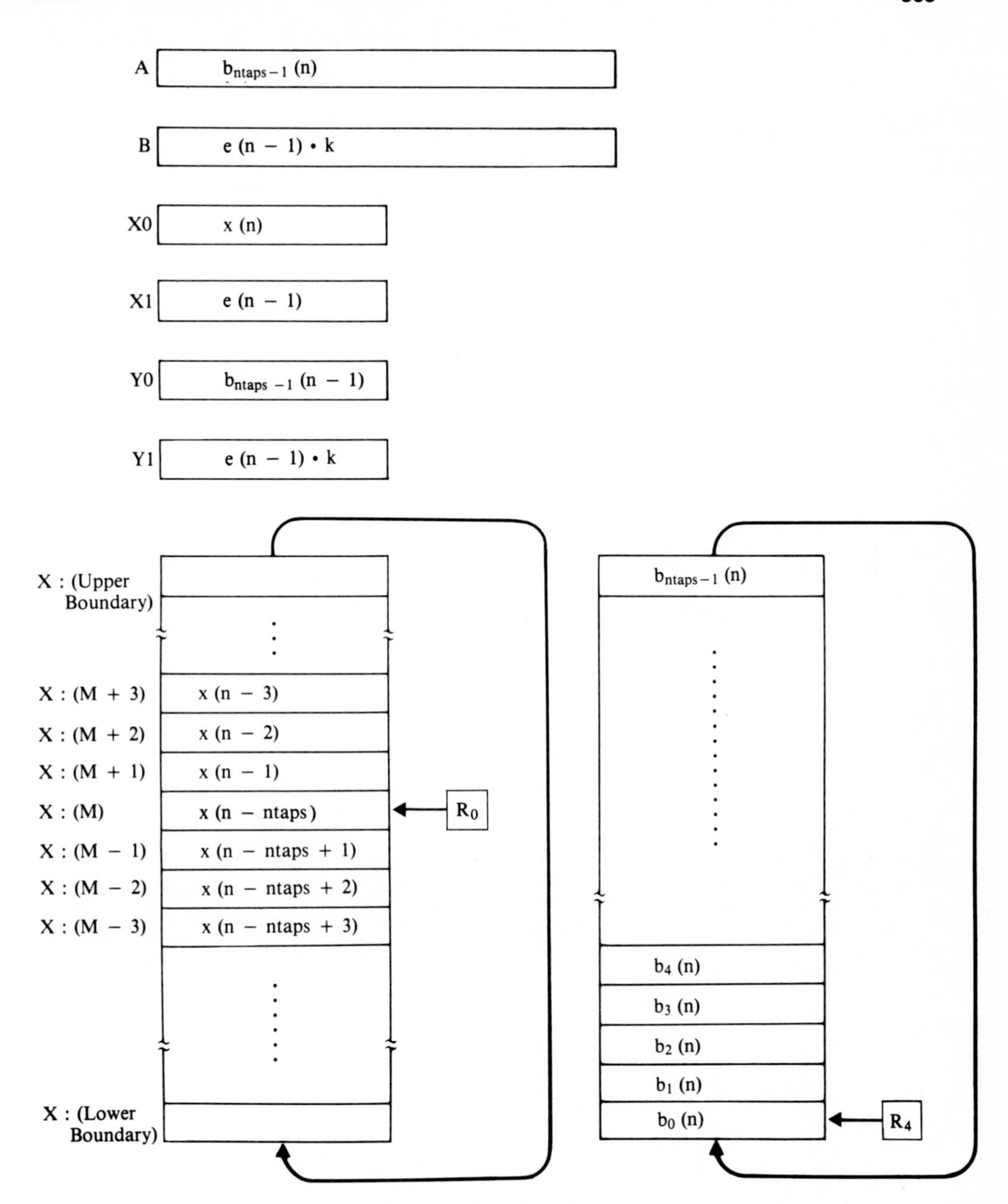

Figure 9-3 Memory Map and Data Registers at the Beginning of the nth Iteration

2. The CLR instruction clears the A-Accumulator and simultaneously:

   a. moves the input sample x(n) from the Data ALU's Input Register X0 to the X-Memory location pointed to by Address Register R0, and

   b. moves the first coefficient from the Y-Memory location pointed to by Address Register R4 to the Data ALU's Input Register Y0.

   Both Address Registers R0 and R4 are automatically incremented by one at the end of the CLR instruction (post incremented).

The X-Memory map for the filter states, the Y-Memory map for the coefficients, and the contents of the A and B Accumulators and Data ALU Input Registers X0, X1, Y0 and Y1 after execution of the CLR instruction are as shown in Figure 9.4.

3. The REP instruction regulates execution of NTAPS-1 iterations of the MAC instruction:

```
mac      x0,y0,a         x:(r0)+,x0        y:(r4)+,y0
```

4. The MAC instruction multiplies the filter state variable in X0 by the coefficient in Y0, adds the product to the A-Accumulator, and simultaneously:

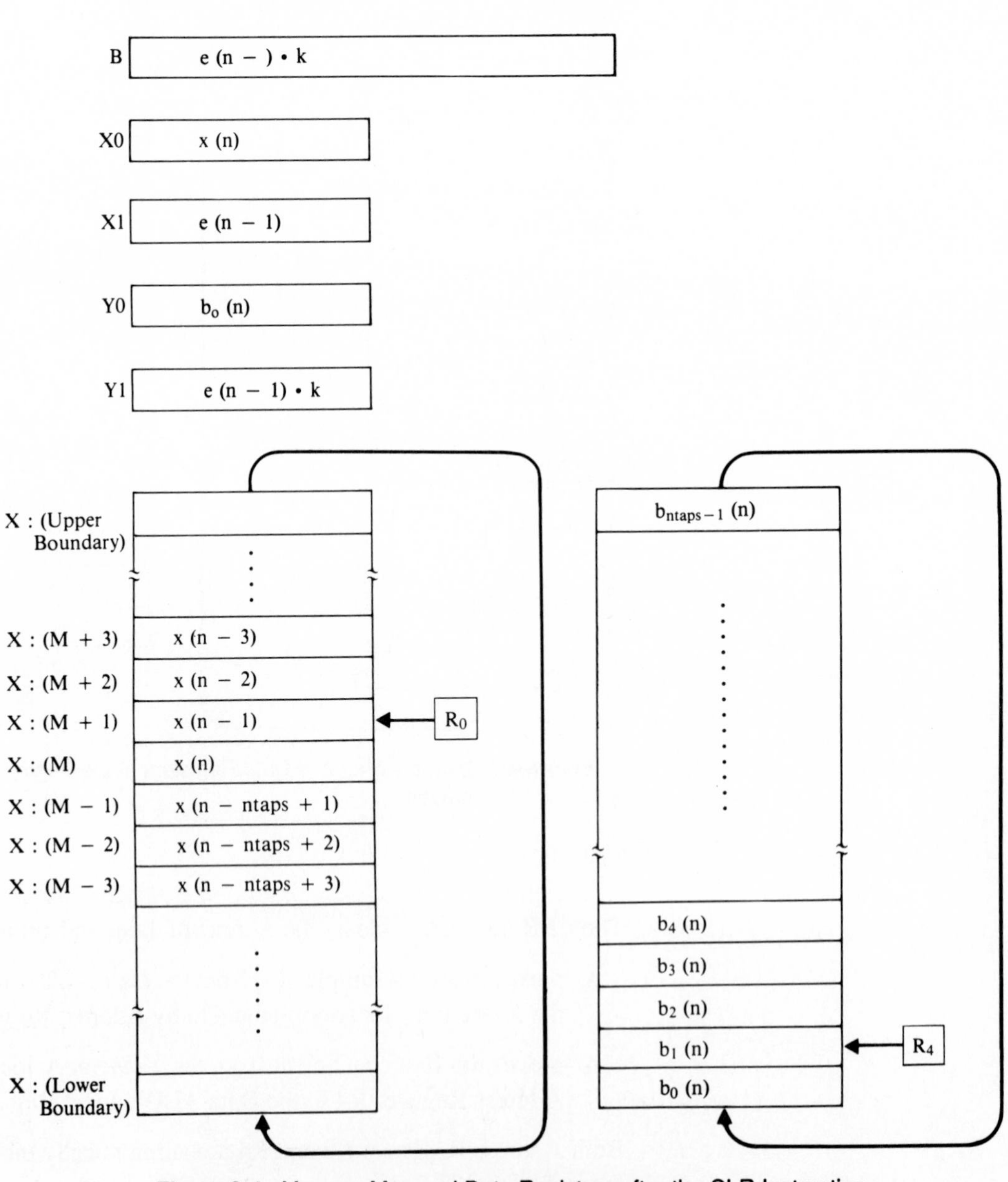

Figure 9-4 Memory Map and Data Registers after the CLR Instruction

a. moves the next state variable from the X-Memory location pointed to by the Address Register R0 to the Input Register X0, and

b. moves the next coefficient from the Y-Memory location pointed to by Address Register R4 to Input Register Y0.

Both Address Registers R0 and R4 are automatically incremented by one at the end of the MAC instruction (post incremented).

The X-Memory map for the filter states, the Y-Memory map for the coefficients, and the contents of the A and B Accumulators and Data ALU Input Registers X0, X1, Y0 and Y1 after execution of the first, second and last MAC instruction are as shown in Figures 9.5, 9.6 nd 9.7 respectively.

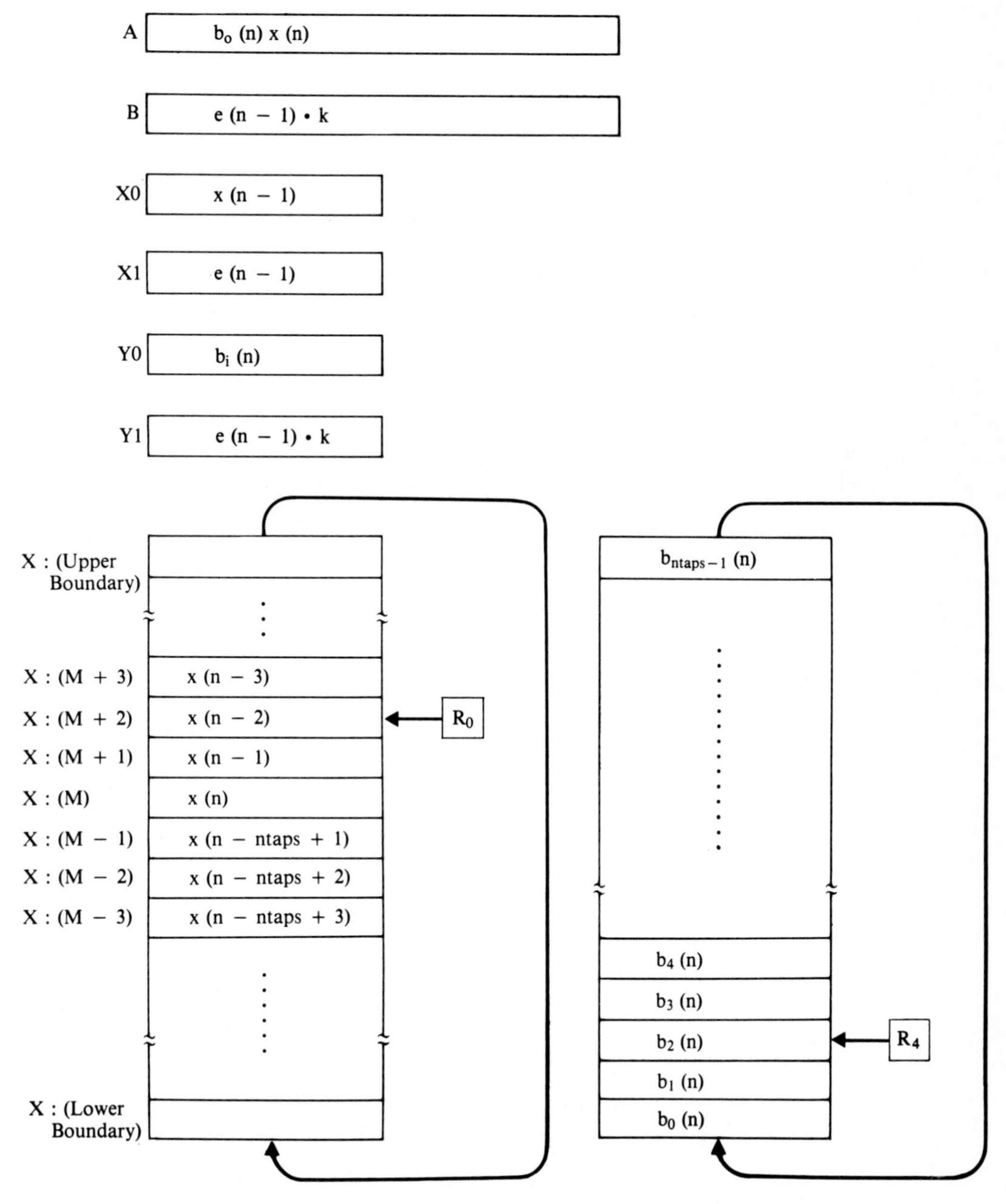

Figure 9-5 Memory Map and Data Registers after First MAC Instruction

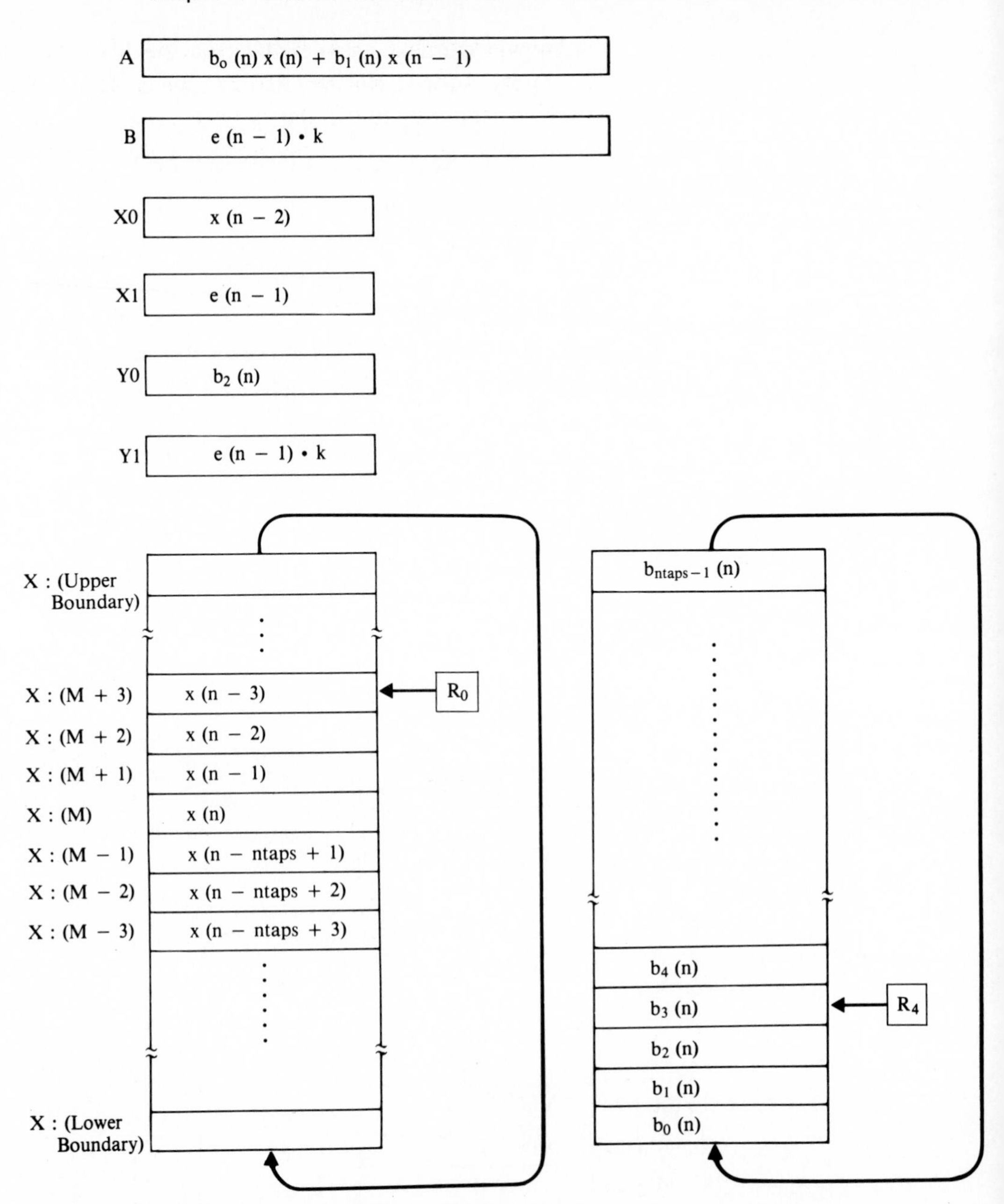

Figure 9-6 Memory Map and Data Registers after Second MAC Instruction

Note that during the execution of the filter algorithm, Address Register R4 is post incremented a total of NTAPS times; once in conjunction with the CLR instruction and NTAPS-1 times (due to the REP instruction) in conjunction with the MAC instruction. Since the modulus for R4 is NTAPS and R4 is incremented NTAPS times, the address value in R4 wraps around and points to the coefficient buffer's lower boundary location.

Note that during the execution of the filter algorithm, Address Register R0 is post incremented at total of NTAPS times; once in conjunction with the CLR instruction and NTAPS-1 times (due to the REP instruction) in conjunction with the MAC instruction. Also note that at the beginning of the algorithm, the input sample x(n) is moved from Data ALU Input Register X0 to the X-Memory location pointed to by R0. Since the modulus for R0 is NTAPS and R0 is incremented NTAPS times, the address value in R0 wraps around and points to the state variable buffer's X-Memory location M.

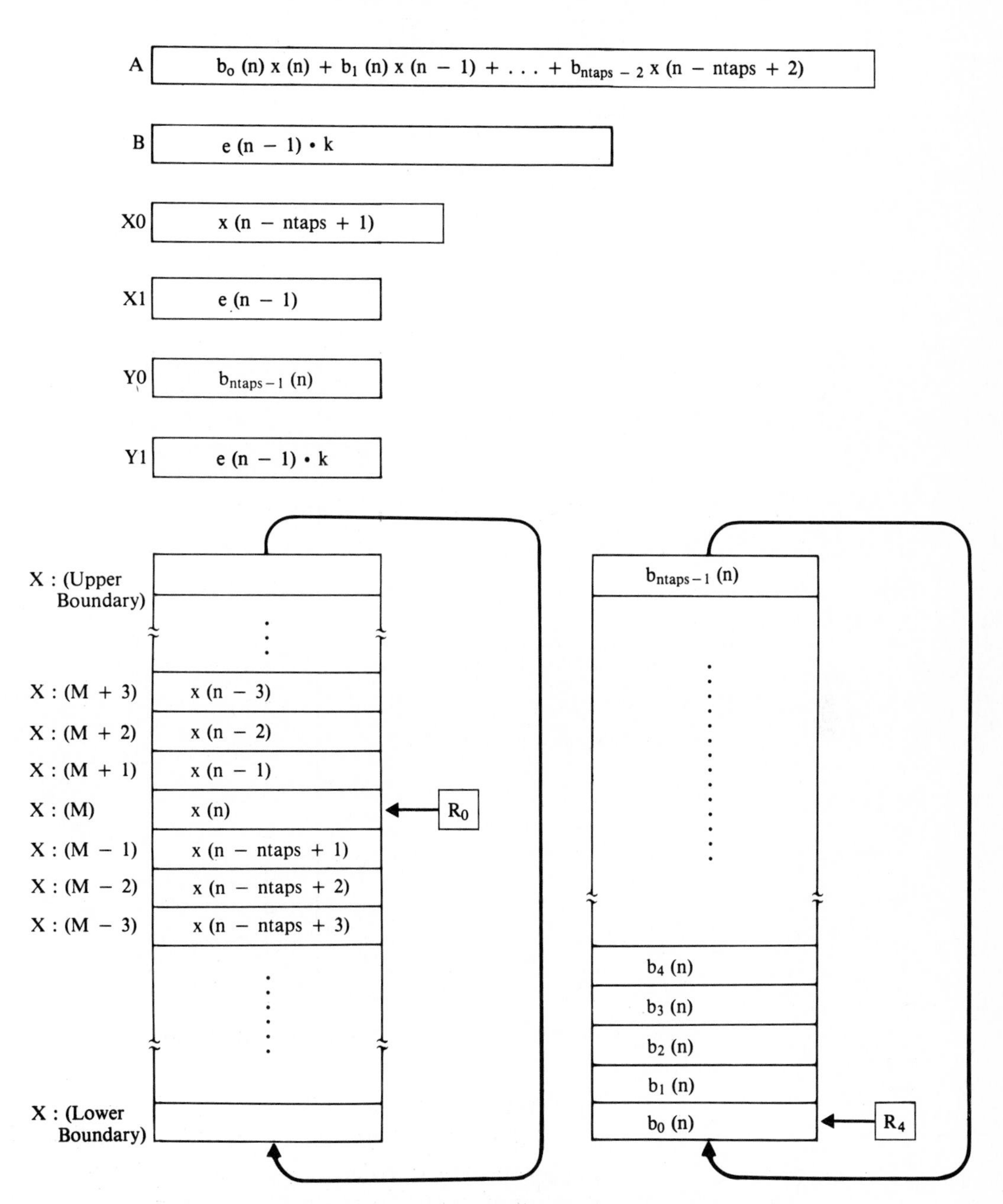

Figure 9-7 Memory Map and Data Registers after Last MAC Instruction

5. The MACR instruction calculates the final tap of the filter algorithm and performs convergent rounding of the result. The data move portion of this instruction loads the input sample x(n) into the B-Accumulator. At the end of the MACR instruction, the A-Accumulator contains the filter output sample y(n).

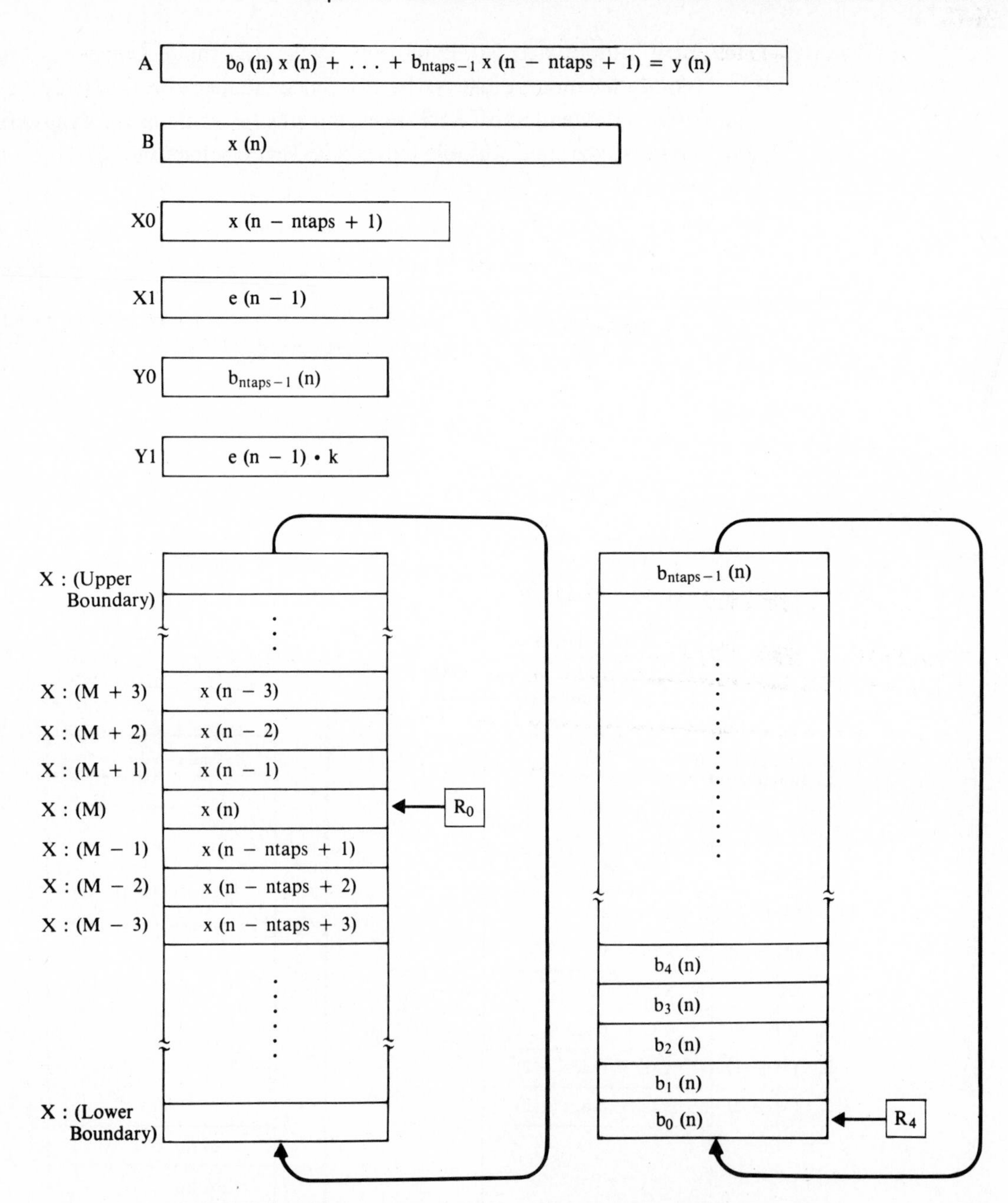

Figure 9-8 Memory Map and Data Registers after MACR Instruction

The X-Memory map for the filter states, the Y-Memory map for the coefficients, and the contents of the A and B Accumulators and Data ALU Input Registers X0, X1, Y0 and Y1 after the MACR instruction are as shown in Figure 9.8.

6. The SUB instruction calculates the error sample e(n) by subtracting the output sample y(n), in the A-Accumulator, from the input sample x(n), in the B-Accumulator. The error sample e(n) is stored in B-Accumulator.

7. The two MOVE instructions **move #lg,y1 and move b,x1** transfer the loop gain K to the Data Register Y1 and the error sample e(n) to the Data Input Register X1, respectively.

8. The MPY instruction multiplies the error sample e(n) by the loop gain K, and stores the result in the B-Accumulator. The X-Memory map for the filter states, the Y-Memory map for the coefficients, and the contents of the A and B Accumulators and Data ALU Input Registers X0, X1, Y0 and Y1 after the MPY instruction are shown in Figure 9.9.

9. The MOVE instruction **move b,y1** transfers the product of the error sample e(n) and the loop gain K to the Data Register Y1.

10. The DO instruction sets up a hardware "do loop" to update the NTAPS filter's parameters. The three instructions following the DO instruction are repeated NTAPS times.

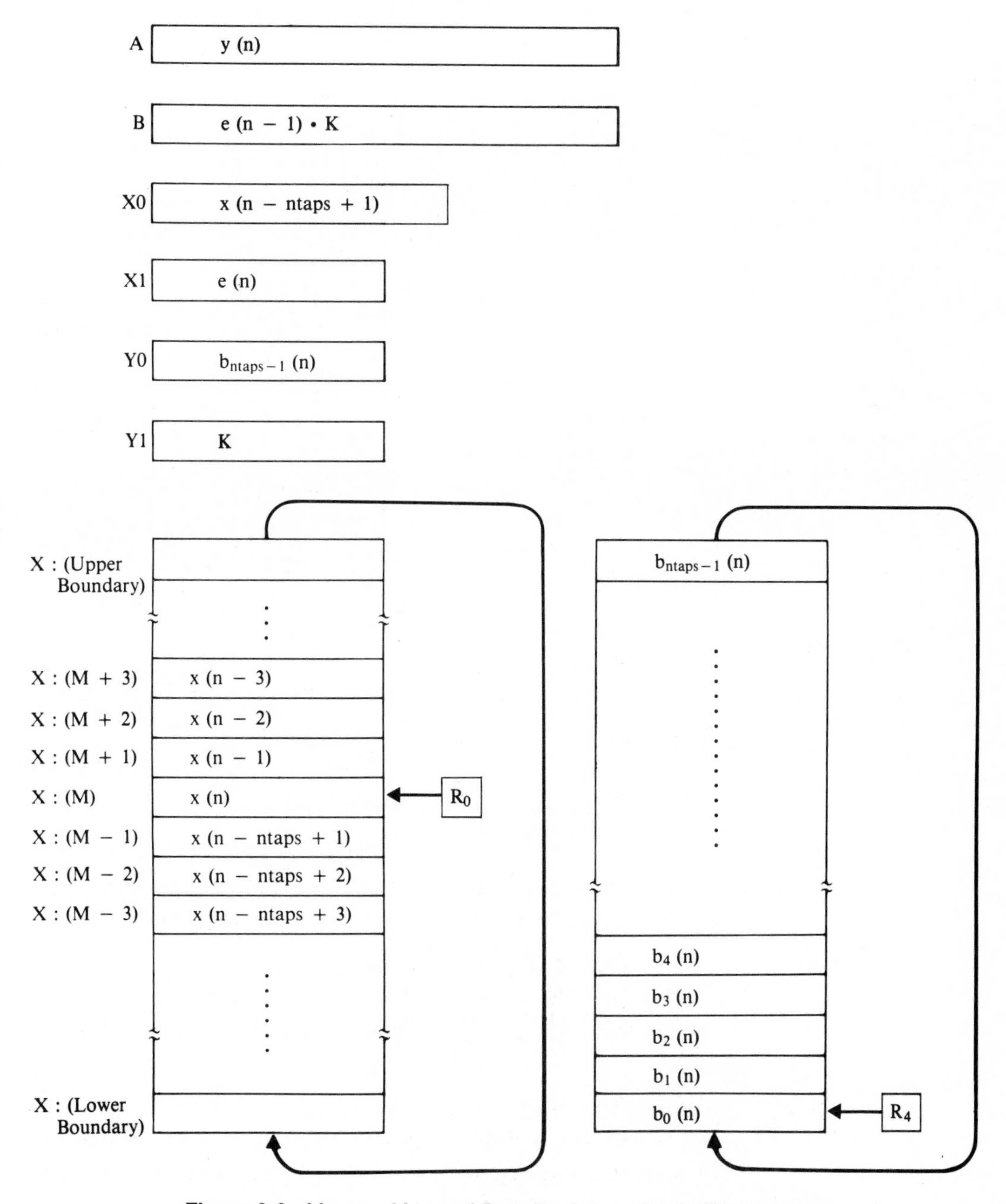

Figure 9-9 Memory Map and Data Registers after MPY Instruction

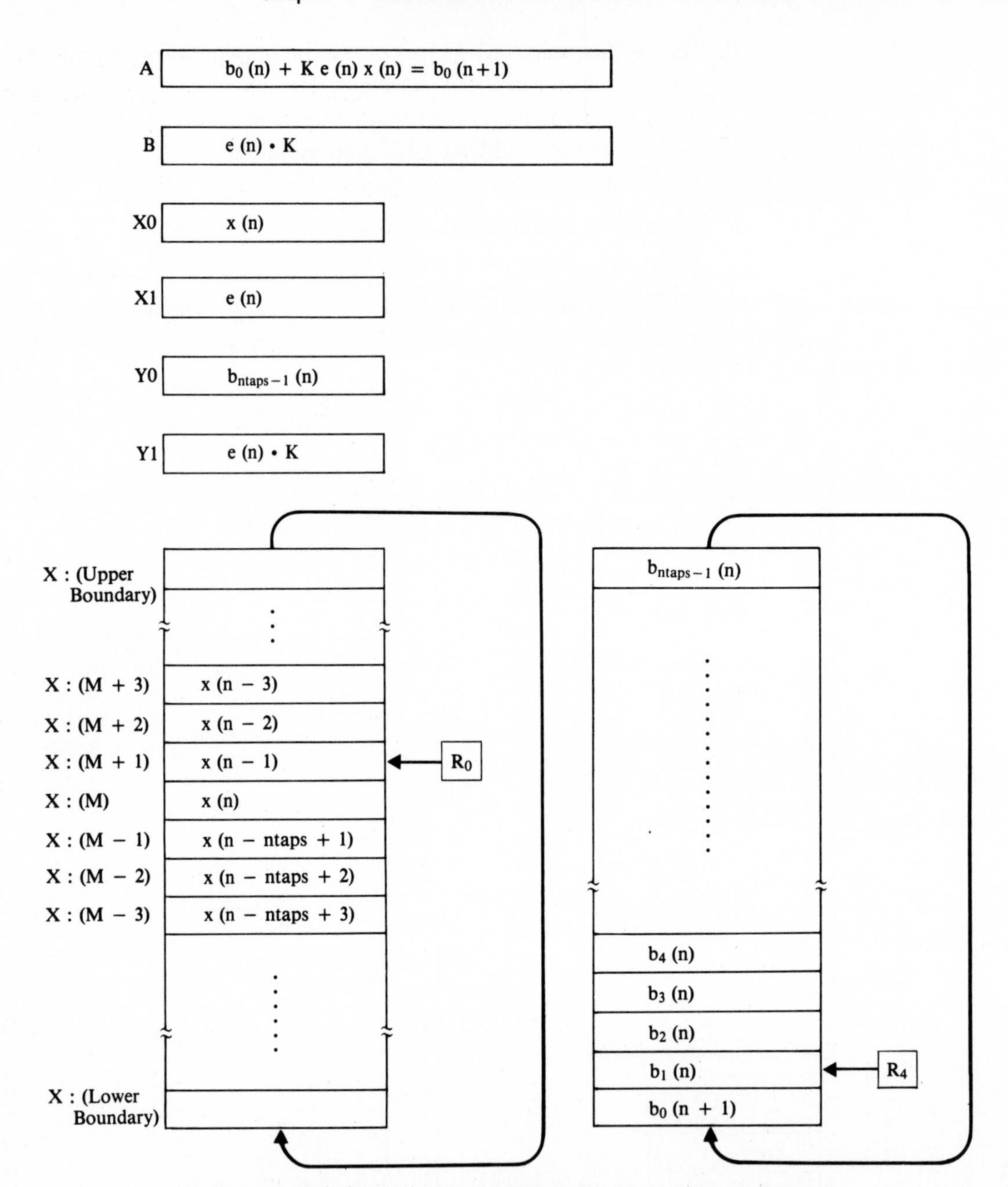

Figure 9-10 Memory Map and Data Registers after First Pass of DO Loop

11. The first MOVE instruction in the "do loop" transfers the parameter $b_i(n)$ to the A-Accumulator and the filter state x(n-i) to the Data Input Register X0. Address Register R0 is incremented by one to point to the next filter state.
12. The MAC instruction multiplies the filter state, in X0 by the product of the loop gain and the error sample, in Y1, and adds the product to the A-accumulator. The result in the A-accumulator is the updated parameter $b_i(n+1)$.
13. The second MOVE instruction in the "do loop" transfers the parameter $b_i(n+1)$ to the Y-Memory location pointed to by the Address Register R4. R4 is incremented by one to point to the next filter parameter. The X-Memory map for the filter states, the Y-Memory map for the coefficients, and the contents

of the A and B Accumulators and Data ALU Input Registers X0, X1, Y0 and Y1 after the first, second and last passes of the " do loop" are as shown in Figures 9.10, 9.11 and 9.12 respectively.

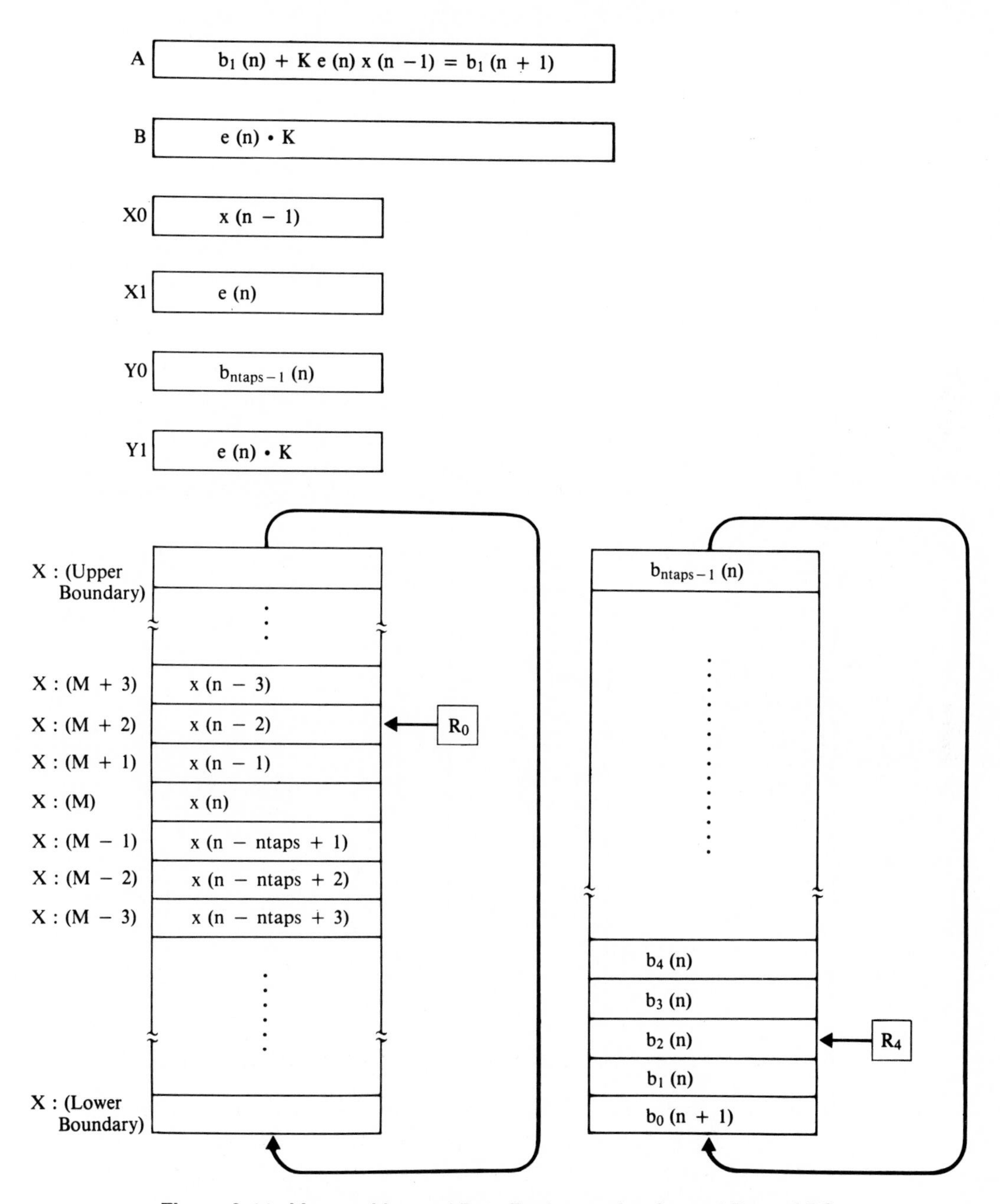

Figure 9-11 Memory Map and Data Registers after Second Pass of DO Loop

14. The LUA instruction decrements R0 by one. As a result, R0 then points to the state variable buffer's X-Memory location M-1. The next time the algorithm is executed, a new (next) input sample x(n+1) will then overwrite the value in X-Memory location M-1. Thus FIFO-like shifting of the filter state variables is accomplished by simply adjusting the R0 address pointer. The

X-Memory map for the filter states, the Y-Memory map for the coefficients, and the contents of the A and B Accumulators and Data ALU Input Registers X0, X1, Y0 and Y1 after the LUA instruction are shown in Figure 9.13.

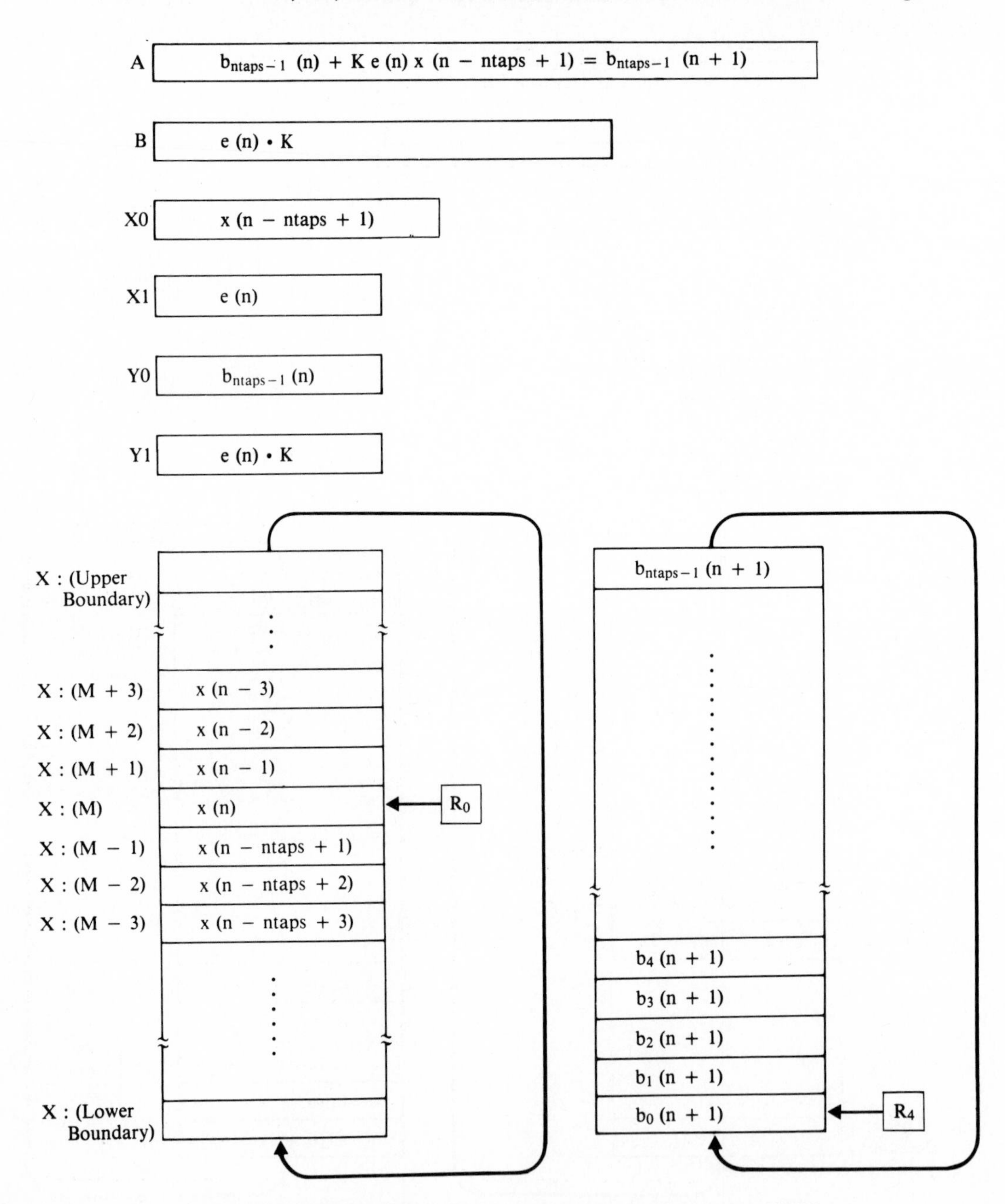

Figure 9-12 Memory Map and Data Registers after Last Pass of DO Loop

9.3 ADAPTIVE FIR FILTER DEMONSTRATION SYSTEM:

The adaptive filtering process can be demonstrated by using the DSP system described in Chapter 5. The DSP system consists of the DSP56000ADS application development system (ADS) and an A/D-D/A evaluation board (EVB). The analog input signal x(t) consists of the desired input signal s(n) plus a white noise signal w(n). The analog input signal x(t) is first digitized using the A/D converter on the EVB. The

DSP56001 processor then executes an adaptive filter algorithm to process the digitized input signal x(n), thereby producing a filtered output signal y(n). The output signal y(n) is an estimate of the desired input signal s(n). The D/A converter chip on the EVB is then used to convert the digital output signal y(n) to the analog output signal y(t).

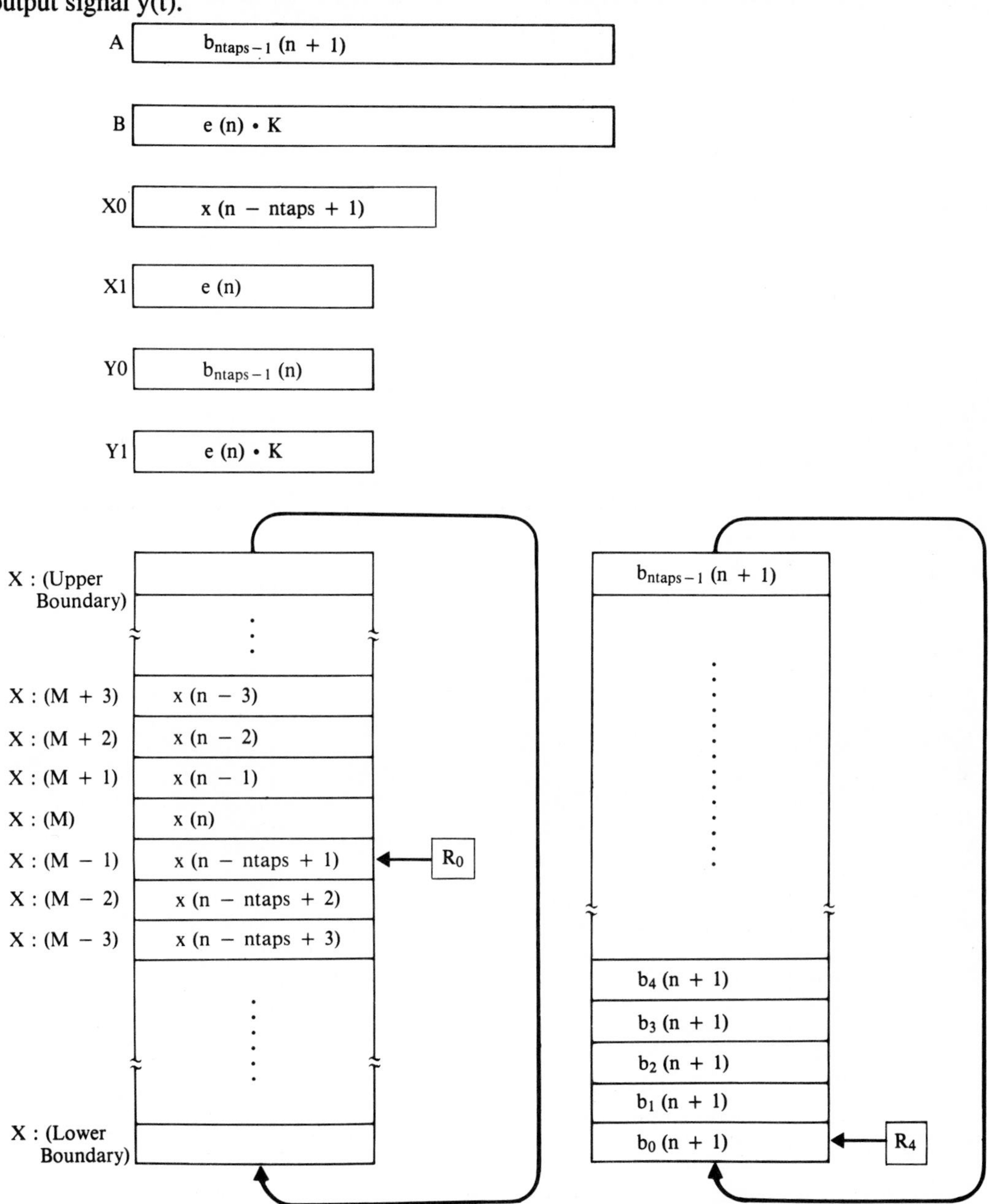

Figure 9-13 Memory Map and Data Registers after LUA Instruction

9.4 IMPLEMENTING AN ADAPTIVE FIR FILTER ON THE DEMONSTRATION SYSTEM:

The DSP56001 instructions shown in Figure 9.2 are combined with the DSP56001 instructions shown in Figure 5.8 to form the AFIR.ASM assembler program macro shown in Figure 9.14. The AFIR.ASM assembler program macro is used to implement the adaptive FIR filter on the demonstration system.

1. POWER-DOWN THE PC.
2. SET-UP THE DSP SYSTEM as described in section 5.2.2 of Chapter 5.
3. POWER-UP AND BOOT THE PC.

```
afir  macro  ntaps,lg

;
;
; ntaps = number of parameter taps in the filter
; lg    = loop gain
;
;

;***********************************************************
; Initialize routine
;***********************************************************

init2   reset                   ; Reset the system

        movep   #0,x:$fffe      ; set bcr to zero
        movep   #0,x:$ffef      ; clear the transfer register
        movep   #0,x:$ffff      ; stop all the interrupt

;************************************************************
; Set up SSI for operation with the EVB
; The following code sets port C to function as SSI
;************************************************************

pcc     equ     $1e0
        movep   #pcc,x:$ffe1    ; write PCC

cra     equ     $4000
        movep   #cra,x:$ffec    ; CRA pattern for word
                                ; length = 16 bits

crb     equ     $3200           ; CRA pattern for continuous
        movep   #crb,x:$ffed    ;ck,synch,normal mode
                                ;word long frame sync:

        move    #states,r0  ;point to filter states
        move    #ntaps-1,m0 ;mod(ntaps)
        move    #para,r4    ;point to filter parameters
        move    #ntaps-1,m4 ;mod(ntaps)

;
;************************************************************
;  Read A/D
;************************************************************
; The following code polls the RDF flag in the SSI-SR and
; waits for RDF=1 and then reads the RX register to
; retrieve the data from the A/D converter.
; Sample rate is controlled by EVB board.
;
wait2   nop
wait1   btst    #7,x:$ffee
        jcc     wait1

        movep   x:$ffef,x0      ; Read input signal

;************************************************************
; Adaptive FIR Filter
;************************************************************

; x0 = Input sample
; a  = Output sample
; b  = Error sample * loop gain
```

```
;    The following code obtains the output sample y(n) and
;    the error sample e(n):

     clr     a            x0,x:(r0)+        y:(r4)+,y0
     rep     #ntaps-1
     mac     x0,y0,a      x:(r0)+,x0        y:(r4)+,y0
     macr    x0,y0,a      x:(r0),b
     move    #$ffff00,x1
     and     x1,a
     sub     a,b
;
; Write to D/A
;

     movep   a,x:$ffef

;    The following code obtains the product of the error
;    sample and the loop gain:

     move    #lg,y1
     move    b,x1
     mpy     x1,y1,b
     move    b,y1

;     The following code updates the parameters:

       do      #ntaps,_update
       move    y:(r4),a   x:(r0)+,x0
       mac     x0,y1,a
       move    a,y:(r4)+
_update

;   The following instruction updates the address register R0

       lua     (r0)-,r0

       jmp     wait2           ; Do again
       nop

       endm
```

Figure 9.14. AFIR.ASM Assembler Macro to Implement the Adaptive FIR Filter

---

EXAMPLE:

The "AFIREX.ASM" assembler program, shown in Figure 9.15, implements an adaptive filter with 7 taps and 0.2 loop gain on the demonstration system. Note that this program has an "include" statement which references the AFIR.ASM assembler program macro shown in Figure 9.14. A copy of the "AFIREX.ASM" assembler program, AFIR.ASM assembler program macro, and the "AFIREX.LOD" absolute load program can be found on the DSP56000/1 Demonstration Software #4 diskette. A white noise generator is used to generate the white noise signal.

1. POWER-DOWN THE PC, FUNCTION, AND WHITE NOISE GENERATORS.

2. BUILD A SUMMING OPERATIONAL AMPLIFIER TO GENERATE THE INPUT SIGNAL X(t) BY ADDING THE FUNCTION GENERATOR'S OUTPUT S(t) TO THE WHITE NOISE GENERATOR'S OUTPUT W(t).
3. CONNECT THE INPUT SIGNAL X(t) TO THE EVB'S "BNC2" INPUT (A/D INPUT) AND TO THE OSCILLOSCOPE'S CHANNEL ONE INPUT.
4. CONNECT THE EVB'S "BNC1" OUTPUT (D/A OUTPUT) TO THE OSCILLOSCOPE'S CHANNEL TWO INPUT.
5. POWER-ON AND BOOT THE PC.
6. POWER-ON THE FUNCTION AND WHITE NOISE GENERATORS.
7. INSERT THE "DSP56000/1 DEMONSTRATION SOFTWARE #4" DISKETTE INTO DRIVE "A".
8. LOAD THE "ADS56000 USER INTERFACE SOFTWARE" PROGRAM.
9. TYPE **load a:afirex.lod<return>** to load the assembled and linked version of the AFIREX.ASM assembler program shown in Figure 9.15.
10 TYPE **go<return>** to begin execution of the program.
11. VARY THE SINUSOIDAL INPUT FREQUENCY FROM 0.5 TO 20 KHZ AND OBSERVE THE INPUT AND OUTPUT SIGNAL ON THE OSCILLOSCOPE. Note the close match between the output signal y(t) and the sinusoidal input signal s(t).
12. TYPE **force b<return>** to halt execution of the program.
13. TYPE **quit<return>** IF YOU WANT TO EXIT THE "ADS56000 USER INTERFACE SOFTWARE" PROGRAM.

```
;***********************************************************
;  Adaptive FIR Example
;***********************************************************
;

        include 'afir'

;

ntaps   equ     7
lg      equ     $199999

        org     x:$0
states  dsm     ntaps   ; filter states

        org     y:$0
para    dsm     ntaps   ; parameters

        org     p:$40   ; program start address

        afir    ntaps,lg

        end
```

Figure 9-15 AFIREX.ASM Assembler Program to Implement the Adaptive FIR Filter in Section 9.4.

9.5 REFERENCES:

1. Bernard Widrow and Samuel Stearns, Adaptive Signal Processing, Prentice-Hall, 1985.
2. C.F.N.Cowan and P.M.Grant, Adaptive Filters, Prentice-Hall, 1985.
3. Samuel D. Stearns and Ruth David, Signal Processing Algorithms, Prentice-Hall, 1988.
4. M. M. Sondhi, D.A. Berkley, " Silencing Echoes on the Telephone Network", Proc. IEEE, Vol. 68, No.8., pp. 948-963, Aug. 1980.

10

RECURSIVE ADAPTIVE LINE ENHANCEMENT WITH THE DSP56001 PROCESSOR

The adaptive filter's environment can be deterministic or stochastic. In a deterministic environment, all the inputs of the adaptive filter are measurable inputs. The adaptive algorithm presented in Chapter 9 is an example of an adaptive nonrecursive algorithm in the deterministic environment. In a stochastic environment, some of the inputs are unmeasurable. An important application of adaptive recursive algorithms in the stochastic environment is the filtering of narrowband signals corrupted with additive noise. In the digital signal processing (DSP) literature, such filters are usually discussed under the name Recursive Adaptive Line Enhancement (RALE). The RALE detects and tracks a moving spectral line in broadband noise while enhancing the signal to noise ratios. Adaptive line enhancers are used in application areas such as communications, radar, speech, sonar, control, and spectral estimation.

The RALE algorithm described in this chapter is based on the adaptive stochastic algorithm developed in References 1 and 2. The adaptive stochastic algorithm consists of three stages as shown in Figure 10.1. The adaptive stochastic algorithm does not require a strict positive real test or a stability test to check the poles and zeros locations. This makes the three stage adaptive stochastic algorithm suitable for implementing the RALE on the DSP56001 signal processor. In the first stage, the residual of an autoregressive model is used as an estimate of the unknown noise. In the second stage, an autoregressive recursive moving average (ARMA) model is fitted using the residual of the first stage. The modified residual obtained from the second stage is then filtered, using a debiasing parameter, to generate an improved estimate of the noise. In the third stage, this improved estimate is used to obtain a better ARMA model.

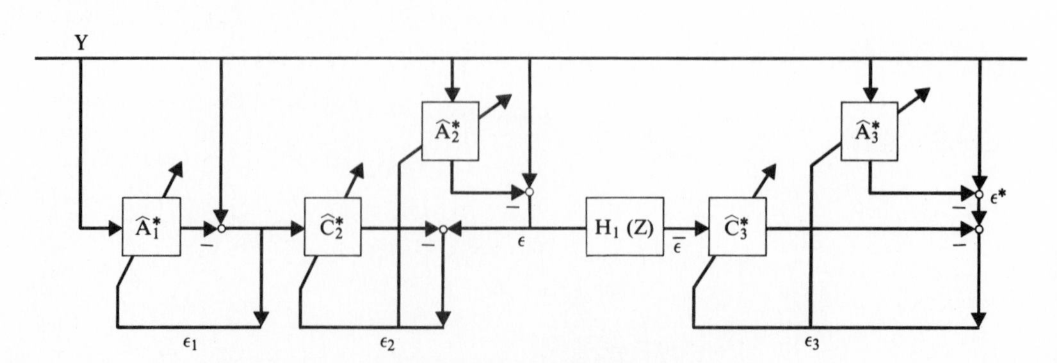

Figure 10-1 Rale Algorithm

This chapter begins with a review of the RALE algorithm. The RALE algorithm is then implemented on the DSP Demonstration System to estimate the RALE parameters. The estimated parameters are used to compute the estimated spectra.

10.1 RALE ALGORITHM:

The particular RALE algorithm addressed in this chapter can be represented by the following ARMA model:

$$A(q^{-1})y(k) = C(q^{-1})e(k) \tag{10.1}$$

where y(k) is the input data and e(k) is the white noise. $A(q^{-1})$ and $C(q^{-1})$ are the nth order polynomials in the delay operator q^{-1} :

$$A(q^{-1}) = 1 + a_1 q^{-1} + \ldots + a_n q^{-n} \tag{10.2}$$

and

$$C(q^{-1}) = 1 + c_1 q^{-1} + \ldots + c_n q^{-n} \tag{10.3}$$

The objective of the proposed adaptive stochastic algorithm is to estimate the parameters a_i and c_i from the input data y(k). The algorithm consists of three stages as shown in Figure 10.1. In the first stage, the following autoregressive representation of Equation 10.1 is considered:

$$A_1(q^{-1})\, y(k) = e(k) \tag{10.4}$$

where

$$A_1(q^{-1}) = A(q^{-1})/C(q^{-1}) \tag{10.5}$$

Another operator, given by

$$A_1^*(q^{-1}) = 1 - A_1(q^{-1}) \tag{10.6}$$

will occasionally be employed instead of $A(q^{-1})$ in the analysis developed below.

The objective of the first stage of the algorithm is to obtain an estimate of the white noise e(k). This is achieved by estimating A_1 by a polynomial $\hat{A}_1$ of degree P, where

$$\hat{A}_1(q^{-1}) = 1 + \hat{a}_{11} q^{-1} + \ldots + \hat{a}_{1p} q^{-p} \tag{10.7}$$

or equivalently,

$$\hat{A}_1(q^{-1}) = 1 - \hat{A}_1^*(q^{-1}) \tag{10.8}$$

where

$$\hat{A}_1^*(q^{-1}) = -(\hat{a}_{11} q^{-1} + \ldots + \hat{a}_{1p} q^{-p}) \tag{10.9}$$

and $\hat{a}_{li}$ are the estimates of a_{1i}. A predicted output of the first stage $\hat{y}_1(k)$ can be defined as:

$$\begin{aligned}\hat{y}_1(k) &= \hat{A}_1^*(q^{-}1)\, y(k) \\ &= \hat{\Theta}_1(k)^T \Phi_1(k-1) \qquad (10.10)\end{aligned}$$

where

$$\begin{aligned}\hat{\Theta}_1(k)^T &= [\hat{a}_{11}(k), \ldots., \hat{a}_{1p}(k)] \\ \Phi_1(k)^T &= [-y(k), \ldots.., -y(k-p+1)] \qquad (10.11)\end{aligned}$$

The residual (error signal) $\varepsilon_1(k)$ of the first stage is defined as the difference between the input data y(k) and the predicted output of the first stage.

$$\varepsilon_1(k) = y(k) - \hat{y}_1(k) \qquad (10.12)$$

which, using Equations 10.8 and 10.10, will become

$$\varepsilon_1(k) = \hat{A}_1(q^{-1})\, y(k) \qquad (10.13)$$

A comparison between Equations 10.13 and 10.4 shows that the residual $\varepsilon_1(k)$ of the first stage can be used as an estimate of the white noise e(k).

Since an exact autoregressive recursive representative of Equation 10.1 has an infinite degree, a bias is introduced into the estimated noise. This bias can be reduced by suitable choice of p, where p = 2n or 3n will usually suffice.

In the second stage of the algorithm, an improved estimate of the white noise e(k) is achieved by using the input data y(k) and the first stage residual $\varepsilon_1(k)$ to estimate the polynomials $\hat{A}_2$, and $\hat{C}_2$ in a manner which best fits in the model given by

$$A(q^{-1})\, y(k) = C(q^{-1})\, \varepsilon_1(k) \qquad (10.14)$$

The polynomials $\hat{A}_2$, and $\hat{C}_2$ can be defined as

$$\hat{A}_2(q^{-1}) = 1 + \hat{a}_{21} q^{-1} + \ldots + \hat{a}_{2n}\, q^{-n}$$

or

$$\hat{A}_2(q^{-1}) = 1 - \hat{A}_2^*(q^{-1}) \qquad (10.15)$$

where

$$\hat{A}_2^*(q^{-1}) = -(\hat{a}_{21}q^{-1} + \ldots + a_{2n}q^{-n})$$

and

$$\hat{C}_2(q^{-1}) = 1 + \hat{c}_{21}q^{-1} + \ldots + \hat{c}_{2n}q^{-n}$$

or

$$\hat{C}_2(q^{-1}) = 1 - \hat{C}_2^*(q^{-1}) \tag{10.16}$$

where

$$\hat{C}_2^*(q^{-1}) = -(\hat{c}_{21}q^{-1} + \ldots + \hat{c}_{2n}q^{-n})$$

A predicted output $y_2(k)$ of the second stage can be defined as

$$\hat{y}_2(k) = \hat{\Theta}_2(k)^T \, \Phi_2(k-1) \tag{10.17}$$

where

$$\hat{\Theta}_2(k)^T = [\hat{a}_{21}(k), \ldots, \hat{a}_{2n}(k), \hat{c}_{21}(k), \ldots, \hat{c}_{2n}(k)]$$

$$\hat{\Phi}_2(k)^T = [-y(k), \ldots, -y(k-n+1), \varepsilon_1(k), \ldots, \varepsilon_1(k-n+1)]$$

The error of the second stage can be defined as

$$\varepsilon_2(k) = y(k) - \hat{y}_2(k) \tag{10.18}$$

A modified error e(k) is defined using the new estimated polynomial A_2 as

$$\varepsilon(k) = \hat{A}_2(q^{-1})y(k) \tag{10.19}$$

The modified error is then filtered using a parameter d to obtain the improved estimate of the noise $\bar{\varepsilon}(k,\delta)$, where

$$\begin{aligned} \bar{\varepsilon}(k,\delta) &= -\sum_{i=1}^{n} \delta^i \hat{a}_{2i} \bar{\varepsilon}(k-i,\delta) + \varepsilon(k) \\ &= H_1(q^{-1},\delta)\, \varepsilon(k) \end{aligned} \tag{10.20}$$

and

$$H_1(q^{-1},\delta) = 1/A_2(\delta q^{-1}) \quad \text{for} \quad 0 \le \delta < 1$$

The reasoning behind using this particular filtered version of the residual $\bar{\varepsilon}(k,d)$ is explained as follows: Using Equation 10.19, Equation 10.20 can be written as:

$$\bar{\varepsilon}(k,\delta) = H(q^{-1},\delta)\, y(k) \tag{10.21}$$

where

$$H(q^{-1},\delta) = \hat{A}_2(q^{-1})/\hat{A}_2(\delta q^{-1}) \tag{10.22}$$

The "tuned" filter $H(q^{-1},\delta)$ selects the noise component e(k) of the input data y(k) and rejects the signal component. Therefore, an improved estimate of e(k) is obtained. This filter, or the equivalent filter $(1-H(q^{-1},\delta))$, is referred to as a "notch" filter. The closer δ is to unity, the flatter will be the response of the notch filter. The flatter the response of the "notch" filter, the smaller will be the bias in the estimate of e(k). The parameter δ is sometimes referred to as a debiasing parameter.

In the third stage of the algorithm, the filtered modified residual $\bar{\varepsilon}(k,\delta)$, and the input data y(k), are used to obtain an improved estimated $\hat{A}_3, \hat{C}_3$ in a manner which fits the model

$$A(q^{-1})\, y(k) = C(q^{-1})\bar{\varepsilon}(k,\delta) \tag{10.23}$$

A predicted output of the third stage $\hat{y}_3(k)$ can be defined as

$$\hat{y}_3(k) = \hat{\Theta}_3(k)^T\, \Phi_3(k-1) \tag{10.24}$$

where

$$\hat{\Theta}_3(k)^T = [\hat{a}_{31}(k), \ldots, \hat{a}_{3n}(k), \hat{c}_{31}(k), \ldots, \hat{c}_{3n}(k)]$$

$$\Theta_2(k)^T = [-y(k), \ldots, -y(k-n+1), \bar{\varepsilon}(k,\delta), \ldots, \bar{\varepsilon}(k-n+1,\delta)]$$

The error of the third stage can be defined as

$$\varepsilon_3(k) = y(k) - \hat{y}_3(k) \tag{10.25}$$

Using the Supermartingale approach, the parameter adaptation algorithm of the three stages can be given as:

$$\hat{\Theta}_i(k+1) = \hat{\Theta}_i(k) + F_i(k)\, \Phi_i(k)\, \varepsilon_i(k+1) \qquad i = 1,2 \text{ and } 3 \tag{10.26}$$

where

$$F_i^{-1}(k+1) = \lambda_1(k)\, F_i^{-1}(k) + \lambda_2(k)\, \Phi_i(k)\, \Phi_i(k)^T$$

$$F_i(0) > 0 \qquad 0 \le \lambda_1(k) < 1 \qquad 0 \le \lambda_2(k) < 2$$

A detailed description of this algorithm and a proof using the Supermartingale approach are given in Reference 1. Unlike other algorithms, the three stage adaptive stochastic algorithm does not require a strict positive real test or a stability test to check the poles and zeros locations.

10.2 IMPLEMENTING THE RALE ON THE DSP DEMONSTRATION SYSTEM:

The RALE can be demonstrated by using the DSP Demonstration System described in Chapter 5. The DSP system consists of the DSP56000ADS application development system (ADS) and the A/D converter on the evaluation module board (EVB). The analog input signal y(t) is obtained by adding a white noise v(t) to a sinusoidal input x(t). The analog input signal y(t) is then digitized using the A/D converter.

The digital input signal y(k) can then be written as:

$$y(k) = x(k) + v(k) \tag{10.27}$$

Since the digital sinusoidal input x(k) can be represented by

$$x(k) = u(k)/A(q^{-1}) \tag{10.28}$$

where

$$A(q^{-1}) = 1 + a_1 q^{-1} + a_2 q^{-2}$$

and u(k) is an unknown white noise process.

By using Equation 10.28, Equation 10.27 can be written as:

$$A(q^{-1})y(k) = u(k) + A(q^{-1})v(k) \tag{10.29}$$

or equivalently,

$$A(q^{-1})\, y(k) = C(q^{-1})\, e(k) \tag{10.30}$$

where

$$C(q^{-1})\, e(k) = u(k) + A(q^{-1})\, v(k) \tag{10.31}$$

Note that Equation 10.30 is exactly the same as Equation 10.1.

Using Equation 10.26, the parameter adaptation algorithm can be rewritten as:

$$\hat{\theta}_{li}(k+1) = \hat{\theta}_{li}(k) + au_{li}(k)\, \varepsilon_i(k+1)/bt_i(k) \tag{10.32}$$

where

$$au_{li}(k) = \sum_{j=1}^{4} f_{lji}(k)\, \hat{\theta}_{ji}(k)$$

$$u_{li}(k) = \sum_{j=i}^{4} f_{jli}(k)\, \hat{\theta}_{ji}(k)$$

$$bt_i(k) = \sum_{l=1}^{4} \hat{\theta}_{li}(k)\, au_{li}(k)$$

and

$$f_{lji}(k+1) = f_{lji}(k) - au_{li}(k)\, u_{ji}(k)/bt_i(k) \tag{10.33}$$

$$l = 1,\ldots,4 \quad j = 1,\ldots,4 \quad \text{and} \quad i = 1,2,3$$

and $\hat{\theta}_{ji}$ and f_{lji} are elements of $\hat{\Theta}_i$ and F_i respectively. The flow chart of the RALE algorithm is shown in Figure 10.2.

The RALE.LOD absolute load program can be used to estimate the coefficients of the polynomials $A(q^{-1})$ and $C(q^{-1})$ from the input data y(k). The estimated parameters are then used to calculate the estimated spectra. The RALE.LOD absolute load program can be found on the DSP56000/1 Demonstration Software #4 diskette.

1. POWER-DOWN THE PC, AND FUNCTION AND WHITE NOISE GENERATORS.
2. ON THE EVB, REMOVE JUMPER JP2, INSTALL JUMPER JP1 AND A 2 MHZ CRYSTAL TO ALLOW OPERATION OF THE EVB AT AN OUTPUT SAMPLE RATE OF 15.625 KHZ. The output sample rate is equal to the input clock frequency divided by 128 (2 MHz/128 = 15.625 Khz).
3. BUILD A SUMMING OPERATIONAL AMPLIFIER TO GENERATE THE INPUT SIGNAL Y(t) BY ADDING THE FUNCTION GENERATOR'S OUTPUT X(t) TO THE WHITE NOISE GENERATOR'S OUTPUT V(t).
4. CONNECT THE INPUT SIGNAL Y(t) TO THE EVB'S "BNC2" INPUT (A/D INPUT) AND TO THE OSCILLOSCOPE'S CHANNEL ONE INPUT.
5. POWER-ON AND BOOT THE PC.
6. POWER-ON THE FUNCTION AND WHITE NOISE GENERATORS.
7. SET THE SINUSOIDAL INPUT FREQUENCY TO 5000 HZ.
8. INSERT THE "DSP56000/1 DEMONSTRATION SOFTWARE #4" DISKETTE INTO DRIVE "A".
9. LOAD THE "ADS56000 USER INTERFACE SOFTWARE" PROGRAM.
10. TYPE **load a:rale.lod<return>** to load the assembled and linked version of the RALE assembler program.
11. TYPE
```
display off<return>
break #1 p:$257<return>
go<return>
```
to execute the RALE program.
12. ALLOW A FEW SECONDS FOR THE RALE PARAMETERS TO CONVERGENCE, THEN TYPE:
```
force b<return>
radix f y:$108..$10b<return>
display y:$108..$10b<return>
```
to display the estimated coefficients a1, a2, c1 and c2.

The display on your screen should be similar to:

```
Y:$0108 -0.4035406 -0.3607159  0.1752567 -0.0798273
ADM#0   P:$0257 0C0251          = JMP <$251
```

Note that the value of the coefficients can vary slightly depending on the signal-to-noise ratio in your system.
13. TYPE **quit**<return> IF YOU WANT TO EXIT THE "ADS56000 USER INTERFACE SOFTWARE" PROGRAM.

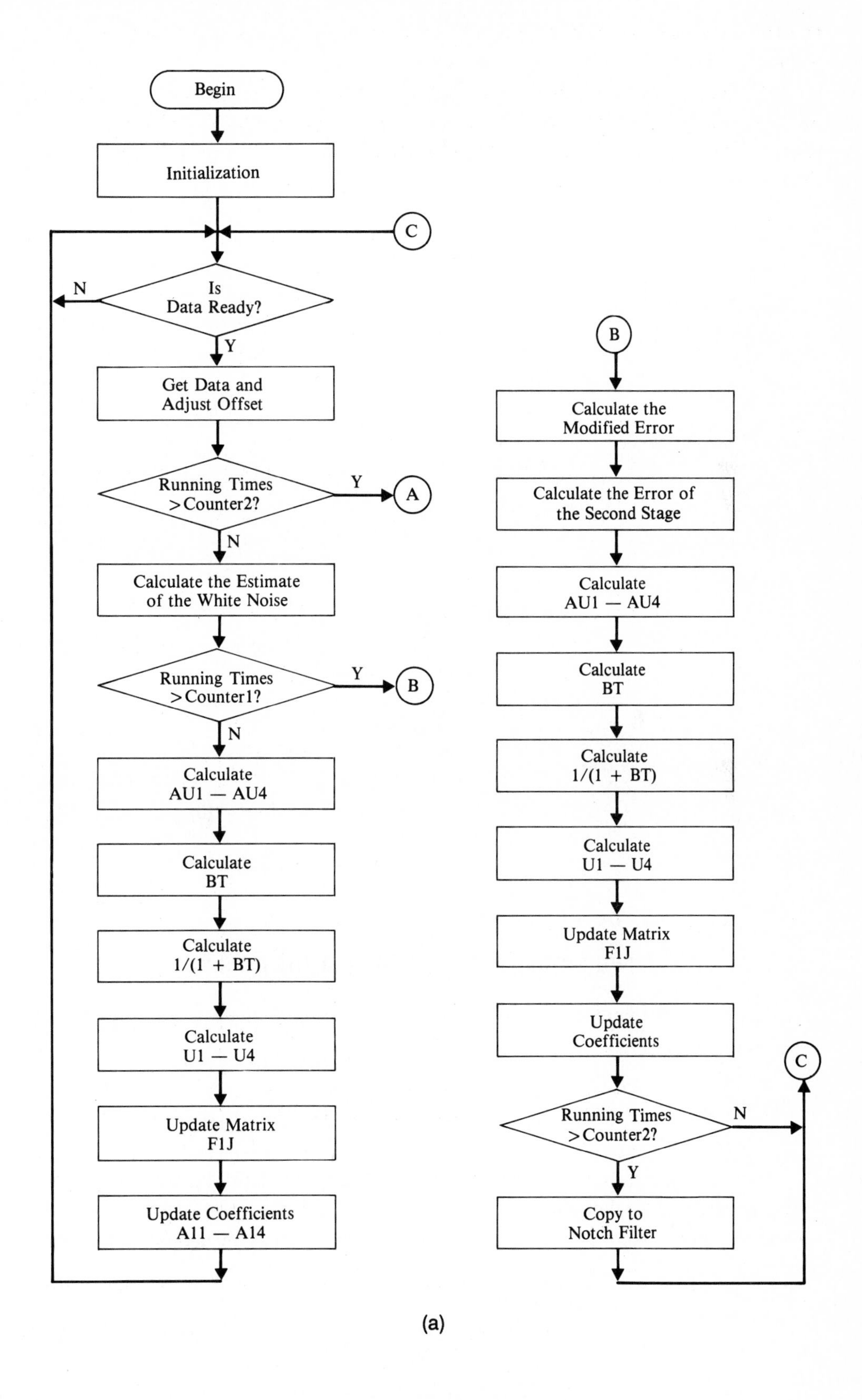

Figure 10-2 Flow Chart of Rale Algorithm

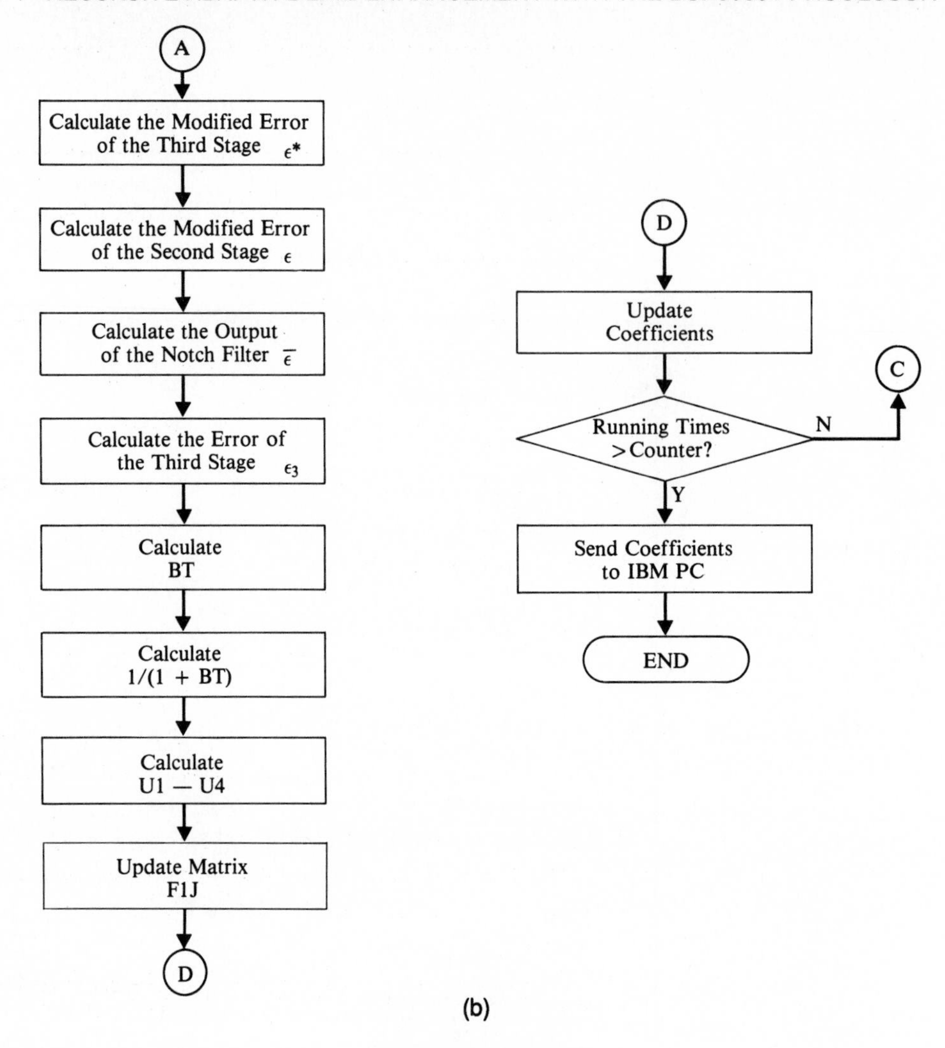

(b)

Figure 10-2 *(cont.)*

10.3 PLOTTING THE ESTIMATED SPECTRA ON THE PC'S SCREEN:

The estimated parameters $\hat{a}_{3i}$ and $\hat{c}_{3i}$ are used to compute the estimated spectra $S_y(w,k)$ according to the relation:

$$S_y(w,k) = \hat{C}_3(q^{-1},k)\,\hat{C}_3(q,k)/\hat{A}_3(q^{-1},k)\,\hat{A}_3(q,k); \qquad q = e^{jw} \tag{10.34}$$

The estimated spectra $S_y(w,k)$ can be computed, and displayed on the PC's screen, by using the SPECT program located on the DSP56000/1 Demonstration Software #4 diskette. Note that the program requires a VGA display adapter. Without the VGA, only the value of the estimated frequency will be appear on the screen.

1. INSERT THE "DSP56000/1 DEMONSTRATION SOFTWARE #4 DISKETTE" IN DRIVE A.
2. TYPE **a:spect<return>**

   The following display will appear on your screen

```
Input Sampling Frequency:
```

3. Type **15625<return>** as the input sampling frequency.
   The following display will appear on your screen

```
Input a1,a2,c1,c2:
```

4. Type **-0.4035406,-0.3607159,0.1752567,-0.0798273<return>** as the inputs for a1, a2, c1 and c2 respectively. A display of the estimated spectra will appear on your screen.

10.4 REFERENCES:

[1] M. El-Sharkawy, and B. Peikari," Adaptive Stochastic Filter With No Strict Positive Real Condition," IEEE Trans on ASSP, Vol ASSP-35, No. 11, PP 1547 - 1556, November 1987.

[2] M. El-Sharkawy, and B. Peikari," Multistage Adaptive Stochastic Filters," IEEE Transactions on Circuits and Systems, Vol.35, August 1988.

[3] I.D.Landau, "A Feedback Approach to Adaptive Filtering," IEEE Trans. on Information Theory, vol. IT-30, No.2, pp 251-262, March 1984.

[4] I.D. Landau, et. al.," Application of Output Error Recursive Estimation Algorithms for Adaptive Signal Processing," IEEE International Conference on ASSP, Vol.2, May 1982.

[5] I.D. Landau, " Near Supermartingales for Convergence Analysis of Recursive Identification and Adaptive Control Schemes," Int.J.Contr., Vol.35, pp 197-226, 1982.

[6] B. Friedlander and J.O.Smith," Analysis and Performance Evaluation of Adaptive Notch Filter," IEEE Trans. Information Theory, Vol. IT-30, pp 283-295, March 1984.

[7] B. Friedlander," Recursive Maximum Likehood Algorithm for ARMA Line Enhancement," IEEE Trans. ASSP, Vol.30, August 1982.

[8] R.A. David et al., "IIR Algorithms for Adaptive Line Enhancement," ICASSP-83, Boston, 1983.

[9] L. Ljung, "On Positive Real Transfer Function and the Convergence of Some Recursive Schemes, " IEEE Trans. Automat. Contr., Vol AC-22, No 4, August 1984.

[10] G.C.Goodwin and K.S.Sin," Adaptive Filtering Prediction and Control," Prentice-Hall 1984.

[11] B.Widrow and S.Stearns, "Adaptive Signal Processing," Prentice-Hall 1985.

[12] C. Cowan and P.Grant, "Adaptive Filters," Prentice-Hall 1985.

[13] D.Graupe, "Identification and Adaptive Filtering," Krieger 1984.

[14] N.B.Jones,"Digital Signal Processing," Peter Peregrinus 1982.

[15] M. Honig and D. Messerschmitt, "Adaptive Filters," Klumer Academic 1984.

11

IMPLEMENTING FIR AND ADAPTIVE FIR FILTERS WITH THE DSP56200 PERIPHERAL

In this chapter, FIR filters and adaptive FIR (AFIR) filters are implemented on a simple DSP56200 Demonstration System. The DSP56200 Demonstration System consists of an A/D converter, a DSP56001 processor, a DSP56200 peripheral, and a D/A converter. An ADS is used to provide and control the DSP56001; an EVB is used to provide and control the A/D converter and the D/A converter. A simple DSP56200 Development Board is designed and used to provide and control the DSP56200 peripheral. The DSP56001 processor's Memory Expansion Port A and the DSP56200 peripheral's Asynchronous Parallel Interface Port are used to interface the DSP56001 processor to the DSP56200 peripheral. The DSP56200 peripheral is initialized and accessed by the DSP56001 processor. The DSP56001 processor's SSI Port is used to accommodate serial data transfers from the A/D converter to the DSP56001 processor and from the DSP56001 processor to the D/A converter.

In the DSP56200 Demonstration System defined above and shown in Figure 11.1, an analog input signal x(t) is digitized by the A/D converter on the EVB. Via the receive channel of the DSP56001 processor's SSI Port, the digitized signal x(n) is received by the DSP56001 processor located on the ADM. The DSP56001 processor initializes the DSP56200 peripheral and provides it with commands and data to process the digital signal x(n) using its hardware-coded FIR or AFIR filtering algorithms. The digital signal produced by the DSP56200 peripheral is either:

1. the output y(n) from the FIR filter algorithm, or
2. the error signal -e(n), the difference between the desired response d(n) and the output y(n) of the AFIR filter algorithm.

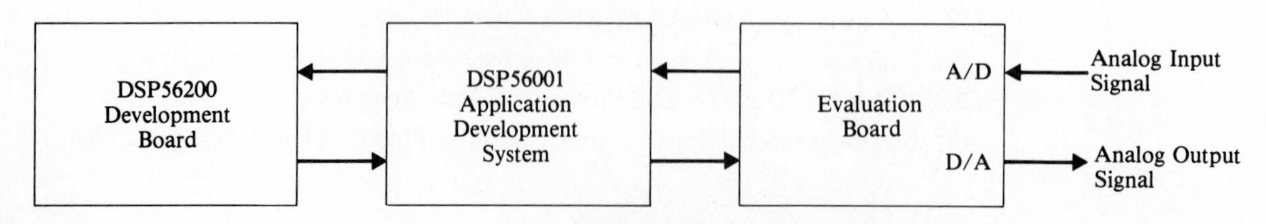

Figure 11-1 DSP56200 Demonstration System

As in Chapter 9, it is assumed that the desired response signal d(n) is equal to the input signal x(n). Via the transmit channel of the DSP56001 processor's SSI Port, the digital output signal y(n) or -e(n) is transferred to the D/A converter where it is converted to the analog output signal y(t) or -e(t).

This chapter begins with a general description of the DSP56200 peripheral. A simple DSP56200 Development Board is then designed to provide the DSP56200 peripheral and control it from the DSP56001 processor on the ADM. The register model of the DSP56200 peripheral and the instructions and macros to implement FIR and AFIR filters on the DSP56200 Demonstration System are also described.

11.1 GENERAL DESCRIPTION OF THE DSP56200 PERIPHERAL:

The DSP56200 peripheral is a 28-pin HCMOS, algorithm-specific device designed to perform computationally intensive digital filtering tasks. It implements both a FIR filtering algorithm and an adaptive AFIR filtering algorithm, where the AFIR algorithm uses the least-mean square (LMS) adaption technique. The DSP56200 peripheral's high performance architecture combined with its low cost and compact 28-pin package, make it an attractive solution for many DSP applications such as conventional filtering, noise cancelling, and echo cancelling. Since the DSP56200 peripheral's FIR and AFIR algorithms are hardware-coded, software development time is minimized, thereby further lowering system cost. The DSP56200 peripheral's high performance features include:

11.1.1 High Performance Architecture:

The DSP56200 peripheral's high performance results from its algorithm-specific architecture where three execution units operate in parallel. The execution units include: the Asynchronous Parallel Interface Unit, the Arithmetic Unit, and the Address Generation Unit. The major architectural components of the DSP56200 peripheral are shown in Figure 11.2 and include:

1. a 256*24-bit Coefficient RAM,
2. a 256*16-bit Data RAM,
3. an Arithmetic Unit with a 16*24 hardware multiplier and a 40-bit accumulator,
4. an Address Generator,
5. an Asynchronous Parallel Interface,
6. a Serial Cascade Interface, and
7. Timing and Control circuitry.

The Computation Unit, shown in Figure 11.3, includes:

1. the 256*24-bit Coefficient RAM,
2. the 256*16-bit Data RAM,
3. the Arithmetic Unit,
4. a Parallel Interface, and
5. two Bus Controllers.

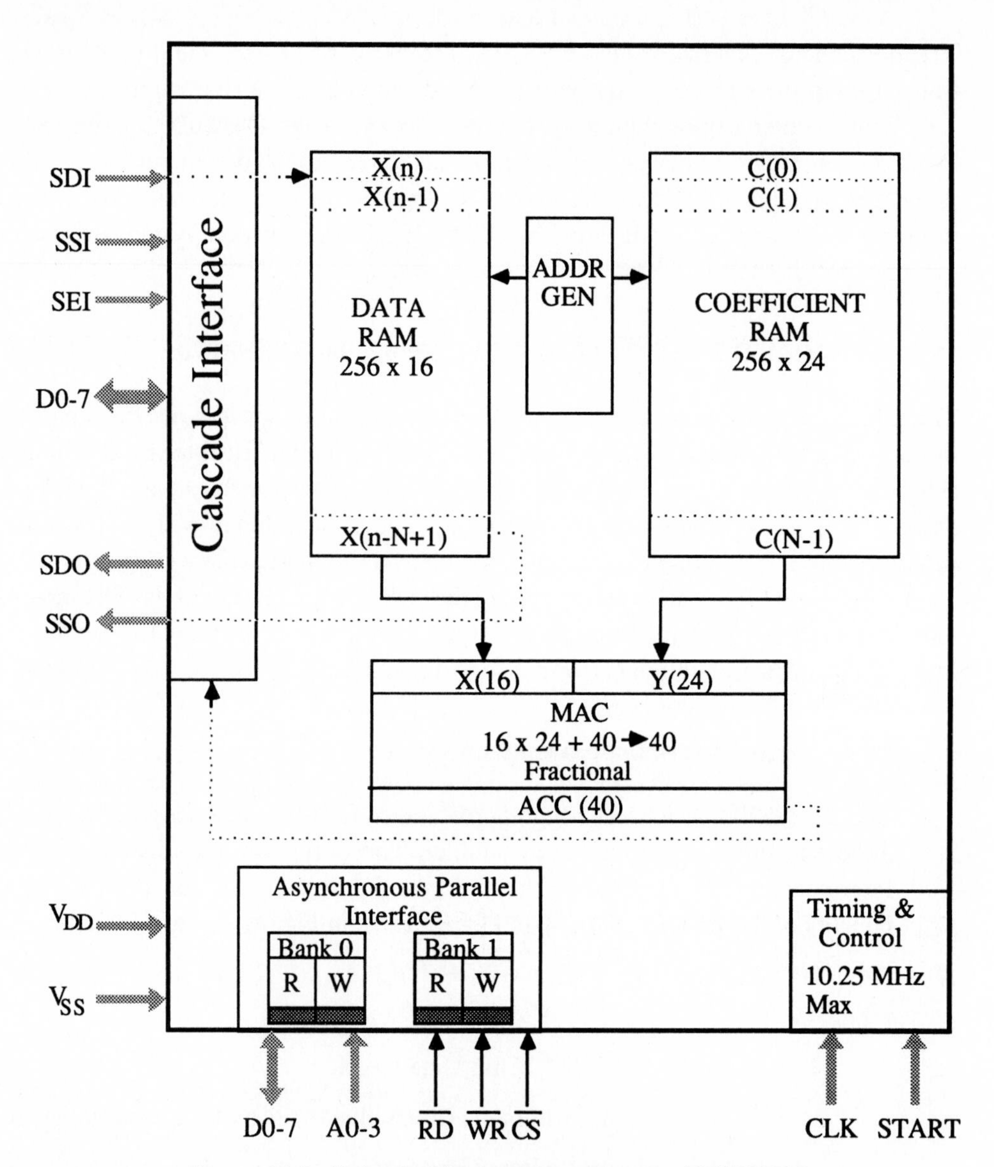

Figure 11-2 Major Architectural Components of DSP56200

The Computation Unit performs the arithmetic necessary to execute the hardware-coded FIR or AFIR filtering algorithms. The Data RAM is configured as a variable length circular queue, allowing it to function as a circular FIFO-type shift register. The Arithmetic Unit is configured to execute either the LMS adaptation algorithm for the AFIR filter or execute the FIR filter algorithm. It performs 24*16-bit multiply-accumulate (MAC) operations, or coefficient update operations, at a rate of 100 nsec per operation. The Parallel Interface resembles that of a fast static RAM, thereby allowing an easy interface to MPUs which have tight timing requirements or to fast general purpose DSP processors. In this chapter, for example, the DSP56200 peripheral is interfaced to the DSP56001 processor.

The DSP56200 peripheral's Cascade Interface performs all of the functions associated with cascading multiple DSP56200 peripherals. Note, the DSP56200 peripheral's Cascade Interface is not used in this chapter.

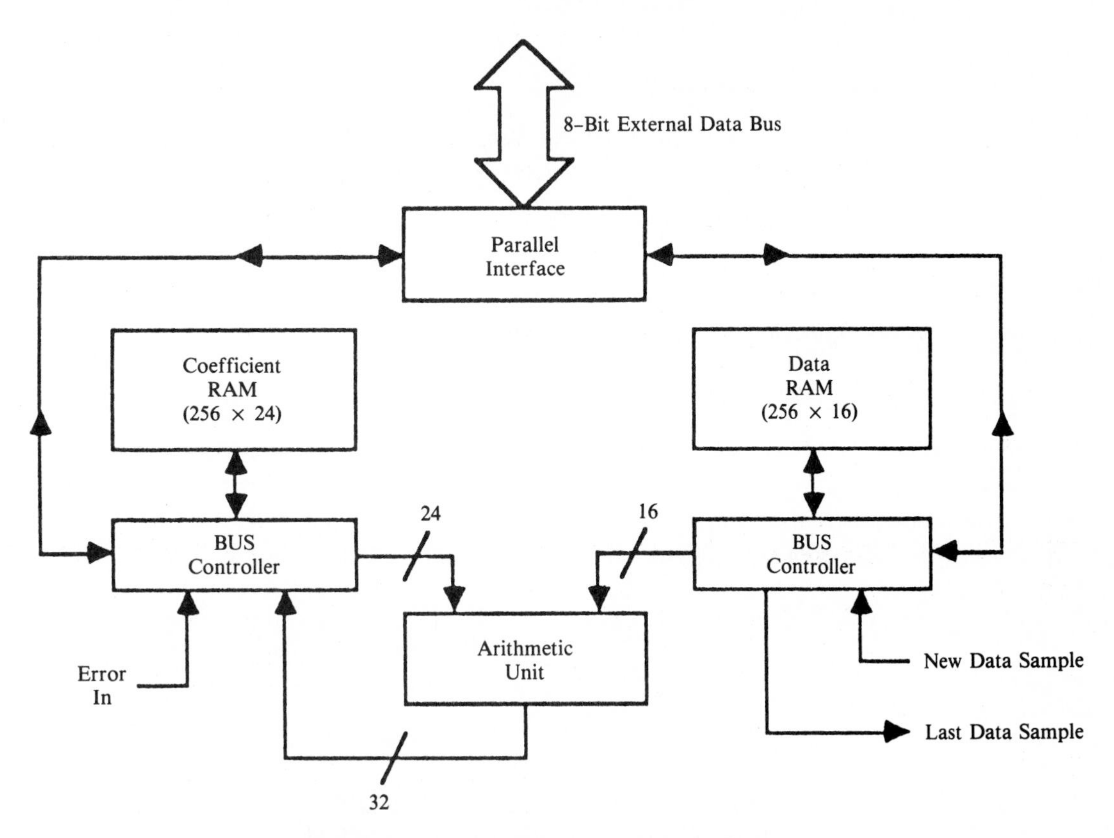

Figure 11-3 Computational Unit Block Diagram

11.1.2 Three Operating Modes:

The DSP56200 peripheral operates in one of three selectable modes:

1. ***Single FIR Filter:***

   The single FIR filter mode is used to implement one FIR filter, either on a single DSP56200 peripheral or on several cascaded DSP56200 peripherals, where each peripheral can have up to 256 filter taps. In this mode, the DSP56200 peripheral is configured to implement the FIR filtering operation described in Chapter 6.

2. ***Dual FIR Filter:***

   The dual FIR mode is an extension of the single FIR filter mode. It allows two independent filters to be implemented using a single DSP56200 peripheral, where both filters must have the same number of taps up to 128. In this mode, the DSP56200 peripheral is not cascadable.

3. ***Single Adaptive FIR Filter:***

   The single adaptive FIR filter mode is used to implement one adaptive filter, either on a single DSP56200 peripheral or on several cascaded DSP56200 peripherals, where each peripheral can have up to 256 filter taps. Adaptive FIR filters perform a multiply-accumulate operation, followed by an adaptation operation which modifies the coefficients. The DSP56200 peripheral uses the LMS adaptation algorithm described in Chapter 9 and updates all filter coefficients once during each sampling period.

11.1.3 High Filtering Rates:

With a 10 MHz external clock frequency, the DSP56200 peripheral(s) can implement, for example:

1. a 227 kHz FIR Filter (32 Taps, 1 DSP56200),
2. a 37 kHz FIR Filter (256 Taps, 1 DSP56200),
3. a 115 kHz Adaptive FIR filter (256 Taps, 8 Cascaded DSP56200s), and
4. a 19 kHz Adaptive FIR filter (256 Taps, 1 DSP56200).

Many other configurations are also possible.

11.1.4 Digital Filtering Options:

The DSP56200 peripheral offers a 16-bit rounding option for the filter's output, and filter tap lengths programmable from four to 256. For an adaptive FIR filter, the DSP56200 peripheral offers a dc tap option, a programmable loop gain, a programmable coefficient leakage term.

The DSP56200 peripheral's LMS adaptation algorithm is described in Chapter 9 and can be rewritten as:

$$\begin{aligned} b_i(n+1) &= b_i(n) + Ke(n)x(n-i) \pm \text{leakage} \\ i &= 0, 1, .., N-1 \end{aligned} \tag{11.1}$$

where the **leakage** term is included to compensate for coefficient drift due to a narrowband input signal. For example, when a narrowband signal such as a sine wave is applied to the input of the AFIR filter, the coefficient values drift because of an insufficient number of frequency components in the input signal. As a result, the coefficient values will erroneously grow in magnitude, creating a larger than expected error term. However, the DSP56200 peripheral's programmable leakage term slowly pushes the coefficient values towards zero, effectively offsetting the slow and erroneous growth in coefficient magnitude. The leakage term is programmed to be small enough that it does not affect the convergence of the filter, but large enough to prevent the occurrence of coefficient drift.

The DSP56200 peripheral's dc tap option can be used in conjuction with the AFIR algorithm to synthesize an internal signal which cancels the effect of a dc offset in the input signal. Such a dc offset could be present, for example, in the digitized output of an A/D converter. The dc tap option can also be used to add a dc offset to the output of the FIR filter algorithm.

11.2 DSP56200 REGISTER MODEL:

The DSP56200 peripheral is initialized and accessed by the DSP56001 processor, or any other host processor, through a set of internal control and data transfer registers. These registers also provide an access path to the contents of the Coefficient RAM and Data RAM, thereby allowing unused memory to be used as auxiliary system storage. All accesses to the DSP56200 peripheral's registers occur through its Asynchronous Parallel Interface Unit.

The DSP56200 peripheral's registers are divided into two banks of sixteen registers each as shown in Figure 11.4. Register Bank 0 contains those registers which are commonly accessed during algorithm execution; Register Bank 1 contains those registers which are used for device initialization. The two register banks share a common Configuration Register at location $0F in each register bank. Selection of a particular register bank for access is accomplished by writing a logic one or a logic zero to the least significant bit (bit 0) of the Configuration Register, where a logic one selects Register Bank 1 and a logic zero selects Register

Bank 0. Once the desired register bank has been selected, the individual registers within the register bank are accessed through the DSP56200 peripheral's package pins A0-A3, /RD or /WR, and /CS. The registers have been ordered so that the host processor can efficiently access them as a group through an autoincrementing address mode.

In this chapter, the DSP56001 processor treats the DSP56200 peripheral as shown in Figure 3.8. In the following sections, the sixteen DSP56200 registers of each register bank are mapped into Y-Memory locations $FFF0-$FFFF. That is, the DSP56001 processor accesses the DSP56200 registers at Y-Memory locations $FFF0-$FFFF.

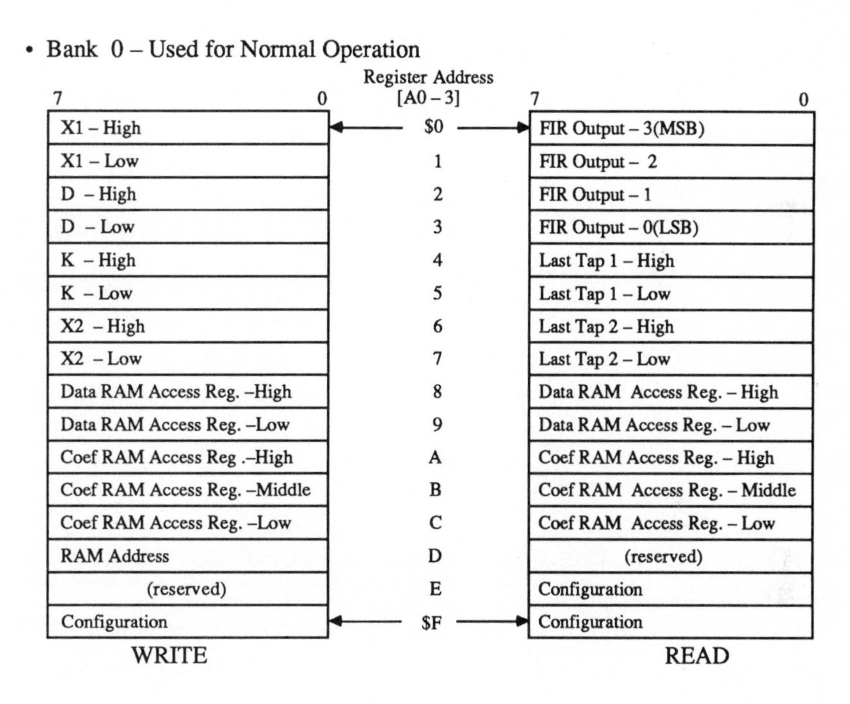

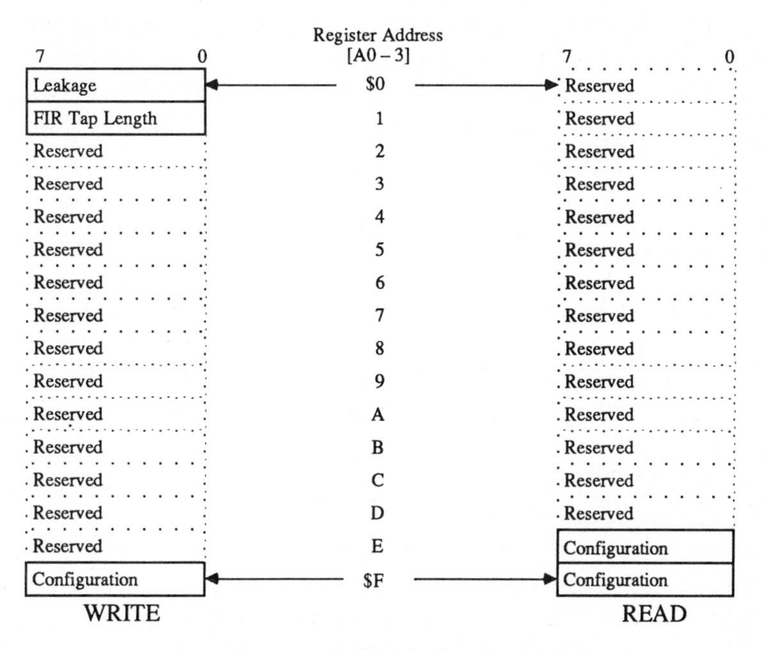

Figure 11-4 DSP56200 Register Model

The DSP56200 peripheral's registers are:

1) *X1 Register, X2 Register*

The 16-bit X1 Register holds the present data input value for both the FIR filter mode and the Adaptive FIR filter mode. The data value in the X1 Register is transferred to the Data RAM once per START cycle. In the dual FIR filter mode, an additional 16-bit X2 Register holds the present data input value for the second FIR filter. The data value in the X2 Register is transferred to the Data RAM once per START cycle.

2) *D Register* (Adaptive FIR Filter Mode Only)

When the DSP56200 peripheral is used in the adaptive FIR filter (noise or echo cancellation) mode, the 16-bit D Register holds a reference (echo) input value d(n). This register is not used in the nonadaptive FIR filter modes.

3) *K Register* (Adaptive FIR Filter Mode Only)

The 16-bit K Register holds the loop gain value K which is used in the LMS adaption process, where K is multiplied by the error term and each tap to generate an updated coefficient value (see Chapter 9). The maximum recommended 16-bit loop gain values, shown in Table 11.2, must always be positive numbers, i.e., Bit 7 of the Most Significant Byte of the K Register should always be zero.

TABLE 11.2. Maximum Recommended Loop Gains.

FIR Tap Length	*Maximum Recommended K*
Up To 32 (20 Hex)	0.75 (6000 Hex)
Up To 64 (40 Hex)	0.375 (3000 Hex)
Up To 128 (80 Hex)	0.125 (1000 Hex)
Up To 192 (C0 Hex)	0.078125 (A00 Hex)
Up To 256 (FF Hex)	0.0703125 (900 Hex)

4) *Configuration Register*

The Configuration Register is used to select Register Bank 0 or 1 for access, and select the operating mode and associated options. The Configuration Register's individual bits are shown in Figure 11.5 and defined as follows:

Bit 7 (FIR/Adaptive-FIR): When set to a logic one, bit 7 selects the DSP56200 peripheral to be configured as a single adaptive FIR filter; when set to a logic zero, the device is configured as either a single or dual, nonadaptive FIR filter.

Bit 6 (Single-FIR/dual-FIR): When set to a logic one, bit 6 selects the DSP56200 peripheral to be configured as a single FIR filter; when set to a logic zero, the device is configured as either a single FIR filter or a single adaptive FIR filter. Table 11.3 summarizes the valid filter modes.

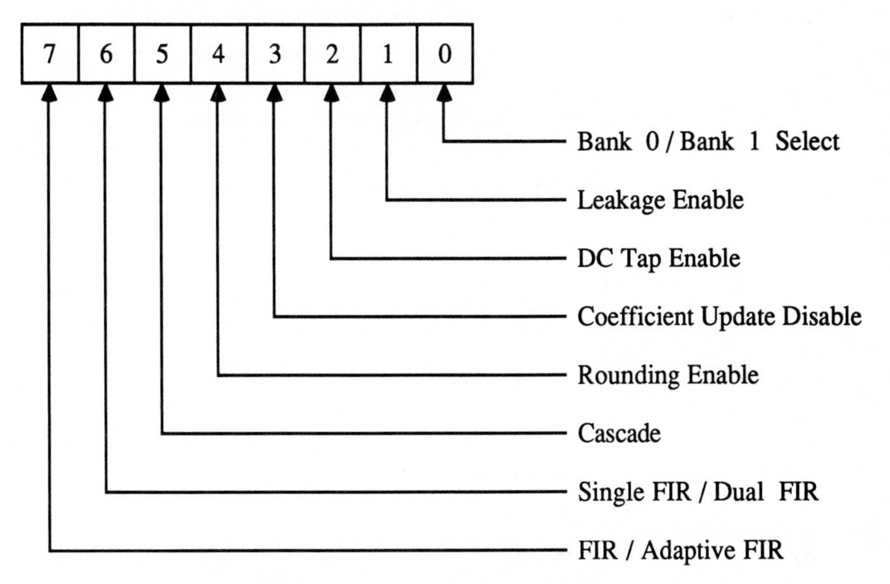

Figure 11-5 Configuration Register

TABLE 11.3. DSP56200 Filter Modes.

Adaptive/ Nonadaptive, Config. Reg. Bit 7	*Single/Dual, Config. Reg. Bit 6*	*Mode*
0	0	Single FIR Filter
0	1	Dual FIR Filter
1	0	Single Adaptive FIR Filter
1	1	(Operation Undefined)

Bit 5 (Position in Cascade): When set to a logic one, Bit 5 selects the DSP56200 peripheral to operate as either a stand-alone device or the first device in a cascade; when set to a logic zero, it selects the device to operate as not the first device in a cascade. Bit 5 must be set to a logic zero whenever the DSP56200 peripheral is used in the dual FIR filter mode, where the device is not cascadable.

TABLE 11.4. DSP56200 Cascade Configurations.

Config. Reg. Bit 5	*System Configuration*
0	Single DSP56200 System (Standalone) or
0	First DSP56200 in a Cascaded System
1	Not First DSP56200 in a Cascaded System

Bit 4 (16-Bit Rounding): When set to a logic one, Bit 4 selects the 16-bit rounding mode; when set to a logic zero, it disables the 16-bit rounding mode. In the single FIR filter mode, the output data can be 32-bits or 16-bits of precision. When the 16-bit rounding mode is selected, the 32-bit output data is rounded to 16-bits of precision, where Data Bytes 3 and 2 of Register Bank 0 contain the valid 16-bit rounded result and Data Bytes 1 and 0 contain invalid data. If the rounding mode is not selected, Data Bytes 3, 2, 1, and 0 contain the full 32-bit result, where Byte 4 is the most significant byte. When the 16-bit rounding mode is selected for the dual FIR filter mode, the output data value of each filter is rounded to 16-bits of precision.

Bit 3 (Adaptive Disable): When set to a logic one, bit 3 selects the DSP56200 peripheral to operate as an adaptive FIR filter; when set to a logic zero, it selects the device to operate as either a single or dual, nonadaptive FIR filter. When bit 3 is set to a logic one, the device will use the LMS algorithm compute error terms from the last set of updated filter coefficients. In echo canceling applications, bit 3 is typically set to a logic one when "double talk" is detected.

Bit 2 (dc Tap Enable): When bit 2 is set to a logic one, the DSP56200 peripheral's dc Tap capability is disabled; when set to a logic zero, the dc Tap capability is enabled. The dc Tap is used as if it is a data tap, where, when enabled, it has a fixed value of $7FFF and it is substituted for the last tap in the Data RAM. Note, upon substitution, the last tap data is not lost, however, and can be read using the Host Data RAM Access Register or the last Tap Register. The dc Tap is multiplied by its corresponding (last) coefficient during the FIR filtering phase, and is also used when updating the last coefficient in the adaptive FIR filter mode. The dc Tap is typically used in the adaptive FIR filter mode to remove differential dc offset in the digitized output of A/D converters. The dc Tap should be selected only in the last device of a cascade.

Bit 1 (Leakage Enable): When bit 1 is set to a logic one, the DSP56200 peripheral's leakage capability is enabled; when set to a logic zero, the leakage capability is disabled.

Bit 0 (Register Bank Select): When bit 0 is set to a logic one, the DSP56200 peripheral's Register Bank 1 is selected for access by the host processor; when set to logic zero, Register Bank 0 is selected for access. Since the Configuration Register is located in both Register Bank 0 and 1, bit-0 is always available for selection of the desired register bank. The Configuration Register can be asychronously accessed from the Parallel Interface, allowing access to both register banks during one sampling period.

5. *Leakage Register* (Adaptive FIR Filter Mode Only)

   The 8-bit Leakage Register contains a "leakage" value which is used to control coefficient drift in the adaptive filter mode. For this mode, Configuration Register bit 3 enables coefficient updating and bit 1 enables use of the leakage value.

6. *FIR Tap Length Register*

   The 8-bit Filter Tap Length Register specifies the quantity (number) of taps used in the FIR filter. The register is loaded with the quantity of taps minus one. For example, if a 256-tap filter is desired, the register is loaded with 255. Valid tap quantities range from 3 to 255 in the single FIR filter mode or the single Adaptive FIR filter mode, and from 3 to 127 in the dual FIR filter mode. Loading this register also resets the DSP56200 peripheral. Typically, this register is loaded by the host processor upon power-up.

7. *RAM Address Register*

   The 8-bit RAM Address Register points to the Coefficient RAM or Data RAM location to be accessed by the host processor. It allows the host to access any taps used in the filter and any RAM location not used in the filter. The RAM Address Register is automatically postincremented by one during each sampling period. Thus, the same Coefficient RAM and Data RAM location can be internally accessed during each sampling period. In the Dual FIR Mode, the data samples of the first FIR filter cannot be accessed using the RAM Address Register. In the single FIR filter modes, the starting address of unused RAM is FTL + 1. In the dual FIR filter mode the starting address of unused RAM is 2 * (FTL + 1). Note that in the dual FIR filter mode, the coefficients for the second filter start at Coefficient RAM location FTL + 1; the coefficients for the first filter start at Coefficient RAM location 0.

8. *Coefficient RAM Access Register*

   The 24-bit Coefficient RAM Access Register allows the host processor to read from or write to any location in the Coefficient RAM. To read a Coefficient RAM location, the host processor:

   1. writes into the RAM Address Register the value which points to the desired location,
   2. waits for two pulses to occur on the START pin (i.e., two sample periods), and
   3. then reads the coefficient values from the Coefficient RAM Access Register.

   To write to a Coefficient RAM location, the host processor writes the desired value to the Coefficient RAM Access Register. The write operation is completed during the next sampling period. Locations in the Coefficient RAM are pointed to by the RAM Address Register.

9. *Data RAM Access Register*

   The 16-bit Data RAM Access Register allows the host processor to read from or write to any location in the Data RAM. To read a Data RAM location, the host processor:

   1. writes into the RAM Address Register the value which points to the desired location,
   2. waits for two pulses to occur on the START pin (i.e., two sample periods), and
   3. reads the data values from the Data RAM Access Register.

   To write to a Data RAM location, the host processor writes the desired value to the Data RAM Access Register. The write operation is completed during the next sampling period. If the addressed Data RAM location resides within the FIR filter structure (a circular buffer in the Data RAM), the DSP56200 peripheral automatically performs a logical-to-physical address conversion, thereby correctly accessing the desired filter tap.

10. *LAST TAP 1, LAST TAP 2*

    The 16-bit Last Tap 1 Register can provide the host processor with a copy of the last data sample in the Data RAM. When in the Dual FIR Filter mode, the Last Tap 2 Register can provide the host processor with a copy of the last data sample of the second FIR filter. Since the last tap data is also transmitted serially to the next device in a cascade, these registers are provided only for convenience and are not required when cascading devices. These registers, in conjunction with the X1 and X2 Input Registers, are useful for signal power calculations.

11. *Output Data Register*

    Output Data Register bytes 3 through 0 contain the output of the filter(s). In the single filter mode, the output will be four bytes (32-bits) unless the 16-bit rounding mode is set. If the 16-bit rounding mode is selected, bytes 3 and 2 contain the rounded 16-bit output and bytes 1 and 0 contain invalid data. Only 16-bits of output are available for each filter in the dual FIR filter mode. Bytes 3 and 2 contain the output of the first FIR filter and bytes 1 and 0 contain the output of the second FIR filter. Both outputs are rounded to 16-bits of precision if the 16-bit rounding mode is enabled. Otherwise, the outputs are truncated to 16-bits. In the adaptive FIR filter mode, the output is the negative of the error e(n).

11.3 USING THE DSP56001 PROCESSOR TO INITIALIZE THE DSP56200 PERIPHERAL

Upon power-up, the DSP56200 peripheral's registers and RAMs must be initialized. In this chapter, the DSP56001 processor initializes the DSP56200 peripheral by:

1. writing to the Configuration Register at Y-Memory location $FFFF to select Register Bank 1;
2. writing to the FIR Tap Length Register at Y-Memory location $FFF1 to specify the quantity of filter taps and thereby also reset the peripheral's timing circuitry;
3. writing to the Configuration Register at Y-Memory location $FFFF to select Register Bank 0 and select the operating mode and its associated options;
4. writing valid coefficient values into the Coefficient RAM; and
5. writing a zero value into each location of the Data RAM.

11.4 DSP56200 DEVELOPMENT BOARD:

The DSP56001 processor's Expansion Port A and the DSP56200 peripheral's Asynchronous Parallel Interface are used to interface the DSP56001 processor to the DSP56200 peripheral. Note, the DSP56001 processor's Expansion Port A, interrupt, power supply, and clock signals are shown in Figure 11.6 and described in Section 11.7; the DSP56200 peripheral's Asynchronous Parallel Interface, Cascade Interface, power supply, and clock signals are shown in Figure 11.7 and described in Section 11.8. The DSP56200 Development Board is used to interface the DSP56200 peripheral to the DSP56001 processor on the ADM so the DSP56200 peripheral can be initialized and accessed by the DSP56001 processor. The DSP56200

Development Board can be build on a wire wrap prototype card which has a male connector that mates with the standard J3 Eurocard prototype female connector on the ADM. The ADM's J3 connector has the DSP56001 signal-to-pin assignments shown in Table 11.1.

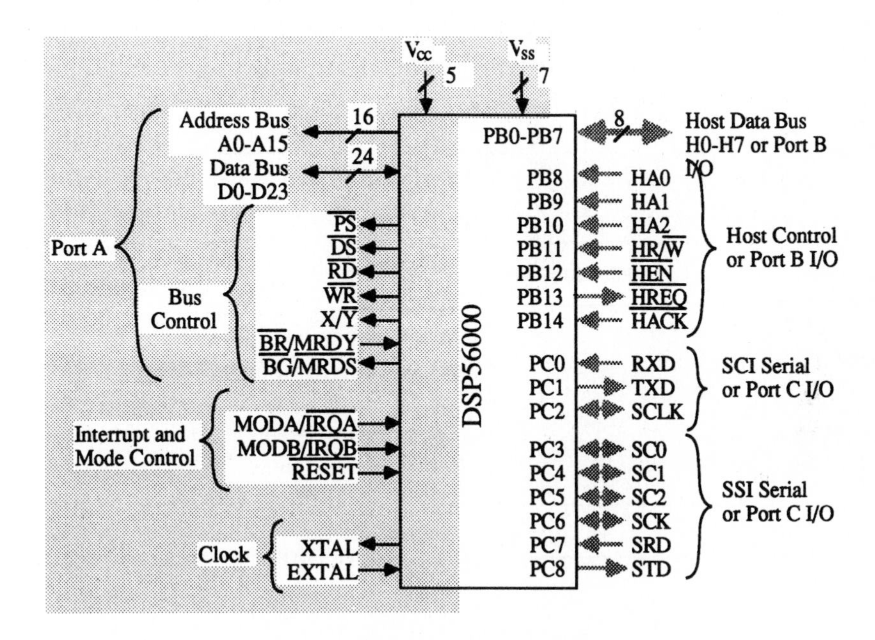

PGA Sockets Available From

Manufacturer	Part Number	Telephone	Comment
SAMTEC	MVAS - 120 - ZSTT - 13	(812) 944-6733	Includes center cutout
	CPAS - 88 - ZSTT - 13BF		No cutout
Advanced Interconnections	4CS088 - 01TG	(401) 823-5200	Includes Center cutout
AMP	55269 - 1	Local Distribution	Zero Insertion Force

Figure 11-6 DSP56001 Signal Groups

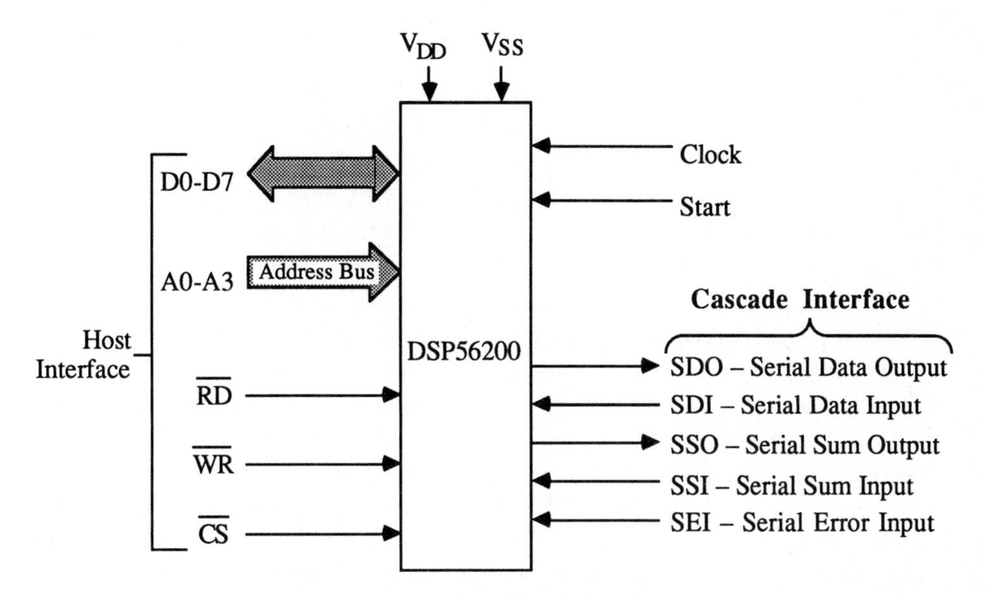

Figure 11-7 DSP56200 Signal Groups

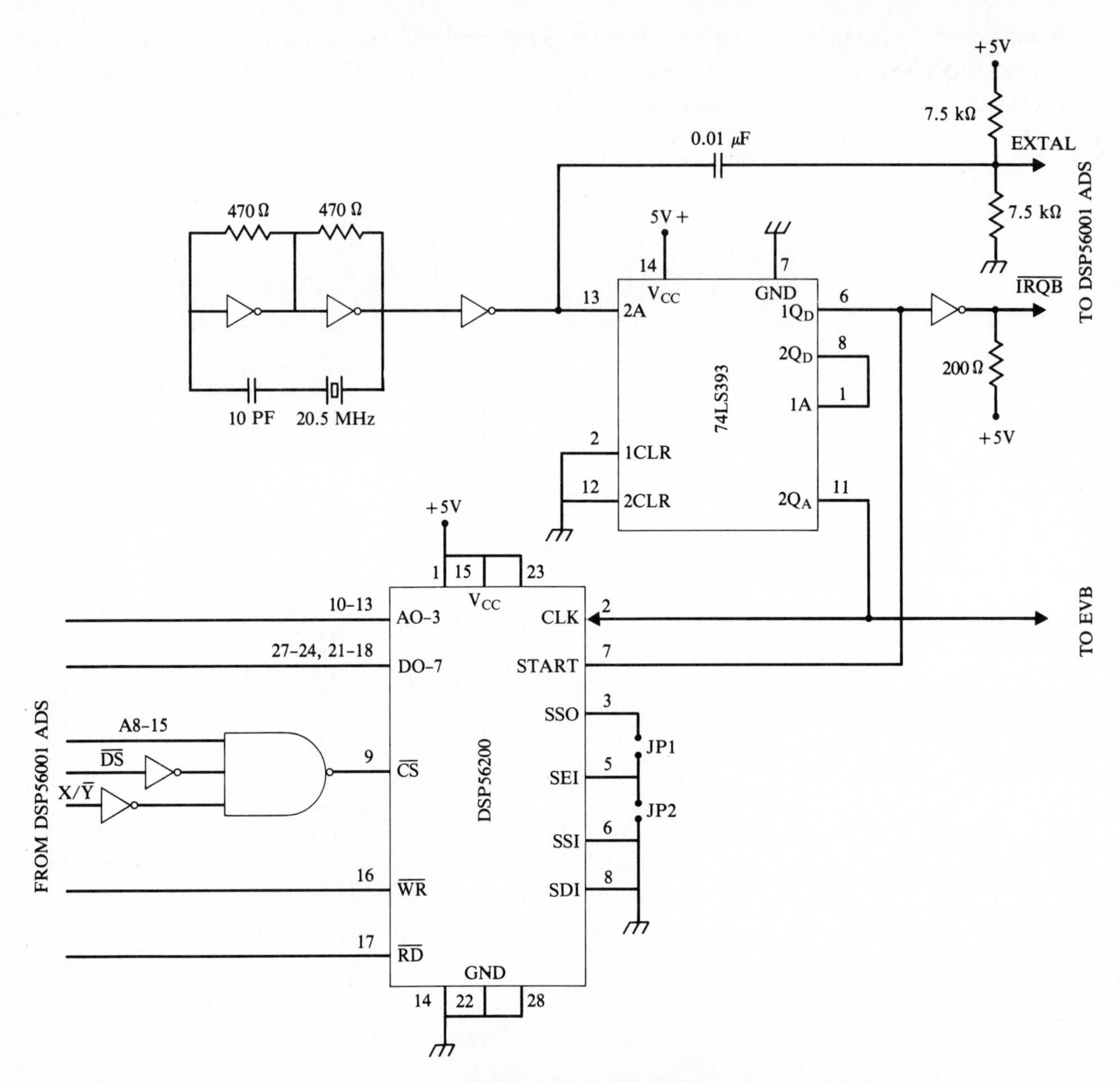

Figure 11-8 DSP56200 Development Board

The schematic diagram for the DSP56200 Development Board is shown in Figure 11.8. A 20.5 MHz crystal provides an external clock to the DSP56001 processor (on the ADM) and the 74LS393 Dual 4-bit Counters. The 74LS393 counters, in turn, provide the clock and START pulse for the DSP56200 peripheral, the clock for the EVB, and the /IRQB interrupt signal for the DSP56001 processor (on the ADM).

The 74LS393 counter device has two independent 4-bit binary counters, each having a CLEAR and a CLOCK input. The two 4-bit counters can be concatenated to form a single 8-bit counter, thereby forming a binary counter with a divide-by-256 capability. The 74LS393 counter has parallel outputs from each counter stage so that any submultiple of the input count frequency is available. In the case of the DSP56200 Development Board:

1. a 80 KHz (20.5 MHz/256) output provides the START pulse for the DSP56200 peripheral, and the /IRQB interrupt signal for the DSP56001 processor; and
2. a 10.25 MHz (20.5 MHz/2) output provides the external clock for the DSP56200 peripheral and the EVB.

TABLE 11.1. DSP56001/J3 Connector Description.

J3 PIN NUMBER	*DSP56001 SIGNAL MNEMONIC*	*DSP56001 PROCESSOR SIGNAL DESCRIPTION*
A1-A24	D23-D0	Data bus bits 23-0
A25	PB0	Port B bits 0,1,3,5,7,9,11,13
A26	PB1	
A27	PB3	
A28	PB5	
A29	PB7	
A30	PB9	
A31	PB11	
A32	PB13	
B1,20,32	+5V	ADM +5V reference.
B2,11,21, 31	GND	ADM ground reference.
B3	RESET_IN~	Reset input. This signal can be an input or an open collector output.
B4,6,8,10, 22,23,		NO CONNECT
B5	CLOCK_IN	External clock input.
B7	IRQB_IN~	IRQB input.
B9	IRQA_IN~	IRQA input.
B12	Y7~	Address block E000-FFFF decode signal.
B13	Y6~	Address block C000-DFFF decode signal.
B14	Y5~	Address block A000-BFFF decode signal.
B15	Y4~	Address block 8000-9FFF decode signal.
B16	Y3~	Address block 6000-7FFF decode signal.
B17	Y2~	Address block 4000-5FFF decode signal.
B18	Y1~	Address block 2000-3FFF decode signal.
B19	Y0~	Address block 0000-1FFF decode signal.
B24	WR~	Write signal output.
B25	RD~	Read signal output.
B26	XY~	X/Y data select signal output.
B27	PS~	Program Select signal output.
B28	DS~	Data Select signal output.
B29	BR~	Bus Request signal input.
B30	BG~	Bus Grant signal output.
C1-C16	A15-A0	Address lines 15 to 0 outputs.
C17-C25	PC8-PC0	Port C bits 8 to 0.
C26	PB2	Port B bits 2,4,6,8,10,12,14.
C27	PB4	
C28	PB6	
C29	PB8	
C30	PB10	
C31	PB12	
C32	PB14	

11.5 IMPLEMENTING FIR FILTERS ON THE DSP56200 DEMONSTRATION SYSTEM:

The DSP56001 instructions listed below configure the DSP56200 peripheral as a stand-alone single FIR filter. Register Bank 1 is first selected, then the quantity (number) of taps minus one is loaded to the FIR Tap Length Register. Register Bank 0 is then selected, the 16-bit rounding mode is enabled and the adaptation, dc tap and leakage options are disabled.

```
movep   #$01,y:$ffff
movep   #ntaps-1,y:$fff1
movep   #$10,y:$ffff
```

The DSP56001 instructions listed below clear the DSP56200 peripheral's data input X1 Register and set up Coefficient RAM and Data RAM locations $0 for access by the DSP56001 processor.

```
move    #$0,x0
movep   x0,y:$fff0
movep   x0,y:$fff1
movep   x0,y:$fffd
```

The DSP56001 instructions listed below transfer coefficient values from the X-Memory to the DSP56200 peripheral's Coefficient RAM via the Coefficient RAM Access Register. DSP56001 Address Register R1 points to the coefficient buffer in X-Memory.

```
set     movep   x:(r1)+,y:$fffa
        movep   x:(r1)+,y:$fffb
        movep   x:(r1)+,y:$fffc

        move    r1,x0
        move    #MTAPS,a            ; MTAPS = 3*NTAPS-1
        cmp     x0,a
        jcc     set
```

The DSP56001 instructions listed below read the A/D converter, transfer the digitized data to the DSP56200 peripheral's data input X1 Register, read the DSP56200 peripheral's Output Data Register, and transfer the output data to the D/A converter. The contents of the data input X1 Register and Output Data Register are assumed to be $FFFF00 and $0000FF, respectively.

```
clr     a
clr     b

movep   x:$FFEF,x0          ; read SSI RX reg. for A/
                            ; data
move    x0,a0               ;x0 and a0 is $XXXX00
rep     #8
asl     a                   ;a1 is $0000XX
movep   a1,y:$fff0
rep     #8
asl     a                   ;a1 is $00XXXX
and     y1,a                ;a1 is $0000XX
movep   a1,y:$fff1

movep   y:$fff0,b1
and     y1,b                ;b1 is $0000XX
movep   y:$fff1,a1
and     y1,a
rep     #8
asl     b
add     a,b                 ;b1 is $00XXXX
rep     #8
asl     b                   ;b1 is $XXXX??
and     x1,b                ;b1 is $XXXX00
movep   b1,x:$FFEF          ;write to SSI TX reg.for D/A
```

The **FIR1.ASM** assembler macro shown in Figure 11.9 is used to implement the single FIR filter on the DSP56200 Demonstration System. The **FIR1.ASM** assembler macro uses the DSP56001 instructions described above with DSP56001 instructions similar to those shown Figure 5.8.

```
fir1  macro ntaps,mtaps

; Program start address

        org     p:$40

; Set up ADS board.

        movep   #0,x:$FFFF
        movep   #0,x:$FFFE          ;set bcr to zero
                                    ;no delay
        move    #0,r0
        move    #coef,r1
        move    #0,r2
        move    #2,r3
        move    #3,r4

        movec   #1,sp               ;init stack pointer
        movec   #0,sr               ;clear loop flag/interrupt
                                    ;mask bits

; Set up the SSI for operation with the EVB
; The following code sets port C to function as SCI/SSI

        movep   #0,x:$FFE1
        movep   #$1ff,x:$FFE1       ;write PCC

; The following code sets the SSI CRA and CRB control
; registers for external cont. synchronous clock, normal
; mode.

        movep   #$4000,x:$FFEC
        movep   #$3200,x:$FFED

; Initialize IPR to allow interrupt 2 & 3

        ori     #$02,mr
        andi    #$fe,mr

; Allow the IRQB interrupt, level

        movep   #$38,x:$FFFF

; Sample rate is controlled by DSP56200 system.

self    jmp     self                ;looping waiting for
                                    ; interrupt

;*********************************************
; Start Initialize                           *
;*********************************************

initial
        lua     (r0)+,r0
        movep   #$01,y:$ffff
        movep   #ntaps-1,y:$fff1
        movep   #$10,y:$ffff
        rti

set
        jset    #0,y:(r2),coeff
```

```
;
        lua     (r2)+,r2
        move    #$0,x0
        movep   x0,y:$fffd
        movep   x0,y:$fff0
        movep   x0,y:$fff1
coeff
        movep   x:(r1)+,y:$fffa
        movep   x:(r1)+,y:$fffb
        movep   x:(r1)+,y:$fffc
;
        rti

;*******************************************
; Interrupt Routine- Read A/D and write D/A *
;*******************************************

ad      jclr    #0,y:(r0),initial   ;Initialization

        move    r1,x0
        move    #mtaps,a        ; mtaps = 3 ntaps -1
        cmp     x0,a
        jcc     set
;
;
;
        clr     a       y:(r4),y1
        clr     b       y:(r3),x1

        movep   x:$FFEF,x0      ;read SSI RX reg. for A/D
                                ; data
        move    x0,a0           ;x0, a0 is $XXXX00
        rep     #8
        asl     a               ;a1 is $0000XX
        movep   a1,y:$fff0
        rep     #8
        asl     a               ;a1 is $00XXXX
        and     y1,a            ;a1 is $0000XX
        movep   a1,y:$fff1

        movep   y:$fff0,b1
        and     y1,b            ;b1 is $0000XX
        movep   y:$fff1,a1
        and     y1,a
        rep     #8
        asl     b
        add     a,b             ;b1 is $00XXXX
        rep     #8
        asl     b               ;b1 is $XXXX??
        and     x1,b            ;b1 is $XXXX00
        movep   b1,x:$FFEF      ;write to SSI TX reg.for D/A
        rti
;
;**********************************************************
; Set up JSR's at the interrupt addresses for SSI interrupts
;**********************************************************

        org     p:$000a         ;IRQB interrupt
        jsr     ad              ;Run program

        endm
```

Figure 11-9 FIR1.ASM Assembler Macro to Implement the FIR Filter on the DSP56200 Demonstration System

EXAMPLE 11.1: Implementing a Low-Pass Filter:

The *Filter Design and Analysis System* (FDAS) software is used to design the filters in Examples 11.1 and 11.2.
Filter Specifications:

Filter Type:	Low-Pass	
Passband Attenuation	.5000	
Stopband Attenuation	50.0000	
Passband Cutoff Frequency	10000.0	HERTZ
Stopband Cutoff frequency	12000.0	HERTZ
Sampling Frequency	80000.0	HERTZ

Design Results:

```
SAMPLING FREQUENCY          .800000E+05 HERTZ
101                   /* number of taps in decimal     */
0065                  /* number of taps in hexadecimal */
24                    /* number of bits in quantized
                         coefficients (dec) */
0018                  /* number of bits in quantized
                         coefficients (hex) */
 0                     /* shift count in decimal     */
0000                   /* shift count in hexadecimal */

      -3983 FFFFF071  /* coefficient of tap     0 */
       6159 0000180F  /* coefficient of tap     1 */
      13435 0000347B  /* coefficient of tap     2 */
      19828 00004D74  /* coefficient of tap     3 */
      19834 00004D7A  /* coefficient of tap     4 */
      11056 00002B30  /* coefficient of tap     5 */
      -3357 FFFFF2E3  /* coefficient of tap     6 */
     -15527 FFFFC359  /* coefficient of tap     7 */
     -17366 FFFFBC2A  /* coefficient of tap     8 */
      -6399 FFFFE701  /* coefficient of tap     9 */
      11262 00002BFE  /* coefficient of tap    10 */
      23669 00005C75  /* coefficient of tap    11 */
      20700 000050DC  /* coefficient of tap    12 */
       1839 0000072F  /* coefficient of tap    13 */
     -21713 FFFFAB2F  /* coefficient of tap    14 */
     -33307 FFFF7DE5  /* coefficient of tap    15 */
     -22273 FFFFA8FF  /* coefficient of tap    16 */
       7087 00001BAF  /* coefficient of tap    17 */
      36143 00008D2F  /* coefficient of tap    18 */
      43340 0000A94C  /* coefficient of tap    19 */
      19742 00004D1E  /* coefficient of tap    20 */
     -22244 FFFFA91C  /* coefficient of tap    21 */
     -54579 FFFF2ACD  /* coefficient of tap    22 */
     -51840 FFFF3580  /* coefficient of tap    23 */
     -10450 FFFFD72E  /* coefficient of tap    24 */
      45189 0000B085  /* coefficient of tap    25 */
      76281 000129F9  /* coefficient of tap    26 */
      55979 0000DAAB  /* coefficient of tap    27 */
      -8906 FFFFDD36  /* coefficient of tap    28 */
     -77598 FFFED0E2  /* coefficient of tap    29 */
     -99971 FFFE797D  /* coefficient of tap    30 */
     -51937 FFFF351F  /* coefficient of tap    31 */
      42663 0000A6A7  /* coefficient of tap    32 */
     121629 0001DB1D  /* coefficient of tap    33 */
     123738 0001E35A  /* coefficient of tap    34 */
      33740 000083CC  /* coefficient of tap    35 */
     -98382 FFFE7FB2  /* coefficient of tap    36 */
```

```
-182216 FFFD3838  /* coefficient of tap   37 */
-145429 FFFDC7EB  /* coefficient of tap   38 */
  10016 00002720  /* coefficient of tap   39 */
 194039 0002F5F7  /* coefficient of tap   40 */
 274586 0004309A  /* coefficient of tap   41 */
 162849 00027C21  /* coefficient of tap   42 */
-112861 FFFE4723  /* coefficient of tap   43 */
-395565 FFF9F6D3  /* coefficient of tap   44 */
-468487 FFF8D9F9  /* coefficient of tap   45 */
-174139 FFFD57C5  /* coefficient of tap   46 */
 481689 00075999  /* coefficient of tap   47 */
1304403 0013E753  /* coefficient of tap   48 */
1986515 001E4FD3  /* coefficient of tap   49 */
2250842 0022585A  /* coefficient of tap   50 */
1986515 001E4FD3  /* coefficient of tap   51 */
1304403 0013E753  /* coefficient of tap   52 */
 481689 00075999  /* coefficient of tap   53 */
-174139 FFFD57C5  /* coefficient of tap   54 */
-468487 FFF8D9F9  /* coefficient of tap   55 */
-395565 FFF9F6D3  /* coefficient of tap   56 */
-112861 FFFE4723  /* coefficient of tap   57 */
 162849 00027C21  /* coefficient of tap   58 */
 274586 0004309A  /* coefficient of tap   59 */
 194039 0002F5F7  /* coefficient of tap   60 */
  10016 00002720  /* coefficient of tap   61 */
-145429 FFFDC7EB  /* coefficient of tap   62 */
-182216 FFFD3838  /* coefficient of tap   63 */
 -98382 FFFE7FB2  /* coefficient of tap   64 */
  33740 000083CC  /* coefficient of tap   65 */
 123738 0001E35A  /* coefficient of tap   66 */
 121629 0001DB1D  /* coefficient of tap   67 */
  42663 0000A6A7  /* coefficient of tap   68 */
 -51937 FFFF351F  /* coefficient of tap   69 */
 -99971 FFFE797D  /* coefficient of tap   70 */
 -77598 FFFED0E2  /* coefficient of tap   71 */
  -8906 FFFFDD36  /* coefficient of tap   72 */
  55979 0000DAAB  /* coefficient of tap   73 */
  76281 000129F9  /* coefficient of tap   74 */
  45189 0000B085  /* coefficient of tap   75 */
 -10450 FFFFD72E  /* coefficient of tap   76 */
 -51840 FFFF3580  /* coefficient of tap   77 */
 -54579 FFFF2ACD  /* coefficient of tap   78 */
 -22244 FFFFA91C  /* coefficient of tap   79 */
  19742 00004D1E  /* coefficient of tap   80 */
  43340 0000A94C  /* coefficient of tap   81 */
  36143 00008D2F  /* coefficient of tap   82 */
   7087 00001BAF  /* coefficient of tap   83 */
 -22273 FFFFA8FF  /* coefficient of tap   84 */
 -33307 FFFF7DE5  /* coefficient of tap   85 */
 -21713 FFFFAB2F  /* coefficient of tap   86 */
   1839 0000072F  /* coefficient of tap   87 */
  20700 000050DC  /* coefficient of tap   88 */
  23669 00005C75  /* coefficient of tap   89 */
  11262 00002BFE  /* coefficient of tap   90 */
  -6399 FFFFE701  /* coefficient of tap   91 */
 -17366 FFFFBC2A  /* coefficient of tap   92 */
 -15527 FFFFC359  /* coefficient of tap   93 */
  -3357 FFFFF2E3  /* coefficient of tap   94 */
  11056 00002B30  /* coefficient of tap   95 */
  19834 00004D7A  /* coefficient of tap   96 */
  19828 00004D74  /* coefficient of tap   97 */
  13435 0000347B  /* coefficient of tap   98 */
   6159 0000180F  /* coefficient of tap   99 */
  -3983 FFFFF071  /* coefficient of tap  100 */
```

```
-.4748106002807617D-03 BF3F1E0000000000/*  tap   0 */
 .7342100143432617D-03 3F480F0000000000/*  tap   1 */
 .1601576805114746D-02 3F5A3D8000000000/*  tap   2 */
 .2363681793212891D-02 3F635D0000000000/*  tap   3 */
 .2364397048950195D-02 3F635E8000000000/*  tap   4 */
 .1317977905273437D-02 3F55980000000000/*  tap   5 */
-.4001855850219727D-03 BF3A3A0000000000/*  tap   6 */
-.1850962638854980D-02 BF5E538000000000/*  tap   7 */
-.2070188522338867D-02 BF60F58000000000/*  tap   8 */
-.7628202438354492D-03 BF48FF0000000000/*  tap   9 */
 .1342535018920898D-02 3F55FF0000000000/*  tap  10 */
 .2821564674377441D-02 3F671D4000000000/*  tap  11 */
 .2467632293701172D-02 3F64370000000000/*  tap  12 */
 .2192258834838867D-03 3F2CBC0000000000/*  tap  13 */
-.2588391304016113D-02 BF65344000000000/*  tap  14 */
-.3970503807067871D-02 BF70436000000000/*  tap  15 */
-.2655148506164551D-02 BF65C04000000000/*  tap  16 */
 .8448362350463867D-03 3F4BAF0000000000/*  tap  17 */
 .4308581352233887D-02 3F71A5E000000000/*  tap  18 */
 .5166530609130859D-02 3F75298000000000/*  tap  19 */
 .2353429794311523D-02 3F63478000000000/*  tap  20 */
-.2651691436767578D-02 BF65B90000000000/*  tap  21 */
-.6506323814392090D-02 BF7AA66000000000/*  tap  22 */
-.6179809570312500D-02 BF79500000000000/*  tap  23 */
-.1245737075805664D-02 BF54690000000000/*  tap  24 */
 .5386948585510254D-02 3F7610A000000000/*  tap  25 */
 .9093403816223145D-02 3F829F9000000000/*  tap  26 */
 .6673216819763184D-02 3F7B556000000000/*  tap  27 */
-.1061677932739258D-02 BF51650000000000/*  tap  28 */
-.9250402450561523D-02 BF82F1E000000000/*  tap  29 */
-.1191747188568115D-01 BF88683000000000/*  tap  30 */
-.6191372871398926D-02 BF795C2000000000/*  tap  31 */
 .5085825920104980D-02 3F74D4E000000000/*  tap  32 */
 .1449930667877197D-01 3F8DB1D000000000/*  tap  33 */
 .1475071907043457D-01 3F8E35A000000000/*  tap  34 */
 .4022121429443359D-02 3F70798000000000/*  tap  35 */
-.1172804832458496D-01 BF8804E000000000/*  tap  36 */
-.2172183990478516D-01 BF963E4000000000/*  tap  37 */
-.1733648777008057D-01 BF91C0A800000000/*  tap  38 */
 .1194000244140625D-02 3F53900000000000/*  tap  39 */
 .2313125133514404D-01 3F97AFB800000000/*  tap  40 */
 .3273320198059082D-01 3FA0C26800000000/*  tap  41 */
 .1941311359405518D-01 3F93E10800000000/*  tap  42 */
-.1345407962799072D-01 BF8B8DD000000000/*  tap  43 */
-.4715502262115479D-01 BFA824B400000000/*  tap  44 */
-.5584800243377686D-01 BFAC981C00000000/*  tap  45 */
-.2075898647308350D-01 BF9541D800000000/*  tap  46 */
 .5742180347442627D-01 3FAD666400000000/*  tap  47 */
 .1554969549179077D+00 3FC3E75300000000/*  tap  48 */
 .2368110418319702D+00 3FCE4FD300000000/*  tap  49 */
 .2683212757110596D+00 3FD12C2D00000000/*  tap  50 */
 .2368110418319702D+00 3FCE4FD300000000/*  tap  51 */
 .1554969549179077D+00 3FC3E75300000000/*  tap  52 */
 .5742180347442627D-01 3FAD666400000000/*  tap  53 */
-.2075898647308350D-01 BF9541D800000000/*  tap  54 */
-.5584800243377686D-01 BFAC981C00000000/*  tap  55 */
-.4715502262115479D-01 BFA824B400000000/*  tap  56 */
-.1345407962799072D-01 BF8B8DD000000000/*  tap  57 */
 .1941311359405518D-01 3F93E10800000000/*  tap  58 */
 .3273320198059082D-01 3FA0C26800000000/*  tap  59 */
 .2313125133514404D-01 3F97AFB800000000/*  tap  60 */
 .1194000244140625D-02 3F53900000000000/*  tap  61 */
-.1733648777008057D-01 BF91C0A800000000/*  tap  62 */
-.2172183990478516D-01 BF963E4000000000/*  tap  63 */
-.1172804832458496D-01 BF8804E000000000/*  tap  64 */
 .4022121429443359D-02 3F70798000000000/*  tap  65 */
 .1475071907043457D-01 3F8E35A000000000/*  tap  66 */
 .1449930667877197D-01 3F8DB1D000000000/*  tap  67 */
 .5085825920104980D-02 3F74D4E000000000/*  tap  68 */
```

```
-.6191372871398926D-02 BF795C2000000000/*  tap   69 */
-.1191747188568115D-01 BF88683000000000/*  tap   70 */
-.9250402450561523D-02 BF82F1E000000000/*  tap   71 */
-.1061677932739258D-02 BF51650000000000/*  tap   72 */
 .6673216819763184D-02 3F7B556000000000/*  tap   73 */
 .9093403816223145D-02 3F829F9000000000/*  tap   74 */
 .5386948585510254D-02 3F7610A000000000/*  tap   75 */
-.1245737075805664D-02 BF54690000000000/*  tap   76 */
-.6179809570312500D-02 BF79500000000000/*  tap   77 */
-.6506323814392090D-02 BF7AA66000000000/*  tap   78 */
-.2651691436767578D-02 BF65B90000000000/*  tap   79 */
 .2353429794311523D-02 3F63478000000000/*  tap   80 */
 .5166530609130859D-02 3F75298000000000/*  tap   81 */
 .4308581352233887D-02 3F71A5E000000000/*  tap   82 */
 .8448362350463867D-03 3F4BAF0000000000/*  tap   83 */
-.2655148506164551D-02 BF65C04000000000/*  tap   84 */
-.3970503807067871D-02 BF70436000000000/*  tap   85 */
-.2588391304016113D-02 BF65344000000000/*  tap   86 */
 .2192258834838867D-03 3F2CBC0000000000/*  tap   87 */
 .2467632293701172D-02 3F64370000000000/*  tap   88 */
 .2821564674377441D-02 3F671D4000000000/*  tap   89 */
 .1342535018920898D-02 3F55FF0000000000/*  tap   90 */
-.7628202438354492D-03 BF48FF0000000000/*  tap   91 */
-.2070188522338867D-02 BF60F58000000000/*  tap   92 */
-.1850962638854980D-02 BF5E538000000000/*  tap   93 */
-.4001855850219727D-03 BF3A3A0000000000/*  tap   94 */
 .1317977905273437D-02 3F55980000000000/*  tap   95 */
 .2364397048950195D-02 3F635E8000000000/*  tap   96 */
 .2363681793212891D-02 3F635D0000000000/*  tap   97 */
 .1601576805114746D-02 3F5A3D8000000000/*  tap   98 */
 .7342100143432617D-03 3F480F0000000000/*  tap   99 */
-.4748106002807617D-03 BF3F1E0000000000/*  tap  100 */
```

The **FIR1LP.ASM** assembler program which implements a low-pass filter on the DSP56200 Demonstration System is shown in Figure 11.10. A copy of the **FIR1LP.ASM** assembler program and the **FIR1LP.LOD** absolute load program can be found on the DSP56000/1 Demonstration Software #4 diskette.

1. POWER-DOWN THE PC, ADS, AND FUNCTION GENERATOR.
2. ON THE DSP56200 DEVELOPMENT BOARD, INSTALL JUMPER JP2 AND REMOVE JUMPER JP1. CONNECT BOARD TO ADM.
3. ON THE EVB, INSTALL JUMPER JP5, REMOVE JUMPERS JP1, JP2,JP3 AND JP4 AND CONNECT THE EXTERNAL CLOCK SUPPLIED BY THE DSP56200 DEVELOPMENT BOARD TO J1-6.
4. CONNECT THE FUNCTION GENERATOR'S OUTPUT TO THE EVB'S "BNC2" INPUT (A/D INPUT).
5. CONNECT THE FUNCTION GENERATOR'S OUTPUT TO THE OSCILLOSCOPE'S CHANNEL ONE INPUT.
6. CONNECT THE EVB'S "BNC1" OUTPUT (D/A OUTPUT) TO THE OSCILLOSCOPE'S CHANNEL TWO INPUT.
7. POWER-ON THE PC AND ADS; BOOT THE PC.
8. POWER-ON THE FUNCTION GENERATOR.
9. INSERT THE "DSP56000/1 DEMONSTRATION SOFTWARE #4" DISKETTE INTO DRIVE "A".

10. Depending on your PC's configuration, INVOKE THE "ADS56000 USER INTERFACE SOFTWARE" PROGRAM from Drive "B" or the hard disk.
11. TYPE **load a:fir1lp.lod<return>** to load the assembled and linked version of the assembler program shown in Figure 11.10.
12. TYPE **go<return>** to begin execution of the program.
13. VARY THE SINUSOIDAL INPUT FREQUENCY FROM 0.5 KHZ TO 20 KHZ AND OBSERVE THE INPUT AND OUTPUT SIGNALS ON THE OSCILLOSCOPE. Note that the filter passes the frequencies from 0.5 KHz up to 10 KHz and stops the frequencies from 12 KHz to 20 KHz.
14. TYPE **force b<return>** to halt execution of the program.
15. TYPE **quit<return>** IF YOU WANT TO EXIT THE "ADS56000 USER INTERFACE SOFTWARE" PROGRAM.

```
;************************************************************
; file:FIR1LP.ASM
;
; Low-Pass FIR Filter
; The interrupt control signal is IRQB
;
;************************************************************

        include 'firl'

ntaps   equ     $65
mtaps   equ     $12e

        org     y:$0        ; store data in RAM
        dc      $000000,$000001,$ffff00,$0000ff

        org     x:$0
coef
        dc      $0000ff,$0000f0,$000071
        dc      $000000,$000018,$00000F
        dc      $000000,$000034,$00007B
        dc      $000000,$00004d,$000074
        dc      $000000,$00004d,$00007A
        dc      $000000,$00002b,$000030
        dc      $0000ff,$0000f2,$0000E3
        dc      $0000ff,$0000c3,$000059
        dc      $0000ff,$0000bc,$00002A
        dc      $0000ff,$0000e7,$000001
        dc      $000000,$00002b,$0000FE
        dc      $000000,$00005c,$000075
        dc      $000000,$000050,$0000DC
        dc      $000000,$000007,$00002F
        dc      $0000ff,$0000ab,$00002F
        dc      $0000ff,$00007d,$0000e5
        dc      $0000ff,$0000a8,$0000FF
        dc      $000000,$00001b,$0000AF
        dc      $000000,$00008d,$00002F
        dc      $000000,$0000a9,$00004C
        dc      $000000,$00004d,$00001E
        dc      $0000ff,$0000a9,$00001C
        dc      $0000ff,$00002a,$0000CD
        dc      $0000ff,$000035,$000080
        dc      $0000ff,$0000d7,$00002E
        dc      $000000,$0000b0,$000085
```

```
        dc      $000001,$000029,$0000F9
        dc      $000000,$0000da,$0000AB
        dc      $0000ff,$0000dd,$000036
        dc      $0000fe,$0000d0,$0000E2
        dc      $0000fe,$000079,$00007D
        dc      $0000ff,$000035,$00001F
        dc      $000000,$0000a6,$0000A7
        dc      $000001,$0000db,$00001D
        dc      $000001,$0000e3,$00005a
        dc      $000000,$000083,$0000CC
        dc      $0000fe,$00007f,$0000B2
        dc      $0000fd,$000038,$000038
        dc      $0000fd,$0000c7,$0000EB
        dc      $000000,$000027,$000020
        dc      $000002,$0000f5,$0000F7
        dc      $000004,$000030,$00009A
        dc      $000002,$00007c,$000021
        dc      $0000fe,$000047,$000023
        dc      $0000f9,$0000f6,$0000D3
        dc      $0000f8,$0000d9,$0000F9
        dc      $0000fd,$000057,$0000C5
        dc      $000007,$000059,$000099
        dc      $000013,$0000e7,$000053
        dc      $00001e,$00004f,$0000D3
        dc      $000022,$000058,$00005A
        dc      $00001e,$00004f,$0000D3
        dc      $000013,$0000e7,$000053
        dc      $000007,$000059,$000099
        dc      $0000fd,$000057,$0000C5
        dc      $0000f8,$0000d9,$0000F9
        dc      $0000f9,$0000f6,$0000D3
        dc      $0000fe,$000047,$000023
        dc      $000002,$00007c,$000021
        dc      $000004,$000030,$00009A
        dc      $000002,$0000f5,$0000F7
        dc      $000000,$000027,$000020
        dc      $0000fd,$0000c7,$0000EB
        dc      $0000fd,$000038,$000038
        dc      $0000fe,$00007f,$0000B2
        dc      $000000,$000083,$0000CC
        dc      $000001,$0000e3,$00005a
        dc      $000001,$0000db,$00001D
        dc      $000000,$0000a6,$0000A7
        dc      $0000ff,$000035,$00001F
        dc      $0000fe,$000079,$00007D
        dc      $0000fe,$0000d0,$0000E2
        dc      $0000ff,$0000dd,$000036
        dc      $000000,$0000da,$0000AB
        dc      $000001,$000029,$0000F9
        dc      $000000,$0000b0,$000085
        dc      $0000ff,$0000d7,$00002E
        dc      $0000ff,$000035,$000080
        dc      $0000ff,$00002a,$0000CD
        dc      $0000ff,$0000a9,$00001C
        dc      $000000,$00004d,$00001E
        dc      $000000,$0000a9,$00004C
        dc      $000000,$00008d,$00002F
        dc      $000000,$00001b,$0000AF
        dc      $0000ff,$0000a8,$0000FF
        dc      $0000ff,$00007d,$0000e5
        dc      $0000ff,$0000ab,$00002F
        dc      $000000,$000007,$00002F
        dc      $000000,$000050,$0000DC
        dc      $000000,$00005c,$000075
        dc      $000000,$00002b,$0000FE
        dc      $0000ff,$0000e7,$000001
        dc      $0000ff,$0000bc,$00002A
        dc      $0000ff,$0000c3,$000059
        dc      $0000ff,$0000f2,$0000E3
```

```
        dc      $000000,$00002b,$000030
        dc      $000000,$00004d,$00007A
        dc      $000000,$00004d,$000074
        dc      $000000,$000034,$00007B
        dc      $000000,$000018,$00000F
        dc      $0000ff,$0000f0,$000071

        firl    ntaps,mtaps

        end
```

Figure 11-10 FIR1LP.ASM Assembler Program to Implement the Low-Pass Filter in Example 11.1 on the DSP56200 Demonstration System

EXAMPLE 11.2: Implementing a Band-Pass Filter:

Filter Specifications:

Filter Type:	Band-Pass		
Passband Attenuation	.5000		
Stopband Attenuation	50.0000		
Passband Cutoff Frequencies	12000.0	17000.0	HERTZ
Stopband Cutoff frequencies	10000.0	19000.0	HERTZ
Sampling Frequency	80000.0	HERTZ	

Design Results:

```
SAMPLING FREQUENCY          .800000E+05 HERTZ
101                   /* number of taps in decimal     */
0065                  /* number of taps in hexadecimal */
24                    /* number of bits in quantized
                         coefficients (dec) */
0018                  /* number of bits in quantized
                         coefficients (hex) */
 0                    /* shift count in decimal     */
0000                  /* shift count in hexadecimal */

        -422 FFFFFE5A  /* coefficient of tap    0 */
       -1326 FFFFFAD2  /* coefficient of tap    1 */
        1219 000004C3  /* coefficient of tap    2 */
       17220 00004344  /* coefficient of tap    3 */
        9915 000026BB  /* coefficient of tap    4 */
      -13825 FFFFC9FF  /* coefficient of tap    5 */
      -30173 FFFF8A23  /* coefficient of tap    6 */
       -9524 FFFFDACC  /* coefficient of tap    7 */
       29022 0000715E  /* coefficient of tap    8 */
       35560 00008AE8  /* coefficient of tap    9 */
           0 00000000  /* coefficient of tap   10 */
      -29843 FFFF8B6D  /* coefficient of tap   11 */
      -18909 FFFFB623  /* coefficient of tap   12 */
        3679 00000E5F  /* coefficient of tap   13 */
        -773 FFFFFCFB  /* coefficient of tap   14 */
       -9756 FFFFD9E4  /* coefficient of tap   15 */
       18729 00004929  /* coefficient of tap   16 */
       54057 0000D329  /* coefficient of tap   17 */
       22207 000056BF  /* coefficient of tap   18 */
      -63556 FFFF07BC  /* coefficient of tap   19 */
      -89452 FFFEA294  /* coefficient of tap   20 */
       -3972 FFFFF07C  /* coefficient of tap   21 */
       87503 000155CF  /* coefficient of tap   22 */
```

```
  68351 00010AFF  /* coefficient of tap   23 */
 -15693 FFFFC2B3  /* coefficient of tap   24 */
 -38163 FFFF6AED  /* coefficient of tap   25 */
  -1761 FFFFF91F  /* coefficient of tap   26 */
 -19018 FFFFB5B6  /* coefficient of tap   27 */
 -84281 FFFEB6C7  /* coefficient of tap   28 */
 -45305 FFFF4F07  /* coefficient of tap   29 */
 123424 0001E220  /* coefficient of tap   30 */
 198590 000307BE  /* coefficient of tap   31 */
  18699 0000490B  /* coefficient of tap   32 */
-218825 FFFCA937  /* coefficient of tap   33 */
-198691 FFFCF7DD  /* coefficient of tap   34 */
  42916 0000A7A4  /* coefficient of tap   35 */
 167270 00028D66  /* coefficient of tap   36 */
  62483 0000F413  /* coefficient of tap   37 */
  -3210 FFFFF376  /* coefficient of tap   38 */
 107800 0001A518  /* coefficient of tap   39 */
  92227 00016843  /* coefficient of tap   40 */
-263494 FFFBFABA  /* coefficient of tap   41 */
-517617 FFF81A0F  /* coefficient of tap   42 */
 -82666 FFFEBD16  /* coefficient of tap   43 */
 732425 000B2D09  /* coefficient of tap   44 */
 835403 000CBF4B  /* coefficient of tap   45 */
-177426 FFFD4AEE  /* coefficient of tap   46 */
-1194948 FFEDC43C /* coefficient of tap   47 */
-858578 FFF2E62E  /* coefficient of tap   48 */
 574330 0008C37A  /* coefficient of tap   49 */
1388699 0015309B  /* coefficient of tap   50 */
 574330 0008C37A  /* coefficient of tap   51 */
-858578 FFF2E62E  /* coefficient of tap   52 */
-1194948 FFEDC43C /* coefficient of tap   53 */
-177426 FFFD4AEE  /* coefficient of tap   54 */
 835403 000CBF4B  /* coefficient of tap   55 */
 732425 000B2D09  /* coefficient of tap   56 */
 -82666 FFFEBD16  /* coefficient of tap   57 */
-517617 FFF81A0F  /* coefficient of tap   58 */
-263494 FFFBFABA  /* coefficient of tap   59 */
  92227 00016843  /* coefficient of tap   60 */
 107800 0001A518  /* coefficient of tap   61 */
  -3210 FFFFF376  /* coefficient of tap   62 */
  62483 0000F413  /* coefficient of tap   63 */
 167270 00028D66  /* coefficient of tap   64 */
  42916 0000A7A4  /* coefficient of tap   65 */
-198691 FFFCF7DD  /* coefficient of tap   66 */
-218825 FFFCA937  /* coefficient of tap   67 */
  18699 0000490B  /* coefficient of tap   68 */
 198590 000307BE  /* coefficient of tap   69 */
 123424 0001E220  /* coefficient of tap   70 */
 -45305 FFFF4F07  /* coefficient of tap   71 */
 -84281 FFFEB6C7  /* coefficient of tap   72 */
 -19018 FFFFB5B6  /* coefficient of tap   73 */
  -1761 FFFFF91F  /* coefficient of tap   74 */
 -38163 FFFF6AED  /* coefficient of tap   75 */
 -15693 FFFFC2B3  /* coefficient of tap   76 */
  68351 00010AFF  /* coefficient of tap   77 */
  87503 000155CF  /* coefficient of tap   78 */
  -3972 FFFFF07C  /* coefficient of tap   79 */
 -89452 FFFEA294  /* coefficient of tap   80 */
 -63556 FFFF07BC  /* coefficient of tap   81 */
  22207 000056BF  /* coefficient of tap   82 */
  54057 0000D329  /* coefficient of tap   83 */
  18729 00004929  /* coefficient of tap   84 */
  -9756 FFFFD9E4  /* coefficient of tap   85 */
   -773 FFFFFCFB  /* coefficient of tap   86 */
   3679 00000E5F  /* coefficient of tap   87 */
 -18909 FFFFB623  /* coefficient of tap   88 */
 -29843 FFFF8B6D  /* coefficient of tap   89 */
      0 00000000  /* coefficient of tap   90 */
  35560 00008AE8  /* coefficient of tap   91 */
```

```
      29022 0000715E  /* coefficient of tap    92 */
      -9524 FFFFDACC  /* coefficient of tap    93 */
     -30173 FFFF8A23  /* coefficient of tap    94 */
     -13825 FFFFC9FF  /* coefficient of tap    95 */
       9915 000026BB  /* coefficient of tap    96 */
      17220 00004344  /* coefficient of tap    97 */
       1219 000004C3  /* coefficient of tap    98 */
      -1326 FFFFFAD2  /* coefficient of tap    99 */
       -422 FFFFFE5A  /* coefficient of tap   100 */

-.5030632019042969D-04 BF0A600000000000/*  tap     0 */
-.1580715179443359D-03 BF24B80000000000/*  tap     1 */
 .1453161239624023D-03 3F230C0000000000/*  tap     2 */
 .2052783966064453D-02 3F60D10000000000/*  tap     3 */
 .1181960105895996D-02 3F535D8000000000/*  tap     4 */
-.1648068428039551D-02 BF5B008000000000/*  tap     5 */
-.3596901893615723D-02 BF6D774000000000/*  tap     6 */
-.1135349273681641D-02 BF529A0000000000/*  tap     7 */
 .3459692001342773D-02 3F6C578000000000/*  tap     8 */
 .4239082336425781D-02 3F715D0000000000/*  tap     9 */
 .0000000000000000D+00 0000000000000000/*  tap    10 */
-.3557562828063965D-02 BF6D24C000000000/*  tap    11 */
-.2254128456115723D-02 BF62774000000000/*  tap    12 */
 .4385709762573242D-03 3F3CBE0000000000/*  tap    13 */
-.9214878082275391D-04 BF18280000000000/*  tap    14 */
-.1163005828857422D-02 BF530E0000000000/*  tap    15 */
 .2232670783996582D-02 3F624A4000000000/*  tap    16 */
 .6444096565246582D-02 3F7A652000000000/*  tap    17 */
 .2647280693054199D-02 3F65AFC000000000/*  tap    18 */
-.7576465606689453D-02 BF7F088000000000/*  tap    19 */
-.1066350936889648D-01 BF85D6C000000000/*  tap    20 */
-.4734992980957031D-03 BF3F080000000000/*  tap    21 */
 .1043117046356201D-01 3F855CF000000000/*  tap    22 */
 .8148074150085449D-02 3F80AFF000000000/*  tap    23 */
-.1870751380920410D-02 BF5EA68000000000/*  tap    24 */
-.4549384117126465D-02 BF72A26000000000/*  tap    25 */
-.2099275588989258D-03 BF2B840000000000/*  tap    26 */
-.2267122268676758D-02 BF62928000000000/*  tap    27 */
-.1004707813262939D-01 BF84939000000000/*  tap    28 */
-.5400776863098145D-02 BF761F2000000000/*  tap    29 */
 .1471328735351562D-01 3F8E220000000000/*  tap    30 */
 .2367377281188965D-01 3F983DF000000000/*  tap    31 */
 .2229094505310059D-02 3F6242C000000000/*  tap    32 */
-.2608597278594971D-01 BF9AB64800000000/*  tap    33 */
-.2368581295013428D-01 BF98411800000000/*  tap    34 */
 .5115985870361328D-02 3F74F48000000000/*  tap    35 */
 .1994013786315918D-01 3F946B3000000000/*  tap    36 */
 .7448554039001465D-02 3F7E826000000000/*  tap    37 */
-.3826618194580078D-03 BF39140000000000/*  tap    38 */
 .1285076141357422D-01 3F8A518000000000/*  tap    39 */
 .1099431514739990D-01 3F86843000000000/*  tap    40 */
-.3141093254089355D-01 BFA0151800000000/*  tap    41 */
-.6170475482940674D-01 BFAF97C400000000/*  tap    42 */
-.9854555130004883D-02 BF842EA000000000/*  tap    43 */
 .8731186389923096D-01 3FB65A1200000000/*  tap    44 */
 .9958779811859131D-01 3FB97E9600000000/*  tap    45 */
-.2115082740783691D-01 BF95A89000000000/*  tap    46 */
-.1424489021301270D+00 BFC23BC400000000/*  tap    47 */
-.1023504734039307D+00 BFBA33A400000000/*  tap    48 */
 .6846547126770020D-01 3FB186F400000000/*  tap    49 */
 .1655458211898804D+00 3FC5309B00000000/*  tap    50 */
 .6846547126770020D-01 3FB186F400000000/*  tap    51 */
-.1023504734039307D+00 BFBA33A400000000/*  tap    52 */
-.1424489021301270D+00 BFC23BC400000000/*  tap    53 */
-.2115082740783691D-01 BF95A89000000000/*  tap    54 */
 .9958779811859131D-01 3FB97E9600000000/*  tap    55 */
 .8731186389923096D-01 3FB65A1200000000/*  tap    56 */
-.9854555130004883D-02 BF842EA000000000/*  tap    57 */
-.6170475482940674D-01 BFAF97C400000000/*  tap    58 */
```

```
-.3141093254089355D-01 BFA0151800000000/*   tap    59 */
 .1099431514739990D-01 3F86843000000000/*   tap    60 */
 .1285076141357422D-01 3F8A518000000000/*   tap    61 */
-.3826618194580078D-03 BF39140000000000/*   tap    62 */
 .7448554039001465D-02 3F7E826000000000/*   tap    63 */
 .1994013786315918D-01 3F946B3000000000/*   tap    64 */
 .5115985870361328D-02 3F74F48000000000/*   tap    65 */
-.2368581295013428D-01 BF98411800000000/*   tap    66 */
-.2608597278594971D-01 BF9AB64800000000/*   tap    67 */
 .2229094505310059D-02 3F6242C000000000/*   tap    68 */
 .2367377281188965D-01 3F983DF000000000/*   tap    69 */
 .1471328735351562D-01 3F8E220000000000/*   tap    70 */
-.5400776863098145D-02 BF761F2000000000/*   tap    71 */
-.1004707813262939D-01 BF84939000000000/*   tap    72 */
-.2267122268676758D-02 BF62928000000000/*   tap    73 */
-.2099275588989258D-03 BF2B840000000000/*   tap    74 */
-.4549384117126465D-02 BF72A26000000000/*   tap    75 */
-.1870751380920410D-02 BF5EA68000000000/*   tap    76 */
 .8148074150085449D-02 3F80AFF000000000/*   tap    77 */
 .1043117046356201D-01 3F855CF000000000/*   tap    78 */
-.4734992980957031D-03 BF3F080000000000/*   tap    79 */
-.1066350936889648D-01 BF85D6C000000000/*   tap    80 */
-.7576465606689453D-02 BF7F088000000000/*   tap    81 */
 .2647280693054199D-02 3F65AFC000000000/*   tap    82 */
 .6444096565246582D-02 3F7A652000000000/*   tap    83 */
 .2232670783996582D-02 3F624A4000000000/*   tap    84 */
-.1163005828857422D-02 BF530E0000000000/*   tap    85 */
-.9214878082275391D-04 BF18280000000000/*   tap    86 */
 .4385709762573242D-03 3F3CBE0000000000/*   tap    87 */
-.2254128456115723D-02 BF62774000000000/*   tap    88 */
-.3557562828063965D-02 BF6D24C000000000/*   tap    89 */
 .0000000000000000D+00 0000000000000000/*   tap    90 */
 .4239082336425781D-02 3F715D0000000000/*   tap    91 */
 .3459692001342773D-02 3F6C578000000000/*   tap    92 */
-.1135349273681641D-02 BF529A0000000000/*   tap    93 */
-.3596901893615723D-02 BF6D774000000000/*   tap    94 */
-.1648068428039551D-02 BF5B008000000000/*   tap    95 */
 .1181960105895996D-02 3F535D8000000000/*   tap    96 */
 .2052783966064453D-02 3F60D10000000000/*   tap    97 */
 .1453161239624023D-03 3F230C0000000000/*   tap    98 */
-.1580715179443359D-03 BF24B80000000000/*   tap    99 */
-.5030632019042969D-04 BF0A600000000000/*   tap   100 */
```

The **FIR1BP.ASM** assembler program which implements a band-pass filter on the DSP56200 Demonstration System is shown in Figure 11.11. A copy of the **FIR1BP.ASM** assembler program and **FIR1BP.LOD** absolute load program can be found on the DSP56000/1 Demonstration Software #4 diskette.

1. POWER-DOWN THE PC, ADS, AND FUNCTION GENERATOR.
2. ON THE DSP56200 DEVELOPMENT BOARD, INSTALL JUMPER JP2 AND REMOVE JUMPER JP1. CONNECT BOARD TO ADM.
3. ON THE EVB, INSTALL JUMPER JP5, REMOVE JUMPERS JP1, JP2,JP3 AND JP4 AND CONNECT THE EXTERNAL CLOCK SUPPLIED BY THE DSP56200 DEVELOPMENT BOARD TO J1-6.
4. CONNECT THE FUNCTION GENERATOR'S OUTPUT TO THE EVB'S "BNC2" INPUT (A/D INPUT).
5. CONNECT THE FUNCTION GENERATOR'S OUTPUT TO THE OSCILLOSCOPE'S CHANNEL ONE INPUT.

6. CONNECT THE EVB'S "BNC1" OUTPUT (D/A OUTPUT) TO THE OSCILLOSCOPE'S CHANNEL TWO INPUT.
7. POWER-ON THE PC AND ADS; BOOT THE PC.
8. POWER-ON THE FUNCTION GENERATOR.
9. INSERT THE "DSP56000/1 DEMONSTRATION SOFTWARE #4" DISKETTE INTO DRIVE "A".
10. Depending on your PC's configuration, INVOKE THE "ADS56000 USER INTERFACE SOFTWARE" PROGRAM from Drive "B" or the hard disk.
11. TYPE **load a:fir1bp.lod<return>** to load the assembled and linked version of the assembler program shown in Figure 11.11.
12. TYPE **go<return>** to begin execution of the program.
13. VARY THE SINUSOIDAL INPUT FREQUENCY FROM 0.5 KHZ TO 20 KHZ AND OBSERVE THE INPUT AND OUTPUT SIGNAL ON THE OSCILLOSCOPE. Note that the filter passes the frequencies from 12 KHz to 17 KHz and stops the frequencies from 0.5 KHz to 10 KHz and from 19 KHz to 20 KHz.
14. TYPE **force b<return>** to halt execution of the program.
15. TYPE **quit<return>** IF YOU WANT TO EXIT THE "ADS56000 USER INTERFACE SOFTWARE" PROGRAM.

```
;***********************************************************
; file:FIR1BP.ASM
;
; Band-Pass FIR filter
; The interrupt control signal is IRQB
;
;***********************************************************

        include 'firl'

ntaps   equ    $65
mtaps   equ    $12e

        org    y:$0        ; store data in RAM
        dc     $000000,$000001,$ffff00,$0000ff

        org    x:$0
coef
        dc      $0000FF,$0000FE,$00005A
        dc      $0000FF,$0000FA,$0000D2
        dc      $000000,$000004,$0000C3
        dc      $000000,$000043,$000044
        dc      $000000,$000026,$0000BB
        dc      $0000FF,$0000C9,$0000FF
        dc      $0000FF,$00008A,$000023
        dc      $0000FF,$0000DA,$0000CC
        dc      $000000,$000071,$00005E
        dc      $000000,$00008A,$0000E8
        dc      $000000,$000000,$000000
        dc      $0000FF,$00008B,$00006D
        dc      $0000FF,$0000B6,$000023
        dc      $000000,$00000E,$00005F
        dc      $0000FF,$0000FC,$0000FB
        dc      $0000FF,$0000D9,$0000E4
        dc      $000000,$000049,$000029
```

```
        dc      $000000,$0000D3,$000029
        dc      $000000,$000056,$0000BF
        dc      $0000FF,$000007,$0000BC
        dc      $0000FE,$0000A2,$000094
        dc      $0000FF,$0000F0,$00007C
        dc      $000001,$000055,$0000CF
        dc      $000001,$00000A,$0000FF
        dc      $0000FF,$0000C2,$0000B3
        dc      $0000FF,$00006A,$0000ED
        dc      $0000FF,$0000F9,$00001F
        dc      $0000FF,$0000B5,$0000B6
        dc      $0000FE,$0000B6,$0000C7
        dc      $0000FF,$00004F,$000007
        dc      $000001,$0000E2,$000020
        dc      $000003,$000007,$0000BE
        dc      $000000,$000049,$00000B
        dc      $0000FC,$0000A9,$000037
        dc      $0000FC,$0000F7,$0000DD
        dc      $000000,$0000A7,$0000A4
        dc      $000002,$00008D,$000066
        dc      $000000,$0000F4,$000013
        dc      $0000FF,$0000F3,$000076
        dc      $000001,$0000A5,$000018
        dc      $000001,$000068,$000043
        dc      $0000FB,$0000FA,$0000BA
        dc      $0000F8,$00001A,$00000F
        dc      $0000FE,$0000BD,$000016
        dc      $00000B,$00002D,$000009
        dc      $00000C,$0000BF,$00004B
        dc      $0000FD,$00004A,$0000EE
        dc      $0000ED,$0000C4,$00003C
        dc      $0000F2,$0000E6,$00002E
        dc      $000008,$0000C3,$00007A
        dc      $000015,$000030,$00009B
        dc      $000008,$0000C3,$00007A
        dc      $0000F2,$0000E6,$00002E
        dc      $0000ED,$0000C4,$00003C
        dc      $0000FD,$00004A,$0000EE
        dc      $00000C,$0000BF,$00004B
        dc      $00000B,$00002D,$000009
        dc      $0000FE,$0000BD,$000016
        dc      $0000F8,$00001A,$00000F
        dc      $0000FB,$0000FA,$0000BA
        dc      $000001,$000068,$000043
        dc      $000001,$0000A5,$000018
        dc      $0000FF,$0000F3,$000076
        dc      $000000,$0000F4,$000013
        dc      $000002,$00008D,$000066
        dc      $000000,$0000A7,$0000A4
        dc      $0000FC,$0000F7,$0000DD
        dc      $0000FC,$0000A9,$000037
        dc      $000000,$000049,$00000B
        dc      $000003,$000007,$0000BE
        dc      $000001,$0000E2,$000020
        dc      $0000FF,$00004F,$000007
        dc      $0000FE,$0000B6,$0000C7
        dc      $0000FF,$0000B5,$0000B6
        dc      $0000FF,$0000F9,$00001F
        dc      $0000FF,$00006A,$0000ED
        dc      $0000FF,$0000C2,$0000B3
        dc      $000001,$00000A,$0000FF
        dc      $000001,$000055,$0000CF
        dc      $0000FF,$0000F0,$00007C
        dc      $0000FE,$0000A2,$000094
        dc      $0000FF,$000007,$0000BC
        dc      $000000,$000056,$0000BF
        dc      $000000,$0000D3,$000029
        dc      $000000,$000049,$000029
        dc      $0000FF,$0000D9,$0000E4
```

```
        dc      $0000FF,$0000FC,$0000FB
        dc      $000000,$00000E,$00005F
        dc      $0000FF,$0000B6,$000023
        dc      $0000FF,$00008B,$00006D
        dc      $000000,$000000,$000000
        dc      $000000,$00008A,$0000E8
        dc      $000000,$000071,$00005E
        dc      $0000FF,$0000DA,$0000CC
        dc      $0000FF,$00008A,$000023
        dc      $0000FF,$0000C9,$0000FF
        dc      $000000,$000026,$0000BB
        dc      $000000,$000043,$000044
        dc      $000000,$000004,$0000C3
        dc      $0000FF,$0000FA,$0000D2
        dc      $0000FF,$0000FE,$00005A

        firl    ntaps,mtaps

        end
```

Figure 11-11 FIR1BP.ASM Assembler Program to Implement the Band-Pass Filter in Example 11.2 on the DSP56200 Demonstration System

11.6 IMPLEMENTING AN ADAPTIVE FIR FILTER ON THE DSP56200 DEMO SYSTEM:

As in Chapter 9, the input signal x(n) is assumed to be equal to the reference input signal d(n). It is also assumed that d(n) consists of the desired signal s(n) and a white noise signal w(n).

The DSP56001 instructions listed below configure the DSP56200 peripheral as a single adaptive FIR filter. Register Bank 1 is first selected, then the quantity (number) of taps minus one is loaded to the FIR Tap Length Register. Register Bank 0 is then selected and the 16-bit rounding mode and coefficient update options are enabled. The value of the loop gain is then loaded into the K Register.

```
movep   #>$99,y:$ffff     ;set the adaptive filter
                          ; mode
movep   #ntaps-1,y:$fff1
movep   #>90,y:$ffff
movep   #kh,y:$fff4     ;set the K factor (loop
movep   #kl,y:$fff5     ;gain)
```

The DSP56001 instructions listed below read the A/D converter, transfer the digitized data to the DSP56200 peripheral's data input X1 Register and D Reference Register, update the coefficients of the adaptive filter, and write the negative of the filter's error signal e(n) to the D/A converter. The contents of the data input X1 Register and Output Data Register are assumed to be $FFFF00 and $0000FF, respectively.

The adaptive FIR filter's output can be obtained by adding the reference input term to the value read from the Output Data Register:

$$\begin{aligned}\text{Adaptive Filter Output} &= \text{Reference Input} - \text{error}\\ &= \text{Reference Input} + \text{DSP56200 Output Data Register} \qquad (11.2)\end{aligned}$$

Note that due to the pipelined structure of the DSP56200 peripheral, the reference input term must be delayed by three sample periods before adding it to the value read from the Output Data Register.

```
clr     a
clr     b

movep   x:$FFEF,a0        ; Read SSI RX reg. for A/D
                          ; data
                          ; a0 is $XXXX00
rep     #8
asl     a                 ; a1 is $0000XX
movep   a1,y:$fff0        ; Write High byte to 56200
movep   a1,y:$fff2
rep     #8
asl     a                 ; a1 is $00XXXX
and     y1,a              ; a1 is $0000XX
movep   a1,y:$fff1        ; Write Low byte to 56200
movep   a1,y:$fff3

clr     a
movep   y:$fff0,b1        ; Read from 56200 High byte
and     y1,b              ; b1 contains the High byte
movep   y:$fff1,a1        ; Low byte
and     y1,a
move    a1,y0             ; y0 contains the Low byte
                          ; y0 is $XXXX00
rep     #8
asl     b
add     y0,b              ; b1 is $00XXXX
rep     #8
asl     b                 ; b1 is $XXXX??
and     x1,b              ; b1 is $XXXX00
clr     a
move    y:(r1),a1
move    b1,x1
add     x1,a
movep   b1,x:$FFEF        ; Write to SSI TX reg. for
                          ; D/A
```

The **AFIR1.ASM** assembler macro shown in Figure 11.12 is used to implement the adaptive FIR filter on the DSP56200 Demonstration System. In the AFIR1.ASM assembler macro, the DSP56001 instructions listed above are combined with DSP56001 instructions similar to those shown in Figure 5.8.

```
;************************************************************
; AFIR1.ASM
;
; Adaptive FIR filters on the DSP56200
; The interrupt control signal is IRQB
;
;************************************************************
;

afirl   macro   ntaps,kh,kl

;
        org     y:$0     ; Store data in Y RAM
        dc      $000000,$000000,$ffff00,$0000ff
```

```
;
        org     p:$40   ; Program start address

;
; Set up ADS board. Stop all the interrupts
;

        movep   #0,x:$FFFF
        movep   #0,x:$FFFE      ; Set bcr to zero and there
                                ; will be waiting time
        move    #0,r0
        move    #-1,m0
        move    #1,r1
        move    r0,r2
        move    #2,r3
        move    #3,r4
        move    m0,m1
        move    m0,m2
        move    m0,m3
        move    m0,m4

;
        movec   #1,sp           ; Init stack pointer
        movec   #0,sr           ; Clear loop flag/interrupt
                                ; mask bits

;
; Set up the SSI for operation with the A/D board
; The following code sets port C to function as SSI
;

        movep   #0,x:$FFE1
        movep   #$1e0,x:$FFE1   ; Write PCC

;
; The following code sets the SSI CRA and CRB control
; registers for external cont. synchronous clock, normal
; mode.

        movep   #$4000,x:$FFEC
        movep   #$3200,x:$FFED

;
; Initialize IPR to allow interrupt 2 & 3
        ori     #$02,mr
        andi    #$fe,mr

;
; Allow the IRQB interrupt, level
;
        movep   #$38,x:$FFFF

;
; Sample rate is controlled by DSP56200 system.

self    jmp     self            ; Looping waiting for
                                ; interrupt

;
;*********************************************
; Start Initialize                           *
;*********************************************
initial
        move    #>$1,x0
```

```
        move    x0,y:(r0)
        movep   #>$99,y:$ffff     ;set the adaptive filter
                                  ;mode
        movep   #ntaps-1,y:$fff1
        movep   #>$90,y:$ffff
        movep   #kh,y:$fff4   ; loop gain
;
        movep   #kl,y:$fff5
        rti

;*********************************************
; Interrupt Routine- Read A/D and write D/A *
;*********************************************
ad      jclr    #0,y:(r0),initial       ; Initialization

        clr     a       y:(r4),y1
        clr     b       y:(r3),x1
        movep   x:$FFEF,a0       ; Read SSI RX reg. for A/D
;                                ; a0 is $XXXX00

;
        rep     #8
        asl     a                ; a1 is $0000XX
        movep   a1,y:$fff0       ; Write High byte to 56200
        movep   a1,y:$fff2
        rep     #8
        asl     a                ; a1 is $00XXXX
        and     y1,a             ; a1 is $0000XX
        movep   a1,y:$fff1       ; Write Low byte to 56200
        movep   a1,y:$fff3
;
        clr     a
        movep   y:$fff0,b1       ; Read from 56200 High byte
        and     y1,b             ; b1 contains the High byte
        movep   y:$fff1,a1       ; Low byte
        and     y1,a
        move    a1,y0            ; y0 contains the Low byte
;                                ; y0 is $XXXX00
        rep     #8
        asl     b
        add     y0,b             ; b1 is $00XXXX
        rep     #8
        asl     b                ; b1 is $XXXX??
        and     x1,b             ; b1 is $XXXX00
        movep   b1,x:$FFEF       ; Write to SSI TX reg. for
                                 ; D/A
;
        rti
;
;*************************************************************
;
; Set up JSR's at the interrupt addresses for SSI interrupts
;
;*************************************************************

        org     p:$000a ; If IRQB interrupt
        jsr     ad      ; Run program
;
        endm
```

Figure 11-12 AFIR1.ASM Assembler Macro to Implement the Adaptive Filter on the DSP56200 Demonstration System

EXAMPLE 11.3: Adaptive Filter

The **AFIR1.ASM** assembler program which implements an adaptive FIR filter, having 7 taps and 0.02 loop gain, on the DSP56200 Demonstration System is shown in Figure 11.13. A copy of the **AFIR1.ASM** assembler program and the **AFIR1.LOD** absolute load program can be found on the DSP56000/1 Demonstration Software #4 diskette.

1. POWER-DOWN THE PC, ADS, FUNCTION AND WHITE NOISE GENERATORS.
2. ON THE DSP56200 DEVELOPMENT BOARD, INSTALL JUMPER JP2 AND REMOVE JUMPER JP1. CONNECT BOARD TO ADM.
3. ON THE EVB, INSTALL JUMPER JP5, REMOVE JUMPERS JP1,JP2, JP3 AND JP4 AND CONNECT THE EXTERNAL CLOCK SUPPLIED BY THE DSP56200 DEVELOPMENT BOARD TO J1-6.
4. BUILD A SUMMING OPERATIONAL AMPLIFIER TO GENERATE THE INPUT SIGNAL X(t) BY ADDING THE FUNCTION GENERATOR'S OUTPUT S(t) TO THE WHITE NOISE GENERATOR'S OUTPUT W(t).
5. CONNECT THE INPUT SIGNAL X(t) TO THE EVB'S "BNC2" INPUT (A/D INPUT) AND TO THE OSCILLOSCOPE'S CHANNEL ONE INPUT.
6. CONNECT THE EVB'S "BNC1" OUTPUT (D/A OUTPUT) TO THE OSCILLOSCOPE'S CHANNEL TWO INPUT.
7. POWER-ON THE PC AND ADS; BOOT THE PC.
8. POWER-ON THE FUNCTION AND WHITE NOISE GENERATORS.
9. INSERT THE "DSP56000/1 DEMONSTRATION SOFTWARE #4" DISKETTE INTO DRIVE "A".
10. Depending on your PC's configuration, INVOKE THE "ADS56000 USER INTERFACE SOFTWARE" PROGRAM from Drive "B" or the hard disk.
11. TYPE **load a:afir1.lod<return>** to load the assembled and linked version of the assembler program shown in Figure 11.13.
12. TYPE **go<return>** to begin execution of the program.
13. VARY THE SINUSOIDAL INPUT FREQUENCY FROM 0.5 KHZ TO 20 KHZ AND OBSERVE THE INPUT AND OUTPUT SIGNAL ON THE OSCILLOSCOPE. Note that the error signal converges quickly to small values.
14. TYPE **force b<return>** to halt execution of the program.
15. TYPE **quit<return>** IF YOU WANT TO EXIT THE "ADS56000 USER INTERFACE SOFTWARE" PROGRAM.

```
;*************************************************************
;
;   Adaptive Filter Example
;
;   AFIR1EX.ASM
;
;*************************************************************

        include 'afir1'

ntaps   equ     $7
kh      equ     $02
kl      equ     $8f

        afir1    ntaps,kh,kl

        end
```

Figure 11-13 AFIR1EX.ASM Assembler Program to Implement the Adaptive FIR Filter in Example 11.3

11.7 PERTINENT DSP56001 SIGNALS:

The DSP56001 processors pertinent Expansion Port A, interrupt, power and clock signals are:

a) *Expansion Port (Port A):*

1) *Address Bus (A0-A15)*

These three-state output pins specify the address for external program and data memory accesses. To minimize power dissipation A0-A15 do not change state when external memory spaces are not being accessed.

2) *Data Bus (D0-D23)*

These pins provide the bidirectional data bus for external program and data memory accesses . D0-D23 are in the high-impedance state when the bus grant signal is asserted.

3) *Bus Control:*

Program Memory Select ($\overline{PS}$)

This three-state output is asserted only when external program memory is referenced.

Data Memory Select ($\overline{DS}$)

This three-state output is asserted only when external data memory is referenced.

X/Y Select ($X/\overline{Y}$)

This three-state output selects which external data memory space (X or Y) is referenced by data memory select ($\overline{DS}$).

Read Enable ($\overline{RD}$)

This three-state output is asserted to read external memory on the data bus D0-D23.

Write Enable ($\overline{WE}$)

This three-state output is asserted to write external memory on the data bus D0-D23

Bus Request ($\overline{BR}$)

The bus request input $\overline{BR}$ allows another device such as a processor or DMA controller to become the master of external data bus D0-D23 and external address bus A0-A15. When $\overline{BR}$ is asserted, the DSP56001 will always release the external data bus D0-D23, address bus A0-A15 and bus Control pins $\overline{PS}$, $\overline{DS}$, X/$\overline{Y}$, $\overline{RD}$, and $\overline{WR}$ (i.e., Port A), by placing these pins in the high-impedance state after the execution of the current instruction has been completed.

Bus Grant ($\overline{BG}$)

This three-state output is asserted to acknowledge an external bus request after Port A has been released.

b) *Interrupt and Mode Control*

1) *Mode Select A/External Interrupt Request A (MODA/$\overline{IRQA}$*
Mode Select B/External Interrupt Request B (MODB/$\overline{IRQB}$)

These two inputs have dual functions: 1) to select the initial chip operating mode and, 2) to receive an interrupt request from an external source. MODA and MODB are read and internally latched in the DSP when the processor exits the RESET state. After leaving the RESET state, the MODA and MODB pins automatically change to external interrupt requests $\overline{IRQA}$ and $\overline{IRQB}$. After leaving the RESET state the chip mode can be changed by software. $\overline{IRQA}$ and *$\overline{IRQB}$* may be programmed to be level sensitive or negative edge triggered. When edge triggered, triggering occurs at a voltage level and is not directly related to the fall time of the interrupt signal, however, the probability of noise on IRQA or IRQB generating multiple interrupts increases with increasing fall time of the interrupt signal.

2) *Reset* ($\overline{RESET}$)

This Schmitt trigger input pin is used to reset the DSP56001. When $\overline{RESET}$ is asserted, the DSP56001 is initialized and placed in the reset state. When $\overline{RESET}$ signal is negated, the initial chip operating mode is latched from the MODA and MODB pins. When coming out of reset, negation occurs at a voltage level and is not directly related to rise time of the reset signal, however, the probability of noise on $\overline{RESET}$ generating multiple resets increases with increasing rise time of the reset signal.

c) *Power and Clock*

1) *Power* (V_{CC}), *Ground (GND)*

There are five sets of power and ground pins. Two pairs for internal logic, one power and two ground for port A address and control pins, one power and two ground for port A data pins, and one pair for peripherals.

2) *External Clock / Crystal Input (EXTAL)*

EXTAL may be used to interface the crystal oscillator input to an external crystal or an external clock.

3) *Crystal Output (XTAL)*

This output connects the internal crystal oscillator output to an external crystal. If an external clock is used, XTAL should not be connected.

11.8 PERTINENT DSP56200 PERIPHERAL SIGNALS

The DSP56200 peripheral's pertinent Asynchronous Parallel Interface, Cascade Interface, power supply and clock signals are:

a) *Asynchronous Parallel Interface:*

1) *D0-D7 (Data Bus)*

These eight pins provide a bidirectional data bus for communication with a host processor. The pins remain in the high-impedance state unless both $\overline{RD}$ and $\overline{CS}$ are asserted.

2) *A0-A3 (Register Address Pins)*

A0-A3 select (in conjunction with the least significant bit of the Configuration Register) which register will be addressed when the Chip Select line is brought low and a read or write operation is performed.

3) *$\overline{CS}$ (Chip Select)*

This pin (active low) enables accesses to the chip operating registers. When not asserted, the D0-D7 lines will go into high-impedance state and all access to the chip will be disabled. $\overline{CS}$ must be high when the START pulse arrives.

4) *$\overline{RD}$ (Read Strobe)*

When $\overline{RD}$ (active low) is asserted, the contents of the register specified by A0-A3 will be driven onto D0-D7. When $\overline{RD}$ is high, pins D0-D7 go into the high-impedance state.

5) *$\overline{WR}$ (Write Strobe)*

This pin (active low) enables host writes to the register specified by A0-A3. Data on D0-D7 must be valid for the specified setup time before the rising edge of $\overline{WR}$.

b) *Cascade Interface*

SDI (Serial Data Input)

This pin is used in the cascade mode to receive data from the last tap of the data shift register in the preceding DSP56200 chip. It connects to the SDO pin of the previous chip cascade. If the chip is first in cascade or is used in standalone mode, this pin should be grounded.

SDO (Serial Data Output)

This pin is used in the cascade mode to pass the last data sample in the data shift register to the next DSP56200 in the cascade. It connects to the SDI pin of the next DSP56200. The output of this pin is not connected if the chip is used in standalone mode or if it is the last chip in a cascaded system.

SSI (Serial Sum Input)

This pin is used in the cascade mode to receive the partial sums from preceding stages. If the chip is first in the cascade or is used in standalone mode, this pin should be grounded.

SSO (Serial Sum Out)

This pin is primarily used in the cascade mode to pass the partial sums to the next DSP56200 in the cascade. The SSO pin is usually connected to the SSI pin of the next chip in the cascade. In the adaptive filter mode, the SSO pin of the last chip in the cascade is connected to the SEI pin on all chips cascaded including itself. This pin should not be connected in standalone FIR modes (single and dual FIR filters).

SEI (Serial Error Input)

This pin is used in the adaptive filter mode. It provides the means of receiving the error term output from the last chip in the cascade. In standalone Adaptive Filter mode, this pin is tied to the SSO pin. In cascade Adaptive Filter mode, this pin is tied to the SSO pin of the last chip in the cascade. This pin should be grounded in the Single FIR Filter mode or the Dual FIR Filter mode.

c) *Clocks and Power*

1) *CLOCK (Clock Input)*

This pin accepts the input clock from the DSP56200. The internal and external clocking frequencies are the same, and the maximum frequency for this input is 10.25 MHz. In cascaded systems, all DSP56200s must be driven from the same clock source. When CLOCK is held low, the device enters a low power mode.

2) *START (Start Processing Command)*

This pin is used to provide a second clock to the chip at the system's sampling rate. This clock must be synchronized with the signal on the CLOCK pin in order to ensure proper operation. The START pulse duty cycle is not critical; however, the start pulse must be low for at least two clock periods prior to going high signalling the start of a new cycle and, it must meet the AC electrical specifications. In most applications the START pin will be cycled at the system data sampling frequency and data will be written and read by the host once per START cycle. Some form of interrupt should be provided to the host processor to indicate when a START transition has occurred so that the host can provide filter data I/O service.The simplest approach is to connect START pin signal to the interrupt input of the host processor. When this approach is used it is guaranteed that the host will not service the DSP56200 until after the START transition has occurred (since interrupt service usually has non-zero latency) and the aforementioned timing restrictions on host accesses during START transition will be met.

3) V_{CC} *(Power) and GND (Ground)*

The DSP56200 provides three VCC pins and three GND pins. As with any high-speed logic family design, careful attention should be paid to bypassing all VCC and GND pins on the DSP56200. 0.1 μF ceramic bypass capacitors with short (less than 0.5 inch) leads should be used. In particular, during DSP56200 read operations care should be taken to assure stable logic low (as per DC specifications) of the START input. Improper supply bypassing or grounding may cause noise to be induced on the START input causing false cycling of the chip.

12

IMPLEMENTING AN ADAPTIVE NOISE CANCELLER WITH THE DSP56200 PERIPHERAL AND/OR THE DSP56001 PROCESSOR

Noise cancelation seeks to reduce noise power while leaving a desired signal relatively unchanged. In this chapter, an adaptive noise canceller (ANC) is implemented on both the DSP56001 Demonstration System and the DSP56200 Demonstration System.

An ANC is shown in Figure 12.1. The first sensor of the ANC receives the signal s(t) plus an uncorrelated noise $n_2(t)$. The combined signal and noise is then digitized to form the primary signal d(k) which is then provided to the primary input of the noise canceller, where

$$d(k) = s(k) + n_2(k) \tag{12.1}$$

The second sensor of the ANC receives noise $n_1(t)$, which is correlated with noise $n_2(t)$. The sensor input is then digitized to provide the reference signal $n_1(k)$ which is then provided to the reference input of the noise canceller. The lumped linear model of the transmission path characteristics between the noise source and the sensors is shown in Figure 12.2. The relationship between $n_1(k)$ and $n_2(k)$ can be described by:

$$n_2(k) = B(q^{-1})n(k) \tag{12.2}$$

$$n_1(k) = A(q^{-1})n(k) \tag{12.3}$$

Therefore

$$A(q^{-1})n_2(k) = B(q^{-1})n_1(k), \tag{12.4}$$

where

$$A(q^{-1}) = 1 - a_1q^{-1} - \dots - a_nq^{-n} \tag{12.5}$$

and

$$B(q^{-1}) = b_0 + b_1q^{-1} + \dots + b_nq^{-n}. \tag{12.6}$$

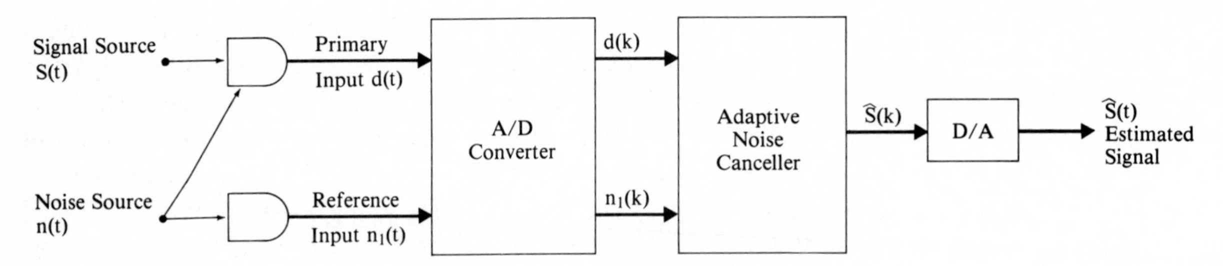

Figure 12-1 Adaptive Noise Canceller

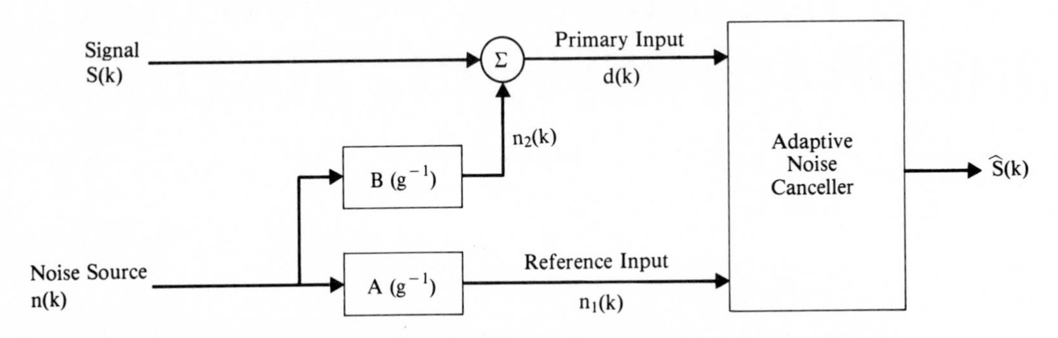

Figure 12-2 Lumped Linear Model of Transmission Path Characteristics between Noise Sources and Sensors

To produce an estimate of the noise which is corrupting a desired signal at the primary input of the ANC, a system identification is performed to estimate a_i and b_i. The system identification is performed by the recursive-like ANC shown in Figure 12.3, where the LMS algorithm described in Chapter 9 is used to estimate the parameters a_i and b_i. The resulting noise estimate is then subtracted from the primary signal to produce the desired signal and a residual noise component.

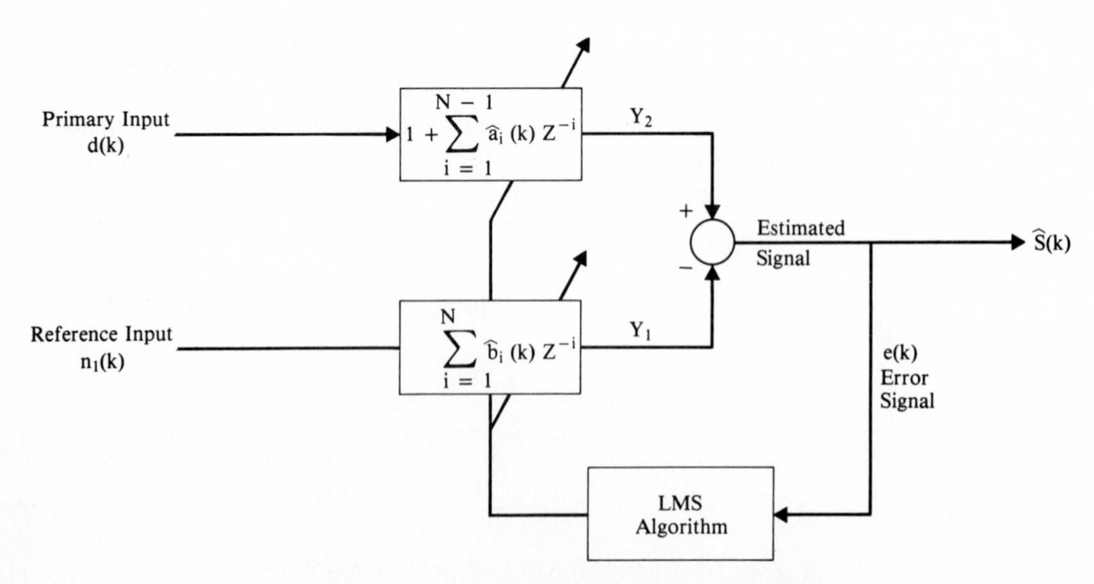

Figure 12-3 Recursive-like Adaptive Noise Canceller

12.1 IMPLEMENTING AN ANC WITH THE DSP56001 PROCESSOR:

The DSP56001 instructions listed below generate the error signal E. The reference signal N1, which is provided to the reference input of the ANC, is transferred to the A-Accumulator and the primary signal D, which is provided to the primary input of the ANC, is transferred to the B-Accumulator. Address Registers R1 and R5 point to the first AFIR filter's state and coefficient buffers, respectively. Address Registers R2 and R6 point to the second AFIR filter's state and coefficient buffers, respectively. The reference signal is processed by the first AFIR filter to produce the output signal Y1. The primary signal is processed by the second AFIR filter to produce an output which is then subtracted from the primary signal to in turn produce the output signal Y2. The output signal Y1 is then subtracted from the output signal Y2 to produce the error signal E. The error signal E is then transferred to the D/A converter.

```
          move    a,x0             ; reference input

; output signal Y1

          clr     a       x0,x:(r1)-       y:(r5)+,y0
          rep     #ntaps-1
          mac     x0,y0,a    x:(r1)-,x0    y:(r5)+,y0
          mac     x0,y0,a
          asr     a

          move    b,x0              ; primary input

; output signal Y2

          asr     b
          neg     b
          move    #0,y0
          move    x0,x:(r2)-       y0,y:(r6)+
          move    x:(r2)-,x0       y:(r6)+,y0
          rep     #ntaps-2
          mac     x0,y0,b    x:(r2)-,x0    y:(r6)+,y0
          mac     x0,y0,b
          neg     b

;  error E

          sub     a,b               ; b-a --> b
          move    b,x0              ; Error = Estimated Signal

;  write D/A

          move    #$ffff00,x1
          and     x1,b
          movep   b,x:$ffef         ; Output error
```

The DSP56001 instructions listed below multiply the loop gain by the error signal, and update the coefficients of both AFIR filters.

```
          move    #lg,y1            ; loop gain
          mpy     x0,y1,b
          move    b,y1              ; product of loop gain and
                                    ; error

;
; update the coefficients of the first filter
;
```

```
        do      #ntaps,_updatex
        move    y:(r5),a      x:(r1)-,x0
        mac     x0,y1,a
        move    a,y:(r5)+
_updatex
        lua     (r1)-,r1          ; Address update

;
; update the coefficients of the second filter
;

        do      #ntaps,_updated
        move    y:(r6),a      x:(r2)-,x0
        mac     x0,y1,a
        move    a,y:(r6)+
_updated
        lua     (r2)-,r2
```

EXAMPLE 12.1:

In this example, a white noise generator is connected to the input of the A/D converter on the EVB to generate the reference signal N1. The primary signal D is simulated by the DSP56001 processor, where the white noise N2 is also simulated by the DSP56001 processor in accordance with the transfer function:

$$\frac{N_2(k)}{N_1(k)} = \frac{1}{1 + a_1 z^{-1} + a_2 z^{-2}} \tag{12.7}$$

where

$$a_1 = -.81 \quad \text{and} \quad a_2 = 1.56$$

The DSP56001 instructions listed below simulate a 26.6 KHz sinewave, where the sinewave's data values are stored in X-Memory.

```
pi1     equ     3.141592654
ratio   equ     3.0
freq1   equ     2.0*pi1/@cvf(ratio)
points  equ     128

        move    #sinew,r3
        move    #points-1,m3

        org     x:sinew
count   set     0

        dup     points
        dc      -0.05*@sin(@cvf(count)*freq1)
count   set     count+1
        endm
```

The DSP56001 instructions listed below read the reference signal from the output of the A/D converter and generate simulated noise signal N2. The simulated noise signal N2 is then added to the simulated 26.6 KHz sinewave to produce the desired primary signal D, where D is then provided to the primary input of the ANC. Address Registers R0 and R4 point to the states and the coefficients, respectively, of the filter used to generate the noise signal N2.

```
        movep   x:$FFEF,a          ; Read SSI RX reg. for A/D
        asr     a                  ; A = N1/2

        move    a,b
        asr     b                  ; B = N1/4

; generating primary input D

        move    #states,r0
        move    #coef,r4
        ori     #$8,mr

        move              x:(r0)+,x0     y:(r4)+,y0
        mac     x0,y0,b   x:(r0)-,x1     y:(r4),y0
        mac     x1,y0,b   x1,x:(r0)+
        move    b,x:(r0)

        move    x:(r3)+,y1
        add     y1,b

        andi    #$f7,mr
        asl     b                  ; B = D/2
```

The **ANC1.ASM** assembler program shown in Figure 12.4 implements Example 12.1 on the DSP56001 DSP Demonstration System, where a 10.25 MHz clock is provided to the EVB by the DSP56200 Demonstration Board. The 10.25 MHz clock is used by the EVB to produce an 80 KHz (10.25 MHz/128) sampling rate. The **ANC1.ASM** assembler program and the **ANC1.LOD** absolute load program can be found on the *DSP56000/1 Demonstration Software #4* diskette.

1. POWER-DOWN THE PC, ADS, AND WHITE NOISE GENERATOR.
2. ON THE EVB, INSTALL JUMPER JP5 AND REMOVE JUMPERS JP1, JP2,JP3 AND JP4 AND CONNECT THE EXTERNAL CLOCK SUPPLIED BY THE DSP56200 DEVELOPMENT BOARD TO J1-6.
3. CONNECT THE WHITE NOISE GENERATOR TO THE EVB'S "BNC2" INPUT (A/D INPUT) AND TO THE OSCILLOSCOPE'S CHANNEL ONE INPUT. CONNECT 56200 BOARD TO ADM.
4. CONNECT THE EVB'S "BNC1" OUTPUT (D/A OUTPUT) TO THE OSCILLOSCOPE'S CHANNEL TWO INPUT.
5. POWER-ON AND BOOT THE PC.
6. POWER-ON THE WHITE NOISE GENERATOR.
7. INSERT THE "DSP56000/1 DEMONSTRATION SOFTWARE #4" DISKETTE INTO DRIVE "A".
8. Depending on your PC's configuration, INVOKE THE "ADS56000 USER INTERFACE SOFTWARE" PROGRAM from Drive B or the hard disk.
9. TYPE **load a:ancl.lod<return>** to load the assembled and linked version of the assembler program shown in Figure 12.4.
10. TYPE **go<return>** to begin execution of the program.
11. VARY THE NOISE LEVEL AND OBSERVE THE INPUT AND OUTPUT SIGNALS ON THE OSCILLOSCOPE. NOTE THAT THE OUTPUT SIGNAL IS THE 26.7 KHZ SINUSOID AND A RESIDUAL NOISE.

12. TYPE **force b<return>** to halt execution of the program.
13. TYPE **quit<return>** IF YOU WANT TO EXIT THE "ADS56000 USER INTERFACE SOFTWARE" PROGRAM.

```
;*************************************************************
;
; file: ANC1.ASM
;
; Noise Canceller with the DSP56001
;
;
;*************************************************************

ntaps   equ     3           ; length of filter = 3
lg      equ     $100000     ; Loop gain = 0.125
pi1     equ     3.141592654
ratio   equ     3.0         ; sinewave freq =  sampling
                            ; freq/ratio

freq1   equ     2.0*pi1/@cvf(ratio)

sinew   equ     $100        ; base address for sinewave in X
                            ; memory

points  equ     128
;

;*************************************************************
; Initialization
;*************************************************************

;
; Filter used to generate primary input
;

        org     x:$0
states  ds      2

;
        org     y:$0                    ; filter coefficient
coef    dc      -.81/2.0,1.56/2.0
;

;
; First adaptive filter
;

        org     x:$4
xx      dsm     ntaps
        org     y:$4
cx      dsm     ntaps
;

;
; Second Adaptive Filter
;

        org     x:$8
dd      dsm     ntaps
        org     y:$8
dx      dsm     ntaps

;
; generate sinewave
;
```

```
        org     x:sinew
count   set     0

        dup     points
        dc      -0.05*@sin(@cvf(count)*freq1)
count   set     count+1
        endm

;
        org     p:$40    ; Program start address
;

        move    #sinew,r3
        move    #points-1,m3

        move    #xx,r1
        move    #cx,r5
        move    #dd,r2
        move    #dx,r6
        move    #ntaps-1,m1
        move    m1,m5
        move    m1,m2
        move    m1,m6
;
; Set up ADS board. Stop all the interrupts
;
        movep   #0,x:$FFFF
        movep   #0,x:$FFFE        ; Set BCR to zero
        movep   #0,x:$FFE1
        movep   #$1e0,x:$FFE1     ; Write PCC
;
        movep   #$4000,x:$FFEC
        movep   #$3200,x:$FFED

;
; Sample rate is controlled by DSP56200 Development board
; Sampling rate = 10MHz/128.

begin
input   btst    #7,x:$ffee
        jcc     input

;**********************************************************
; Read reference input
; Generate primary input
;**********************************************************

;
; read reference input
;

        movep   x:$FFEF,a         ; Read SSI RX reg. for A/D
                                  ; data
        asr     a                 ; half the input value to
                                  ; eliminate overflow
        move    a,b               ; B = A
        asr     b                 ; B = input / 4.

;
; generate primary input
;

        move    #states,r0
        move    #coef,r4
        ori     #$8,mr
;
        move             x:(r0)+,x0      y:(r4)+,y0
        mac     x0,y0,b  x:(r0)-,x1      y:(r4),y0
```

```
        mac     x1,y0,b  x1,x:(r0)+
        move    b,x:(r0)
        move    x:(r3)+,y1
        add     y1,b
;
        andi    #$f7,mr          ; Return scaling mode
        asl     b                ; B = primary input / 2

;**********************************************************
;
; adaptive filter
;
;**********************************************************

start

        move    a,x0    ; reference input

; output signal Y1

        clr     a        x0,x:(r1)-       y:(r5)+,y0
        rep     #ntaps-1
        mac     x0,y0,a x:(r1)-,x0       y:(r5)+,y0
        mac     x0,y0,a
        asr     a                ; sum /2
;

        move    b,x0             ; primary input

; output signal Y2

        asr     b
        neg     b                ; d/2 - sum
;
        move    #0,y0
        move    x0,x:(r2)-       y0,y:(r6)+
        move    x:(r2)-,x0       y:(r6)+,y0
        rep     #ntaps-2
        mac     x0,y0,b x:(r2)-,x0       y:(r6)+,y0
        mac     x0,y0,b
;
        neg     b

;  error

        sub     a,b              ; b-a --> b
        move    b,x0             ; error = estimated signal

;  Write D/A

        move    #$ffff00,x1
        and     x1,b
        movep   b,x:$ffef        ; Output error

;
; product of loop gain and error
;

        move    #lg,y1           ; loop gain
        mpy     x0,y1,b
        move    b,y1             ; Error * loop gain

;
; update the coefficients of the first filter
;

        do      #ntaps,_updatex
        move    y:(r5),a  x:(r1)-,x0
        mac     x0,y1,a
```

```
        move    a,y:(r5)+
_updatex
        lua     (r1)-,r1         ; Address update

;
; update the coefficients of the second filter
;

        do      #ntaps,_updated
        move    y:(r6),a  x:(r2)-,x0
        mac     x0,y1,a
        move    a,y:(r6)+
_updated
        lua     (r2)-,r2         ; Address update

;

        jmp     begin

        end
```

Figure 12-4 ANC1.ASM Assembler Program to Implement the ANC on the DSP56001

12.2 IMPLEMENTING AN ANC WITH THE DSP56200 PERIPHERAL:

To implement an ANC with the DSP56200 peripheral, the device must be configured, as shown in Figure 12.5, as a multi-channel AFIR filter where both the primary input signal and the reference input signal are passed through the AFIR filter structure. Note that the multi-channel AFIR filter structure is composed of two AFIR filters. The DSP56200 peripheral uses its Data RAM Access Register to provide the primary input signal D to the multi-channel AFIR filter. Note that the primary input signal D is delayed by one sample period before it is also provided to the input of the second AFIR filter structure.

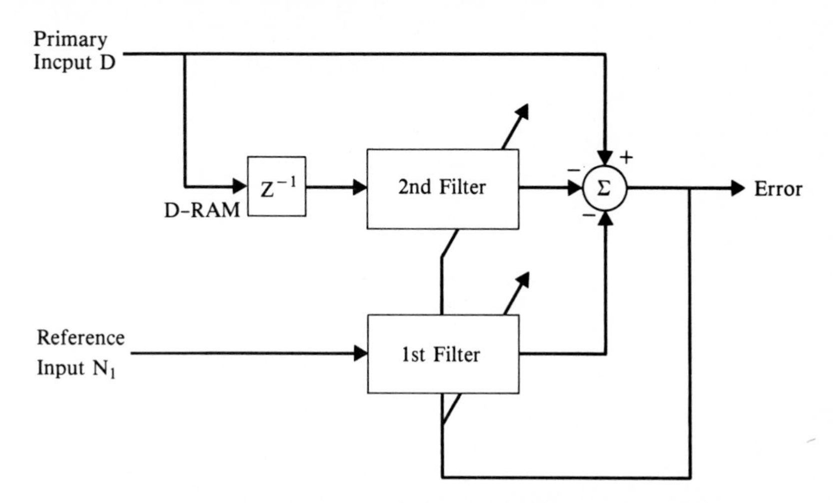

Figure 12-5 Multi-Channel Adaptive Filter with DSP56200

The DSP56001 instructions listed below configure the DSP56200 peripheral as a multi-channel AFIR filter. Register Bank 1 is first selected, then the sum of the quantity (number) of taps in each of the two AFIR filter structures is loaded into the Tap Length Register. The sum of the quantity of taps must be less than or equal to 255 and must be greater than or equal to four. Note that this is different than the normal operating mode(s) of the DSP56200 peripheral where the Tap Length Register is loaded with the total quantity of taps minus one. Also note that the quantity of taps in the first AFIR filter structure can differ from the quantity of taps in the second AFIR filter structure. Register Bank 0 is then selected, and the Configuration Register is programmed to select the single AFIR filter mode and enable the 16-bit rounding option. The value of the loop gain is then loaded to the K Register.

```
movep   #>$01,y:$ffff
movep   #ntaps-1,y:$fff1
movep   #>$90,y:$ffff

movep   #>$kh,y:$fff4    ; loop gain
movep   #>$kl,y:$fff5
```

The following two instructions write the X1 Register with the reference input for the first AFIR filter structure. The reference input is stored in the XY-Memory location pointed to by the R2 Address Register. The higher 8 bits of the reference input stored in the X-Memory location pointed to by the R2 Address Register are transferred to the Y-Memory location $FFF0 (X1-HIGH), while the lower 8 bits of the reference input stored in the Y-Memory location pointed to by the R2 Address Register are transferred to the Y-Memory location $FFF1 (X1-LOW).

```
movep   x:(r2),y:$fff0
movep   y:(r2),y:$fff1
```

The following two instructions write the DATA RAM ACCESS Register with the primary input for the second AFIR filter structure. The primary input is stored in the XY-Memory location pointed to by the R5 Address Register. The higher 8 bits of the primary input stored in the X-Memory location pointed to by the R5 Address Register are transferred to the Y-Memory location $FFF8 (DATA RAM-HIGH), while the lower 8 bits of the primary input stored in the Y-Memory location pointed to by the R5 Address Register are transferred to the Y-Memory location $FFF9 (DATA RAM-LOW).

```
movep   x:(r5),y:$fff8
movep   y:(r5)+,y:$fff9
```

The following instructions write the D Register with the primary input stored in the P-Memory pointed to by the R2 Address Register. The higher 8 bits of the primary input are transferred to the Y-Memory location $FFF2 (D-HIGH), while the lower 8 bits are transferred to the Y-Memory location $FFF3 (D-LOW).

```
clr     a
movem   p:(r2)+,a0
rep     #8
asl     a
movep   a1,y:$fff2

rep     #8
asl     a
and     y1,a
movep   a1,y:$fff3
```

The following instructions load the RAM Address Register with the quantity of taps (MTAPS) in the first AFIR filter structure and load the three bytes of the Coefficient RAM Access Register with logic zeros.

```
movep   #>$mtaps,y:$fffd

move    #0,x0
movep   x0,y:$fffa
movep   x0,y:$fffb
movep   x0,y:$fffc
```

The DSP56001 instructions listed below transfer the error signal, produced by the multi-channel AFIR filter, to the D/A converter; where, the error signal is an estimate of the desired signal.

```
clr     a             y:(r4),y1
clr     b

movep   y:$fff0,b1
and     y1,b              ; b1 is 0000XX
movep   y:$fff1,a1
and     y1,a
move    a1,y0             ; y0 is 0000XX

rep     #8
asl     b
add     y0,b          ; b1 is $00XXXX
rep     #8
asl     b             ; b1 is $XXXX??
move    y:(r3),x1     ; x1 is $ffff00
and     x1,b          ; b1 is $XXXX00
movep   b1,x:$FFEF    ; Write to SSI TX reg. for D/A
```

EXAMPLE 12.2:

In this example, a white noise generator is connected to the input of the A/D converter on the EVB to generate the reference signal N1. The white noise signal N2 is simulated by the DSP56001 processor using Equation 12.7. The DSP56001 processor simulates the primary signal D by summing the simulated noise signal N2 with the simulated 26.6 KHz sinewave.

The **DATA.ASM** assembler program shown in Figure 12.6 generates and stores the two primary signal data and the reference signal data. The reference signal data, which will be transferred to the X1 Register is stored in the XY memory starting at location $400. The primary signal data, which will be transferred to the Data RAM Access Register, is stored in XY memory starting at location $10. The primary signal data, which will be transferred to the D Register, is stored in P-Memory starting at location $400. The **ANC2.ASM** assembler program shown in Figure 12.7 implements the ANC on the DSP56200 Demonstration System. The sampling frequency is 80 KHz. The **DATA.ASM** and **ANC2.ASM** assembler programs and the **DATA.LOD** and **ANC2.LOD** absolute load programs can be found on the *DSP56000/1 Demonstration Software #4* diskette.

1. POWER-DOWN THE PC, ADS, AND WHITE NOISE GENERATOR.
2. ON THE DSP56200 DEVELOPMENT BOARD, INSTALL JUMPER JP2 AND REMOVE JUMPER JP1.
3. ON THE EVB, INSTALL JUMPER JP5, REMOVE JUMPERS JP1,JP2, JP3, AND JP4 AND CONNECT THE EXTERNAL CLOCK SUPPLIED BY THE DSP56200 DEVELOPMENT BOARD TO J1-6.
4. CONNECT THE WHITE NOISE GENERATOR TO THE EVB'S "BNC2" INPUT (A/D INPUT) AND TO THE OSCILLOSCOPE'S CHANNEL ONE INPUT. CONNECT 56200 BOARD TO ADM.
5. CONNECT THE EVB'S "BNC1" OUTPUT (D/A OUTPUT) TO THE OSCILLOSCOPE'S CHANNEL TWO INPUT.
6. POWER-ON THE PC AND ADS; BOOT THE PC.
7. POWER-ON THE WHITE NOISE GENERATOR.

8. INSERT THE "DSP56000/1 DEMONSTRATION SOFTWARE #4" DISKETTE INTO DRIVE "A".
9. Depending on your PC's configuration, INVOKE THE "ADS56000 USER INTERFACE SOFTWARE" PROGRAM from Drive B or the hard disk.
10. To load and execute the assembled and linked versions of the assembler programs shown in Figures 12.6 and 12.7, TYPE:

```
change x:$0..$7ff 0  y:$0..$7ff 0<return>
change x 0 y 0 a 0 b 0 r0..r7 0 n0..n7 0<return>
load a:data<return>
break p:$83<return>
go #1<return>
break off<return>
wait<return>
load a:anc2<return>
go<return>
```

11. VARY THE NOISE LEVEL AND OBSERVE THE INPUT AND OUTPUT SIGNAL ON THE OSCILLOSCOPE. NOTE THAT THE OUTPUT SIGNAL IS THE 26.7 KHZ SINUSOID AND A RESIDUAL NOISE.
12. TYPE **force b<return>** to halt execution of the program.
13. TYPE **quit<return>** IF YOU WANT TO EXIT THE "ADS56000 USER INTERFACE SOFTWARE" PROGRAM.

```
;************************************************************
; file: DATA.ASM
;
; Generate ANC2 input signals
;
;************************************************************

pil     equ     3.141592654
ratio   equu    3.0        ; sinewave freq =  sampling
                           ; freq/ratio

freql   equ     2.0*pil/@cvf(ratio)

sinew   equ     $100       ; base address for sinewave in X
                           ; memory

points  equ     128

;
; generate sinewave
;

        org     x:sinew
count   set     0

        dup     points
        dc      -0.05*@sin(@cvf(count)*freql)
count   set     count+1
        endm

; FIR filter states and coefficients

;
```

```
        org     x:$0            ; Reserve memory places
states  ds      2               ; for the IIR filter
;
        org     y:$0
coef    dc      -.81/2.0,1.56/2.0
;
; program start

        org     p:$40           ; Program start address

        move    #sinew,r1
        move    #points-1,m1
        move    #$400,r7
        move    r7,r6
        move    #$10,r3
        move    #-1,m7
        move    m7,m6
        move    m7,m3

;
; Set up ADS board. Stop all the interrupts
;

        movep   #0,x:$FFFF
        movep   #0,x:$FFFE      ;set bcr to zero
        movep   #0,x:$FFE1
        movep   #$1e0,x:$FFE1   ;write PCC
;
        movep   #$4000,x:$FFEC
        movep   #$3200,x:$FFED

;
; Sample rate is controlled by DSP56200 development board.
;

;  read and store reference input

;
        do      #1001,_endin
;

input   btst    #7,x:$ffee
        jcc     input
;
        clr     a
        movep   x:$FFEF,b       ; Read SSI RX reg. for A/D
                                ;
        asr     b               ; reduce input value to
;                               ; avoid overflow

;       asr     b

        move    b,a0
;
        move    #>$ff,x1
        rep     #8
        asl     a
        move    a1,x:(r7)       ; reference input
        rep     #8              ; High byte
        asl     a
        and     x1,a
        move    a1,y:(r7)+      ; Low byte

;
; generate primary input
;

        move    #states,r0
        move    #coef,r4
```

```
        ori     #$8,mr

;
;  IIR Filter used to generate N2
;

        move              x:(r0)+,x0     y:(r4)+,y0
        mac     x0,y0,b   x:(r0)-,x1     y:(r4),y0
        mac     x1,y0,b   x1,x:(r0)+
        move    x:(r1)+,y1

; generating primary input that is loaded to D register

        add     y1,b
        move    b,x:(r0)
        andi    #$f7,mr          ; Return the mode
        movem   b,p:(r6)+        ; half of the primary input

;
; Generating primary input that is loaded to the DATA
; RAM ACCESS register
;

        move    #>$ff,x1
        clr     a
        move    x:(r0),a0        ; primary input
        rep     #8
        asl     a                ; primary input
        move    a1,x:(r3)        ; High Byte
        rep     #8
        asl     a
        and     x1,a
        move    a1,y:(r3)+       ; Low Byte
;
_endin
;
        nop
        end
```

Figure 12-6 DATA.ASM assembler program used to generate and store the three inputs for the ANC

```
;***********************************************************
; file: ANC2.ASM
;
; Perform the Noise Cancelling by using the DSP56200
; The interrupt control signal is IRQB
;
;***********************************************************

ntaps   equ     6       ; ntaps = sum of two filters + 1
kh      equ     $06     ; high byte of loop gain
kl      equ     $66     ; low byte of loop gain
mtaps   equ     3       ; mtaps = number of first filter taps

;
        org     y:$2     ; Store data in RAM
        dc      $ffff00,$0000ff
;
        org     p:$40    ; Program start address
```

```
;
; Set up ADS board. Stop all the interrupts
;

        movep   #0,x:$FFFF
        movep   #0,x:$FFFE       ; set bcr to zero
                                 ; no waiting time
        move    #0,r0
        move    #-1,m0
        move    r0,r1
        move    #2,r3
        move    #3,r4
        move    m0,m1
        move    m0,m3
        move    m0,m4

        move    #$400,r2
        move    #1000,m2         ; points - 1
        move    #$10,r5
        move    m2,m5            ; points - 1
;
        movec   #1,sp            ; Init stack pointer
        movec   #0,sr            ; Clear loop flag/interrupt
                                 ; mask bits

;
; Set up the SSI for operation with the A/D of Motorola
; The following code sets port C to function as SSI
;

        movep   #0,x:$FFE1
        movep   #$1e0,x:$FFE1    ; Write PCC

;
; The following code sets the SSI CRA and CRB control
; registers for external cont. synchronous clock, normal
; mode.
;

        movep   #$4000,x:$FFEC
        movep   #$3200,x:$FFED

;
; Initialize IPR to allow interrupt 2 & 3 to occur
;
        ori     #$02,mr
        andi    #$fe,mr

;
; Allow the IRQB interrupt, level
;
        movep   #$38,x:$FFFF

;
; Sample rate is controlled by DSP56200 EVB.
;

self    jmp     self    ; Looping waiting for interrupt

;
;********************************************
; Start Initialize                          *
;********************************************

initial
        lua     (r0)+,r0

        movep   #>$01,y:$ffff
        movep   #ntaps-1,y:$fff1
```

```
        movep   #>$90,y:$ffff

        movep   #kh,y:$fff4    ; loop gain
        movep   #kl,y:$fff5

        rti

;
;*******************************************
; Interrupt Routine- Read A/D and write D/A *
;*******************************************

ad
        jclr    #0,y:(r0),initial       ; Initialization
;
        clr     a           y:(r4),y1
        clr     b
;
        movep   y:$fff0,b1
        and     y1,b            ; b1 is 0000XX
        movep   y:$fff1,a1
        and     y1,a
        move    a1,y0           ; y0 is 0000XX
;
;

        move    #0,x0
        movep   x0,y:$fffa
        movep   x0,y:$fffb
        movep   x0,y:$fffc

        movep   x:(r2),y:$fff0
        movep   y:(r2),y:$fff1

        movep   #mtaps,y:$fffd

        movep   x:(r5),y:$fff8  ; The value should be
                                ; delayed by 1
        movep   y:(r5)+,y:$fff9

        clr     a
        movem   p:(r2)+,a0
        rep     #8
        asl     a
        movep   a1,y:$fff2

        rep     #8
        asl     a
        and     y1,a
        movep   a1,y:$fff3

;

        rep     #8
        asl     b
        add     y0,b        ; b1 is $00XXXX
        rep     #8
        asl     b           ; b1 is $XXXX??
        move    y:(r3),x1   ; x1 is $ffff00
        and     x1,b        ; b1 is $XXXX00
        movep   b1,x:$FFEF  ; Write to SSI TX reg. for D/A
;
        rti

;**********************************************************
; Set up JSR's at the interrupt addresses for SSI interrupts
;**********************************************************
;
```

```
        org     p:$000a        ;IRQB interrupt
        jsr     ad             ;Run program
;
        end
```

Figure 12-7 ANC2.ASM Assembler Program Used to Implement the ANC on the DSP56200 Demonstration System

13

STUDENT PROJECTS: A Spectrum Analyzer and an Adaptive Differential Pulse Code Modulator (ADPCM) using the DSP56001 Processor

In the first project, a simple Spectrum Analyzer is implemented using the DSP56001 processor. Although the A/D on the EVB can produce 16-bit digitized data samples at a maximum rate of 100 KHz, the *maximum* sample precision (number of bits) which can be used in the implementation of the Spectrum Analyzer is related to the number of data samples N, where N must be a power of two. The calculation of the *maximum* sample precision is described. The N data samples, with an adjusted precision equal to the calculated maximum precision, are multiplied by a Hamming window to improve their spectral content. The spectrum is then calculated using an N-point FFT. The DSP56001 instructions which implement the Spectrum Analyzer are described and included on the *DSP56000/1 Demonstration Software #3* diskette.

In the second project, an Adaptive Differential Pulse Code Modulation (ADPCM) system, useful in telephony applications to reduce (compact) the number of bits needed to transmit a voice signal, is described for implementation on an ADPCM Evaluation System. The major components of the ADPCM Evaluation Board are the DSP56001 processor and the MC145503 PCM codec. The schematic diagram for the ADPCM Evaluation Board is shown in Figure 13.6. Completion of the ADPCM project is left as an exercise for the student.

13.1 A SPECTRUM ANALYZER PROJECT USING THE DSP56001 PROCESSOR:

This project uses the DSP56001 processor to compute and display the spectrum of 1024 samples of input data.

13.1.1 Using the Maximum Precision of the Input Data Samples:

The digitized input samples can be divided by a factor of 2^G to limit the magnitude of the output spectrum to a value less than or equal to one, where G is less than or equal to $\log_2$(# of FFT points). Since the A/D converter on the EVB offers 15-bits of precision and the DSP56001 processor offers 24-bits of precision, G can be chosen to be less than or equal to 24-15=9. In this project, G=7.

13.1.2 Windowing:

To calculate the spectrum, the Spectrum Analyzer algorithm uses only N samples of the digitized input signal, where limiting the number of input signal samples to N is referred to as windowing. This limiting of samples is equivalent to multiplying the input signal by a rectangular window sequence w(n), where w(n) = 1 for $0 \leq n \leq N-1$ and where w(n) = 0 otherwise. Sample limiting, however, introduces error into the spectrum. To reduce the effect of this error on the spectrum, one of several types of window functions can be used. The Hamming window is considered here. The causal Hamming window function can be written as:

$$w(n) \;=\; 0.54 \;-\; 0.46 * \cos(2\pi n/N) \tag{13.1}$$

The **HAMMING.ASM** assembler macro shown in Figure 13.1 is used to generate the N-point Hamming window.

```
;********************
; Hamming Window
;********************
;
hamming   macro   points,coef
;
;    points - number of points
;    coef   - base address for window table
;
;
;
pi2     equu    3.141592654
freq2   equ     2.0*pi2/@cvf(points)     ;freq = 2 pi/points

        org     p:coef
count   set     0
        dup     points           ; count = 0, ..., points-1
        dc      0.54-0.46*@cos(@cvf(count)*freq2)
count   set     count+1
        endm

        endm                 ; end of hamming window macro
```

Figure 13-1 HAMMING.ASM Assembler Macro to Generate an N-point Hamming Window

13.1.3 Implementing a Spectrum Analyzer with the DSP5601 Processor:

1. The instructions listed below read real-part data from the A/D converter and down-scale the real-part data by a factor equal to 2**scldw. Since imaginary-part data is not acquired by the A/D converter in this application, but the real-part and imaginary-part data FFT developed in Chapter 8 is used in this application, the imaginary-part data expected by the FFT is cleared to zero. The real-part and "zeroed" imaginary-part data are stored in the X-Memory and Y-Memory locations pointed to by Address Register R0.

```
getl      do       #points,end_ld

; read A/D

wait      jclr     #$7,x:$FFEE,wait
          movep    x:$FFEF,a
```

```
;
; scale the real part of data down
;

           dup       scldw
           asr       a
           endm
           move      a1,x:(r0)+
end_ld     nop

;
; Clear the imaginary part of data
;

clri       move      #$0,a
           move      #imagm,r0
           do        #points,end_clr
           move      a1,y:(r0)+
end_clr    nop
```

2. The **SINCOS1.ASM** assembler macro listed below generates the sine and cosine tables for the FFT.

```
;
; Sine-Cosine Table Generator for FFTs
;
;
sincos1  macro   points,coef
;
;   points - number of points
;   coef   - base address of sine/cosine table
;          - cosine value in X memory
;          - negative sine value in Y memory
;
;
pi1     equ     3.141592654
freq1   equ     2.0*pi1/@cvf(points)  ;freq = 2 pi/points

        org     x:coef          ; Calculation of
count   set     0               ; cos(freq*count) for
        dup     points/2        ; count =0,..,(points/2)-1
        dc      @cos(@cvf(count)*freq1)
count   set     count+1
        endm

        org     y:coef          ; Calculation of
count   set     0               ; -sin(freq*count) for
        dup     points/2        ; count =0,..,(points/2)-1
        dc      -@sin(@cvf(count)*freq1)
count   set     count+1
        endm

        endm                   ; end of sincos macro
```

3. The instructions listed below multiply the N input data samples by the Hamming window, where the Hamming window's coefficients are stored in the P-Memory locations pointed to by Address Register R1. The real-part data is stored in the X-Memory locations pointed to by Address Register R0.

```
ham        move       #realm,r0
           move       #coef,r1
           move       #$FFFFFF,m0
           move       m0,m1
           do         #points,end_ham
```

```
          move      x:(r0),x0
          movem     p:(r1)+1,x1
          mpy       x1,x0,a
          move      a,x:(r0)+
end_ham   nop
```

4. The instructions listed below compute the FFT, where its input sequence is in normal order and its output sequence is in bit-reversed order.

```
stage     do        #nstage,end_stage
          move      #$0,r0
          move      #coef,r5
          move      #$0,r2
          move      n1,n3
          move      n1,n4

tfs       do        n2,end_tfs
          move      r2,r3
          lua       (r2)+n2,r4

tf        do        n0,end_tf
          move      x:(r3),a          y:(r4),y1
          move      x:(r4),x1         y:(r3),b
          sub       x1,a
          sub       y1,b              a,x0
          move      x:(r3),a          b,y0
          add       x1,a              y:(r3),b
          add       y1,b              a,x:(r3)
          move      x:(r5),x1         b,y:(r3)+n3
          mpy       x0,x1,a           y:(r5),y1
          macr      -y0,y1,a
          mpy       x0,y1,b           a,x:(r4)
          macr      y0,x1,b
          move      b,y:(r4)+n4
end_tf
          lua       (r0)+n0,r0
          lua       (r2)+,r2
          move      r0,r5
          nop
          lua       (r5)+n5,r5
end_tfs
          move      n2,n1
          move      n2,b
          lsr       b                 n0,a1
          lsl       a                 b1,n2
          move      a1,n0
end_stage nop
```

5. The instructions listed below compute the square of the magnitude of the FFT and up-scale it by a factor equal to 2**sclup.

```
          move      #realm,r0
          move      #imagm,r1
          move      #$0,n0
          move      #$0,n1
          move      #$FFFFFF,m0
          move      #$FFFFFF,m1
mag       do        #points,end_mag
          move      x:(r0),x0
          mpy       x0,x0,a
          move      y:(r1)+,y0
          mac       y0,y0,a
```

```
            dup         sclup
            asl         a
            endm
            rnd         a

            move        a,x:(r0)+
end_mag     nop
```

6. The instructions listed below transfer the FFT's output values, in a normal order, to the D/A converter, where the DSP56001 processor's reverse-carry addressing capability is used to automatically address the output sequence elements in normal order.

```
write       move        #realm,r0
            move        #points/2,n0
            move        #0,m0
            do          #points,end_wr
wait2       jclr        #$6,x:$FFEE,wait2
            movep       x:(r0)+n0,x:$FFEF
            rep         #delay
            nop
end_wr      nop
```

13.1.4 Example: 1024-Point Spectrum Analyzer

The **SPECTRUM.ASM** assembler program shown in Figure 13.2 is used to implement a 1024-point spectrum analyzer, where the up-scale factor is 4 and the down-scale is 7. A copy of the **SPECTRUM.ASM** assembler program and the **SPECTRUM.LOD** absolute load program can be found on the *DSP56000/1 Demonstration Software #3* diskette.

1. POWER-DOWN THE PC AND FUNCTION GENERATOR.
2. CONNECT THE FUNCTION GENERATOR'S OUTPUT TO THE EVB'S "BNC2" INPUT (A/D INPUT).
3. CONNECT THE FUNCTION GENERATOR'S OUTPUT TO THE OSCILLOSCOPE'S CHANNEL ONE INPUT.
4. CONNECT THE EVB'S "BNC1" OUTPUT (D/A OUTPUT) TO THE OSCILLOSCOPE'S CHANNEL TWO INPUT.
5. POWER-ON AND BOOT THE PC.
6. POWER-ON THE FUNCTION GENERATOR.
7. INSERT THE "DSP56000/1 DEMONSTRATION SOFTWARE #3" DISKETTE INTO DRIVE "A".
8. Depending on your PC's configuration, INVOKE THE "ADS56000 USER INTERFACE SOFTWARE" PROGRAM from Drive B or the hard disk.
9. TYPE **load a:spectrum.lod<return>** to load the assembled and linked version of the assembler program shown in Figure 13.2.
10. TYPE **change pc $800<return>** and **go<return>** to begin execution of the program.
11. VARY THE SINUSOIDAL INPUT FREQUENCY FROM 0.5 TO 20 KHZ AND OBSERVE THE INPUT SIGNAL, AND OUTPUT SIGNAL SPECTRUM, ON THE OSCILLOSCOPE.
12. TYPE **force b<return>** to halt execution of the program.
13. TYPE **quit<return>** IF YOU WANT TO EXIT THE "ADS56000 USER INTERFACE SOFTWARE" PROGRAM.

```
;***************************************
;
; Spectrum Analyzer
;
;***************************************

points    equ   1024     ;fft size
nstage    equ   10       ;number of stages = log2 (number of
                         ;points)
coef      equ   $400     ;base address for sine, cos and
                         ;hamming

scldw     set    $7      ;scale input down
sclup     set    $4      ;scale output up
delay     equu   $d2     ;delay in output between pts

          org    x:$0
realm     dsm    points

          org    y:$0
imagm     dsm    points

          include  'sincos1.asm'
          include  'hamming.asm'

; Program start address

          org        p:$100

          sincos1    points,coef
          hamming    points,coef

;***************************************
; Initialize routine
;***************************************

START     reset

init2     movep     #0,x:$FFEF
          movep     #0,x:$FFEF
          movep     #0,x:$FFFF

;**************************************************
; Set up the SSI for operation with EVB
; The following code sets port C to function as SSI
;**************************************************

pcc       equ       $0001e0
          movep     #pcc,x:$FFE1  ; write PCC

;***************************************
; The following code sets the SSI CRA
; and CRB control registers
;***************************************

cra        equ       $004000
           movep     #cra,x:$FFEC

crb        equ       $003200
           movep     #crb,x:$FFED

;***************************************
; Read A/D and scale the data down
;***************************************

; read A/D
```

```
get2        move       #realm,r0
            move       #$0,n0
            move       #$FFFFFF,m0

get1        do         #points,end_ld

wait        jclr       #$7,x:$FFEE,wait
            movep      x:$FFEF,a

;
; scale the real part of data down
;

            dup        scldw
            asr        a
            endm
            move       a1,x:(r0)+
end_ld      nop

;***************************************
; Clear the imaginary part of data
;***************************************

clri        move        #$0,a
            move        #imagm,r0
            do          #points,end_clr
            move        a1,y:(r0)+
end_clr     nop

;***************************************
; Multiply by hamming window
;***************************************

ham         move         #realm,r0
            move         #coef,r1
            move         #$FFFFFF,m0
            move         m0,m1
            move         #1,n1
            do           #points,end_ham
            move         x:(r0),x0
            movem        p:(r1)+n1,x1
            mpy          x1,x0,a
            move         a,x:(r0)+
end_ham     nop

;***************************************
; Init registers for fft
;***************************************

           move          #$200,m0
           move          #$FFFFFF,m1
           move          m1,m2
           move          #1023,m3
           move          m3,m4
           move          m1,m5
           move          #coef,n5
           move          #(points/2),n2
           move          #points,n1
           move          #$1,n0

;***************************************
; Compute fft
;***************************************

stage      do          #nstage,end_stage
           move        #$0,r0
           move        #coef,r5
           move        #$0,r2
           move        n1,n3
```

```
          move      n1,n4

tfs       do        n2,end_tfs
          move      r2,r3
          lua       (r2)+n2,r4

tf        do        n0,end_tf
          move      x:(r3),a          y:(r4),y1
          move      x:(r4),x1         y:(r3),b
          sub       x1,a
          sub       y1,b              a,x0
          move      x:(r3),a          b,y0
          add       x1,a              y:(r3),b
          add       y1,b              a,x:(r3)
          move      x:(r5),x1         b,y:(r3)+n3
          mpy       x0,x1,a           y:(r5),y1
          macr      -y0,y1,a
          mpy       x0,y1,b           a,x:(r4)
          macr      y0,x1,b
          move      b,y:(r4)+n4
end_tf
          lua       (r0)+n0,r0
          lua       (r2)+,r2
          move      r0,r5
          nop
          lua       (r5)+n5,r5
end_tfs
          move      n2,n1
          move      n2,b
          lsr       b                  n0,a1
          lsl       a                  b1,n2
          move      a1,n0
end_stage nop

;****************************************
; Compute fft Magnitude
;****************************************

          move      #realm,r0
          move      #imagm,r1
          move      #$0,n0
          move      #$0,n1
          move      #$FFFFFF,m0
          move      #$FFFFFF,m1
mag       do        #points,end_mag
          move      x:(r0),x0
          mpy       x0,x0,a
          move      y:(r1)+,y0
          mac       y0,y0,a

          dup       sclup
          asl       a
          endm
          rnd       a

          move      a,x:(r0)+
end_mag   nop

;****************************************
; Pulse output to trigger scope
;****************************************

pulse     move      #$0,a
          do        #$4,l_pls
w_pls     jclr      #$6,x:$FFEE,w_pls
          movep     a1,x:$FFEF
l_pls     move      #$27cd00,a
          neg       a
          do        #$4,l2pls
```

```
w2pls       jclr        #$6,x:$FFEE,w2pls
            movep       a1,x:$FFEF
l2pls       nop

;*************************************
; Write to D/A
;*************************************

write       move        #realm,r0
            move        #points/2,n0
            move        #0,m0
            do          #points,end_wr
wait2       jclr        #$6,x:$FFEE,wait2
            movep       x:(r0)+n0,x:$FFEF
            rep         #delay
            nop
end_wr      nop

;*************************************
; Clear output
;*************************************

clout       move        #$0,a
w_clout     jclr        #$6,x:$FFEE,w_clout
            movep       a1,x:$FFEF
;*************************************
; Loop Indefinitely
;*************************************

again       jmp         get2
            reset
            end
```

Figure 13-2 SPECTRUM.ASM Assembler Program

13.2 AN ADAPTIVE DIFFERENTIAL PULSE CODE MODULATION (ADPCM) PROJECT USING THE DSP56001 PROCESSOR:

In this project, an Adaptive Differential Pulse Code Modulation (ADPCM) system, useful in telephony applications to reduce (compact) the number of bits needed to transmit a voice signal, is described for implementation on an ADPCM Evaluation System. Unlike the Spectrum Analyzer project which was taken to completion in Section 13.1, this section presents only enough information to get the ADPCM project started. Completion of the ADPCM project is left as an exercise for the student.

ADPCM is a method whereby a highly correlated signal is decorrelated to reduce redundancies in the signal. Because a voice signal is usually highly correlated, use of an ADPCM system can decrease its dynamic range. Therefore, a reduced number of bits can be used to transmit a voice signal without reducing its level of resolution. For example, with the same number of telephone lines, the bandwidth allocated to each phone call can be decreased such that the total number of phone calls can be increased. The ADPCM system described herein decreases the bandwidth of a phone call by one-half.

In the ADPCM Evaluation System shown in Figure 13.3, the coder (A/D converter) portion of the MC145503 CODEC is used to digitize a "transmitted" voice signal and the decoder (D/A converter) portion of the MC145503 CODEC is used to reconstruct a "received" voice signal. The name "codec" is an acronym constructed from the names "coder" and "decoder" and the method of signal coding used by the MC145503 CODEC is Pulse-Code Modulation (PCM).

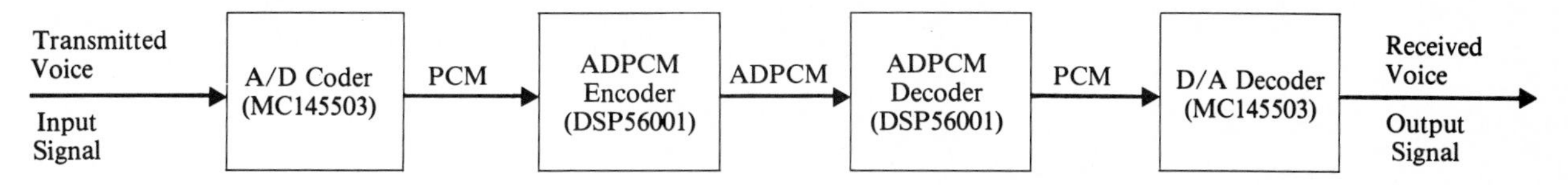

Figure 13-3 ADPCM Evaluation System

To maintain industry standard voice signal quality, the digitized voice signal must have a signal-to-distortion level of at least 30 dB over a 40 dB dynamic range. These specifications can be achieved with linear 13-bit A/D and D/A codecs, but at amplitudes greater than 40 dB below the peak amplitude 13-bit linear codecs far exceed the signal-to-distortion specification. Since this excess performance is not needed, two methods of data reduction (companding) can be implemented to compress the 13-bits of data needed to meet specifications using linear codecs to 8-bits of data needed to meet specifications with "companding" codecs. The Mu-255 Law companding technique is the standard used in North America; the A-Law companding technique is the standard used in Europe. The MC145503 CODEC offers both the Mu-255 Law and A-Law companding techniques. The Mu-255 Law companding technique is considered herein. Samples from the coder portion of the MC145503 CODEC are transferred to the ADPCM encoder in bit-serial format at a bit rate of $8*f_s$ bits/sec, where f_s is the sampling frequency.

The flow chart of the ADPCM encoder system is shown in Figure 13.4. The received companded-PCM signal is first converted to a linear digitized signal by using the factory programmed Mu-Law expansion table ROM located in the DSP56001 processor's on-chip X-Memory space. An estimate of the signal is then subtracted from the linear signal to obtain an error signal. The error signal is up-scaled by an adaptive gain and the result is quantized to 4-bits. The ADPCM encoder's output bit rate is thereby reduced to $4*f_s$. The magnitude of this 4-bit sample is then used in the adaptive gain adjustment to prevent saturation. A feedback loop that includes the inverse quantization operations to reconstruct a quantized version of the error, is used by the adaptive filter to obtain an estimate of the input voice signal.

The same quantizer and adaptive filter hardware used in this feedback loop serves in the ADPCM decoder system as well. The flow chart of the ADPCM decoder is shown in Figure 13.5.

The ADPCM Evaluation Board schematic diagram is shown in Figure 13.6. There, the MC145503 PCM codec is used to digitize the input voice signal and to reconstruct the output voice signal. Three 8K*8-bit "fast" static MCM6264 RAMs are used as external P-Memory to avoid running the ADPCM program from "slow" EPROMs. Hardware reset circuitry invokes the DSP56001 processor's on-chip bootstrap loader to load the monitor program stored in one of the 2764 EPROMs. The DSP56001 processor executes the monitor program to download the ADPCM application program from the "slow" EPROMs to the "fast" external static RAMs. The DSP56001 processor then executes the ADPCM application program. For a detailed description of the ADPCM algorithm and the ADPCM Evaluation Board, obtain a copy of References 1 and 2 (see Appendix E).

13.3 REFERENCES:

1. J.Wang, "Adaptive Differential Pulse Code Modulation with the DSP56001 Processor", Master Thesis, Bucknell University, 1989.

2. Full-Duplex 32-Kbit/s CCITT ADPCM Speech Coding on the DSP56001, Application Note, Motorola 1989.

3. John R. Treichler, C.Johnson,Jr. and Michael Larimore, "Theory and Design of Adaptive Filters", John Wiley, 1987.

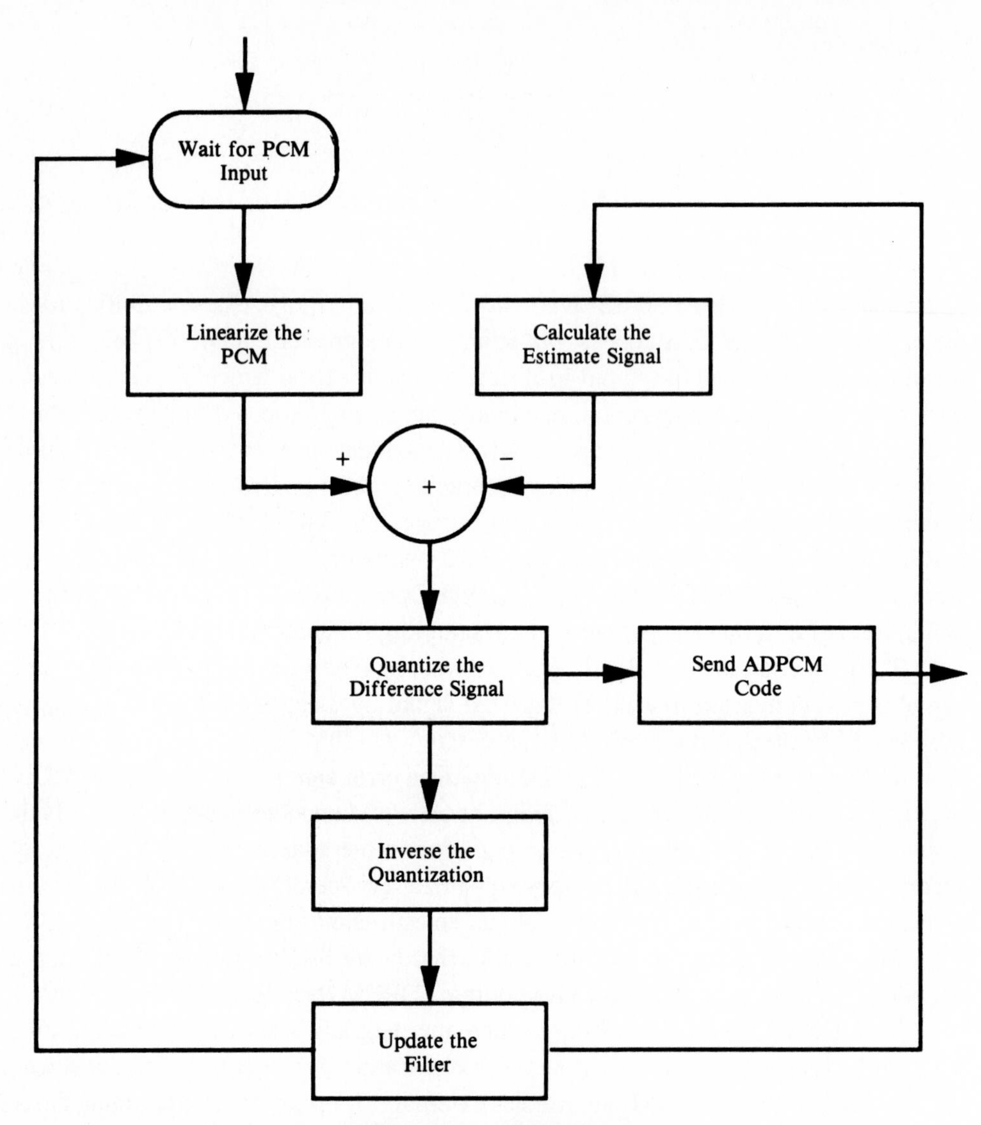

Figure 13-4 ADPCM Encoder

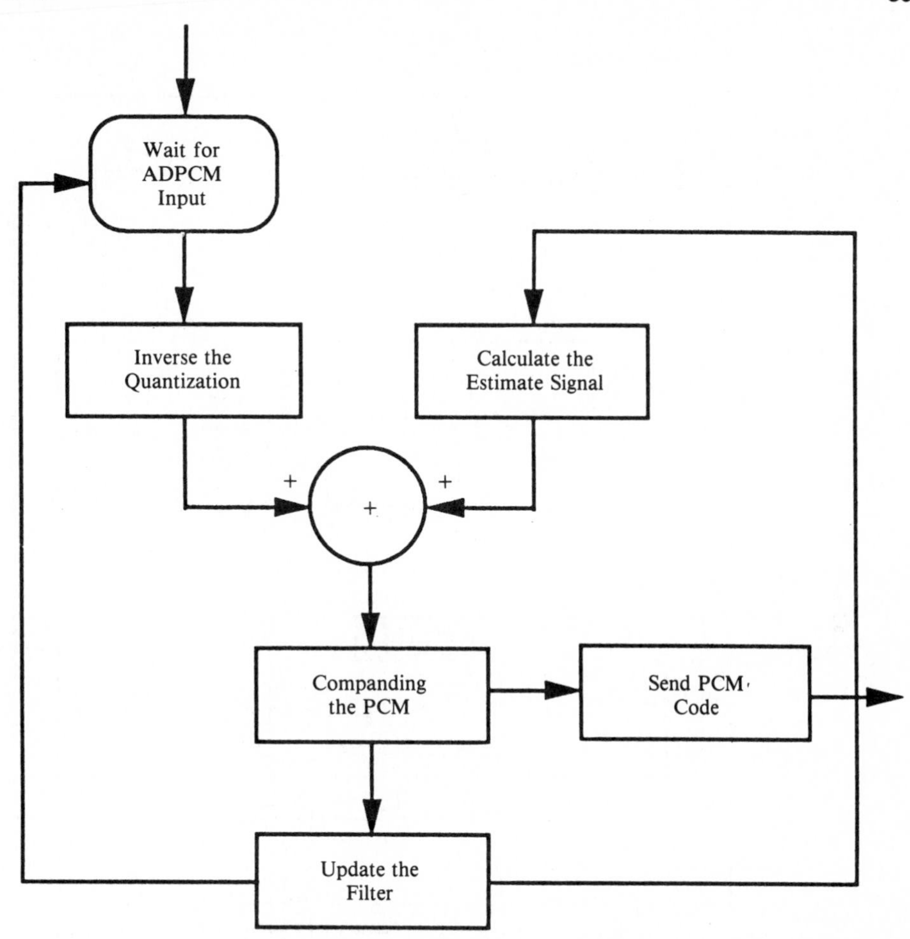

Figure 13-5 ADPCM Decoder

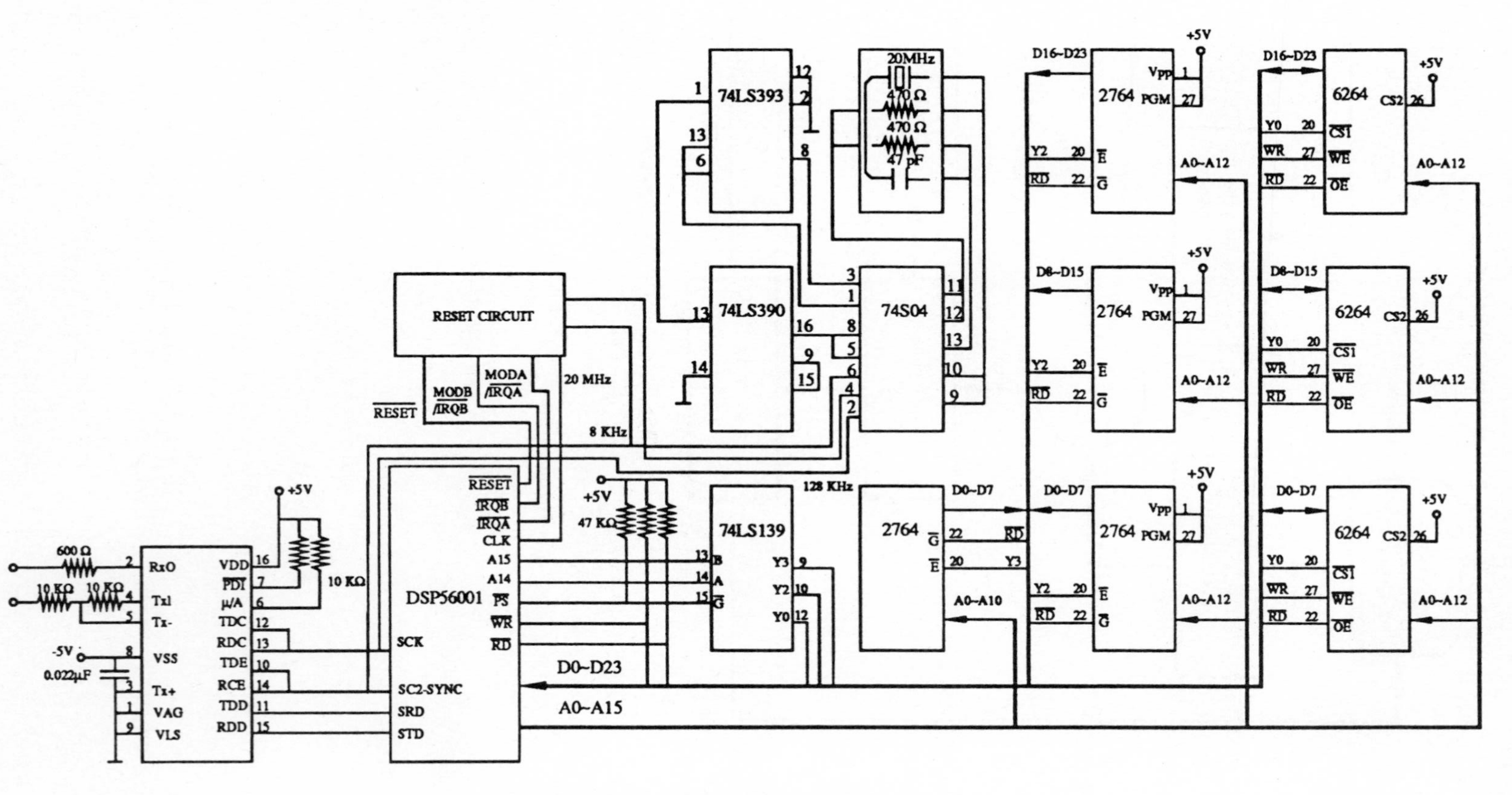

Figure 13-6 ADPCM Evaluation Board

Appendix A

Fractional and Integer Arithmetic Using the DSP56000 Family of General-Purpose Digital Signal Processors

by

Andreas Chrysafis and Steve Lansdowne
Strategic Applications
Digital Signal Processor Operation
Telephone: (512) 440-3035

INTRODUCTION

The DSP5600 Family of general-purpose digital signal processors (DSPs) is distinctive in that the on-chip multiplier directly supports fractional data formats and indirectly supports integer data formats. This application note discusses using the DSP56000/DSP56001 processors to perform arithmetic operations on data represented as integers, fractions, and combinations thereof, variously called mixed numbers, real numbers, or floating-point numbers.

A fractional data representation was chosen for the DSP56000/DSP56001 for many reasons including the following:

- The most significant product (MSP) of a multiplication has the same format as the input and can be immediately used as an input to the multiplier without a shifting operation.
- The least significant product (LSP) of a multiplication can be rounded into the MSP naturally: i.e., without having to handle a changing exponent.
- All floating-point formats use fractional mantissas.
- Coefficients in digital filters are output as fractions by high-level filter-design software packages.

From a hardware point of view, this decision had its primary impact on the design of the multiplier (see **SIGNED MULTIPLICATION**). Since the format of the resultant operands from addition or subtraction operations is unchanged from the format of the input operands, the choice of integer or fractional formats does not impact the design of the arithmetic logic unit (ALU).

Many data representations will be defined, including twos-complement fractional and integer numbers, real and mixed numbers, and double-precision (48-bit) numbers (see **DATA REPRESENTATIONS**). **SIGNED MULTIPLICATION** and **SIGNED DIVISION** discuss multiplication and division using these data representations.

Data representations are not as difficult as they may first appear. The basic difference between data representations is a shifting or scaling operation. Performing shifting operations with the DSP56000 Family of processors is discussed in **DATA SHIFTING. MIXED- AND REAL-NUMBER ADDITION AND SUBTRACTION** discusses operations involving more than simple ADD and SUB instructions. Division-yielding quotients and remainders that are not word (24-bit) multiples are described in **DIVIDE ROUTINES WITH N ≤ 24 BITS.**

DATA REPRESENTATIONS

Different data representations are introduced and discussed in the following paragraphs. Where required, routines are given that convert numbers from one data representation to another.

TWOS-COMPLEMENT FRACTION

A fraction, F, is any number whose magnitude satisfies the inequality:

$$0.0 \leq \mathrm{mag}(F) < 1.0$$

Examples of fractions are 0.25 and −0.87. The twos-complement fractional data representation is shown in Figure 1. The binary word is interpreted as having a binary point after the most significant bit (MSB). The range of numbers that can be represented using N-bit twos-complement fractional data is

$$-1.0 \leq F \leq 1-2^{-(N-1)}$$

DSP56000/DSP56001 processors use a fractional data representation for all arithmetic operations. Figure 2 shows the bit weighting and register names for words, long words, and accumulator operands in the DSP56000/DSP56001 processors.

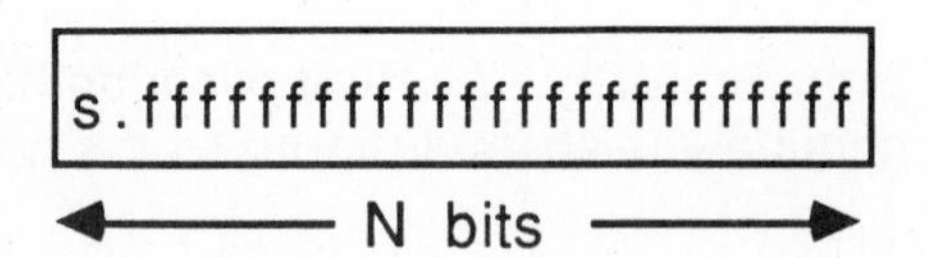

where

s = the sign bit
f = a fractional bit
. = the implied binary point and
N = the number of bits

For words and long words, the most negative number that can be represented is −1.0 whose internal representation is \$800000 and \$800000000000, respectively. The "\$" sign denotes a hexadecimal value. The most positive word is \$7FFFFF or $1-2^{-23}=0.9999998$, and the most positive long word is \$7FFFFFFFFFFF or $1-2^{-47}=0.999999999999993$. These limits apply to data stored in memory and to data stored in the data ALU input pipeline

Word Operand (24 bits)

Weighting = -2^{0} 2^{-23}

Registers: X1 X0 Y1 Y0 A1 A0 B1 B0

Long Word Operand (48 bits)

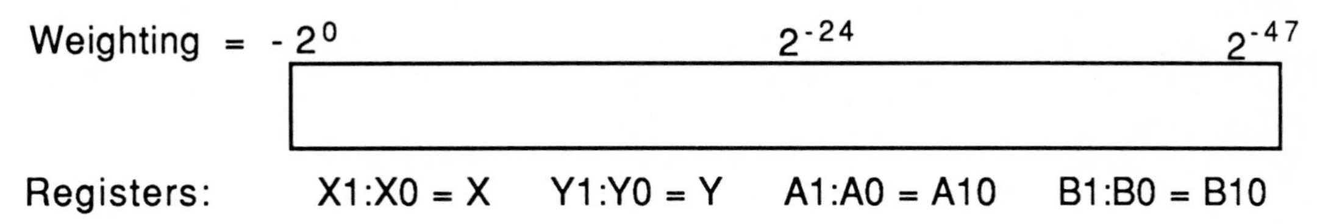

Accumulator (56 bits)

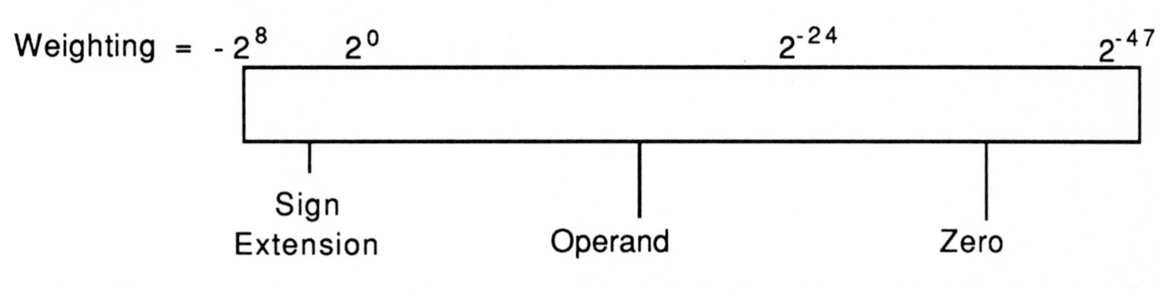

Registers: A = A2:A1:A0 or B = B2:B1:B0

Figure 2. DSP56000 Operands

registers. The accumulators, A and B, have 8-bit extension registers, A2 and B2, respectively. This extension allows word growth so that the most positive and negative numbers that can be represented in the accumulators are +255.999999999999993 and −250.0, respectively.

An immediate fractional number can be stored in a general-purpose register (for example, X0) by simply using the MOVE immediate instruction. For example, execution of

```
MOVE  #.5,X0
```

with result in $400000 (0.5) being stored in X0.

Twos-Complement Integer

An integer, I, is a number that contains no decimal point and its magnitude satisfies the inequality:

$$1 \leq \text{mag}(I)$$

Examples of integer numbers are 1, 256, and −67. The twos complement integer data representation is shown in Figure 3. The binary word is interpreted as having a binary point after the least significant bit (LSB) — that is, the data is right-hand justified. The range of number that can be represented using N-bit twos-complement integers is

$$-2^{N-1} \leq I \leq 2^{N-1}-1$$

The MSBs are sign-extension bits. Caution must be exercised when moving integer data on the DSP56000/DP56001 because the DSP56000/DSP56001 will naturally tend to left-hand justify data (i.e., assuming the data is a fraction). The DSP56000 cross assembler provides forcing functions that can be used to facilitate handling integer data, which is best described by way of example.

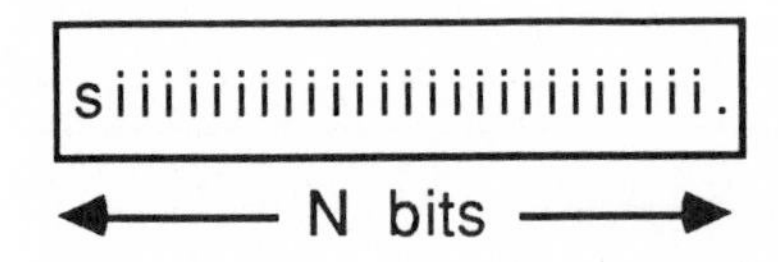

where

s = the sign bit
i = an integer bit or sign extension bit
. = the implied binary point and
N = the number of bits

Figure 3. Twos-Complement Integer

To store the integer value 56 ($38) in X1, the programmer would be tempted to use

```
MOVE   #56,X1
```

The value stored in X1 would be $380000 (3670016), which is clearly incorrect. The error occurred because the assembler will always pick the shortest form of instruction encoding. Data less than eight bits can be encoded in the MOVE instruction without the use of an extension word. The DSP56000/DSP56001 interprets this data as fractional and therefore stores it left-hand justified as demonstrated in this example. If the immediate long force operator, >, is used, an extension word will be used, and the data will be right-hand justified as the following example shows:

```
MOVE   #>56,X1
```

The content of X1 will be $000038.

For integers of magnitude greater than 128, the short addressing is not applicable since the number will occupy more than eight bits. The value will therefore be treated as a 24-bit number by the assembler and encoded into the LSBs of the extension word. As an example, execution of

```
MOVE   #1234,X1
```

will result in X1 = $0004D2 (1234).

DOUBLE-PRECISION NUMBERS

A double-precision number is a 48-bit twos-complement number, fraction or integer, that is stored as a long-word operand. The range of a double-precision twos-complement fractional number is

$$-1 \leq \text{double-precision fraction} \leq 1-2^{-47}$$

The range for a double-precision integer is

$$-140737488355328 \leq \text{double-precision integer} \leq 140737488355327$$

REAL NUMBERS

A real number, R, consists of an integer part and a fractional part. The decimal point separates the two parts. Only the integer part has a sign bit, and it may assume the value zero. The real-number representation discussed in this document consists of a 24-bit integer portion and a 24-bit fractional portion (see Figure 4). For long-word operands, like X (X1:X0), the integer part of the real number will occupy the upper 24 bits of X, X1, and the fractional part with occupy the lower 24 bits, X0. The binary point is assumed to occupy an imaginary place between bit 23 and bit 24; whereas, the sign bit occupies the leftmost bit of the integer portion. The range of a real number, R, is

$$-8388608.0 \leq R < 8388607.9999999$$

Examples of real numbers are 56.789, 0.345, and −789.123.

```
47                      24  23                      0
 siiiiiiiiiiiiiiiiiiiiiii . ffffffffffffffffffffffff
```

where

s = the sign bit
i = integer bit or sign extension bit
f = fractional bit
. = binary point

Figure 4. Real-Number Format

The convert to real macro, CONVR, presented in Figure 5, performs the conversion of a real, positive decimal number, xr, to the real-number format for storage. This macro uses the convert to integer function, CVI, built into the assembler to convert real numbers to integers by simply truncating the fractional part of the number. When the fractional portion is moved into A0, it will be signed. A left shift is subsequently performed to eliminate this sign bit.

Converting a negative real number to the real-number format involves two steps. First, the absolute value of the number has to be stored into an accumulator in the real-number format; then the stored value is negated.

The convert real general macro, CONVRG, depicted in Figure 6, handles both positive and negative operands. The ASL instruction will eliminate the sign bit in the fractional part.

```
;CONVR.ASM
;This macro converts a real positive decimal number,
;xr (0.0 < xr < 8388607.9999999), to the real number format.
;The signed integer part is stored in the upper part of the
;A accumulator (A1), and the unsigned fractional
;part is stored in A0.
CONVR macro xr                  ;macro definition
      clr a                     ;clear the accumulator
      move #(xr-@cvi(xr)),a0 ;store fractional part in A0
      asl a                     ;eliminate sign bit in fract. part
      move #@cvi(xr),a1         ;store the integer part in A1
      endm                      ;end macro definition
```

Figure 5. CONVR Macro Definition

If the number is positive, the number in register A has been correctly converted. If xr is negative, the sign bit of the fractional part will propagate from A0 to A1 due to the ASL instruction. This sign bit is used to subtract out the one that was added to the integer portion when it was converted to a twos-complement number. The single case in which one should not be subtracted from the integer is when the fraction portion is zero. In this case, since CVI always returns a zero as positive, zero is subtracted from the integer, and the result is correct.

```
;CONVRG.ASM
;This macro converts a real decimal number,
;xr (8388607.9999999 > xr > -8388608.0), to the real number format.
;The signed integer part is stored in the upper part
;of the A accumulator (A1), and the unsigned fractional part
;is stored in A0.
CONVRG macro xr                 ;macro definition
      clr a                     ;clear accumulator a
      move #(xr-@cvi(xr)),a0    ;move the fraction into A0
      asl  a  #>@cvi(xr),x1     ;shift the fraction's sign bit into A1
      move a1,x0                ;move the int. into X1, move A1 to X0
      move x1,a1                ;move the int. to A1
      sub x0,a                  ;convert the integer to one's complement
      endm
```

Figure 6. CONVRG Macro Definition

A good example to demonstrate the CONVRG macro is the case where xr = −1.5. After the SUB X0,A instruction has been executed, A = $00:FFFFFE:800000 (−1.5 in the real-number format). The extension register contains all zeros even though the real number is negative. That is, bit 47 is the true sign bit for the 48-bit real number. It is not immediately apparent that $00:FFFFFE:800000 represents −1.5. It becomes apparent after the absolute value of A1:A0 is taken, which is accomplished by moving the long word, A1:A0, into accumulator B (so that bit 47 is properly sign extended), and then taking the absolute value of B. This yields B = $00:000001:800000 (+1.5 in the real-number format).

In summary, real numbers must be treated as 48-bit entities — for example, do not take the absolute value of the integer portion only.

MIXED NUMBERS

A mixed number, MN, is a special case of a real number in that the number occupies 24 bits instead of 48 bits and satisfies the inequality

$$-128.0 \leq MN < 128.0$$

Examples of mixed numbers are −67.875 and 89.567. As seen in the MN format given in Figure 7, the 24-bit word is divided into two parts. The 8-bit part contains the integer portion of the mixed number, including the sign bit related to the whole 24-bit number. The second part, which is 16-bits, contains the unsigned fractional portion of the number.

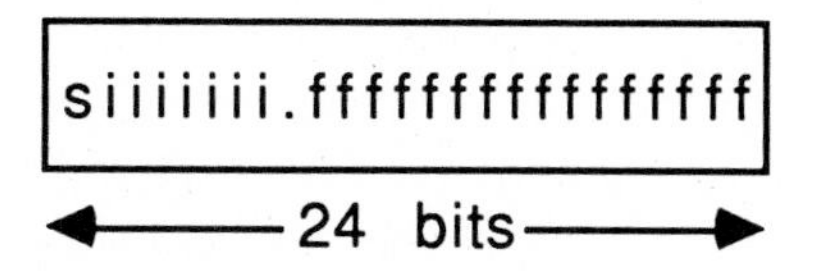

where

s = the sign bit
i = an integer bit or a sign extension bit
. = the binary point and
f = a fractional bit

Figure 7. Mixed-Number Format

The virtue of the MN format is that MN-formatted data can be treated as 24-bit signed data by the machine — that is, the integer and fractional portions do not have to be treated separately. Two macros have been prepared to show how real numbers having a magnitude less than 128 can be stored in 24 bits and set up in the MN format. The first macro, convert to MN, CONVMN, converts any positive mixed number to the MN format (see Figure 8). The macro does not check the sign or the magnitude of the input; it is the user's responsibility to make the necessary checks. The fractional part of the number is shifted to the right by seven bits, using a shift constant, $010000, and the MPYR instruction discussed in **DATA SHIFTING**; therefore, the first eight bits of the 24-bit word are zero (see Figure 8). This instruction ensures that the fraction is stored in the upper part of the accumulator (A1 in this case) and that it is 16 bits (rounded). In parallel with the MPYR instruction, the integer is moved to X1 by using the move immediate short instruction that places the 8-bit signed integer in the upper eight bits of the 24-bit register, X1. The ADD X1,A instruction concatenates the integer and fractional parts to form the mixed number and sign extends the mixed number so that the number can be immediately used in the data ALU.

```
;CONVMN.ASM
;This macro converts a positive decimal mixed number to the MN
;format. The mixed number is stored in A1. 0.0 ≤ xmn < 128.0
CONVMN  macro xmn                        ;macro definition
        move #(xmn-@cvi(xmn)),x0         ;fractional part to X0
        move #$010000,y1                 ;shift constant in Y1
        mpyr y1,x0,a   #@cvi(xmn),x1     ;shift X0; integer part in X1
        add  x1,a                        ;concatenate int. and fract.
        endm                             ;end macro definition
```

Figure 8. CONVMN Macro

A second macro, convert to MN general, CONVMNG, handles signed real numbers having a magnitude less than 128 (see Figure 9). When a negative number is detected, it is made positive, then transformed into the MN format, and finally negated again. As shown in Figure 9, the detection of the sign of the number is done using the EOR instruction. The EOR is performed with accumulator B, which is already zero. The integer will reside in the most significant byte of B1, and the N bit in the condition code register (CCR) will be set if the integer was negative. The negation is now performed by using the NEG instruction. When the negative B1 is negated to turn its contents into a positive number, the integer occupies the lowest eight bits of B1. To move the 8-bit number to the upper eight bits of a 24-bit register, a left shift by 16 bits must be performed by utilizing the ideas presented in **DATA SHIFTING**. A 16-bit left shift by the use of a shift constant will force the 8-bit number to reside in the lower 24 bits, B0, of the destination accumulator B. To concatenate the 8-bit signed integer with the 16-bit fraction, the number is moved to X1 and then added to accumulator A containing the unsigned fraction.

```
;CONVMNG.ASM
;This macro will convert a decimal number xmn,
;where  -128.0 < xmn < 128.0, into the mixed number, MN, format.
;
CONVMNG     macro xmn               ;macro definition
            move #(xmn - @cvi(xmn)),x0 ;obtain signed fract. part
            clr b        #$010000,y1;clear b,shift constant in Y1
            mpyr y1,x0,a #@cvi(xmn),x1;shift fract.; integer in X1
            eor x1,b                ;set sign
            jeq _endf               ;finished if integer = 0
            jpl _endit              ;jump if positive
            neg b        #$008000,y1 ;int. positive;left shift const.
            neg a        b1,x1      ;fract. positive;integer in X1
            mpy y1,x1,b             ;shift integer and store in B
            move b0,x1              ;obtain the shifted integer
            add x1,a                ;add the int. to the fract.
            neg a                   ;negate mixed number entity
            jmp _endf               ;jump to the end
_endit      add b,a                 ;concatenate positive int. and fr.
_endf                               ;finished
         endm                       ;end macro definition
```

Figure 9. CONVMNG Macro Definition

DATA SHIFTING

Data shifting is used in converting one data representation into another data representation. Four distinct ways exist for performing data shifts on DSP56000 Family processors, which include 1) 1-bit shifts/rotates, 2) multi-bit shifts/rotates, 3) fast multi-bit shifts, and 4) dynamic scaling. These approaches to shifting are described in the following paragraphs.

1-Bit Shifts/Rotates

For 1-bit shifts/rotates of either 56-bit accumulator A (A = A2:A1:A0 = 8:24:24 bits) or 56-bit accumulator B (B = B2:B1:B0 = 8:24:24 bits), use the arithmetic shift right (ASR) and arithmetic shift left (ASL) instructions. If 1-bit shifts on only the most significant 24-bit word of accumulator A, A1, or accumulator B, B1 are required, use the rotate right (ROR), rotate left (ROL), logical shift right (LSR), or logical shift left (LSL) instructions.

Multi-Bit Shifts/Rotates

The most straightforward approach for multi-bit shifts/rotates of the accumulator is to use ASR, ASL, LSR, LSL, ROR, or ROL instructions with the repeat instruction, REP, or the hardware DO loop instruction, with the loop consisting of a single instruction as the examples in Figure 10 show. The repeat instruction is not interruptible; whereas, the DO instruction is interruptible.

The DSP56000 macro cross assembler supports macros with the MACRO DEFINITION and MACLIB directives (see Reference 1). Two accumulator shift macros, one for left shifts, SHLAC, and the other for right shifs, SHRAC, can be defined as shown in Figure 11.

```
        REP #n
        ASL A
or
        DO #n,END1
        ASR A
        END1
```

where n is the number of positions to be shifted/rotated.

Figure 10. Multi-Bit Shifts Using REPeat, DO

```
;Macros for performing multi-bit shifts right and left.
;For the two given macros
;      Let      acc = accumulator A or B
;                n  = the number of bits to be shifted
;
SHRAC   macro   acc,n     ;macro definition for shifting the
        rep #n            ;accumulator right n bits.
        asr acc           ;shift right
        endm              ;end macro definition
;
SHLAC   macro acc,n       ;macro definition for shifting the
        rep #n            ;accumulator left n bits.
        asl acc           ;shift left
        endm              ;end macro definition
```

Figure 11. Multi-Bit Shift Macros

Fast Multi-Bit Shifts

The fastest way to do multi-bit shifting is to multiply the operand by a shift constant. In the case of a right shift, the constant KR is a fraction given by $KR = 2^{-n}$. The example in Table 1 shows how to shift the content of X0 right by four bits. The shifted result resides in the upper part of accumulator A, A1. The code executed to implement the 4-bit right shift shown in Table 1 is

```
MOVE  #KR,X1
MPY   X0,X1,A
```

Table 1. Fast 4-Bit Right Shift

Register	Hexadecimal Value	Comments
X0	060000	Value To Be Shifted
X1	080000	Shift Constant, KR
A1	006000	Shifted Result

Similarly, the example in Table 2 shows how to shift the content of X0 left by four bits. In the case of a left shift, the constant KL is an integer given by $KL = 2^{n-1}$. KL is 2^{n-1}, not 2^n, because the DSP56000 multiplier is fractional, thereby automatically implementing a 1-bit left shift. The result for left shifts resides in A0, the least significant word of A.

The code executed to implement the 4-bit left shift is

```
MOVE  #>KL,X1
MPY   X0,X1,A
```

Table 2. Fast 4-Bit Left Shift

Register	Hexadecimal Value	Comments
X0	060000	Value To Be Shifted
X1	000008	Shift Constant, KL
A0	600000	Shifted Result

Generating the constants for the shifts is made easy by using the POW and CVI functions built into the DPS56000 macro cross assembler. The raise to the power function, POW, returns a real number for any base raised to a real number. For example,

K=@pow(2,−4) returns 0.0625, and
K=@pow(2,+4) returns 16.0

However, because DPS56000 is a fractional machine, the assembler will limit real numbers unless precautions are taken. In the previous example, the object code for 16.0 will be limited to $7FFFFF or +0.999998 decimal by the assembler, which is incorrect. To obtain the integer form of the the real number, the assembler provides a convert to integer function, CVI. The CVI function converts real numbers to integers by truncating the fractional portions. For example,

```
@CVI (@pow(2,+4))
```

returns 16 (not 16.0), which will be assembled as $000010 in object code.

The previous instruction sequences can be put in two distinct macros for programming ease. The two macros, MSHR for multi-bit shifts right and MSHL for multi-bit shifts left, are listed in Figure 12.

The immediate long-data move (note the greater than sign, >, in the move instruction) must be used in the MSHL example to prevent the data from being treated as a fraction and shifted accordingly. In the following example, for

```
MOVE  #2,X1
```

X1 will be $020000 because the immediate short data (i.e., data which can be represented as eight bits) is treated as an 8-bit fraction occupying the two most significant bytes of the destination register; whereas, for

```
MOVE  #>2,X1
```

X1 will be $000002 because the data is treated as an integer occupying the least significant bytes of the destination register with sign extension.

The immediate long-data move is a two-word instruction that executes in two instruction cycles; whereas, the immediate short-data move is a one-word instruction that executes

```
;Macro definitions for generating right and left shift constants,
;KR and KL, and performing the right and left shifts.
;
;              Let     s   = the source register
;                      m   = the multiplier register
;                      n   = the number of bits to be shifted
;                      acc = the destination accumulator
;
;          where  s,m can be one of X0,X1,Y0,Y1
;           and   acc can be A or B
;
;
MSHR   macro s,m,n,acc           ;four input variables.
       move  #@pow(2,-n),m       ;load the multiplier register
       mpy   s,m,acc             ;shift right n bits
       endm                      ;end macro definition
;
MSHL   macro s,m,n,acc           ;four input variables
       move  #>@cvi(@pow(2,n-1)),m ;load the mult. reg.
       mpy   s,m,acc             ;shift left n bits
       endm                      ;end macro definition
```

Figure 12. Constant Generation and Multi-Bus Shifts

in one instruction cycle. The immediate short-data move can be used for multi-bit right shifts of less than or equal to eight bits. For right shifts of more than eight bits, the immediate long-data move must be used with the appropriate 24-bit fraction, 2**(−n), utilizing the POW directive.

No-Overhead, Dynamic Scaling (1-Bit Shifts)

For no-overhead 1-bit shifts of either accumulator A or B, the scaling mode is used. In this mode the shift occurs automatically when transferring data from either of the 56-bit data ALU accumulators, A or B (not A2, A1, A0, A10 or B2, B1, B0, B10), to the XD or YD buses. This shift function is activated by appropriately setting the scaling-mode bits in the status register. This mode is primarily intended for adding a scaling operation to existing code without modifying the code (simply setting the scaling-mode bits).

For more details on the scaling mode, consult the **DSP56000 Digital Signal Processor User's Manual** (see Reference 2) and the **DATA ALU** subsection of ADI1290 (see Reference 3).

MIXED- AND REAL-NUMBER ADDITION AND SUBTRACTION

The arithmetic operations of addition and subtraction performed on mixed and real numbers are discussed in the following paragraphs.

MIXED NUMBERS

Mixed numbers can be represented in a 24-bit word using the MN format discussed in **REAL NUMBERS**. To better understand addition and subtraction of mixed numbers, consider the examples in the following paragraphs.

Addition

Two examples of mixed-number addition are considered.

Example One — The simplest case is the addition of two positive numbers as shown in Table 3. The instruction executed is

```
ADD X1,A
```

Table 3. Positive Mixed Numbers with Sum Less Than 128

Register	Hexadecimal Value	Mixed-Number Value
X1	43C000	67.75
A (before)	00 : 178000 : 000000	23.50
A (after)	00 : 5B4000 : 000000	91.25

Example Two — In this example, the result of the addition will be greater than 128, which is the limit for 24-bit MN-formatted mixed numbers. However, the status register will signify the use of the extension part of the accumulator; thus, the exact representation of the mixed number having a magnitude greater than 128 can be contained in the accumulator. It cannot be stored as a 24-bit word, however, since it requires more than 24 bits to represent it.

Consider the example shown in Table 4. The value of the status register, SR, is $0320, signifying the E bit, bit 5, has been set. The hexadecimal value of B1 represents the decimal number 131.0 because the status register indicates the extension bits of accumulator B are in use; thus, bit 47 of B is not a sign bit but part of the mixed number.

Table 4. Positive Mixed Numbers with Sum Greater Than 128

Register	Hexadecimal Value	Mixed-Number Value
Y1	464000	70.25
B (before)	00 : 3CC000 : 000000	60.75
B (after)	00 : 830000 : 000000	131.00

Subtraction

Mixed-number subtraction is as straightforward as the mixed-number addition.

Example Three — Consider the case shown in Table 5. The instruction executed is

```
SUB   X0,B
```

The status register value remains the same.

Table 5. Mixed-Number Subtraction

Register	Hexadecimal Value	Mixed-Number Value
X0	178000	23.50
B (before)	00 : 43C000 : 000000	67.75
B (after)	00 : 2C4000 : 000000	44.25

Example Four — Consider the case depicted in Table 6 where the result is negative. The N bit, bit 3, and the borrow (carry) bit, bit 0, in the status register are set, which indicates that the result is negative and that a borrow has occurred. The magnitude of the negative mixed number can be easily found by executing the ABS B instruction.

Table 6. Mixed-Number Subtraction with Negative Result

Register	Hexadecimal Value	Mixed-Number Value
X1	464000	70.25
B (before)	00 : 3CC000 : 000000	60.75
B (after)	FF : F68000 : 000000	−9.50

REAL NUMBERS

Consider real numbers having the format discussed in **REAL NUMBERS** in which the signed integer occupies the most significant 24-bit word, and the unsigned fraction occupies the least significant 24-bit word of a 48-bit-long word.

Addition

The numbers to be added should be moved into any of the acceptable source registers (X, Y, A, B) and destination registers (A, B) for the 48-bit addition (see Reference 2). If the sum of the fractional parts is greater than unity, a carry is propagated into the integer part. If, after adding the two real numbers, the integer result cannot be represented in 24 bits, then the extension part of the accumulator will be used. Bit 5 in the status register will indicate whether the extension bits are in use.

Example One — Consider the case shown in Table 7. The instruction executed is

```
ADD   X,A
```

Although the real-number source may have been saved in A10, bit 47 and A2 must represent proper sign extension if the C bit in the status register is to be set correctly, which is necessary when doing multiple-precision arithmetic. The fraction parts in X0 and A0 are unsigned.

Table 7. Real-Number Addition

Register	Hexadecimal Value	Real-Number Value
X	000237 : C00000	567.750
A (before)	00 : 0003DB : A00000	987.625
A (after)	00 : 000613 : 600000	1555.375

Example Two — Consider the case shown in Table 8. If the first bit of the result is interpreted as a sign bit, the decimal value of A1 is not 8389160 but is −8388056. The reason that the hexadecimal value in A represents the correct result (i.e., +8389160) is because the extension bit, bit 5 in the status register, is set. This fact indicates that the extension bits of accumulator A are in use; therefore, the sign bit is not the leftmost bit (bit 47) of A1 but is the leftmost bit (bit 55) of the extension register, A2.

Table 8. Real-Number Addition Using the Extension Bit

Register	Hexadecimal Value	Real-Number Value
X	000237 : C00000	567.750
A (before)	00 : 7FFFF0 : A00000	8388592.625
A (after)	00 : 800228 : 600000	8389160.375

Subtraction

The subtraction of real numbers is similar to the addition of real numbers.

Example Three — The case shown in Table 9 generates a positive result. The instruction executed is

```
SUB   X,A
```

Table 9. Real-Number Subtraction with a Positive Result

Register	Hexadecimal Value	Real-Number Value
X	000138 : C00000	312.750
A (before)	00 : 00037A : 400000	890.250
A (after)	00 : 000241 : 800000	577.500

Example Four — The case depicted in Table 10 generates a negative result. The N bit, bit 3, and the C bit, bit 0, in the status register are set, indicating the result is negative and a borrow has occurred. The magnitude of the negative real number can be easily found by executing the ABS A instruction.

Table 10. Real-Number Subtraction with a Negative Result

Register	Hexadecimal Value	Real-Number Value
X	00037A : 400000	890.250
A (before)	00 : 000138 : C00000	312.750
A (after)	FF : FFFDBE : 800000	−577.500

SIGNED MULTIPLICATION

Consider the multiplication of two signed-integer numbers (see Figure 13). The product of the signed multiplier is 2^{N-1} bits long. To keep the product properly normalized and to

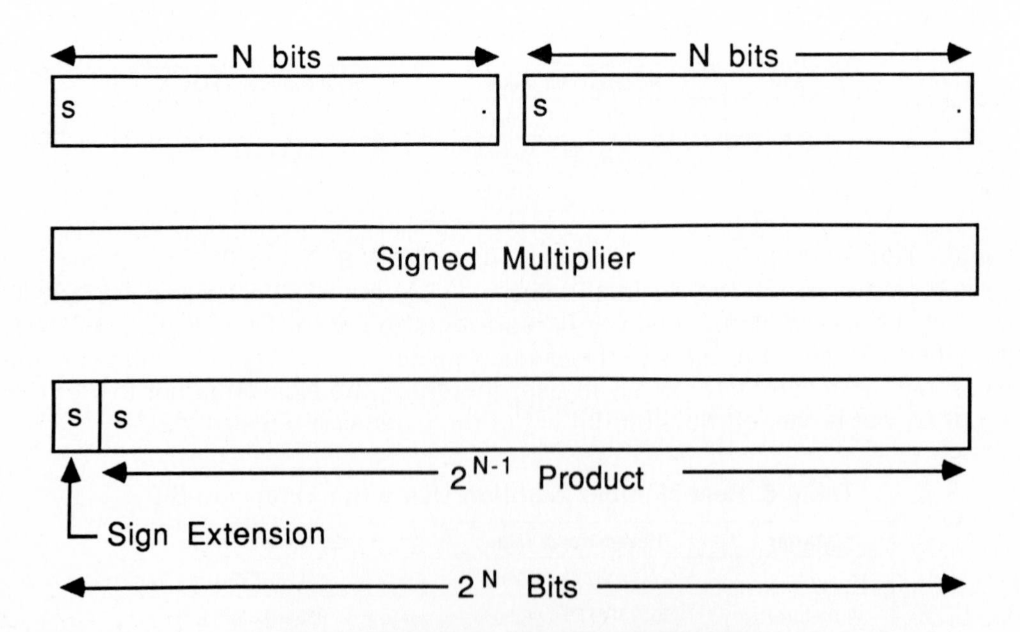

Figure 13. Signed-Integer Multiplication

further process the product, it is advantageous to format the product as multiple operand words. Therefore, there is an extra bit because two sign bits exist before multiplication and only one exists after the multiplication. Integer multipliers use the extra bit as a sign-extension bit.

Multiplication of signed fractions is shown in Figure 14. As is the case for signed-integer multiplication, the result of the multiplication is a 2^{N-1} bit word, including the sign bit. In this case, the extra bit is appended to the LSP as a zero in the LSB position. This bit is called the zero-fill bit.

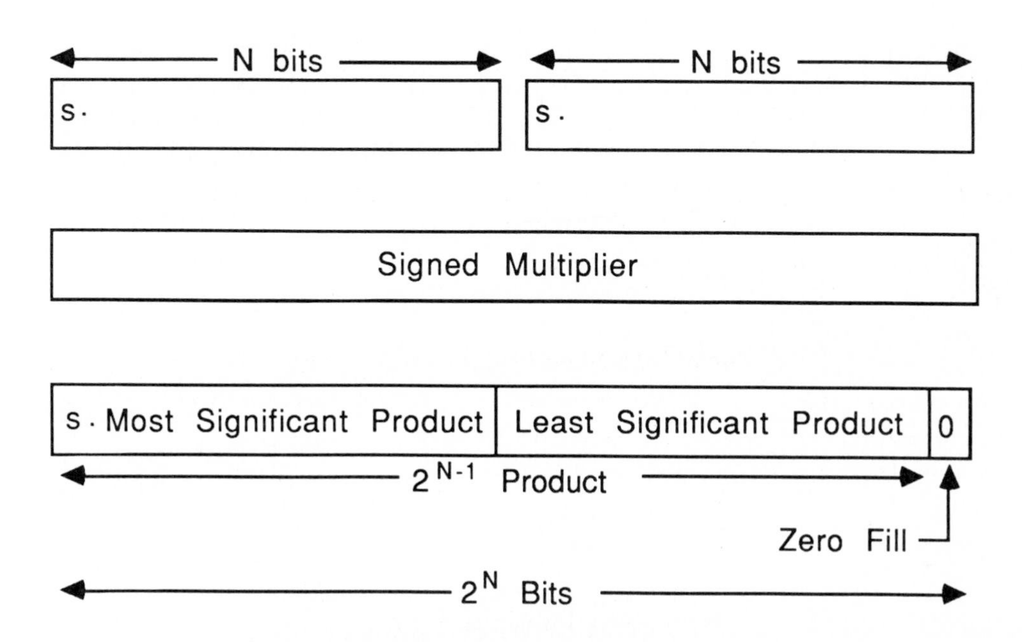

Figure 14. Signed-Fraction Multiplication

In summary, signed-integer and signed-fractional multipliers differ only in the way in which they treat the extra bit. In the integer case, the bit is used for sign extension; whereas, in the fractional case, it is used as zero fill. By shifting appropriately, integer or fractional multiplication can be accomplished on any signed hardware multiplier. The following paragraphs discuss performing fractional, integer, mixed-number, and real-number multiplications using the DSP56000 Family of processor.

MULTIPLICATION OF A SIGNED FRACTION WITH A SIGNED FRACTION

Let the values of the 24-bit general-purpose registers X1 and X0 be as shown in Table 11. After executing MPY X0,X1,A on the DSP56000/DSP56001, the content of accumulator A is as shown in Table 11. The last bit of the accumulator is zero, and the first bit carries the sign of the product. When accumulator A is rounded to 24 bits using the instructions RND or MPYR X0,X1,A, the value in A is $00:009D99:000000 (see Table 11). The lower 24 bits, A0, are zeros, and the eight sign-extension bits, A2, of the 56-bit accumulator are zeros, indicating a positive number.

MULTIPLICATION OF A SIGNED INTEGER WITH A SIGNED INTEGER

Consider the case represented in Table 12 in which two signed integers in X0 and X1 are multiplied, and the result is stored in accumulator A. It can be seen from Table 12 that if the contents of X0, X1, and A are interpreted as fractions, the result is correct. However,

Table 11. Signed-Fraction Multiplication

Register	Hexadecimal Value	Real-Number Value
X0	0647D9	+0.049067616462708
X1	0C8BD3	+0.098017096519470
A	00 : 009D98 : B815B6	+0.004809465298806
A (RND)	00 : 009D99 : 000000	+0.004809498786926

if the contents of X0, X1, and A are interpreted as integers, then a shift is required immediately after the multiplication to obtain the correct results. This shift moves the LSB out of the accumulator and adds a sign-extension bit in the MSB position. Therefore, the instruction sequence to perform integer multiplication on DSP56000/DSP56001 processors is a multiplication followed by a right shift, namely,

```
MPY  X0,X1,A
ASR  A
```

Table 12. Signed-Integer Multiplication

Register	Hexadecimal Value	Interger Value	Fractional Value
X0	000002	2	2.3841858E−07
X1	000138	312	3.7193298E−05
A	00 : 000000 : 0004E0	1248	8.8675733E−11

MULTIPLICATION OF A SIGNED INTEGER WITH A SIGNED FRACTION

Multiplication of an integer with a fractional number is a unique case since the result will be a real number — i.e., it will consist of an integer and a fractional part. When the contents of X0 and X1 are as shown in Table 13, execution of the instruction MPY X0,X1,A will result in A1 = $00003F and A0 = $800000.

Table 13. Signed-Integer and Signed-Fraction Multiplication

Register	Hexadecimal Value	Interger Value	Fractional Value
X0	400000		0.5000000
X1	00007F	127	
A10	00003F:800000		
A1	00003F	63	
A0	800000		0.5 (unsigned)

The integer part will be stored in the upper 24 bits, A1, of the 48-bit result, and the fractional part will reside in the lower 24 bits, A0, of the result. A0 is being interpreted as an unsigned fraction.

When performing multiple-precision arithmetic on real numbers, it is necessary to convert real numbers into a signed-integer operand and a signed-fraction operating (see **DOUBLE-PRECISION MULTIPLICATION**). To format A0 as a twos-complement positive fraction, two shift operations must be performed, LSL A followed by ASR A. The execution of LSL

A shifts the MSP, A1, one bit left and inserts a zero in the LSB position of A1. The execution of ASR A shifts the full 56-bit accumulator A one bit right, thereby restoring A1 and forming a positive twos-complement fraction in A0.

If the product of the multiplication is negative, then introducing the sign bit in the fractional part involves three steps. First, the absolute value of the number must be obtained. Second, the shift LSL A followed by ASR A should be performed to generate a signed twos-complement fraction. Finally, the negative values of both parts, integer and fractional, must be obtained separately. The convert to signed integer and signed fraction routine, CONVSISF, given in Figure 15, implements these three steps.

```
;CONVSISF.ASM
;This routine will convert a negative 56-bit number in the real number
;format (with a signed integer in A2:A1 and an unsigned fraction in A0)
;to a signed integer in A1 and a signed fraction in B1
;
        abs a          ;obtain the absolute value of the result
        lsl a          ;shift left to introduce sign bit
        asr a          ;introduce positive sign in fractional part
        move a0,b      ;move positive fraction to B1.
        neg b  #0,a0   ;negate fraction, clear A0
        neg a          ;negate integer
```

Figure 15. CONVSISF Routine

DOUBLE-PRECISION MULTIPLICATION

In double-precision multiplication, two 48-bit numbers are multiplied together to generate a 96-bit signed product. The concept of double-precision multiplication is depicted in Figure 16. When two 48-bit numbers P and Q (where P1 and P0 are the most significant and least significant 24-bit words, respectively, of P and, similary, Q1 and Q0 for Q) are multiplied, four single-precision products, P0Q0, P1Q0, P0Q1, and P1Q1, are generated. These products must be added with the proper weighting to yield the correct result, R3:R2:R1:R0.

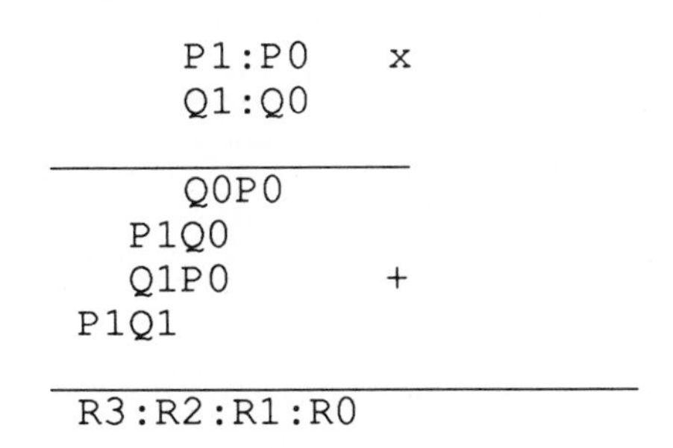

Figure 16. Double-Precision Concept

DOUBLE-PRECISION MULTIPLICATION OF FRACTIONS

The flowchart for the 48-bit general fraction multiplication routine, MULT48FG, is given in Figure 17. To compensate for the fact that signed multiplications are performed, a trick is used. The trick is to force bits 23 of P0 (P0(#23)) and Q0 (Q0(#23)) to zero before performing the Q0P0, P1Q0, and Q1P0 multiplications and then to adjust the result if P0(#23) and/or Q0(#23) were set. The least significant word of the adjusted intermediate product, LS(IP), is concatenated with the least significant 48 bits of the result, R1:R0. IP is then shifted right by 24 bits to weight the MSB of IP correctly before performing the P1Q1 multiplication and accumulation. When A2 is moved in A1, A1 is sign extended. After the P1Q1 multiplication, A contains the sign-extended result, R3:R2. This routine executes in 27 cycles if both P0(#23) and Q0(#23) are set and in 26 cycles if both are zero.

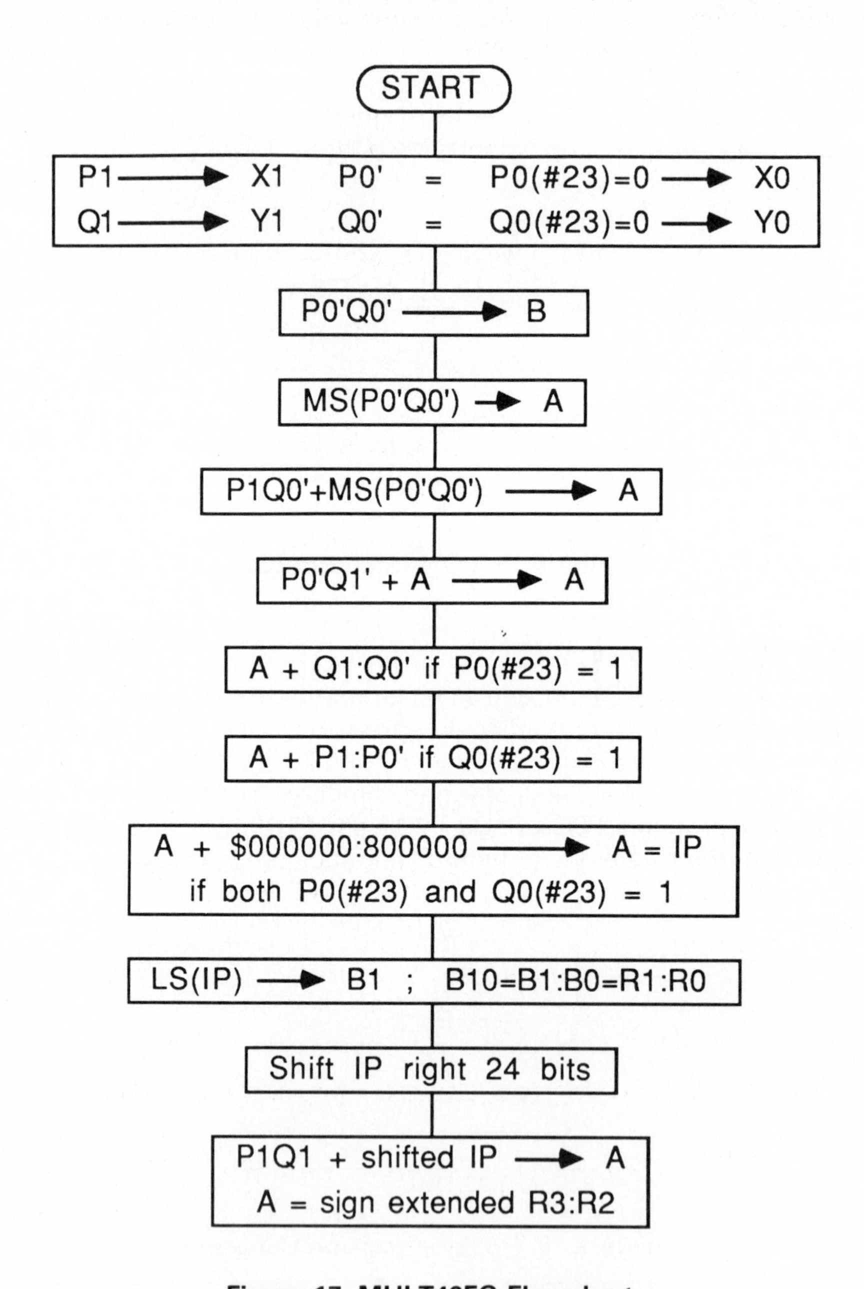

Figure 17. MULT48FG Flowchart

Listed in Figure 18, MULT48FG performs double-precision signed multiplication of fractions. Consider the multiplication of two 48-bit fractions stored in X and Y as shown in Table 14. The result is stored in accumulators A and B. The upper 48 bits of the 96-bit result are stored in A10, and the lower 48 bits are stored in B10 (see Table 14). A2 contains the sign extension of A10. The fractional result in decimal form is obtained after concatenating the two results, A0:B10, as indicated.

```
;MULT48FG.ASM
;This routine will execute the multiplication of two 48-bit
;FRACTIONAL numbers that are already stored in memory as follows.
;           x:$Paddr  P1                 y:$Paddr  P0
;           x:$Qaddr  Q1                 y:$Qaddr  Q0
;The initial 48-bit numbers are:
;             P = P1:P0  (24:24 bits)
;             Q = Q1:Q0  (24:24 bits)
;P0 with bit #23 forced to zero is P0'
;Q0 with bit #23 forced to zero is Q0'
;The result, R, is a 96 bit number that is stored in the two
;accumulators A and B as follows:
;                 R = R3:R2:R1:R0
;                   = A10:B10  (48:48 bits)
;                   = A:B10  ( sign extended )
    move #paddr,r4            ;initialize pointer for P
    move #qaddr,r5            ;initialize pointer for Q
    move #$7fffff,x1          ;load x1 with mask value
    move y:(r4),a             ;load A with P0
    and x1,a   y:(r5),b       ;create P0'; Q0 into B
    and x1,b   a1,x0          ;create Q0'; P0' into x0
    clr  a    b1,y0           ;clear A, Q0' into y0
    mpy x0,y0,b  x:(r4),x1    ;mpy P0' with Q0', P1 into x1
    move b1,a0                ;most significant word MS(P0'Q0') to a0
    mac x1,y0,a  x:(r5),y1    ;P1 * Q0' + a into a, Q1 into Y1
    mac x0,y1,a               ;P0' * Q1 + a into a
    jset #23,y:(r4),one       ;P0(#23) = 1 ?
    jset #23,y:(r5),two       ;Q0(#23) = 1 ?
    jmp thr                   ;both P0(#23) and Q0(#23) = 0
one add y,a   #$000800,y0     ;adjust for P0(#23) = 1; load y0
    jclr #23,y:(r5),thr       ;both P0(#23) and Q0(#23) = 1 ?
    mac y0,y0,a               ;generate cross term ($400000) and adj
two add x,a                   ;adjust for Q0(#23) = 1 product
thr move a0,b1                ;concatenate R1:R0 in B10
    move a1,x0                ;shift accumulator A 24 bits right
    move a2,a                 ;and sign extend
    move x0,a0                ;interm. product (IP) weighted properly
    mac x1,y1,a               ;R3:R2 sign extended in A
    end                       ;end of routine
```

Figure 18. MULT48FG Routine

Table 14. Double-Precision Fractional Multiplication

Register	Hexadecimal Value	Fractional Value
X	345678 : FFFFFF	0.408888936042779
Y	006789 : 7FFFFF	0.003159701824181
A10	002A55 : CE41FA	
B10	9683FB : 000002	
A10:B10		0.001291967117102

DOUBLE-PRECISION MULTIPLICATION OF INTEGERS

Double-precision integer multiplication is the same as double-precision fractional multiplication except that an ASR instruction needs to be introduced in the routine. The ASR eliminates the zero-fill bit and adds a sign-extension bit, thereby converting the fractional multiplier into an integer multiplier as discussed in **SIGNED MULTIPLICATION**. The shift right is done in two stages since the result is 96 bits. The lower 48 bits are shifted first, which results in a zero in bit 47. The upper 48 bits are subsequently shifted right with bit 0 going to the carry bit. If the carry is set, a one is loaded into bit 47 of the lower 48 bits of the result.

The double-precision multiplication is performed by the 48-bit general integer multiplication routine, MULT48IG, listed in Figure 19. An example is given in Table 15. The result of the multiplication is stored in the two accumulators. The 96-bit result can be obtained by concatenating A10 with B10 (see Table 15).

```
;MULT48IG.ASM
;This routine will execute the multiplication of two 48-bit
;INTEGER numbers that are already stored in memory as follows.
;          x:$Paddr  P1               y:$Paddr  P0
;          x:$Qaddr  Q1               y:$Qaddr  Q0
;The initial 48-bit numbers are:
;           P = P1:P0  (24:24 bits)
;           Q = Q1:Q0  (24:24 bits)
;P0 with bit #23 forced to zero is P0'
;Q0 with bit #23 forced to zero is Q0'
;The result, R, is a 96 bit number that is stored in the two
;accumulators A and B as follows:
;              R = R3:R2:R1:R0
;                = A10:B10  (48:48 bits)
;                = A:B10 ( sign extended )
    move #paddr,r4             ;initialize pointer for P
    move #qaddr,r5             ;initialize pointer for Q
    move #$7fffff,x1           ;load x1 with mask value
    move y:(r4),a              ;load A with P0
    and x1,a  y:(r5),b         ;create P0'; Q0 into B
    and x1,b  a1,x0            ;create Q0'; P0' into x0
    clr  a   b1,y0             ;clear A, Q0' into y0
    mpy x0,y0,b  x:(r4),x1     ;mpy P0' with Q0', P1 into x1
    move b1,a0                 ;most significant word MS(P0'Q0') to a0
    mac x1,y0,a  x:(r5),y1     ;P1 * Q0' + a into a, Q1 into Y1
    mac x0,y1,a                ;P0' * Q1 + a into a
    jset #23,y:(r4),one        ;P0(#23) = 1 ?
    jset #23,y:(r5),two        ;Q0(#23) = 1 ?
    jmp thr                    ;both P0(#23) and Q0(#23) = 0
one add y,a  #$000800,y0       ;adjust for P0(#23) = 1; load y0
    jclr #23,y:(r5),thr        ;both P0(#23) and Q0(#23) = 1 ?
    mac y0,y0,a                ;generate cross term ($400000) and adj.
two add x,a                    ;adjust for Q0(#23) = 1 product
thr move a0,b1                 ;concatenate R1:R0 in B10
    move #0,b2                 ;clear the extension register B2
    asr b                      ;start adjusting the product to integer
    move a1,x0                 ;shift accumulator A 24 bits right
    move a2,a                  ;and sign extend
    move x0,a0                 ;interm. product (IP) weighted properly
    mac x1,y1,a                ;R3:R2 sign extended in A
    asr a                      ;finish adjusting the product to integ.
    jcc end                    ;finished if A(#0) is 0
    move #$800000,x0           ;if A(#0) is 1
    or x0,b                    ;set B10 (#47)
    end                        ;end of routine
```

Figure 19. MULT48IG Routine

Table 15. Double-Precision Integer Multiplication

Register	Hexadecimal Value	Integer Value
X	000006 : 123456	101856342
Y	006789 : 7FFFFF	444688498687
A10	000000 : 027495	
B10	D5B62A : EDCBAA	

NOTE: The A10:B10 concatenated result is 4.52943438572962E+19.

MULTIPLICATION OF A REAL NUMBER WITH A REAL NUMBER

When two real numbers are multiplied together, four 24-bit multiplications must be performed: one integer with an integer, one fraction with a fraction, and two fraction with an integer. Both the integer and the fractional parts must be in the signed twos-complement format. The result will be 96 bits long; the most significant 48 bits will be the integer part, and the least significant 48 bits will be the fractional portion.

To perform a real-number multiplication using the real multiply routine, REALMULT, the multiplicand, P, is stored in register X and the multiplier, Q, is stored in register Y (see Figure 20). The signed-integer portion of the real-number result, Ri, is stored in A10, and the unsigned fractional part, Rf, is stored in B10. The data ALU programmer's model for REALMULT is shown in Figure 21. An example is given in Table 16.

This routine is similar to MULT48FG or MULT48IG in that the interim products must be properly weighted to yield the correct result. Unlike the MULT48FG and MULT48IG routines, however, there are no adjustment terms to consider because the fractions, Pf and Qf, are assumed to be signed. The CONVSISF macro in Figure 15 will perform the conversion of a real number into a signed integer and signed fraction.

The 96-bit result from the REALMULT routine should be treated as an entity. If the positive value of a negative result is required, then the absolute value of the whole 96 bits should be obtained before the integer part and fractional part can be separated.

Table 17 shows a second example using two negative numbers that produce a positive result.

```
;REALMULT.ASM
;
;This routine multiplies two signed real numbers P and Q. It
;assumes the signed integer part of P (Pi), is in X1 and the
;signed fractional part (Pf), is in X0, the signed integer part of
;Q (Qi), is in Y1 and the signed fractional part of Q (Qf), is in
;Y0. The signed integer part of the result, Ri, is stored in A10
;and the unsigned fractional part, Rf, is stored in B10.
;
        mpy  x0,y0,b        ;Pf * Qf
        asl  b              ;remove the sign bit from the product
        move b2,a           ;shift PfQf product 24 bits right
        move b1,a0          ;and preload accumulator A
        mac  x1,y0,a        ;mult. Pi with Qf and accumulate in A
        mac  x0,y1,a        ;mult. Pf with Qi and accumulate in A
        move a0,b1          ;concatenate MS(Rf) with LS(Rf)
        move a1,a0          ;adjust weighting by shifting interm.
        move a2,a1          ;product right 24 bits in prep. for
        mac  x1,y1,a        ;final product Pi * Qi + A into A
        asr  a              ;eliminate the zero fill bit
```

Figure 20. REALMULT Routine

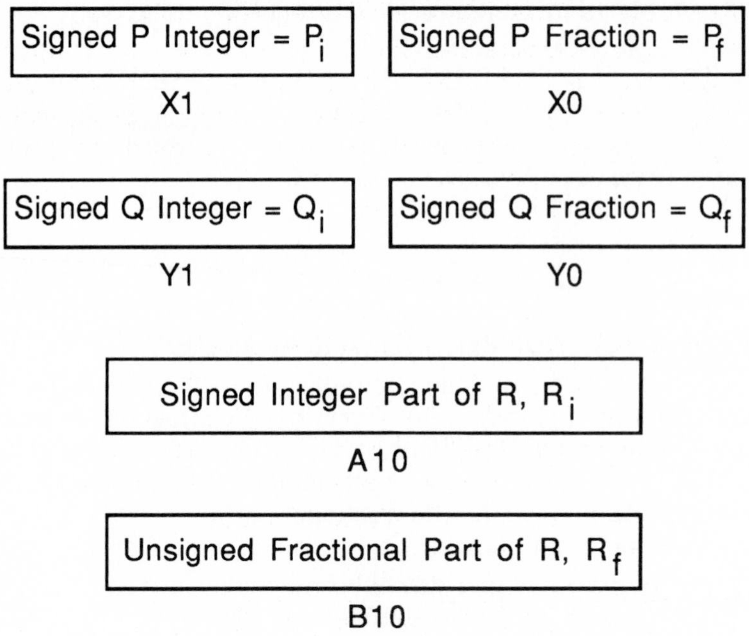

Signed P Integer = P_i	Signed P Fraction = P_f
X1	X0
Signed Q Integer = Q_i	Signed Q Fraction = Q_f
Y1	Y0

Signed Integer Part of R, R_i

A10

Unsigned Fractional Part of R, R_f

B10

Figure 21. REALMULT Data ALU Programmer's Model

Table 16. Real-Number Multiplication

Register	Hexadecimal Value	Integer Value	Fractional Value
X1	00007B	123	
X0	600000		0.75
Y1	FFFFB1	−79	
Y0	B00000		−0.625
A10	FF : FFFFFF : FFD982		
B10	68 : 000000 : 000000		

NOTE: The A10:B10 concatenated result is −9853.59375.

Table 17. Multiplication of Two Negative Real Numbers

Register	Hexadecimal Value	Integer Value	Fractional Value
X1	FFFFBF	−65	
X0	933334		−0.85
Y1	FFFFE9	−23	
Y0	ECCCCD		−0.15
A10	000000 : 0005F4		
B10	6D7064 : 75C290		

NOTE: The A10:B10 concatenated result is 1524.427496222257616.

MULTIPLICATION OF A MIXED NUMBER WITH A MIXED NUMBER

Assume that the mixed numbers are stored in the MN format. Multiplying two mixed numbers is simply a multiplication using the MPY instruction or the MAC instruction for a multiply and accumulate. The multiplication will be a 48-bit result, which will be in the format shown in Figure 22 only after a one-bit right shift to compensate for the zero-fill bit introduced by the fractional multiplication. After this shift, the most significant 16 bits will be the signed integer part, and the least significant 32 bits will be the unsigned fractional part.

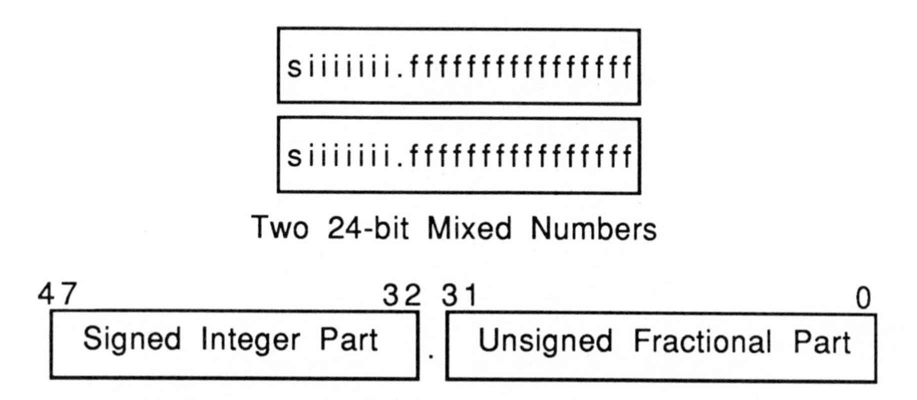

Figure 22. Multiplication of Two Mixed Numbers

Consider the example given in Table 18. If the result is desired in the real-number format, where the integer part is separated from the unsigned fractional part, then an additional right shift by eight bits must be performed on the product. This shift can be performed using the REP instruction or the appropriate shift multiplier as shown in **DATA SHIFTING**. By performing the shift, the integer part of the product is stored in the most significant word of the accumulator, and the unsigned fractional part is stored in the least significant word of the accumulator.

Table 18. Multiplication of Two Mixed Numbers

Register	Hexadecimal Value	Mixed-Number Value
X1	068000	6.5
X0	044000	4.25
A	00 : 003740 : 000000	
A10[1]	1BA000 : 000000	27.625
A10[2]	0001BA : 000000	27.625

NOTES:
1. In the MN format after the left shift by seven bits net.
2. In the real-number format after the right shift by nine bits net.

If the magnitude of the result is less than 128, it can be stored back in a 24-bit register in the MN format, which is performed by a left shift by eight bits on the result in accumulator A (already shifted one bit to the right, implying a net shift of seven bits left). The result will reside in A1 as shown in Table 18.

SIGNED DIVISION

Even though division is the inverse mathematical process of multiplication, it differs from multiplication in many aspects. Division is a shift and subtract divisor operation in contrast to multiplication, which is a shift and add multiplicand operation. In division, the results of one subtraction determine the next operation in the sequence; thus, division is inherently iterative and data dependent. The answer consists of a quotient and a remainder, both of which can have variable word lengths. In multiplication, the number of bits in the product is know a priori, which means that division is not a deterministic process, but rather a trial-and-error process.

This fact makes implementing divide routines a challenge. There are, however, additional data- and hardware-related factors that must be considered, such as:

- Input Data
  - Signed/Unsigned
  - Integer
  - Fractional Normalized
  - Fractional Unnormalized
- Output Requirements
  - Quotient
  - Remainder
  - Quotient with Remainder
  - Magnitude Only
  - Signed
  - Number of Bits of Accuracy
- Machine Architecture
  - Fractional
  - Integer
  - Register Structure

The instruction set of the DSP56000/DSP56001 processors includes a divide iteration instruction, DIV. Execution of a DIV generates one quotient bit using a nonrestoring algorithm on signed fractional operands. The original dividend must occupy the low-order 48 bits of the destination accumulator and must be a positive number. Also, the divisor must be larger than the dividend so that a fractional quotient is generated. After the first DIV execution, the destination accumulator holds both the partial remainder and the formed quotient. The partial remainder, which occupies the high-order portion of the destination operand, is a signed fraction. The partial remainder is not a true remainder and must be corrected before it may be used because of the nonrestoring nature of the division algorithm. Therefore, once the divide is complete, it is necessary to reverse the last DIV operation to restore the remainder, if the true remainder is desired. The formed quotient, which occupies the low-order portion of the destination accumulator, is a signed fraction. One bit of the formed quotient is shifted into the LSB of the destination accumulator for each DIV execution. Thus, portions of the destination accumulator allocated to the remainder and to formed quotients depend on the number of DIV executions.

In summary, for the division to produce the correct results on the DSP56000/DSP56001, two conditions must be satisfied: 1) the dividend must be positive and sign extended, and 2) the magnitude of the divisor must be greater than the magnitude of the dividend so that a fractional quotient is generated except for integer division.

DIVISION OF A SIGNED FRACTION BY A SIGNED FRACTION

The signed 24-bit divide routine for a four-quadrant divide (i.e., a signed divisor and a signed dividend) that generates a 24-bit signed quotient and a 48-bit signed remainder,

SIG24DIV, is given in Figure 23. The dividend is assumed to be in X0, but could have been in X1, Y1, or Y0.

The first three instructions save the appropriate sign bits and ensure that the dividend is positive. The first instruction copies A1 to B1 so that the sign bit of the dividend, bit 47 of A, is saved in B1 prior to taking the absolute value of the dividend. The exclusive OR in the second instruction will result in the N bit in the status register being set if the signs of the divisor and the dividend are different. Since the DIV instruction does not affect the N bit, the N bit represents the sign of the final quotient. The sign of the remainder is the sign of the dividend; therefore, bit 23 of B1 contains the sign of the remainder. B1 is stored in data memory as well as in the second instruction so that the sign bit can be tested using the bit manipulation instructions later in the routine. The third instruction clears the carry bit, C, in the condition code register. This fact ensures the quotient will be positive because the C bit is always the next quotient bit and because the C bit is shifted into the accumulator at the beginning of the execution of the DIV instruction. Execution of the next two instructions, REP and DIV, generates the 24-bit quotient and 48-bit remainder (the first 24 bits of which will be zero). The transfer instruction, TFR A,B, copies the quotient A0 into B0 (also A1 into B1 and A2 into B2) so that the sign of the quotient (i.e., bit 23 of B0) can be corrected. If the N bit in the status register was set, B is complemented. The only purpose is to complement B0 at this point; bit 23 of B0 is zero prior to the negation. Therefore, B0 is a valid, signed quotient that is saved in X1. The divisor in X0 is copied into B1 so that its absolute value can be generated and used to restore the remainder. If the remainder needs negating, B0 must be cleared first to prevent an unwanted carry from propagating into B1, the true remainder.

In Figure 24 the contents of the A accumulator are shown after each iteration of the DIV instruction for the case shown in Table 19; the repeat instruction is not interruptible. **DIVIDE**

```
;SIG24DIV.ASM
;This is a routine for a 4 quadrant divide (i.e., a signed
;divisor and a signed dividend ) which generates a 24-bit signed
;quotient and a 48-bit signed remainder. The quotient is stored
;in the lower 24 bits of accumulator A, A0, and the remainder in
;the upper 24 bits, A1. The true (restored) remainder is stored
;in B1. The original dividend must occupy the low order 48 bits
;of the destination accumulator, A, and must be a POSITIVE
;number. The divisor (x0) must be larger than the dividend so that a
;fractional quotient is generated. The quotient will be in x1 and
;the remainder will be in B1.
;
        abs a   a,b         ;make dividend positive,copy A1 to B1
        eor x0,b  b,x:$0    ;save rem. sign in X:$0, quo. sign in N
        and #$fe,ccr        ;clear carry bit C ( quotient sign bit )
        rep #$18            ;form a 24-bit quotient
        div x0,a            ;form quotient in A0, remainder in A1
        tfr a,b             ;save remainder and quotient in B1, B0
        jpl savequo         ;go to SAVEQUO if quotient is positive
        neg b               ;complement quotient if N bit is set
savequo tfr x0,b  b0,x1     ;save quo. in X1, get signed divisor
        abs b               ;get absolute value of signed divisor
        add a,b             ;restore remainder in B1
        jclr #23,x:$0,done ; go to DONE if remainder is positive
        move #$0,b0         ;prevent unwanted carry
        neg b               ;complement remainder
done                        ;end of routine.
```

Figure 23. SIG24DIV Routine

ROUTINES WITH N ⩽ 24 BITS contains four divide routines that generate quotients having less than 24 bits of precision. In Table 19 the true remainder in B1 is shifted right by 24 bits because the quotient is 24 bits, and the remainder is always smaller than the quotient. The 24 bits are implied and are not in a register.

```
DIV ITERATION    CONTENTS OF ACCUMULATOR A (in HEX)

      1      A2 = 00    A1 = 000000      A0 = 000000
      2      A2 = FF    A1 = A00000      A0 = 000001
      3      A2 = FF    A1 = A00000      A0 = 000002
      4      A2 = FF    A1 = A00000      A0 = 000004
      5      A2 = FF    A1 = A00000      A0 = 000008
      6      A2 = FF    A1 = A00000      A0 = 000010
      7      A2 = FF    A1 = A00000      A0 = 000020
      8      A2 = FF    A1 = A00000      A0 = 000040
      9      A2 = FF    A1 = A00000      A0 = 000080
      10     A2 = FF    A1 = A00000      A0 = 000100
      11     A2 = FF    A1 = A00000      A0 = 000200
      12     A2 = FF    A1 = A00000      A0 = 000400
      13     A2 = FF    A1 = A00000      A0 = 000800
      14     A2 = FF    A1 = A00000      A0 = 001000
      15     A2 = FF    A1 = A00000      A0 = 002000
      16     A2 = FF    A1 = A00000      A0 = 004000
      17     A2 = FF    A1 = A00000      A0 = 008000
      18     A2 = FF    A1 = A00000      A0 = 010000
      19     A2 = FF    A1 = A00000      A0 = 020000
      20     A2 = FF    A1 = A00000      A0 = 040000
      21     A2 = FF    A1 = A00000      A0 = 080000
      22     A2 = FF    A1 = A00000      A0 = 100000
      23     A2 = FF    A1 = A00000      A0 = 200000
      24     A2 = FF    A1 = A00000      A0 = 400000
```

Figure 24. Contents of Accumulator after Signed-Fraction Division Iterations

Table 19. Signed-Fraction Division

Register	Hexadecimal Value	Fractional Value
X0 (Divisor)	600000	0.75
A (Dividend)	00 : 300000 : 000000	0.375
A0 (Quotient)	400000	0.500
A1 (Remainder)	A00000	
$000000:B1 (True Remainder)	000000 : 000000	0.0

If the dividend is greater than the divisor, the dividend must be scaled down to be smaller than the divisor. The quotient must then be scaled up before being output, or it must be output directly and interpreted correctly. This interpretation involves assuming the binary point has moved to compensate for the original downscaling — that is, the quotient will now be a real number.

If it can be guaranteed that the divisor and dividend are normalized, then faster quadratic convergence and reciprocal methods can be used to calculate the quotient.

DIVISION OF A SIGNED INTEGER WITH A SIGNED INTEGER

Integer division can be treated in the same manner as fractional division. That is, if the dividend is positive and smaller in magnitude than the divisor, executing the SIG24DIV routine will generate the correct results. However, since the remainder is not used in integer division (the remainder is truncated), SIG24DIV can be shortened for use with integer division. Consider the example using SIG24DIV shown in Table 20. The contents of the A accumulator after each DIV iteration are shown in Figure 25.

Table 20. Signed-Integer Division

Register	Hexadecimal Value	Decimal Value
X0 (Divisor)	00162E	5678
A (Dividend)	00 : 0004D2 : 000000	1234
A0 (Quotient)	1BD178	0.21732998
A1 (Remainder)	0002B2	
$000000:B1 (True Remainder)	000000 : 0018E0	0.000000000045247

```
DIV ITERATION    CONTENTS OF ACCUMULATOR A (in HEX)
      1       A2 = FF    A1 = FFF376      A0 = 000000
      2       A2 = FF    A1 = FFFD1A      A0 = 000000
      3       A2 = 00    A1 = 001062      A0 = 000000
      4       A2 = 00    A1 = 000A96      A0 = 000001
      5       A2 = FF    A1 = FFFEFE      A0 = 000003
      6       A2 = 00    A1 = 00142A      A0 = 000006
      7       A2 = 00    A1 = 001226      A0 = 00000D
      8       A2 = 00    A1 = 000E1E      A0 = 00001B
      9       A2 = 00    A1 = 00060E      A0 = 000037
     10       A2 = FF    A1 = FFF5EE      A0 = 00006F
     11       A2 = 00    A1 = 00020A      A0 = 0000DE
     12       A2 = FF    A1 = FFEDE6      A0 = 0001BD
     13       A2 = FF    A1 = FFF1FA      A0 = 00037A
     14       A2 = FF    A1 = FFFA22      A0 = 0006F4
     15       A2 = 00    A1 = 000A72      A0 = 000DE8
     16       A2 = FF    A1 = FFFEB6      A0 = 001BD1
     17       A2 = 00    A1 = 00139A      A0 = 0037A2
     18       A2 = 00    A1 = 001106      A0 = 006F45
     19       A2 = 00    A1 = 000BDE      A0 = 00DE8B
     20       A2 = 00    A1 = 00018E      A0 = 01BD17
     21       A2 = FF    A1 = FFECEE      A0 = 037A2F
     22       A2 = FF    A1 = FFF00A      A0 = 06F45E
     23       A2 = FF    A1 = FFF642      A0 = 0DE8BC
     24       A2 = FF    A1 = 0002B2      A0 = 1BD178
```

Figure 25. Contents of Accumulator after Each Signed-Integer Division Iteration

The quotient is stored in the lower 24 bits, A0, of accumulator A. The value $1BD178 (0.21732998 decimal) is the quotient. A1 will contain the lower 24 bits of the 48-bit true remainder after the addition of the absolute value of the divisor. In this example, B1 = $0018E0 after the remainder has been restored. Therefore, the true remainder is $000000:0018E0 or 0.000000000045247 decimal.

Shown in Figure 26, the INTDIV macro performs a signed-integer divide without the extra instructions that SIG24DIV uses to generate the remainder. This routine reduces the number of operative instructions by about one-half.

```
; Signed integer divide macro.
;         Registers used: a,b,x0
;         Input: macro pass parameters "dividend, divisor"
;         Output: Quotient --> a0

INTDIV  macro dividend,divisor
        move #>dividend,a       ;sign extend a2
        move a2,a1              ; and A1
        move #>dividend,a0      ;move the dividend into A
        asl  a  #>divisor,x0    ;prepare for divide, and
                                ;move divisor into x0 (24 bit)
        abs a    a,b            ;make dividend positive, save in B
        and #$fe,ccr            ;clear the carry flag
        rep #$18                ;form a 24-bit quotient
        div x0,a                ;for quotient in a0, remainder in a1
        eor x0,b                ;save quotient sign in N
        jpl done                ;go to done if quotient is positive
        neg a                   ;complement quotient if N bit is set
done    nop                     ;finished, the quotient is in a0
        endm
```

Figure 26. INTDIV Routine

DOUBLE-PRECISION DIVISION

Division of a 48-bit number by another 48-bit number is not possible using the REP,DIV instruction sequence because the divisor is restricted to be a 24-bit fraction. Therefore, to perform double-precision division producing a 48-bit quotient and 96-bit remainder, the DIV48 routine is introduced. This routine implements the nonrestoring divide algorithm that the DIV instruction implements in hardware. The flowchart for this algorithm is shown in Figure 27.

The DIV48 dividend is stored in B10; the divisor is stored in X; and the quotient is developed in A10. After iterating the loop 48 times, B10 will contain the least significant 48 bits of the 96-bit remainder. The sign of the remainder must be made the same as that of the dividend. The quotient will be negative if the signs of the dividend and the divisor are different.

The borrow bit generated as the result of a subtraction (i.e., $N1-N2$) is the complement of the carry bit generated as the result of the addition of the negative (i.e., $N1+(-N2)$) of an operand. Therefore, the carry bit must be inverted if the divisor is subtracted from the dividend. The carry is introduced into the quotient using the add long with carry (ADC) instruction with one of the addends, Y, equal to zero. Consider the example in Table 21. The result, which is stored in accumulator A, is obtained using the DIV48 routine shown in Figure 28.

REAL-NUMBER DIVISION

This type division is not possible by simply using the DIV instruction repeatedly because a real number is a 48-bit number, and the DIV operation only operates on 24-bit operands. A real-number division cannot be broken down into four divisions and subtractions like the real-number multiplication because division is nondeterministic.

One way to divide two real numbers is to scale the numbers to form integers, fractions, or mixed numbers by multiplying them by the appropriate constant. Consider the following example:

$$\frac{123.750}{837.875}=0.14769506191257$$

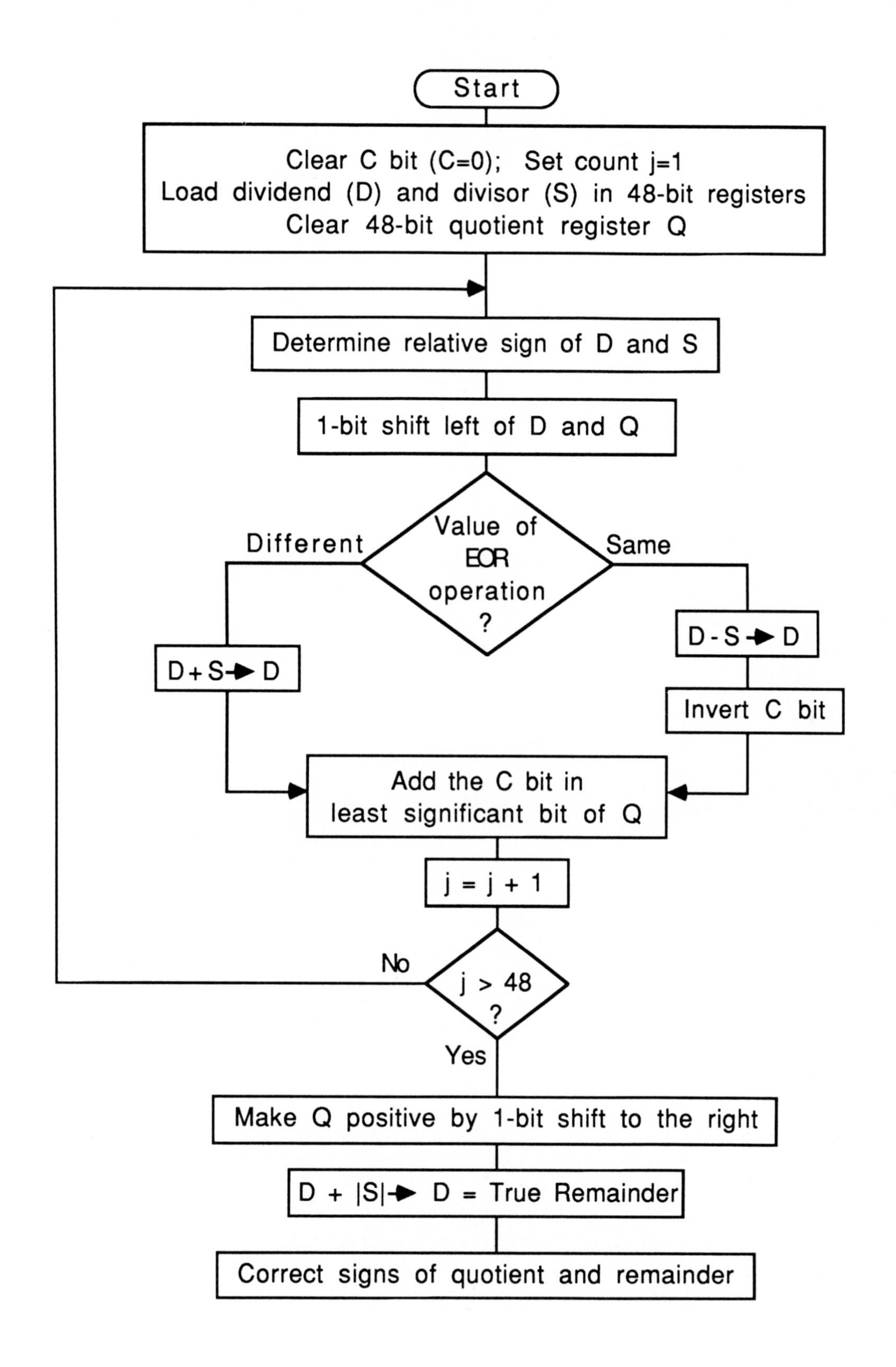

Figure 27. Double-Precision Divide Flowchart

Table 21. Double-Precision Division

Register	Decimal Hexadecimal Value	Integer Value	Decimal Fractional Value
X (Divisor)	000078 : 123450	+2014458960	0.000014313591805
B10 (Dividend)	00000F : 02468A	+251807370	0.000001789198976
A10 (Quotient)	100000 : 000000		0.125
B10 (Remainder before Correction)	FFFF87 : EDCBB0		
B10 (True Remainder after Correction)	000000 : 000000 000000 : 000000	0	0.0

```
;DIV48.ASM
;This routine performs double precision division on two 48-bit
;operands. The operands can be either fractions or integers. The
;dividend must be positive; the magnitude of the divisor must be
;greater than the magnitude of the dividend.
;The dividend and divisor are assumed to be in memory as
;      L:$divaddr     dividend
;      L:$divaddr+1   divisor
;      L:$divaddr+2   quotient
;The dividend is loaded in the long word operand B10 and the
;divisor is loaded in the long operand word X. The 48-bit true
;remainder is stored in B.
;
     move #divaddr,r0          ;initialize r0
     move #0,y0                ;clear y0
     move r0,r1                ;initialize r1
     and #$fe,ccr              ;clear carry bit C
     move   l:(r0)+,b          ;load dividend into B
     abs b         l:(r0)+,x   ;make the dividend positive, divisor in X
     clr a         #0,y1       ;clear a and y1
     do #48,endloop            ;execute the loop 48 times
     eor x1,b                  ;do operands have the same sign?
     jmi opp                   ;if opposite sign jump to location one
     eor x1,b                  ;restore the value of the operand in B
     asl a                     ;prepare A to receive a quotient bit
     asl b                     ;multiply the dividend by 2
     sub x,b                   ;subtract the divisor from the dividend
     move sr,x:$0              ;process to invert the carry bit
     bchg #0,x:$0              ;invert the carry bit
     move x:$0,sr              ;restore the SR with inverted carry bit
     jmp quo                   ;jump to location "quo"
opp  eor x1,b                  ;operands have opposite sign
     asl a                     ;prepare A to receive a quotient bit
     asl b                     ;multiply the dividend by 2
     add x,b                   ;add the divisor to the dividend
quo  adc y,a                   ;add the carry bit to develop quotient
endloop                        ;end of the 48 iteration loop
     asr a                     ;introd. the positive sign bit in quotient
     tfr x1,a  a,l:(r0)        ;divisor in A, save quotient
     move x0,a0                ;lower part of divisor in A
     abs a                     ;get the absolute value of divisor.
     add a,b                   ;generate lower 48 bits of true remainder
     jclr #23,x:(r1),done      ;if dividend is positive, finished
     neg b                     ;if dividend is negative, negate remainder
done nop                       ;end of routine
```

Figure 28. DIV48 Routine

To perform the previous division using the DSP56000/DSP56001, multiply both numbers by 1000 to change them to integers. The new values with the results after executing the SIG24DIV routine are shown in Table 22.

Table 22. Result of Real-Number Division Using DIV24 Routine

Register	Hexadecimal Value	Integer Value	Fractional Value
X1 (Divisor)	0CC8F3	837875	
A (Dividend)	00 : 01E366 : 000000	123750	
A0 (Quotient)	12E7AB		0.1476949
A1 (Remainder)	0C3868		
$000000:B1 (True Remainder)	000000 : 19015E		0.00000001164

The second and most accurate method of real-number division is by using the DIV48 routine shown in Figure 28. The real number is kept in the real-number format and stored according to the requirements of the routine. The values used are shown in Table 23. Although the result is more accurate than the result obtained by the first method, it is slower than the first method.

Table 23. Result of Real-Number Division Using DIV48 Routine

Register	Hexadecimal Value	Real Number	Fractional Value
X (Divisor)	000345 : E00000	+837.875	
B10 (Dividend)	00007B : C00000	+123.750	
A10 (Quotient)	12E7AB : FA58FC		0.147695061912572
B10 (Remainder)	000129 : 200000		
True Remainder	000000 : 000000 00046F : 000000		4.8069E−19

MIXED-NUMBER DIVISION

The mixed number is stored in the appropriate register or accumulator in the MN format. Since the number is represented in 24 bits, the SIG24DIV routine can be used. The mixed-number division must satisfy the same two conditions as the fractional division. Since the magnitude of the divisor must be greater than the magnitude of the dividend, the quotient will be a fraction represented in the signed-fraction format.

Consider the example shown in Table 24. After executing the SIG24DIV routine, the quotient will be in A0, and the lower part of the true remainder will be in B1. The results obtained after the execution of each DIV instruction are given in Figure 29.

Table 24. Result of Mixed-Number Division Using SIG24DIV Routine

Register	Hexadecimal Value	Decimal Value
X0 (Divisor)	3FC000	63.75
A (Dividend)	00 : 188000 : 000000	24.50
A0 (Quotient)	313131	0.3843137
A1 (Remainder)	D8C000	
$000000:B1 (True Remainder)	000000 : 188000	0.000000011408702

```
DIV ITERATION    CONTENTS OF ACCUMULATOR A ( in HEX)
      1      A2 = FF    A1 = F14000      A0 = 000000
      2      A2 = 00    A1 = 224000      A0 = 000000
      3      A2 = 00    A1 = 04C000      A0 = 000001
      4      A2 = FF    A1 = C9C000      A0 = 000003
      5      A2 = FF    A1 = D34000      A0 = 000006
      6      A2 = FF    A1 = E64000      A0 = 00000C
      7      A2 = 00    A1 = 0C4000      A0 = 000018
      8      A2 = FF    A1 = D8C000      A0 = 000031
      9      A2 = FF    A1 = F14000      A0 = 000062
      10     A2 = 00    A1 = 224000      A0 = 0000C4
      11     A2 = 00    A1 = 04C000      A0 = 000189
      12     A2 = FF    A1 = C9C000      A0 = 000313
      13     A2 = FF    A1 = D34000      A0 = 000626
      14     A2 = FF    A1 = E64000      A0 = 000C4C
      15     A2 = 00    A1 = 0C4000      A0 = 001898
      16     A2 = FF    A1 = D8C000      A0 = 003131
      17     A2 = FF    A1 = F14000      A0 = 006262
      18     A2 = 00    A1 = 224000      A0 = 00C4C4
      19     A2 = 00    A1 = 04C000      A0 = 018989
      20     A2 = FF    A1 = C9C000      A0 = 031313
      21     A2 = FF    A1 = D34000      A0 = 062626
      22     A2 = FF    A1 = E64000      A0 = 0C4C4C
      23     A2 = 00    A1 = 0C4000      A0 = 189898
      24     A2 = FF    A1 = D8C000      A0 = 313131
```

Figure 29. Contents of Accumulator After Each Mixed-Number Division Iteration

DIVIDE ROUTINES WITH N ≤ 24 BITS

Four distinct routines for the division of fractional numbers where an N-bit (N<24) quotient is required are given in the following paragraphs.

Positive Operands with Remainder Where N Is Variable

For positive fractional operands, the code in Figure 30 may be used to perform a divide operation, which generates an N-bit quotient and a 48-bit remainder having 48 N bits of precision for N<24.

In this routine, the quotient is built up by rotating the C bit into B. The correct C bit is generated by executing the DIV instruction. The remainder is built up in A. The REP Y1 and ASL B instruction sequence sets the signed-fraction format for the N-bit quotient into a signed fraction. Similarly, the REP Y0 and ASR B instruction sequence formats the 48-N-bit true remainder into a signed fraction.

Positive Operands without Remainder Where N Is Fixed

For positive fractional operands, the code in Figure 31 (or similar code) may be used to perform a divide operation yielding only an N-bit quotient without a remainder for N<24.

The quotient bits must be extracted out of the accumulator that contains both remainder and quotient bits after the execution of the DIV instructions. These bits must then be formatted as a positive fraction.

Signed Operands with Remainder Where N Is Variable

For signed fractional operands, the code shown in Figure 32 may be used to perform a divide operation yielding an N-bit quotient and a 48-bit remainder having 48 N bits of precision for N<24. Bits 0 and 1 in location X:$0 are used to save the quotient and remainder

```
;This routine assumes that the 48-bit positive fractional
;dividend is stored in the A accumulator, the 24-bit  positive
;fractional divisor is stored in the X0 register, the value N is
;stored in the Y0 register and the value 24-N is stored in the Y1
;register. This routine stores the N-bit positive fractional
;quotient in the X1 register and the 48-bit positive fractional
;remainder with 48-N bits of precision in the B accumulator. Note
;that in this routine the value of N and 24-N may be changed at
;run time without reassembling since they are stored in
;registers.
;
            clr b               ;initialize B1 for quotient
            and #$fe,ccr        ;clear carry, C, (quotient sign bit)
            do y0,loop1         ;compute N-bit quotient (Y0=N)
            rol b               ;build up N-bit quotient in B1
            div x0,a            ;build up 48-N bit remainder in A
    loop1   rep y1              ;repeat 24-N times ( Y1=24-N )
            asl b               ;format quotient as positive fraction
            tfr x0,b    b1,x1   ;save N-bit quotient, move divisor
            add a,b             ;recover 48-N bit remainder in B
            rep y0              ;repeat N times ( Y0 = N )
            asr b               ;format remainder as positive fraction
```

Figure 30. Positive Divide: 48-Bit Operand and Remainder

```
;This routine assumes that the 48-bit positive fractional
;dividend is stored in the A accumulator and that the 24-bit
;positive fractional divisor is stored in the X0 register. This
;routine stores the N-bit positive fractional quotient in the A
;accumulator, A1. The value of N is not stored in a register and
;is specified at the time of compilation using the CVI and POW
;functions built into the DSP56000 Cross Assembler. The quotient
;is stored in A1 and the remainder is destroyed.
;
      and #$fe,ccr                  ;clear carry, C, (quotient sign bit)
      rep #n                        ;form an N-bit quotient
      div x0,a                      ;perform divide iteration N times
      move a0,a1                    ;move quotient to A1, destroy remainder
      move #>(@cvi(@pow(2,n))-1,x1  ;store N-bit quotient bit mask in X1
      and x1,a                      ;extract N-bit quotient in A1
      rep #(24-n)                   ;repeat (24-N) times
      lsl a                         ;format quotient as positive fraction
```

Figure 31. Positive Divide: N-Bit Quotient without Remainder

sign flags, respectively. The first function of this routine is to set up these flags. The quotient is built up by rotating the C bit into B. The correct C bit is generated by executing the DIV instruction. The remainder is built up in A. The REP Y1 and ASL B instruction sequence formats the N-bit quotient as a signed fraction. Similarly, the REP Y0 and ASR B instruction sequence formats the 48 N-bit true remainder into a signed fraction.

Signed Operands without Remainder Where N Is Fixed

For signed fractional operands, the code given in Figure 33 may be used to perform a divide operation, yielding only an N-bit quotient without a remainder for $N<24$.

The NEG A instruction is guaranteed to negate the quotient because the first bit of the quotient has been set to zero, making it a positive fraction. Quotient bits must be extracted out of the accumulator, which contains both remainder and quotient bits after the execution of the DIV instructions. The quotient bits must then be formatted as a positive fraction.

```
;This routine assumes that the sign extended 48-bit fractional
;dividend is stored in the A accumulator, the 24-bit signed
;divisor is stored in the X0 register, the value N is stored in
;the Y0 register and that the value 24-N is stored in the Y1
;register. This routine stores the N-bit signed fractional
;quotient in the X1 register and the 48-bit positive fractional
;remainder with 48-N bits of precision in the B accumulator. In
;this routine the values of N and 24-N may be changed at runtime.
;
        bclr #0,x:$0           ;clear quotient sign flag (bit 0,x:$0)
        bclr #1,x:$0           ;clear remainder sign flag (bit 1,x:$0)
        tst  a                 ;determine the sign of the dividend
        jpl  signquo           ;go to SIGNQUO if the dividend is positive
        bset #1,x:$0           ;set remainder sign flag if negative
signquo abs a    a,b           ;make dividend positive, copy a1 to b1
        eor x0,b               ;get sign of quotient  ( N bit )
        jpl start              ;go to START if the sign is positive
        bset #0,x:$0           ;set quotient sign flag if negative
start   clr b                  ;initialize B1 for quotient
        and #$fe,ccr           ;clear carry, C, (quotient sign bit=0)
        do y0,loop1            ;compute N-bit quotient ( Y0=N )
        rol b                  ;build up N-bit quotient in B1
        div x0,a               ;build up 48-N bit remainder in A
loop1   rep y1                 ;repeat 24-N times ( Y1=24-N )
        asl b                  ;format quotient as positive fraction
        jclr #0,x:$0,savequo   ;go to SAVEQUO if the quot. is positive
        neg b                  ;complement quot. if sign flag is set
savequo tfr x0,b b1,x1         ;save N-bit quotient, divisor into B
        abs b                  ;get the absolute value of divisor
        add a,b                ;recover 48-N bit remainder in B
        rep y0                 ;repeat N times ( Y0=N )
        asr b                  ;format remainder as signed fraction
        jclr #1,x:$0,done      ;go to DONE if the remainder is positive
        neg b                  ;complement remainder if negative
done
```

Figure 32. Signed Divide: 48-Bit Operand and Remainder

```
;This routine assumes that the sign extended 48-bit fractional
;dividend is stored in the A accumulator and that the 24-bit
;signed fractional divisor is stored in the X0 register. This
;routine stores the N-bit signed fractional quotient in the A
;accumulator. The value of N is not stored in a register and is
;specified at the time of compilation using the CVI and POW
;functions built into the DSP56000 Cross Assembler. The quotient
;is stored in A1 and the remainder has been destroyed.
;
     abs a       a,b      ;make dividend positive, copy signed div. to B1
     eor x0,b             ;get sign of quotient and save in N bit
     and #$fe,ccr         ;clear carry (quotient sign bit = 0 )
     rep #n               ;form an N-bit quotient by executing
     div x0,a             ;the divide iteration N times
     jpl mask             ;go to MASK if quotient is positive
     neg a                ;negate quotient
mask move a0,a1           ;destroy remainder, move quotient in A1
     move #>(@cvi(@pow(2,n))-1),x1 ;N-bit quot.  mask in X1
     and x1,a             ;recover signed N-bit quotient in A1
     rep #24-n            ;repeat 24-N times
     lsl a                ;format quotient as signed fraction.
```

Figure 33. Signed Divide: N-Bit Quotient without Remainder

CONCLUSION

For the case in which only positive fractional operands are used to compute both a full-precision (i.e., 24-bit) unsigned quotient and its 48-bit remainder, both the quotient and the remainder are correctly aligned with their respective register boundaries after the 24 divide iterations. Neither the quotient nor the remainder must be shifted to produce the correct result (see **DIVISION OF A SIGNED FRACTION BY A SIGNED FRACTION**).

In general, unless an application specifically requires the number of bits of the precision (N) in the quotient to be variable, using a fixed value N, declared at the time of assembling, results in a significantly faster division execution time. Similarly, unless an application specifically requires that the remainder has to be computed, using a routine that computes only the quotient results in a significantly faster division execution time. Finally, unless and application specifically demands using less than 24 bits of precision, the use of an unsigned full-precision division routine will probably result in the fastest division execution time.

REFERENCES

1. *DSP56000 Macro Assembler Reference Manual* (available from DSP, Oak Hill, Texas, (512) 440-2030), Motorola Inc., 1986.

2. *DSP56000 Digital Signal Processor User's Manual* (DSP56000UM/AD), Motorola Inc., 1986.

3. *DSP56001 56-Bit General-Purpose Digital Signal Processor* (ADI1290), Motorola Inc., 1988.

Appendix B

DIGITAL SINE-WAVE SYNTHESIS USING THE DSP56001

by
Andreas Chrysafis
Strategic Applications
Digital Signal Processing Operation

1.0 INTRODUCTION

Sine-wave generators are used in communications and control applications.[1,2] With the introduction of high-speed high-precision digital signal processors, stable and low distortion sine waves of any frequency can be produced digitally using some form of table look-up with interpolation to reduce distortion.[3,4,5] This document describes three table look-up methods for sine-wave generation and provides the Total Harmonic Distortion (THD) performance and Maximum Synthesizable Frequency (MSF) for each case.

A routine for synthesizing sine waves having frequencies limited to integer multiples of the Fundamental Table Frequency (FTF) is described in Section 3.1. The MSF is highest using this approach. In Section 3.2 a routine using only direct table look-up for synthesizing sine waves having frequencies which are fractional multiples of the FTF is described. This approach can be used to synthesize sine waves with frequencies which are not integer multiples of the FTF but they have substantially higher THD. A routine for synthesizing sine waves using table look-up with interpolation is described in Section 4. Sine waves with frequencies which are not limited to multiples of the FTF and yet have low THD are possible using this synthesis approach.

2.0 BACKGROUND

The values used for approximating a sine wave are stored in a table in memory as follows:

i = N − 1	sin[(N − 1)•360/N]
	•
	•
i →	sin[(i)•360/N]
	•
	•
	sin[(2)•360/N]
	sin[(1)•360/N]
BASE ADDRESS (i = 0)	sin[(0)•360/N]

where

N = the table length

i = the index into the table; $0 \leq i \leq N-1$

sin[i•360/N] = the value stored at the ith location in the table. i•360/N is the angle, in degrees, for which the sine function is calculated. Throughout the remainder of this document, the abbreviation S[i] will be used to represent this function.

Figure 1. Sine-Wave Table Values

Note that the length of the table can be traded off against software in that, except for the sign, only one quarter of the table values are unique. The frequency of the digital sine wave generated depends upon the time interval and the phase angle increment (delta, Δ) between successive table accesses.

If delta is unity (i.e., the entries are read sequentially) and the table is accessed every T seconds (T is referred to as the sampling interval), the FTF of the sine wave synthesized will be

$$\text{FTF} = 1/NT \text{ Hz}$$

On the other hand, if delta is greater than unity, e.g., every second (Δ = 2) or third (Δ = 3) entry is read, (see Figure 2) and the table is still accessed every T seconds, then the frequency of the sine wave synthesized will be

$$f = \Delta \cdot \text{FTF Hz}\ ; \quad \Delta \leq N/2$$

If delta is an integer value, only multiples of the FTF can be generated; whereas, if delta is allowed to be fractional, any frequency up to the MSF can be generated. The maximum value delta can assume is N/2 since at least two samples per cycle are required to synthesize a sine wave without aliasing.

The value of the sample output, x(n), will depend on the initial phase angle, phi (ϕ), and the time or sample index, n, as follows:

$$x(n) = \sin[\phi + n \cdot \Delta \cdot 360/N]\ ; \quad n = 0, 1, 2, \ldots$$

The THD of the synthesized sine wave depends upon the length of the table, N, the accuracy (number of bits of precision) of the data stored in the table and the value of delta.

3.0 DIRECT TABLE LOOK-UP

3.1 Integer Delta Implementation

This implementation is a direct table look-up method with delta being a positive integer number. Because delta is limited to being an integer all the required samples are contained within the table; no approximations are necessary.

Figure 2 illustrates the method discussed in this section. The illustration is done for N = 8 and Δ = 2.

The assembler listing for the SINe-Wave Generation Integer Delta ("SINWGID") routine is presented in Figure 3a. The corresponding memory map is shown in Figure 3b.

Although no assembler options are indicated in this listing, a number of options are available. Refer to the Macro Assembler Reference Manual for information.[6]

The memory locations can also be changed to suit the user's needs. In this routine the sine table (N = 256) is in internal Y ROM starting at address HEX 100 to minimize external accesses and preserve RAM for data variables. The actual 256 sine table values stored are given in Appendix B. The output location is chosen to be an address in external I/O space, Y:$FFE0.

The part of the routine that generates the sine samples is given in the form of a subroutine to facilitate calculation of the MSF. Only a MOVEP and a JMP instruction must be executed to output a sample point. This MOVEP instruction utilizes the modulo addressing capability of the DSP56001. Three address arithmetic unit registers are used in this routine; the address register, R1, contains the index into the sine table, the offset register, N1, contains the value of delta, and the modifier register, M1, contains the value N − 1 to set up the modulo buffer.

NOTE

The numbers in the parentheses at the far right of the subroutine portion of the listing indicate the number of instruction cycles required to complete the specific instruction. These numbers have been manually included as comments to ease the calculation of MSF in equation 1.

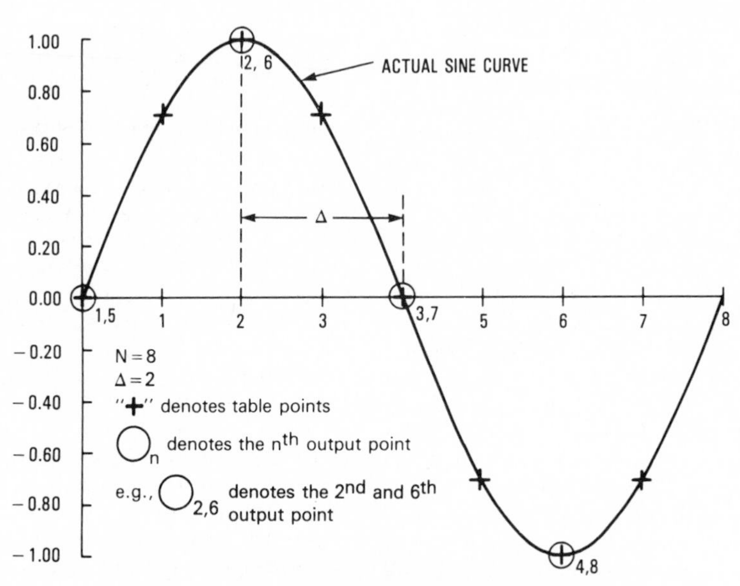

Figure 2. Integer Delta

```
Motorola DSP56000 Macro Cross Assembler  Version 1.10  06-18-87  13:03:01  sinwgid.asm  Page 1

1                                page    132,50,0,10
2                       ;SINWGID is a direct table look-up routine for SINe Wave Generation using Integer Delta
3                       ;values.  The sine table is in on-chip ROM starting at address Y:$100 and contains 256 (N)
4                       ;entries which correspond to the sines of 256 equally spaced angles between 0 and 360 degrees.
5                       ;Integer delta values between 1 and N/2 can be used to step through the table.  Here Delta=2
6                       ;and is saved in N1.  The table size is saved in M1.
7                       ;
8         00000060      start    equ     $60                               ;starting address.
9         00000100      sinep    equ     $100                              ;sine table starting address.
10        000000FF      mask     equ     255                               ;set mask = table size - 1.
11        00000002      delta    equ     2                                 ;set delta = 2
12        0000FFE0      sinea    equ     $ffe0                             ;address of output device.
13        P:0060                 org     p:start
14        P:0060 61F400          move            #sinep,r1                 ;initialize sine table pointer.
                 000100
15        P:0062 05FFA1          movec           #mask,m1                  ;set up modulo N addressing.
16        P:0063 390200          move            #delta,n1                 ;offset equals Delta.
17        P:0064 0BF080          jsr     sineg                             ;jump to subroutine.
                 000066
18        P:0066 09C9E0 sineg    movep           y:(r1)+n1,y:sinea         ;output sample.              (2)
19        P:0067 00000C          rts                                       ;return from subroutine.     (2)
20                               end                                       ;the end of listing.
0    Errors
0    Warnings
```

Figure 3a. The "SINWGID" Routine

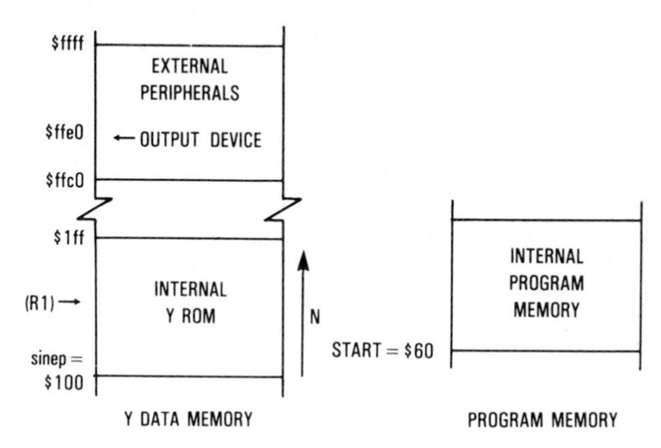

Figure 3b. "SINWGID" Memory Map

Resources required for the "SINWGID" routine:

Data ROM:	256 24-bit words for the sine table
Program Memory:	6 24-bit words for initialization
	2 24-bit words for the subroutine

The MSF for the given sine-wave subroutine would be achieved by replacing the RTS instruction with the JMP SINEG instruction.

After performing the above replacement, MSF is given by,

$$MSF = \frac{1}{2 \cdot Icyc \cdot SIC} \quad (1)$$

where SIC = the total number of Subroutine Instruction Cycles in "sineg". SIC can be obtained by adding the numbers in the brackets incorporated in the subroutine, and Icyc = the instruction cycle execution time.

For Icyc = 100 ns and SIC = 4 cycles, in the example above,

$$MSF = 1/(2 \cdot 100 \cdot 4) \text{ MHz}$$
$$= 1.25 \text{ MHz}$$

The results above compare very favorably in terms of both program memory requirements and execution speed with those for other competitive products.[7]

3.2 Real Delta Implementation

This implementation is a direct table look-up method with delta being a positive real number, that is, a number consisting of an integer and a fractional part. When fractional values of delta are used, samples of points between table entries must be estimated using the table values. The most straightforward estimation is to use the previous table entry. This approach is described in this section and illustrated in Figure 4 for N = 8 and $\Delta = 2.5$.

The assembler listing for the SINe-Wave Generation Real Delta ("SINWGRD") routine is presented in Figure 5a. The memory map (see Figure 5b) is the same as that used for the SINWGID routine except that R4 is used to point to the output device. This saves one cycle since the previous sample can be output while the accumulator is being updated using the parallel move feature of the DSP56001. The programmer's model for the Data ALU is presented in Figure 5c.

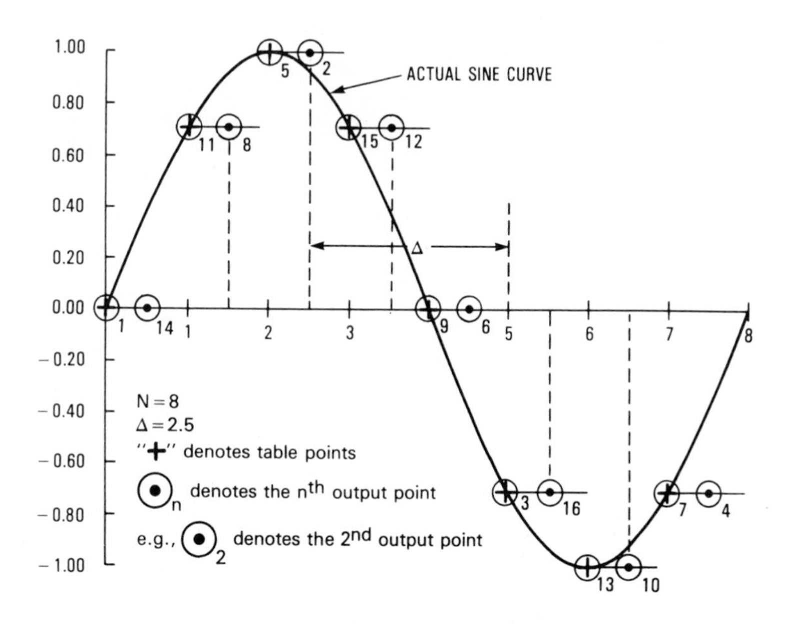

Figure 4. Real Delta Without Interpolation

```
1                                 page    132,50,0,10
2                         ;SINWGRD is a direct table look-up routine for SINe Wave Generation using positive Real
3                         ;numbers for Delta.  The sine table is in on-chip ROM starting at address Y:$100 and
4                         ;contains 256 (N) entries corresponding to the sine values of 256 equally spaced angles
5                         ;between 0 and 360 degrees.  N must be a power of two.  Delta can be between 0.0 and N/2.
6                         ;Here delta = 2.5 and is saved in B1 and B0.  The table size minus one, N-1, is saved in X1.
7                         ;
8       00000060          start   equ     $60                                 ;program starting address.
9       00000100          sinep   equ     $100                                ;sine table starting address
10      000000FF          mask    equ     $ff                                 ;N - 1.
11      2.500000          delta   equ     2.5                                 ;delta = 2.5
12      0000FFE0          sinea   equ     $ffe0                               ;address of output device.
13      P:0060                    org     p:start
14      P:0060 64F400             move            #sinea,r4                   ;set up output pointer.
               00FFE0
15      P:0062 51F400             move            #(delta-@cvi(delta)),b0     ;store the fract. part in b0.
               400000
16      P:0064 61F43A             asl     b       #sinep,r1                   ;eliminate sign bit,init. r1.
               000100
17      P:0066 2D0200             move            #@cvi(delta),b1             ;integer part of delta in b1.
18      P:0067 390013             clr     a       #0,n1                       ;initialize acc. a and n1.
19      P:0068 45F400             move            #>mask,x1                   ;set modulo mask.
               0000FF
20      P:006A 0BF080             jsr     sineg                               ;jump to subroutine.
               00006C
21      P:006C 4EE910     sineg   add     b,a               y:(r1+n1),y0 ;update a, get sine value.      (2)
22      P:006D 4E6466             and     x1,a              y0,y:(r4)    ;mask a, output sample.         (1)
23      P:006E 219900             move            a1,n1                       ;update offset register n1.     (1)
24      P:006F 00000C             rts                                         ;return from the subroutine.    (2)
25                                end                                         ;end of listing.
0       Errors
0       Warnings
```

Figure 5a. The "SINWGRD" Routine

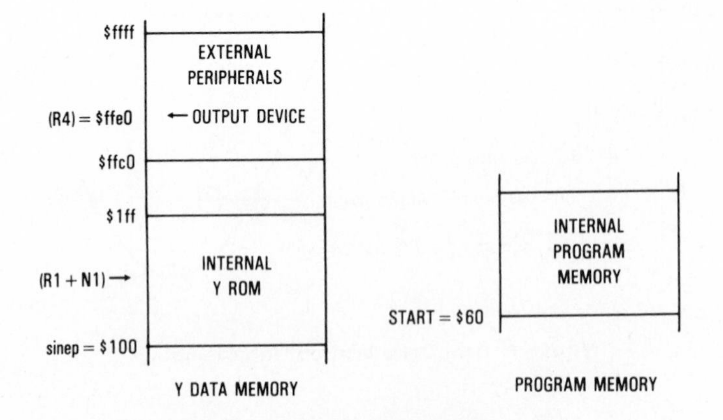

Figure 5b. "SINWGRD" Memory Map

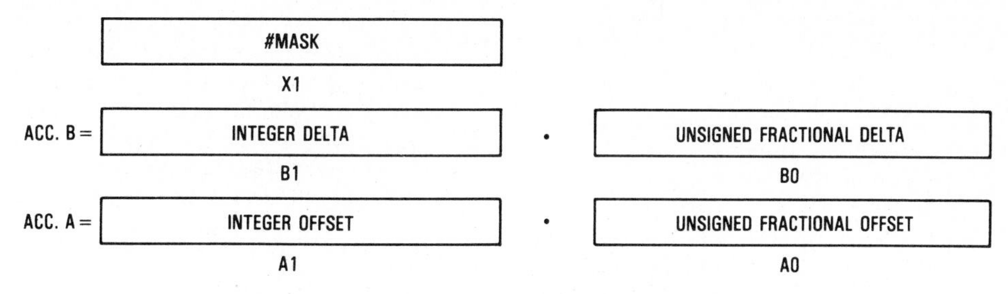

Figure 5c. "SINWGRD" Data ALU Programmer's Model

Given delta is now a real number, the inherent modulo addressing mode in the DSP56001 cannot be used and therefore a modulo addressing scheme must be implemented in software. The following is a description of one approach to implementing such a scheme. This approach uses both accumulators.

In this routine the integer part of delta is stored in the upper portion of an accumulator (B1 for the example given), and the fractional part is stored in the lower portion of the same accumulator (B0). Care should be taken to ensure that the fractional part is stored correctly. The number moved into B0 as a result of subtracting the integer portion from delta, i.e. $\Delta - @cvi(\Delta)$, is a signed, i.e. positive, fraction. This is undesirable since the sign bit should be associated with the integer portion. Therefore a shift to the left must be performed to eliminate the sign bit leaving the unsigned fraction in B0.

The separation of integer and fractional portions is done by using the assembler "ConVert to Integer" built-in function (@cvi). The CVI function converts real numbers to integers by simply truncating the fractional part of the number. Therefore

$@cvi(\Delta)$ returns the integer portion of delta
and $\Delta - @cvi(\Delta)$ returns the signed fractional part of delta.

This assumes that delta is not a runtime variable. If delta is required to be a runtime variable, it can be passed by separating it into a signed integer and unsigned fraction which are loaded into B1 and B0 respectively.

A second accumulator, A in this example, contains the positive real number value used to load the offset register, N1, which is used to offset the pointer, R1. R1 is used to address the correct location in the sine table. Since the table index must be an integer, only A1 is moved to N1. Incrementing by delta, however, is done on both the integer and fractional parts of the A accumulator. It should be noted that the increment of the register R1 is done by indexing the register with the offset register N1, thus leaving the original value of R1 (base address) unchanged. Note that the binary point is at an imaginary point between the two 24-bit parts of the accumulators. Initially we have:

ACC. A =	0	•	0
	A1		A0

After the first addition we get:

ACC. A =	0 + INTEGER DELTA	•	UNSIGNED FRACTIONAL DELTA
	A1		A0

To wrap the pointer, A1, around when it is incremented past the length of the table, a masking operation is performed on A. The mask, contained in X1, is the table length minus one, $N-1$, where N is restricted to be a power of two. This restriction is necessary to ensure that the mask consists of k least significant ones where k is defined by $2^k = N$. Using the above masking operation, the value of A1 and consequently N1, is restricted to be between 0 and $N-1$.

Resources required for the "SINWGRD" routine:

Data ROM:	256 24-bit words for sine table
Program Memory:	11 24-bit words for initialization
	4 24-bit words for subroutine memory
MSF:	1/(2•100•6) MHz
	=0.833 MHz

3.3 Harmonic Distortion

Due to the fact that the sine wave generated is an approximation, not all of the energy is at the fundamental frequency; a certain amount of the energy of the generated samples falls into frequencies other than the fundamental. Those frequencies are:

1. Harmonic frequencies, hf, i.e. integer multiples of the fundamental frequency, f.
2. Subharmonic frequencies, sf, where $s = h/d$ and h,d are integers.

The resulting noise is measured in terms of THD, given by the following equation:

$$THD = \frac{\text{spurious harmonic energy}}{\text{total energy of the waveform}} \qquad (2)$$

When using the table look-up algorithms, harmonic distortion occurs from two distinct sources—

1. Quantization Error: Since the sine table values stored in memory are of finite word length (24 bits in this case) the sine values cannot be represented exactly. Quantization error is directly proportional to the word length. For example the sine of 45 degrees in decimal will be

   0.7070707 using 24 bits
   and 0.7071 using 16 bits

2. Sampling Error: When points between table entries are sampled, i.e., delta is not an integer, then large errors are introduced because these points must be estimated from the table values. Since the sampling errors are derived from the table values, sampling errors are always greater than quantization errors.

The THD for different deltas and different table sizes is shown in Tables 1a, 1b, and 2.

Table 1a. Total Harmonic Distortion—Integer Delta

N	$\Delta=2$	$\Delta=3$
8	$3.5527141 \cdot 10^{-15}$	$5.7949068 \cdot 10^{-15}$
16	$5.7949068 \cdot 10^{-15}$	$3.6274700 \cdot 10^{-15}$
32	$3.6274700 \cdot 10^{-15}$	$2.8356370 \cdot 10^{-15}$
64	$2.8356370 \cdot 10^{-15}$	$3.4157912 \cdot 10^{-15}$
128	$3.4157912 \cdot 10^{-15}$	$2.8659804 \cdot 10^{-15}$
256	$2.8659804 \cdot 10^{-15}$	$2.6423040 \cdot 10^{-15}$
512	$2.6423040 \cdot 10^{-15}$	$2.6142553 \cdot 10^{-15}$
1024	$2.6142553 \cdot 10^{-15}$	$2.4857831 \cdot 10^{-15}$

Table 1b. Total Harmonic Distortion—Integer Delta

Δ	N = 64	N = 256
1	$3.4157912 \cdot 10^{-15}$	$2.6423040 \cdot 10^{-15}$
2	$2.8356370 \cdot 10^{-15}$	$2.8659804 \cdot 10^{-15}$
3	$3.4157912 \cdot 10^{-15}$	$2.6423040 \cdot 10^{-15}$
4	$3.6274702 \cdot 10^{-15}$	$3.4157912 \cdot 10^{-15}$
5	$3.4157912 \cdot 10^{-15}$	$2.6423040 \cdot 10^{-15}$
6	$2.8356370 \cdot 10^{-15}$	$2.8659804 \cdot 10^{-15}$
7	$3.4157912 \cdot 10^{-15}$	$2.6423040 \cdot 10^{-15}$
8	$5.7949069 \cdot 10^{-15}$	$2.8356370 \cdot 10^{-15}$
9	$3.4157912 \cdot 10^{-15}$	$2.6423040 \cdot 10^{-15}$
10	$2.8356370 \cdot 10^{-15}$	$2.8659804 \cdot 10^{-15}$

Table 2. Total Harmonic Distortion—Real Delta

Δ	N = 64	N = 128	N = 256
2.00	$2.8356370 \cdot 10^{-15}$	$3.4157912 \cdot 10^{-15}$	$2.8659804 \cdot 10^{-15}$
2.25	$7.5141324 \cdot 10^{-04}$	$1.8539830 \cdot 10^{-04}$	$4.7061084 \cdot 10^{-05}$
2.50	$6.0107522 \cdot 10^{-04}$	$1.4805426 \cdot 10^{-04}$	$3.7649080 \cdot 10^{-05}$
2.75	$7.5141324 \cdot 10^{-04}$	$1.8539830 \cdot 10^{-04}$	$4.7061084 \cdot 10^{-05}$
3.00	$3.4157912 \cdot 10^{-15}$	$2.8659804 \cdot 10^{-15}$	$2.6423040 \cdot 10^{-15}$
8.25	$7.5141324 \cdot 10^{-04}$	$1.8539830 \cdot 10^{-04}$	$4.7061084 \cdot 10^{-05}$
11.625	$7.9041085 \cdot 10^{-04}$	$1.9724369 \cdot 10^{-04}$	$4.9414069 \cdot 10^{-05}$

The equations as well as the FORTRAN code used for calculating the THD in the above tables are included in Appendix A. With respect to Tables 1a, 1b, and 2 some important observations and conclusions can be drawn:

1. For integer delta the THD for $N \leq 1024$ is of the same order of magnitude. This can be seen in Tables 1a and 1b. For integer deltas the only errors which cause distortion are quantization errors.
2. For integer delta and small N the THD is nonuniform. This is evident for $\Delta=2$ and $N \leq 128$ or $\Delta=3$ and $N \leq 64$ in Table 1a. For small N, the quantization error distribution is not uniformly distributed between $\pm 1/2$ Least Significant Bit (LSB) thereby causing the THD values to be nonmonotonic.
3. For integer delta and large N the THD decreases monotonically with increasing N. This is evident for $\Delta=2$ and $N \geq 128$ or $\Delta=3$ and $N \geq 64$ in Table 1a. For large N, the quantization error distribution will tend to be uniform between $\pm 1/2$ LSB resulting in the monotonic behavior.
4. THD for odd deltas and any N is constant. Similarly THD for even deltas and any N is constant with the exception of delta equal to a power of two greater than 2. This interesting result is evident in Table 1b. The explanation rests with the observation that all N points are used for generating sine waves using odd deltas. The N/2 even points are used twice for generating sine waves using even deltas with the exception of delta being a power of two greater than 2. For this exception case the points used in generating the sine waves are separated by delta and are used delta times. As noted in Appendix A, for integer deltas the total number of points used in the THD calculations will be independent of delta and equal to N.
5. THD depends on the fractional part of delta. From Table 2 is can be seen that the THD for $\Delta=2.50$ is less than either 2.25 or 2.75. This is because the fractional part generates an integer every other access for $\Delta=2.50$ and every other fourth access for 2.25 or 2.75. Therefore every other sample is free of sampling error for $\Delta=2.5$ but only every fourth sample is free of sampling error for the other two cases. Observe that the THD for 2.25, 2.75, and 8.25 is the same. The THD for 11.625 is slightly higher because the fractional part will form an integer only every eighth access.
6. The THD for non-integer deltas decreases with increasing table length. This can be verified by considering the same delta entries for different N in Table 2. By increasing the table length the difference between table entries decreases, resulting in better approximations to the non-integer samples, hence reduced sampling errors.

4.0 TABLE LOOK-UP WITH LINEAR INTERPOLATION

In order to synthesize a sine wave of any frequency with low distortion, an interpolation method must be used together with table look-up. By using interpolation, sine values between table entries can be represented more accurately. The easiest interpolation technique to implement is linear interpolation. For linear interpolation the sine value for a point between successive table entries is assumed to lie on the straight line between the two values. This is illustrated in Figure 6 where $\Delta=2.5$ and $N=8$. This figure should be contrasted with Figure 4.

The algorithm used for linear interpolation is based on the equation of a straight line, namely

$$y = m \cdot x + b$$

where m = the slope of the line
x = the x-coordinate value
b = the initial y value
y = the new y value

For linear interpolation:

$m = S[i+1] - S[i]$; i.e., it is the slope of the line segment between successive table entries i and i+1.
$b = S[i]$, the value of the sine table at the base address plus the ith offset location.
x = the fractional part of the pointer, $0 < x < 1.0$.
$y = S[i+x]$, the approximated sample value.

Therefore, we have

$$S[i+x] = S[i] + x \cdot \{S[i+1] - S[i]\}.$$

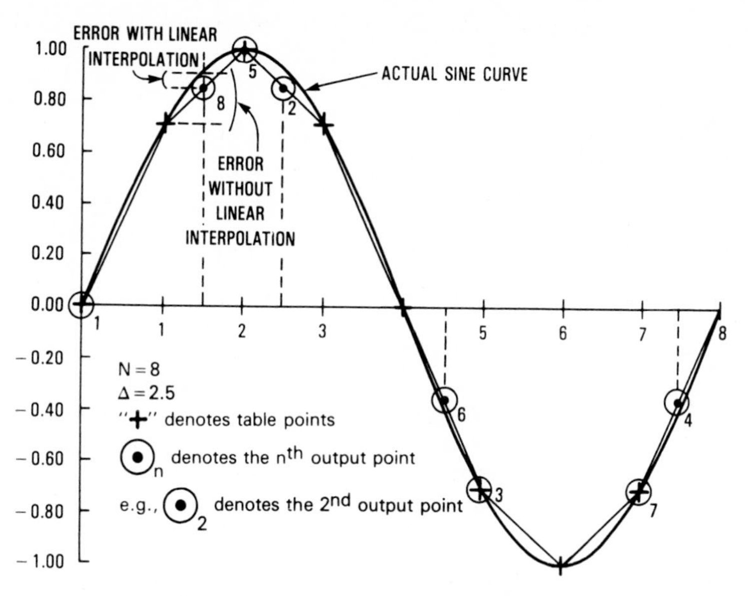

Figure 6. Real Delta Interpolation

The program flow diagram for the SINe-Wave Generation Linear Interpolation ("SINWGLI") routine is presented in Figure 7.

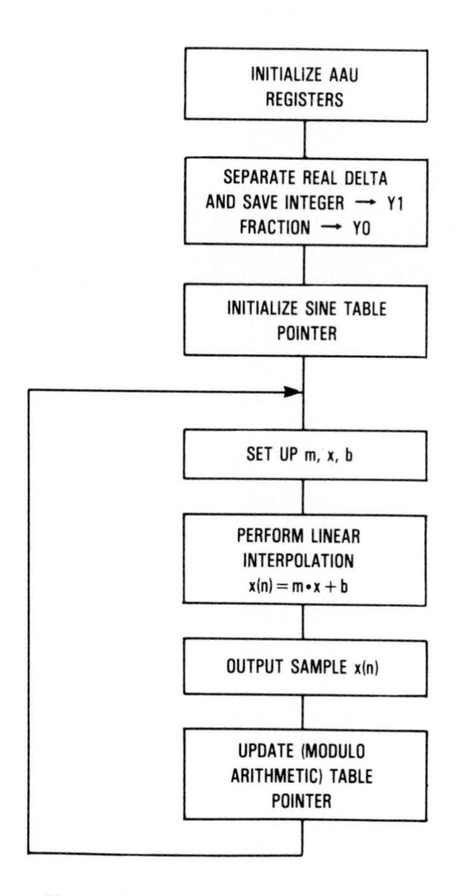

Figure 7. "SINWGLI" Flow Diagram

The assembler listing is given in Figure 8a; the memory map is presented in Figure 8b; and the Data ALU programmer's model is presented in Figure 8c.

The sine table is stored in Y ROM starting at HEX 100. The address pointer, R1, points to the current sine value while address pointer, R2, points to the current plus one sine value, R2 = R1 + 1. The slope between successive points is determined by subtracting the table entry pointed to by R2 from the table entry pointed to by R1.

For the "SINWGLI" routine, the fractional part of the linear approximation, x, is stored in the accumulator, A0, and the integer portion is stored in A1. A1 is "moved" into N1 which is used to index the R1 register. The same value that is moved in N1 is moved in N2 which is used to index R2. Note that in contrast to the SINWGRD routine, the content of A0 has to be right shifted so that it is in the correct positive signed fractional format prior to performing the multiplication to generate the linear approximation. Delta can be any positive real number between 0.0 and N/2. Accumulator B is first used to separate delta into a signed integer and an unsigned fraction (see "SINWGRD" routine), and then it is used to calculate the interpolated sample value. Note that an immediate long move (i.e., using the ">" sign) must be specified when saving the integer part of delta in Y1 because (@CVI(Δ)) would be interpreted as a signed fraction otherwise.[8] The part of the routine that generates the sine samples is given in the form of a subroutine.

Six address arithmetic unit registers are used in this routine; besides R1, R2, R4, N1, and N2, the modifier register M2 is used. The use of M2 is twofold. First, M2 is used to store the mask value before it is used in the AND masking operation. (This masking operation constrains R1 + N1 within the desired region in memory, i.e., implements modulo addressing.) Second, it is used to activate modulo arithmetic updates of R2 + N2. This is necessary to take care of the special case where R1 + N1 points to \$1FF. Since R2 + N2 equals R1 + N1 + 1 it would point to \$200 instead of \$100 without modulo arithmetic.

```
1                               page    132,50,0,10
2                       ;SINWGLI is a direct table look-up routine for SINe Wave Generation using Linear
3                       ;Interpolation with any positive real delta greater than 0.0 and less than N/2.0.  Here
4                       ;delta = 2.5 and is saved in Y1 and Y0.  The sine table contains 256 (N) entries corres-
5                       ;ponding to the sines of 256 equally spaced angles between 0 and 360 degrees.  N must be
6                       ;a power of two.  The table size minus one, N-1, is saved in N0.
7                       ;
8    00000060           start   equ     $60                                     ;program starting address.
9    00000100           sinep   equ     $100                                    ;sine table starting address.
10   000000FF           mask    equ     $ff                                     ;base address plus N minus one.
11   2.500000           delta   equ     2.5                                     ;delta = 2.5
12   0000FFE0           sinea   equ     $ffe0                                   ;address of output device.
13   P:0060                     org     p:start
14   P:0060 64F400              move            #sinea,r4                       ;set up output pointer
            00FFE0
15   P:0062 51F400              move            #(delta-@cvi(delta)),b0 ;store the fraction part in b0
            400000
16   P:0064 61F43A              asl     b       #sinep,r1                       ;eliminate sign bit, inititialize r1
            000100
17   P:0066 47F400              move            #>@cvi(delta),y1                ;integer part of Delta in y1
            000002
18   P:0068 212600              move            b0,y0                           ;unsigned fractional Delta in y0
19   P:0069 390013              clr     a       #0,n1                           ;initialize a and n1
20   P:006A 38FF00              move            #mask,n0                        ;set up mask value
21   P:006B 3A0000              move            #0,n2                           ;initialize n2
22   P:006C 62F400              move            #sinep+1,r2                     ;initialize r2 with the value r1+1
            000101
23   P:006E 0BF080              jsr     sineg                                   ;jump to subroutine
            000070
24   P:0070 210F00      sineg   move            a0,b                            ;fraction part of ptr in b1, b0=0     (1)
25   P:0071 4CE92B              lsr     b                    y:(r1+n1),x0 ;b1=signed fr. Delta, x0=sine           (2)
26   P:0072 1FEA00              move            b,x1         y:(r2+n2),b ;sign. fract. in x1, sine in b1          (2)
27   P:0073 20004C              sub     x0,b                                    ;subtr sines to obtain slope          (1)
28   P:0074 1BE900              move            b,x0         y:(r1+n1),b ;slope value to x0, sine in b1           (2)
29   P:0075 2000AB              macr    x0,x1,b                                 ;sample approximation = m*x+b         (1)
30   P:0076 5F6430              add     y,a                  b,y:(r4)    ;update calculation, output sample       (1)
31   P:0077 230500              move            n0,x1                           ;set up mask value in x1              (1)
32   P:0078 200066              and     x1,a                                    ;mask a1 = modulo arithmetic          (1)
33   P:0079 219900              move            a1,n1                           ;update offset register n1            (1)
34   P:007A 219A00              move            a1,n2                           ;update offset register n2            (1)
35   P:007B 00000C              rts                                             ;return from the subroutine.          (2)
36                              end                                             ;the end of the routine.
0    Errors
0    Warnings
```

Figure 8a. The "SINWGLI" Routine

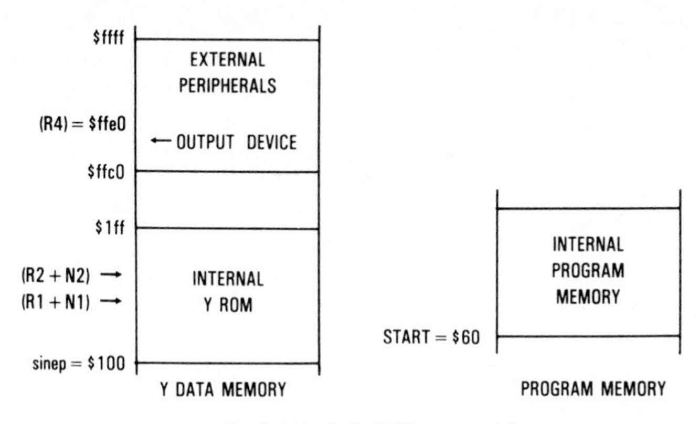

Figure 8b. "SINWGLI" Memory Map

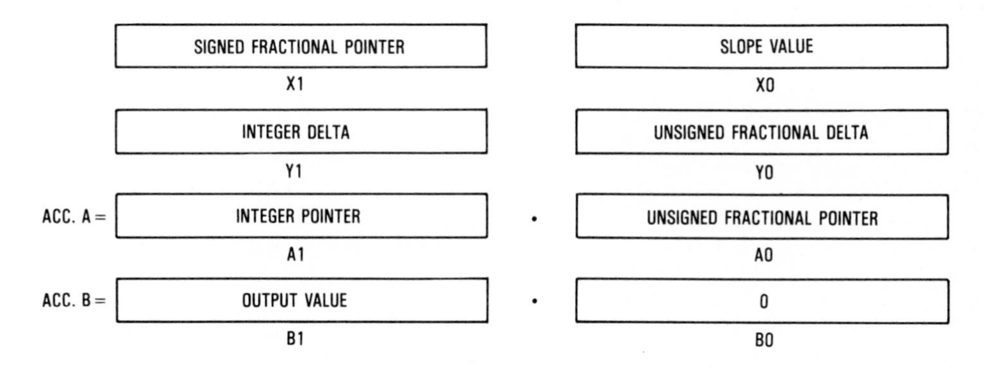

Figure 8c. "SINWGLI" Data ALU Programmer's Model

Resources required for the "SINWGLI" routine:

Data ROM:	256 24-bit words for sine table
Program Memory:	17 24-bit words for initialization
	12 24-bit words for program memory
MSF:	1/(2•100•16) MHz =0.3125 MHz

Harmonic Distortion

As expected the THD improves dramatically for non-integer values of delta when using linear interpolation. This is of course because the intermediate points are better approximated. The performance for integer deltas remains the same as that for the direct table look-up methods already discussed. The results for the THD using linear interpolation are shown in Table 3 and these results should be directly compared with the results in Table 2.

With respect to Table 3 some important observations and conclusions can be drawn:

1. THD is reduced by 10^3 by using linear interpolation with direct table look-up. This can be verified by comparing Table 2 with Table 3. This is because an extra correctional value is added to the sine value obtained from the sine table. This correctional value is the product of the slope at the specific point and the fractional part of the pointer (i.e., m•x).
2. THD depends only on the fractional part of delta when linear interpolation is used. It is a maximum for the fractional part = 0.5 and symmetrical about this midpoint. This is evident in Table 3. This is because the linear approximation is poorest at the midpoint.

Table 3. Total Harmonic Distortion—Real Delta

Δ	N = 64	N = 128	N = 256
2.00	$2.8356370 \cdot 10^{-15}$	$3.4157912 \cdot 10^{-15}$	$2.8659804 \cdot 10^{-15}$
2.25	$2.0443282 \cdot 10^{-07}$	$1.2762312 \cdot 10^{-08}$	$7.9741605 \cdot 10^{-10}$
2.50	$3.6316863 \cdot 10^{-07}$	$2.2683957 \cdot 10^{-08}$	$1.4175620 \cdot 10^{-09}$
2.75	$2.0443282 \cdot 10^{-07}$	$1.2762312 \cdot 10^{-08}$	$7.9741605 \cdot 10^{-10}$
3.00	$3.4157912 \cdot 10^{-15}$	$2.8659804 \cdot 10^{-15}$	$2.6423040 \cdot 10^{-15}$
8.25	$2.0443282 \cdot 10^{-07}$	$1.2762312 \cdot 10^{-08}$	$7.9741605 \cdot 10^{-10}$
11.625	$1.4913862 \cdot 10^{-07}$	$9.3069933 \cdot 10^{-09}$	$5.8146748 \cdot 10^{-10}$

APPENDIX A
COMPUTATION OF TOTAL HARMONIC DISTORTION (THD)

The equation for calculating THD is given by equation 2 which can be rewritten as

$$THD = \frac{ET - EF}{ET} \quad (A1)$$

where ET is the total energy of the wave and
EF is the energy of the fundamental frequency.

For accurate and correct results, the above energy terms must be calculated over a full cycle of the synthesized sine wave. In the case of direct table look-up, a full cycle may require several passes through the sine table. The number of passes depends on the value of delta used.

A full cycle will be synthesized for the smallest n for which $n \bullet \Delta$ is evenly divisible by N. For example, if the table length, N = 128, and the step size, $\Delta = 2.5 = 5/2$, then a complete cycle occurs for n = 256, since $256 \bullet 2.5/128 = 5$. Figure 4 illustrates an example for N = 8 and $\Delta = 2.5$. The $(2 \bullet 8 =)$ 16 points required for the THD calculation are shown on the figure.

In general, if $\Delta = A/B$, where A and B are relatively prime numbers, then the minimum number of samples, N′, which must be output to synthesize a full cycle is $N' = B \bullet N$.

The total energy, ET, in a cycle of length BN ($BN = B \bullet N$) is given by

$$ET = \sum_{i=0}^{BN-1} x(i) \bullet x(i) \quad (A2)$$

where x(i) is the ith sample of the sine-wave sequence.

The amount of energy in the fundamental frequency, EF, over the same period is given by

$$\begin{aligned} EF &= (1/BN) \bullet (|X(A)|^2 + |X(BN - A)|^2) \quad (A3) \\ &= (2/BN) \bullet (|X(A)|^2) \quad \text{for a real sequence} \end{aligned}$$

where the X(k) are terms of the Discrete Fourier Transform (DFT) defined by the following equation:

$$X(k) = \sum_{n=0}^{BN-1} x(n) e^{[-j(2 \bullet \pi / BN) \bullet n \bullet k]} \quad (A4)$$

The x(n) values used to calculate THD in Tables 1, 2, and 3 are based on actual values computed by the DSP56001 for the 3 sample sine-wave generator programs described in this document. The THD computation was carried out using VAX VMS FORTRAN and by using the formulas given above with double precision floating point arithmetic.[9]

The VAX VMS FORTRAN source code used for the computation of THD is given in Figure A1.

Some important details concerning the computation of the THD should be noted. ENERG3 and (ENERG2-EFUND) ideally should have the same value. However, a difference may occur because ENERG2 is much greater than EFUND. The result of the subtraction will be inaccurate due to numerical precision limitations. To determine the precision needed to calculate THD, consider the inherent symmetry in the DFT. For a real sequence x(n), the DFT sequence, X(k), exhibits the following symmetries:[10]

1. $Re\{X(k)\} = Re\{X(N-k)\}$
2. $Im\{X(k)\} = -Im\{X(N-k)\}$
3. $Mag\{X(k)\} = Mag\{X(N-k)\}$
4. $Arg\{X(k)\} = Arg\{X(N-k)\}$

where N is the length of the DFT. Therefore, using symmetry 3 as a check for the correct results for N = 128, we should have

$$\begin{gathered} Mag\{X(1)\} = Mag\{X(127)\} \\ Mag\{X(2)\} = Mag\{X(126)\} \\ \bullet \\ \bullet \\ \bullet \\ Mag\{X(n)\} = Mag\{X(128-n)\} \end{gathered}$$

This, in turn, means that the following equalities should hold:

$$\cos(2 \bullet \pi \bullet k \bullet n/N) = \cos(2 \bullet \pi \bullet (N-k) \bullet n/N)$$

and (A5)

$$\sin(2 \bullet \pi \bullet k \bullet n/N) = -\sin(2 \bullet \pi \bullet (N-k) \bullet n/N)$$

where n = 0,1,2 . . . BN − 1
k = 0,1,2 . . . BN − 1

If single precision accuracy is used in calculating the THD, the above equalities do NOT hold. For example,

let k = 1
N = 128
$ARG1 = 2 \bullet \pi \bullet n \bullet 1/128$
$ARG2 = 2 \bullet \pi \bullet n \bullet 127/128.$

Table A1. Single Precision Accuracy

n	cos (ARG1)	cos (ARG2)	sin (ARG1)	− sin (ARG2)
125	0.9891766	0.9891724	0.1467302	−0.1467580
126	0.9951848	0.9951788	0.0980167	−0.0980776
127	0.9987954	0.9987938	0.0490676	−0.0491002

Then, as shown in Table A1, cos (ARG1) does not equal cos (ARG2) when truncating to only 8 digits. If, however, double precision is used, the equalities of equation A5 are satisifed, as shown in Table A2 even when truncating to 10 digits.

Table A2. Double Precision Accuracy

n	cos (ARG1)	cos (ARG2)	sin (ARG1)	− sin (ARG2)
125	0.989176509	0.989176509	0.146730474	−0.146730474
126	0.995147266	0.995147266	0.098017140	−0.098017140
127	0.998795456	0.998795456	0.049067674	−0.049067674

```
C       Program THD.FOR
C       Using double precision real arithmetic
C       Program to calculate THD for digital sine-wave
C       generation without interpolation.
C
      IMPLICIT COMPLEX*16(X)
      REAL*8 ENERG1,ENERG2,ENERG3,S(2048),TWOPI,ISS(2048)
      DIMENSION X(0:2048)
      INTEGER*4 IS(2048)
      TYPE *,' ENTER TABLE SIZE (UP TO 2048)'
      ACCEPT *,NN
      TYPE *,'ENTER INPUT HEX FILENAME'
C     the hex values are assumed to be 11 on each line, each
C     consisting of 6 characters and having a blank space in
C     between them. This is the format of the "LOD" file that is
C     generated by the assembler [5].
      READ(1,110)(IS(I),I=1,NN)
  110 FORMAT(11(Z6,X))
      DO 120 I=1,NN
      IF(BJTEST(IS(I),23)) THEN
           IS(I)=JIOR(IS(I),'FF000000'X)
      ENDIF
  120 CONTINUE

      DO 100 I=2,NN
      ISS(I)=IS(I)
  100 CONTINUE
      TYPE *,' ENTER DELTA, BN, A'
      ACCEPT *,DELTA,BN,A
      AINDEX=0.0
      DO 200 IABC=1,BN
      INDEX=INT(AINDEX)
      S(IABC)=DBLE(ISS(INDEX+1))
      AINDEX=AINDEX+DELTA
      IF (AINDEX.GT.FLOAT(N)) AINDEX=AINDEX-FLOAT(N)
  200 CONTINUE
      TWOPI=8.0*DATAN(1.D0)
      DO 400 IK=0,BN-1
      XSUM=DCMPLX(0.D0,0D0)
      DO 300 IN=0,BN-1
      XARG=DCMPLX(0.D0,-TWOPI*DBLE(IN)*DBLE(IK)/DBLE(BN))
  300 XSUM=XSUM+DCMPLX(S(IN+1),0.D0)*CDEXP(XARG))
      X(IK)=XSUM
  400 CONTINUE
      DO 500 IK=0,BN-1
  500 ENERG1=ENERG1+CDABS(X(IK))**2.D0)
      ENERG1=ENERG1/DBLE(BN)
      DO 505 IK=0,BN-1
      IF((IK.EQ.(A)).OR.(IK.EQ.(BN-INT(A)))) GOTO 505
      ENERG3=ENERG3+(CDABS(X(IK))**2.D0)
  505 CONTINUE
      ENERG3=ENERG3/DBLE(BN)
C     energ3 should be the same as EFUND
      DO 600 I=1,BN
  600 ENERG2=ENERG2+(S(I)*S(I))
      TYPE *,' ENERG1=',ENERG1,' ENERG2=',ENERG2,'
      $ENERG3=',ENERG3
      EFUND=((CDABS(X(A))**2)+(CDABS(X(BN-A))**2))/DBLE(BN)
      TYPE *,' THD=',ENERG3/ENERG2,'EFUND=',EFUND
      END
```

Figure A1. VAX VMS FORTRAN Source Code for THD Computation

APPENDIX B
THE SINE TABLE

Table B1 gives the sine entries for N = 256. The hexadecimal values are given in the signed fractional format used for the DSP56001. "$" denotes a hexadecimal value.

Table B1. Sine Table

Address	Value	Address	Value	Address	Value	Address	Value
S_00	$000000	S_40	$7FFFFF	S_80	$000000	S_C0	$800000
S_01	$03242B	S_41	$7FF622	S_81	$FCDBD5	S_C1	$8009DE
S_02	$0647D9	S_42	$7FD888	S_82	$F9B827	S_C2	$802778
S_03	$096A90	S_43	$7FA737	S_83	$F69570	S_C3	$8058C9
S_04	$0C8BD3	S_44	$7F6237	S_84	$F3742D	S_C4	$809DC9
S_05	$0FAB27	S_45	$7F0992	S_85	$F054D9	S_C5	$80F66E
S_06	$12C810	S_46	$7E9D56	S_86	$ED37F0	S_C6	$8162AA
S_07	$15E214	S_47	$7E1D94	S_87	$EA1DEC	S_C7	$81E26C
S_08	$18F8B8	S_48	$7D8A5F	S_88	$E70748	S_C8	$8275A1
S_09	$1C0B82	S_49	$7CE3CF	S_89	$E3F47E	S_C9	$831C31
S_0A	$1F19F9	S_4A	$7C29FC	S_8A	$E0E607	S_CA	$83D604
S_0B	$2223A5	S_4B	$7B5D04	S_8B	$DDDC5B	S_CB	$84A2FC
S_0C	$25280C	S_4C	$7A7D05	S_8C	$DAD7F4	S_CC	$8582FB
S_0D	$2826B9	S_4D	$798A24	S_8D	$D7D947	S_CD	$8675DC
S_0E	$2B1F35	S_4E	$788484	S_8E	$D4E0CB	S_CE	$877B7C
S_0F	$2E110A	S_4F	$776C4F	S_8F	$D1EEF6	S_CF	$8893B1
S_10	$30FBC5	S_50	$7641AF	S_90	$CF043B	S_D0	$89BE51
S_11	$33DEF3	S_51	$7504D3	S_91	$CC210D	S_D1	$8AFB2D
S_12	$36BA20	S_52	$73B5EC	S_92	$C945E0	S_D2	$8C4A14
S_13	$398CDD	S_53	$72552D	S_93	$C67323	S_D3	$8DAAD3
S_14	$3C56BA	S_54	$70E2CC	S_94	$C3A946	S_D4	$8F1D34
S_15	$3F174A	S_55	$6F5F03	S_95	$C0E8B6	S_D5	$90A0FD
S_16	$41CE1E	S_56	$6DCA0D	S_96	$BE31E2	S_D6	$9235F3
S_17	$447ACD	S_57	$6C2429	S_97	$BB8533	S_D7	$93DBD7
S_18	$471CED	S_58	$6A6D99	S_98	$B8E313	S_D8	$959267
S_19	$49B415	S_59	$68A69F	S_99	$B64BEB	S_D9	$975961
S_1A	$4C3FE0	S_5A	$66CF81	S_9A	$B3C020	S_DA	$99307F
S_1B	$4EBFE9	S_5B	$64E889	S_9B	$B14017	S_DB	$9B1777
S_1C	$5133CD	S_5C	$62F202	S_9C	$AECC33	S_DC	$9D0DFE
S_1D	$539B2B	S_5D	$60EC38	S_9D	$AC64D5	S_DD	$9F13C8
S_1E	$55F5A5	S_5E	$5ED77D	S_9E	$AA0A5B	S_DE	$A12883
S_1F	$5842DD	S_5F	$5CB421	S_9F	$A7BD23	S_DF	$A34BDF
S_20	$5A827A	S_60	$5A827A	S_A0	$A57D86	S_E0	$A57D86
S_21	$5CB421	S_61	$5842DD	S_A1	$A34BDF	S_E1	$A7BD23
S_22	$5ED77D	S_62	$55F5A5	S_A2	$A12883	S_E2	$AA0A5B
S_23	$60EC38	S_63	$539B2B	S_A3	$9F13C8	S_E3	$AC64D5
S_24	$62F202	S_64	$5133CD	S_A4	$9D0DFE	S_E4	$AECC33
S_25	$64E889	S_65	$4EBFE9	S_A5	$9B1777	S_E5	$B14017
S_26	$66CF81	S_66	$4C3FE0	S_A6	$99307F	S_E6	$B3C020
S_27	$68A69F	S_67	$49B415	S_A7	$975961	S_E7	$B64BEB
S_28	$6A6D99	S_68	$471CED	S_A8	$959267	S_E8	$B8E313
S_29	$6C2429	S_69	$447ACD	S_A9	$93DBD7	S_E9	$BB8533
S_2A	$6DCA0D	S_6A	$41CE1E	S_AA	$9235F3	S_EA	$BE31E2
S_2B	$6F5F03	S_6B	$3F174A	S_AB	$90A0FD	S_EB	$C0E8B6
S_2C	$70E2CC	S_6C	$3C56BA	S_AC	$8F1D34	S_EC	$C3A946
S_2D	$72552D	S_6D	$398CDD	S_AD	$8DAAD3	S_ED	$C67323
S_2E	$73B5EC	S_6E	$36BA20	S_AE	$8C4A14	S_EE	$C945E0
S_2F	$7504D3	S_6F	$33DEF3	S_AF	$8AFB2D	S_EF	$CC210D
S_30	$7641AF	S_70	$30FBC5	S_B0	$89BE51	S_F0	$CF043B
S_31	$776C4F	S_71	$2E110A	S_B1	$8893B1	S_F1	$D1EEF6
S_32	$788484	S_72	$2B1F35	S_B2	$877B7C	S_F2	$D4E0CB
S_33	$798A24	S_73	$2826B9	S_B3	$8675DC	S_F3	$D7D947
S_34	$7A7D05	S_74	$25280C	S_B4	$8582FB	S_F4	$DAD7F4
S_35	$7B5D04	S_75	$2223A5	S_B5	$84A2FC	S_F5	$DDDC5B
S_36	$7C29FC	S_76	$1F19F9	S_B6	$83D604	S_F6	$E0E607
S_37	$7CE3CF	S_77	$1C0B82	S_B7	$831C31	S_F7	$E3F47E
S_38	$7D8A5F	S_78	$18F8B8	S_B8	$8275A1	S_F8	$E70748
S_39	$7E1D94	S_79	$15E214	S_B9	$81E26C	S_F9	$EA1DEC
S_3A	$7E9D56	S_7A	$12C810	S_BA	$8162AA	S_FA	$ED37F0
S_3B	$7F0992	S_7B	$0FAB27	S_BB	$80F66E	S_FB	$F054D9
S_3C	$7F6237	S_7C	$0C8BD3	S_BC	$809DC9	S_FC	$F3742D
S_3D	$7FA737	S_7D	$096A90	S_BD	$8058C9	S_FD	$F69570
S_3E	$7FD888	S_7E	$0647D9	S_BE	$802778	S_FE	$F9B827
S_3F	$7FF622	S_7F	$03242B	S_BF	$8009DE	S_FF	$FCDBD5

Appendix C

Implementation of Fast Fourier Transforms on Motorola's DSP56000/DSP56001 and DSP96002 Digital Signal Processors

by
Guy R. L. Sohie
Digital Signal Processor Operation

PREFACE

The human body has inherently slow perception mechanisms. This is illustrated, for instance, when listening to music, or speech: we do not hear individual pressure variations of the sound as they occur very fast in time. Instead, we hear a changing pitch, or frequency. Similarly, our eyes do not "see" individual oscillations of electromagnetic fields (light); rather, we see colors. In fact, we do not directly perceive any fluctuations (or oscillations) which change faster than approximately 20 times per second. Any faster changes manifest themselves in terms of the frequency or rate of change, rather than the change itself. Thus, the concept of frequency is as important and fundamental as the concept of time.

TABLE OF CONTENTS

TABLE OF CONTENTS (Continued)

LIST OF ILLUSTRATIONS

LIST OF TABLES

SECTION 1 — INTRODUCTION TO THE FOURIER INTEGRAL

1.1 — DEFINITION AND HISTORY

The scientific and engineering communities have attempted to represent changing signals in two fundamental domains: time and frequency. Temporal changes are easily shown on oscilloscopes, for instance, where change in time is directly proportional to distance on a screen. Representation of signals in terms of frequencies falls under the general category of "spectrum analysis", and has generated a lot of attention in the more recent past, due to increased availability of hardware which makes such representations possible. The first formal approach to spectrum analysis probably dates back to the work of Fourier, who showed how to represent a general class of time-varying phenomena in terms of sine and cosine functions of particular frequencies. His work is best known in terms of the *Fourier integral* (inverse Fourier transform) (see Reference 1):

$$x(t) = \int_{-\infty}^{+\infty} X(f)e^{j2\pi ft}dt \qquad (1)$$

where $j = \sqrt{-1}$ and $e^{j2\pi ft} = \cos(2\pi ft) + j\sin(2\pi ft)$. When interpreted as an infinite summation, the previous integral is simply a linear combination of a number of sine and cosine functions (expressed by the complex exponential), each one of which is "weighted" by the (complex) amplitude X(f). Conversely, the complex frequency function ("amplitude") X(f) can be derived from the time-varying signal x(t) by the *Fourier transform*:

$$X(f) = \int_{-\infty}^{+\infty} x(t)e^{-j2\pi ft}dt \qquad (2)$$

The two expressions shown in equations (1) and (2) define a *Fourier transform pair* x(t) and X(f). The Fourier transform X(f) determines the frequency content of the signal in question, while x(t) shows the way the signal varies as a function of time. Note that, in general, x(t) can be directly measured (for instance, displayed on an oscilloscope). X(f) remains a mathematical expression which attempts to express our intuitive perception of "frequency". Unfortunately, it is not always true that the theoretical concept of "frequency", as defined by the Fourier transform in (2), and the intuitive concept of frequency as we perceive it, are identical. For instance, music consists of tones (frequencies) which vary over time. Although we can clearly perceive time-varying frequencies, equation (2) does not allow for Fourier's concept of frequency to have any time-varying character: X(f) is a function of frequency only.

1.2 — USE OF THE FOURIER TRANSFORM

Because of the basic nature of the frequency concept, practical applications of the Fourier transform are abundant. As more cost-efficient methods become available to compute the Fourier transform, the number of practical solutions to frequency-based problems will grow even larger. In these frequency-based applications, a digital signal processor can be used to efficiently compute the Fourier transform (as defined in **1.1 DEFINITION AND HISTORY**), and to perform specific frequency-domain tasks such as elimination of certain frequency components, etc.

One can distinguish three general types of Fourier transform applications as shown below:

1. *Number-Based* — Most spectrum analysis applications require the direct evaluation of the Fourier transform as in equation (2). Since the Fourier transform is a mathematical expression, these applications are based on numerical computations, and can be termed "number based". Examples range from spectrum analysis laboratory instrumentation and professional audio equipment to velocity estimation in radar. Note that in number-based applications the accuracy of the computed numbers is of vital importance to the performance of the overall system. For instance, the quality-conscious audio industry requires full 16-bit result precision in order to provide no audible distortion.
2. *Pattern-Based* — Many problems involve the recognition and detection of signals with a specific frequency content (a predefined spectral pattern). For instance, speech consists of segments of sound with very specific frequency characteristics. In this type of application, the conversion to the "frequency domain" is often only a small step in the overall task. It is important that this conversion process be as fast as possible, to allow for enough time to perform computationally intensive pattern matching techniques. In addition to providing fast Fourier transform computations, the processor in question needs to retain a general-purpose nature such that a variety of frequency-based calculations for pattern matching can be done.
3. *Convolution-Based* — The third class of applications of Fourier transforms uses the transform as a simple mathematical tool to perform general filtering in a very efficient manner. This concept is based on the property that the Fourier transform of the *convolution* of two time-signals:

$$y(t) = \int_{-\infty}^{+\infty} x(t-\tau)h(\tau)d\tau \qquad (3)$$

is equal to the *product* of the individual transforms:

$$Y(f) = X(f)H(f) \qquad (4)$$

Equation (3) (better known as the *convolution integral*) represents the output of a linear filter with *impulse response* h(t) and input signal x(t). Clearly, in the frequency domain, the output of a filter can be obtained by a simple multiplication, whereas in the time-domain, a more complicated convolution integral needs to be solved. The amount of computation involved in evaluating the integral in equation (3) becomes particularly large when the impulse response h(t) has a long time duration. This sometimes prevents real-time implementation. Clearly, if the Fourier transform X(f) of the signal can be computed efficiently, the filtering operation itself can be achieved by simple multiplications. The combined number of computations (for computing the Fourier transform, for filtering in the frequency domain, and for obtaining the inverse Fourier Transform) is often less than the total number of calculations required to compute equation (3) directly. This is especially true when the filter in question performs a simple frequency discrimination function (lowpass, bandpass, highpass, bandreject, etc.). In this case, the multiplications in the frequency domain can be replaced by a simple "masking" operation, which deletes the stopbands and leaves the passband(s) unchanged.

Although no direct frequency information is extracted from the signal, the Fourier transform is used as a mathematical tool for fast-filtering applications. Note that again, fast Fourier transform and inverse Fourier transform "engines" are required to provide the real-time filtering operation.

In summary, the basic nature of the concept of frequency indicates that the number of possible frequency domain applications is as large as more conventional time domain applications. In the past, these applications were either impossible to implement or could not be realized in a cost-efficient manner because of the lack of low-cost, high-performance hardware. In this report, it will be demonstrated that the DSP56000/DSP56001 and DSP96002 Families of digital signal processors fulfill the demanding requirements imposed by frequency-domain problems. In addition to providing a *fast* implementation of *high-precision* Fourier transform computations, the *general-purpose nature of the instruction set* allows for a **complete, single-chip**, **low-cost**, integrated solution to a wide variety of frequency domain problems.

SECTION 2 — THE DISCRETE FOURIER TRANSFORM

2.1 — THE DISCRETE-TIME FOURIER TRANSFORM (DTFT)

In order to compute the Fourier transform using digital hardware, equation (2) needs to be approximated by a form which makes machine computation feasible. The first step in this process consists of eliminating the theoretical integral symbol, and replacing it by a computable sum:

$$X(f) \approx \tilde{X}(f) = T \sum_{n=-\infty}^{+\infty} x(nT) e^{-j2\pi fnT} \tag{5}$$

The previous expression uses a *sampled signal* x(nT), where the sampling period T is taken as small as possible to reduce approximation errors. $\tilde{X}(f)$ is appropriately called the *discrete-time Fourier transform (DTFT)*. As T (the sampling period) becomes infinitely small, the previous summation approaches the original Fourier transform in equation (2). In order to assess the accuracy of this approximation, it suffices to realize that the resulting expression for $\tilde{X}(f)$ is a *periodic* function of frequency:

$$\tilde{X}(f) = \tilde{X}\left(f + \frac{1}{T}\right) \tag{6}$$

due to the fact that

$$e^{-\left(j2\pi fnT + j2\pi n \frac{T}{T}\right)} = e^{-j2\pi fnT} e^{-j2\pi n} = e^{-j2\pi fnT} \tag{7}$$

In general, the original spectrum X(f) is not periodic, and the approximation is only justified for a range of small values of f. In Figure 2-1, the DTFT magnitude and the Fourier transform magnitude of a simple rectangular function are shown for several values of the sample rate $f_s = 1/T$. Note the periodicity of the resulting function, as well as the approximation errors due to the sampling process.

A well-accepted criterion for the sampling rate is given by the *Nyquist sampling theorem*, which says that a signal needs to be sampled faster than twice its highest frequency. In other words, if

$$X(f) = 0 \tag{8}$$

for $|f| \geq B$ (B is referred to as the *bandwidth* of the signal), then the sampling frequency needs to satisfy:

$$f_s \geq 2B \tag{9}$$

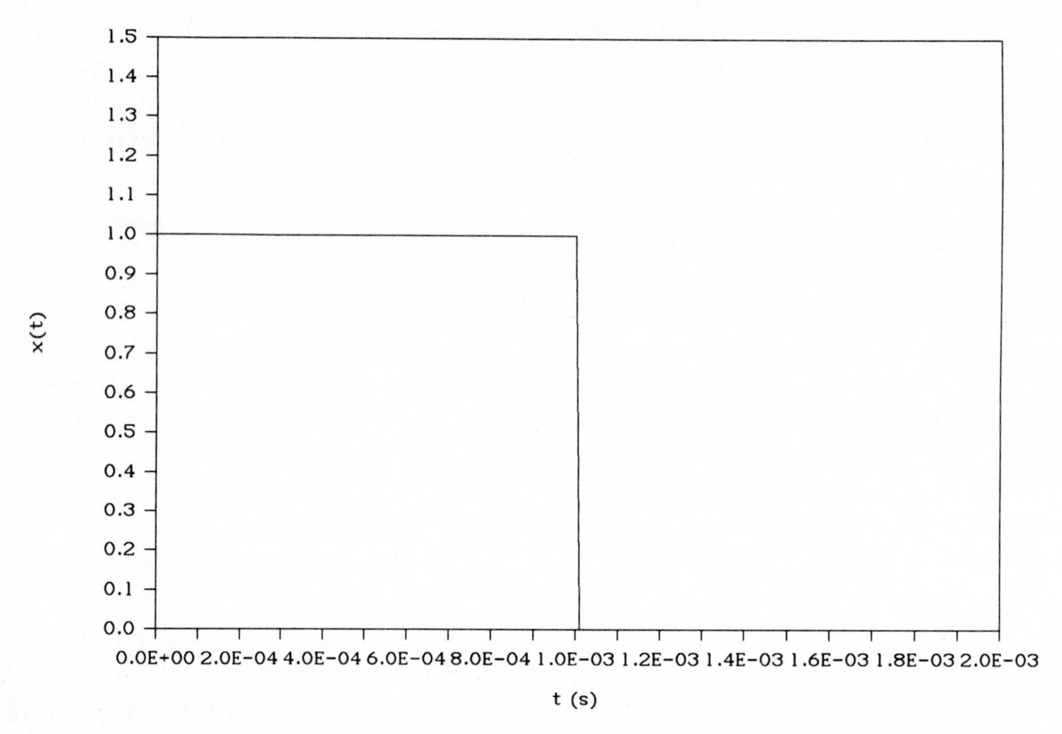

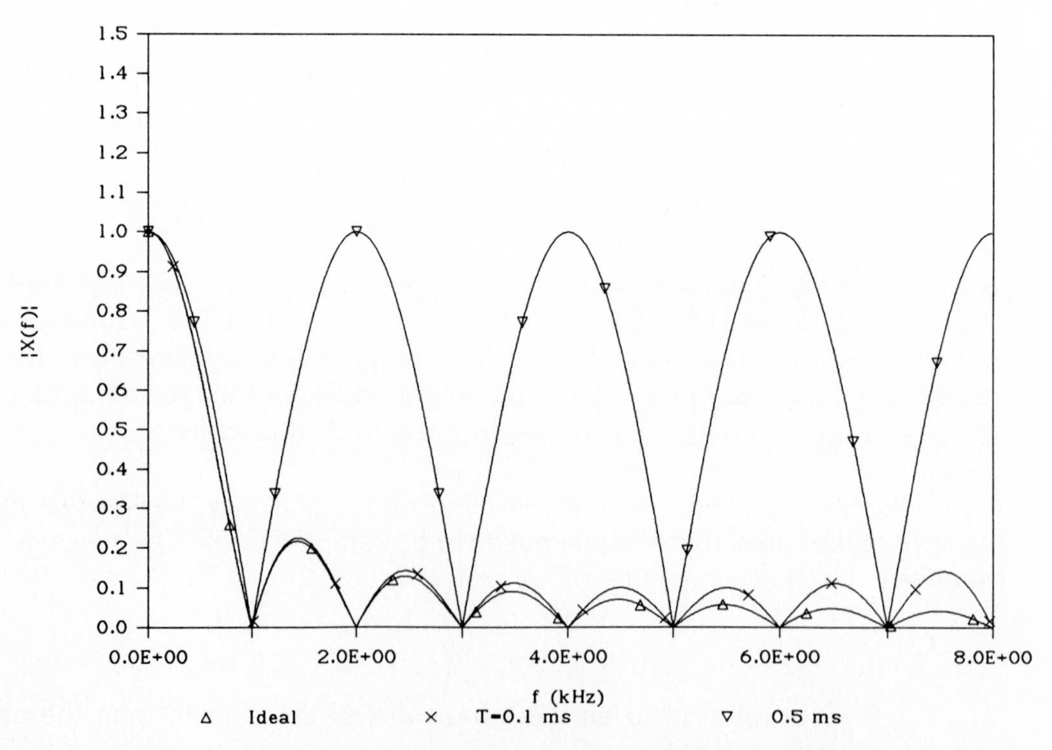

Figure 2-1. Fourier Transform of a Rectangular Function

In practice, signals rarely satisfy equation (9), and some error can be expected in the evaluation of X(f). This error is called the *aliasing error*. It is generated by frequency components at higher frequencies, which manifest themselves at lower frequencies because of the periodicity of $\tilde{X}(f)$ (aliases). The aliasing error can be reduced by filtering out the higher-frequency components of the signal using a low-pass filter ("anti-aliasing" filter) and/or by increasing the sampling rate.

2.2 — WINDOWING AND WINDOWING EFFECTS

The discussion in the previous paragraph illustrates how the Fourier transform can be approximated by an infinite summation. In practice, the results need to be available within a finite time-period, and the infinite summation needs to be somehow reduced to a finite summation. One obvious way of accomplishing this is by simply truncating the sum in equation (6) to N terms as:

$$\tilde{X}_w(f) = T \sum_{n=0}^{N-1} x(nT)e^{-j2\pi fnT} \qquad (10)$$

This truncation is frequently referred to as windowing: it appears as if the signal is "looked" at through a finite window. The resulting transform is called the *windowed discrete-time Fourier transform* (WDTFT). In mathematical terms, windowing is nothing but the multiplication of the signal by a "window" sequence of finite-length, w(n). In the simple case above, $w(n)=1$ for $0 \leq n \leq N-1$; otherwise, $w(n)=0$. Because of its rectangular shape, this window is called the *rectangular* window.

Unless the signal in question is of finite duration, this truncation will introduce other errors, resulting in a number of artifacts in the spectrum. In order to assess the effect of the windowing operation, a simple sine wave of the form:

$$x(t) = \sin(2\pi 1000t) \qquad (11)$$

is sampled with a sampling frequency of 4000 Hz, and the windowed DTFT is computed with $N=20$.

Figure 2-2 shows the result of windowing a sine wave by a rectangular window. Several errors due to windowing can be noticed as shown below:

1. *Leakage* — Even though the input signal consists of a single-frequency component at 1000 Hz, the result clearly shows components at other frequencies than 1000 Hz. This is called the leakage effect: it appears as if energy has "leaked" from 1000 Hz to the rest of the spectrum.
2. *Smoothing* — Although the theoretical transform exhibits an infinitely narrow, and infinitely large peak at 1000 Hz, the actual peak has finite magnitude and exhibits finite width. It appears that the narrow peak has been "smeared" out in the frequency domain as a result of the windowing function in the time domain. This effect is appropriately termed the smoothing effect.

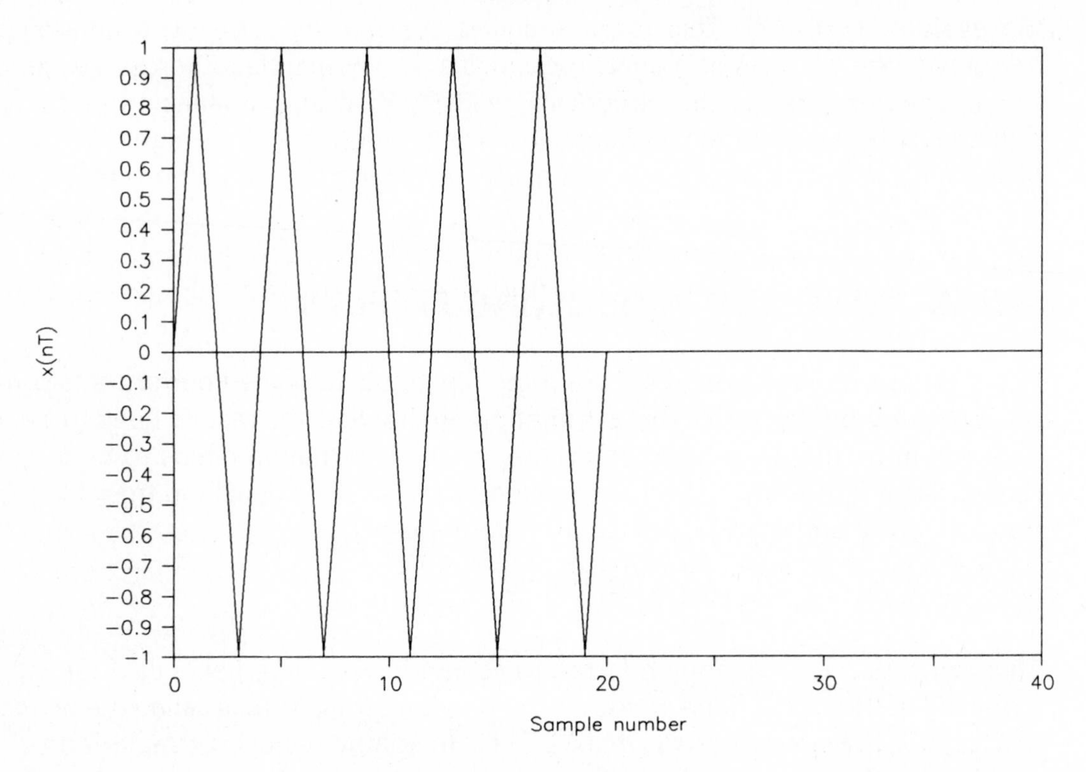

Fourier Transform Magnitude

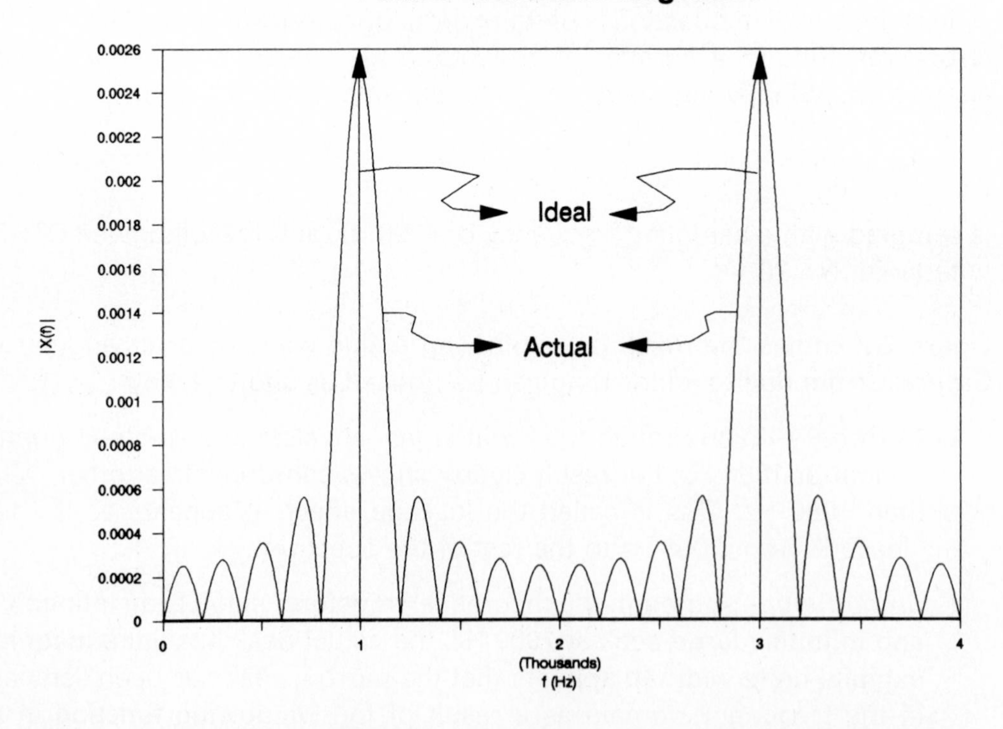

Figure 2-2. Windowing Effects when Windowing a Single Sine Wave

3. *Ripple* — The overall magnitude plot in Figure 2-2 shows an oscillatory character not present in the original Fourier transform: this is called the "ripple" effect. The origin of the ripple effect lies in the discontinuities (abrupt start and end) introduced in the signal by the window. Windows with "smoother" transitions generally have lower "sidelobes", or less ripple.

In general, a tradeoff exists between these different effects, and an appropriate windowing function can be chosen for a specific application. For an excellent summary of existing windowing functions and their properties see Reference 2.

2.3 — SAMPLING THE FREQUENCY FUNCTION

The windowed DTFT is now ready for machine computation, with one exception: the independent frequency variable f is still a continuous variable, and needs to be discretized, or sampled. Since the DTFT is periodic in the frequency domain with period f_S, only values of f from 0 to f_S (the sampling frequency) need to be computed. Although one could go through similar arguments concerning the distance between successive frequency samples as in the case of time-sampling, it turns out that when the WDTFT is sampled every f_s/N Hz, fast algorithms for computing the transform can be derived. Note that in this case, the number of samples in the window (= N) and the number of samples in the frequency domain (= N) are equal. The resulting transform is called the discrete-time Fourier series (DTFS):

$$\tilde{X}_N(k) = T \sum_{n=0}^{N-1} x(nT) e^{-j\frac{2\pi}{N}nk} \quad (12)$$

The inverse DTFS is given by:

$$x_N(nT) = \frac{1}{NT} \sum_{k=0}^{N-1} \tilde{X}_N(k) e^{j\frac{2\pi}{N}nk} \quad (13)$$

Keep in mind that the values of the frequency samples (f_k are equal to f_s/N k.

Note that many textbooks simply define the discrete Fourier transform (DFT) $X_N(k)$:

$$X_N(k) = \sum_{n=0}^{N-1} x(nT) e^{-j\frac{2\pi}{N}nk} \quad (14)$$

with inverse transform:

$$x_N(n) = \frac{1}{N} \sum_{n=0}^{N-1} X_N(k) e^{j\frac{2\pi}{N}nk} \quad (15)$$

Obviously, the DFT and DTFS differ only by a scaling factor of T, making the spectrum independent of the sampling period. Consequently, explicit T dependence is dropped in the previous expression.

Although the sequence $x_N(n)$ corresponds to the original sampled and windowed sequence $x(nT)$ for sampling instants 0 through $N-1$, the complete sampled sequence $x(nT)$ for any n cannot necessarily be recovered from it. Indeed, $x_N(n)$ appears to be periodic with period N due to the periodicity of $e^{j\frac{2\pi}{N}nk}$, whereas the original sampled signal was not assumed to be periodic.[1] This must be kept in mind in "convolution-based" applications, where the forward as well as inverse transforms are used: the incoming signal stream needs to be segmented, and the computed signal segments need to be "pieced together" to construct the complete output stream. Techniques for accomplishing this are discussed in most basic textbooks on digital signal processing (see, for instance, Reference 3).

[1]The error introduced in the time-domain by sampling a frequency function is termed "aliasing in time". This is completely analogous to the "aliasing in frequency" caused by sampling a time function, as discussed in **2.1 THE DISCRETE-TIME FOURIER TRANSFORM (DTFT)**. That is, if a frequency spectrum is not sampled sufficiently densely, the signal constructed in the time domain through the inverse "discrete-frequency Fourier transform" will show distorion.

SECTION 3 — THE FAST FOURIER TRANSFORM

3.1 — MOTIVATION

Upon closer examination of equation (15), it becomes clear that for every frequency point, $N-1$ complex summations and N complex multiplications need to be evaluated. Since there are N frequency points to be evaluated, this gives a total of $N(N-1)$ complex sums, and N^2 complex multiplications. Counting two real sums for every complex one, and four real multiplications plus two real summations for every complex multiplication, this is a total of $4\,N^2-2N$ real summations and $4\,N^2$ real multiplications.

The above numbers grow rapidly for increasing N. In fact, for $N=1024$ (1024-point DFT), 4,194,304 real multiplications are required. If this is computed on a DSP56000/DSP56001 with a 27-MHz clock, it takes 0.31 seconds just to execute that many real multiplications. Since the DFT computation needs to be completed by the time the next 1024 data points are collected for real-time performance, this limits the sampling rate to a maximum of 3.3 kHz. Obviously, faster solutions need to be sought.

3.2 — DIVIDE AND CONQUER

A faster algorithm for computing the DFT can easily be derived. The principle behind this is very basic; in fact, it is one with which we are all very familiar, as shown in Figure 3-1: a square of only half the dimensions of a larger square only has one-fourth the surface area.

This is because the surface area is quadratically proportional to the dimensions of the square. Similarly, the number of multiplications needed to compute the DFT is proportional to the square of the DFT's length (N). Thus, if we could replace the DFT over N points by two DFTs over N/2 points, computations would be reduced in order of magnitude of 0.5 ($=0.25+0.25$).

Since there are two independent variables (time and frequency) in the Fourier transform, dividing (or *decimating*) the DFT into smaller ones can be done in two ways. We can attempt to represent an N-point transform in terms of DFTs over half the number (N/2) of time-samples. This approach is appropriately called the decimation-in-time or DIT approach. Alternatively, the N-point DFT can be represented in terms of DFTs with N/2 frequency samples. This approach is called the decimation-in-frequency or DIF approach.

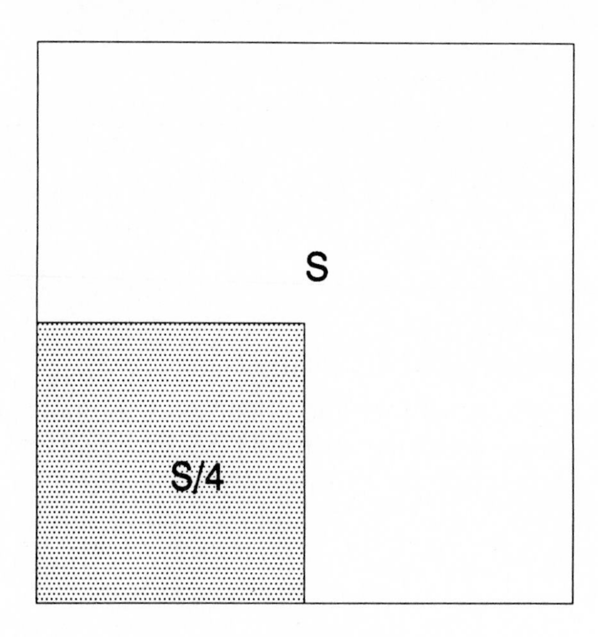

Figure 3-1. The FFT Principle in Layman's Terms

3.3 — THE DECIMATION-IN-TIME AND DECIMATION-IN-FREQUENCY, RADIX-2, FAST FOURIER TRANSFORMS

It is easily shown that equation (15) can be rewritten when N is even as:

$$X_N(k) = \sum_{r=0}^{(N/2)-1} x(2rT)e^{-j\frac{2\pi}{(N/2)}rk} + e^{-j\frac{2\pi}{N}} \sum_{r=0}^{(N/2)-1} X[(2r+1)T]e^{-j\frac{2\pi}{(N/2)}rk} \qquad (16)$$

As illustrated in Figure 3-2, this expression shows how two N/2-point DFTs can be combined to obtain one N-point DFT. If N is an integer power of 2, this process can be repeated, as shown in Figures 3-3 and 3-4, until a simple, two-point DFT is obtained. This gives rise to the flow diagram of a DIT fast Fourier transform (FFT) as shown in Figure 3-5, which represents a complete 8-point FFT computation.

The basic flow diagram of Figure 3-5 can be further simplified by rearranging the terms in the basic building block (the butterfly) as in Figure 3-6. Also, it is seen from Figure 3-5 that input samples no longer occur in normal, sequential order. When the indices are represented in their binary equivalent, however, the input samples appear in "bit-reversed" order. Figure 3-7 shows how the diagram can be rearranged for normally-ordered inputs and bit-reversed outputs.

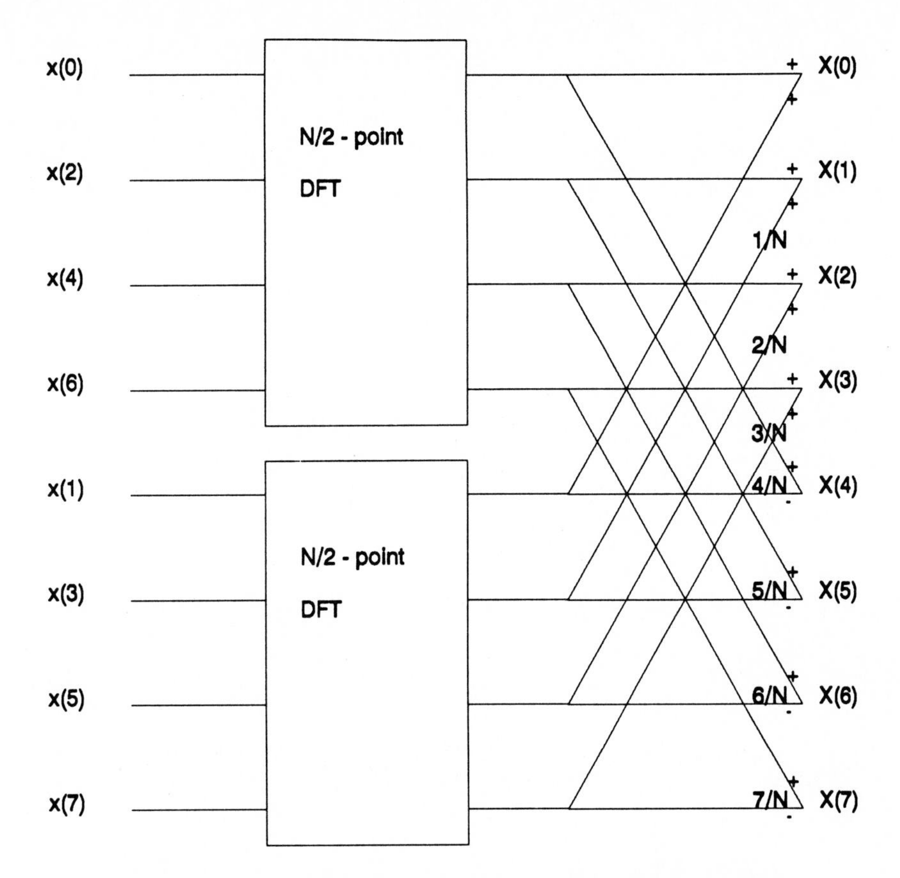

Figure 3-2. Decimation-in-Time of an N-Point FFT

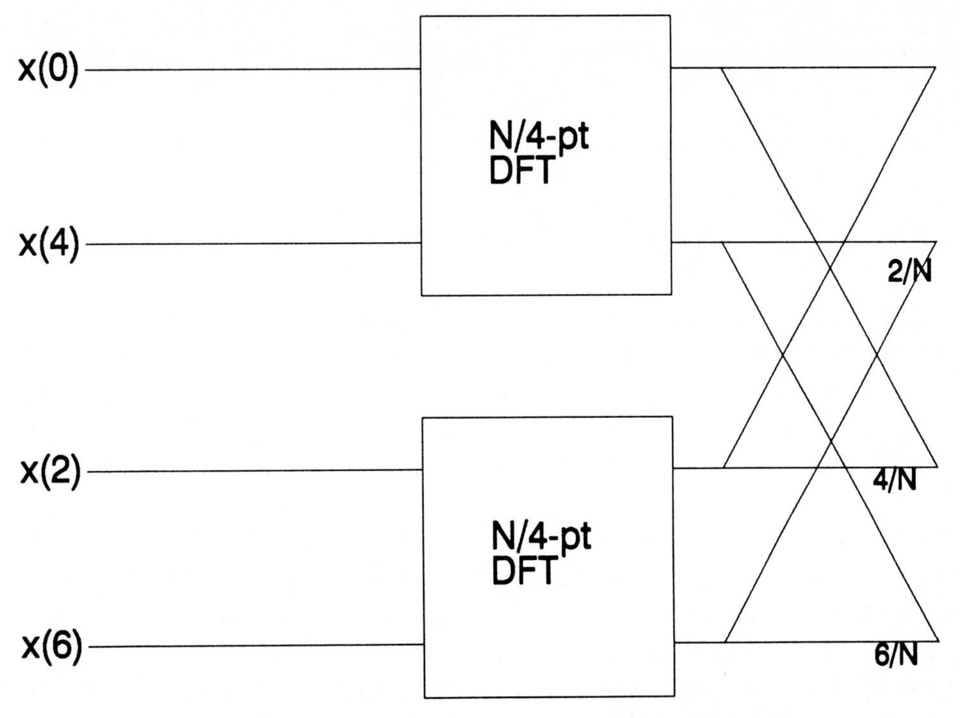

Figure 3-3. Decimation-in-Time FFT: Step Two

k/N denotes multiplication by the "twiddle factors" $e^{-j\frac{2\pi}{N}k}$ throughout this document

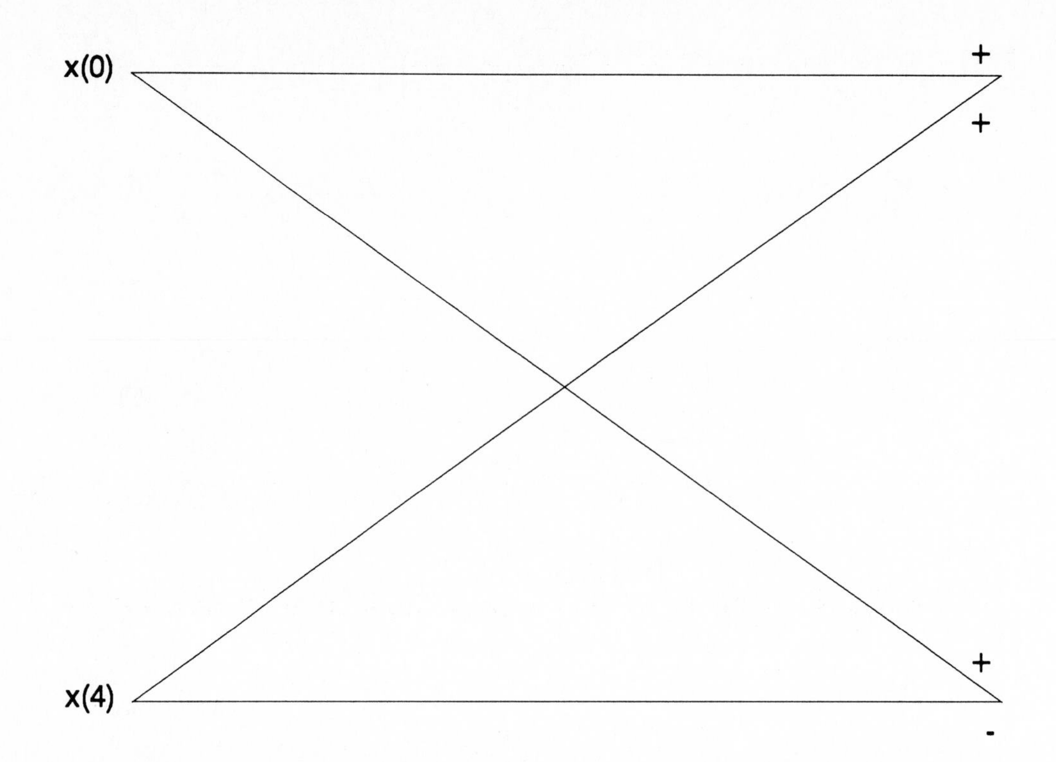

Figure 3-4. Decimation-in-Time FFT: Final Step (2-Point DFT)

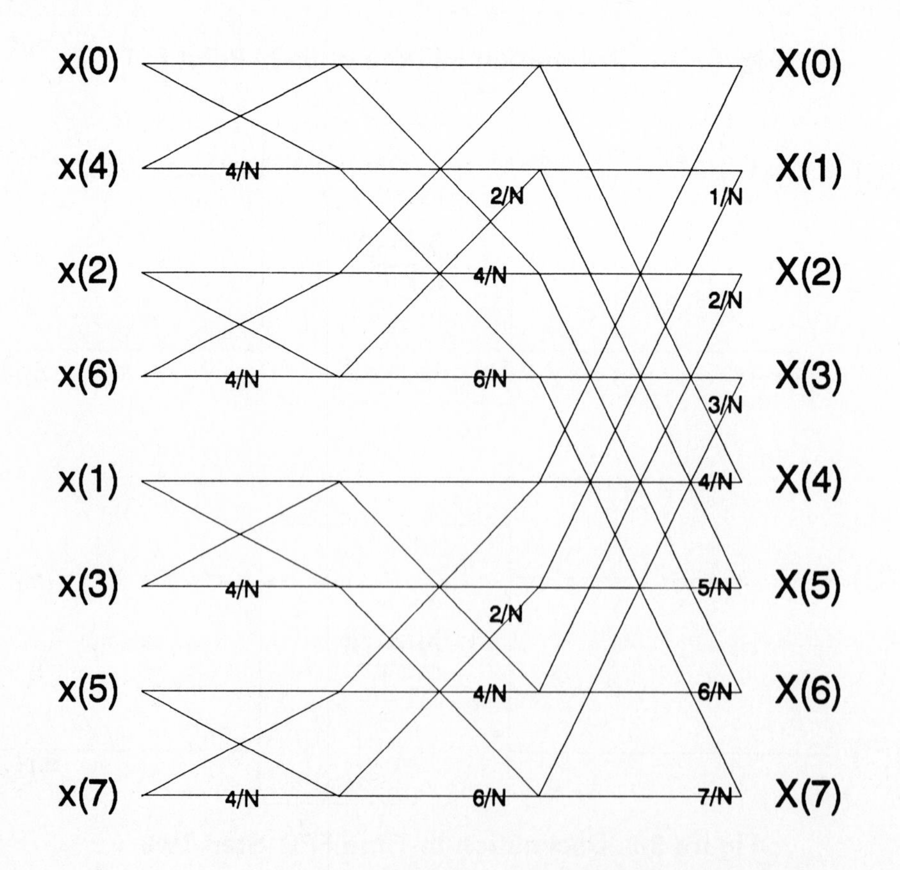

Figure 3-5. An 8-Point, Radix-2, Decimation-in-Time FFT

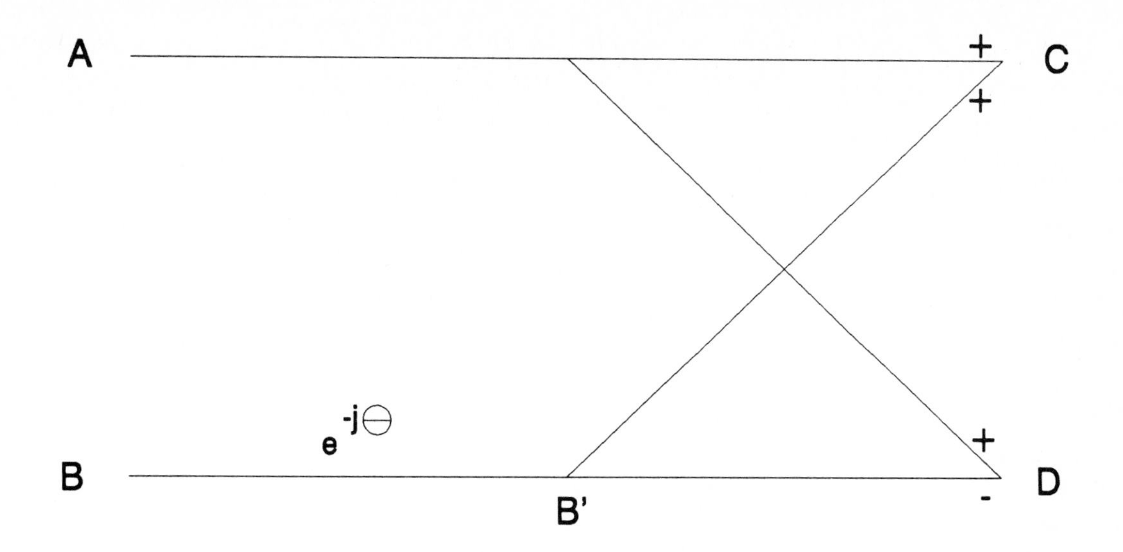

Figure 3-6. Rearrangement of the "Butterfly" Building Block of the DIT FFT

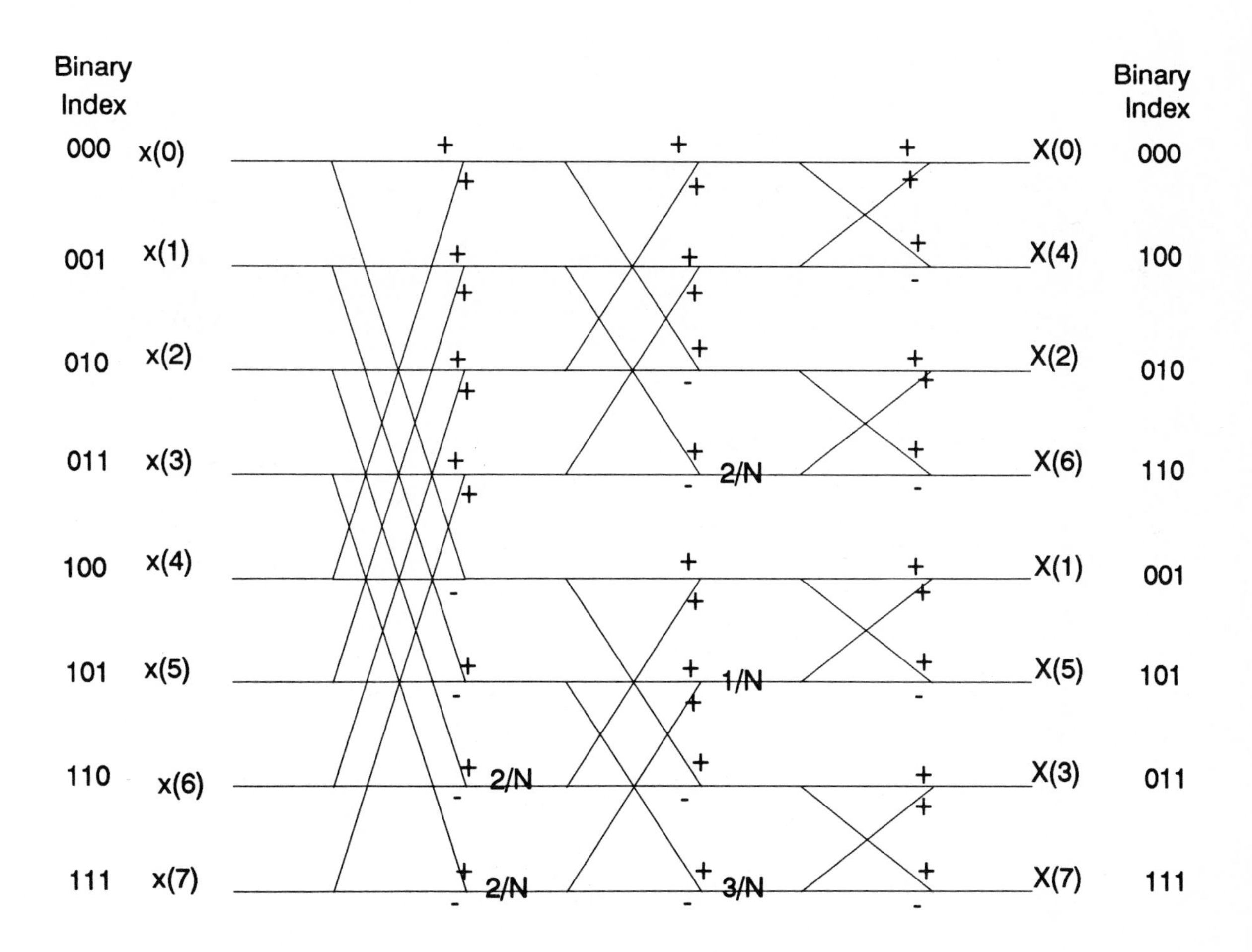

Figure 3-7. Rearrangement of the DIT Computation of Figure 3-6

Figures 3-8 and 3-9 show how the DFT with N frequency points can be obtained in terms of DFTs with a smaller number of frequency samples (decimation-in-frequency FFT). Note that the basic building block (butterfly) is different than for the DIT case (see Figure 3-9).

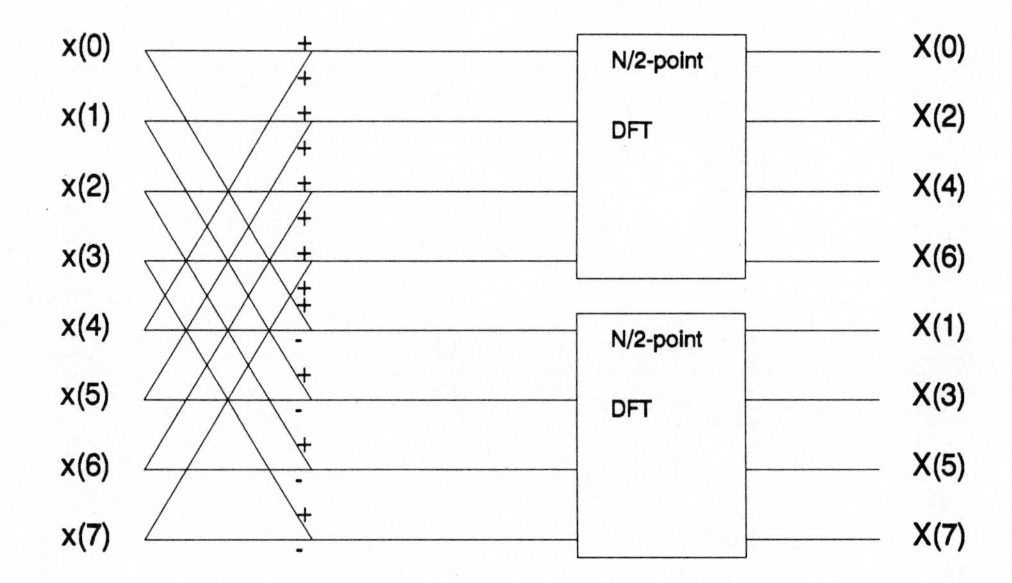

Figure 3-8. Decimation-in-Frequency Concept

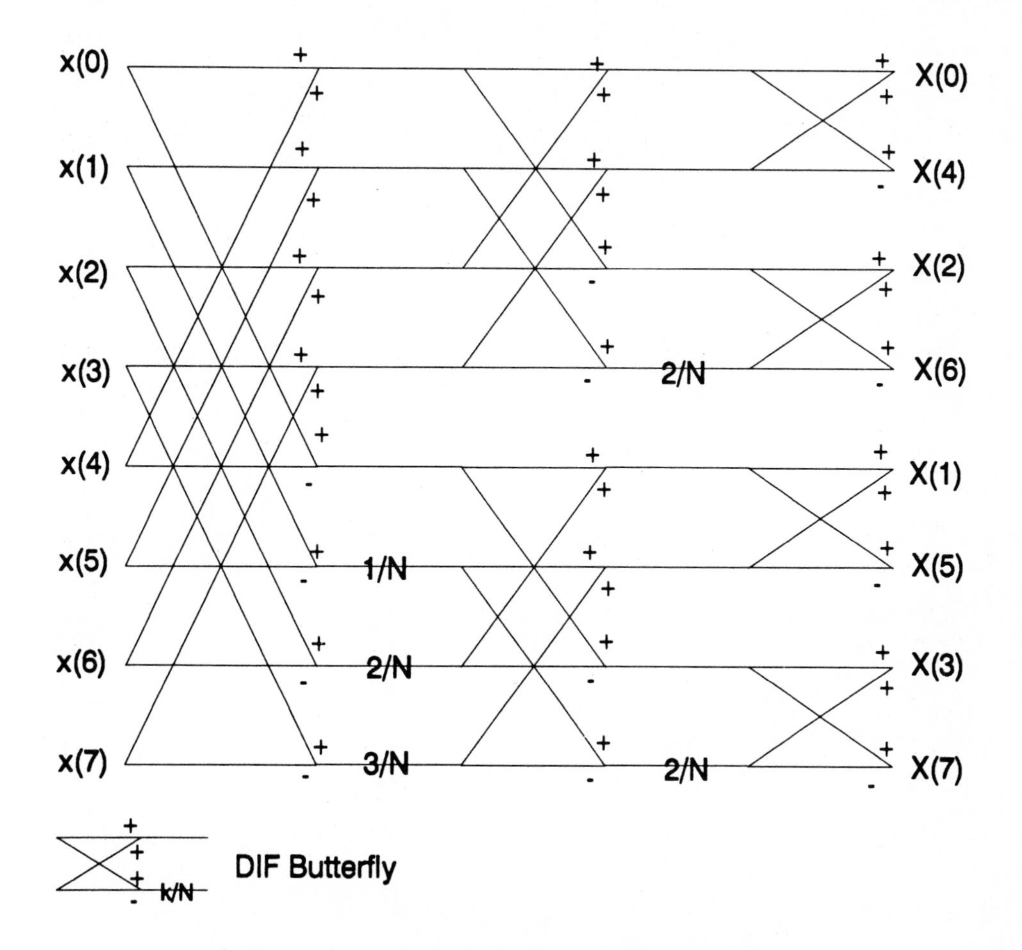

Figure 3-9. Complete 8-Point Radix-2 DIF FFT

SECTION 4 — IMPLEMENTATION OF THE FFT ON THE DSP56000/DSP56001 AND DSP96002

4.1 — REQUIREMENTS FOR IMPLEMENTATION

The basic building block of the DIT[2] FFT routine is the butterfly computation of Figure 3-6. Consequently, the architecture and instruction set of a DSP device should allow efficient computation of this basic butterfly. Since the butterfly consists of additions and multiplications, a hardware adder/subtractor and multiplier are paramount. Since the butterfly data are complex, the architecture must easily support complex arithmetic. The input and output data to the butterflies are moved between the processor's arithmetic unit and memory; consequently, efficient moves are needed, which impact the execution time of the butterfly as little as possible.

The overall DIT FFT algorithm is a collection of many such butterflies, the number of which depends upon the number of points (N) in the FFT. In order to write general FFT routines (for any N), efficient means are required for repetitive execution of the basic butterfly element. Although this could be accomplished in software, a hardware solution is preferred which has minimum impact on execution speed and program length.

In real-life applications, time as well as frequency data is used in normal order, even though the diagram of Figure 3-7 delivers the frequency data in bit-reversed order. Thus, an efficient method for bit-reverse addressing is needed. Again, a time-consuming software solution is to be avoided.

The input data (time samples) of the FFT are in practice obtained from an external source (A/D converter). This data collection must occur in parallel with the FFT computation to make real-time performance possible. Consequently, a DSP device must provide easy interface with a variety of A/D converters, and must support low-overhead interrupt schemes which can load data from an external device with minimal effect on the FFT computation.

[2]The DIT FFT is the main one discussed here, since the DSP56000/DSP56001 instruction set lends itself particularly well for its implementation. DIF FFTs can also be implemented.

4.2 — IMPLEMENTATION ON MOTOROLA'S DSP56000/DSP56001

4.2.1 — Minimum Program Length

The parallel architecture and instruction set of Motorola's DSP56000/DSP56001 (see Reference 4) lends itself particularly well to the radix-2, DIT FFT computation. The architecture is shown in Figure 4-1.

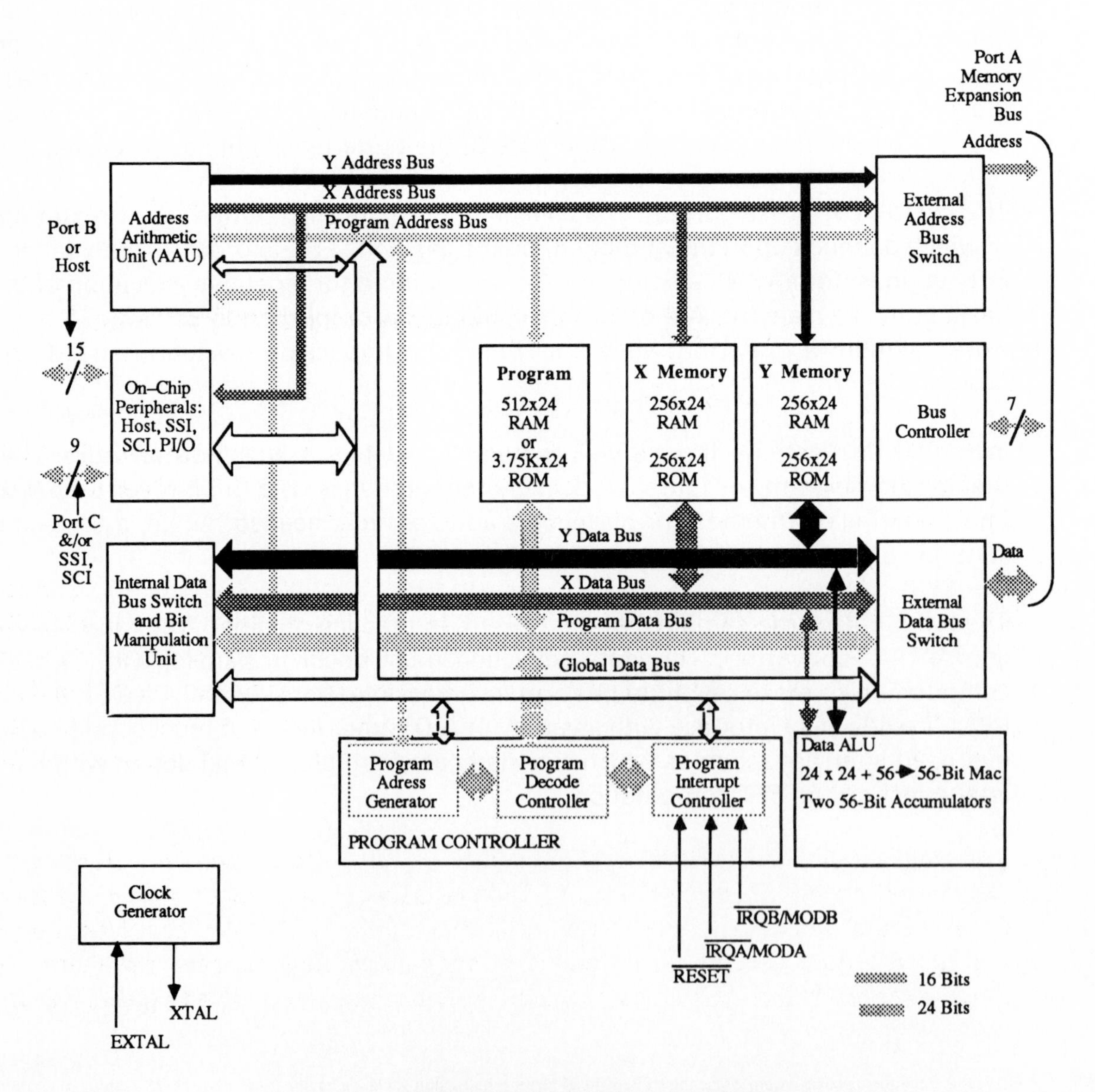

Figure 4-1. DSP56000/DSP56001 Architecture

The DIT butterfly equations are programmed on Motorola's DSP56000/DSP56001 as given below:

$$C_r = A_r - B_r(-\cos(\theta)) - B_i(-\sin(\theta)) \quad (17)$$
$$C_i = A_i - B_i(-\cos(\theta)) + B_r(-\sin(\theta))$$
$$D_r = 2A_r - C_r$$
$$D_i = 2A_i - C_i$$

where the variables refer back to Figure 3-6. The basic butterfly "kernel" is implemented in assembly language in Figure 4-2.

```
;r0 → A
;r1 → B
;r4 → C
;r5 → D

mac     x1,y0,b     y:(r1)+,y1               ;Ai+Br(-sin) → b, Bi → y1
macr    -x0,y1,b    a,x:(r5)+    y:(r0),a    ;Ai+Br(-sin)-Bi(-cos) → b,Ai → a
subl    b,a         x:(r0),b     b,y:(r4)    ;2Ai-b → a, Ar → b
mac     -x1,x0,b    x:(r0)+,a    a,y:(r5)    ;Ar-Br(-cos) → b, Ar → a
macr    -y1,y0,b    x:(r1),x1                ;Ar-Br(-cos)-Bi(-sin) → b,Br → x1
subl    b,a         b,x:(r4)+    y:(r0),b    ;2Ar-b → a, Ai → b
```

Figure 4-2. The Radix-2, DIT Butterfly Kernel on the DSP56000/DSP56001

Note that the previous equations are written in this particular form such that the SUBL instruction can be used. This instruction allows efficient implementation of the DIT butterfly in a two-accumulator ALU.

The kernel of Figure 4-2 executes in six instruction cycles, or a total of 12 clock cycles[3]. This is made possible because of the parallel architecture of the DSP56000/DSP56001, which allows up to two data ALU operations (multiply/accumulate) in parallel with two data moves to/from memory in a single instruction cycle. The dual data spaces X and Y with the appropriate X and Y buses are ideally suited for complex arithmetic: the real parts are stored in X memory, and imaginary parts are stored in Y memory.

The simplest way of combining all of the butterflies into a complete program is shown in Figure 4-3. The FFT diagram (Figure 3-7) is first divided into FFT passes. On each pass, the data is fetched from memory, the butterfly calculations are done, and the results are moved back out to memory. It is easily shown that there are $\log_2 N$ passes. Within each pass, the butterflies are seen to cluster in groups. From

[3]These figures assume that the input and output data points to the butterfly are stored in internal memory.

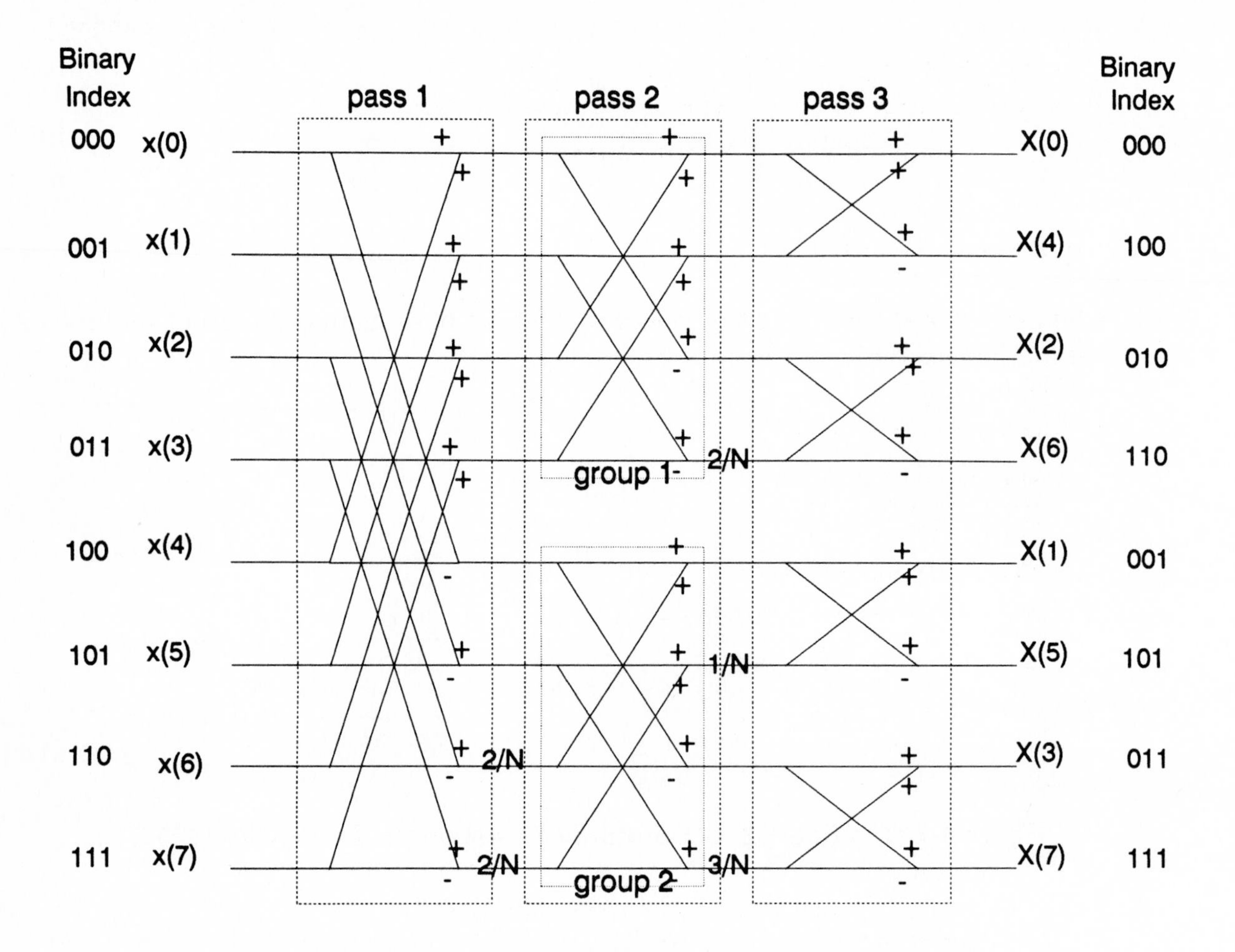

Figure 4-3. Grouping of Butterflies in the FFT Calculation

one pass to the next, the number of groups doubles, while the number of butterflies per group is divided by two. Note that the "twiddle factors" are the same for all butterflies within each group, and that the order of the twiddle factors from one group to the next is bit-reversed. This is easily implemented on the DSP56000/DSP56001 by setting the appropriate modifier register (m6) equal to zero and the offset register (n6) equal to N/4 (=coefficient table size/2), such that the twiddle factors are addressed in bit-reversed manner. This gives rise to the simple, triple-nested DO-loop program of Figure 4-4: the outer DO loop steps through passes, the middle loop goes through all the groups within a pass, and the inner loop cycles through all the butterflies inside a group. The DSP56000/DSP56001 is particularly well suited for looped program execution, as it has hardware DO loop capability. Once a loop is entered via the DO instruction, this loop is executed without any time penalty. The resulting program takes 40 words in program memory. This is the most compact implementation of the radix-2 DIT FFT. A 1024-point complex FFT using this code executes in 4.72 ms when using a 27-MHz clock.

```
;
;This program originally available on the Motorola DSP bulletin board.
;It is provided under a DISCLAIMER OF WARRANTY available from
;Motorola DSP Operation, 6501 Wm. Cannon Drive W., Austin, Tx., 78735.
;
;Radix 2, In-Place, Decimation-In-Time FFT (smallest code size).
;
;Last Update 30 Sept 86     Version 1.1
;
fftr2a   macro    points, data, coef
fftr2a   ident    1,1
;
;Radix 2 Decimation in Time In-Place Fast Fourier Transform Routine
;
;  Complex input and output data
;     Real data in X memory
;     Imaginary data in Y memory
;  Normally ordered input data
;  Bit reversed output data
;     Coefficient lookup table
;      -Cosine values in X memory
;      -Sine values in Y memory
;
;Macro Call — fftr2a      points,data,coef
;
;     points     number of points (2-32768, power of 2)
;     data       start of data buffer
;     coef       start of sine/cosine table
;
;Alters Data ALU Registers
;     x1    x0    y1    y0
;     a2    a1    a0    a
;     b2    b1    b0    b
;
;Alters Address Registers
;     r0    n0    m0
;     r1    n1    m1
;           n2
;
;     r4    n4    m4
;     r5    n6    m5
;     r6    n6    m6
;
;Alters Program Control Registers
;     pc    sr
;
;Uses 6 locations or System Stack
;
;Latest Revision — September 30, 1986
;
;r0 points to A
;r1 points to B
;r4 points to C
;r5 points to D
;r6 points to twiddle factor
        move      #points/2,n0                    ;initialize butterflies per group
        move      #1,n2                           ;initialize groups per pass
        move      #points/4,n6                    ;initialize C pointer offset
        move      #-1,m0                          ;initialize A and B address modifiers
        move      m0,m1                           ;for linear addressing
        move      m0,m4
        move      m0,m5
        move      #0,m6                           ;initialize C address modifier for
                                                  ;reverse carry (bit-reversed) addressing
```

Figure 4-4. A Simple, Triple-Nested DO Loop Radix-2 DIT FFT on DSP56000/DSP56001 (Sheet 1 of 2)

```
;
;Perform all FFT passes with triple nested DO loop
;
        d0      #@cvi(@log(points)/@log(2)+0.5),_end_pass
        move    #data,r0                                        ;initialize A input pointer
        move    r0,r4                                           ;initialize A output pointer
        lua     (r0)+n0,r1                                      ;initialize B input pointer
        move    #coef,r6                                        ;initialize C input pointer
        lua     (r1)-,r5                                        ;initialize B output pointer
        move    n0,n1                                           ;initialize pointer offsets
        move    n0,n4
        move    n0,n5
        d0      n2,_end_grp
        move    x:(r1),x1       y:(r6),y0                       ;lookup -sine and
                                                                ; -cosine values
        move    x:(r5),a        y:(r0),b                        ;preload data
        move    x:(r6)+n6,x0                                    ;update C pointer

        do      n0,_end_bfy
        mac     x1,y0,b         y:(r1)+,y1                      ;Radix 2 DIT
                                                                ;butterfly kernel
        macr    -x0,y1,b        a,x:(r5)+       y:(r0),a
        subl    b,a             x:(r0),b        b,y:(r4)
        mac     -x1,x0,b        x:(r0)+,a       a,y:(r5)
        macr    -y1,y0,b        x:(r1),x1
        subl    b,a             b,x:(r4)+       y:(r0),b
_end_bfy
        move    a,x:(r5)+n5     y:(r1)+n1,y1                    ;update A and B pointers
        move    x:(r0)+n0,x1    y:(r4)+n4,y1
_end_grp
        move    n0,b1
        lsr     b               n2,a1                           ;divide butterflies per group by two
        lsl     a               b1,n0                           ;multiply groups per pass by two
        move    a1,n2
_end_pass
        endm
```

Figure 4-4. A Simple, Triple-Nested DO Loop Radix-2 DIT FFT on DSP56000/DSP56001 (Sheet 2 of 2)

4.2.2 — Optimization for Faster Execution

Although the previously discussed program executes extremely efficiently, some applications may exist which impose less stringent requirements on program size, but which demand even faster execution. Faster execution can be obtained by further optimizing the above algorithm. Several steps can be taken to achieve this optimization as shown below:

1. The twiddle factors for the first stage are all equal to one. Consequently, the multiplications in pass one are trivial, and need not be calculated. The first two passes can be combined into one, and are actually computed as four-point butterflies.

2. The groups in the last pass all consist of one single butterfly each. A triple-nested DO loop is thus no longer required in this pass: it can be "split out" and handled by a single DO loop.

3. For longer FFTs (>256 points), internal memory in the DSP56000/DSP56001 is no longer sufficient to contain the complete data set. Consequently, the butterflies execute more slowly when the processor needs to fetch a data value in external X and in external Y memory in the same instruction cycle. This causes the instruction cycle to be "stretched", resulting in slower execution time. Through intelligent memory usage, however, this effect can be minimized. In a further optimized routine (see Reference 5), the first two passes, combined into one pass as above, are executed. Next, separate 256-point FFTs are computed, whereby the data is moved into internal memory, and the results are not moved to external memory until the final pass. This avoids stretching the instruction cycle on the middle passes, and makes optimal use of the available internal memory.

With these optimizations, a significantly faster routine is obtained. For instance, a 1024-point optimized complex FFT routine is available for DSP56000/DSP56001 which executes in 2.49 ms (see Reference 6) (using a 27-MHz clock). Note, however, that more "straight-line" code always results in longer programs: this routine takes up 105 words in program memory.

4.2.3 — FFTs with Real Inputs

In most practical situations, the data to be analyzed by the FFT is real: it is usually obtained from a single analog-to-digital (A/D) converter.[4] This knowledge can be exploited in several ways to increase the speed of the FFT calculation even further:

1. Since the input data is real, there is no need to multiply, add, or subtract the imaginary parts.

2. Use can be made of symmetries within the FFT:

$$X_N(k) = X^*_N(N-k) \tag{18}$$

when x(nT) is real, * denotes complex conjugate. Clearly, not all of the frequency points need to be calculated, as many of them can be obtained by taking a simple complex conjugate of other, previously computed points. Taking a complex conjugate can be easily achieved by moving the same values to different memory locations, after taking the negative of the value which goes to Y memory (imaginary part). Figure 4-5 shows the procedure for a 16-point,

[4]Complex data is obtained when sampling in-phase and quadrature components of a bandpass signal.

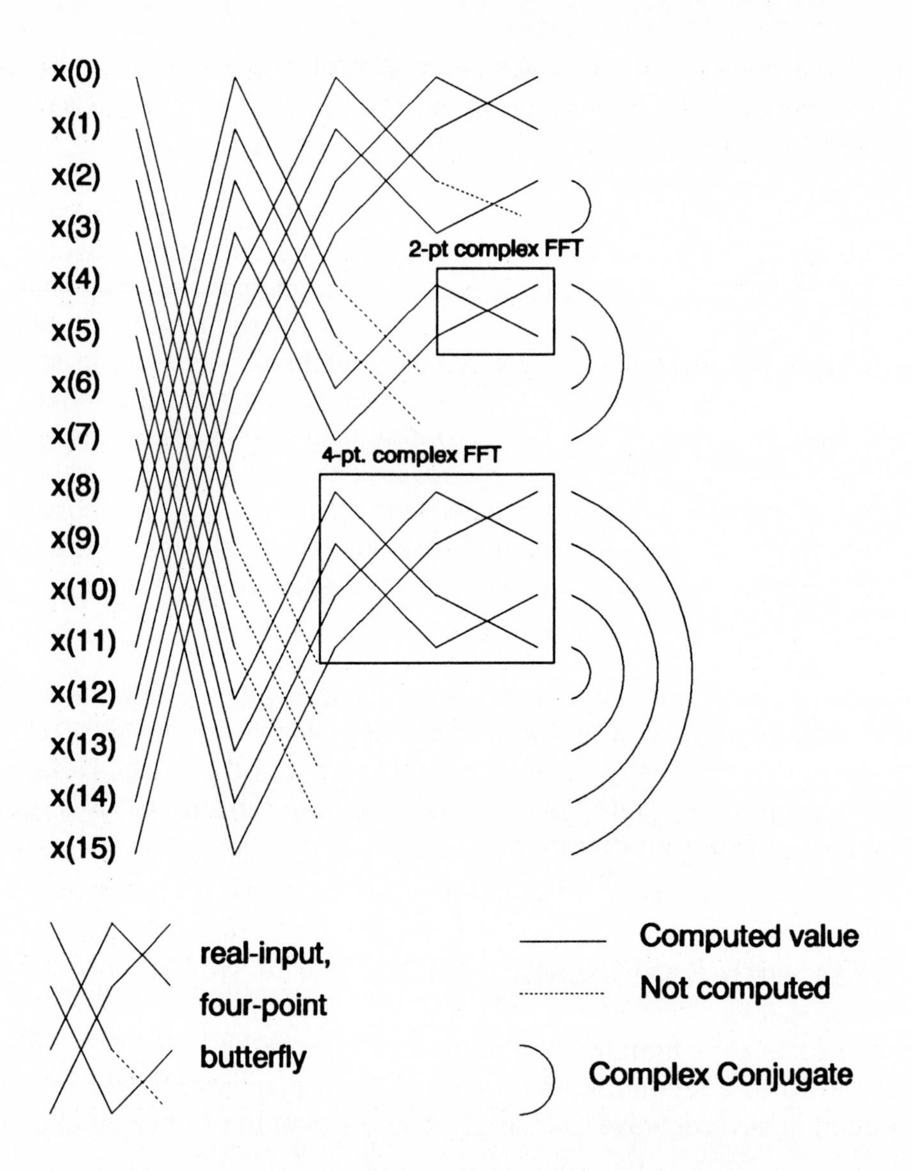

Figure 4-5. Computation of the Real-Input, DIT FFT

real FFT in greater detail. A real-input FFT routine is available for DSP56000/ DSP56001, which executes in 1.49 ms using a 27-MHz clock. This also includes the amount of time necessary to bring in 1024 sampled data points from an external A/D converter. Because of the fast interrupt capability of the DSP56000/ DSP56001, data sampling creates very little overhead. As a result, the maximum sampling rate at which a 1024-point real FFT can be executed equals:

$$f_{smax} = \frac{1024}{(1.49 \cdot 10^{-3})} = 687 \text{ kHz}$$

Comparing this with the sampling rate of 3.3 kHz mentioned in **3.1 MOTIVATION,** a more than 200-fold improvement is obtained by carefully optimizing the Fourier transform algorithm!

4.3 — IMPLEMENTATION ON MOTOROLA'S DSP96002

4.3.1 — Minimum Program Length

The butterfly equations implemented in the radix-2, DIT FFT on DSP96002 are the following:

$$B_r' = B_r\cos\theta + B_i\sin\theta \qquad (19)$$
$$B_i' = B_i\cos\theta - B_r\sin\theta$$
$$D_r = A_r - B_r'$$
$$D_i = A_i - B_i'$$
$$C_r = A_r + B_r'$$
$$C_i = A_i + B_i'$$

where the variables again refer back to Figure 3-6. The implementation of this basic butterfly in DSP96002 assembly is shown in Figure 4-6. This kernel executes in only four instruction cycles, or eight clock cycles.[5] Since four real multiplications are needed, and only one real multiplier is available, this is the most efficient implementation possible. In addition to the features available on the DSP56000/DSP56001 (discussed in **4.2.1 Minimum Program Length**), this efficient execution is obtained by the FADDSUB instruction which delivers the sum and the difference of two operands, in parallel with a multiplication and two data moves. With this feature, a total of three floating-point operations can be executed in one instruction cycle, resulting in a peak performance of 40.5 million floating-point operations per second (MFLOPS) with a 27-MHz clock.

The triple-nested DO loop routine, which computes the radix-2, DIT FFT on the DSP96002 takes only 30 words in program memory. A 1024-point complex FFT is executed in only 2.31 ms, assuming a 27-MHz clock.

[5]Assumming no additional clock cycles are used for treatment of unnormalized numbers, memory wait states, etc.

```
;r0 → A
;r1 → B
;r4 → C
;r5 → D

fmpy   d8,d6,d2   fadd.s      d3,d0   x:(r0),d4.s     d2.s,y:(r5)+    ;Br*sin → d2
                                                                      ;Bi*sin+Br*cos → d0
                                                                      ;Ar → d4,Di → mem.

fmpy   d8,d7,d3   faddsub.s   d4,d0   x:(r1)+,d6.s    d5.s,y:(r4)+    ;Bi*sin → d3
                                                                      ;Ar+Br1 → d0
                                                                      ;Ar-Br1 → d4
                                                                      ;Br → d6
                                                                      ;Ci → mem.

fmpy   d9,d6,d0   fsub.s      d1,d2   d0.s,x:(r4)     y:(r0)+,d5.s    ;Br*cos → d0
                                                                      ;Br*sin-Bi*cos → d2
                                                                      ;Cr → mem.
                                                                      ;Ai → d5

fmpy   d9,d7,d1   faddsub.s   d5,d2   d4.s,x:(r5)     y:(r1),d7.s     ;Bi*cos → d1
                                                                      ;Ai+Bi1 → d2
                                                                      ;Ai-Bi1 → d5
                                                                      ;Dr → mem.
                                                                      ;Bi → d7
```

Figure 4-6. The Radix-2, DIT FFT Butterfly Kernel on the DSP96002

4.3.2 — Optimization for Faster Execution

The techniques employed to optimize execution speed of DSP56000/DSP56001 FFTs discussed in **4.2.2 Optimization for Faster Execution** can be invariably applied to the DSP96002 case. Note that the dual external buses available on the DSP96002 avoid the "stretching" of instruction cycles when a parallel move to/from external X and Y space is attempted in one instruction (assuming zero wait state external memory): external X memory and external Y memory can be mapped to different external buses, and FFTs of virtually unlimited length can be executed without any penalty in execution time.

In addition to the methods described in **4.2.2 Optimization for Faster Execution**, the large number of internal registers in the DSP96002's data ALU allow the two butterflies per group in the next to the last pass to be combined into one, four-point butterfly. A routine is available on the DSP96002 which executes a 1024-point complex FFT in 1.55 ms with a 27-MHz clock. This routine takes 137 words in program memory.

4.3.3 — FFTs with Real Inputs

The ideas explained in **4.2.3 FFTs with Real Inputs** can also be applied to the DSP96002. In addition to the fast interrupts available on its fixed-point counterpart, the DSP96002 has a two-channel DMA controller, which operates unobtrusively in parallel with the ALU. Consequently, the DSP96002 can collect a block of data from an external location, such as another processor or an A/D converter, while the FFT is being computed and without adding any execution time to this FFT. An FFT for real inputs is available on the DSP96002, which runs in 905 microseconds with a 27-MHz clock, and takes up a total of 317 words in program memory. This limits the sampling rate for real-time performance to a maximum of 1.13 MHz. Table 4-1 gives a summary of execution speeds and program memory requirements for the different routines discussed.

Table 4-1. Required Program Memory and Execution Times for Several FFTs

1024-Point FFT	DSP56000/DSP56001		DSP96002	
	Program Size	Execution Time	Program Size	Execution Time
Triple-Nested DO Loop	40 Words	4.72 ms	30 Words	2.31 ms
Minimum Execution Time	105 Words	2.49 ms	137 Words	1.55 ms
Real-Input FFT	254 Words	1.49 ms	317 Words	0.905 ms

SECTION 5 — FIXED POINT, BLOCK FLOATING POINT, AND IEEE FLOATING POINT

Whenever mathematical algorithms are implemented in digital hardware, one must realize that results are only obtained with finite precision. The precision is generally limited by the number of bits used in the number representation, and depends on how the arithmetic limits its results to these bits (truncation, rounding).

When analyzing the effects of finite-precision arithmetic relative to FFT results, it is important to understand how the magnitude of the complex FFT data changes throughout the FFT calculation. The easiest way to characterize behavior of numbers in FFTs is achieved using vector notation. Figure 5-1 shows how the two complex numbers (vectors) at the input of the DIT butterfly of Figure 3-6 are combined to give the two outputs.

First, vector B is multiplied by the twiddle factor. Since all twiddle factors in the FFT have the form $e^{-j\theta}$, which has unit magnitude, the magnitude of B′ is the same as that of B. As shown in Figure 5-1, the multiplication by the twiddle factor can be

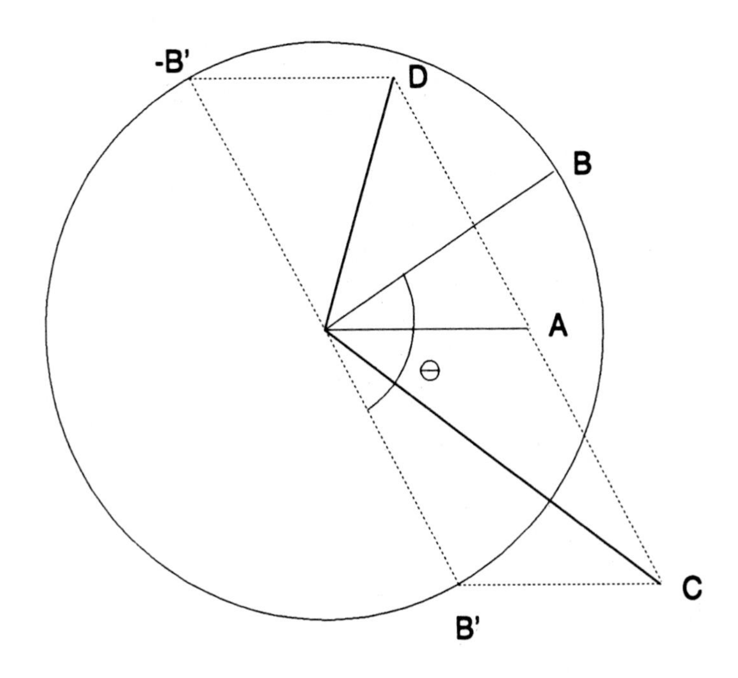

Figure 5-1. Vector Representation of the DIT Butterfly

interpreted as a simple rotation of the vector B over an angle θ. The vectors B′ and A are next added and subtracted to give the butterfly outputs C and D, respectively.

Limits on the output vectors' magnitudes are derived in Figure 5-2. Obviously, the largest magnitude is achieved if the vectors B′ and A line up, such that the total magnitude of either C (B′ and A point in the same direction), or D (B′ and A point in opposite directions) is the sum of the magnitudes of A and B′. It is also clear from Figure 5-2 that either the magnitude of C or the magnitude of D is at least equal to the magnitude of the larger of the two vectors A and B. This leads to the relationships:

$$\max(|A|,|B|) \leq \max(|C|,|D|) \leq 2\max(|A|,|B|) \quad (20)$$

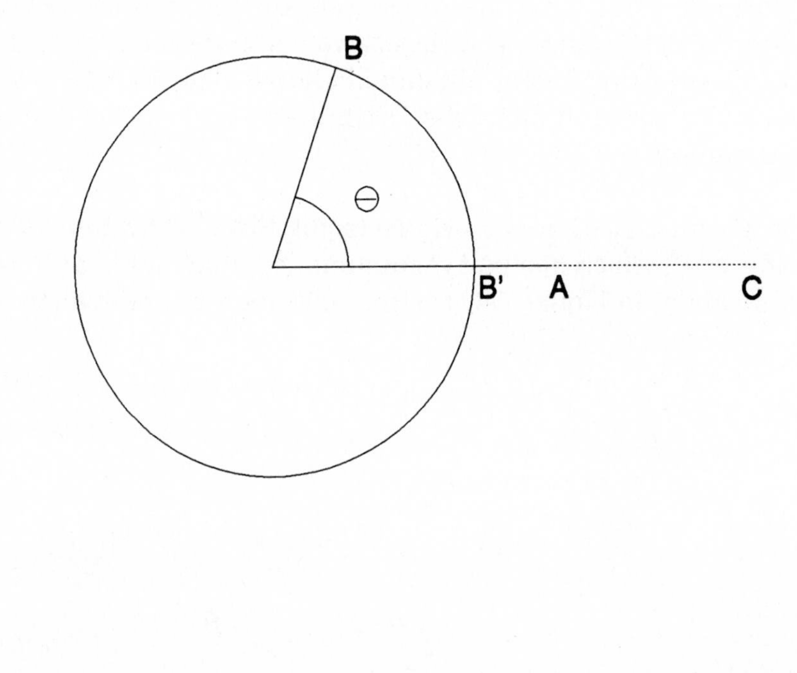

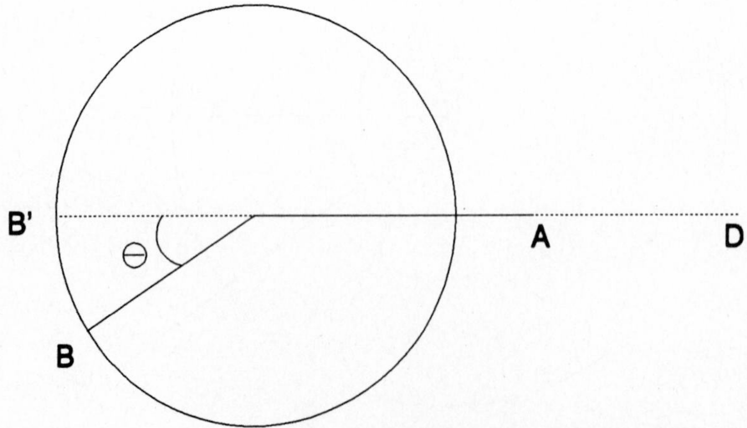

Figure 5-2. Bounds on the Butterfly's Output Magnitude

It is apparent from the previous discussion, that the complex numbers at the output of the butterflies "grow" in magnitude from stage to stage. The maximum growth of the magnitude, as shown by expression (20), is a factor of two per stage. In digital terms, one can conclude that the maximum *word growth* of the magnitude in the FFT that can occur is one bit per pass.

5.1 — FIXED-POINT IMPLEMENTATION

5.1.1 — Input Bounds

The DSP56000/DSP56001 Family implements 24-bit, fixed-point, fractional arithmetic (see Reference 7). This means that in practice, 24-bit numbers $\varkappa$ stored in memory are limited as $-1.0 \leq \varkappa < +1.0$. Numbers α in the 56-bit accumulators are limited by $-256.0 \leq \alpha < +256.0$. Since complex numbers are moved out to memory at the output of each butterfly in the FFT computation, the real and imaginary parts must be kept less than one (in absolute value) at all times.

The previous discussion indicates that the magnitude of the complex numbers can grow by a total factor of N, or $\log_2 N$ bits, in an N-point FFT. Note that, because of the twiddle factor "rotation", real and imaginary parts of complex numbers can at any time equal the magnitude (i.e., when they become either purely real or purely imaginary). Thus, the initial magnitude of the FFT input data needs to be limited by:

$$M < \frac{1.0}{N} \tag{21}$$

Examples:

1. For a real-input, 1024-point FFT, the absolute value of the input data equals the magnitude, and therefore needs to satisfy:

   $$|\, x(nT) \,| < \frac{1}{1024} \tag{22}$$

   This bound on the input is achieved by placing the input data in the lower 14 bits of the 24-bit word. As such, 10 "guard-bits" are kept to accomodate the worst-case word growth.

2. For a complex-input, 1024-point FFT, the absolute value of the input data's magnitude needs to be limited as in equation (21). Keeping 10 guard-bits is no longer sufficient, since now the imaginary part contributes to the magnitude

as well. The simplest way to place a bound on the data such that equation (21) is satisfied, is by forcing the inputs to satisfy:

$$-\frac{0.5}{1024} \leq R\{x(nT)\} \leq \frac{0.5}{1024} \tag{23}$$

$$-\frac{0.5}{1024} \leq I\{x(nT)\} \leq \frac{0.5}{1024}$$

This bound forces the real and imaginary parts to the lower 13 bits of the input word, and results in an input magnitude bound of $(\sqrt{2}/2)/1024$, which definitely satisfies equation (21).

The previous examples indicate that the following general rules must be obeyed for an N-point FFT:

$$|x(nT)| < \frac{1}{N} \tag{24}$$

for real-input FFTs, and:

$$|R\{x(nT)\}| \leq \frac{1}{2N} \tag{25}$$

$$|I\{x(nT)\}| \leq \frac{1}{2N}$$

for complex-input FFTs. This can be directly translated into the number of guard-bits needed. For a real-input FFT, a total of

$$G = \log_2 N \tag{26}$$

guard-bits are needed, while for a complex-input:

$$G = \log_2 N + 1 \tag{27}$$

The number of guard-bits, in turn, determines the maximum number of bits in the A/D converter used to digitize the input signal. Clearly, for a fixed-point DSP with P-bit memory storage, the maximum number of A/D bits (A) equals

$$A_{max} = P - \log_2 N \tag{28}$$

for real-input FFTs, and

$$A_{max} = P - \log_2 N - 1 \tag{29}$$

Example:

For a 24-bit processor, like the DSP56000/DSP56001, the theoretical maximum A/D size for a 1024-point real-input FFT is 14 bits. For a 16-bit processor, this is reduced to six bits.

In general, the bounds above are quite strict and are affected by windowing parameters, the type of signal that is analyzed (harmonic signal versus broadband signal), etc. For instance, it is found that a 1024-point FFT of a Blackman-Harris windowed sine wave only requires eight guard-bits. The reason for this apparent contradiction with equation (26) lies in the fact that the windowing operation "smooths" out the peak energy (as discussed previously in **2.2 WINDOWING AND WINDOWING EFFECTS**). Consequently, when using a Blackman-Harris window, a 16-bit A/D converter can be used with the DSP56000/DSP56001, where the A/D bits are placed in the 16 least significant bits of the 24-bit word.

5.1.2 — Roundoff Errors

When mathematical algorithms are implemented using finite-precision arithmetic, roundoff errors or truncation errors occur throughout the computations. These errors become especially important if the number of computations is large, and if the results are not appropriately scaled relative to the errors. This may happen in fixed-point implementations.

Roundoff errors are caused when the results of multiplications are "reduced" to the number of bits used in the processor's data storage. The mantissa size of the result of a multiplication is twice the mantissa size of the operands when the multiplier is implemented with "infinite precision", i.e., all bits of the multiplication result are computed. In the DSP56000/DSP56001 Family, other arithmetic operations (add and subtract) can use a total of 56 bits, and thus accomodate the complete multiplication result. Arithmetic results are "rounded" (using convergent rounding or "round to nearest" (see Reference 4)) only before moving these results out to 24-bit memory locations. Thus, a total error of at most plus or minus one-half of one least significant bit (LSB) of the 24-bit word occurs. In the next FFT pass, results are moved from memory to the arithmetic unit, new calculations are performed, the new results are rounded again and moved back to memory. The successive rounding errors tend to accumulate and appear as roundoff noise in the final results. Because of the accumulative effect, not only the LSB becomes unreliable, but other bits as well. Obviously, one needs to keep as many reliable bits as possible in the final result, and thus the rounding error per operation needs to be kept as small as possible. This is achieved by making the number of bits in the number representation as large as possible, and by making sure the rounding errors do not add coherently. The 24-bit number representation in DSP56000/DSP56001 has proved sufficient for most applications requiring high-precision results. In addition, the convergent rounding scheme used in the DSP56000/DSP56001 assures that rounding errors occur in a truly random fashion, and do not add coherently (hence the name "convergent" rounding: the mean rounding error converges to zero over a large number of operations).

Expressions for the signal-to-roundoff noise ratio (SRNR) are derived in **APPENDIX A**. The SRNR on a digital signal processor (DSP) with infinite-precision internal arithmetic and convergent rounding, such as the DSP56000/DSP56001, is shown to be:

$$SRNR_{dB} = 10\ \log_{10}[3 \cdot 2^{2P-1}] - 10\ \log_{10}(N-1) \tag{30}$$

The following example illustrates how the SRNR shows up in the FFT results.

Example:

For a 1024-point FFT on the DSP56000/DSP56001, $P = 24$ and $N = 1024$. This results in a theoretical SRNR of 116 dB. In Figure 5-3, two Blackman-Harris windowed sine waves of the form:

$$x(nT) = \{\sin(2\pi \cdot 0.1255 \cdot n) + j \cdot 2^{-15} \cdot \sin(2\pi \cdot 0.02505 \cdot n)\} \cdot w(n) \cdot K \tag{31}$$

are digitized using a 16-bit A/D converter, and the magnitude of the FFT is plotted. w(n) denotes the Blackman-Harris window (see Reference 2). The scale factor K is taken such that the output peak magnitude is as close to one as possible ($K \approx 2^{-8}$). The resulting noise floor around −120 dB is clearly visible.

It is important to realize that the signal-to-quantization noise ratio (SQNR) and the SRNR are separate, independent quantities. The quantization noise is due to the finite-precision representation of the input signal, and does not change because of the FFT calculations: it can be adequately represented as (white) noise added to the ideal input signal. The roundoff noise, on the other hand, is independent of how the input signal is quantized, and increases with every butterfly calculation. If the DSP makes full use of the available number of bits by using the maximum-sized A/D converter (LSB of the A/D = LSB of the processor), this LSB is soon affected by the growing roundoff error, and the roundoff error is dominant in the final results. This is illustrated in Figure 5-4.

The quantization noise and the roundoff noise error affect the same bits in the FFT results (become of the same order of magnitude) when (as is derived in **APPENDIX A**, equation A-6):

$$A = P - 0.5 \cdot \log_2(N-1) - G \tag{32}$$

for complex-input FFTs, and (equation A-8)

$$A = P - 0.5 - 0.5 \cdot \log_2(N-1) - G \tag{33}$$

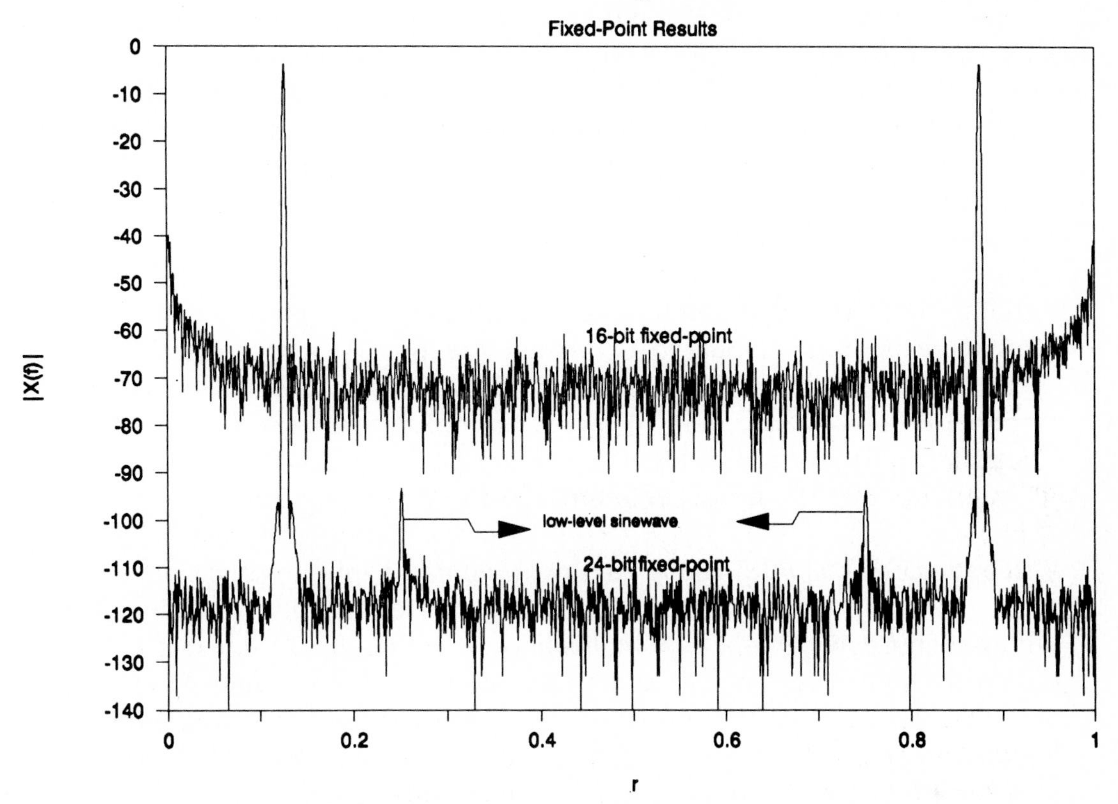

Figure 5-3. Fixed-Point FFT Results: 16-Bit Truncation Arithmetic versus 24-Bit DSP56000/DSP56001 ($r = f/f_s$)

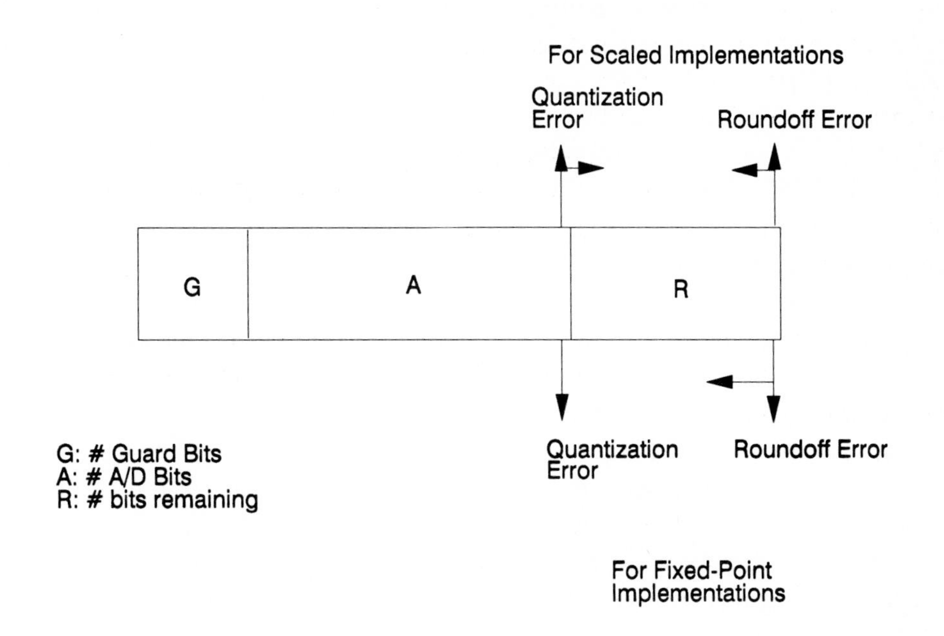

Figure 5-4. Quantization Noise versus Roundoff Noise

for real-input FFTs. A is the number of bits in the A/D converter, and G is the number of guard bits used to avoid overflow or limiting of the FFT output. This A/D size may well be called "optimal": a larger A/D precision does not result in more precise results, as the roundoff noise becomes dominant. A smaller A/D size does not offer the maximum precision obtainable with the processor, since the quantization noise dominates.

Example:

The "cutoff" point at which the roundoff noise and the quantization noise become of the same order of magnitude, assuming complex-input FFTs as in the previous example, is roughly equal to 11, using equation (31). Consequently, a 1024-point FFT which uses more than 11-bit A/D converter inputs, is limited in precision by roundoff noise. Fewer than 11 bits at the input result in more quantization than roundoff noise. Alternatively, if P=16 (16-bit DSP), a 1024-point complex FFT is limited by roundoff noise up to an A/D size of only A=3.

While the DSP56000/DSP56001 Family implements infinite-precision internal arithmetic, with one final convergent rounding operation, many existing DSPs use a different approach. Often, a multiplication result is simply truncated to the number of bits P used in storage of variables in memory. This results in introduction of errors after every real multiplication in the FFT flow diagram. Clearly, the noise sources are located in different locations than is the case for a DSP56000/DSP56001 type machine. The signal-to-truncation noise ratio (STNR) is derived in **APPENDIX A** (equation A-12).

$$STNR_{dB} = 10\ \log_{10}[3 \cdot 2^{2P-2}] - 10\ \log_{10}(N-1) \tag{34}$$

In addition, it is shown in **APPENDIX A** (equation A-9) that truncation arithmetic results in a bias in the FFT magnitude, which is prevalent at DC.

Example:

The signal of equation (31) is digitized and the 1024-point FFT computed on a 16-bit DSP, which implements truncation arithmetic after every multiplication. The same number of guard-bits needs to be kept as in the 24-bit implementation to avoid overflow problems (G=8). According to equation (28), the maximum number of A/D bits A_{max} is consequently equal to eight. Figure 5-3 shows the resulting FFT magnitude. Note the bias at DC, which is predicted in **APPENDIX A** (equation A-7) as a result of the truncation arithmetic. The STNR, given in equation (34) for N=1024 is predicted to be 65 dB. The actual STNR obtained in Figure 5-3 is seen to be roughly equal to 70 dB. This discrepancy is due to the fact that in the calculations of **APPENDIX A**, it is assumed that every multiplication contributes to truncation noise. In practice, trivial multiplications (such as in pass one) do not contribute to the noise, and the STNR is actually larger. Note that the smaller harmonic component, which was clearly visible in the 16-bit input, DSP56000/DSP56001 computation, is now completely lost in the truncation noise.

5.2 — AUTOMATIC SCALING

5.2.1 — Input Bounds

The main problem in the previous fixed-point implementation lies in the requirement of a number of guard-bits to avoid overflow or limiting when the magnitude of the butterfly results grows every stage. This growth can be avoided, however, by automatically scaling down the outputs of all FFT butterflies by a factor of 0.5. Since this scaling is uniform throughout the FFT (all points are scaled equally), it can be easily taken into account by keeping in mind that a common scale factor of $2^{\log_2 N} = N$ needs to be applied to the FFT results if absolute results are required. This automatic scaling mode is implemented on the DSP56000/DSP56001 in hardware, with no penalty in execution speed.[6]

Since the real and imaginary outputs, and thus the magnitudes, are automatically divided by two at every stage, the bounds on the magnitude at the butterfly outputs are directly obtained from equation (20) as:

$$\frac{\max(|A|,|B|)}{2} \leq \max(|C|,|D|) \leq \max(|A|,|B|) \qquad (35)$$

Thus, with a two's complement, fixed-point notation, it is now sufficient to make sure that the magnitude of the butterfly inputs is less than one at all times. Since there is no more growth of the magnitude of butterfly results, it is sufficient to require that the input magnitude is less than one.

For real-input FFTs, the magnitude of the input is equal to the absolute value of the input signal, and thus:

$$|x(nT)| < 1.0 \qquad (36)$$

Thus, no guard-bits are needed, and the maximum number of bits of the A/D converter A_{max} becomes equal to the number of bits used in the DSP's memory storage, i.e.:

$$A_{max} = P \qquad (37)$$

For complex-input FFTs, the previous limit is no longer sufficient, since both real and imaginary parts affect the magnitude. However, by keeping one guard bit:

$$\begin{aligned} |R\{x(nT)\}| &\leq 0.5 \\ |I\{x(nT)\}| &\leq 0.5 \end{aligned} \qquad (38)$$

[6]The user can set the scaling bits S0 and S1 in the status register prior to executing the FFT.

the magnitude of the input is limited to $\sqrt{2}/2 \approx 0.707$, and is definitely less than one. In this case, the maximum number of bits that can be used for digitizing the input signal equals:

$$A_{max} = P - 1 \tag{39}$$

In addition to allowing a larger number of A/D bits at the input, the main advantage of the automatic scaling mode lies in the fact that FFTs of virtually unlimited size can be computed: in the fixed-point (unscaled) version, the maximum number of points in the FFT is obviously limited by the required A/D bits and the number of bits P used for number storage.

5.2.2 — Roundoff Errors

The roundoff errors which occur in the FFT butterfly calculation using automatic scaling are analyzed in **APPENDIX B**. The total SRNR in decibels is given by:

$$SRNR_{dB} = 10 \log_{10}[3 \cdot 2^{2P-1}] - 10 \log_{10}[\log_2 N] \tag{40}$$

The SRNR in this case is seen to grow more slowly ($\sim \log_2 N$) than the SRNR for the fixed-point case ($\sim N$) . As is shown in **APPENDIX B**, this is due to the fact that the roundoff errors in early FFT passes are attenuated throughout the calculation because of the division by two at every successive stage. If the maximum-length A/D converter is used, the roundoff noise again exceeds the quantization noise. The cutoff point at which the quantization and roundoff errors are of equal order of magnitude is derived in **APPENDIX B** to be:

$$A = P - 1 - \log_2 N - 0.5 \cdot \log_2(\log_2 N) \tag{41}$$

for complex-input FFTs and

$$A = P - 0.5 - \log_2 N - 0.5 \cdot \log_2(\log_2 N) \tag{42}$$

for real-input FFTs.

Example:

For the complex-input, 1024-point FFT of the previous examples, equation (40) suggests a total SRNR of approximately 145 dB. This is right at the limit of the 24-bit representation. Using this technique, a theoretical input A/D size of 23 bits is possible. The roundoff noise in this case exceeds the quantization noise in the final results. Figure 5-5 shows the magnitude of the complex FFT of the same signal as in the previous examples, but with automatic scaling turned on. The cutoff point at which the quantization noise becomes of the same order of magnitude as the roundoff noise is calculated to occur for an input A/D size of approximately 11 bits.

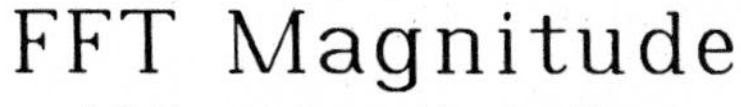

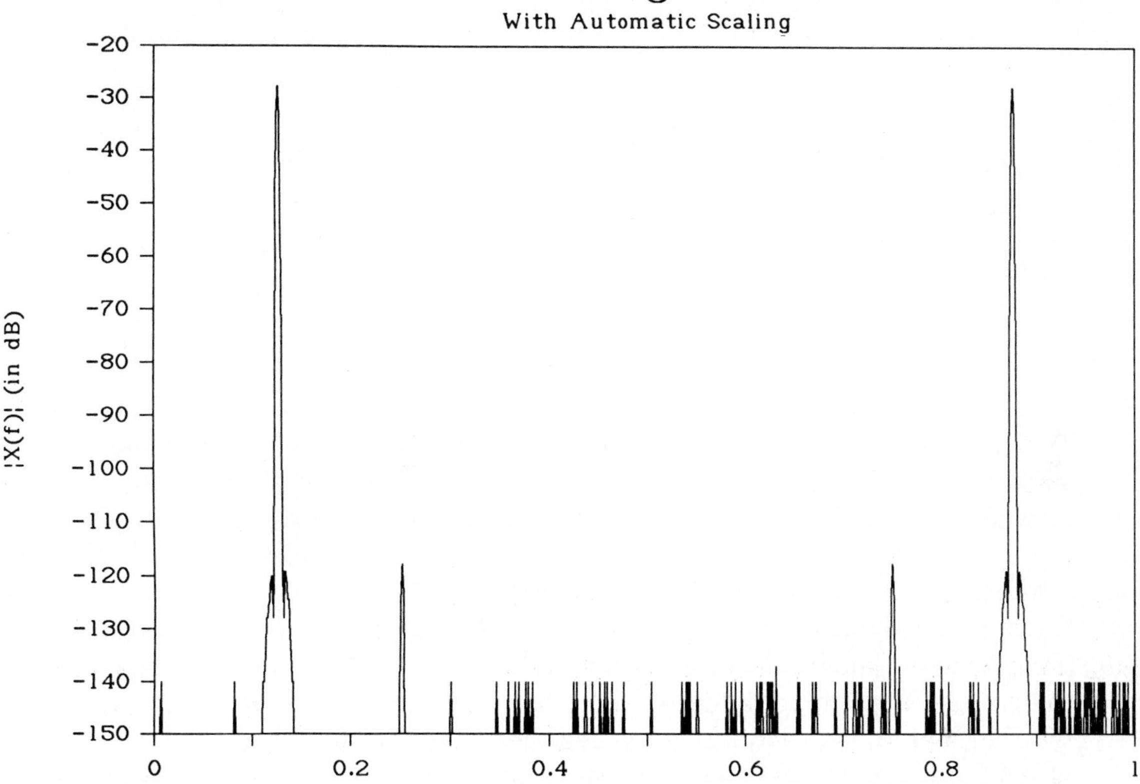

Figure 5-5. Magnitude of the FFT of the Same Signal as in Figure 5-3, with Automatic Scaling Turned On

Note from Figure 5-5 that the noise appears very irregular. The reason for this is that the roundoff error is so small (because of indiscriminate down shifts, − 145 dB) that it only sporadically affects the least significant bit of the result. The resulting FFT has mainly zero values except for at the signal peak. The low-level side-lobes of the Blackman-Harris window are not at all visible, as they "fell off the edge" due to the down shifts. Also note that the unconditional down shifts in the FFT result in a peak magnitude which is less than one. This means that full 1-bit word growth of the magnitude did not occur in every stage. Consequently, the SRNR is in practice worse than predicted: an effective SRNR roughly equal to the fixed-point case is obtained.

Although automatic scaling does offer the ability of implementing FFTs of virtually unlimited size, it is clear from the previous example that a better SRNR is not necessarily obtained in practice. Also, the A/D size at which quantization noise "overtakes" roundoff noise (the "optimal" A/D precision) is roughly equal to that for fixed-point results, unless the number of points N is very large. Consequently, for FFTs of sizes 1024 and less, automatic scaling does not offer any main advantages. However, the advantages do become more pronounced for larger FFTs. For example, when N=4096, the optimal A/D size for a fixed-point realization is seven, whereas the size for an automatically scaled version is nine.

5.3 — BLOCK FLOATING POINT

5.3.1 — General

Although it becomes clear from the previous discussions that scaling each FFT butterfly improves the maximum SRNR in some cases, this improvement is often not fully used in practical applications. For instance, as is seen in equation (35), automatic scaling also decreases the lower bound on the FFT butterfly output magnitude. Although all of the discussions above are based on a scaled input such that the output peak magnitude is equal to one, practical situations arise where the input is not *a priori* known. For signals without pronounced harmonic components, the maximum output magnitude will be closer to its lower bound. Hence, there is no word growth in the magnitude for unscaled, fixed-point FFTs, and the peak output magnitude is much less than one. As a result, the previously derived expressions are no longer correct, and the SRNR and SQNR are actually found to be much worse: the same noise components now need to be scaled according to the actual peak signal magnitude. In the case of automatic scaling, the peak magnitude of the output may even be much smaller than the maximum magnitude of the input. As becomes evident from Figure 5-5, even with pronounced harmonic components the peak magnitude is considerably smaller than 1 (0 dB). Consequently, the SQNR and SRNR for FFTs of signals without pronounced harmonic components is much worse for the automatically scaled FFT than for the fixed-point FFT. For signals with a combined strong harmonic input as well as an underlying broad-band signal, the harmonic peak is generally reliably computed, while the broadband component is not retained: it "falls off the edge" (is shifted below the least significant bit) of the P-bit representation.

Clearly, some decision mechanism is needed which selectively scales the output of butterflies when needed, and doesn't scale the outputs when there is no appropriate growth in the magnitude of the results.

The simplest way of accomplishing conditional scaling is by a technique termed *block floating point*. With this technique, complete passes of the FFT calculation are selectively scaled, depending on whether sufficient growth in magnitude occurred in the previous pass. This method is especially of interest on fixed-point machines with a hardware scaling mode, such as the DSP56000/DSP56001. When implemented correctly, minimal additional hardware and software is required for implementing block floating-point algorithms with a negligible penalty in speed performance.

5.3.2 — Implementation of Block Floating Point on DSP56000/DSP56001[7]

On the DSP56000/DSP56001 Family, block floating point is easily implemented by using the "scaling bit" S in the status register (bit 7). This bit is set upon moving a result to memory which is larger than 0.25 in absolute value. Upon completion of each FFT pass, this bit is tested. If S is set, the automatic scaling mode is turned on[8] before the next FFT pass, indicating that in at least one butterfly the magnitude has increased. If S is not set, no growth has occurred and the scale-down mode is turned off; butterflies on the next pass are computed as in **5.1 FIXED-POINT IMPLEMENTATON**. Testing S also resets it, such that it is ready for the next pass. Since this testing of S only occurs once for every FFT pass, only a few instruction cycles are added. Compared to the total number of instructions required to calculate the FFT, these few cycles per pass are negligible in the overall execution time, as is evident from Table 4-1.

The operation of this algorithm is graphically depicted in Figure 5-6. For complex-input FFTs, the input data is limited to be less than or equal to 0.25 in absolute value, while for real-input FFTs, a maximum input magnitude of 0.5 is retained. This should avoid all limiting at the output of the first pass, where scaling is always off. The grey area depicts all real and imaginary values less than 0.25, where scaling down is turned off. Note that any results moved to memory, which fall outside the grey area, will result in scale-down mode being turned on for the next FFT pass. Thus, at all times the results in memory are forced to lie inside the outer circle, and thus the real and imaginary values are always less than or equal to $\sqrt{2}/2 \approx 0.707$, and no overflow or limiting occurs.

Since scaling is implemented for complete "blocks" of memory, a common scaling factor can be used if absolute numbers are required. This scaling factor can be constructed in terms of an "exponent" by adding up the number of passes for which scale-down was turned on. This common exponent is easily obtained by incrementing a counter or memory location each time S is detected to be set. The name *block floating point* indicates this common exponent for a complete memory block.

Example:

The signal of the previous examples is used as input to the block floating-point FFT on the DSP56000/DSP56001. The results are shown in Figure 5-7.

[7]The technique discussed here is only available on the full CMOS version of the DSP56000/DSP56001.
[8]By setting $S0 = 1$ and $S1 = 0$ in the status register (scale-down mode).

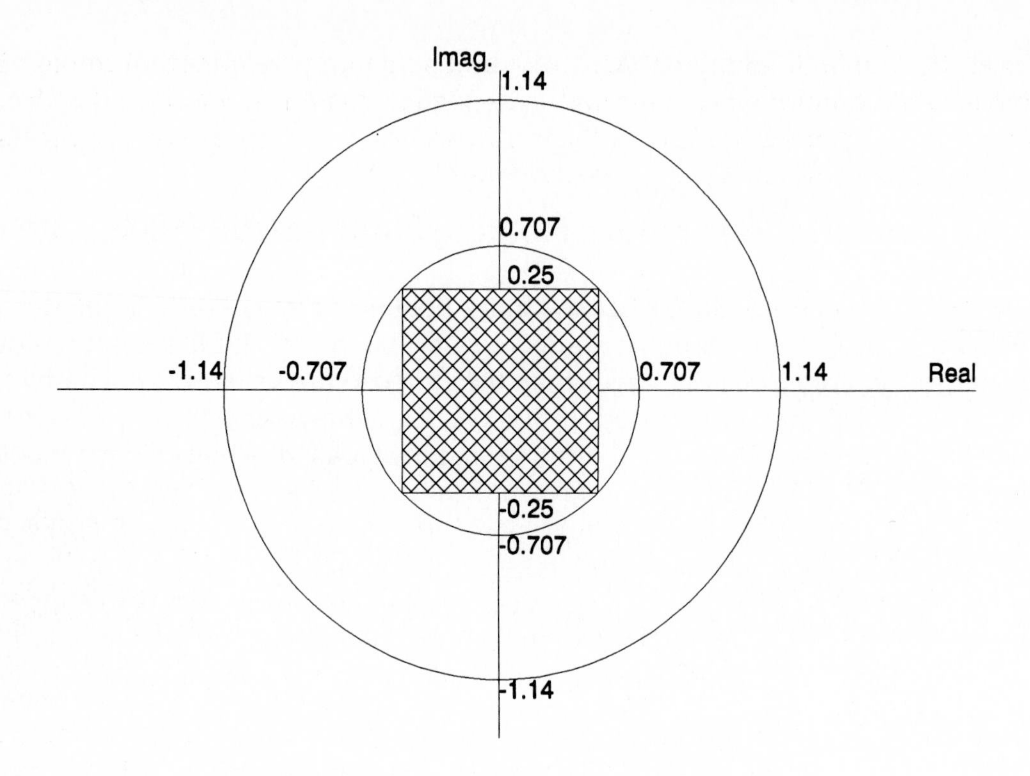

Figure 5-6. Block Floating-Point Bounds

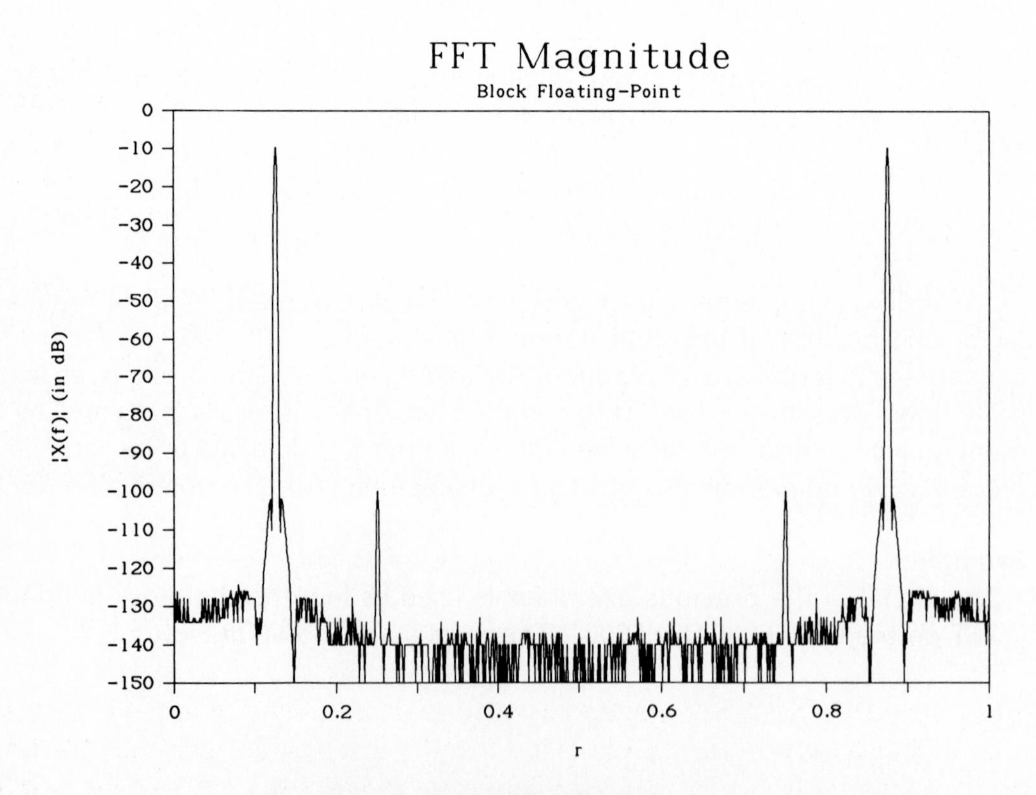

Figure 5-7. Magnitude of the Block Floating-Point FFT of the Same Signal as in Figure 5-3

Note that the outline of the Blackman-Harris window's side lobes becomes visible, with a small amount of roundoff noise imposed. This result is obtained with full 22-bit input A/D converter. Clearly, this result shows much more accuracy than any of the previously discussed implementations.

5.4 — IEEE FLOATING POINT

5.4.1 — Introduction

The previous discussion shows that the key element to reduce roundoff noise effects is the scaling of the complex results, such that the magnitude of the result is always as large as possible compared to the roundoff error introduced in the computation. This scaling operation is obviously data dependent. Although block floating point is an effective way of implementing conditional scaling on fixed-point DSPs, the fact that one common scaling factor is used for a complete memory block indicates that individual butterfly outputs may not be scaled optimally. For instance, if a harmonic component in the input increases the magnitude of one butterfly output, other butterfly outputs may not experience any word growth and thus should not be scaled, since unnecessary scaling invariably reduces the SRNR at the butterfly output.

Optimal scaling of all arithmetic results is obtained by implementation of floating-point arithmetic. Floating point uses a different scaling factor in terms of an exponent of two for every number. The exponent is determined such that the most significant bit of the remaining number (the mantissa) is always normalized to one. As such, the representation of the mantissa always uses the total number of bits available, and the SRNR is always optimal for *every* number in the total computation.

Floating-point arithmetic can be implemented on fixed-point DSPs in software. For instance, a complete floating-point library for DSP56000/DSP56001 is available (see Reference 8). Even though this approach gives satisfactory performance for various applications (the DSP56000/DSP56001 floating-point library gives an average performance of twice the speed of Motorola's MC68881 Floating-Point Coprocessor, see Reference 9), fast real-time requirements and compatibility issues with standard floating-point formats often necessitate another approach.

5.4.2 — IEEE Floating-Point on Motorola's DSP96002

Motorola's DSP96002 provides a complete hardware implementation of the *IEEE Standard 754-1985 for Binary Floating-Point Arithmetic* (see Reference 10). Because of the similarity between the DSP96002's architecture with that of the DSP56000/DSP56001, compatibility between the two processors is maintained. Consequently, implementation of FFTs is very similar on both processors, as illustrated in **4.3 IMPLEMENTATON ON MOTOROLA'S DSP96002**. The result clearly shows both harmonic components, as well as the spectral shape of the Blackman-Harris window. Because of the hardware IEEE floating-point arithmetic, FFTs execute as fast as or even faster than their fixed-point counterpart, as illustrated in Table 4-1.

In addition to the optimal scaling offered by floating-point arithmetic, conformance to the IEEE standard provides several other advantages. These advantages are discussed in greater detail in Reference 11, and include guaranteed error bounds, standard error indication, and portability across many different implementations. This last aspect is of particular importance in spectrum analysis algorithm development: adherence to the IEEE standard guarantees results which are bit-for-bit equal with results obtained in simulations using high-level languages that conform with the standard. Since much time is spent in algorithm development for spectrum-based applications, compatibility of results often saves an additional step in application development, where the use of fixed-point arithmetic or other floating-point formats may require considerable further effort in comparison with simulated results.

Example:

The signal of the previous examples is generated in IEEE standard floating point, and the magnitude of the 1024-point complex FFT is shown in Figure 5-8. Clearly, the optimal scaling of the floating-point numbers has reduced the roundoff noise to a negligeable level, and the figure clearly displays the smooth outline of the Blackman-Harris window. Bit-per-bit compatibility is obtained with high-level language implementations which use the standard. However, one must keep in mind that sine and cosine tables, generated in high-level languages using single-precision sine and cosine calls, are not obtained in full IEEE single precision (the standard only defines addition, multiplication, division, and square root). The sine and cosine tables in the DSP96002 were generated in double precision, and rounded to single precision to obtain correct results for all 32 bits. Consequently, FFTs generated in high-level languages may actually demonstrate additional "noise", created by the lack of precision of the sine and cosine tables. This is illustrated in Figure 5-9. The results do become bit-for-bit identical with the DSP96002 when the sine and cosine tables are generated in the same way. This is illustrated in Figure 5-10.

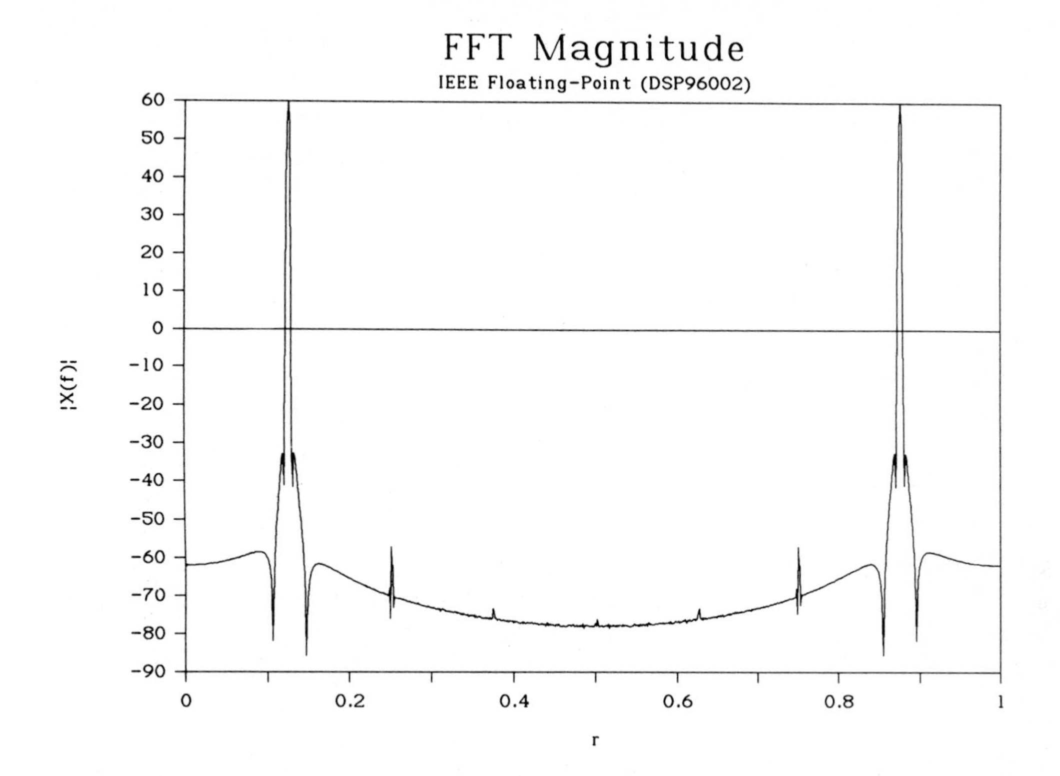

Figure 5-8. Magnitude of the IEEE Standard Floating-Point FFT of the Same Signal as in Figure 5-3

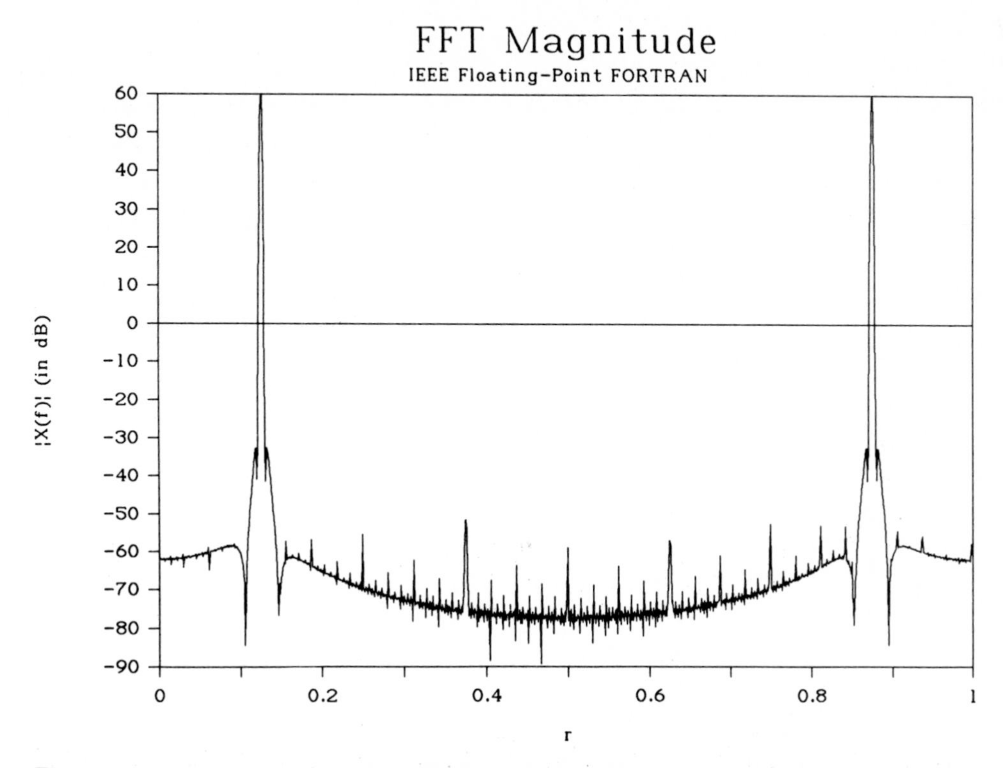

Figure 5-9. Magnitude of an IEEE Floating-Point FORTRAN FFT of the Same Signal as in Figure 5-3

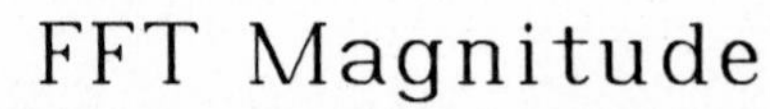
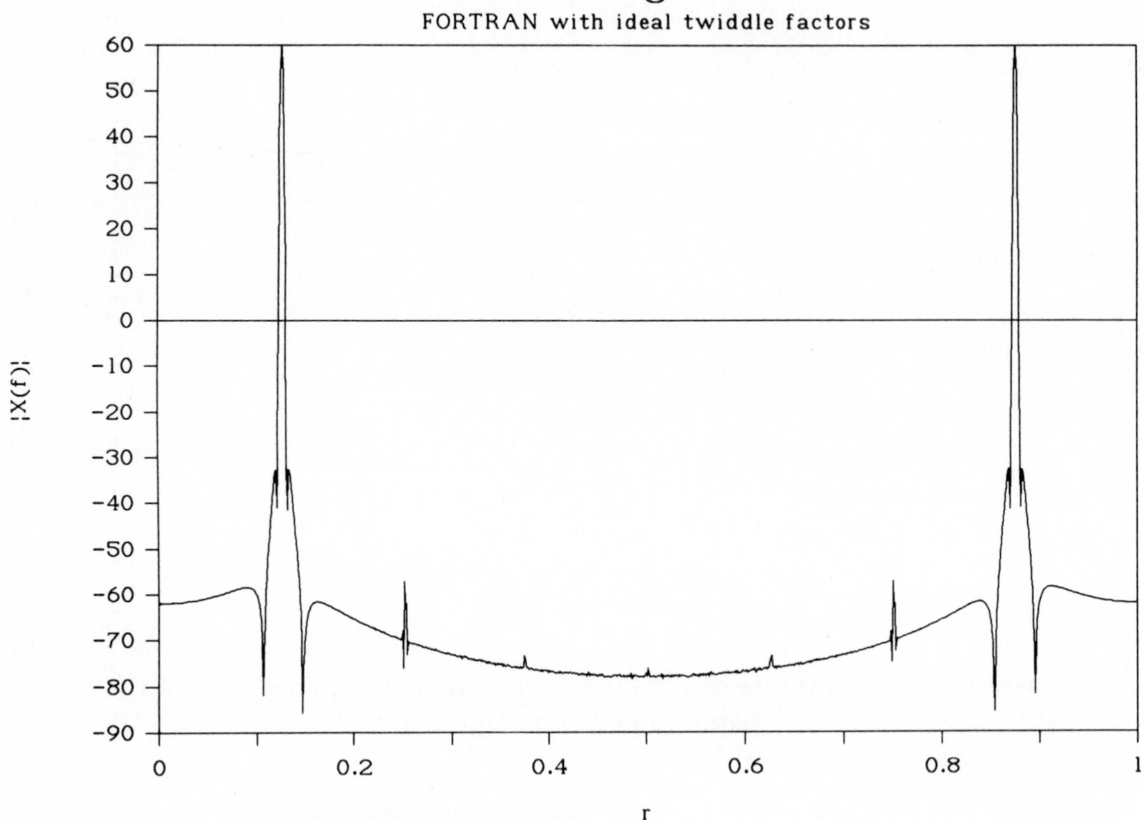

Twiddle factors are generated in double precision and rounded to signal precision.

Figure 5-10. Magnitude of an IEEE Floating-Point FORTRAN FFT of the Same Signal as in Figure 5-3

SECTION 6 — CONCLUSIONS

Frequency domain applications are becoming more important as inexpensive hardware solutions become more readily available. Motorola's Family of DSP56000/DSP56001 and DSP96002 digital signal processors (DSP) provide particularly effective solutions to frequency domain problems. A highly parallel architecture, combined with an instruction set particularly suited for implementation of fast Fourier transforms allow real-time computation of high-resolution FFTs up to very high sampling rates. Fast interrupts of DSP56000/DSP56001 and parallel DMA over a separate bus in DSP96002 provide for data I/O with hardly any penalty in speed. Furthermore, the dual external buses on DSP96002 allow fast calculation of FFTs of virtually unlimited size, with no performance penalty on external data access.

The large, 24-bit data representation of DSP56000/DSP56001, together with infinite-precision internal arithmetic and convergent rounding, lead to numerically superior results over 16-bit DSPs with truncation arithmetic. Special hardware provided in the DSP56000/DSP56001 allows no-overhead automatic scaling and block floating-point implementations of FFTs of virtually unlimited size, with result precision rivaling that of true floating point, for a fixed-point price.

For high-end applications, the DSP96002 provides full IEEE standard floating-point arithmetic for negligible roundoff errors. In addition to providing standard IEEE exception handling capabilities, the results obtained in the DSP96002 are portable across many applications that use the standard, such as high-level language simulations, data bases, etc. Motorola's family of digital signal processors, combined with Motorola's data conversion parts (see Reference 12), provide a complete, cost-efficient solution to frequency domain problems; from low-end small-size FFT applications, to high-end instrumentation and computer workstations for scientific computing.

APPENDIX A — NOISE EXPRESSIONS FOR FIXED-POINT FFTs

A-1 — ROUNDOFF NOISE EXPRESSIONS FOR FIXED-POINT FFTs ON DSP56000/DSP56001

Although a detailed analysis of roundoff errors is beyond the scope of this work (for a more detailed tutorial on roundoff errors in FFTs, see Reference 13[9]), the source of roundoff noise within butterfly calculations is depicted in Figure A-1.

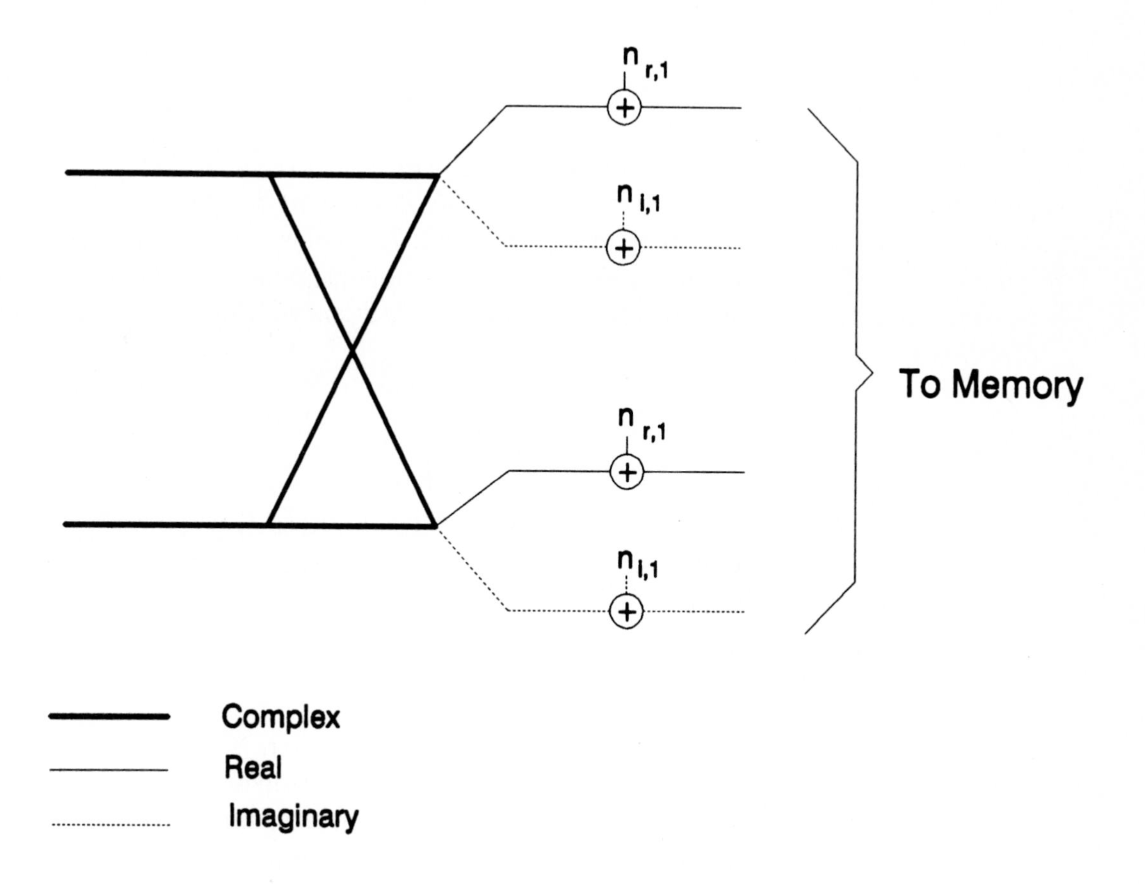

Figure A-1. Roundoff Noise Sources in DSP56000/DSP56001 Butterfly Calculation

[9]The reader should be cautioned that most roundoff noise analysis in the literature holds for "generic" fixed-point problems. Specific features of DSPs such as infinite internal precision, different multiplier and accumulator sizes, number of available registers, etc. are usually not taken into account.

Figure A-1 shows roundoff noise in the FFT butterfly when implemented on the DSP56000/DSP56001. In this particular case, the four sources of roundoff noise occur prior to moving each of the butterfly's four outputs to memory. Note that the errors on the computations for D (bottom butterfly output) are the same as the errors on the computation of C because of the use of the SUBL instruction. These are the only sources of roundoff noise. Internal to the CPU, infinite-precision arithmetic is obtained by the 56-bit accumulators, followed by only one final rounding operation before the result is moved to memory. The noise sources $n_{i,r}$ thereby have zero mean because of the convergent rounding ("round to nearest even") which is implemented in the DSP56000/DSP56001, and have variance (or average "noise power") of:

$$\sigma^2_{i,r} = \frac{2^{-2p}}{3} \tag{A-1}$$

in real and imaginary parts (where P is the number of bits).[10] Refer to Reference 14. This corresponds to average errors on the squared magnitude of butterfly outputs as:

$$\sigma^2 = \sigma^2_i + \sigma^2_r = 2 \bullet \frac{2^{-2P}}{3} \tag{A-2}$$

For $P = 24$ (DSP56000/DSP56001), an average noise power of $\approx 2.4 \bullet 10^{-15}$ is added to the real and imaginary parts at the output of each of the butterflies. Closer examination of the FFT butterfly diagram (Figure 3-7) reveals that every output point of the N-point FFT is connected to exactly N-1 (complex) noise sources (as in Figure A-2). Assuming the roundoff noise adds incoherently throughout the FFT calculation, a total noise power of:

$$\sigma^2_N = 2 \bullet \frac{2^{-2p}}{3}(N-1) \tag{A-3}$$

is generated throughout the FFT. Assuming that the input is appropriately normalized as in **5.2.1 Input Bounds** such that the peak signal output magnitude is one, a peak signal-to-roundoff-noise ratio (SRNR) is obtained as:

$$SRNR_{dB} = 10 \log_{10}[3 \bullet 2^{2P-1}] - 10 \log_{10}(N-1) \tag{A-4}$$

The quantization noise in fixed-point FFTs is the error which is caused by the representation of the *input signal* by a finite number of bits. This error affects the least significant bit of the A/D converter. When the A/D converter bits are loaded in the DSP memory, G guard bits are kept. Consequently, the quantization noise and the

[10]Note that here P is the total number of bits and not the number of mantissa bits only, as in Reference 13.

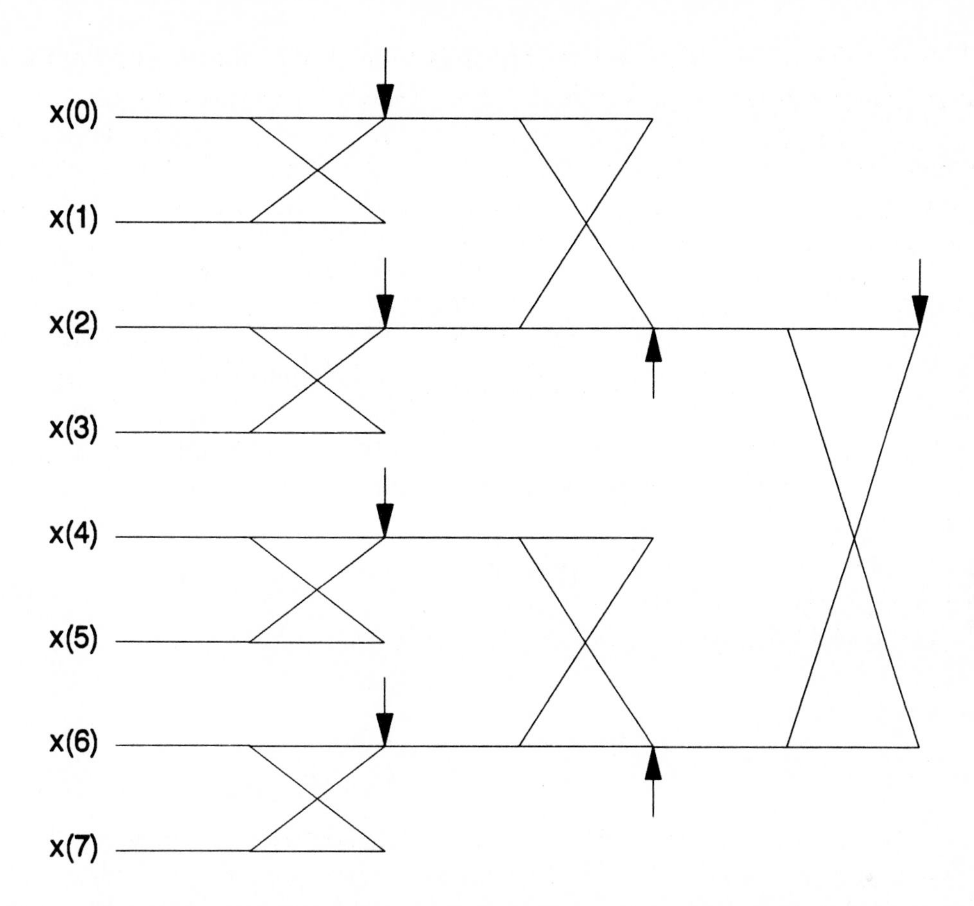

Figure A-2. Combination of Noise Sources to the FFT Result

roundoff noise at the output of the FFT affect the same bits (become of the same order of magnitude) when the associated errors become equal (assuming all noise sources are "white"):

$$\frac{2^{1-2(A+G)}}{3} = \frac{2^{1-2P}}{3}(N-1) \tag{A-5}$$

or:

$$A = P - G - 0.5 \bullet \log_2(N-1) \tag{A-6}$$

for complex-input FFTs and

$$\frac{2^{-2(A+G)}}{3} = \frac{2^{1-2P}}{3} \bullet (N-1) \tag{A-7}$$

or:

$$A = P - G - 0.5 - 0.5 \bullet \log_2(N-1) \tag{A-8}$$

for real-input FFTs. A equals the number of bits in the A/D converter and G is the number of guard bits.

A-2 — TRUNCATION NOISE EXPRESSIONS FOR IMPLEMENTATION OF FIXED-POINT FFTs ON DSPs WITH TRUNCATION ARITHMETIC.

Figure A-3 shows sources of errors for processors which do not implement internal arithmetic with infinite precision. In this case, the multiplier result is simply truncated to the resulting number of bits P. This results in the four noise sources as in Figure A-3, which now occur at the output of the twiddle factor multiplication. Assuming simple truncation to P bits as opposed to the convergent rounding used previously, the noise sources $n_{i,r}$ no longer have zero mean, which can result in a bias in the FFT calculation. The error sources in Figure A-3 result in total average error values in real and imaginary parts at the output of the complex multiplication:

$$\mu_i = 0 \tag{A-9}$$

$$\mu_r = 2 \cdot 2^{-P}$$

and variances of:

$$\sigma^2_{i,r} = \frac{2 \cdot 2^{-2p}}{3} \tag{A-10}$$

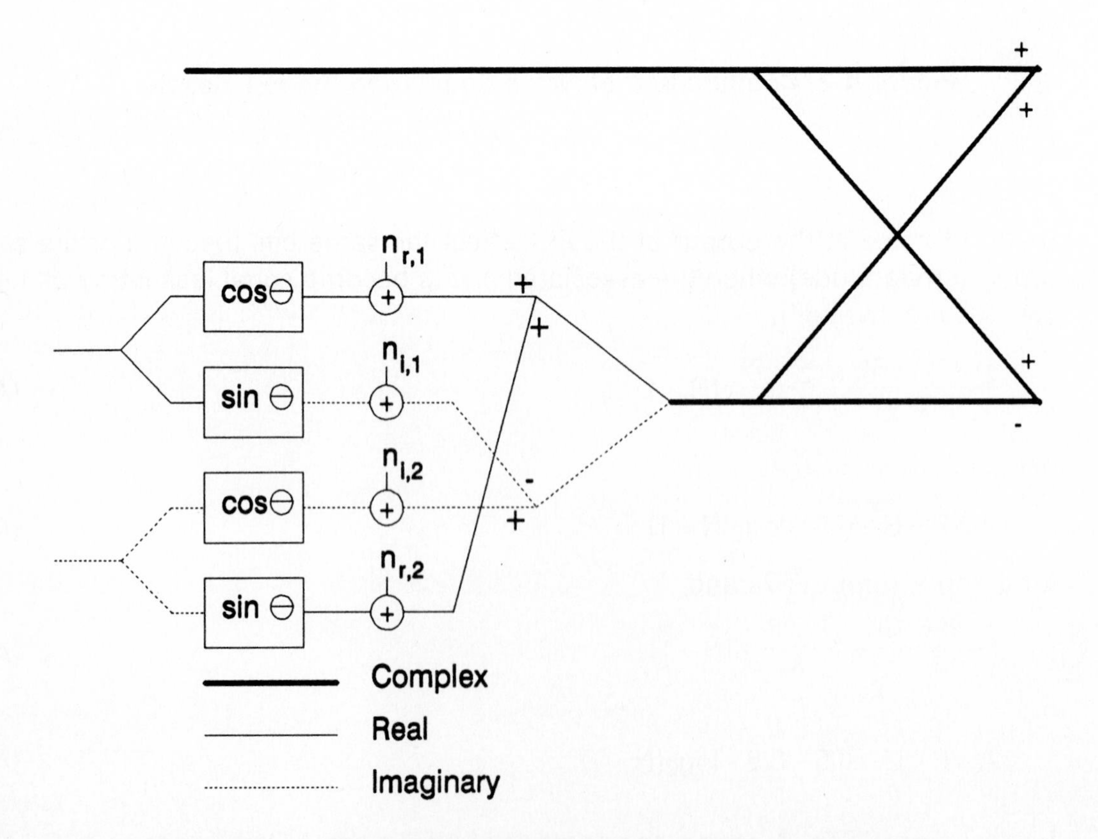

Figure A-3. Truncation Error Sources in the DIT Butterfly

on real and imaginary parts, since the imaginary part is obtained by the difference of two products, while the real part is obtained by the sum. Note that this variance is twice the value obtained when using infinite-precision internal arithmetic as in **A-1 ROUNDOFF NOISE EXPRESSIONS FOR FIXED-POINT FFTs ON DSP56000/ DSP56001**. The nonzero average error in the calculation of the imaginary part of equation (A-7) results in a bias in the FFT calculation. Since this value is constant across the FFT flow diagram, a low-frequency bias will result. The variance, expressed in equation (A-6), propagates as truncation noise variance to the output of the FFT: as before, each output point of the flow-diagram is connected to exactly N-1 butterflies. Assuming that the noises add incoherently, a total variance on the complex outputs is given by:

$$\sigma^2_N = (N-1)\frac{2^{-2(P-1)}}{3} \qquad \text{(A-11)}$$

This past expression can again be used to determine the total Signal-to-Truncation noise ratio (STNR) as:[11]

$$STNR_{dB} = 10 \log_{10}[3 \cdot 2^{2P-2}] - 10 \log_{10}(N-1) \qquad \text{(A-12)}$$

[11]It is again assumed that the input is appropriately scaled such that the peak output signal magnitude is one.

APPENDIX B — ROUNDOFF NOISE EXPRESSIONS FOR FFTs ON DSP56000/DSP56001 WITH AUTOMATIC SCALING

Figure B-1 shows the appropriate butterfly when automatic scaling is introduced at every pass. The noise sources introduced have the same characteristics as the ones in the fixed-point representation (equation (A-1)) in **APPENDIX A**. These noise sources still add incoherently throughout the FFT. However, they are no longer of equal strength when they add: noise which is created in the first stage of the FFT shows up scaled down $\log_2 N - 1$ times before contributing to the final output noise, while

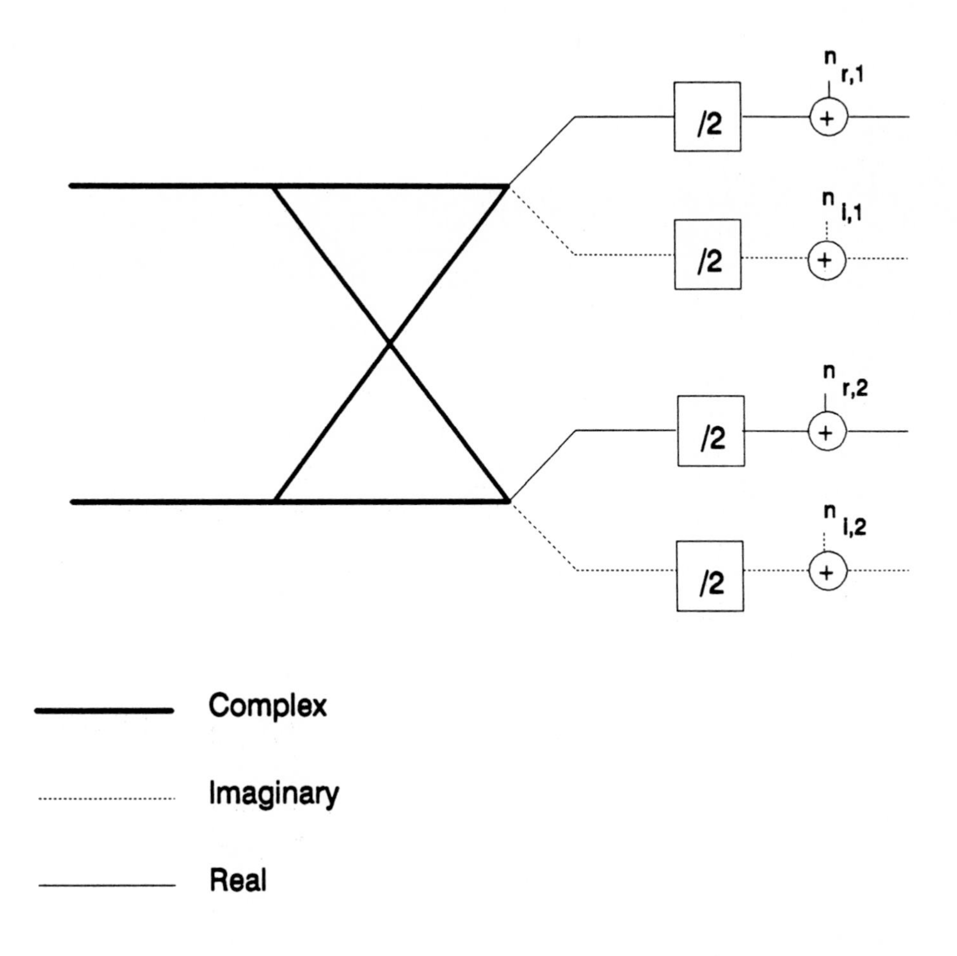

Figure B-1. Roundoff Noise Sources with Automatic Scaling

the noise source in the final stage contributes fully. Thus, an equation for the total noise variance on the complex output is given by:

$$\sigma^2_N = \sigma^2 \left\{ \frac{N}{2}\left(\frac{1}{2}\right)^{\log_2 N - 1} + \frac{N}{4}\left(\frac{1}{2}\right)^{\log_2 N - 2} + \cdots + 1 \right\} \quad \text{(B-1)}$$

$$= \sigma^2 \sum_{k=1}^{\log_2 N} \frac{N}{2^k}\left(\frac{1}{2}\right)^{\log_2 N - k}$$

$$= \sigma^2 \log_2 N$$

where σ^2 is given in equation (A-2). The total SRNR in this case is thus given by:[12]

$$\mathrm{SRNR}_{dB} = 10\ \log_{10}[3 \cdot 2^{2P-1}] - \log_{10}[\log_2 N] \quad \text{(B-2)}$$

The bit affected by the quantization noise when automatic scaling is turned on, is now shifted down after every FFT butterfly. The point at which the quantization error and the roundoff error affect the same bits can be obtained as in **A-1 ROUND-OFF NOISE EXPRESSION FOR FIXED-POINT FFTs ON DSP56000/DSP56001**. This gives:

$$\frac{2^{1-2(A+G)}}{3 \cdot N^2} = \frac{2^{1-2P}}{3} \cdot \log_2 N \quad \text{(B-3)}$$

or:

$$A = P - 1 - \log_2 N - 0.5 \cdot \log_2(\log_2 N) \quad \text{(B-4)}$$

for complex-input FFTs (G = 1). For real-input FFTs, G = 0 as explained in **5.2.1 Input Bounds**, and the result:

$$\frac{2^{-2(A+G)}}{3 \cdot N^2} = \frac{2^{1-2P}}{3} \cdot \log_2 N \quad \text{(B-5)}$$

$$A = P - 0.5 - \log_2 N - 0.5 \cdot \log_2(\log_2 N) \quad \text{(B-6)}$$

is obtained.

[12]Assuming that the input is appropriately scaled such that the peak output magnitude is equal to one.

REFERENCES

1. A. Papoulis, *The Fourier Integral and its Applications*

2. F. J. Harris, "On the Use of Windows for Harmonic Analysis with the Discrete Fourier Transform" *Proc. IEEE*, Vol. 66, No. 1, Jan. 1978, pp. 57-84

3. A. V. Oppenheim and R. W. Schafer, *Digital Signal Processing* Englewood Cliffs, NJ: Prentice-Hall, 1975

4. Kevin L. Kloker, "The Motorola DSP56000 Digital Signal Processor" *IEEE Micro*, December 1986

5. Motorola Electronic Bulletin Board (Dr. BuB) FFTR2C Fast Fourier Transform Routine

6. EDN's DSP Benchmarks EDN, September 29, 1988

7. *Fractional and Integer Arithmetic Using the DSP56000 Family of General-Purpose Digital Signal Processors* Motorola, Inc., Application Report APR3/D

8. Motorola's DSP Bulletin Board (Dr. BuB) DSP56000 Floating-Point Library

9. *MC68881/MC68882 Floating-Point Coprocessor User's Manual* Publication No. MC68881 UM/AD, Motorola, Inc., Austin, TX 1987

10. *IEEE Standard for Binary Floating-Point Arithmetic* ANSI/IEEE Std. 754-1985 New York, NY: IEEE, 1985

11. Guy R.L. Sohie and K. Kloker, "A Digital Signal Processor with IEEE Floating-Point Arithmetic" *IEEE Micro*, Vol. 8, No. 6, December 1988

12. Motorola's DSP56ADC16 Data Sheet

13. A. V. Oppenheim and C. J. Weinstein, "Effects of Finite Register Length in Digital Filtering and the Fast Fourier Transform" *Proc. IEEE*, Aug. 1972, pp. 957-976

14. See ref. [3], pp. 404 - 418

INDEX

— T —

— W —

BR522/D

MOTOROLA
SEMICONDUCTOR
TECHNICAL DATA

DSP320to56001

Appendix D

Software Summary

DSP320to56001 Translator Software

The DSP320to56001 translator software will convert any 32010 code into code for Motorola's powerful new digital signal processor chip, the DSP56001. The primary features of DSP320to56001 are:

- Translation of any 32010 applications software into DSP56001 source code
- Two modes of operation:
  - — Translates to 56001 source code for potential optimization and assembly with the DSP56000SASMA or DSP56000CLASA software
  - — Translates and runs 32010 Code "as is" directly and immediately on the DSP56000ADS, Motorola's DSP560001 Applications Development System
- Runs on IBM®-PC under MS™-DOS or PC-DOS
- C source code of DSP320to56001 program is provided on diskette
  - — User may modify for 32020 and 320C25 translation
  - — Third party vendors may contact Motorola for licensing details
- Registration card provided so users can obtain future optimized versions of DSP320to56001 software, hand-coded macro routines, etc.

MOTOROLA DSP DEVELOPMENT SOFTWARE 32010 TO 56000/1 CODE CONVERSION

HARDWARE REQUIREMENTS

The conversion programs are delivered on one double-sided, double-density 5 1/4 inch floppy disk and may be run from either a floppy disk or a hard disk. They require only enough disk space to hold the output of the converted source file.

The minimum hardware requirements for the conversion programs are:
IBM-PC, XT, AT, or compatible with 256K bytes of RAM and one 5 1/4 inch floppy disk drive.
PC-DOS/MS-DOS v2.0 or later.

The DSP56000 Application Development System (DSP56000ADS) is recommended as a development tool for designing real-time DSP56000/1 signal processing systems.

IBM is a registered trademark of International Business Machines Corporation.
MS™-DOS is a trademark of Microsoft, Inc.

EMULATION METHODS

Two methods of 32010 emulation are provided. One provides support for execution of code that is modified at run-time. This method is supported by the **xasm/lodled** combination. The **xasm** code has a built-in compiler that allows the DSP56000/1 to recompile 32010 code if it is modified during execution. The **xasm** code has a DSP56000/1 subroutine for each of the 32010 instructions. The built-in compiler selects the proper subroutine based on the 32010 opcode rather than generating in-line code.

The second emulation method, **xmac/lodasm** combination, generates in-line code in the DSP56000/1 assembler source format. The in-line code execution is fast compared with the subroutine method used in **xasm**. Execution of self-modified code cannot be emulated with this in-line method.

Note that both conversion routines, **lodasm** and **lodled**, operate only on 32010 **.lod** files. These are object code files which have been absolutely located.

DSP320to56001 SOFTWARE

The DSP320to56001 software includes the following programs:

lodasm.exe
- Converts 32010 **.lod** file to DSP56000/1 **.asm** file
- Generates in-line DSP56000/1 macro assembler mnemonics
- Output combines with **xmac.asm** at assembly time
- Does not support execution of self-modified 32010 code
- Relatively faster code compared with **lodled.exe** output

lodled.exe
- Converts 32010 **.lod** file to DSP56000/1 **.lod** file
- Transfers 32010 opcodes to a file loadable by DSP56000ADS, DSP56000/1
- The opcodes are compiled by the DSP56000/1 at runtime
- Supports execution of self-modified 32010 code
- Combines with **xasm.asm** at runtime
- 32010 emulation is executed in subroutine calls

xmac.asm
- DSP56000/1 macro definitions used with **lodasm.exe** output
- Combines with DSP56000/1 **.asm** file produced by **lodasm.exe**

xasm.asm
- 32010 emulation subroutines used with **lodled.exe** output
- Compiles and executes 32010 code at runtime
- Recompiles code if modified by 32010 **TBLW** instruction

xasm.lod
- DSP56000/1 load module created from **xasm.asm**

lodasm.c
- C language source file for **lodasm.exe**

lodled.c
- C language source file for **lodled.exe**

Literature Distribution Centers:
USA: Motorola Literature Distribution; P.O. Box 20912; Phoenix, Arizona 85036.
EUROPE: Motorola Ltd.; European Literature Center; 88 Tanners Drive, Blakelands Milton Keynes, MK145BP, England.
ASIA PACIFIC: Motorola Semiconductors H.K. Ltd.; P.O. Box 80300; Cheung Sha Wan Post Office; Kowloon Hong Kong.

Appendix E

Full-Duplex 32-kbit/s CCITT ADPCM Speech Coding on the Motorola DSP56001

by Kent Terry
Motorola Digital Signal Processor Operation

Section 1 - Introduction

This application report describes the implementation of an Adaptive Differential Pulse Code Modulation (ADPCM) speech coder on the Motorola DSP56001 digital signal processor. The algorithm described in this document has been standardized by the International Telegraph and Telephone Consultative Committee (CCITT) in Recommendation G.721 [1] for digital speech coding in a telecommunications environment. The standard, as defined by the CCITT, specifies the translation of μ-law or A-law PCM encoded speech at 64-kbit/s to ADPCM encoded speech at 32-kbit/s to provide a 2 to 1 compression of the speech signal with very little perceptual loss of speech quality. The algorithm also has added complexity to handle non-speech signals such as modem signals. The block diagrams of the CCITT ADPCM encoder and decoder are shown in Figures 1-1 and 1-2.

The implementation on the DSP56001 adheres completely with the CCITT Recommendation G.721 (revised version dated August 1986). It provides bit-for-bit compatibility with the test vectors described in G.721, meaning that the code correctly passes all the digital test sequences defined by the CCITT. In addition to providing compatibility with the standard, the DSP56001 implementation provides full-duplex operation. This means that one DSP56001 is able to perform both an encode and a decode in real-time.

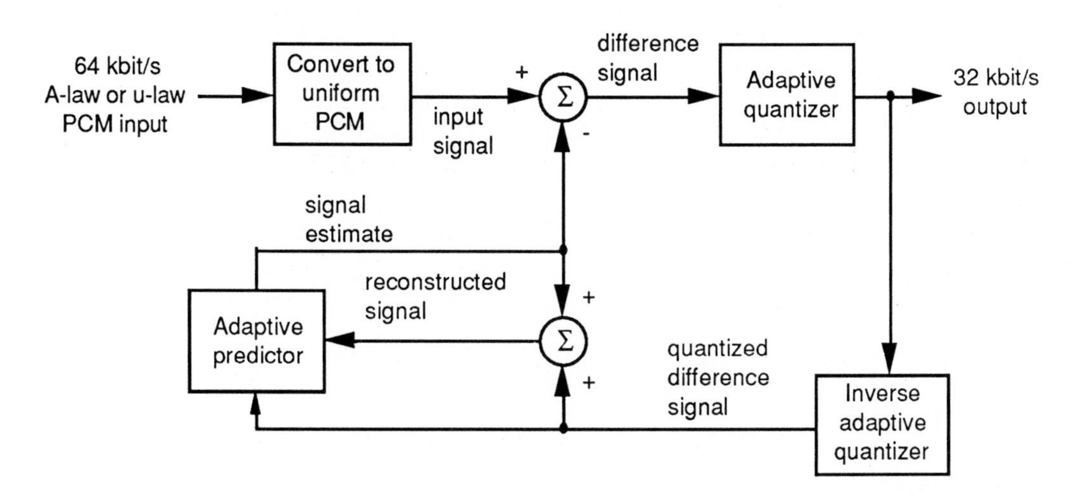

Fig. 1-1. CCITT ADPCM Encoder Block Diagram

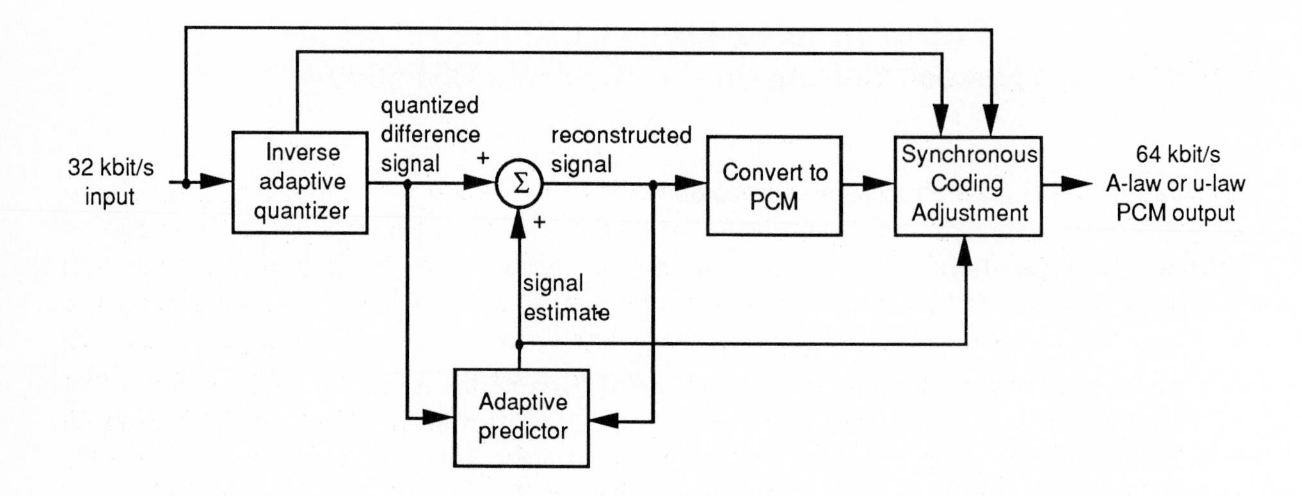

Fig. 1-2. CCITT ADPCM Decoder Block Diagram

This application report will first point out some of the advantages of speech coding in general and then some of the particular advantages of the CCITT standard. Then the basic concepts of ADPCM coding will be presented followed by a detailed description of the actual CCITT algorithm. After describing the algorithm, the implementation of the algorithm on the DSP56001 will be discussed including various techniques used to improve performance. Finally the real-time performance and other technical characteristics of the DSP56001 implementation will be described.

Section 2 - Speech Coding

Speech coding, or speech compression, is one of the major application areas of digital signal processing in the field of speech processing. The goal of speech coding is to digitally code speech in an efficient manner for either storage or transmission. In telecommunications applications the goal is typically to code an analog speech signal into a digital format, transmit the digital signal, and then decode the digital signal back into an analog waveform, all in real time. While the goal is very simple there are many trade-offs to be considered in a practical application. These include decoded speech quality, transmission bandwidth, coder complexity, overall system cost, and real-time considerations. Different applications may have very different requirements. For example, military communications applications often sacrifice coder complexity and speech quality to achieve very low transmission bandwidths. Other applications, such as some voice mail systems, may not require real-time performance.

Digital speech coding has other advantages beyond the compression savings. Digital signals in general have many desirable properties. They are less sensitive to transmission noise and are easier to multiplex, error protect, and encrypt for security than analog signals. Since coding algorithms can be implemented in software, modifications and improvements to algorithms are much easier than with dedicated hardware. In many applications, such as computer workstations, a common DSP may be used for many functions. These applications can add speech coding to a system without adding additional hardware.

As noted above, the ADPCM algorithm defined by the CCITT is intended for use in a telecommunications environment, although it may be applied in other areas. The CCITT algorithm actually implements a PCM/ADPCM/PCM conversion process and is also called a transcoder for this reason. The standard is intended for use on digital channels that contain the digital equivalent of analog signals on analog telephone channels, so the CCITT algorithm has added complexity to handle non-speech signals (such as modem and DTMF signals) that may be present on analog telephone channels. It provides 2 to 1 compression allowing two ADPCM coded signals to be easily multiplexed into one basic 64-kbit/s digital channel.

Section 3 - Types of Speech Coders

Although there are many different methods for speech coding they can generally be classified into two main categories: waveform coders and source coders. Waveform coders deal with speech on a sample by sample basis. Their goal is to have the output waveform of the decoder match the original speech waveform as closely as possible. Source coders (also called vocoders) on the other hand, attempt to describe a speech signal in terms of parameters of a speech production model. These models typically estimate vocal tract shape and vocal tract excitation. Vocoders can operate at much lower bit rates than waveform coders but generally the output speech quality is not as good. ADPCM is classified as a waveform coder and is actually a combination of several basic waveform coding techniques. It should also be noted that many speech coding algorithms are hybrids that combine a variety of techniques. For instance, ADPCM techniques are used in portions of many other coder algorithms, such as subband coders and LPC coders.

As noted above ADPCM is a combination of several basic techniques. The following discussion is a brief introduction to these techniques. Reference [2] also provides a good introduction to the topics discussed here as well as other types of speech coders.

3.1 - Uniform PCM

In any digital speech coding system an analog speech signal must be converted into a digital representation before it can be processed. In practice most digital coders use a form of pulse-code modulation (PCM) for the A/D conversion. The A/D conversion process can be represented by a combination of two processes, a continuous-to-discrete conversion (C/D) and a quantization (Q[ ]), as shown in Fig. 3-1 [3]. The C/D conversion changes a continuous time waveform into a discrete time waveform with continuous amplitude. Mathematically this will not introduce any error into the input signal as long as the sampling rate is at least twice the highest frequency of the input signal. The quantizer maps each continuous amplitude sample into a digital codeword. The mapping process represented by the function Q[ ] is called the quantizer characteristic. The quantization process will introduce an error called the quantization error into the signal. This error can be represented as an additive white noise source in the quantizer as shown in Fig. 3-2. In speech applications this error will be evident as audible noise at the D/A output. In many speech coding applications this quantization noise is expressed in terms of Signal-to-Quantization Noise Ratio (SNR).

An example of a uniform quantizer is shown in Fig. 3-3. Uniform PCM quantizers have the property that the step size (Δ) of the quantizer is constant [3]. To prevent large errors the range of the quantizer must be such that the maximum amplitude that may occur in the input signal can be represented in this range. The amount of noise the quantizer introduces into the signal is also directly related to the step size. The quantizer range and the step size together determine the number of bits required to adequately represent a signal of given quality. Telephone quality speech signals ("toll" quality) require about 35 dB of SNR in a frequency range of about 200-3200 Hz. It has been found in practice that 11-12 bits quantization at an 8 kHz sampling rate are needed to ensure this SNR over a typical range of speech signals [2].

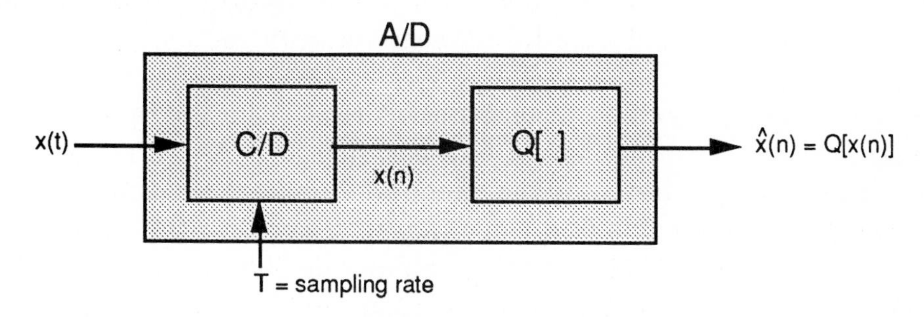

Fig. 3-1. A/D Conversion Process

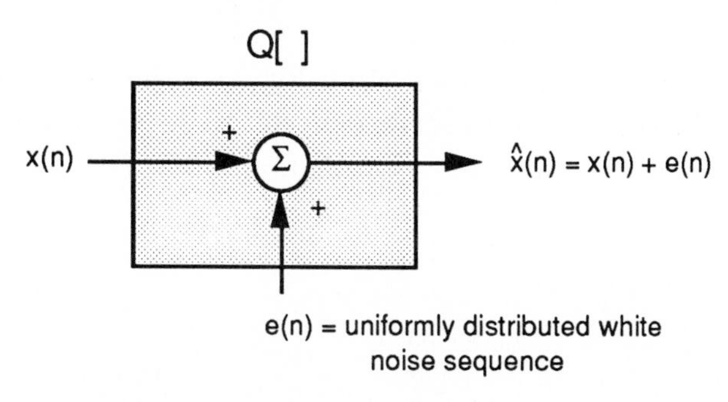

Fig. 3-2. Quantization Noise Model

3.2 - Logarithmic PCM

One of the disadvantages of uniform PCM coding is that the SNR is not constant. For instance, during "voiced" segments of speech (vowels for example) the SNR may be high but during "unvoiced" segments of speech (consonants for example) the SNR may be low. The result is that bits are wasted to make sure the maximum SNR is always less than an acceptable level. If one can reduce the maximum SNR then fewer bits will be needed to ensure this. This is the goal of non-uniform quantizers. They have the property that the step size varies according to the input signal, or alternatively, the input is scaled non-uniformly before it is quantized [4]. With this approach the step size is larger for large input amplitudes than it is for small amplitudes. The goal is to obtain a uniform SNR over all input ranges so that the SNR is independent of the input.

A basic approach to achieve this goal with speech signals is to logarithmically space the quantization levels. This approach, called logarithmic PCM (log PCM) coding, effectively compresses the input signal at the transmission end and expands it at the receiving end. For this reason log PCM coding is also referred to as companding [4]. True logarithmic quantization is not practical in reality, but two methods that approximate this technique have been standardized [5]. They are referred to as μ-law and A-law companding. They achieve a SNR that is constant enough for practical purposes. Compared with uniform PCM these techniques need about four fewer bits per sample for equivalent speech quality, so only 8-bits are needed per sample rather than the 11 or 12 bits that are needed with uniform PCM. 8-bit log-PCM companding is one of the simplest forms of speech coding and has also been standardized for digital telecommunications.

Speech coders are typically specified in terms of bit rate. The bit rate is the number of samples per second times the number of bits per sample. As noted above, telephone quality voice signals are primarily in the range of 200-3200 Hz and they are typically sampled at 8 kHz to maintain this frequency range. Therefore the basic standard data rate for μ-law and A-law PCM data is 8-bits/sample at 8000 samples/second or 64,000 bits/second (64-kbit/s). This is the source of the data rate for a basic digital transmission channel. The process of converting log PCM signals to/from analog signals is often done using CODEC devices (such as the MC145503). In the CCITT algorithm the log PCM data is converted to linear PCM data before the ADPCM encoding itself is performed and the decoded linear signal is converted back into log PCM form after the decoding process is completed. The process for converting between log and linear PCM data, including the implementation on the DSP56001, is described in detail in the Motorola applications brief "Logarithmic/Linear Conversion Routines For DSP56000/1" [6].

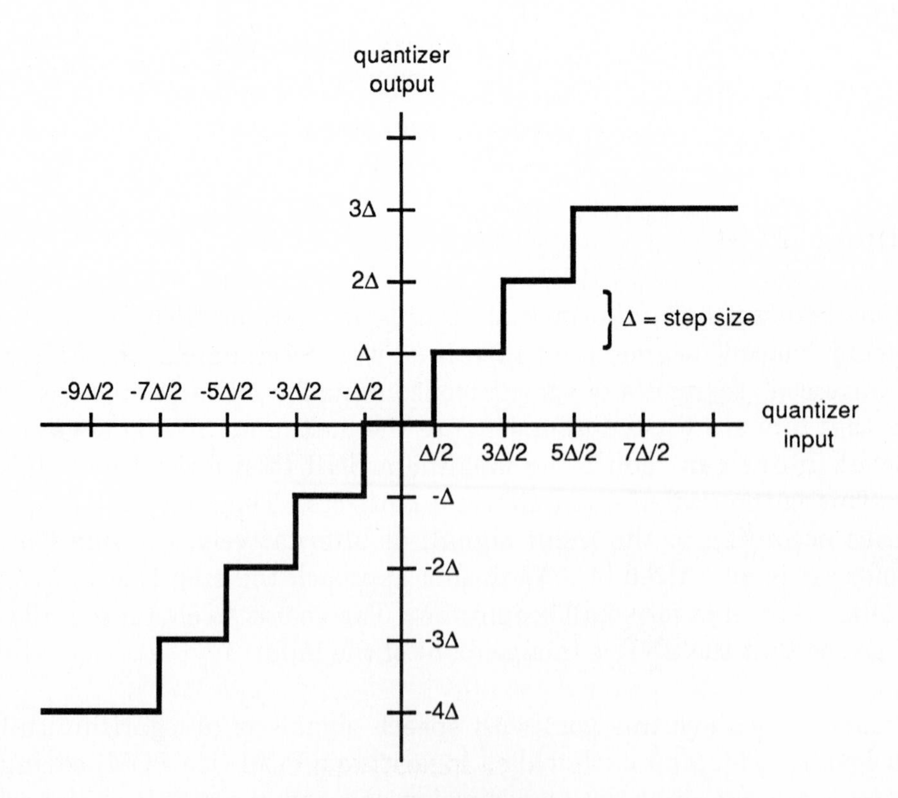

Fig. 3-3. Uniform Quantizer

3.3 - ADPCM Coding

To further reduce the data rate of speech signals from that obtained by PCM methods, speech coders attempt to remove "redundancy" in speech signals. They do this by taking advantage of known characteristics of speech signals. ADPCM is a technique of speech compression based on a combination of two basic speech coding techniques, Adaptive PCM (APCM) and Differential PCM (DPCM). The first basic difference between these techniques and uniform and log PCM methods is that they require previous samples to be remembered while uniform and log PCM methods do not. A second key difference is that uniform and log PCM coders have fixed compression and expansion curves (fixed step sizes), while most adaptive PCM methods change their compression and expansion curves over time (adaptive step sizes).

APCM coders take advantage of the tendency for speech signals to vary relatively slowly. [2] They exploit this property by changing the characteristics of the coder adaptively over time. One way of doing this is by changing the quantizer step size in proportion to the average speech amplitude. The step size modification can be achieved in two ways, by either directly scaling the step size or by scaling the input signal by a gain factor. Updating the step size or gain factor can also be done in one of two ways. The feedforward approach, shown in Fig. 3-4, actually sends the update information to the decoder over the transmission channel [7]. The feedback approach, shown in Fig. 3-5, determines the information from the previously coded samples so that no extra information needs to be transmitted to the decoder [7]. APCM in general provides better SNR performance and speech quality at a given bit rate compared to uniform or log PCM but it does require more computation.

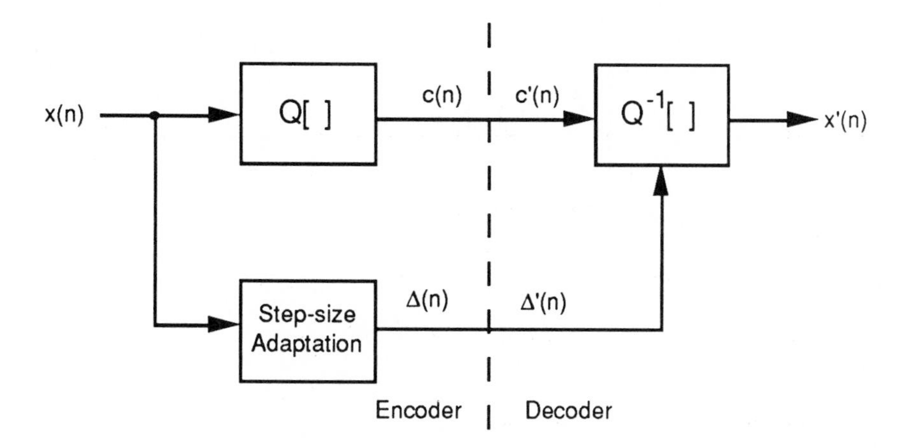

Fig. 3-4. Feedforward APCM Coder

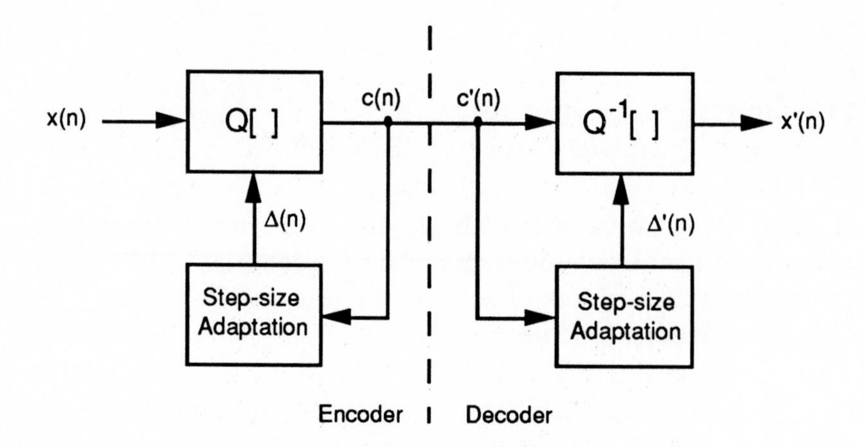

Fig. 3-5. Feedback APCM Coder

Other properties that are unique to speech signals relate to the spectral envelope. A typical short-time magnitude spectrum of a speech signal shows a slowly varying envelope whose shape is primarily determined by the vocal tract response. This causes speech samples to have a high degree of "sample-to-sample correlation" [2]. DPCM techniques try to take advantage of this characteristic. DPCM coders are characterized by the use of a predictor that forms an estimate of each incoming speech sample. This estimate is subtracted from the actual sample and the difference between them is coded instead of the actual input sample. Most DPCM coders use a form of linear prediction where the estimate is based on a linear combination of previous input samples. A block diagram of a DPCM coder is shown in Fig. 3-6 [2].

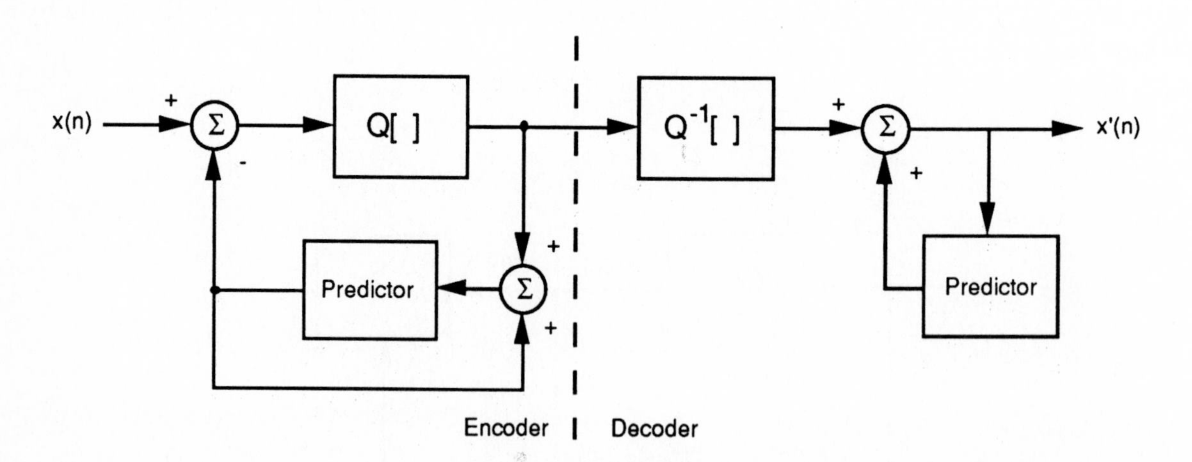

Fig. 3-6. DPCM Coder

ADPCM techniques use a combination of APCM and DPCM techniques. There are many different ways to implement the concepts of APCM and DPCM, so the term ADPCM can justifiably refer to a broad range of speech coders that may have widely varying characteristics. It should be pointed out that ADPCM techniques, as well as APCM and

DPCM techniques, may also be applied to non-speech signals, such as high-fidelity audio signals or video images. These implementations may not exploit properties that are specific to speech, however. The term ADPCM as used in this discussion refers specifically to the algorithm defined by the CCITT for telephone quality speech signals. Therefore the scope of this algorithm may not apply to all applications requiring signal compression. A general block diagram of the ADPCM configuration used in the CCITT algorithm is shown in Fig. 3-7 [2].

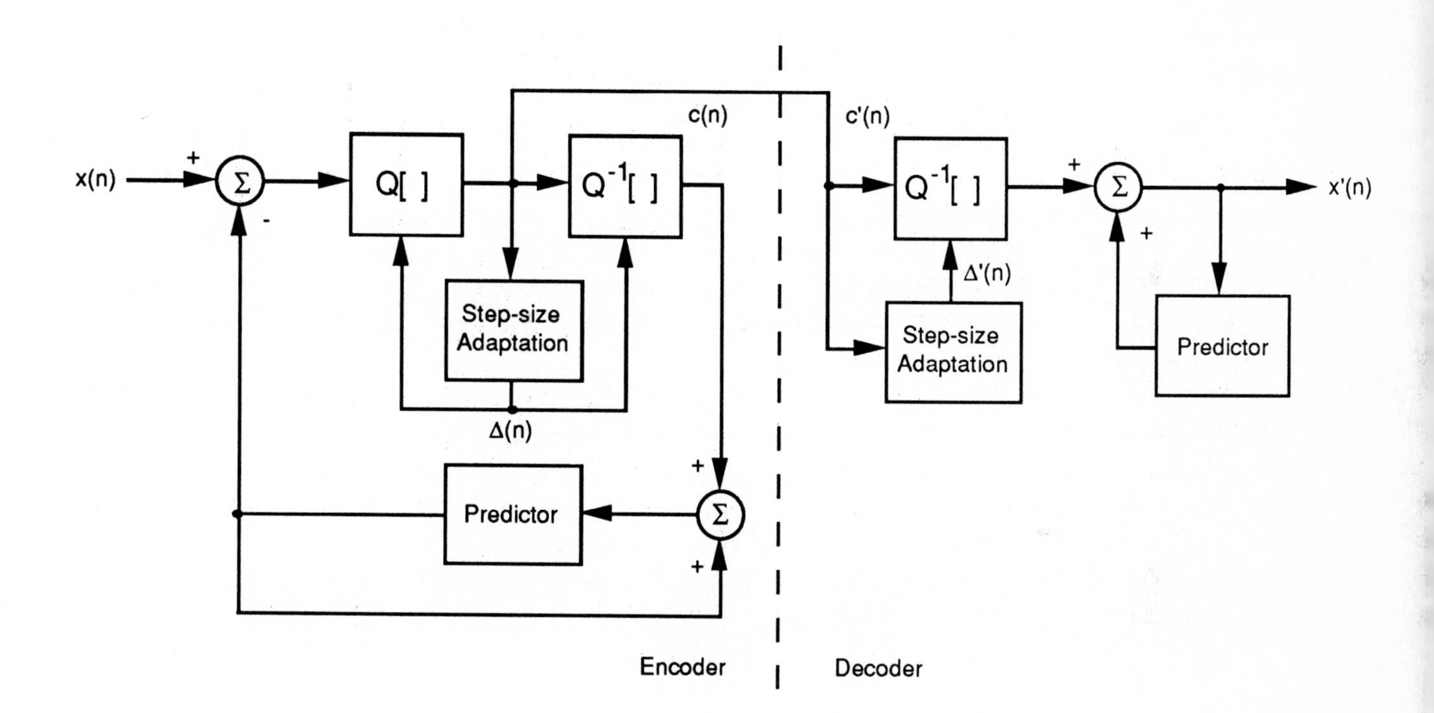

Fig. 3-7. ADPCM Coder

Section 4 - The CCITT ADPCM Algorithm

The CCITT ADPCM coder has several particular requirements that it is designed to meet [8]. The first goal is to provide compression while satisfying the objective signal quality requirements specified in CCITT Recommendation G.712 [9]. It must retain a high enough level of subjective quality (defined by listening tests) even after a series of encodings and decodings. Compatibility with the existing μ-law and A-law PCM formats is required. It must have stable operation in the presence of high bit error rates in transmission and must also operate properly in the presence of voiceband data at up to 4.8 kbit/s. These requirements and others have led to the standardization of this particular algorithm that meets these requirements. The following description first provides an overview of the algorithm and then gives a detailed description of each part of the algorithm. A complete development of the theory behind this algorithm is beyond the scope of this application report so the emphasis will be placed on the algorithm implementation.

A detailed block diagram of the encoder portion of the CCITT ADPCM algorithm is shown in Fig. 4-1. The decoder portion is shown in Fig. 4-12. The algorithm uses the feedback method where the encoder uses only the coded ADPCM signal I(k) for feedback to the prediction and adaptation sections. This is the same information that the decoder uses for adaptation so no update parameters need to be sent over the transmission channel. This structure has two key properties. The first is that the encoder and the decoder are almost identical in terms of function. The second property is that for a given sample the decoder will be in the same "state" as the encoder (assuming no transmission errors); all common internal signals will be identical. This is how the decoder is able to keep track of the encoder's adaptive process without explicitly receiving information from the encoder.

The APCM portion of the algorithm uses the gain factor approach for quantization. The difference signal to be coded is first scaled by the adaptive scale factor y(k) before it is quantized according to a fixed quantization curve. The smallest step size allowed by the overall quantization is equivalent to the smallest step size defined in μ-law or A-law PCM. The largest step size is 1024 times the smallest step size. It should be noted that the gain factor itself and other signals used in its computation are in base 2 logarithmic form. The adaptation of the scale factor y(k) is based on a "bimodal" adaptation technique (the quantizer is also called a dynamic locking quantizer). The scaling adaptation rate is "fast" for signals with large fluctuations, like speech, and "slow" for signals with small fluctuations, like voiceband data and tones. A purely stationary input such as a single tone causes the quantizer to stop adapting or to "lock". The overall speed of adaptation is a combination of the fast (unlocked) and slow (locked) scale factors.

The DPCM portion of the algorithm uses a linear predictor that is based on an autoregressive moving average (ARMA) process which has a combination of poles and zeros in its transfer function [4]. The structure for the predictor is based on several factors including stability in the presence of errors and the ability to track both speech and voiceband data signals. The adaptation of the predictor coefficients is based on a gradient search or steepest descent method and all coefficients are updated for each input sample. The output of the linear predictor is the signal estimate $s_e(k)$. This signal is subtracted from the input signal to form the difference signal that is actually coded and sent to the decoder.

The additional blocks added to the CCITT coder are for PCM format conversion, tone transition detection, and synchronization. The tone transition detection is added to determine when transitions between stationary tone signals occur. When a transition is detected all of the coefficients in the predictor are set to zero and the quantizer is set to the fast (unlocked) mode. The synchronization block is added to help prevent noise accumulation when multiple PCM/ADPCM/PCM conversions (synchronous tandem codings) are performed on a signal. This synchronization block does not affect the internal state of the decoder and has a minimal effect on the output quality of a single PCM/ADPCM/PCM conversion.

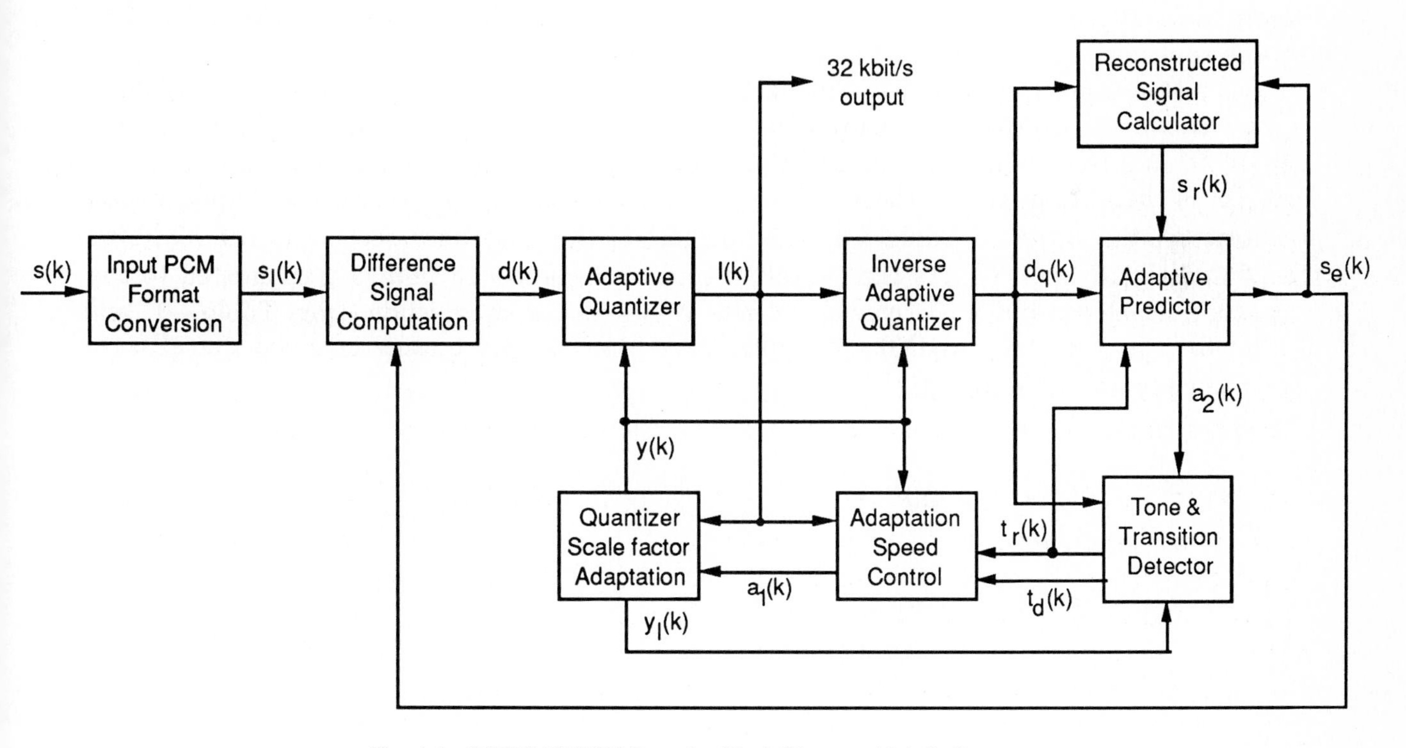

Fig. 4-1. CCITT ADPCM Encoder Block Diagram (detailed)

4.1 - The Encoder Algorithm

Fig. 4-1 shows a block diagram of the major portions of the CCITT ADPCM encoder. This section gives a detailed description of each block. The DSP56001 assembly code routines associated with each block are described in detail in Section 5.

The first stage of the encoder is shown in Fig. 4-2. The input to the encoder is the A-law or μ-law PCM signal s(k). This signal is converted to a uniform (linear) PCM signal $s_l(k)$ in the routine EXPAND. In the next routine, SUBTA, the difference signal d(k) is calculated by subtracting the signal estimate $s_e(k)$ from $s_l(k)$. As in all DPCM type coders it is the difference signal that is actually encoded and transmitted rather than a compressed version of the input signal.

$$d(k) = s_l(k) - s_e(k) \qquad (1)$$

In the next stage of the encoder, shown in Fig. 4-3, the difference signal d(k) is quantized by a 15-level non-uniform adaptive quantizer. The quantization process is performed by three routines. In the routine LOG the linear difference signal d(k) is converted to a base 2 logarithmic form, $d_l(k)$ representing the magnitude and $d_s(k)$ representing the sign. The scale factor y(k) (also in logarithmic form) is then subtracted from $d_l(k)$ in the routine SUBTB, in effect dividing the linear signal d(k) by a gain factor. This normalized signal $d_{ln}(k)$ is then quantized in the routine QUAN according to the normalized input/output characteristic shown in Table 4-1. The output of the quantizer section is the 32-kbit/s ADPCM signal I(k). This is the overall output of the encoder that is transmitted to the decoder. Each sample of I(k) contains 4-bits, 3 bits for the magnitude (from Table 4-1) and 1 bit for the sign (from $d_s(k)$). The quantizer is a 15-level quantizer since the all-zero codeword is not allowed.

INPUT		OUTPUT
$\log_2\lvert d(k)\rvert - y(k)$	$\lvert I(k)\rvert$	$\log_2\lvert d_q(k)\rvert - y(k)$
[3.12,+∞)	7	3.32
[2.72, 3.12)	6	2.91
[2.34, 2.72)	5	2.52
[1.91, 2.34)	4	2.13
[1.38, 1.91)	3	1.66
[0.62, 1.38)	2	1.05
[-0.98, 0.62)	1	0.031
(-∞, -0.98)	0	-∞

Table 4-1. Quantizer Normalized Input/Output Characteristic

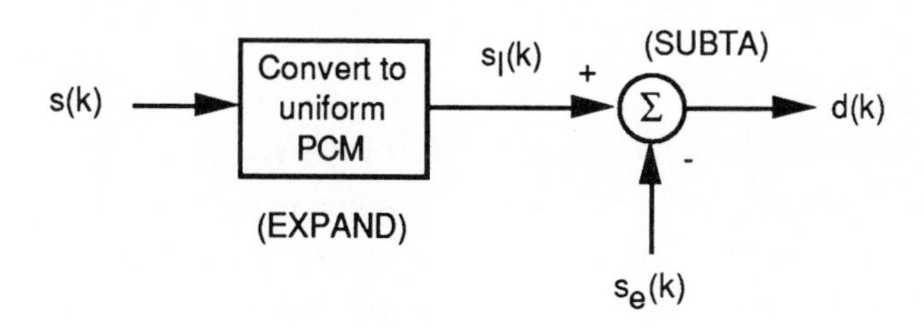

Fig. 4-2. PCM Conversion and Difference Signal Computation

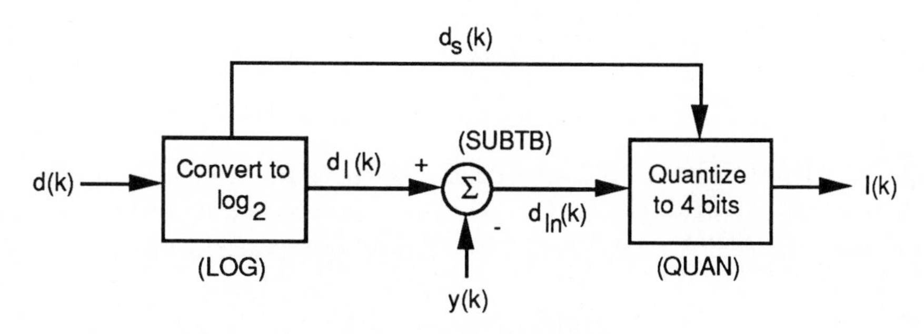

Fig. 4-3. Adaptive Quantizer

The inverse adaptive quantizer, shown in Fig. 4-4, converts the ADPCM signal I(k) into the signal $d_q(k)$, the quantized version of the difference signal. The inverse quantization is performed in three routines that are effectively the inverse of the three quantizer routines. The routine RECONST maps the magnitude of I(k) into one of 8 normalized output values shown in Table 4-1. The routine ADDA adds the scale factor y(k) (the same value as in the quantizer) to the normalized output value $d_{qln}(k)$, in effect multiplying by a gain factor. The routine ANTILOG then converts this logarithmic value $d_{ql}(k)$, along with the sign of I(k) ($d_{qs}(k)$), to the linear quantized difference signal $d_q(k)$.

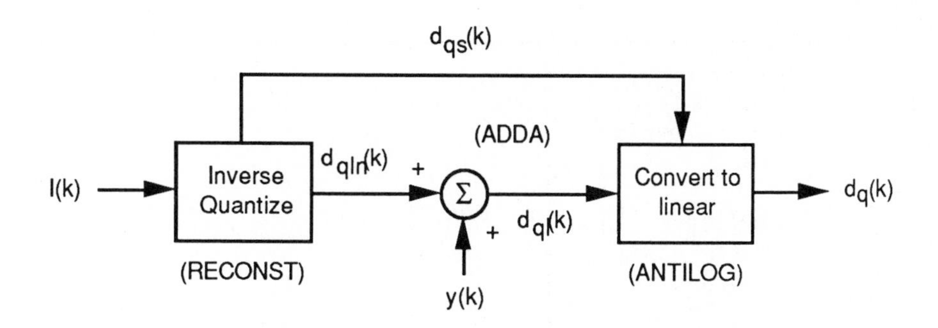

Fig. 4-4. Inverse Adaptive Quantizer

The remaining portion of the encoder performs three main functions: calculating the new signal estimate, performing the adaptation functions, and detecting tones.

The adaptive predictor's primary function is to use the past history of the quantized difference signal $d_q(k)$ to update the signal estimate $s_e(k)$. The linear predictor model used for the prediction consists of a sixth order section that models zeroes and a second order section that models poles. The prediction filter, shown in Fig. 4-5, is implemented in the routines FMULT and ACCUM using the equations:

$$s_{ez}(k) = \sum_{i=1}^{6} b_i(k-1) \bullet d_q(k-i) \qquad (2)$$

$$s_e(k) = \sum_{i=1}^{2} a_i(k-1) \bullet s_r(k-i) + s_{ez}(k) \qquad (3)$$

The CCITT standard specifies that the multiplies in equations (2) and (3) be done in floating point so the values of $d_q(k)$ and $s_r(k)$ must be converted to floating point. This is done in the routines FLOATA and FLOATB. The signal $s_r(k)$ in equations (2) and (3) is the reconstructed signal. It is calculated in the routine ADDB by adding the quantized difference signal $d_q(k)$ to the signal estimate $s_e(k)$, as shown in equation (4). The reconstructed signal represents the overall output of the ADPCM algorithm. The encoder does not output this signal but uses it as feedback for the prediction.

$$s_r(k-i) = s_e(k-i) + d_q(k-i) \qquad (4)$$

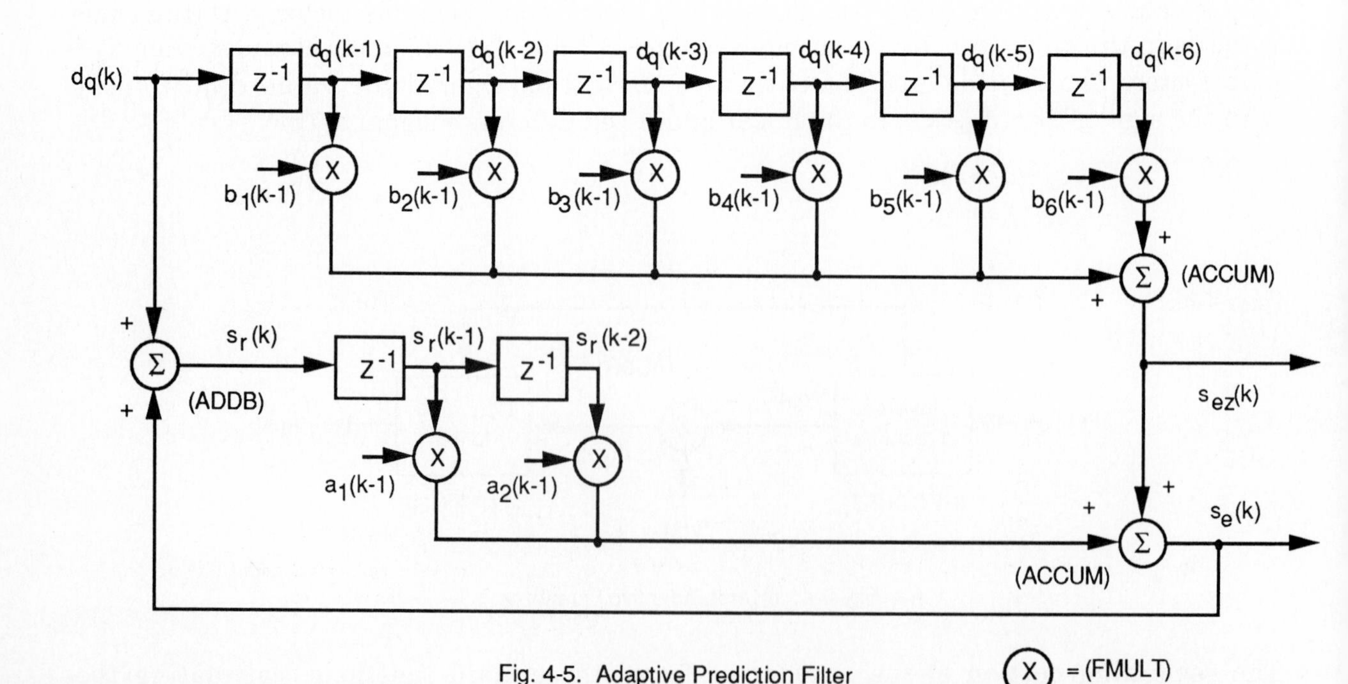

Fig. 4-5. Adaptive Prediction Filter

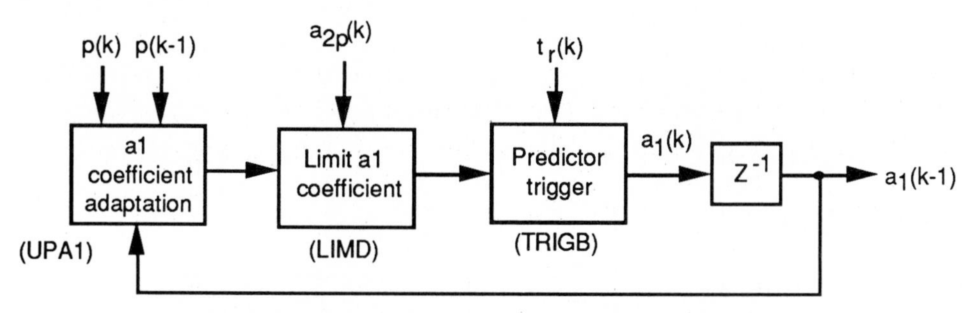

Fig. 4-6. Predictor Pole Coefficient Adaptation for $a_1(k)$

The predictor coefficients $a_i(k)$ and $b_i(k)$ are updated for each sample using a gradient search algorithm. The adaptation of the two pole coefficients, $a_1(k)$ and $a_2(k)$, is shown in Figs. 4-6 and 4-7 respectively. These coefficients are updated according to the following equations:

$$a_1(k) = (1 - 2^{-8})a_1(k-1) + (3 \bullet 2^{-8}) \operatorname{sgn}[p(k)] \operatorname{sgn}[p(k-1)] \quad (5)$$

$$a_2(k) = (1 - 2^{-7})a_2(k-1) + 2^{-7} (\operatorname{sgn}[p(k)] \operatorname{sgn}[p(k-2)] - f[a_1(k-1)] \operatorname{sgn}[p(k)] \operatorname{sgn}[p(k-1)]) \quad (6)$$

where

$$p(k) = d_q(k) + s_{ez}(k) \quad (7)$$

and

$$f(a_1) = 4 a_1 \qquad |a1| \leq 2^{-1}$$
$$= 2 \operatorname{sgn}(a_1) \qquad |a_1| > 2^{-1} \quad (8)$$

Equation (5) is implemented in the routine UPA1 while equations (6) and (7) are implemented in the routine UPA2. The calculation of equation (7) is performed in the routine ADDC. The coefficients $a_1(k)$ and $a_2(k)$ are constrained for stability reasons to the following limits:

$$|a_2(k)| \leq 0.75 \quad (9)$$

$$|a_1(k)| \leq 1 - 2^{-4} - a_2(k) \quad (10)$$

Equation (9) is calculated in the routine LIMC and equation (10) is calculated in the routine LIMD.

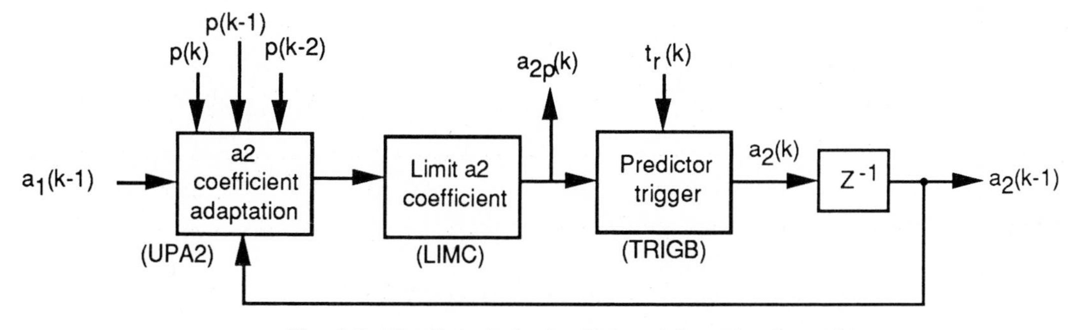

Fig. 4-7. Predictor Pole Coefficient Adaptation for $a_2(k)$

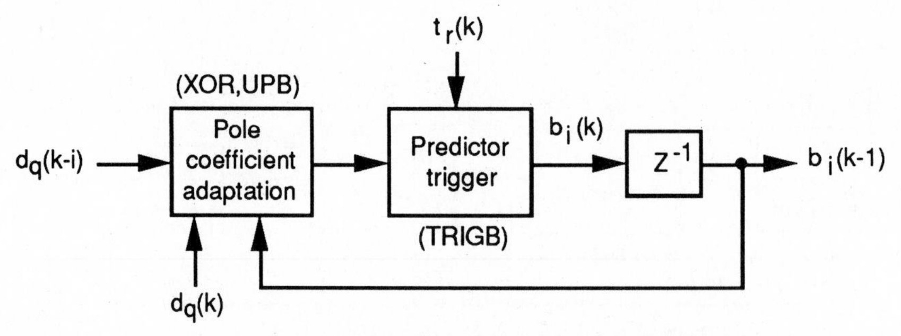

Fig. 4-8. Predictor Zero Coefficient Adaptation

The adaptation of the $b_i(k)$ zero coefficients is shown in Fig. 4-8. They are updated in the routines XOR and UPB according to:

$$b_i(k) = (1 - 2^{-8})\, b_i(k\text{-}1) + 2^{-7}\, \mathrm{sgn}[d_q(k)]\, \mathrm{sgn}[d_q(k\text{-}i)] \qquad (11)$$

$$\text{for } i = 1,2,..,6$$

The $b_i(k)$ coefficients are implicitly limited to $\pm$ 2.

The function sgn[x] in equations (5) - (11) represents the sign of x. It is defined as:

$$\begin{aligned} \mathrm{sgn}[x] &= +1 \quad \text{if} \quad x > 0 \\ &= -1 \quad \text{if} \quad x < 0 \\ &= +1 \quad \text{if} \quad x = 0 \text{ and } i \neq 0 \quad [\text{where } x = p(k\text{-}i) \text{ or } x = d_q(k\text{-}i)] \\ &= 0 \quad \text{if} \quad x = p(k) = 0 \text{ or } x = d_q(k) = 0 \end{aligned} \qquad (12)$$

The predictor coefficients may be further modified by the tone transition signal $t_r(k)$. The routine TRIGB tests $t_r(k)$ for a transition detection and sets the predictor coefficients to 0 if a transition is detected.

$$\text{If} \quad t_r(k) = 1 \quad \text{then} \quad a_i(k) = b_i(k) = t_d(k) = 0 \qquad (13)$$

The adaptation of the scale factor y(k) is based on information from past values of I(k) and the speed control parameter $a_l(k)$. The overall speed of adaptation is a combination of the fast and slow scale factors $y_u(k)$ and $y_l(k)$. The speed control parameter $a_l(k)$ determines how the fast and slow scale factors are combined. A diagram of this process is shown in Fig. 4-9.

The fast (unlocked) scale factor $y_u(k)$ is calculated in the routine FILTD by the equation:

$$y_u(k) = (1 - 2^{-5})\, y(k) + 2^{-5}\ W[I(k)] \qquad (14)$$

The routine LIMB constrains $y_u(k)$ to the limits:

$$1.06 \leq y_u(k) \leq 10.00 \qquad (15)$$

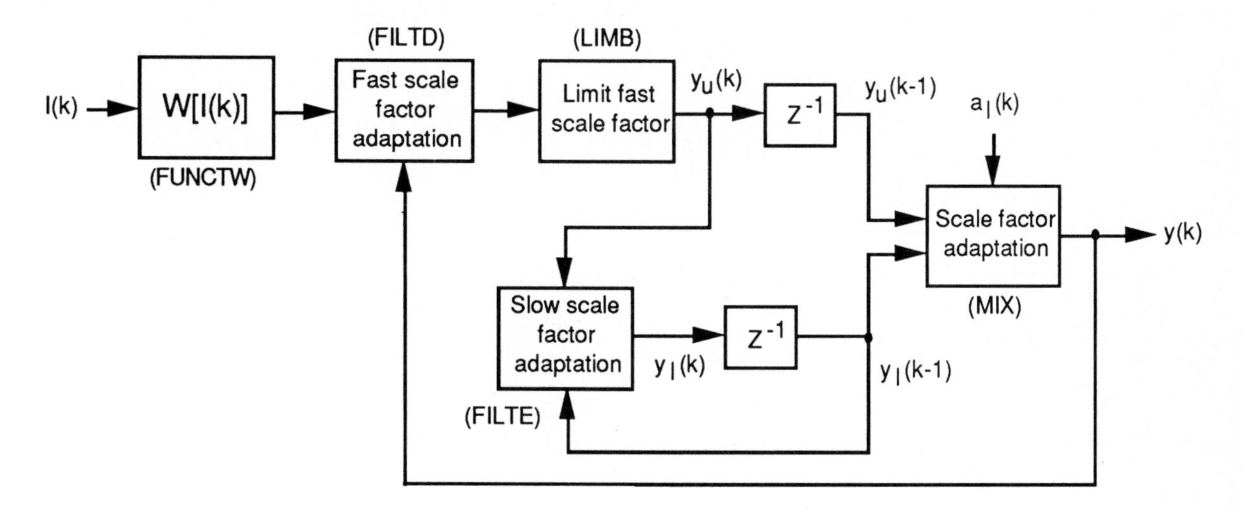

Fig. 4-9. Scale Factor Adaptation

The function W(I) in equation (14) is calculated in the routine FUNCTW according to Table 4-2.

|I(k)|	7	6	5	4	3	2	1	0
W(I)	70.13	22.19	12.38	7.00	4.00	2.56	1.13	-0.75

Table 4-2. W(I) Lookup Table

The slow (locked) scale factor $y_l(k)$ is calculated in the routine FILTE by the equation:

$$y_l(k) = (1 - 2^{-6})\, y_l(k-1) + 2^{-6}\, y_u(k) \tag{16}$$

The overall scale factor y(k) is determined in the routine MIX by the equation:

$$y(k) = a_l(k)\, y_u(k-1) + [1 - a_l(k)]\, y_l(k-1) \tag{17}$$

The speed control parameter $a_l(k)$ is limited to the range $0 \le a_l(k) \le 1$. The value of $a_l(k)$ approaches 1 for speech signals in which case the fast (unlocked) scale factor $y_u(k)$ dominates in equation (17). For data signals and tones however, the value of $a_l(k)$ approaches 0 in which case the slow (locked) scale factor $y_l(k)$ dominates in equation (17). The value of $a_l(k)$ is determined primarily by the rate-of-change of the difference signal, which is encoded in I(k). The update of $a_l(k)$ is shown in Fig. 4-10.

Two measures of the difference signal value are used to update the speed control parameter, $d_{ms}(k)$ and $d_{ml}(k)$. The signal $d_{ms}(k)$ represents the "short term" average of the function F[I(k)], while $d_{ml}(k)$ represents the "long term" average of F[I(k)]. The value of F[I(k)] is a weighted function of I(k) and is determined in the routine FUNCTF according to Table 4-3. The difference signals $d_{ms}(k)$ and $d_{ml}(k)$ are calculated in the routine FILTA and FILTB respectively according to the equations:

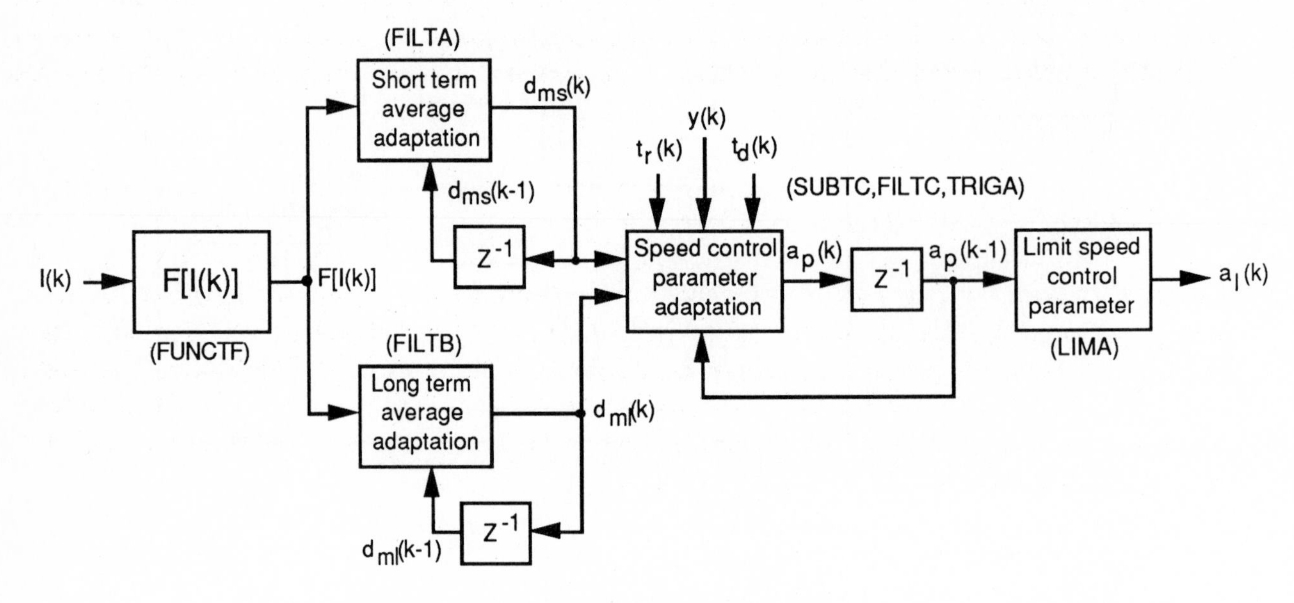

Fig. 4-10. Speed Control Parameter Adaptation

$$d_{ms}(k) = (1 - 2^{-5})\, d_{ms}(k-1) + 2^{-5}\, F[I(k)] \tag{18}$$

$$d_{ml}(k) = (1 - 2^{-7})\, d_{ml}(k-1) + 2^{-7}\, F[I(k)] \tag{19}$$

The values of $d_{ms}(k)$ and $d_{ml}(k)$ are used to determine the "unlimited" speed control parameter $a_p(k)$. The value of $a_p(k)$ is calculated in the routines SUBTC, FILTC, and TRIGA by the equation:

$$\begin{aligned} a_p(k) &= (1 - 2^{-4})a_p(k-1) + 2^{-3} && \text{if} \quad |d_{ms}(k) - d_{ml}(k)| \geq 2^{-3} d_{ml}(k) \\ &= (1 - 2^{-4})a_p(k-1) + 2^{-3} && \text{if} \quad y(k) < 3 \\ &= (1 - 2^{-4})a_p(k-1) + 2^{-3} && \text{if} \quad t_d(k) = 1 \\ &= 1 && \text{if} \quad t_r(k) = 1 \\ &= (1 - 2^{-4})a_p(k-1) && \text{otherwise} \end{aligned} \tag{20}$$

The value of $a_p(k)$ tends towards 2 if the difference between $d_{ms}(k)$ and $d_{ml}(k)$ is large, indicating the rate-of-change of the difference signal is fast, but it tends towards 0 if the difference is small, indicating the rate-of-change is slow. $a_p(k)$ also tends towards 2 when tones are detected ($t_d(k) = 1$) or in idle channel conditions ($y(k) < 3$). When a tone transition is detected ($t_r(k) = 1$), $a_p(k)$ is explicitly set to 1.

|I(k)|	7	6	5	4	3	2	1	0
F[I(k)]	7	3	1	1	1	0	0	0

Table 4-3. F[I(k)] Lookup Table

To form the desired speed control parameter $a_l(k)$, the parameter $a_p(k)$ is constrained in the routine LIMA to the limits given in equation (21). This limiting has the effect of delaying a state transition start until the magnitude of the difference signal become relatively constant.

$$a_l(k) = 1 \quad \text{if} \quad a_p(k\text{-}1) > 1 \qquad\qquad (21)$$
$$= a_p(k\text{-}1) \quad \text{if} \quad a_p(k\text{-}1) \le 1$$

The final step of the encoder is tone detection, shown in Fig. 4-11. It is included to improve the performance of the coder in the presence of non-speech voice-band data signals that may be present on a typical analog phone line (e.g. DTMF tones and data modems). The tone detect signal $t_d(k)$ indicates the presence of a tone. When a tone is detected $t_d(k)$ causes the quantizer to be driven into the fast mode of adaptation. $t_d(k)$ is calculated in the routine TONE by the equation:

$$t_d(k) = 1 \quad \text{if } a_2(k) < -0.71875 \qquad\qquad (22)$$
$$= 0 \quad \text{otherwise}$$

The tone detect signal $t_d(k)$ also causes the tone transition detect signal $t_r(k)$ to be set when a transition between tones occurs. This signal sets the predictor coefficients to 0 and the tone detect signal to 0 (in the TRIGB routine) so that the fast adaptation mode will take effect immediately. $t_r(k)$ is determined in the routine TRANS by the equation:

$$t_r(k) = 1 \quad \text{if} \quad a_2(k) < -0.71875 \text{ and } |d_q(k)| > 24 \bullet 2^{y_l(k)} \qquad\qquad (23)$$
$$= 0 \quad \text{otherwise}$$

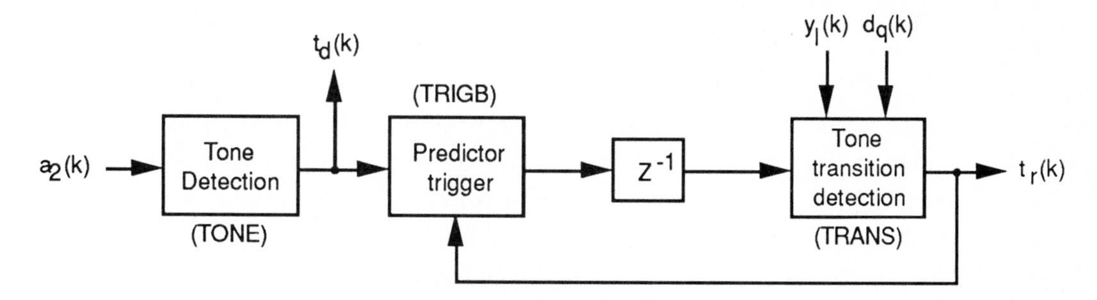

Fig. 4-11. Tone Detection

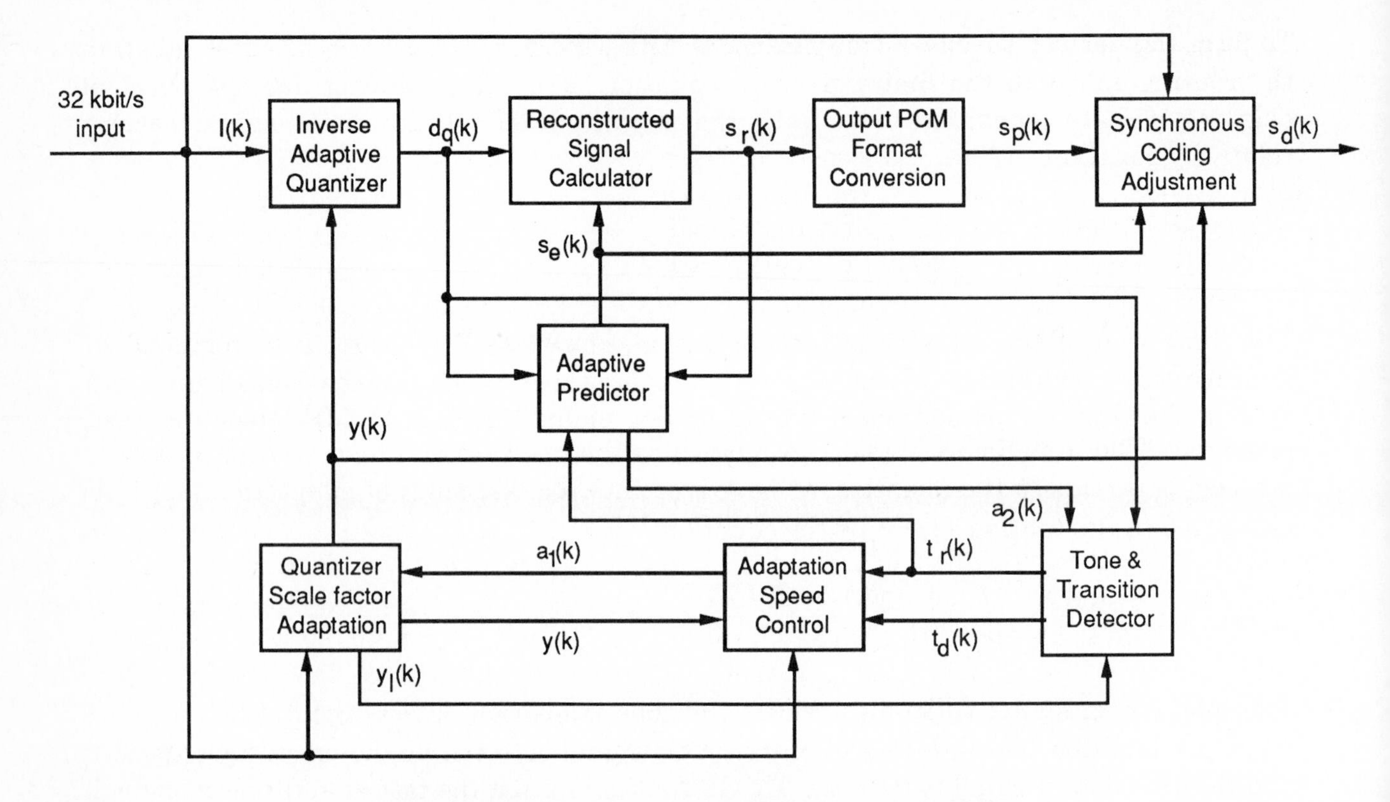

Fig. 4-12. CCITT ADPCM Decoder Block Diagram (detailed)

4.2 - The CCITT Decoder Algorithm

A detailed block diagram of the decoder process is shown in Fig. 4-12. As mentioned previously the CCITT ADPCM coder uses the feedback scheme and one of the properties of this scheme is that the encoder and decoder are almost identical in terms of function. The decoder portion of the CCITT algorithm uses the same routines as the encoder for the inverse quantization, linear prediction, tone detection, and adaptation functions. The input to the decoder, I(k), is the same I(k) used in the encoder for adaptation and prediction. The decoder executes all of the same routines as the encoder (and in the same order) except for the input PCM conversion (EXPAND), the difference signal computation (SUBTA), and the adaptive quantization (LOG, SUBTB, and QUAN). These sections are normally not needed in the decoder but they are used in the CCITT algorithm however, as explained below.

The linear output of the decoder is the reconstructed signal $s_r(k)$ as calculated in the routine ADDB. This is converted to the A-law or μ-law PCM signal $s_p(k)$ in the routine COMPRESS. This would normally be the final output but instead this signal is passed through a synchronous coding adjustment block, shown in Fig. 4-13. The purpose of this block is to prevent cumulative distortions that may occur with "synchronous tandem codings" - multiple ADPCM/PCM/ADPCM conversions on a transmission path. These distortions can only be prevented when the transmission paths are error-free and when no extra digital signal processing functions are performed on intermediate PCM and ADPCM signals.

As noted previously, the encoder and decoder will be in the same internal "state" (all internal variables the same) if there are no transmission errors. Assuming this is the case the decoder estimates the quantization that occurred in the encoder and forces the ADPCM sequence that it reconstructs to match the ADPCM sequence that it received. The decoder does so by converting the PCM signal $s_p(k)$ back to a linear signal $s_{lx}(k)$ in the EXPAND routine. Then a new difference signal $d_x(k)$ is calculated in the SUBTA routine by the equation:

$$d_x(k) = s_{lx}(k) - s_e(k) \tag{24}$$

The new difference signal $d_x(k)$ is then converted to the normalized logarithmic signal $d_{lnx}(k)$ in the routines LOG and SUBTB. The same quantization as in the encoder then occurs in the routine SYNC. But this routine also does a comparison of the new coded ADPCM signal to the received ADPCM signal I(k). The final PCM output of the decoder, $s_d(k)$, is determined by this comparison, defined by:

$$\begin{aligned} s_d(k) &= s_p(k)^+ \quad \text{if} \quad d_x(k) < \text{lower interval boundary} \\ &= s_p(k)^- \quad \text{if} \quad d_x(k) \geq \text{upper interval boundary} \\ &= s_p(k) \quad \text{otherwise} \end{aligned} \tag{25}$$

where

$s_p(k)^+$ = the PCM code word that represents the next more positive PCM output level

$s_p(k)^-$ = the PCM code word that represents the next more negative PCM output level

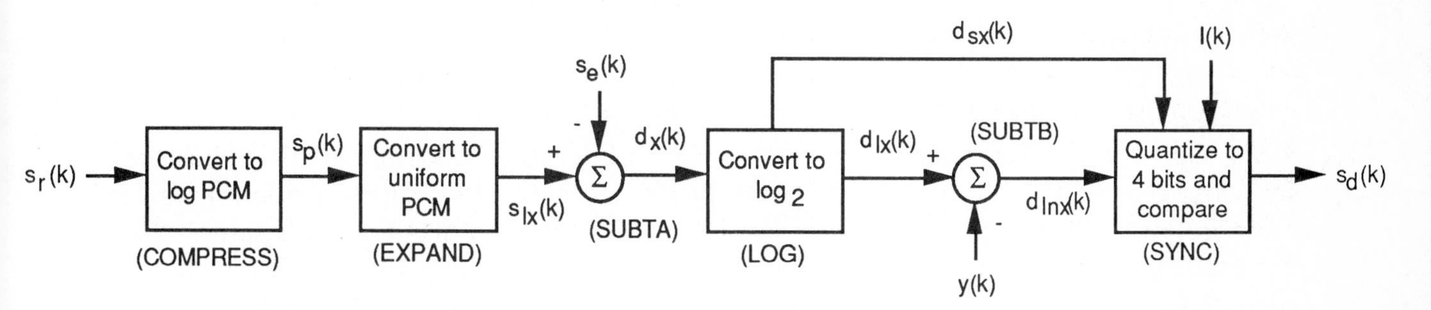

Fig. 4-13. Synchronous Coding Adjustment

Section 5 - ADPCM Implementation on the DSP56001

The CCITT ADPCM algorithm described above has been implemented on the DSP56001 in a full-duplex configuration. The assembly source code described herein is available on the Motorola DSP bulletin board (Dr. Bub) under the name ADPCM.ASM. It has been optimized so that it is able to perform both the encode and the decode portions of the algorithm in real-time on a single DSP56001 running at 27 MHz with external program memory. The source code is set up to run the CCITT test sequences specified in Appendix II of Recommendation G.721 (1986 version) [1]. The code is a bit-for-bit compatible implementation of the CCITT specification and correctly passes all μ-law and A-law test sequences provided by the CCITT. The encoder and decoder portions of the source code are designed to be independent of the I/O sources so that the code can be easily modified to interface to real-time hardware. This has been done to demonstrate real-time performance. The code can also be easily modified for half-duplex configurations. This will permit real-time performance on a slower speed DSP56001 or will allow other tasks to be performed on the 27 MHz DSP56001.

This section discusses the details of the implementation of the CCITT ADPCM algorithm on the DSP56001. This application report by itself provides a basic description of the source code but for a complete understanding of the DSP56001 code it is essential to refer to the CCITT document. Many of the details in Recommendation G.721 are not included in this document but have a significant impact on the assembly implementation. In may cases the code does not implement the equations given in this document in a straightforward manner due to the way the specification is written. Also, many of the comments in the source code refer to the notations used in Recommendation G.721. (Some of the basic terminology is included in Appendix A of this document).

5.1 - I/O Interface

The DSP56001 ADPCM source code is set up to run the CCITT test sequences on the Motorola DSP56000ADS board. The program simulates the PCM and ADPCM interfaces by using the file I/O routines on the ADS board. The file I/O routines allow programs running on the ADS board to access data in ASCII files on the host computer. These routines provide a convenient method for accessing the CCITT test files which are distributed in ASCII format. The ADS does require that the data in the test files be in a slightly different format than that provided by the CCITT. The details of this format can be found in the file ADPCM.HLP located with the ADPCM source code on the Motorola DSP bulletin board. The source code is set up to process a PCM input file and an ADPCM input file simultaneously. Two output files are written, one for the encoded ADPCM output and one for the decoded PCM output. When running the CCITT test sequences these output files can be compared to the CCITT files to verify correct operation. This procedure is also discussed in the ADPCM.HLP file. The data in the two files being processed does not have to be related in any way since the encoder and decoder are designed to operate on two independent signals. Additionally, any file containing PCM data in ASCII hex characters may be used as input to the encoder, and likewise any file containing ADPCM data can be used as input to the decoder. It should be noted however that the file I/O routines on the ADS are not designed for high-speed data transfer so that processing data files with the DSP56001 ADPCM program will not be in real-time.

As mentioned previously the I/O interface of the ADPCM source code is designed to be flexible for a variety of configurations. To test the real-time operation of the implementation a version of the source code was modified to interface to a Motorola MC145503 CODEC though the Synchronous Serial Interface (SSI) of the DSP56001. The only portion of the code that was changed was the file I/O interface sections. In the test configuration the encoder received real-time PCM data from the CODEC, each sample was coded, then decoded, and sent back out to the same CODEC. The algorithm portion of the code did not need to be modified.

5.2 - Standard Implementation

The standard version of the DSP56001 ADPCM source code implements the ADPCM algorithm exactly as specified by the CCITT. The advantage of using this version is that the user can have confidence that the DSP56001 implementation will perform exactly as specified by the CCITT. This includes performance with non-speech signals and in special operating conditions. The disadvantage of this version is that the specification does not always allow the equations to be implemented in an efficient manner on a general purpose digital signal processor. An example is the multiply and accumulation portion of the linear predictor. The standard specifies that the multiply be done in a floating-point format while the accumulation be done in 16-bit fixed point format. Not only is the floating-point multiply less efficient than a native 24-bit fixed point multiply on the DSP56001, but several conversions between fixed and floating-point formats are required for each sample (this is discussed in Section 5.2.5). This is the most time consuming part of the algorithm but other parts of the specification also do not permit efficient implementation on a programmable microprocessor.

The standard implementation has been written with two main objectives that have priority over other considerations. The first objective is the requirement of complete adherence with the CCITT specification in order to maintain bit-for-bit compatibility with the CCITT test sequences. The second objective is the desire to obtain a full-duplex solution with real-time operation on a single DSP56001. These 2 goals may help to clarify the way the algorithm has been implemented.

5.2.1 - Code Structure

The assembly program for the standard version of the ADPCM algorithm is structured as two main routines, the encoder (transmit) and the decoder (receive), plus one subroutine for initialization. After the initialization routine, the encoder routine is executed, followed by the decoder. The code then alternates between the encoder and decoder indefinitely. A flow diagram of the encoder and decoder is shown in Fig. 5-1. This shows the order in which the various portions of the algorithm are executed. The order of execution of the individual CCITT routines along with their execution speed is given in Section 5.4.

The encoder and decoder routines are designed to operate as independent code segments. They do assume that appropriate variables are stored in data memory and that appropriate pointers have been set. In particular, address registers r1, r2, r6, and r7 should contain the appropriate memory addresses prior to executing the encoder and decoder sections. Registers r3 and r5 should contain constant address values used for table lookup. Registers r0 and r4 do not need to be initialized since they are used as general purpose registers. The encoder and decoder are not set up as subroutines in the program. If interrupts are used for data I/O they can easily be made into subroutines or interrupt routines. If this is done, it should be noted that one routine should not interrupt the other routine until it is completely finished executing.

No subroutines are used within the encoder or the decoder so that optimal speed can be obtained. The code has also been optimized to take advantage of the DSP56001's architecture as much as possible. This causes the various CCITT routines in the code to "overlap" in many cases, meaning that variables and data values for one routine may be read from memory while the previous routine is still executing. In one case, the XOR and UPB routines are actually combined into a single section of code. This optimization makes the code more difficult to follow in some cases but extra comments have been added to try and clarify most of these cases. Further discussion of this optimization technique can be found in Section 5.3.

5.2.2 - Initialization

The initialization subroutine accomplishes three main tasks: initializing the DSP56001, initializing program variables and lookup tables, and initializing data buffer pointers and modulo registers. This section can also include any I/O interface configuration that is necessary. The memory map of the internal data RAM is shown in Fig. 5-2. The encoder and decoder algorithms each require several variables to be stored in memory and the DSP56001 code also requires several other temporary storage locations. Most of the variables used by the algorithm are stored in data memory below address $40 so that the short immediate addressing mode can be used when accessing them. Data locations that are accessed with an addressing register are stored at higher locations in data memory since they do not need to use immediate addressing modes.

The DSP56001 initialization consists of enabling the on-chip, factory programmed data ROM tables and setting the DSP56001's Bus Control Register (BCR) for zero wait state external program memory access. The μ-law and A-law ROM table in X memory is needed for the log PCM conversion routine. The zero wait states for external program memory are needed so the algorithm will run in real time.

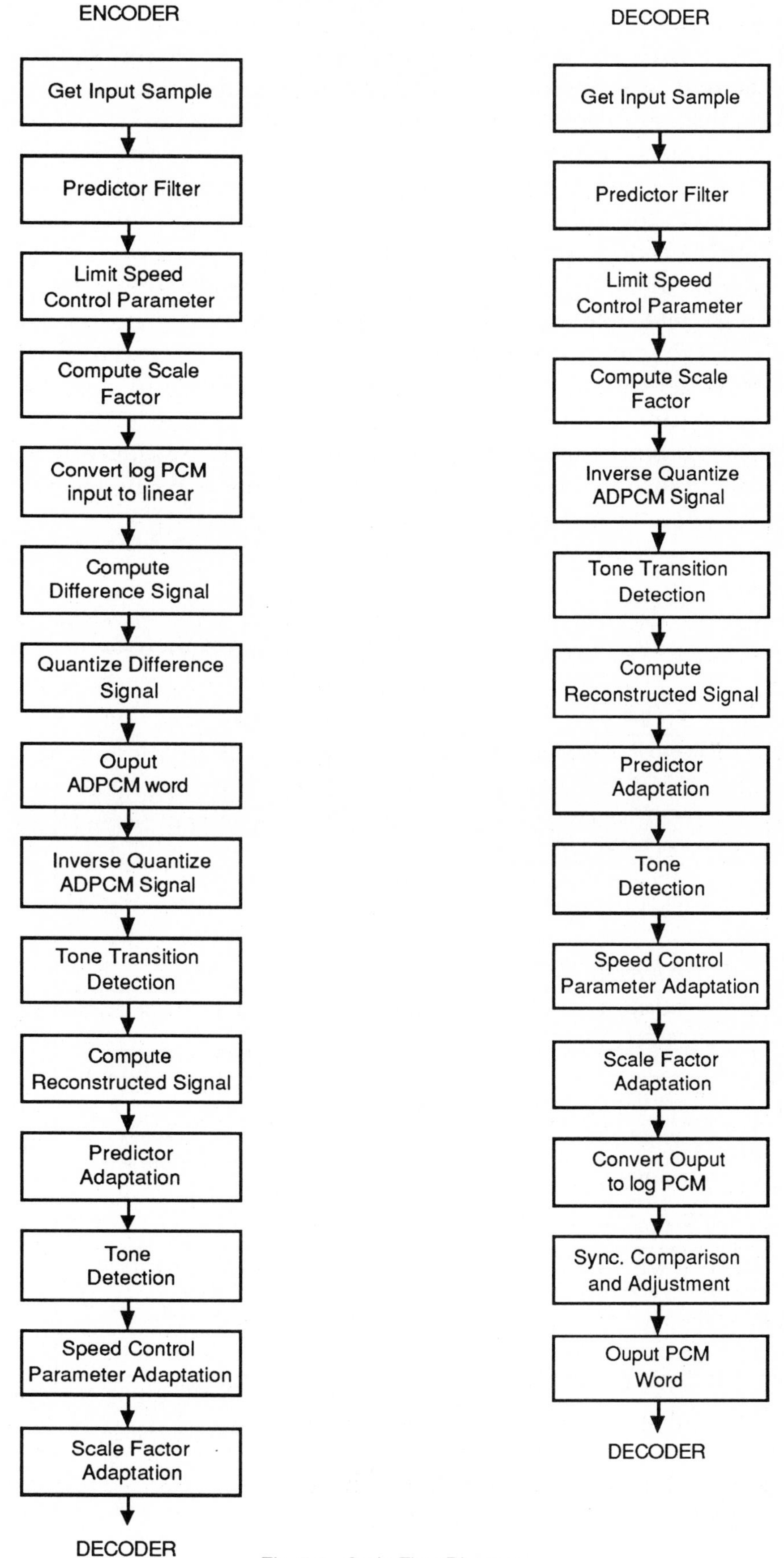

Fig. 5-1. Code Flow Diagram

The program initialization is done in the subroutine INIT. First all internal X and Y data RAM is cleared, then all variables that require specific values are initialized, and then all lookup tables are copied from their load-time locations in program memory to their run-time locations in data RAM. Next the pointers to the receive (decode) and transmit (encode) data buffers are initialized. These buffers hold the delayed values of $d_q(k)$ and $s_r(k)$ used in the linear predictor filter. These are the only true modulo buffers used in the assembly code in the sense that the newest delayed values replace the oldest delayed values without actually moving the other delayed values. The INIT routine initializes the sign and mantissa locations in these buffers since the code assumes a certain range of legal values in these locations. The INIT routine also initializes other variables including the variable LAW. It determines whether the μ-law or A-law format is chosen. The program defaults to setting LAW to zero to select μ-law for the PCM format. The code can be changed to set LAW to any non-zero value which will select the A-law format. It can also be easily modified to select μ-law or A-law based on an external input.

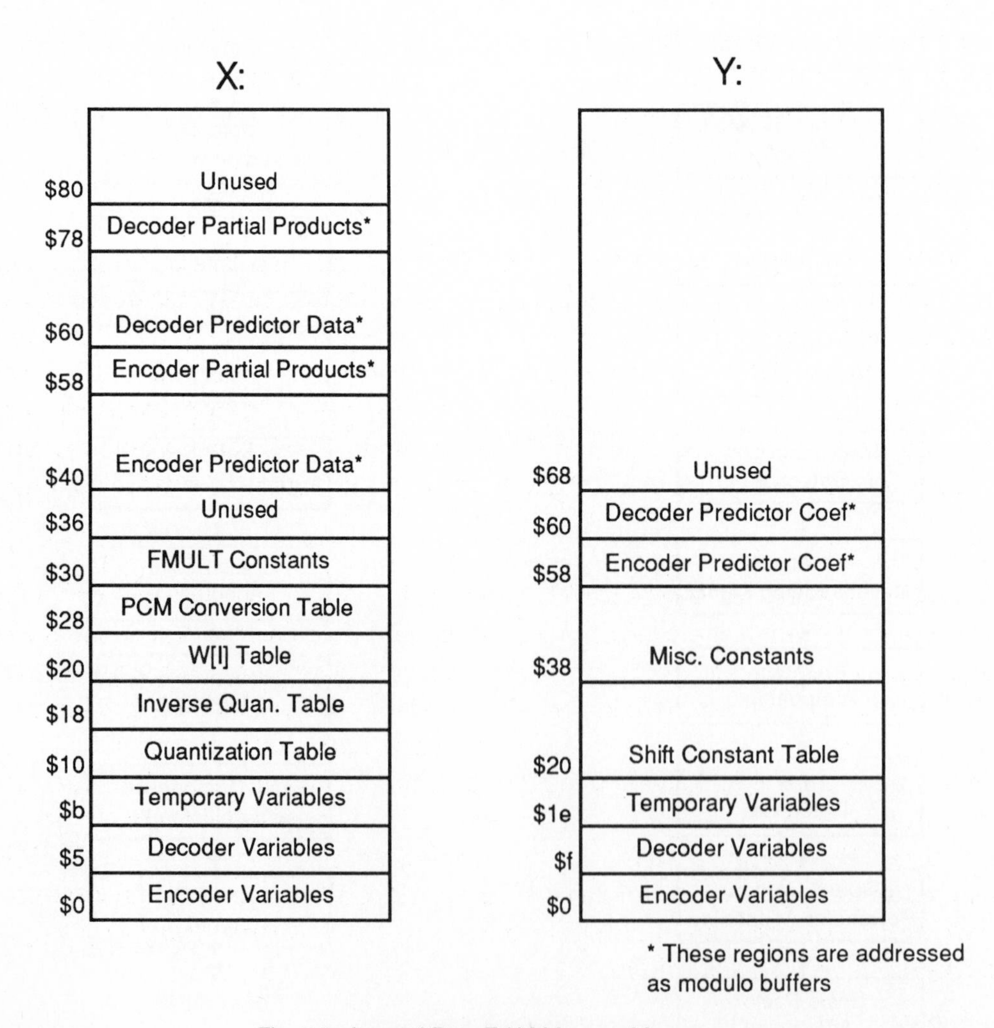

Fig. 5-2. Internal Data RAM Memory Map

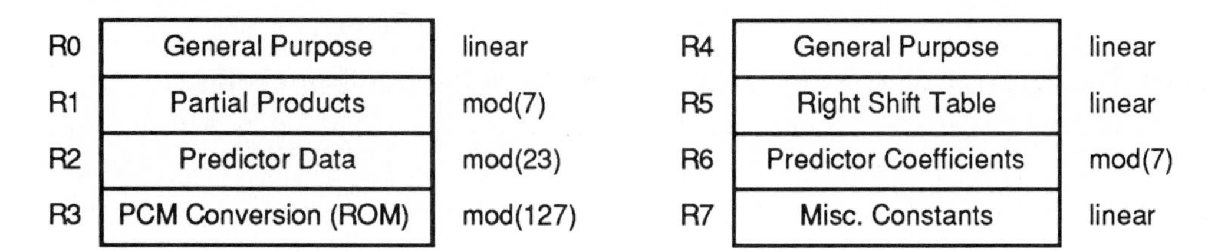

Fig. 5-3. Address Register Usage

The remaining portions of the INIT routine initialize the addressing registers and modifier registers that are used to access the data memory. Six of the variable storage areas (indicated by * in Fig. 5-2) are addressed using modulo pointers. These six areas, three each for the encoder and decoder, are used by the linear predictor filter. The complete addressing register assignments along with the associated addressing modes are shown in Fig. 5-3. Six of the addressing registers are reserved for particular functions, and the remaining two are used for other general purpose tasks requiring addressing registers. The INIT routine can also contain any initialization needed for a hardware interface such as the SSI port. No initialization is needed to use the file I/O routines on the ADS so I/O initialization is not included in the standard code.

5.2.3 - PCM Format Conversion

There are several different types of numeric formats used in the ADPCM algorithm. Conversion between these formats is required in several places in the assembly code. The log PCM format conversion is one of these. Two different routines are used for converting between μ-law or A-law PCM samples and linear (uniform) samples. EXPAND converts an 8-bit log PCM sample to a linear 14-bit two's-complement representation that is suitable for numeric operations. COMPRESS performs the opposite conversion. The DSP56001 ADPCM implementation supports both μ-law and A-law format conversion.

The routine EXPAND performs the conversion by use of the internal μ-law and A-law ROM tables on the DSP56001. Register r3 is used as the pointer into the lookup table. It is set during program initialization to either the μ-law table base or the A-law table base and remains set to this value while the program is running. Register n3 is used as an offset into the ROM table. The assembly code for EXPAND is identical for both formats and is only dependent on the base pointer stored in r3 so separate routines are not needed. To obtain optimal speed the COMPRESS routine requires separate code sections for the μ-law and A-law conversion. In this routine the variable LAW is tested for each sample. If only one format is used the variable LAW may be eliminated and one section of COMPRESS may be removed to save program memory. The SYNC routine discussed in Section 5.2.10 also requires separate code segments for each log PCM format, so similar program memory savings can be obtained there.

The EXPAND and COMPRESS routines are modified versions of routines given in the Motorola applications brief "Logarithmic/Linear Conversion Routines For The

DSP56000/1" [6]. They were chosen based on maximum execution speed. Complete descriptions of each routine can be found in the above applications brief.

5.2.4 - Logarithmic Conversion

Another conversion required in the ADPCM algorithm is between a linear and a base 2 logarithmic format. The quantizer and the inverse quantizer achieve their adaptive characteristic by use of a scale factor. The scale factor itself and many of the variables used to calculate it are in a base 2 logarithmic form. The total number of bits used for these variables differs but they all share a common form of a mixed number. The numeric operations are performed on these numbers assuming an integer exponent portion combined with a fractional mantissa portion in one mixed number. The scale factor adaptation does not require a specific conversion to this format but a conversion is required in the quantizer and inverse quantizer. The routine LOG in the quantizer converts the difference signal to the base 2 logarithmic form so that it can be modified by the scale factor. In the inverse quantizer the routine ANTILOG converts the quantized difference signal back to the linear form after it has been readjusted by the scale factor. In fact the ADPCM codeword is based on the log of the difference signal d(k) rather than the difference signal itself.

A version of the LOG routine is shown in Fig. 5-4 to illustrate the conversion process. The input D is in a 16-bit two's complement format. The first step is to convert this number to a sign magnitude representation saving the sign in register y1. After checking the magnitude for a non-zero value it is then normalized to determine the exponent and mantissa. The iterative NORM instruction is used for this conversion. This instruction will shift the magnitude left one bit for each iteration until a 0 is in bit 23 and a 1 is in bit 22 (this is the normalized fraction format on the DSP56001). For each left shift the value in r0 is decremented. Once the magnitude has been normalized successive iterations will do no further adjustments. Fourteen iterations are performed since the maximum that the magnitude can be shifted is 14 bits, assuming the magnitude is non-zero. Fourteen iterations of the NORM is not needed in all cases but taking time to test after each iteration would cause the worst case delay to be longer. After the normalization process is finished the normalized mantissa will be in accumulator a and the associated exponent will be in register r0.

The remaining instructions combine the exponent and mantissa into a mixed number. The truncation of the mantissa to seven bits is performed by using a mask instead of actually shifting. This technique is common throughout the code. The process of combining the exponent and mantissa also shows the technique of shifting by multiplication. The exponent is moved from r0 to x1 where it will be in the four LSB's but it needs to be left justified to bits 22-19 which are the four MSB's of the DSP56001's fractional format. A shift constant is read from the shift constant table in Y memory and is multiplied with the exponent. The result is that the exponent is effectively shifted left nineteen bits. A shift constant is also used to shift the mantissa right by three bits. In this example the mantissa is shifted right and combined with the exponent in a single

MAC instruction. This shift technique is described in further detail in Section 5.3. The resulting log signal DL is a mixed number with four integer bits and seven fractional bits with an implied radix point. Note that this logarithmic format is similar to the mixed number format discussed in [9].

The routine ANTILOG in the inverse quantizer performs the opposite conversion. It does so by splitting the exponent and mantissa apart and then shifting the mantissa right again according to the exponent.

```
;**************************************************************************
;       LOG
;
; Convert difference signal from the linear to the log
;   domain
;
; Input:
;   D = siii iiii | iiii iiii. | 0000 0000  (16TC) in accum A
;
; Outputs:
;   DL = 0iii i.fff | ffff 0000 | 0000 0000  (11SM) in accum A
;   DS = sXXX XXXX | XXXX XXXX | 0000 0000  (1TC) in Y1
;
;**************************************************************************

        MOVE    #$000E,R0              ;Get exp bias (14)
        MOVE    X:Y_T,B                ;Get Y
        ABS     A   A,Y1               ;Find DQM=|D|, save DS to Y1
        JNE     <NORMEXP_T             ;Check for DQM=0
        CLR     A         (R7)+        ;If DQM=0 set DL=0
        JMP     <SUBTB_T
NORMEXP_T
        REP     #14                    ;If DQM!=0, do norm iteration
        NORM    R0,A                   ; 14 times to find MSB of DQM

;   A1 = 01?? ???? | ???? ???? | 0000 0000 = normalized DQM  (A2=A0=0)
;   R0 = 0000 0000 | 0000 eeee = exponent of normalized DQM

;   Get rid of leading "1" in normalized DQM
;   Truncate mantissa to 7 bits and combine with exponent

        MOVE    Y:(R7)+,X1              ;Get mask K6 ($3F8000)
        AND     X1,A    Y:LSHFT-19,X0   ;Truncate MANT, get EXP shift

;   A1 = 00mm mmmm | m000 0000 | 0000 0000 (A2=A0=0)

        MOVE    R0,X1                  ;Move EXP to X1
        MPY     X0,X1,A    A,X1        Y:(R7)+,Y0 ;Shift EXP<<19, save MANT to X1,
                                       ; get mask K7 ($100000)
        MOVE    A0,A                   ;Move EXP to A1

;   X1 = 00mm mmmm | m000 0000 | 0000 0000
;   A1 = 0eee e000 | 0000 0000 | 0000 0000 (A2=A0=0)

        MAC     Y0,X1,A                ;Shift MANT>>3 & combine with EXP

;   A1 = 0eee e.mmm | mmmm 0000 | 0000 0000  (A2=A0=0)
;      = 0iii i.fff | ffff 0000 | 0000 0000  (A2=A0=0)
```

Fig. 5-4. Linear to Log Conversion Routine

5.2.5 - Floating-Point Conversion

The other type of conversion required in the ADPCM algorithm is a floating-point conversion. The CCITT specifies that the multiplications in the linear predictor filter be done in a specific floating-point format. After each multiply the result must be converted back into fixed point before it is accumulated with the other partial products. The data inputs to the predictor filter are the delayed values of $d_q(k)$ and $s_r(k)$. For each input sample these values are converted to floating-point and then stored in this form so they do not have to be converted again. The coefficients of the predictor filter $a_i(k)$ and $b_i(k)$ are updated in fixed point form for each sample so they must be converted from linear to floating-point form each sample. Since there are six zeros and two poles in the predictor filter, the overall requirement is ten fixed point to floating-point conversions and eight floating-point to fixed point conversions. Clearly this conversion process has a major impact on the execution speed of the overall algorithm so it is desired that it be as fast as possible. Sixteen of these conversions are performed in the FMULT routine and the other two are performed in the FLOATA and FLOATB routines so much of the speed emphasis is placed on the FMULT routine. Again the overall goal is the minimum worst case execution time.

The floating-point format used in this algorithm consists of four exponent bits, six mantissa bits (with an explicit leading 1), and one sign bit for a total of eleven bits (11FL). As mentioned, the values of $d_q(k)$ and $s_r(k)$ are stored in the floating-point format. The coefficients do not need to be stored in this format since they are used immediately after they are converted in the FMULT routine. Further details of the FMULT routine are given in the adaptive predictor section but to illustrate the conversion process a version of FLOATA is shown in Fig. 5-5. It can be seen that the process of normalizing the input value is very similar to that in the logarithmic conversion routine LOG. The main difference is that the sign, exponent, and mantissa components are not combined but are stored separately in the data buffer using register r2 as a pointer. Notice that the form in which they are stored was chosen because it requires less data manipulation and shifting. Also notice that the complete value of the two's complement number is stored as the sign. Only the sign of the value is important so there is no need to separate the sign from the rest of the number.

5.2.6 - Difference Signal Quantization

After the EXPAND conversion routine converts the input signal s(k) to the two's-complement signal $s_l(k)$, the routine SUBTA subtracts the signal estimate $s_e(k)$ from this value to form the difference signal d(k). The computation in SUBTA only requires aligning radix points and subtracting. The adaptive quantization of d(k) is not as straightforward. To perform the quantization d(k) must be normalized by the scale factor y(k). As noted previously the scaling and quantization is performed in base 2 log format. This conversion in the routine LOG is described in the Section 5.2.4.

```
;*************************************************************************
;        FLOATA
;
; Converts the quantized difference signal from 15-bit
;   signed magnitude to floating pt. format (11FL - sign,
;   exp, and mant stored separately)
;
; Inputs:
;   DQ = siii iiii | iiii iii.0 | 0000 0000 (15SM) in accum A
;
; Outputs:
;   DQ0 = (11FL)
;    DQ0EXP = X:(R2) = 0000 0000 | 0000 0000 | 0000 eeee
;    DQ0MANT = X:(R2+1) = 01mm mmm0 | 0000 0000 | 0000 0000
;    DQ0S = X:(R2+2) = sXXX XXXX | XXXX XXXX | 0000 0000
;
;*************************************************************************
;
; R2 points to predictor data buffer - DQ0 will overwrite previous SR2

        MOVE    X:DQ_T,Y0             ;Get DQS
        MOVE    Y:DQMAG,A             ;Get MAG=DQMAG
        TST     A   #$000E,R0         ;Check MAG, get exponent bias (14)
        JNE     <NORMDQ_T             ;Test MAG
        MOVE    #<$40,A               ;If MAG=0 set MANT=100000,
        MOVE    #0,R0                 ; and EXP=0
        JMP     <TRUNCDQ_T
NORMDQ_T
        REP     #13                   ;If MAG!=0 do NORM iteration 13
        NORM    R0,A                  ; times to find MSB of MAG

;   A1 = 01?? ???? | ???? ???0 | 0000 0000 = normalized MAG (A2=A0=0)
;   R0 = 0000 0000 | 0000 eeee = exponent of normalized MAG

TRUNCDQ_T   MOVE    #<$7E,X0          ;Get mask
        AND     X0,A     R0,X:(R2)+   ;Truncate MANT to 6 bits,
                                      ; save EXP to DQ1EXP
;   A1 = 01mm mmm0 | 0000 0000 | 0000 0000  (A2=A0=0)
        MOVE    A1,X:(R2)+            ;Save MANT to DQ1MANT
        MOVE    Y0,X:(R2)+            ;Save DQ to DQ1S
```

Fig. 5-5. Linear to Floating-Point Conversion Routine

After conversion the log signal $d_l(k)$ is scaled in the routine SUBTB and quantized in the routine QUAN. These two routines are shown in Fig. 5-6. SUBTB simply truncates the scale factor y(k) and then subtracts this value from $d_l(k)$. The quantization of this normalized value $d_{ln}(k)$ in the QUAN routine is done by a table search. The boundary values of the eight quantization regions shown in Table 4-1 are stored in the table QUANTAB in data memory. These values are read from the table using register r0 as a pointer and then compared with $d_{ln}(k)$ until the correct range is found. When the range is found an offset from the starting address of QUANTAB is subtracted from the last value in r0. This produces that correct magnitude of I(k) given in Table 4-1.

The magnitude of I(k) is shifted to the MSB's of register a1 and is then combined with the sign value $d_s(k)$ which was previously saved in register y1 in the LOG routine. The ADPCM word I(k) is in a sign magnitude type format. If $d_s(k)$ is negative then the 4 MSB's of the accumulator are inverted, setting the sign bit to one and inverting the magnitude bits. If the sign is positive the magnitude of I(k) is not changed leaving the sign set to 0. A special case occurs when the magnitude of I(k) is 0. In this case the bits are inverted even if the sign is positive. This means that an all zero word is not legal and will never be transmitted. This is why the quantizer is referred to as a 15-level quantizer. It should be noted however that transmission errors can cause an all zero word to be received by the decoder so this case must be taken into account in the inverse quantization.

```
;*************************************************************************
;       SUBTB
;
; Scale log version of difference signal by subtracting
;   scale factor
;
; DLN = DL - Y
;
; Inputs:
;   DL = 0iii i.fff | ffff 0000 | 0000 0000  (11SM) in accum B
;   Y = 0iii i.fff | ffff ff00 | 0000 0000  (13SM) in accum A
;
; Output:
;   DLN = siii i.fff | ffff 0000 | 0000 0000 (12TC) in accum A
;
;*************************************************************************

SUBTB_T
        MOVE    Y:(R7),X0                 ;Get mask K8 ($7FF000)
        AND     X0,B    #QUANTAB,R0       ;Truncate Y to 11 bits (Y>>2),
                                          ; get quantization table base
        SUB     B,A     #>QUANTAB+2,X1    ;Find DLN=DL-Y,
                                          ; get offset for quan. conversion

;*************************************************************************
;       QUAN
;
; Quantize difference signal in log domain
;
;    log2|D(k)| - Y(k)  |  |I(k)|
;    -------------------+--------
;      [3.12, + inf)    |   7
;      [2.72, 3.12)     |   6
;      [2.34, 2.72)     |   5
;      [1.91, 2.34)     |   4
;      [1.38, 1.91)     |   3
;      [0.62, 1.38)     |   2
;      [-0.98, 0.62)    |   1
;      (- inf, -0.98)   |   0
;
; Inputs:
;   DLN = siii i.fff | ffff 0000 | 0000 0000 (12TC) in accum A
;   DS = sXXX XXXX | XXXX XXXX | 0000 0000  (1TC) in reg Y1
;
; Output:
;   I = siii 0000 | 0000 0000 | 0000 0000 (ADPCM format) in accum A
;
;*************************************************************************
;
;    Quantization table in X memory
;
;QUANTAB     DC        $F84000        ;-0.98
;            DC        $050000        ;0.62
;            DC        $0B2000        ;1.38
;            DC        $0F6000        ;1.91
;            DC        $12C000        ;2.34
;            DC        $15D000        ;2.72
;            DC        $190000        ;3.12
;            DC        $7FFFFF        ;15.99
```

Fig. 5-6. Difference Signal Scaling and Quantization

```
        MOVE      #QUANTAB,R0            ;Get quantization table base
        MOVE      #>QUANTAB+2,X1         ;Get offset for quan. conversion
        MOVE      X:(R0)+,X0             ;Get 1st quan table value
TSTDLN_T
        CMP       X0,A     X:(R0)+,X0    ;Compare to DLN, get next value
        JGE       <TSTDLN_T              ;If value<DLN try next range
        MOVE      R0,A
        SUB       X1,A     Y:LSHFT-20,X0  ;When range found subtract pointer
                                         ; from base to get IMAG=|I|
;    A1 = 0000 0000 | 0000 0000 | 0000 0iii (A2=A0=0)
        MOVE      A1,X1
        MPY       X0,X1,A       Y1,B     ;Shift IMAG <<20, result is
                                         ; in A0, move DS into B
        MOVE      A0,A
;    A1 = 0iii 0000 | 0000 0000 | 0000 0000 (A2=A0=0)
        MOVE      A1,X:IMAG              ;Save IMAG
        TST       A        #<$F0,X0      ;Check IMAG, get invert mask
        JEQ       <INVERT_T              ;If IMAG=0 invert bits
        TST       B                      ; else check DS
        JPL       <IOUT_T                ;If DS=1 don't invert IMAG
INVERT_T    EOR      X0,A                ;If DS=0 or IMAG=0 invert IMAG
IOUT_T  MOVE      A1,A                   ;Adjust sign extension
```

Fig. 5-6. Difference Signal Scaling and Quantization (cont.)

5.2.7 - Inverse Quantization

The inverse quantization of the ADPCM sample I(k) is performed in the routines RECONST and ADDA. These routines are shown in Fig. 5-7. The RECONST routine uses a table lookup to find $d_{lnq}(k)$, the quantized version of $d_{ln}(k)$. After removing the sign of I(k) the magnitude is inverted if necessary and is then shifted to the 3 LSB's of the 24-bit word. This magnitude is then moved to the offset register n4 where it is used as an offset to find one of 8 values stored in the lookup table IQUANTAB (defined in Table 4-1). The scale factor y(k) is added to the result to find the denormalized value $d_{ql}(k)$. This logarithmic value is converted back into linear form in the routine ANTILOG to find the result of the overall inverse quantization procedure $d_q(k)$, the quantized version of the difference signal.

The quantization and inverse quantization procedures can serve as illustrations of one way of implementing an adaptive quantization in a waveform coder. The adaptation of y(k), discussed in Section 5.2.10, addresses the adaptive characteristic of the scale factor but the scaling and quantization process described here can still be used no matter how the adaptation is performed.

5.2.8 - Adaptive Predictor

The adaptive predictor portion of the ADPCM algorithm is implemented in two main sections. The first section is the prediction filter itself, shown in Fig. 4-5. This section consists of the routines FMULT and ACCUM. The filter uses delayed data and coefficient values so FMULT and ACCUM are the first two routines executed in the encoder and the decoder. The second section consists of the reconstructed signal calculation and the

```
;************************************************************************
;       RECONST
;
; Reconstruct quantized difference signal in the log domain
;
;     |I(k)|  | log2|DQ(k)| - Y(k)
;    ---------+--------------------
;       7     |       3.32
;       6     |       2.91
;       5     |       2.52
;       4     |       2.13
;       3     |       1.66
;       2     |       1.05
;       1     |       0.031
;       0     |       - inf
;
; Inputs:
;   I = iiii 0000 | 0000 0000 | 0000 0000  (ADPCM format) in accum A
;
; Output:
;   DQLN = siii i.fff | ffff 0000 | 0000 0000  (12TC) in accum A
;   DQS = sXXX 0000 | 0000 0000 | 0000 0000  (1TC) in reg Y1
;
;************************************************************************
;
;    Inverse quantization table in X memory
;
;IQUANTAB    DC      $800000         ;-16    |I|=0
;            DC      $004000         ;0.031  |I|=1
;            DC      $087000         ;1.05   |I|=2
;            DC      $0D5000         ;1.66   |I|=3
;            DC      $111000         ;2.13   |I|=4
;            DC      $143000         ;2.52   |I|=5
;            DC      $175000         ;2.91   |I|=6
;            DC      $1A9000         ;3.32   |I|=7

        MOVE    #<$F0,X1
        MOVE    A,Y1    A,X:I_R     ;Save DQS (sign of I) to Y1
        EOR     X1,A    Y:RSHFT+20,Y0   ;Invert bits of I
        TMI     Y1,A                ;If ^IS=1 use I, else use ^I
;   A1 = 0iii 0000 | 0000 0000 | 0000 0000
        MOVE    A1,X0
        MOVE    A1,X:IMAG           ;Save |I|
        MPY     X0,Y0,A     #IQUANTAB,R4    ;Shift IMAG>>20
;   A1 = 0000 0000 | 0000 0000 | 0000 0iii (A2=A0=0)
        MOVE    A1,N4               ;Load IMAG as offset into IQUAN table
        MOVE    X:Y_R,B             ;Get Y
        MOVE    X:(R4+N4),A         ;Lookup DQLN

;************************************************************************
;       ADDA
;
; Add scale factor to log version of quantized difference
;   signal
;
; DQL = DQLN + Y
;
; Inputs:
;   Y = 0iii i.fff | ffff ff00 | 0000 0000  (13SM) in accum B
;   DQLN = siii i.fff | ffff 0000 | 0000 0000  (12TC) in accum A
;
; Output:
;   DQL = siii i.fff | ffff 0000 | 0000 0000  (12TC) in accum A
;
;************************************************************************

        MOVE    Y:(R7)+,Y0          ;Get mask K8 ($7FF000)
        AND     Y0,B                ;Truncate Y to 11 bits (Y<<2)
        ADD     B,A                 ;Find DQL=DQLN+(Y<<2)
```

Fig. 5-7. Inverse Quantization and Scaling of ADPCM Codeword

adaptation of the predictor coefficients. These routines are executed after the inverse quantization in both the encoder and the decoder.

The adaptive predictor is the most computationally intensive portion of the ADPCM implementation on the DSP56001. One of the main reasons for this is the floating-point multiplies that are required in the FMULT routine, as was discussed in the floating-point conversion section. The FMULT routine is set up as a hardware DO loop that is executed eight times, two for the poles and six for the zeros. For each tap of the filter the two's-complement coefficient must be converted to the floating-point format before it is multiplied with the delayed data value. After the multiplication each partial product must be converted to the fixed point format. Overall eight fixed point to floating-point conversions and eight floating-point to fixed point conversions are required in the FMULT routine. The flow description within each loop of FMULT is as follows:

1. Convert the 16-bit two's complement coefficient to a 13-bit magnitude and a 1-bit sign
2. Convert the 13-bit magnitude to a 4-bit exponent and a 6-bit mantissa
3. Add the exponents of the coefficient and the data to find the 5-bit exponent of the partial product
4. Multiply the mantissas of the coefficient and the data and truncate the results to find the 8-bit mantissa of the partial product
5. Convert the exponent and the mantissa of the partial product to a 15-bit magnitude
6. Exclusive-OR the signs of the coefficient and the data to find the sign of the partial product
7. Convert the 15-bit magnitude and 1-bit sign of the partial product to a 16-bit two's complement number

Since the FMULT routine requires a large percentage of the processing time of the overall algorithm it is desired to have the worst case execution speed of this routine as short as possible. To reduce data movement overhead, several constants that are needed by the routine are stored in the table CONST in data memory. They are addressed using register r0 so parallel moves can be taken advantage of whenever possible. This technique is discussed in Section 5.3. Another aspect of this routine that has a significant effect on the speed is the way the coefficients, data, and partial products are stored in data memory. A description of this buffer structure is shown in Fig. 5-8. The variables $wa_i(k)$ and $wb_i(k)$ in this figure represent the partial products for the pole and zero taps of the filter. To eliminate extra overhead the predictor data buffer that contains the delayed floating point values of $d_q(k)$ and $s_r(k)$ is set up so that the sign, mantissa, and exponent are stored separately. Storing the data in this form uses more data memory but the routine requires much less computation when using these values than if they were combined into one word. The address register r2 is used as a pointer to this buffer so when an element is needed an address register addressing mode can be used. These modes allow the parallel bus structure of the DSP56001 to be taken advantage of whenever possible.

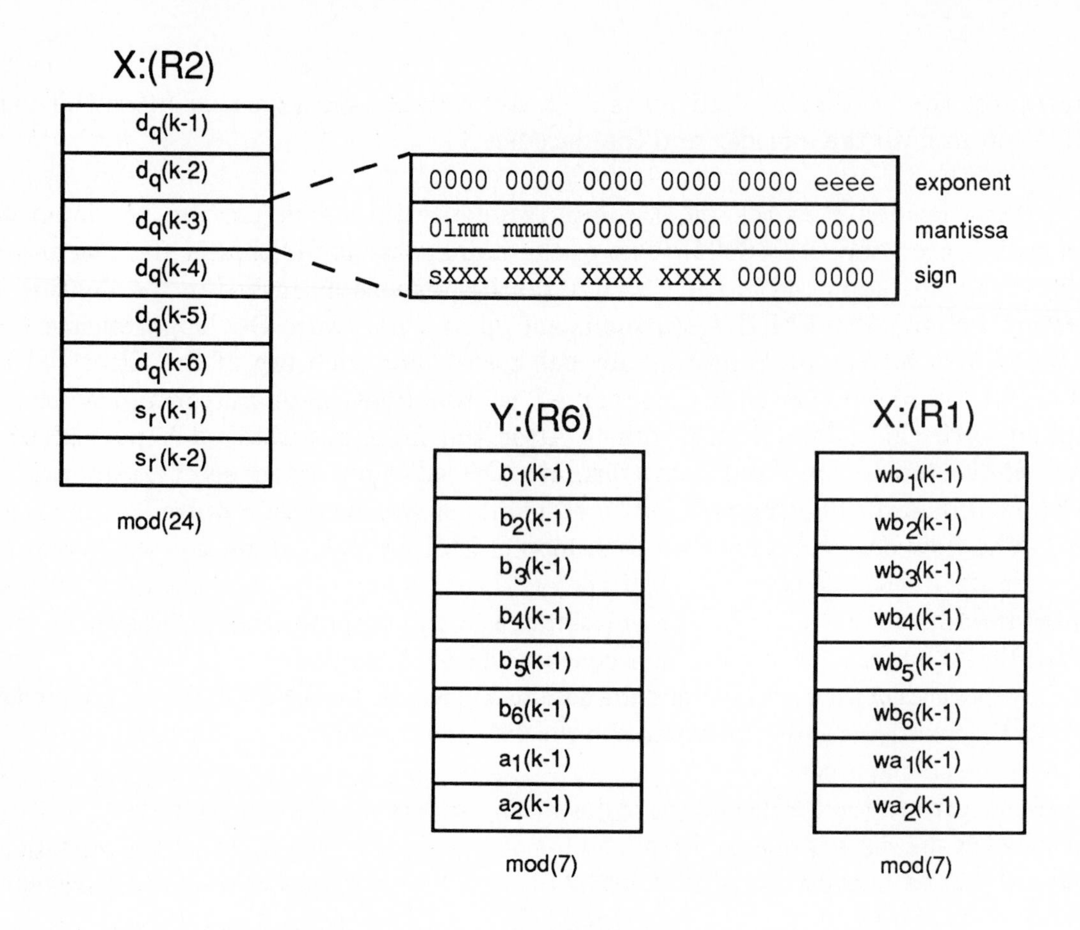

Fig. 5-8. Adaptive Predictor Data Structure

The data buffer structure also allows the efficient offset addressing modes of the DSP56001 to be used when consecutive values of one component are required. For instance, the signs of each delayed $d_q(k)$ value are needed in the XOR/UPB routine. In this routine the offset register n3 is set to 3 so that when one sign is read from the buffer, register r3 is automatically post-incremented by 3 to point to the next sign. The modulo pointer feature of the DSP56001 also reduces execution speed since less software overhead is required to update the pointer. The data buffer for the prediction filter is actually a modified form of the usual modulo buffer structure on the DSP56001 since two new values, $s_r(k\text{-}1)$ and $d_q(k\text{-}1)$, are added to the buffer for each sample. The new $s_r(k\text{-}1)$ overwrites the current $d_q(k\text{-}6)$ and the new $d_q(k\text{-}1)$ overwrites the current $s_r(k\text{-}2)$. The coefficient buffer and partial product buffer are addressed using modulo pointers for efficiency but the physical locations of each component do not change.

The implementation of the ACCUM routine that adds the partial products is more straightforward than the FMULT routine. The $wb_i(k)$ partial products for the six zeros are accumulated first to obtain the partial signal estimate $s_{ez}(k)$. Then the $wa_i(k)$ partial products for the two poles are accumulated with the zeros to form the final signal estimate $s_e(k)$. Even though FMULT and ACCUM account for a large percentage of the overall execution speed they only implement equations (2) and (3) of the CCITT algorithm.

The second section of the adaptive predictor code is executed after the inverse quantizer. The calculation of equation (4) to determine the reconstructed signal $s_r(k)$ is done in the routine ADDB. This routine requires a format conversion since $d_q(k)$ is in a sign magnitude format. Following the ADDB routine is the adaptation of the predictor coefficients which includes equations (5) - (12). Due to the way the specification is written the implementation of equations (5) - (8) to update $a_1(k)$ and $a_2(k)$ is slightly more complicated than Figs. 4-6 and 4-7 indicate. The ADDC routine basically implements equation (7) but it also calculates part of equations (5) and (6) that use the signal p(k). The routine first calculates p(k) and then saves it to memory after delaying the previous values of p(k) and p(k-1). Then the routine calculates the variables PKS1 and PKS2 that represent the multiplication of the sgn[x] functions in equations (5) and (6). The sgn[x] function that is calculated in this routine does not have the values shown in equation (12). Instead a 0 in the sign bit represents a positive number while a 1 in the sign bit represents a negative number. The variable SIGPK is used to distinguish the special case of sgn[p(k)] = 0. The routine UPA1 calculates equation (5) in a slightly rearranged form:

$$a_1(k) = a_1(k-1) - [2^{-8} \bullet a_1(k-1)] + \text{gain} \tag{26}$$

The gain portion of equation (26) represents the second half of equation (5) and is determined by testing the variables PKS1 and SIGPK. The routine UPA2 calculates equation (6) to update $a_2(k)$ in a similar manner. This routine is more complicated since it also has to calculate the value of $f(a_1)$ in equation (8). After updating $a_1(k)$ and $a_2(k)$, the routines LIMD and LIMC limit these values. The routine LIMC uses a constant upper and lower limit defined in equation (9) for the comparison. It makes use of the conditional transfer instructions of the DSP56001 as discussed in Section 5.3. The LIMD routine is very similar but it must calculate the upper and lower limits before the comparison.

The $b_i(k)$ coefficients are updated in the routines XOR and UPB which are combined in a single code segment. The XOR routine accounts for the sgn[x] multiplication in equation (11) that uses the delayed values of $d_q(k)$. The calculation of equation (11) is similar to that of equations (5) and (6) except that the calculation must be done six times for each of the six $b_i(k)$ coefficients. In this routine the special case of $\text{sgn}[d_q(k)] = 0$ is checked only once since it applies to all six calculations. If this case is found a DO loop that does not add the second half of equation (11) is executed. If this case is not present a DO loop that does the full calculation is executed, but execution time is saved since the test for the special case does not have to be performed for each stage of the loop.

After the coefficients have been updated they are passed through the predictor trigger routine. Since this routine also affects the tone detect signal $t_r(k)$, they are all updated at once in the tone detection section of the algorithm. The final step of the adaptive prediction section is the conversion of the quantized difference signal $d_q(k)$ and the reconstructed signal $s_r(k)$ to the floating-point format. This is done in the routines FLOATA and FLOATB respectively. After these values have been converted they are stored in the predictor data buffer. Further details of the floating-point conversion are found in Section 5.2.5.

5.2.9 - Tone Detection

The tone detection portion of the ADPCM algorithm is implemented in 3 routines, TONE, TRIGB, and TRANS. The TONE routine, executed after the predictor adaptation, checks the pre-limited version of the coefficient $a_2(k)$ for a threshold. If it is below that threshold the tone detection variable $t_d(k)$ is set to 1. The TRIGB routine affects the signal $t_d(k)$ as well as the predictor coefficients. TRIGB tests the transition detect signal $t_d(k)$, and if it is set then the tone detect signal and all of the predictor coefficients are set to zero. The transition detect signal $t_d(k)$ is set in the TRANS routine. This routine occurs after a delay block so it is executed after the inverse quantizer (since it needs the value of $d_q(k)$) but before the predictor adaptation. The TRANS routine first checks the delayed signal of $t_d(k)$. If $t_d(k)$ is set indicating a tone is present, TRANS compares the values of $d_q(k)$ and $y_l(k)$ according to equation (23) to detect a transition from a tone to a non-stationary signal. The comparison requires a format conversion since $d_q(k)$ is in sign magnitude format and $y_l(k)$ is in logarithmic format. If the comparison threshold is met then $t_d(k)$ will be set to drive the adaptation to the fast mode immediately.

5.2.10 - Scale Factor Adaptation

The scale factor adaptation consists of two main sections, the speed control parameter adaptation, shown in Fig. 4-10, and the adaptation of the scale factor, shown in Fig. 4-9. The adaptation of the speed control parameter comprises seven routines. The first routine, FUNCTF, maps the magnitude of the ADPCM sample to one of 4 values specified in Table 4-3. The next 2 routines, FILTA and FILTB, use the result of this mapping, F[I(k)], to track a short term and a long term average of the difference signal. As was the case with the adaptation of the predictor coefficients, the FILTA routine implements equation (18) in a slightly different form:

$$d_{ms}(k) = d_{ms}(k-1) + 2^{-5} [F[I(k)] - d_{ms}(k-1)] \qquad (27)$$

The FILTB routine implements equation (19) in a similar manner using $d_{ml}(k)$. The next 3 routines implement equation (20) that determines the unlimited speed control parameter $a_p(k)$ from these averages and other inputs. Equation (20) sets $a_p(k)$ to one of 3 values. The SUBTC routine checks for the three cases in which the factor of 2^{-3} is added. SUBTC sets the variable AX to 1 if either of these three conditions is met, otherwise it is set to 0. The routine FILTC calculates the factor $(1-2^{-4})a_p(k-1)$ and then adds AX to this value. Notice that AX is actually set to a value that can be added directly so that extra shifting is not required. The following routine, TRIGA, tests the final condition based on a transition detection. If $t_r(k)$ is equal to 1 then the value of $a_p(k)$ is set to 1, otherwise TRIGA leaves $a_p(k)$ as the value set by FILTC. The final routine, LIMA, limits the delayed value of $a_p(k)$ according to equation (21) to form the final speed control parameter

$a_l(k)$. Since LIMA uses the delayed value of $a_p(k)$ this routine is executed after FMULT and ACCUM routines near the beginning of the algorithm. The output of LIMA, $a_l(k)$ is fed directly to the scale factor adaptation routine MIX.

The update of the scale factor y(k) is performed in 5 routines that are executed immediately following the speed control parameter update. The first of these routines, FUNCTW, performs a mapping of I(k) that is based on Table 4-2. It is similar to the FUNCTF routine. The output of FUNCTW is the signal W[I(k)] that is used to update the unlocked scale factor $y_u(k)$. This update is performed in the FILTD routine. It implements equation (14) in a rearranged form that is similar to that of equation (27). This value is limited in the routine LIMB according to equation (15). The limited value of $y_u(k)$ is also used in equation (16) to update the locked scale factor yl(k). This is done in the routine FILTE. The final update of the scale factor y(k) is done in the routine MIX. Since this follows a delay block, it is performed immediately after LIMA near the beginning of the algorithm. The MIX routine combine the 3 inputs $y_u(k-1)$, $y_l(k-1)$, and $a_l(k)$ according to equation (16). Like previous equations it is executed in a slightly different form:

$$y(k) = a_l(k)\,[\,y_u(k-1) - y_l(k-1)] + y_l(k-1) \qquad (28)$$

5.2.11 - Decoder Synchronization

As mentioned in Section 4.2 the decoder algorithm is almost identical to the encoder algorithm. The sections added to the decoder that are not included in the encoder are the log PCM conversion and the synchronization block shown in Fig. 4-13. The log PCM conversion routine COMPRESS is discussed in Section 5.2.3. Most of the synchronization section is identical to the first section of the encoder. The routines EXPAND, SUBTA, LOG, and SUBTB are the same as those discussed previously. The SYNC routine is similar to the QUAN routine but instead of transmitting the resulting ADPCM word, it compares it to the received ADPCM word. If they are the same then $s_p(k)$, the log PCM value of the reconstructed signal, becomes the output of the decoder, $s_d(k)$. If they are not the same then the next more positive or negative PCM word becomes the output as described in equation (25). Most of the computation in the SYNC routine is used to check the boundary cases of the PCM word. Also, it has two separate code sections, one for the μ-law case and the other for the A-law case.

5.3 - Optimization Techniques

The implementation of the standard CCITT ADPCM coder on the DSP56001 uses several optimization techniques to obtain real-time performance. The goal of these techniques is to obtain the minimum worst case execution speed for the entire algorithm. A description of them can serve as an aid to understanding the assembly code since many of these techniques make the code harder to follow. Also, many of these techniques may be in other applications. A key factor in optimizing any assembly code is a complete knowledge of the architecture of the part, both from a hardware and software standpoint, and an equal knowledge of the algorithm being implemented. This is also true for this ADPCM example.

1) As noted in Section 5.2.1, subroutines are not used in the algorithm itself. It was found that the overhead of passing parameters and calling subroutines required a significant portion of the total execution time, enough to prohibit real-time performance. Eliminating subroutines also allows the parallelism of the DSP56001 to be exploited as much as possible since less data movement is required. The disadvantage of this approach is that more program memory is required to duplicate common routines. This can be a problem if more external memory is required but in the ADPCM code it was found that even with subroutines the amount of required code would not fit in the on-chip program RAM .

2) In this application memory is sacrificed whenever gains in execution speed can be obtained. This may not work in all cases depending on the memory configuration. A key factor in using this technique is the multiplexed external bus of the DSP56001. The external bus will allow one external access per instruction cycle (assuming zero wait states) with no penalty in execution speed. In this application external data memory is not used so there will not be a speed delay when accessing external program memory.

3) REP and DO instructions are very useful for saving program memory locations and in general are very efficient instructions , however they do take extra cycles to set up the loop registers. In many cases in the ADPCM code instructions that could be performed with a REP or DO loop are instead repeated in the code multiple times. An example is the iterative NORM instruction used for the logarithmic and floating-point conversions. Instead of using a "REP #14" preceding the NORM instruction, 14 separate NORM instructions are used. For a single case the savings are very little but when adding up the savings throughout the algorithm this allows several cycles of execution time to be saved for each sample. In the case of FMULT which is executed 8 times, this allows a total of 16 instruction cycles to be saved for each sample.

4) In many cases gains can be obtained by examining instruction encoding. For instance, immediate operands that are greater than 8 bits (or 12 bits in some cases) require an extra word of storage in program memory and also cause the instruction to take an extra cycle to fetch the operand. This is also the case for immediate data that is less than eight bits but not left-justified. The ADPCM algorithm needs several operands of this type, 32 to be exact. Instead of addressing them as immediate operands they are moved to a constant table in data memory and addressed with register r7. This does not cost extra memory since these constants require an extension word in program memory in any case. The advantage of this technique is that an extra instruction cycle is saved by not having to fetch an instruction extension word. For the data in the table in Y memory this results in a savings of 32 cycles per sample for the entire algorithm. This technique is also used in the FMULT routine that uses its own constant table so even more savings per sample are added. In addition to the instruction fetch savings, this allows more

parallelism to be exploited since the most efficient parallel addressing modes on the DSP56001 require the use of addressing registers. The disadvantages of this technique are that a dedicated register is required to address them and code is harder to follow and modify because of the added complexity. To help alleviate the complexity of the ADPCM code these constants are referred to as Kxx in the comments so that what is actually being read from the constant table is easier to identify.

5) Since the DSP56001 is a fractional-based architecture, many operations are more efficient when the data is left-justified. An example of this is the 8-bit immediate operand storage mentioned above. Another example is the NORM instruction that deals with the left-most bits in an accumulator. Data that is already left-justified as much as possible will require less shifting for normalization and therefore fewer NORM iterations. For this reason data is kept left-justified throughout the code as much as possible.

6) Parallel moves are taken advantage of whenever possible. This makes the assembly code more difficult to understand and modify since values may be fetched from memory long before they are actually used, in many cases while the previous routine is being executed. In many applications this can save considerable time especially when multiple loops of a code segment are executed. Dual parallel moves on the DSP56001 usually require the use of an addressing register for accessing memory. Whenever possible pointers are uses instead of immediate addresses.

7) Knowledge of efficient test instructions can result in savings when decisions are necessary. For example, many cases in the CCITT specification refer to single bit variables. Whenever possible the single bit is stored in the sign bit of the 24-bit word whether it represents a sign or not. This allows instructions to test the sign of the word to determine if the variable is set or not.

8) Some of the most efficient instructions on the DSP56001 are the conditional transfer instructions. These are taken advantage of in several places in the ADPCM source code. These include the routines used for limiting such as the LIMC routine. In this routine a value is tested for both an upper and a lower limit. When this is required the following code segment can be used:

```
; Lower limit in x0
; Upper limit in x1
; Value to be limited in a

        cmp     x0,a    <parallel move>
        tlt     x0,a
        cmp     x1,a    <parallel move>
        tgt     x1,a
```

This code segment can be very efficient since each instruction executes in a single cycle and the CMP instructions can have parallel moves associated with them. When it is possible to use these transfer instructions the savings over using branching instructions can be great.

9) As with virtually all microprocessors, JMP instructions on the DSP56001 should be avoided whenever possible. These instructions take a minimum of two instruction cycles to execute due to the instruction pipeline and do not allow parallel moves. Conditional jump instructions also take a minimum of two instruction cycles even if the jump is not taken. As mentioned above, conditional jumps can be avoided in many cases by using conditional transfer instructions instead.

10) Many different data formats are encountered in the ADPCM assembly code. Some of these have been discussed previously. Trying to adjust these different formats to the DSP56001's fractional data format is not practical. Instead these formats are allowed to "float" freely within the 24-bit data word, or the 56-bit data word in the accumulators. The goal is to find the most efficient format that requires the least amount of data manipulation. An example is the many shifts are specified in the CCITT standard. In many of these cases the shift is not actually performed. Instead, the value is truncated by using a mask constant and an AND instruction.

11) When shifting data cannot be avoided the lack of a single-cycle, multi-bit hardware shifter on the DSP56001 can be viewed as a problem. This is another example of a case where a complete knowledge of the architecture of the part can be a key factor in code optimization. The DSP56001 does not have a hardware shifter but it does have a single-cycle hardware multiplier. Since shift operations are actually just a multiplication by a power of two, the multiplier can be used as a shifter.

A basic description of this technique is described in the Motorola DSP applications report "Fractional and Integer Arithmetic Using the DSP56000 ..." [9]. This applications report describes two techniques for doing multi-bit shifts on the 56000/1. The most straightforward approach uses a REP or DO instruction followed by a shift instruction. As noted previously this is an iterative process and also requires overhead for the DO or REP. A faster way calls for multiplying the operand by a shift constant. If the amount of shift is always the same, the constant can be explicitly coded in the instruction sequence. If however, the shift amount is not always the same there is a convenient method of getting the appropriate shift constant. This method uses a lookup table for determining the shift constant. The table should be of the following form:

```
                org     x:

rshift          equ     *-1
                dc      $400000          ;>>1 or <<23
                dc      $200000          ;>>2 or <<22
                dc      $100000          ;>>3 or <<21
                        •••
                dc      $000002          ;>>22 or <<2
                dc      $000001          ;>>23 or <<1
lshift          equ     *
```

This table can be put in either X or Y memory. To perform an arbitrary right shift the value of rshift should be loaded into one of the address registers as a table base. The amount of the shift will then be loaded into the corresponding offset register so that the appropriate shift constant can be read from the table in memory. The technique is similar for an arbitrary left shift although the offset will be negative. For either a left or right shift the integer shift amount should be between 1 and 23. The following is an example of a right shift:

```
move    #rshift,r0          ;Set r0=table base
move    b0,n0               ;Load shift amount as offset
move    a1,x1               ;Set data up for shift
move    x:(r0+n0),x0        ;Lookup shift constant
mpy     x0,x1,a             ;Shift data right
                            ;Result is shifted into a0
```

In this example the 24-bit number to be shifted is in a1 and the amount of the shift is in the LSBs of b0. The base of the right shift lookup table rshift is loaded into r0 and the amount of the shift is loaded into n0 as an offset into the table. The shift constant is found by using the Indexed by Offset addressing mode. The table base in r0 is not changed so the r0 does not have to be loaded again for another shift unless r0 is used elsewhere. To accomplish the shift the shift constant is put in x0 and the 24-bit number that is to be shifted is put in x1. The MPY instruction performs the shift. The result is found in a1 and a0. It is also sign extended into a2. This method is faster than the REP or DO method because the actual shift takes only 1 instruction cycle. The only overhead is data movement which can often be done in parallel with other operations. The only other extra time needed is the 1 extra instruction cycle required by the Indexed by Offset mode.

An example of a left shift is the following:

```
move    #lshift,r0          ;Set r0=table base
neg     b       a1,x1       ;Find negative shift amount,
                            ; set data up for shift
move    b0,n0               ;Load s.a. as negative offset
move    x:(r0+n0),x0        ;Lookup shift constant
mpy     x0,x1,a             ;Shift data left
                            ;Result is shifted into a1
```

This example is very similar to the right shift except a negative offset is used. Since the Indexed by Offset mode can only use a positive offset the shift amount is negated before it is loaded as an offset. Again the result is found in a0 and a1 and is sign extended into a2. The left shift requires the same amount of data movement as the right shift but also needs an extra ALU operation to negate the shift amount. In the example shown the execution time is the same as the right shift.

This technique is used in the ADPCM code in several forms. The shift table is in Y memory and uses the labels RSHFT and LSHFT. In some cases the table is address like the above examples but using register r5 as the pointer. In other cases where the immediate short addressing mode can be used the table is addressed using this mode instead of the address pointer. In addition, some shift constants are addressed as immediate operands in the instruction word. In all cases though the actual shift is performed in the same manner as the above examples.

5.4 - Performance Specifications

The memory usage of the DSP56001 ADPCM source code is shown in Fig. 5-9. The program uses only internal X and Y memory so no external data memory is required. External program memory is required however. As noted previously, the algorithm must have zero-wait state RAM for a 27 MHz DSP56001 in order to run in real-time.

Fig. 5-10 shows the order of execution of the routines and the worst-case processing time in instruction cycles for each routine. A routine is defined as the code between commented sections even though some processing for that function may not be included in this code. Note that an instruction cycle (Icycle) is defined as two clock cycles on the DSP56001. For a 27 MHz DSP56001 an instruction cycle is 74 ns. Sampling at a rate of 8 kHz gives 125 microseconds to do both an encode and a decode. This translates into 1719 instruction cycles on 27.5 MHz DSP56001. The worst-case calculations in Fig. 5-10 indicate a total of 1719 instruction cycles for doing both the encoder and decoder. These worst-case times were calculated based on worst-case branches and delays in each routine. It has been found, however, that there is very little possibility of the maximum delay occuring in all of the routines in the encoder and decoder simultaneously. No samples of the CCITT test sequences were observed to generate the worst-case condition.

It should be noted that the synchronization block in the decoder is not necessary for the ADPCM algorithm itself. This includes the routines EXPAND, SUBTA, LOG, SUBTB, and SYNC. This block is included for synchronization of multiple PCM/ADPCM/PCM conversions on a single channel. If only one PCM/ADPCM/PCM conversion is used the deletion of this block should not significantly affect the output speech quality. In this case the worst case execution time will be 1589 instruction cycles. Note that these routines are necessary to correctly pass the CCITT test sequences.

	Total	Internal	External
Program memory	1230	447	783
Data memory	222	118 X, 104 Y	

*(all values for 24-bit words)

Fig. 5-9. Memory Usage

Encoder	
FMULT (x8)	341
ACCUM	13
LIMA	4
MIX	14
EXPAND	10
SUBTA	3
LOG	22
SUBTB	3
QUAN	36
RECONST	7
ADDA	3
ANTILOG	25
TRANS	34
ADDB	8
ADDC	19
XOR	
UPB (x8)	76
UPA2	27
LIMC	6
UPA1	12
LIMD	8
FLOATA	22
FLOATB	27
TONE	5
TRIGB	13
FUNCTF	14
FILTA	6
FILTB	5
SUBTC	11
FILTC	5
TRIGA	3
FUNCTW	7
FILTD	5
LIMB	4
FILTE	8
misc.	4
TOTAL	810 Icycles

Decoder	
FMULT (x8)	341
ACCUM	13
LIMA	4
MIX	14
RECONST	13
ADDA	3
ANTILOG	25
TRANS	34
ADDB	8
ADDC	19
XOR	
UPB (x8)	76
UPA2	27
LIMC	6
UPA1	12
LIMD	8
FLOATA	22
FLOATB	27
TONE	5
TRIGB	13
FUNCTF	14
FILTA	6
FILTB	5
SUBTC	11
FILTC	5
TRIGA	3
FUNCTW	7
FILTD	5
LIMB	4
FILTE	9
COMPRESS	33
EXPAND	10
SUBTA	3
LOG	22
SUBTB	3
SYNC	80
misc.	7
TOTAL	897 Icycles

Fig. 5-10. Code Execution Times (worst case)

Appendix A - Terminology

Several different symbols are used in the DSP56001 assembly code and in this document for describing the ADPCM implementation. Most of these symbols are derived from the G.721 specification. The following description may clarify some of these terms.

The following symbols correspond to the variable types defined in the G.721 specification:

SM = signed magnitude value
TC = two's complement value
FL = floating point value

A number preceding one of these symbols shows the number of total bits in a particular variable (e.g. 14TC represents a 14-bit two's complement number). The full binary representation of each variable, including the location of the radix point, is given in Table 3 of Recommendation G.721.

The contents of registers or memory locations at certain points in the code are detailed bit for bit. The terminology used is the following:

. = location of implied radix point
i = integer bit
f = fraction bit
s = sign bit
m = mantissa bit
e = exponent bit
1 = bit is always 1
0 = bit is always 0
X = bit value is unknown
but is not significant

An exception to the above is the PCM word where:

p = sign bit
s = segment bit
q = quantization level bit

Note that when labels are used to refer to variables or program locations in the assembly code the suffix "_T" refers to those labels associated with the encoder (transmit) and the suffix "_R" refers to those associated with the decoder (receive).

Example:

```
;   Y = 0iii i.fff | ffff ff00 | 0000 0000  (13SM)
Y_T                     DS          1                      ;Quantizer scale factor
```

This shows the memory allocation for a variable, the scale factor y(k). It is defined as a 13-bit signed magnitude number with four integer bits and nine fractional bits. The value stored in memory is always stored in the 24-bit format shown above with the implied radix point between bits 18 and 19.

Example:

```
;    A1 = 01mm mmm0 | 0000 0000 | 0000 0000 (A2=A0=0)
;    B1 = 0000 0000 | 0000 0000 | 0000 eeee (B2=B0=0)
```

At the point where these comments appear in the code the register a1 always contains a 1 in bit 22, five other mantissa bits, and all other bits set to 0. Register b1 always contains four exponent bits in bits 0 through 3 with all other bits set to 0. Registers a0,a2,b0, and b2 are always set to 0.

References

[1] CCITT Recommendation G.721, "32 kbit/s Adaptive Differential Pulse Code Modulation (ADPCM)", Study Group XVIII - Report R 26(C), August 1986.

[2] D. O'Shaughnessy, *Speech Communication - Human and Machine*, Addison-Wesley, 1897.

[3] A.V. Oppenheim and R.W. Schafer, *Discrete-Time Signal Processing*, Prentice-Hall, Englewood Cliffs, NJ, 1989.

[4] N.S. Jayant and P. Noll, *Digital Coding of Waveforms*, Prentice-Hall, Englewood Cliffs, NJ, 1984.

[5] CCITT Recommendation G.711, "Pulse Code Modulation (PCM) of Voice Frequencies", *CCITT Red Book*, October, 1984.

[6] "Logarithmic/Linear Conversion Routines for DSP56000/1", Motorola, Inc., DSP Operation Technical Brief

[7] L.R. Rabiner and R.W. Schafer, *Digital Processing of Speech Signals*, Prentice-Hall, Englewood Cliffs, NJ, 1978.

[8] N. Benvenuto et al., "The 32-kb/s ADPCM Coding Standard", *AT&T Technical Journal*, Vol. 65, No. 5, September/October 1986, pp. 12-22.

[9] "Fractional and Integer Arithmetic Using the DSP56000 Family of General Purpose Digital Signal Processors", Motorola, Inc., DSP Operation Technical Report APR3/D

Appendix F

Implementing IIR/FIR Filters With Motorola's DSP56000/DSP56001 Digital Signal Processors

INTRODUCTION

This report considers the design of digital frequency-selective filters which modify the frequency content and phase of input signals according to some specification. Two classes of frequency-selective digital filters are considered - *Infinite Impulse Response* (IIR) and *Finite Impulse Response* (FIR) filters. The design process consists of determining the coefficients of the IIR or FIR filters which results in the desired magnitude and phase response being closely approximated.

The purpose of this report is therefore two-fold:
(1) To provide some intuitive insight into digital filters, in particular how the coefficients are calculated in the digital domain so that a desired frequency response is obtained and;
(2) To show how to implement both classes of digital filters (IIR and FIR) on the DSP56001 digital signal processor.

It is assumed that most readers are analog designers learning *Digital Signal Processing* (DSP). The approach used reflects this assumption in that digital filters are initially presented from a very "analog" point of view. This may be less than traditional, but hopefully it will help speed up bridging the gap for those familiar with analog s-domain transfer functions and faced with realizing the equivalent behavior in the digital z-domain. In keeping with this "analog perspective" IIR filters will be discussed first since the equivalent of FIR filters are infrequently encountered in the analog world.

Section 1 of the report is a brief review of lowpass, highpass, bandpass and stopband analog filters. The s-domain formulas governing the key characteristics, magnitude-frequency response $G(\Omega)$ and phase-frequency response $\phi(\Omega)$ are derived from first principals. Damping factor d, cut-off frequency Ω_c (for low and highpass filters), center frequency Ω_0 (for bandpass and bandstop filters), and quality factor Q are defined for the various filter types.

In *Section 2* the bilinear transformation is introduced so that analog s-domain designs can be transformed into the digital z-domain and the correct coefficients thereby determined. The form of the formulas for the z-domain filter coefficients thus determined are generalized in terms of the key filter characteristics in the z-domain. It is important to emphasize this because the formulas derived are functions of the filter characteristics in the z-domain so that the engineer will be able to design digital filters directly without the necessity of designing the analog equivalent and transforming the design back into the digital domain.

In the analog domain the performance of the filter depends on the tolerance of the components. Similarly, in the digital domain the filter performance is limited by the precision of the arithmetic used to implement the filters. In particular the performance of digital filters is extremely sensitive to overflow which occurs when the accumulator width is insufficient to represent all the bits resulting from many consecutive additions. This is similar to the condition in the analog world where the signal output is larger than the amplifier power supply so that saturation occurs. In any case a short analysis of the gain at critical nodes in the filters is given in *Section 3 and 4* to provide some insight into the scaling requirements for different forms of IIR filters. For this reason the signal flow graphs developed are centralized about the accumulator nodes.

The analysis of IIR filters focuses on second order sections. Clearly higher order filters are often required. Therefore a brief discussion of how second order sections can be cascaded to yield higher order filters is given in *Section 5*. Because the analysis becomes complex quickly the discussion naturally leads to using commercially available filter design software such as *Filter Design and Analysis System* (FDAS) from Momentum Data Systems, Inc. *Section 6* concludes by showing how the filter coefficients just discussed can be used in DSP56000 code to implement practical digital filters. Examples of complete filter designs are given including the code, coefficients, frequency response and maximum sample frequency.

In *Section 7* FIR filters are discussed. Initially, FIR filters are contrasted with IIR filters to show that in many ways they are complimentary, each satisfying weaknesses of the other. An intuitive approach is taken to calculating the filter coefficients by starting from a desired arbitrary frequency response. The importance of and constraint imposed by linear phase is emphasized. Having developed an intuitive appreciation what FIR filter coefficients are, the use of FDAS to accelerate the design process is described. *Section 7* concludes by showing how the filter coefficients just determined can be used in DSP56000 code to implement practical digital filters. An example of a passband digital filter using a Kaiser window design approach is given.

1.0 ANALOG RCL FILTER TYPES

In the following sections, the analog RCL filter network will be analyzed for the four basic filter types: *lowpass, highpass, bandpass, and bandstop.* This is a very important step leading to the analysis of digital filters for it will be seen that the difference between the basic RCL filter and the corresponding digital filter is far less than most may believe. In fact, the reader may soon realize that designing digital IIR filters is in many cases much simpler than designing analog filters.

In this analysis, as in all of the following cases, the input is assumed to be a steady state signal containing a linear combination of sinusoidal components whose rms (or peak) amplitudes are constant in time. This assumption allows simple analytic techniques to be used in determining the network response. Even though these results will then be applied to real-world signals which may not satisfy the original steady state assumption, the deviation of the actual response from the predicted response in most cases is small enough to neglect. General analysis techniques consist of a linear combination of steady state and transient response solutions to the differential equations describing the network.

1.1 ANALOG LOWPASS FILTER

The passive RCL circuit forming a lowpass filter network is shown in Figure 1.1-1 where the transfer function *H(s)* is derived from a voltage divider analysis of the RCL network. This approach is valid since the effect of *C* and *L* can be described as a complex impedance (or reactance, X_C and X_L) under steady state conditions. *s* is a complex variable of the complex transfer function *H(s)*. The filter frequency response is found by evaluating *H(s)* with $s = j\Omega$ where $\Omega = 2\pi f$ and f is the frequency of a sinusoidal component of the input signal. The output signal is calculated from the product of the input signal and $H(j\Omega)$. To facilitate analysis the input and output signal components are described by the complex value $e^{j\Omega t} = \cos\Omega t + j\sin\Omega t$. The actual physical input and output signal components are found by taking the real part of this. The input is $\Re\{e^{j\Omega t}\} = \cos\Omega t$; the output is $\Re\{H(j\Omega)\, e^{j\Omega t}\} = G(\Omega)\cos[\Omega t + \phi(\Omega)]$. The above technique is based upon the solution of the differential equations describing the network when the input is steady state. Describing the circuit response by *H(s)* instead of solving the differential equation is a common simplification used in this type of analysis.

The magnitude of *H(s)* is defined as the gain $G(\Omega)$ of the system, while the ratio of the imaginary part to real part of *H(s)*, $\Im\{H(j\Omega)\}/\Re\{H(j\Omega)\}$ is the tangent of the phase $\phi(\Omega)$ introduced by the filter. If the input signal is $A_k\sin(\Omega_k t + \phi_k)$ then the output signal is $A_k\, G(\Omega_k)\sin[\Omega_k t + \phi_k + \phi(\Omega_k)]$. Figure 1.1-2 shows the gain $G(\Omega)$ and phase $\phi(\Omega)$ plots for the second order lowpass network of Figure 1.1-1 for various values of damping factor d. d also controls the amplitude and position of the peak of the normalized response curve.

The frequency corresponding to the peak amplitude can be easily found by taking the derivative of $G(\Omega)$ (from the equation for $G(\Omega)$ in Figure 1.1-1) with respect to Ω and setting it equal to zero. Solving the resultant equation for Ω then defines Ω_M as the frequency where the peak amplitude occurs. The peak amplitude is then $G_M = G(\Omega_M)$:

$$\Omega_M = \Omega_c\sqrt{(1 - d^2/2)} \qquad (1.1-1)$$

$$G_M = \frac{1}{d\sqrt{(1 - d^2/4)}} \qquad (1.1-2)$$

for $d < \sqrt{2}$. For $d > \sqrt{2}$, $\Omega_m = 0$ is the position of the peak amplitude where $G_M = 1$. When $d = \sqrt{2}$, $G_M = 1$ which gives the *maximally flat* response curve which is used in the Butterworth filter design (usually applies only to a set of *cascaded sections*). Note that Ω_C for a lowpass filter is that frequency where the gain is $G(\Omega_C) = 1/d$, and the phase is $\phi(\Omega_C) = -\pi/2$.

1.2 ANALOG HIGHPASS FILTER

The passive RCL circuit forming a highpass filter network is shown in Figure 1.2-1 where the transfer function $H(s)$ is again derived from a voltage divider analysis of the RCL network. The gain and phase response are plotted in Figure 1.2-2 for different values of damping coefficient. As can be seen, the response of the highpass filter is the mirror image of the lowpass filter response.

1.3 ANALOG BANDSTOP FILTER

The analog RCL network for a bandstop filter network is simply the sum of the lowpass and highpass transfer functions as shown in Figure 1.3-1 where the transfer function $H(s)$ is again derived from a voltage divider analysis of the RCL network. The gain and phase response are plotted in Figure 1.3-2 for different values of quality factor Q (where $Q = 1/d$). Neglecting the departure of real RCL components values from the ideal case, the attenuation at the center frequency f_0 is infinite. Also note that the phase undergoes a 180 degree shift when passing through the center frequency (zero in the s-plane).

Q for bandpass and bandstop filters is a measure of the width $\Delta\Omega$ of the stopband with respect to the center frequency Ω_0, i.e., $\Delta\Omega = Q^{-1}\Omega_0$. $\Delta\Omega$ is measured at the points where $G(\Omega) = 1/\sqrt{2}$.

1.4 ANALOG BANDPASS FILTER

The passive RCL circuit forming a bandpass filter network is shown in Figure 1.4-1 where the transfer function $H(s)$ is again derived from a voltage divider analysis of the RCL network. The gain and phase response are plotted in Figure 1.4-2 for different values of Q. The lowpass gain approaches an asymptotic function of $G = (f_c/f)^2$ for $f/f_c \gg 1$. The highpass asymptotic gain is $G = (f/f_c)^2$ for $f/f_c \ll 1$ whereas the bandstop approaches unity at zero and infinity with a true zero at the center frequency. The bandpass gain, on the other hand, approaches $G = Q^{-1}f/f_c$ for $f/f_c \ll 1$ and $G = Q^{-1}f_c/f$ for $f/f_c \gg 1$. The primary differences to note in the bandpass response are: (1) the stopband attenuation is 6 dB/octave or 20 dB/decade (since it goes as $1/f$) whereas the lowpass and highpass go as $1/f^2$ (12 dB/octave, 40 dB/decade); (2) the stopband attenuation asymptote is dependent on the quality factor whereas for the lowpass and highpass cases the stopband attenuation asymptote is independent of damping factor d; and (3) the maximum value of gain is unity regardless of the filter Q.

The specific features characterizing the bandpass, as well as the lowpass, highpass, and bandstop analog networks, are found to be nearly identical in the digital IIR filter equivalents when the sampling frequency is very high as compared to the frequencies of interest. For this reason it is important to understand the basic properties of the four filter types before proceeding to the digital domain.

SECOND ORDER LOWPASS RCL FILTER ANALYSIS

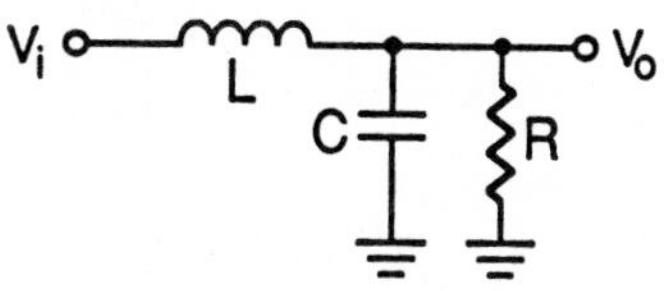

$$\frac{V_o}{V_i} = \frac{X_C || R}{X_L + X_C || R}$$

$$= \frac{(R/j\Omega C)/(R + 1/j\Omega C)}{j\Omega L + (R/j\Omega C)/(R + 1/j\Omega C)}$$

$$= \frac{1/LC}{-\Omega^2 + j\Omega/RC + 1/LC}$$

$$= \frac{\Omega_C^2}{-\Omega^2 + jd\Omega_C\Omega + \Omega_C^2}$$

V_i — L — C — R — V_o

$$X_C = 1/j\Omega C$$

$$X_L = j\Omega L$$

$$\Omega_C = \frac{1}{\sqrt{LC}}$$

$$d = \sqrt{\frac{L}{R^2 C}}$$

Let $s = j\Omega$ and define $H(s) = V_o/V_i$

then,

$$H(s) = \frac{1}{(s/\Omega_C)^2 + d(s/\Omega_C) + 1}$$

which is the *s-Domain Transfer Function.* The *gain* $G(\Omega)$ of the filter is:

$$G(\Omega) \equiv \sqrt{H(s)\, H^*(s)}\Big|_{s=j\Omega}$$

$$= \frac{1}{\sqrt{(1 - \Omega^2/\Omega_C^2)^2 + (d\Omega/\Omega_C)^2}}$$

where * denotes *complex conjugate.* The phase angle $\phi(\Omega)$ is the angle between the imaginary and real components of H(s):

$$\phi(\Omega) \equiv \tan^{-1}[\Im\{H(s)\}/\Re\{H(s)\}]$$

$$= -\tan^{-1}\left[\frac{d(\Omega/\Omega_C)}{1 - (\Omega/\Omega_C)^2}\right] \qquad for\ \Omega \le \Omega_C$$

$$= -\pi - \tan^{-1}\left[\frac{d(\Omega/\Omega_C)}{1 - (\Omega/\Omega_C)^2}\right] \qquad for\ \Omega > \Omega_C$$

Figure 1.1-1. s-Domain Analysis of Second Order Lowpass Analog Filter

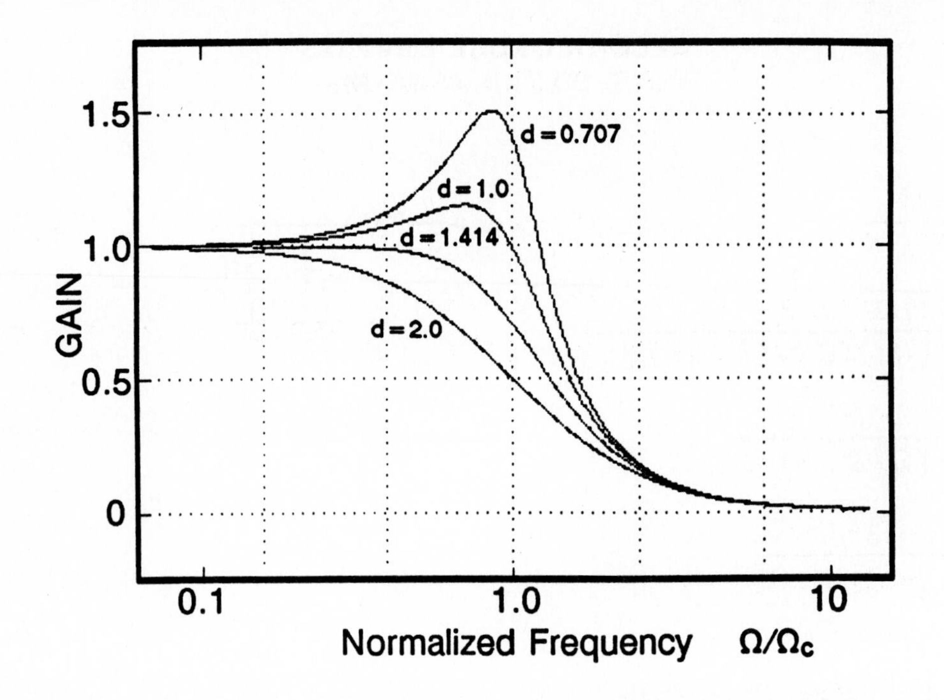

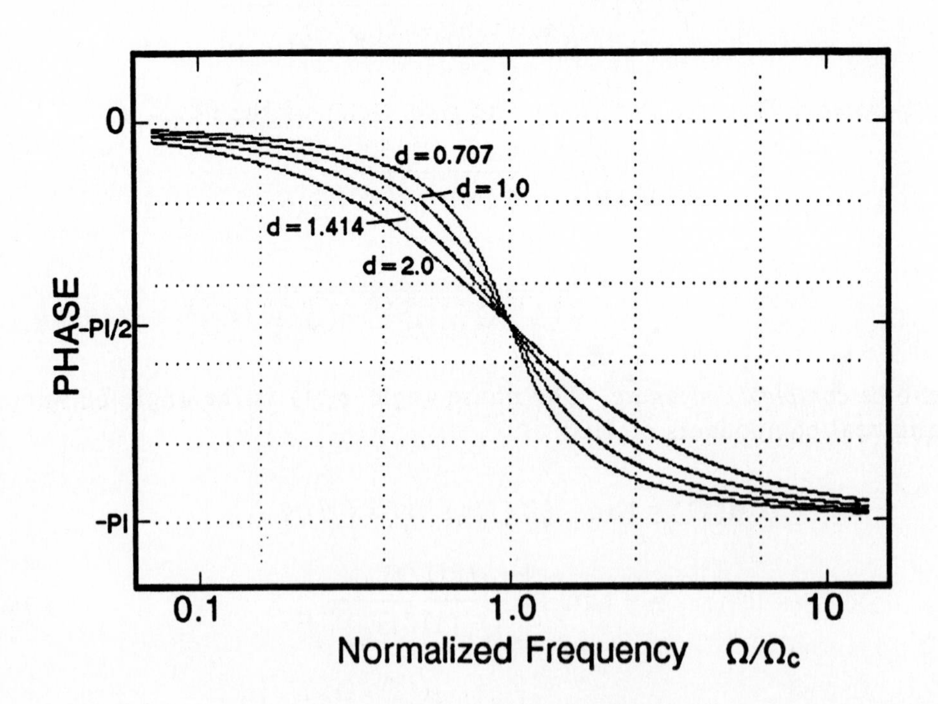

Figure 1.1-2. Gain and Phase Response of Second Order Lowpass Analog Filter at Various Values of Damping Factor, d.

SECOND ORDER HIGHPASS
RCL FILTER ANALYSIS

$$\frac{V_o}{V_i} = \frac{X_L||R}{X_C + X_L||R}$$

$$= \frac{j\Omega LR/(j\Omega L + R)}{1/j\Omega C + j\Omega LR/(j\Omega L + R)}$$

$$= \frac{-\Omega^2}{-\Omega^2 + j\Omega/RC + 1/LC}$$

$$= \frac{-\Omega^2}{-\Omega^2 + jd\Omega_C\Omega + \Omega_C^2}$$

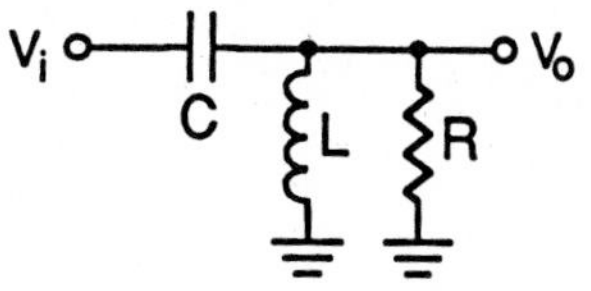

V_i — C — L — R — V_o

$$X_C = 1/j\Omega C$$

$$X_L = j\Omega L$$

$$\Omega_C = \frac{1}{\sqrt{LC}}$$

$$d = \sqrt{\frac{L}{R^2C}}$$

Let $s = j\Omega$ **and define** $H(s) = V_o/V_i$

then,

$$H(s) = \frac{(s/\Omega_C)^2}{(s/\Omega_C)^2 + d(s/\Omega_C) + 1}$$

which is the ***s-Domain Transfer Function.*** **The** ***gain*** $G(\Omega)$ **of the filter is:**

$$G(\Omega) \equiv \sqrt{H(s)\,H^*(s)}\Big|_{s=j\Omega}$$

$$= \frac{(\Omega/\Omega_C)^2}{\sqrt{(1-\Omega^2/\Omega_C^2)^2 + (d\Omega/\Omega_C)^2}}$$

where * denotes ***complex conjugate.*** **The phase angle** $\phi(\Omega)$ **is the angle between the imaginary and real components of H(s):**

$$\phi(\Omega) \equiv \tan^{-1}[\Im\{H(s)\}/\Re\{H(s)\}]$$

$$= \pi - \tan^{-1}\left[\frac{d(\Omega/\Omega_C)}{1-(\Omega/\Omega_C)^2}\right] \qquad \text{for } \Omega \le \Omega_C$$

$$= -\tan^{-1}\left[\frac{d(\Omega/\Omega_C)}{1-(\Omega/\Omega_C)^2}\right] \qquad \text{for } \Omega > \Omega_C$$

Figure 1.2-1. s-Domain Analysis of Second Order Highpass Analog Filter

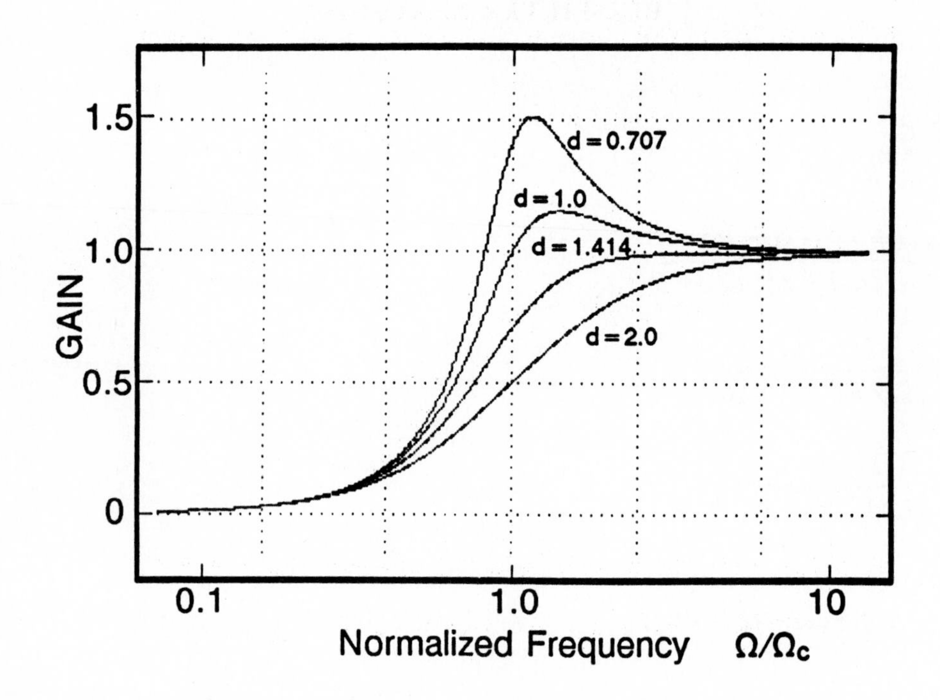

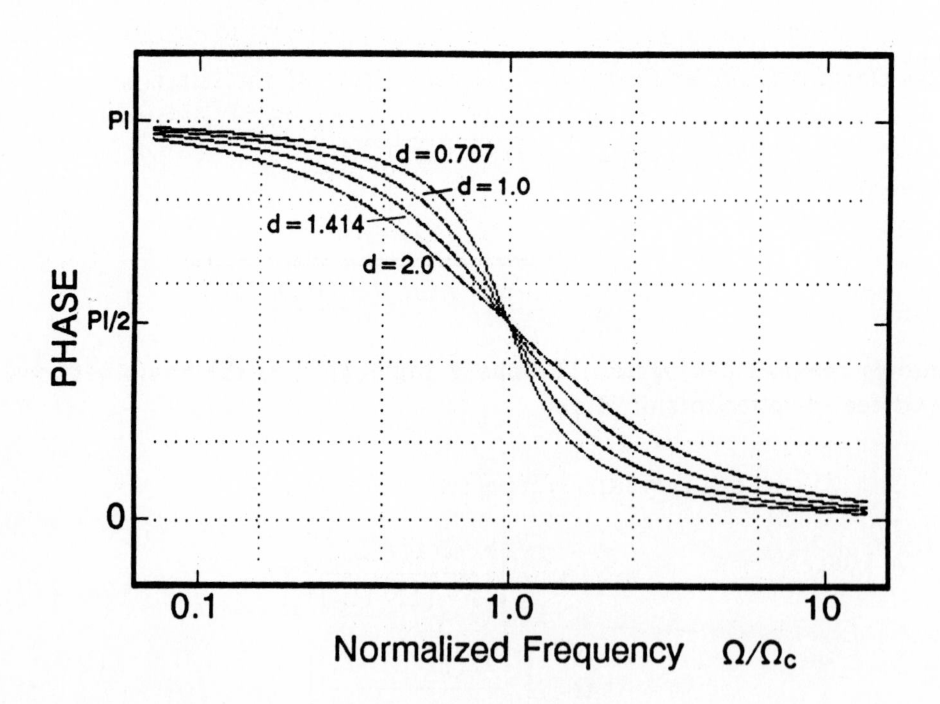

Figure 1.2-2. Gain and Phase Response of Second Order Highpass Analog Filter at Various Values of Damping Factor, d.

SECOND ORDER BANDSTOP RCL FILTER ANALYSIS

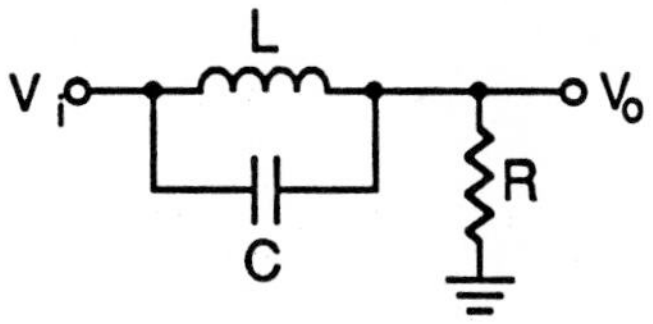

$$\frac{V_o}{V_i} = \frac{R}{X_C||X_L + R}$$

$$= \frac{R}{R + (j\Omega L/j\Omega C)/(j\Omega L + 1/j\Omega C)}$$

$$= \frac{-\Omega^2 + 1/LC}{-\Omega^2 + j\Omega/RC + 1/LC}$$

$$= \frac{-\Omega^2 + \Omega_o^2}{-\Omega^2 + j\Omega_o\Omega/Q + \Omega_o^2}$$

V i, L, C, R, Vo

$$X_C = 1/j\Omega C$$

$$X_L = j\Omega L$$

$$\Omega_C = \frac{1}{\sqrt{LC}}$$

$$Q = 1/d = \sqrt{R^2C/L}$$

Let $s = j\Omega$ and define $H(s) = V_o/V_i$

then,

$$H(s) = \frac{(s/\Omega_o)^2 + 1}{(s/\Omega_o)^2 + s/\Omega_o Q + 1}$$

which is the *s-Domain Transfer Function.* The *gain* $G(\Omega)$ of the filter is:

$$G(\Omega) \equiv \sqrt{H(s)\ H^*(s)}\Big|_{s=j\Omega}$$

$$= \frac{|1 - (\Omega/\Omega_o)^2|}{\sqrt{(1 - \Omega^2/\Omega_o^2)^2 + (\Omega/\Omega_o Q)^2}}$$

where * denotes *complex conjugate.* The phase angle $\phi(\Omega)$ is the angle between the imaginary and real components of H(s):

$$\phi(\Omega) \equiv \tan^{-1}[\Im\{H(s)\}/\Re\{H(s)\}]$$

$$= -\tan^{-1}\left[\frac{\Omega/\Omega_o Q}{1 - (\Omega/\Omega_o)^2}\right]$$

Figure 1.3-1. s-Domain Analysis of Second Order Bandstop Analog Filter

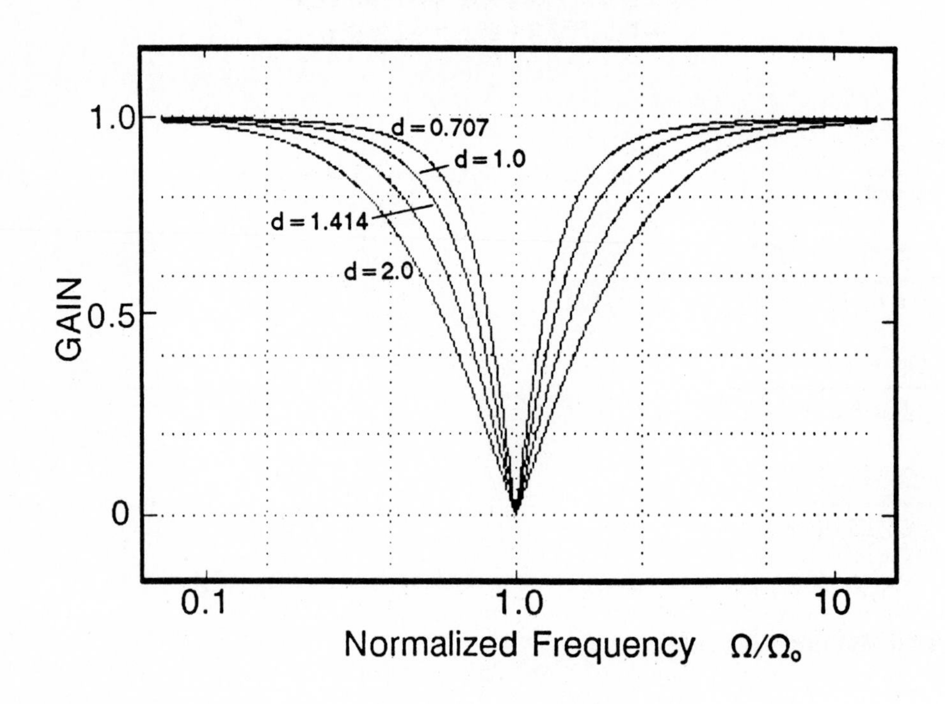

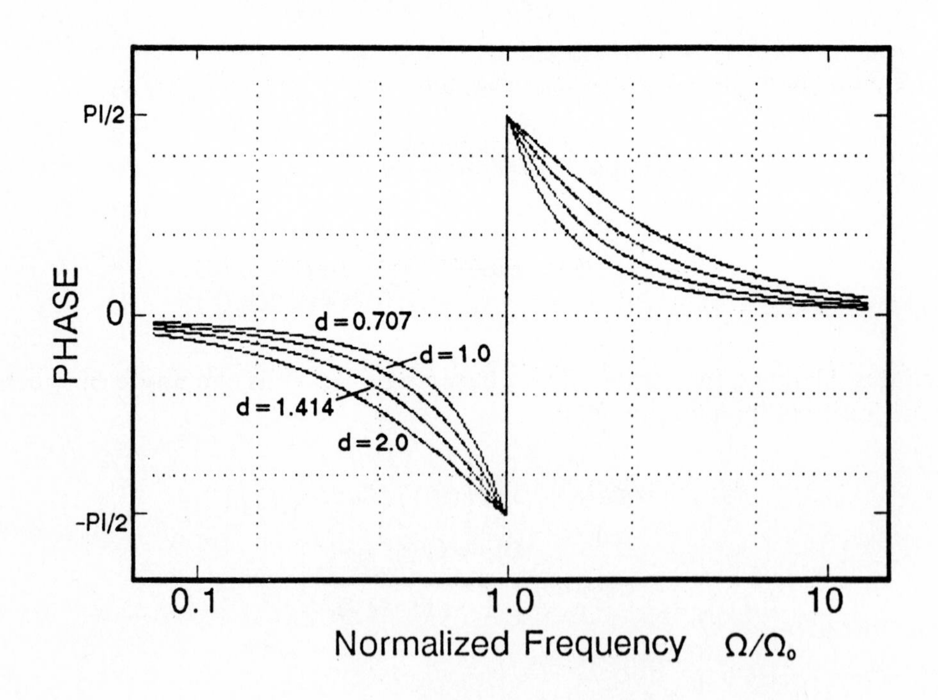

Figure 1.3-2. Gain and Phase Response of Second Order Bandstop Analog Filter at Various Values of Damping Factor, d.

SECOND ORDER BANDPASS RCL FILTER ANALYSIS

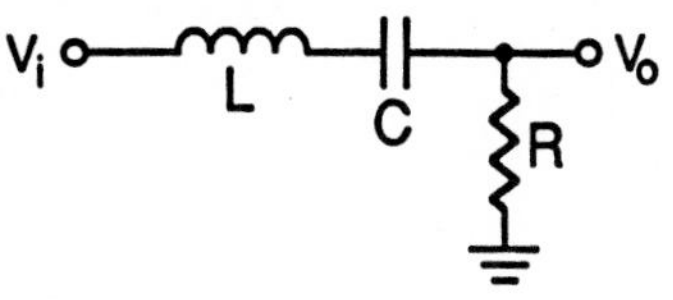

$$\frac{V_o}{V_i} = \frac{R}{X_L + X_C + R}$$

$$= \frac{R}{j\Omega L + 1/j\Omega C + R}$$

$$= \frac{j\Omega R/L}{-\Omega^2 + j\Omega R/L + 1/LC}$$

$$= \frac{j\Omega_o \Omega/Q}{-\Omega^2 + j\Omega_o \Omega/Q + \Omega_o^2}$$

$$X_C = 1/j\Omega C$$

$$X_L = j\Omega L$$

$$\Omega_C = \frac{1}{\sqrt{LC}}$$

$$Q = 1/d = \sqrt{L/R^2 C}$$

Let $s = j\Omega$ **and define** $H(s) = V_o/V_i$

then,

$$H(s) = \frac{s/\Omega_o Q}{(s/\Omega_o)^2 + s/\Omega_o Q + 1}$$

which is the ***s-Domain Transfer Function.*** **The** ***gain*** $G(\Omega)$ **of the filter is:**

$$G(\Omega) \equiv \sqrt{H(s)\, H^*(s)}\Big|_{s=j\Omega}$$

$$= \frac{\Omega/\Omega_o Q}{\sqrt{(1 - \Omega^2/\Omega_o^2)^2 + (\Omega/\Omega_o Q)^2}}$$

where * denotes ***complex conjugate.*** **The phase angle** $\phi(\Omega)$ **is the angle between the imaginary and real components of H(s):**

$$\phi(\Omega) \equiv \tan^{-1}[\Im\{H(s)\}/\Re\{H(s)\}]$$

$$= \tan^{-1}\left[\frac{\Omega/\Omega_o Q}{1 - (\Omega/\Omega_o)^2}\right]$$

Figure 1.4-1. s-Domain Analysis of Second Order Bandpass Analog Filter

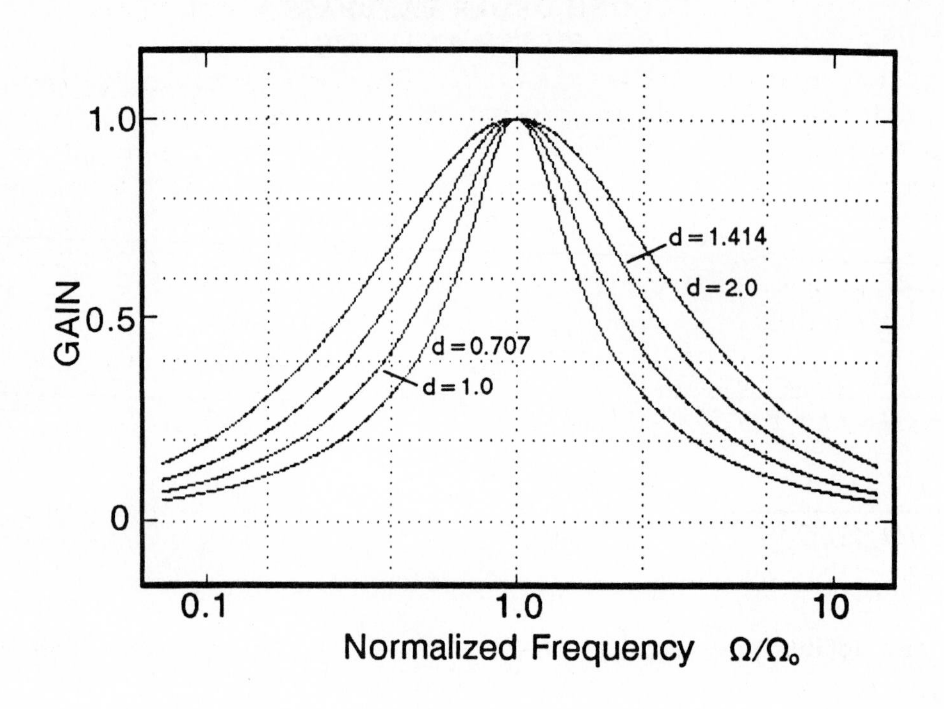

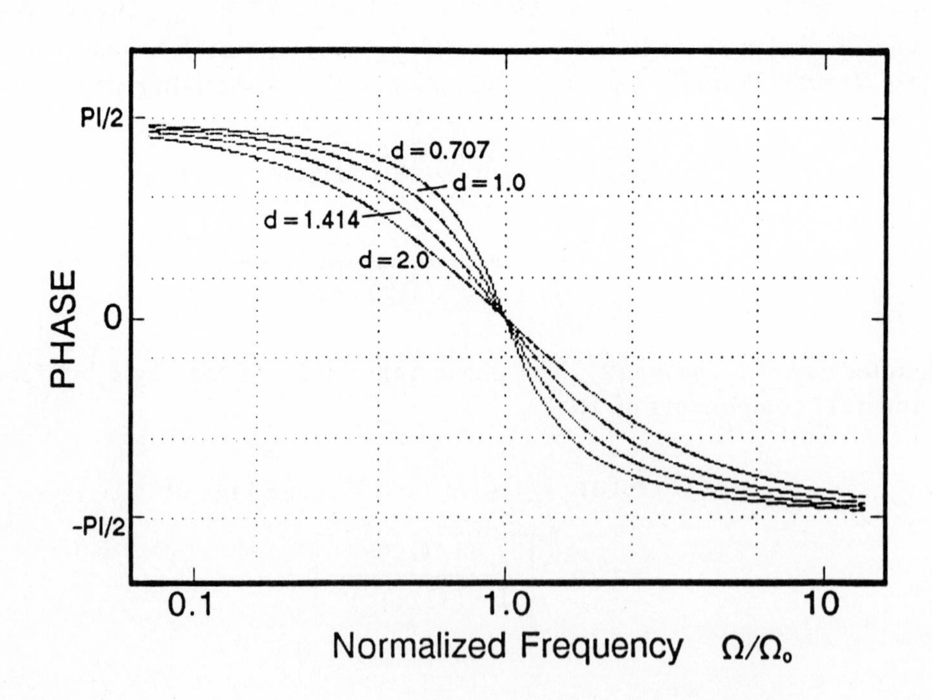

Figure 1.4-2. Gain and Phase Response of Second Order Bandpass Analog Filter at Various Values of Damping Factor, d.

2.0 SECOND ORDER DIRECT FORM IIR DIGITAL FILTER SECTIONS

The traditional approach to deriving the digital filter coefficients has been to start with the digital z-domain description, transform to the analog s-domain where we understand how to design filters, then transform back again to the digital domain in order to implement the filter. This approach will not be used in this report. Instead, we will develope the formulas relating the s-domain filter to the z-domain filter so that the transformations to and from one domain to the other are no longer necessary.

To begin, recall that the Laplace or s-transform in the analog domain was developed to facilitate the analysis of continuous time signals and systems. For example, using Laplace transforms the concepts of poles and zeros made system analysis much faster and more systematic. The Laplace transform of a continuous time signal is:

$$\begin{aligned} X(s) &= L\{x(t)\} \\ &= \int_0^{\infty} x(t)\, e^{-st}\, dt \end{aligned} \tag{2.0-1}$$

where L is the *Laplace transform operator* and implies the operation described in Equation (2.0-1).

In the digital domain the continuous signal $x(t)$ is first sampled and then quantized by an A/D converter before being processed. That is, the signal is only known at discrete points in time which are at multiples of the sampling interval $T = 1/f_S$ where f_S is the sampling frequency. Because of the sampled characteristic of a digital signal its z-transform is given by a summation (as opposed to an integral):

$$\begin{aligned} X(z) &= Z\{x(n)\} \\ &= \sum_{n=-\infty}^{\infty} x(n)\, z^{-n} \end{aligned} \tag{2.0-2}$$

where Z is the *z-transform operator* as described by the operation of Equation (2.0-2) and $x(n)$ are the quantized values from the A/D converter of the continuous time signal $x(t)$ at discrete times $t = nT$.

One of the properties of the z-transform which will be used later in this report is its *time shifting* property. The *time shifting* property states:

$$x(n-k) = Z^{-1}\{z^{-k} X(z)\} \tag{2.0-3}$$

The proof of this property follows directly from the definition of the z-transform.

An obvious question which arises is, "If the s-domain of a signal $x(t)$ is known and that same signal is digitized, what is the relationship between the s-domain transform and the z-transform?" The relationship or mapping is not unique and depends on your point of view.

It is obvious that the trivial mapping $s = z$ is inappropriate, the signal has been sampled. It is widely known that when a bandlimited signal is sampled (i.e., multiplied by a periodic impulse function) the spectrum of the resulting signal is repetitive as shown in Figure 2.0-1 [10].

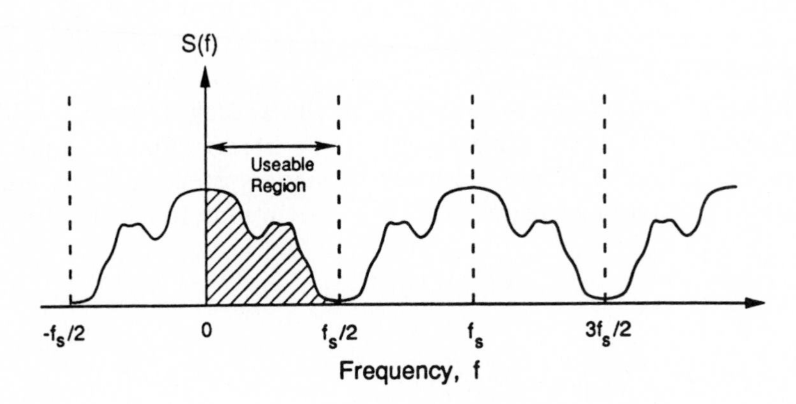

Figure 2.0-1. Spectrum of Bandlimited Signal Repeated at Multiples of the Sampling Frequency f_S.

Clearly the spectrum consists of the spectrum of the bandlimited signal repeated at multiples of the sampling frequency $\omega_S = 2\pi f_S$. That is the resulting spectrum is unique only between 0 and $\omega_S/2$ or multiples thereof whereas, before the signal was sampled the energy at frequencies greater than those in the signal was independent of the signal. Therefore acceptable mappings would either reflect the cyclic nature of the spectrum of the sampled signal or at least be linear over the frequencies of interests.

One mapping or transformation from the s-domain to the z-domain which is often used and will be used later in this report is the *bilinear transformation.* To gain some feeling as to the origin of this transformation consider the simple first order linear analog filter with the system function:

$$H(s) = \frac{Y(s)}{X(s)} = \frac{b}{s+a} \tag{2.0-4}$$

and recall the differentiation property of the s-transform when $x(t) = L^{-1}\{X(s)\}$ (where L^{-1} is the inverse Laplace transform operator), then the time derivative of $x(t)$ is:

$$\frac{d}{dt}x(t) = L^{-1}\{s\,X(s)\} \tag{2.0-5}$$

where $\frac{d}{dt}x(t)$ is the time derivative of *x(t)*.

Using the differentiation property of Equation (2.0-5), the linear system described by Equation (2.0-4) can be expressed as:

$$\frac{d}{dt}y(t) + a\,y(t) = b\,x(t) \qquad (2.0\text{-}6)$$

Now, if this differential equation is solved by expressing *y(t)* with the trapezoidal integration formula:

$$y(t) = \int_{t_0}^{t} \frac{d}{d\tau}y(\tau)\,d\tau + y(t_0) \qquad (2.0\text{-}7)$$

where the approximate solution is given by:

$$y(t) = \frac{1}{2}\left[\frac{d}{dt}y(t) + \frac{d}{dt}y(t_0)\right](t - t_0) + y(t_0) \qquad (2.0\text{-}8)$$

then using $\frac{d}{dt}y(t)$ from Equation (2.0-6) with $t = nT$ and $t_0 = (n-1)T$, Equation (2.0-8) can be expressed as:

$$(2 + aT)\,y(n) - (2 - aT)\,y(n-1) = bT[x(n) + x(n-1)] \qquad (2.0\text{-}9)$$

Now taking the z-transform of this difference equation and using the *time shifting property* of the z-transform, Equation (2.0-3), results in the z-domain system function:

$$H(z) = \frac{Y(z)}{X(z)} = \frac{b}{\frac{2}{T}\left(\frac{1-z^{-1}}{1+z^{-1}}\right) + a} \qquad (2.0\text{-}10)$$

Clearly, the mapping between the s-plane and the z-plane is:

$$s = \frac{2}{T}\left(\frac{1 - z^{-1}}{1 + z^{-1}}\right) \qquad (2.0\text{-}11)$$

This mapping is called the *bilinear transformation.*

Although this transformation was developed using a first order system it holds, in general, for an Nth-order system [14]. It can be shown by letting $s = \sigma + j\Omega$ and $z = re^{j\theta}$ that the left half plane in the s-domain is mapped inside the unit $r = 1$ circle in the z-domain under the *bilinear transformation.* Also, and this is the important point for what follows, when $r = 1$ and $\sigma = 0$, the frequencies in the s-domain and the z-domain are related by:

$$\Omega = \frac{2}{T}\tan\frac{\theta}{2} \qquad (2.0\text{-}12)$$

or equivalently:

$$\theta = 2\tan^{-1}\frac{\Omega T}{2} \qquad (2.0\text{-}13)$$

where θ is the digital domain normalized frequency equal to $2\pi f/f_s$ and Ω is the analog domain frequency used in the analysis of the previous section.

It can be seen that on the $j\Omega$ axis or equivalently along the frequency axis the scale has been changed non-linearly. The gain and phase values depicted on the vertical axis of Figures 1.1-2, 1.2-2, 1.3-2, and 1.4-2 remain exactly the same in the digital domain (or z-plane). The horizontal (frequency) axis is modified so that an infinite frequency in the analog domain maps to one half of the sample frequency, $f_s/2$ in the digital domain whereas for frequencies much less that $f_s/2$ the mapping is approximately 1:1 with $\theta = \Omega$. In summary the *bilinear transformation* is a one to one nonlinear mapping from the s-domain into the z-domain in which high frequencies ($\Omega > 2\pi f_s/4$) in the s-domain are compressed into a small interval in the z-domain. Therefore the gain and phase expressions of the previous section can be directly transformed into the digital domain by simply substituting Equation (2.0-12) into the corresponding expressions. This will be done for each filter type in the following subsections.

First, it is appropriate to introduce the *Direct Form* implementation of a digital filter by noting that, in general, if the *bilinear transformation*, Equation (2.0-11), is substituted into the system functions H(s) of the previous section the resulting H(z) will have the following generalized form:

$$H(z) = \frac{b_0 + b_1 z^{-1} + b_2 z^{-2}}{1 + a_1 z^{-1} + a_2 z^{-2}} \qquad (2.0-14)$$

where the digital domain coefficients a_i and b_i are exactly related to the s-domain characteristics of the system such as the center frequency and bandwidth, etc. In the *Direct Form* implementation the a_i and b_i are used directly in the difference equation which can be easily programmed on a high speed digital signal processor (DSP) such as the DSP56001. The time domain difference equation is derived from the z-domain transfer functions by applying the inverse z-transform in general and the inverse time shifting property in particular as follows:

$$Z^{-1}\{H(z)\} = Z^{-1}\{Y(z)/X(z)\} \qquad (2.0-15)$$

$$= Z^{-1}\{[b_0 + b_1 z^{-1} + b_2 z^{-2}]/[1 + a_1 z^{-1} + a_2 z^{-2}]\}$$

so that:

$$Z^{-1}\{Y(z)[1 + a_1 z^{-1} + a_2 z^{-2}]\} = Z^{-1}\{X(z)[b_0 + b_1 z^{-1} + b_2 z^{-2}]\}$$

and, therefore using the *inverse time shifting* property of Equation (2.0-3):

$$Z^{-1}\{X(z)z^{-k}\} = x(n-k)\}$$

and

$$Z^{-1}\{Y(z)z^{-k}\} = y(n-k)\}$$

Equation (2.0-15) becomes:

$$y(n) = b_0 x(n) + b_1 x(n-1) + b_2 x(n-2) - a_1 y(n-1) - a_2 y(n-2) \qquad (2.0-16)$$

Equation (2.0-16) can be directly implemented in software where $x(n)$ is the sample input and $y(n)$ is the corresponding filtered digital output. When the filter output is calculated using Equation (2.0-16) it is said that $y(n)$ is calculated using the *Direct Form* implementation of the digital filter.

There are other implementations which can be used for the same system (filter) function H(z). Two of these which will be discussed in this report are the *Canonic Form* implementation and the *Transpose Form* implementation. First the *Direct Form* implementation will be applied to the system transfer functions H(s), developed in *Section 1.*

2.1 DIGITAL LOWPASS FILTER

Using the analog transfer function H(s) from Figure 1.1-A and Equations (2.0-11) and (2.0-12), the digital transfer function H(z) becomes that shown in Figure 2.1-1 where the coefficients α, β, and γ are expressed in terms of the digital cutoff frequency, θ_c, and the damping factor d. The value of the transfer function at $\theta = \theta_c$ in the digital domain is identical to the value of the s-domain transfer function at $\Omega = \Omega_c$:

$$H_z(e^{j\theta_c}) = H_s(j\Omega_c) \qquad (2.1-1)$$

As can be seen in Figure 2.1-2 the digital gain and phase response calculated from the equations of Figure 2.1-1 are similar to the analog plots shown in Figure 1.1-2 except for the asymmetry introduced by the zero at $f_s/2$. That is, the frequency axis is modified so that a gain of zero at $f = \infty$ in the s-domain corresponds to a gain of zero at $f = f_S/2$ in z-domain. The fact that the magnitude of the transfer functions H(s) and H(z) are identical once the proper frequency transformation is made, is very useful for understanding the digital filter and its relationship to the analog equivalent. This fact is also useful for purposes of scaling the gain since the maximum magnitude of $G_s(\Omega_M) = G_z(\theta_M)$ where Ω_M and θ_M are related by Equation (2.0-12). In other words, scaling analysis of the digital transfer function H(z) can be done in the s-domain (not that it may necessarily be simpler there but the algebra is often easier to manage). Scaling of the gain is an essential part of digital filter implementation since the region of numeric calculations on fixed point DSP's such as the DSP56001 are usually restricted to a range of -1 to 1.

Using the formulas in Figure 2.1-1 along with Equation (1.1-2) guarantees the behavior of the digital filter. Since the gain is scaled to unity at $f = 0$ (DC), the input data x(n) in Figure 2.1-1 must be scaled down by a factor of $1/G_M$ from Equation (1.1-2) if the entire dynamic range of the digital network is to be utilized. The alternative is automatic gain control to insure that $x(n)$ is smaller than $1/G_M$ before it arrives at the filter input. Note that for $d \geq \sqrt{2}$, the input does not require scaling since the gain of the filter will never exceed unity.

The DSP56001 code to implement the second order lowpass filter section is shown in Figure 2.1-3. Note that the address register modifiers are initially set to M0=4, M4=1, and M5=1 to allow use of the circular buffer or modulo addressing in this particular implementation [8]. Typically, this code would be an interrupt routine driven by the input data (ADC for example) sample rate clock. The basic filter code along with the interrupt overhead and data I/O moves for this second order filter could be executed at a sample rate of nearly 1 MHz on the DSP56001.

2.2 DIGITAL HIGHPASS FILTER

The highpass filter is nearly identical to the lowpass as can be seen by the formulas in Figure 2.2-1. As with the analog case, the digital highpass filter is just the mirror image of the lowpass digital filter, as can be seen from Figure 2.2-2. The frequency transformation from high to low in the analog case is , $\Omega \rightarrow 1/\Omega$, while in the digital case it is $\theta \rightarrow \pi - \theta$.

The DSP56001 code is shown in Figure 2.2-3 and as can be seen by comparison to Figure 2.1-3, exactly the same instruction sequence is used. The only difference is the coefficient data which is calculated by the formulas in Figure 2.2-1. Note that the *scaling mode* is turned on so that a move from the A or B accumulator to the X or Y registers or memory results in an automatic multiply by two (represented by a "2" enclosed by a diamond in the network diagrams to be discussed later in this report). The *scaling mode* is used not only in the code for the lowpass case but also in the code for the highpass, bandstop, and bandpass cases.

2.3 DIGITAL BANDSTOP FILTER

The formulas and network diagram for the digital bandstop filter are presented in Figure 2.3-1. The DSP56001 code from Figure 2.3-3 is identical to that for the lowpass and highpass cases except for the coefficient data calculated from the equations of Figure 2.3-1. Note that scaling of this filter is not a problem for the single section case since the gain from the equation in Figure 2.3-1 (as is true in the analog case as seen by the gain equation from Figure 1.3-1) never exceeds unity. This is also true of the bandpass filter described in the next section. Figure 2.3-2 is the calculated gain and phase of the digital filter which should compare to the response curves of the equivalent analog filter plotted in Figure 1.3-2.

2.4 DIGITAL BANDPASS FILTERS

Because there is one less coefficient in the bandpass network (as shown in Figure 2.4-1), one instruction can be saved in the DSP56001 code implementation shown in Figure 2.4-3. Otherwise, the instructions are identical to those in the other three filter routines. As with the second order bandstop network of the previous section, the maximum response at the center frequency θ_0 is unity for any value of Q so that scaling does not need to be considered in the implementation of a single section bandpass filter. This is true when the formulas for α, β, and γ from Figure 2.4-1 are used in the *Direct Form* implementation in Figure 2.4-3. Figure 2.4-2 is the calculated gain and phase of the digital filter which should compare to the response curves of the equivalent analog filter plotted in Figure 1.4-2.

2.5 SUMMARY OF DIGITAL COEFFICIENTS

Figure 2.5-1 gives a table summary of the coefficient values for the four basic filter types. Note that the coefficient β has the same form for all four filter types and that it can only assume values between 0 and 1/2 for practical filters. β is bounded by 1/2 because Q (or d) and θ_0 are not independent. For $Q \gg 1$ $\beta \to 1/2$; whereas for $\theta_0 = f_S/4$ and $Q = 1/2$, $\beta \to 0$. These properties are independent of the form of implementation; they are only dependent on the form of the transfer function. Alternate implementations (difference equations) will be described in the following sections.

Note that the Q described in Figure 2.5-1 meets the traditional requirements, i.e., Q is the ratio of the bandwidth at the -3 dB points divided by the center frequency. The formula for β can be modified in the case of the bandpass or bandstop filter by replacing the damping coefficient d with the formula for Q. When the coefficients are described in this manner, a constant Q filter results. This relationship between d and Q makes it possible to implement a bandpass or bandstop filter where the bandwidth is any function of center frequency by replacing Q with the desired function of bandwidth and center frequency.

Figure 2.5-2 shows the relationship between the *pole* of the second order section and the center frequency. Note that the *pole* is on the real axis for $d>2$, where d is also constrained by the relation $d < 2/\sin\theta_0$.

Second Order Lowpass Filter

NETWORK DIAGRAM

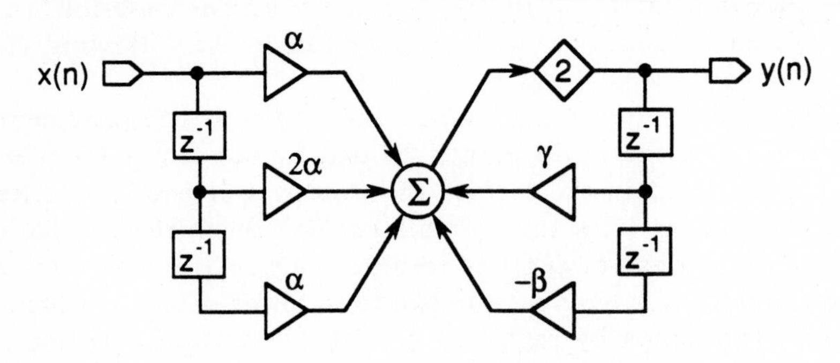

TRANSFER FUNCTION

$$H(z) = \frac{\alpha(1+2z^{-1}+z^{-2})}{1/2-\gamma z^{-1}+\beta z^{-2}}$$

GAIN

$$G(\theta) = \frac{(1+\cos\theta)(1-\cos\theta_c)}{[(d\sin\theta\sin\theta_c)^2+4(\cos\theta-\cos\theta_c)^2]^{1/2}}$$

PHASE

$$\phi(\theta) = \begin{cases} \tan^{-1}\left[\dfrac{2(\cos\theta-\cos\theta_c)}{d\sin\theta\sin\theta_c}\right] & \textit{for}\ \theta \le \theta_c \\ -\pi+\tan^{-1}\left[\dfrac{2(\cos\theta-\cos\theta_c)}{d\sin\theta\sin\theta_c}\right] & \textit{for}\ \theta > \theta_c \end{cases}$$

COEFFICIENTS

$$\beta = \frac{1}{2}\,\frac{1-d/2\sin\theta_c}{1+d/2\sin\theta_c}$$

$$\gamma = (1/2+\beta)\cos\theta_c$$

$$\alpha = (1/2+\beta-\gamma)/4$$

Figure 2.1-1. Direct Form Implementation of Second Order Lowpass IIR Digital Filter and Analytical Formulas Relating Desired Response to Filter Coefficients.

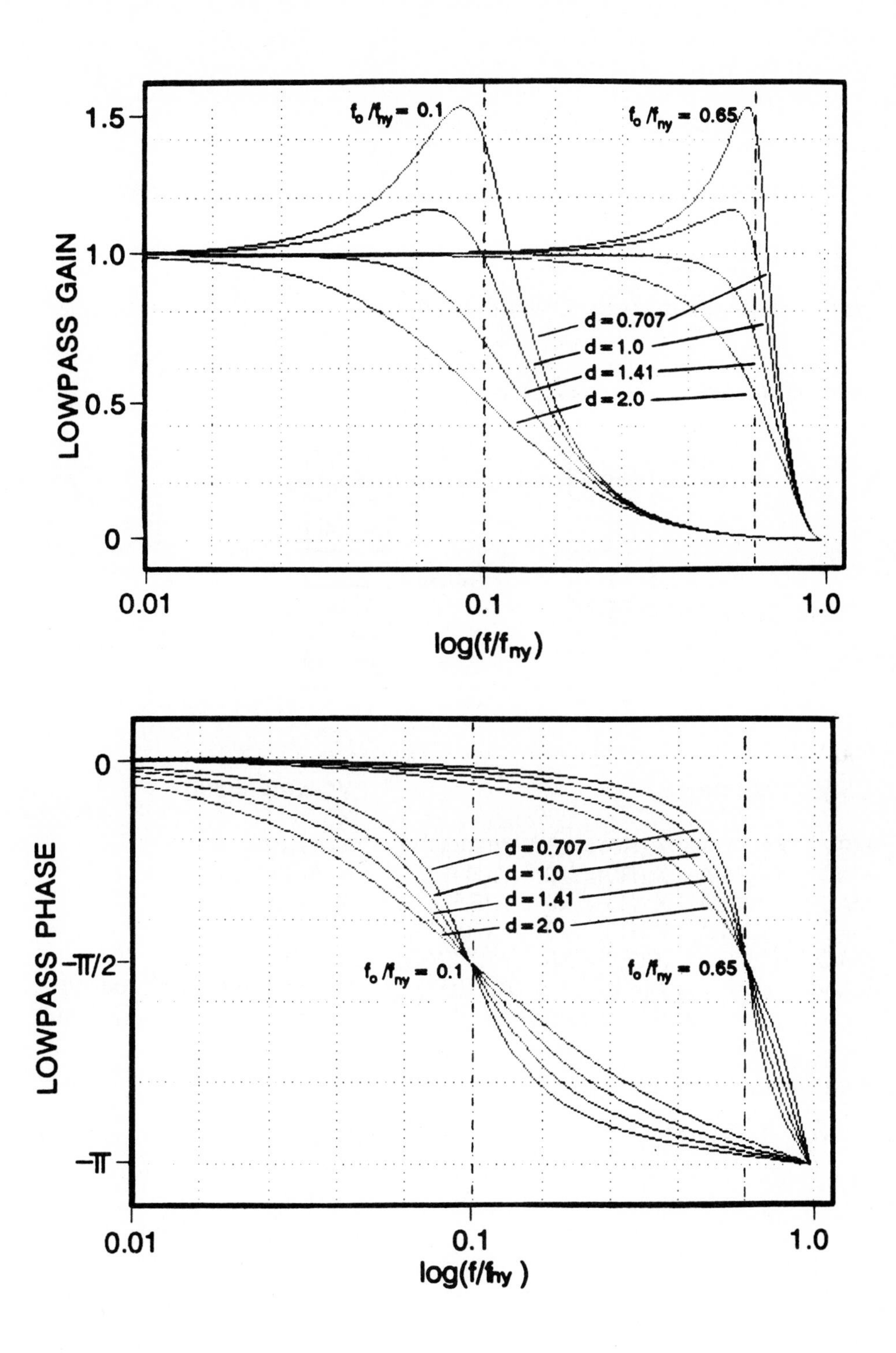

Figure 2.1-2. Gain and Phase Response of Second Order Lowpass IIR Filter. Nyquist Frequency f_{ny} is Equal to One Half of the Sample Frequency f_S.

SECOND ORDER LOWPASS DIRECT FORM IMPLEMENTATION

Difference Equation:

$$y(n)=2\{\alpha[x(n)+2x(n-1)+x(n-2)]+\gamma y(n-1)-\beta y(n-2)\}$$

Data Structures:

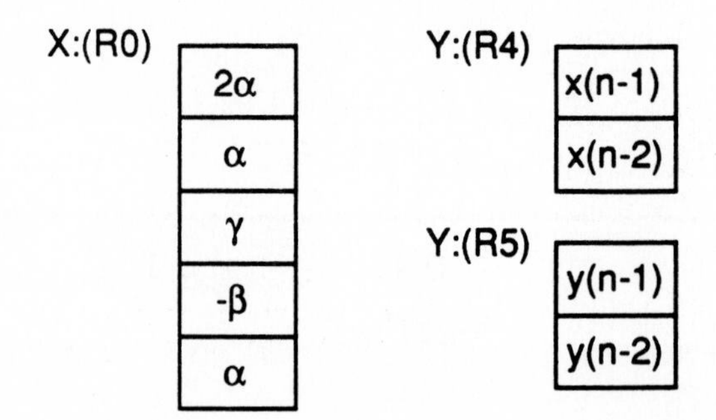

DSP56001 Code:

```
                                              ;Y1=x(n) (Input)
                                              ;X0=α
MPY   X0,Y1,A   X:(R0)+,X0   Y:(R4)+,Y0   ;A=αx(n)
MAC   X0,Y0,A   X:(R0)+,X0   Y:(R4),Y0    ;A=A+2αx(n-1)
MAC   X0,Y0,A   X:(R0)+,X0   Y:(R5)+,Y0   ;A=A+αx(n-2)
MAC   X0,Y0,A   X:(R0)+,Y0   Y:(R5),Y0    ;A=A+γy(n-1)
MAC   X0,Y0,A   X:(R0)+,X0   Y1,Y:(R4)    ;A=A-βy(n-2)
MOVE            A,X1         A,Y:(R5)     ;y(n)=2A (assumes scaling
                                          ;mode is set). X1 is Output.
```

Total Instruction Cycles:

6 Icyc @20 MHz=600ns

Figure 2.1-3. DSP56001 Code and Data Structures for Second Order Direct Form Implementation of a Lowpass Digital IIR Filter.

Second Order Highpass Filter

NETWORK DIAGRAM

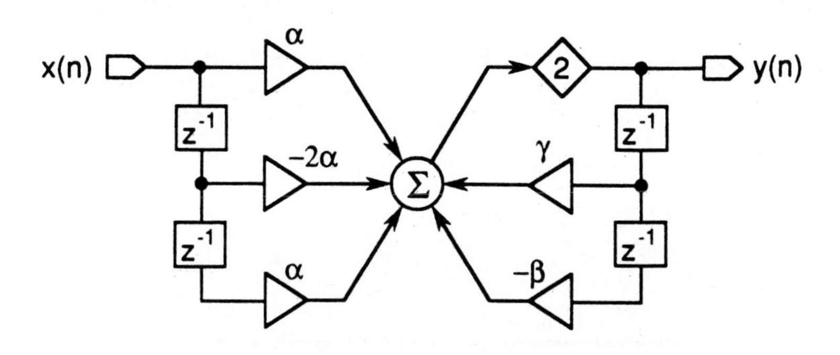

TRANSFER FUNCTION

$$H(z) = \frac{\alpha(1-2z^{-1}+z^{-2})}{1/2-\gamma z^{-1}+\beta z^{-2}}$$

GAIN

$$G(\theta) = \frac{(1-\cos\theta)(1+\cos\theta_c)}{[(d\sin\theta\sin\theta_c)^2+4(\cos\theta-\cos\theta_c)^2]^{1/2}}$$

PHASE

$$\phi(\theta) = \begin{cases} \pi + \tan^{-1}\left[\frac{2(\cos\theta-\cos\theta_c)}{d\sin\theta\sin\theta_c}\right] & \textit{for } \theta \le \theta_c \\ \tan^{-1}\left[\frac{2(\cos\theta-\cos\theta_c)}{d\sin\theta\sin\theta_c}\right] & \textit{for } \theta > \theta_c \end{cases}$$

COEFFICIENTS

$$\beta = \frac{1}{2}\,\frac{1-d/2\sin\theta_c}{1+d/2\sin\theta_c}$$

$$\gamma = (1/2+\beta)\cos\theta_c$$

$$\alpha = (1/2+\beta+\gamma)/4$$

Figure 2.2-1. Direct Form Implementation of Second Order Highpass IIR Digital Filter and Analytical Formulas Relating Desired Response to Filter Coefficients.

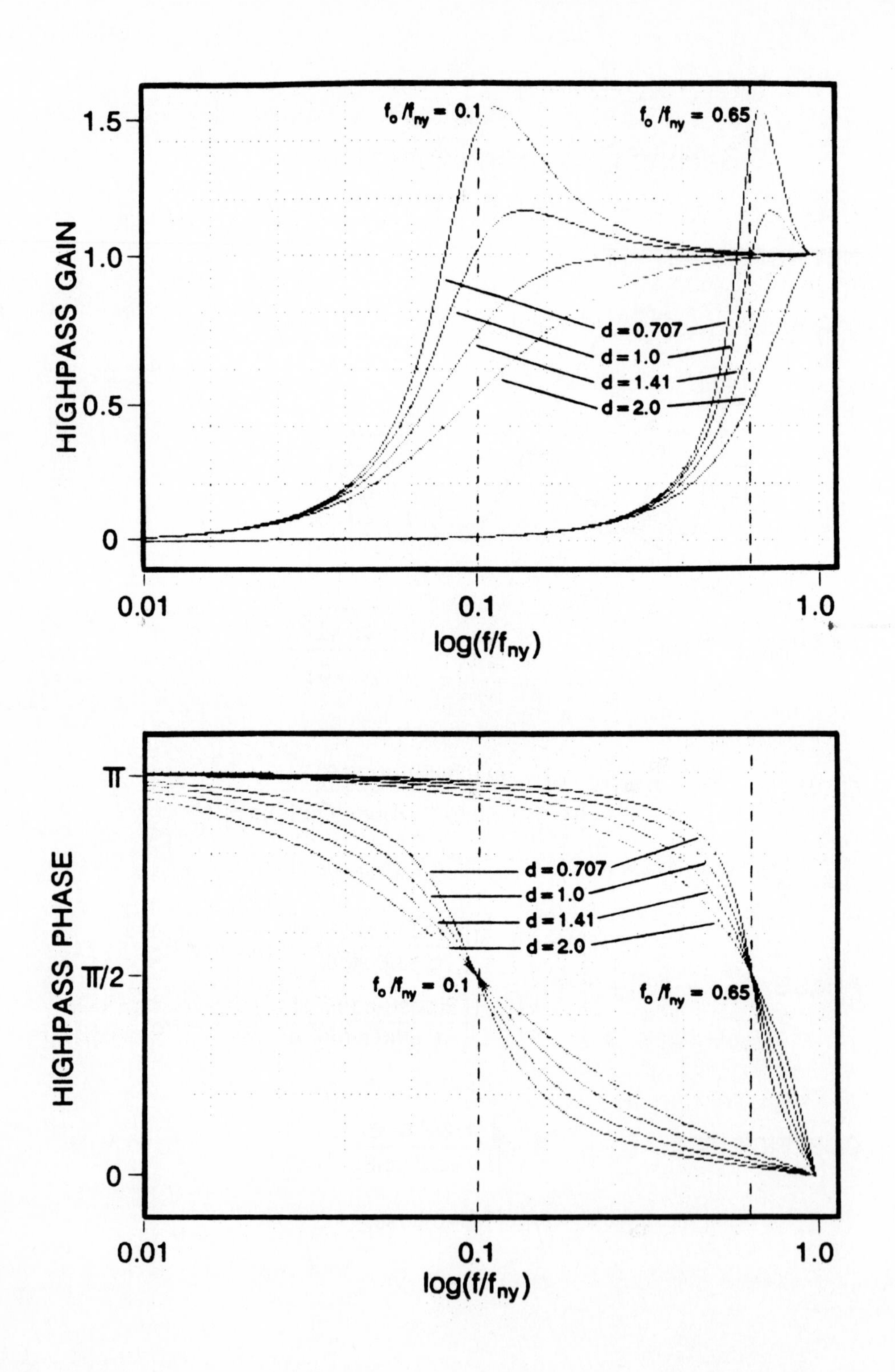

Figure 2.2-2. Gain and Phase Response of Second Order Highpass IIR Filter. Nyquist Frequency f_{ny} is Equal to One Half of the Sample Frequency f_S.

SECOND ORDER HIGHPASS DIRECT FORM IMPLEMENTATION

Difference Equation:

$$y(n)=2\{\alpha[x(n)-2x(n-1)+x(n-2)]+\gamma y(n-1)-\beta y(n-2)\}$$

Data Structures:

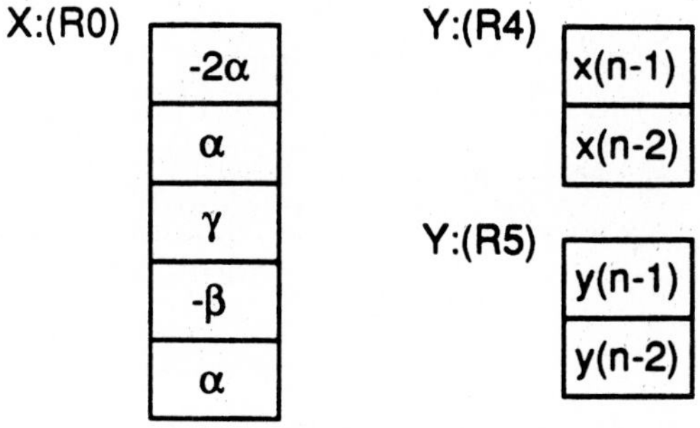

X:(R0)
-2α
α
γ
$-\beta$
α

Y:(R4)
x(n-1)
x(n-2)

Y:(R5)
y(n-1)
y(n-2)

DSP56001 Code:

```
                                                 ;Y1=x(n) (Input)
                                                 ;X0=α
MPY   X0,Y1,A   X:(R0)+,X0   Y:(R4)+,Y0   ;A=αx(n)
MAC   X0,Y0,A   X:(R0)+,X0   Y:(R4),Y0    ;A=A-2αx(n-1)
MAC   X0,Y0,A   X:(R0)+,X0   Y:(R5)+,Y0   ;A=A+αx(n-2)
MAC   X0,Y0,A   X:(R0)+,Y0   Y:(R5),Y0    ;A=A+γy(n-1)
MAC   X0,Y0,A   X:(R0)+,X0   Y1,Y:(R4)    ;A=A-βy(n-2)
MOVE            A,X1         A,Y:(R5)     ;y(n)=2A (assumes scaling
                                          ;mode is set).  X1 is Output.
```

Total Instruction Cycles:

6 Icyc @20 MHz=600ns

Figure 2.2-3. DSP56001 Code and Data Structures for Second Order Direct Form Implementation of a Highpass Digital IIR Filter.

Second Order Bandstop Filter

NETWORK DIAGRAM

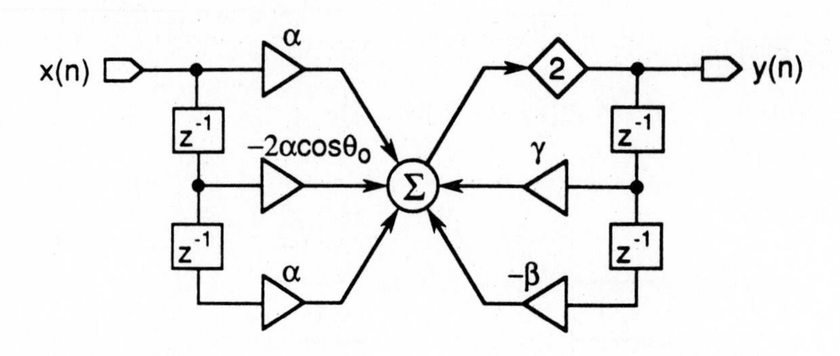

TRANSFER FUNCTION
$$H(z) = \frac{\alpha(1-2\cos\theta_0 z^{-1}+z^{-2})}{1/2-\gamma z^{-1}+\beta z^{-2}}$$

GAIN
$$G(\theta) = \frac{|\cos\theta-\cos\theta_0|}{[(d \sin\theta \sin\theta_0)^2+4(\cos\theta-\cos\theta_0)^2]^{1/2}}$$

PHASE
$$\phi(\theta) = \tan^{-1}\left[\frac{2(\cos\theta-\cos\theta_0)}{d \sin\theta \sin\theta_0}\right]$$

COEFFICIENTS
$$d = \frac{2\tan(\theta_0/2Q)}{\sin\theta_0}$$

$$\beta = \frac{1}{2}\,\frac{1-\tan(\theta_0/2Q)}{1+\tan(\theta_0/2Q)}$$

$$\gamma = (1/2+\beta)\cos\theta_0$$

$$\alpha = (1/2+\beta)/2$$

Figure 2.3-1. Direct Form Implementation of Second Order Bandstop IIR Digital Filter and Analytical Formulas Relating Desired Response to Filter Coefficients.

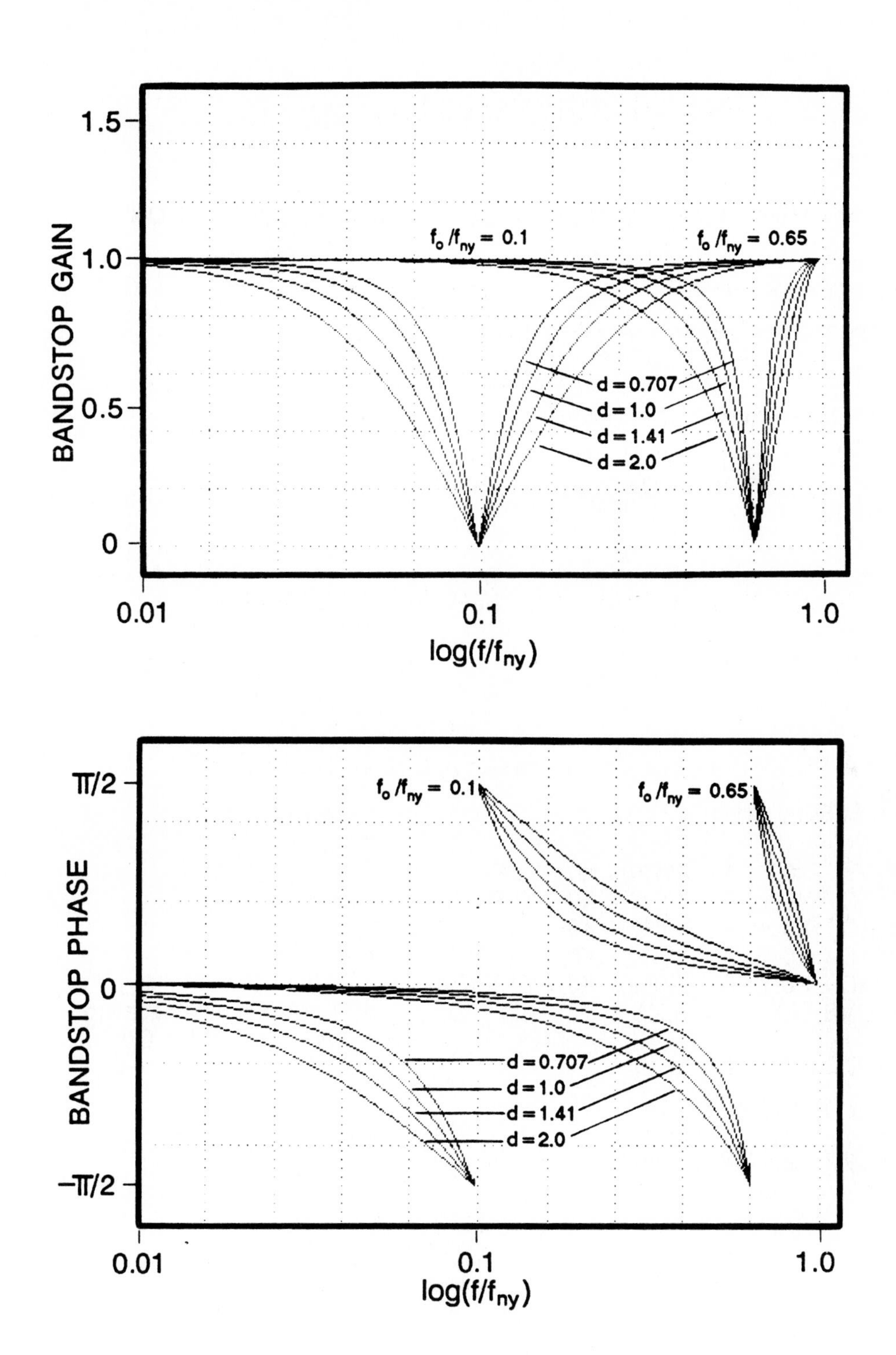

Figure 2.3-2. Gain and Phase Response of Second Order Bandstop IIR Filter. Nyquist Frequency f_{ny} is Equal to One Half of the Sample Frequency f_S.

SECOND ORDER BANDSTOP DIRECT FORM IMPLEMENTATION

Difference Equation:

$$y(n)=2\{\alpha[x(n)-2\cos\theta_0 x(n-1)+x(n-2)]+\gamma y(n-1)-\beta y(n-2)\}$$

Data Structures:

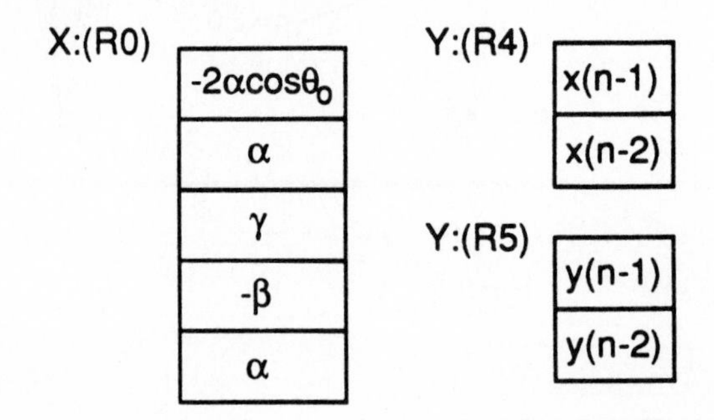

DSP56001 Code:

```
                                              ;Y1=x(n) (Input)
                                              ;X0=α
MPY   X0,Y1,A   X:(R0)+,X0   Y:(R4)+,Y0   ;A=αx(n)
MAC   X0,Y0,A   X:(R0)+,X0   Y:(R4),Y0    ;A=A-2αcosθ0x(n-1)
MAC   X0,Y0,A   X:(R0)+,X0   Y:(R5)+,Y0   ;A=A+αx(n-2)
MAC   X0,Y0,A   X:(R0)+,Y0   Y:(R5),Y0    ;A=A+γy(n-1)
MAC   X0,Y0,A   X:(R0)+,X0   Y1,Y:(R4)    ;A=A-βy(n-2)
MOVE            A,X1         A,Y:(R5)     ;y(n)=2A (assumes scaling
                                              ;mode is set).  X1 is Output.
```

Total Instruction Cycles:

6 Icyc @20 MHz=600ns

Figure 2.3-3. DSP56001 Code and Data Structures for Second Order Direct Form Implementation of a Bandstop Digital IIR Filter.

Second Order Bandpass Filter

NETWORK DIAGRAM

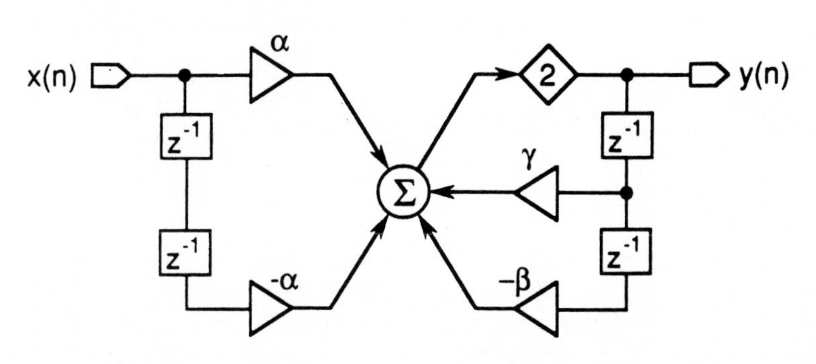

TRANSFER FUNCTION $$H(z) = \frac{\alpha(1-z^{-2})}{1/2-\gamma z^{-1}+\beta z^{-2}}$$

GAIN $$G(\theta) = \frac{d \sin\theta_o \sin\theta}{[(d \sin\theta \sin\theta_o)^2+4(\cos\theta-\cos\theta_o)^2]^{1/2}}$$

PHASE $$\phi(\theta) = -\tan^{-1}\left[\frac{2(\cos\theta-\cos\theta_o)}{d \sin\theta \sin\theta_o}\right]$$

COEFFICIENTS $$d = \frac{2\tan(\theta_o/2Q)}{\sin\theta_o}$$

$$\beta = \frac{1}{2}\,\frac{1-\tan(\theta_o/2Q)}{1+\tan(\theta_o/2Q)}$$

$$\gamma = (1/2+\beta)\cos\theta_o$$

$$\alpha = (1/2-\beta)/2$$

Figure 2.4-1. Direct Form Implementation of Second Order Bandpass IIR Digital Filter and Analytical Formulas Relating Desired Response to Filter Coefficients.

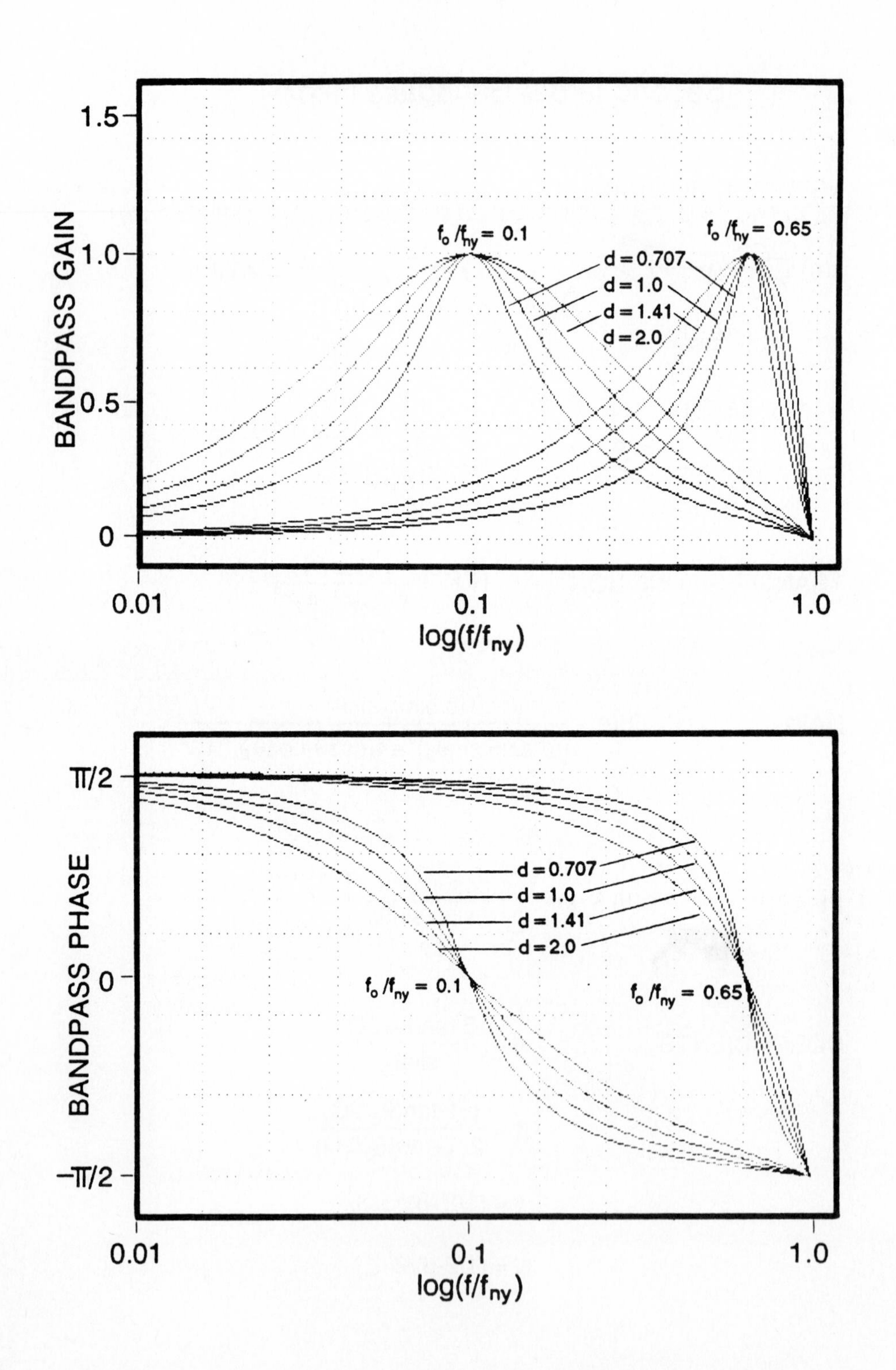

Figure 2.4-2. Gain and Phase Response of Second Order Bandpass IIR Filter. Nyquist Frequency f_{ny} *is Equal to One Half of the Sample Frequency* f_S.

SECOND ORDER BANDPASS DIRECT FORM IMPLEMENTATION

Difference Equation:

$y(n)=2\{\alpha[x(n)-x(n-2)]+\gamma y(n-1)-\beta y(n-2)\}$

Data Structures:

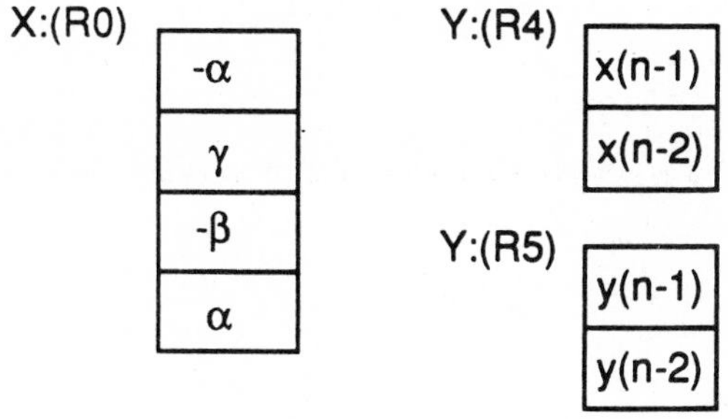

X:(R0)	Y:(R4)	Y:(R5)
-α	x(n-1)	y(n-1)
γ	x(n-2)	y(n-2)
-β		
α		

DSP56001 Code:

```
                                                  ;Y1=x(n) (Input)
                                                  ;X0=α
MPY   X0,Y1,A   X:(R0)+,X0   Y:(R4),Y0           ;A=αx(n)
MAC   X0,Y0,A   X:(R0)+,X0   Y:(R5)+,Y0          ;A=A-αx(n-2)
MAC   X0,Y0,A   X:(R0)+,Y0   Y:(R5),Y0           ;A=A+γy(n-1)
MAC   X0,Y0,A   X:(R0)+,X0   Y1,Y:(R4) +         ;A=A-βy(n-2)
MOVE            A,X1         A,Y:(R5)            ;y(n)=2A (assumes scaling
                                                  ;mode is set). X1 is Output.
```

Total Instruction Cycles:

5 Icyc @20 MHz=500ns

Figure 2.4-3. DSP56001 Code and Data Structures for Second Order Direct Form Implementation of a Bandpass Digital IIR Filter.

z-DOMAIN TRANSFER FUNCTION

$$H(z) = \frac{\alpha(1 + \mu z^{-1} + \sigma z^{-2})}{\frac{1}{2} - \gamma z^{-1} + \beta z^{-2}}$$

DIFFERENCE EQUATION (DIRECT FORM)

$$y(n) = 2\{\alpha[x(n) + \mu x(n-1) + \sigma x(n-2)] + \gamma y(n-1) - \beta y(n-2)\}$$

COEFFICIENTS

$$\beta = \frac{1}{2}\left[\frac{1 - \frac{1}{2}d \sin \theta_0}{1 + \frac{1}{2}d \sin \theta_0}\right] \qquad d = \frac{2 \tan(\theta_0/2Q)}{\sin\theta_0} \qquad \gamma = (1/2 + \beta)\cos\theta_0$$

where $0 < \beta < 1/2$ *and,*

$$Q = \frac{\theta_0}{\Delta\theta} = \frac{2\pi f_0/f_S}{2\pi(f_2 - f_1)/f_S} = \frac{f_0}{f_2 - f_1}$$

where f_0 *is the center frequency of the bandpass or bandstop filter,* f_1 *and* f_2 *are the half-power points (where gain is equal to* $1/\sqrt{2}$*), and* f_S *is the sample frequency. Note that* f_0 *is replaced with* f_C *in the lowpass and highpass cases.*

Numerator Coefficients

TYPE	α	μ	σ	Unity Gain at...
Lowpass	$(1/2+\beta-\gamma)/4$	2	1	$f = 0$
Highpass	$(1/2+\beta+\gamma)/4$	-2	1	$f = f_S/2$
Bandpass	$(1/2-\beta)/2$	0	-1	$f = f_0$
Bandstop	$(1/2+\beta)/2$	$-2\cos\theta_0$	1	$f = 0$ and $f = f_S/2$

where $\theta_0 = 2\pi f_0/f_S$.

Figure 2.5-1. Summary of Digital Coefficients for the Four Basic Filter Types.

POLE EQUATION OF H(z)

$$z_p = r\cos\theta_p + j\,r\sin\theta_p$$

$$= \gamma \pm j\sqrt{2\beta - \gamma^2}$$

$$= \frac{\cos\theta_0 \pm j\sin\theta_0\sqrt{1-\left(\frac{1}{2}d\right)^2}}{1+\frac{1}{2}d\sin\theta_0} \qquad \text{for} \quad d<2$$

where $\beta = \frac{1}{2}(2 - d\sin\theta_0)/(2 + d\sin\theta_0)$ and $\gamma = (1/2 + \beta)\cos\theta_0$. This result applies to $H(z)$ as well as $H_w(z)$. Distance from origin to pole is $|z_p| = \sqrt{2\beta}$.

For $d > 2$...

$$z_p = \gamma - \sqrt{\gamma^2 - 2\beta}$$

$$= \frac{\cos\theta_0 - \sin\theta_0\sqrt{\left(\frac{1}{2}d\right)^2 - 1}}{1+\frac{1}{2}d\sin\theta_0}$$

where $\theta_p = 0$. Note that in order to satisfy requirement $0 < \beta < \frac{1}{2}$ results in $\frac{1}{2}d\sin\theta_0 < 1$.

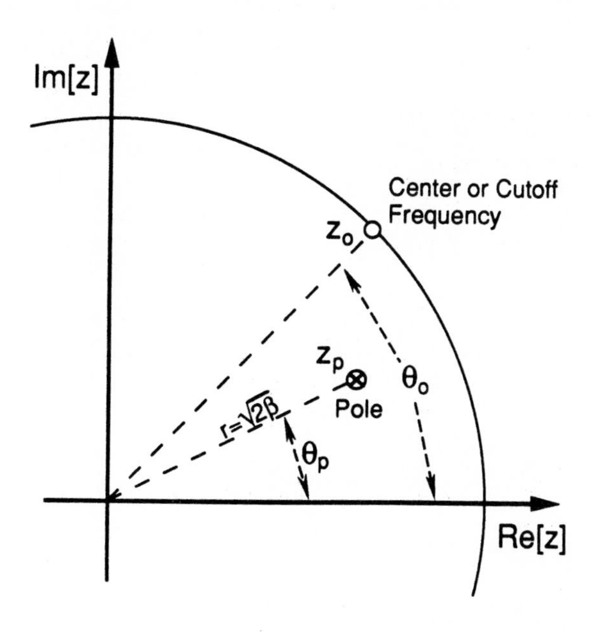

Figure 2.5-2. Pole Location and Analysis of Second Order Section.

3.0 SINGLE SECTION CANONIC FORM (DIRECT FORM II)

3.1 THE CANONIC FORM DIFFERENCE EQUATION

The direct form difference equation, rewritten from Equation (2.0-16), is:

$$y(n) = \sum_{i=0}^{2} b_i\, x(n-i) - \sum_{j=1}^{2} a_j\, y(n-j) \tag{3.1-1}$$

Equation (3.1-1) can be represented by the flow graph of Figure 3.1-1a. This is the same as those shown in Figures 2.1-1, 2.2-2, 2.3-2, and 2.4-1 except that the summations have been separated to highlight the correspondence with the *direct form* difference equation (3.1-1a). From this flow graph it is clear that the direct form implementation requires four delay elements or, equivalently four internal memory locations.

The flow graph of Figure 3.1-1b represents the same transfer function as that implemented by the difference equation (3.1-1), but now the delay variable is $w(n)$. Comparison of this network with that of the *direct form* network of Figure 3.1-1a shows that interchanging the order of the left and right half's does not change the overall system response [ ]. The delay elements can then be collapsed to produce the final *canonic form* network shown in Figure 3.1-1c. As a result the memory requirements for the system are reduced to the minimum, two locations, and therefore this realization of the IIR filter is often referred to as the *canonic form.* The system difference equations for the canonic realization are:

$$y(n) = \sum_{i=0}^{2} b_i\, w(n-i) \tag{3.1-2a}$$

where,

$$w(n) = x(n) - \sum_{j=1}^{2} a_j\, w(n-j) \tag{3.1-2b}$$

In order to prove that Equations (3.1-2a) and (3.1-2b) are equivalent to the original direct form difference equation (3.1-1), the following procedure can be used. First, combine Equations (3.1-2a) and (3.1-2b):

$$y(n) = \sum_{i=0}^{2} b_i \left[x(n-i) - \sum_{j=1}^{2} a_j\, w(n-j-i) \right] \tag{3.1-3}$$

$$= \sum_{i=0}^{2} b_i\, x(n-i) - \sum_{j=1}^{2} a_j \sum_{i=0}^{2} b_i\, w(n-i-j)$$

$$= \sum_{i=0}^{2} b_i\, x(n-i) - \sum_{j=1}^{2} a_j\, y(n-j)$$

The last step uses the definition for $y(n)$ from Equation (3.1-2a). The result is exactly equivalent to Equation (3.1-1), the *direct form* difference equation.

In order to take advantage of the scaling mode on the DSP56001 it is advantageous to write the canonic realization equations as:

$$y(n) = 2\left\{\frac{1}{2}\,w(n) + \frac{\mu}{2}\,w(n-1) + \frac{\alpha}{2}\,w(n-2)\right\} \tag{3.1-4a}$$

and,

$$w(n) = 2\{\alpha\,x(n) + \gamma\,w(n-1) - \beta\,w(n-2)\} \tag{3.1-4b}$$

where the coefficients have been substituted for the a_i and b_j.

From these equations it can be seen that both $y(n)$ and $w(n)$ depend on a sum of products and therefore they are the output of an accumulator. The accumulators on the DSP56001 have eight extension bits so the sum can exceed unity by 255 without overflowing. However, whenever the contents of a 56-bit accumulator are transferred to a 24-bit register or memory location (i.e., delay element) an overflow error may occur. It is the digital filter designer's responsibility to insure that only values less than unity are stored. Of course even if memory had the precision of the accumulators, overflow may still occur if the accumulator itself experiences overflow. Note however, that intermediate sums are allowed to exceed the capacity of the accumulator providing the final result can be represented in the accumulator. (This is a result of the circular nature of two's complement arithmetic.)

In order to insure that overflow does not occur, it is necessary to calculate the gain at the internal nodes (accumulator output) of a filter network. As will be seen in the next section, the canonic realization is susceptible to overflow. It's advantages are the minimum storage requirements and that it can be implemented in fewer instruction cycles.

3.2 ANALYSIS OF INTERNAL NODE GAIN

Calculating the gain at any internal node is no different then calculating the total network gain. In Figure 3.2-1 the transfer function of node $w(n)$ is $H_w(z)$. $H_w(z)$ represents the transfer function from the input node $x(n)$ to the internal node $w(n)$. The gain $G_w(\theta)$ at $w(n)$ is found in the standard manner by evaluating the magnitude of $H_w(z)$ as shown in Figure 3.2-3. Note that the network flow diagram of Figure 3.2-1 uses the automatic *scaling mode* (multiply by a factor of two when transferring data from the accumulator to memory) so that the coefficients are by definition (as seen in section 2.0) less than one.

The peak gain g_o of $G_w(\theta)$ is found by taking the derivative of $G_w(\theta)$ with respect to θ and setting the result equal to zero. A simple expression for g_0 is derived as:

$$g_0 = \frac{\alpha}{\left(\frac{1}{2} - \beta\right)\sin\theta_p} \quad \text{for} \quad \left[\frac{\left(\frac{1}{2}+\beta\right)^2}{2\beta}\right]\cos\theta_0 \le 1 \tag{3.2-1}$$

where θ_p is the angle (see Figure 2.5-2) to the pole of the filter. Note that this result does not depend on the numerator of the filter transfer function $H(z)$ so that the result is valid for all four of the basic filter types. However, as can be seen by the example given for a bandpass network in Figure 3.2-2, g_0 may exceed unity by a large amount especially for filters having poles at frequencies much less than $f_S/2$ where $\sin\theta_p \ll 1$. In order to compensate, the input must be scaled down by an amount equal to $1/g_0$ or guaranteed not to exceed $1/g_0$ before arriving at the filter. This scaling aspect of the *canonic form* network is a definite disadvantage, but this network has the significant advantage that it can be implemented in one less instruction than the other filter realizations for the lowpass, highpass, and bandstop filters and, of course, less memory is required since only two intermediate variables are stored.

In general, the behavior of systems at internal nodes can be unexpected. For instance, it is interesting to note that the frequency at which the gain at the internal node of the canonic IIR filter section peaks is not the same as the frequency at which the gain of the filter peaks as given by the poles of the filter. From Figure 2.5-2 this behavior is expressed by:

$$\frac{\sqrt{2\beta}}{\frac{1}{2}+\beta} = \frac{\cos\theta_0}{\cos\theta_p} \tag{3.2-2}$$

The *canonic (direct form II)* network has trade-offs that must be carefully understood and analyzed for the particular application.

3.3 IMPLEMENTATION ON THE DSP56001

Figure 3.3-1 shows the DSP56001 code and data structures for implementation of the canonic realization network. Note that the modifier register M4 is set equal to 4 in order to allow circular operation for addressing coefficient data. M0 is set to FFFF to turn off circular addressing for $w(n-1)$ and $w(n-2)$.

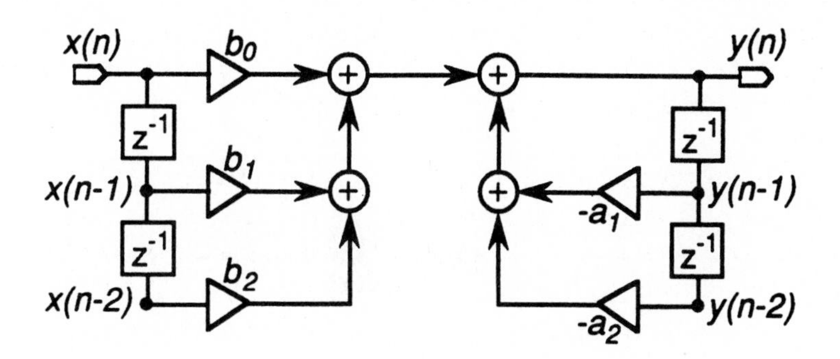

Figure 3.1-1a. Direct Form Network Flow Diagram of Equation (3.1-1).

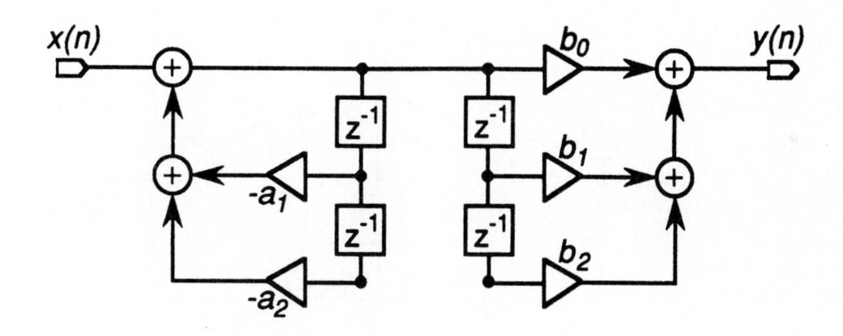

Figure 3.1-1b. Left and Right Halves of Direct Form Network Interchanged.

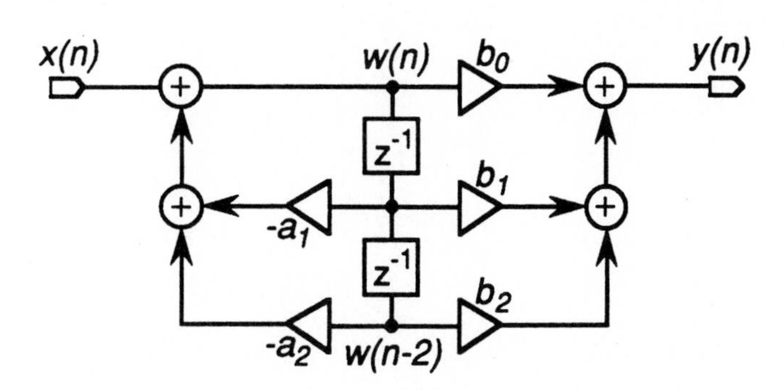

Figure 3.1-1c. Collapsing Delay Terms from Above Diagram Resulting in Canonic Form Network Diagram.

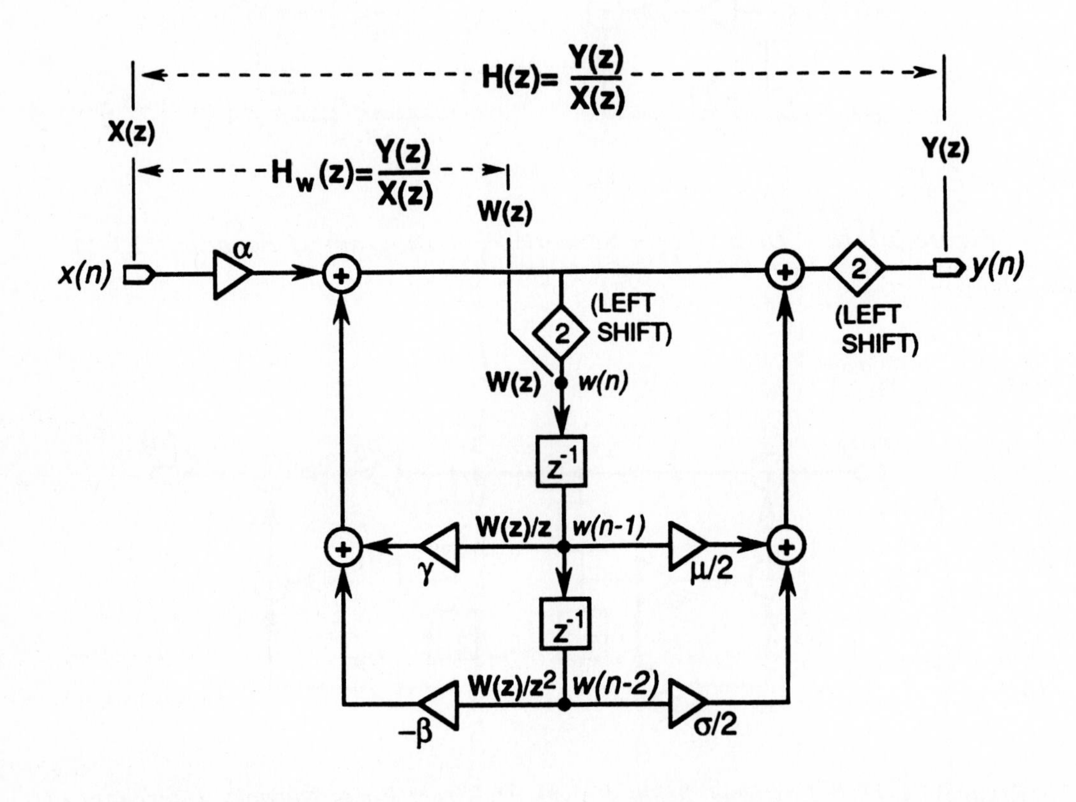

$$H(z) = \frac{\alpha(1+\mu z^{-1} + \sigma z^{-2})}{1/2 - \gamma z^{-1} + \beta z^{-2}} \qquad H_W(z) = \frac{\alpha}{1/2 - \gamma z^{-1} + \beta z^{-2}}$$

Figure 3.2-1. Internal Node Transfer Function $H_w(z)$ *of Canonic (Direct Form II) Network.*

INTERNAL NODE GAIN

Gain at node $w(n)$**...**

$$H_w(z) = \frac{W(z)}{X(z)} = \frac{\alpha}{\frac{1}{2} - \gamma z^{-1} + \beta z^{-2}}$$

$$G_w(\theta) = \sqrt{H_w(e^{j\theta})\, H_w(e^{-j\theta})}$$

$$= \frac{\alpha}{\sqrt{\left(\frac{1}{2}-\beta\right)^2 \sin^2\theta + \left(\frac{1}{2}+\beta\right)^2 (\cos\theta - \cos\theta_o)^2}}$$

where $\gamma = (1/2+\beta)\cos\theta_0$ has been used.

Peak Gain ...

$$\left.\frac{d}{d\theta} G_w(\theta)\right|_{\theta=\theta_m} = 0$$

Frequency of Peak Gain ...

$$\cos\theta_m = \xi\cos\theta_0 \qquad \text{for } \xi\cos\theta_0 \le 1$$

$$= 1 \qquad \text{otherwise}$$

where $\xi = \dfrac{\left(\frac{1}{2}+\beta\right)^2}{2\beta}$

$$g_0 = G_w(\theta_m) = \frac{\alpha}{\left(\frac{1}{2}-\beta\right)\sqrt{1-\gamma^2/2\beta}} \qquad \text{for } \xi\cos\theta_0 \le 1$$

$$= \frac{\alpha}{\left(\frac{1}{2}+\beta\right)(1-\cos\theta_0)} \qquad \text{otherwise}$$

EXAMPLE: Maximum Internal Node Gain for Bandpass

$$g_0 = \frac{1}{2\sin\theta_p} \qquad \text{for } \xi\cos\theta_0 \le 1$$

where $\gamma^2 = 2\beta\cos^2\theta_p$ has been used, and where $\alpha = (1/2-\beta)/2$ for a bandpass filter.

If $\sin\theta_p > 1/2$, then $g_0 < 1$, otherwise an overflow (i.e. $g_0 > 1$) may occur at the internal node $w(n)$ unless the input is scaled down by $1/g_0$.

Figure 3.2-2. Internal Node Gain Analysis of Canonic Second Order Section.

SECOND ORDER DIRECT FORM II (CANONIC) IMPLEMENTATION

Difference Equation:

$$y(n)=2\{\tfrac{1}{2}w(n)+\tfrac{\mu}{2}w(n-1)+\tfrac{\sigma}{2}w(n-2)\}$$

$$w(n)=2\{\alpha\, x(n)+\gamma\, w(n-1)-\beta\, w(n-2)\}$$

Data Structures:

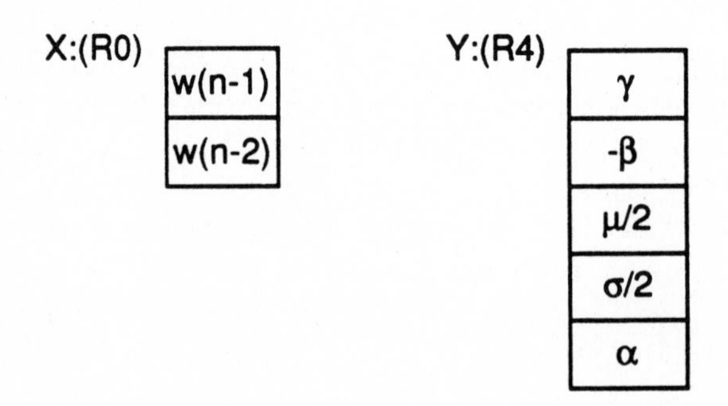

DSP56001 Code:

```
                                                  ;Y1=x(n) (Input)
                                                  ;Y0=α
MPY   Y0,Y1,A   X:(R0)+,X0   Y:(R4)+,Y0   ;A=αx(n)
MAC   X0,Y0,A   X:(R0),X1    Y:(R4)+,Y0   ;A=A+γw(n-1)
MACR  X1,Y0,A   X0,X:(R0)-   Y:(R4)+,Y0   ;A=A-βw(n-2)
MAC   X0,Y0,A   A,X:(R0)     Y:(R4)+,Y0   ;A=½w(n)+μ/2w(n-1)
MACR  X1,Y0,A   A,X0         Y:(R4)+,Y0   ;A=A+σ/2w(n-2). X0=2A
                                          ;assumes scaling mode is
                                          ;set. X0 is Output.
```

Total Instruction Cycles:

5 Icyc @20 MHz=500ns

Figure 3.3-1. Single Second Order Canonic Section (Direct Form II) DSP56001 Code Implementation and Data Structures.

4.0 SINGLE SECTION TRANSPOSE FORM

A third realization of IIR filters is the *transposed form (direct form I)* as shown in Figure 4.0-1. This network implementation can be derived directly from the *direct form* difference equation, as shown in this figure, or by taking the transpose of the canonic network [10,11].

The transpose realization is characterized by three accumulator operations. One reason for popularity of this realization is that like the canonic realization, it only requires two memory locations, however, unlike the canonic realization, it is much less prone to overflow at internal node. The disadvantage is that this realization requires more instructions to implememt.

4.1 GAIN EVALUATION OF INTERNAL NODES

With the same techniques used to calculate $H_w(z)$ for the canonic realization, $H_u(z)$ and $H_v(z)$ are found as shown in Figures 4.1-1 and 4.1-2. The resulting expressions, unlike the *canonic form* results, depend on the numerator of the transfer function, so that the internal gains $G_u(\theta)$ and $G_v(\theta)$ have different forms for the different filter types. In the case of the bandpass filter, these results simplifiy signifcantly such that a closed form expression for the maximum gain at the internal nodes can be derived by calculating the maxima of the gain functions. For the bandpass and bandstop networks, the maximum of $G_u(\theta)$ (and $G_v(\theta)$) is $g_m = \beta + 1/2$. Since $\beta < 1/2$ then $g_m < 1$ so that there is no overflow at these nodes in the bandpass or bandstop case.

Figure 4.1-3 contain example plots of $G_u(\theta)$ and $G_v(\theta)$ for the second order *transpose form* lowpass filter with various values of cutoff frequency θ_c and damping factor d. In most cases $G_u(\theta)$ and $G_v(\theta)$ never exceed the maximum value of $G(\theta)$, so that if the total gain does not exceed unity, then the internal nodes will not exceed unity (i.e. no overflow).

4.2 IMPLEMENTATION ON THE DSP56001

The code for a single second order section of the *transposed form* network is shown in Figure 4.2-1, together with diagrams of the data structures. Referring back to the network diagram of Figure 4.0-1, the diamond blocks enclosing a *1/2* is represented in the code by an ASR *(Accumulator Shift Right)* instruction. As in the previous code implementation, the *2* enclosed by a diamond can be implemented by the automatic *scaling mode* (equivalent to a left shift of the accumulator) feature of the DSP56001. The modifier register M4 is again set to a value of 4 initially so that a circular buffer can be conveniently used to address the coefficient data. Note that this network requires more instructions than either of the previous two network forms.

DIRECT FORM DIFFERENCE EQUATION:

$$y(n) = 2\{\alpha[x(n) + \mu x(n-1) + \sigma x(n-2)] + \gamma y(n-1) - \beta y(n-2)\}$$
$$= 2\{\alpha x(n) + \alpha\mu x(n-1) + \gamma y(n-1) + [\alpha\sigma x(n-2) - \beta y(n-2)]\}$$
$$= 2\left\{\alpha x(n) + \left[\alpha\mu x(n-1) + \gamma y(n-1) + \frac{1}{2}u(n-2)\right]\right\}$$
$$= 2\left\{\alpha x(n) + \frac{1}{2}v(n-1)\right\}$$

TRANSPOSE FORM DIFFERENCE EQUATION:

$$y(n) = 2\left\{\alpha x(n) + \frac{1}{2}v(n-1)\right\}$$

where $$v(n) = 2\left\{\alpha\mu x(n) + \gamma y(n) + \frac{1}{2}u(n-1)\right\}$$

and $$u(n) = 2\{\alpha\sigma x(n) - \beta y(n)\}$$

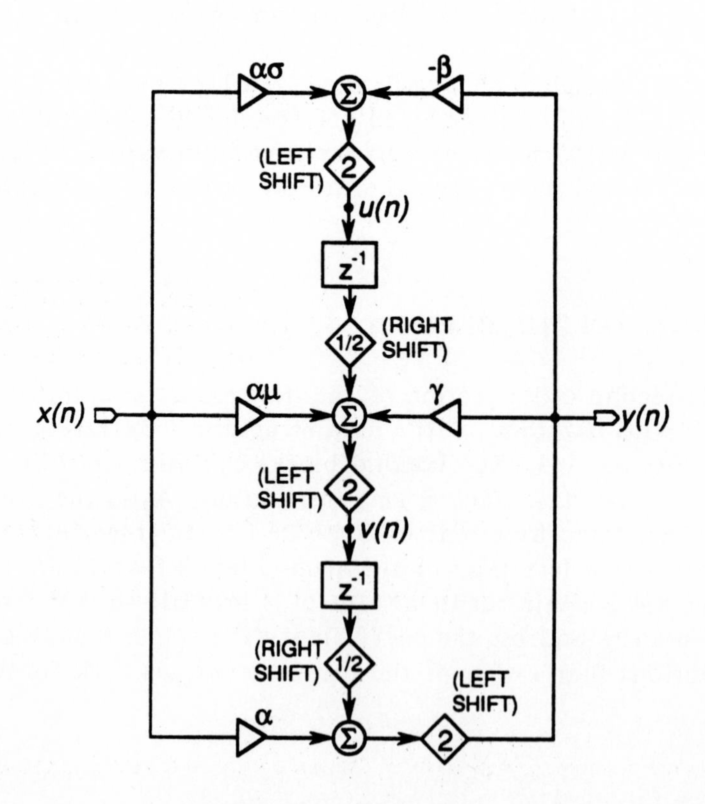

Figure 4.0-1. Network Diagram for Transpose Form (Direct Form I) Implementation of Single Second Order Section Filter.

GAIN AT INTERNAL NODE u(n)

Transfer Function of u(n) ...

$$U(z) = 2[\alpha\sigma X(z) - \beta Y(z)]$$
$$= H_U(z)\, X(z)$$

$$H_U(z) = 2[\alpha\sigma - \beta H(z)]$$
$$= 2\left[\frac{\alpha\sigma\left(\frac{1}{2} - \gamma z^{-1} + \beta z^{-2}\right) - \alpha\beta(1 + \mu z^{-1} + \sigma z^{-2})}{\frac{1}{2} - \gamma z^{-1} + \beta z^{-2}}\right]$$
$$= \frac{2\alpha(A - Bz^{-1})}{\frac{1}{2} - \gamma z^{-1} + \beta z^{-2}}$$

where $H_U(z)$ is the transfer function from input node $x(n)$ to internal node $u(n)$ and $A = \sigma/2 - \beta$ and $B = \sigma\gamma + \mu\beta$

Gain of u(n) ...

$$G_U(\theta) = |H_U(z)|_{z=e^{j\theta}}$$
$$= \frac{2\alpha\sqrt{A^2 + B^2 - 2AB\cos\theta}}{\sqrt{\left(\frac{1}{2} - \beta\right)^2 \sin^2\theta + \left(\frac{1}{2} + \beta\right)^2 (\cos\theta - \cos\theta_0)^2}}$$

BANDPASS EXAMPLE

Bandpass Coefficients ...

$\sigma = -1 \quad \mu = 0 \quad \alpha = (1/2 - \beta)/2$ so that, $A = -(1/2 + \beta) \quad B = -\gamma$

$$= -(1/2 + \beta)\cos\theta_0$$

and,
$$G_U(\theta) = \frac{\left(\frac{1}{2} - \beta\right)\left(\frac{1}{2} + \beta\right)\sqrt{\sin^2\theta + (\cos\theta - \cos\theta_0)^2}}{\sqrt{\left(\frac{1}{2} - \beta\right)^2 \sin^2\theta + \left(\frac{1}{2} + \beta\right)^2 (\cos\theta - \cos\theta_0)^2}}$$

Peak Gain ...

$g_m = G_U(\theta_m)$ is found by evaluating $\left.\frac{d}{d\theta} G_U(\theta)\right|_{\theta=\theta_m} = 0$

$\theta_m = \theta_0$ after evaluating derivative, so that,

$$g_m = \frac{(1/2 - \beta)(1/2 + \beta)\sin\theta_0}{(1/2 - \beta)\sin\theta_0} = 1/2 + \beta \quad < 1 \quad (\text{since } \beta < 1/2)$$

Figure 4.1-1. Gain Evaluation at First Internal Node u(n) of Transpose Network.

GAIN AT INTERNAL NODE v(n)

Transfer Function of v(n) ...

$$V(z) = [Y(z) - 2\alpha X(z)]/z^{-1}$$
$$= H_V(z)X(z)$$

where $Y(z) = 2\left[\frac{1}{2}V(z)z^{-1} + \alpha X(z)\right]$

$$H_V(z) = \frac{H(z) - 2\alpha}{z^{-1}}$$

$$= \frac{\alpha\left[(1 + \mu z^{-1} + \sigma z^{-2}) - 2\left(\frac{1}{2} + \gamma z^{-1} + \beta z^{-2}\right)\right]}{z^{-1}\left(\frac{1}{2} - \gamma z^{-1} + \beta z^{-2}\right)}$$

$$= \frac{2\alpha(C + Az^{-1})}{\frac{1}{2} - \gamma z^{-1} + \beta z^{-2}}$$

where $H_V(z)$ is the transfer function from input node $x(n)$ to internal node $v(n)$ and $A = \sigma/2 - \beta$ and $C = \mu/2 + \gamma$.

Gain of v(n) ...

$$G_V(\theta) = |H_V(z)|_{z=e^{j\theta}}$$

$$= \frac{2\alpha\sqrt{A^2 + C^2 + 2AC\cos\theta}}{\sqrt{\left(\frac{1}{2} - \beta\right)^2 \sin^2\theta + \left(\frac{1}{2} + \beta\right)^2 (\cos\theta - \cos\theta_0)^2}}$$

BANDPASS EXAMPLE

Bandpass Coefficients ...

$\sigma = -1$ $\quad \mu = 0$ $\quad \alpha = (1/2 - \beta)/2$ $\quad$ so that, $\quad A = -(1/2 + \beta)$ $\quad C = \gamma$

$$= (1/2 + \beta)\cos\theta_0$$

and,
$$G_V(\theta) = \frac{\left(\frac{1}{2} - \beta\right)\left(\frac{1}{2} + \beta\right)\sqrt{\sin^2\theta + (\cos\theta - \cos\theta_0)^2}}{\sqrt{\left(\frac{1}{2} - \beta\right)^2 \sin^2\theta + \left(\frac{1}{2} + \beta\right)^2 (\cos\theta - \cos\theta_0)^2}}$$

Peak Gain ...

$g_m = G_V(\theta_m)$ is found by evaluating $\left.\frac{d}{d\theta}G_V(\theta)\right|_{\theta=\theta_m} = 0$

$\theta_m = \theta_0$ after evaluating derivative, so that as before,

$$g_m = \frac{(1/2 - \beta)(1/2 + \beta)\sin\theta_0}{(1/2 - \beta)\sin\theta_0} = 1/2 + \beta \quad < 1 \quad (\text{since } \beta < 1/2)$$

Figure 4.1-2. Gain Evaluation at Second Internal Node v(n) of Transpose Network.

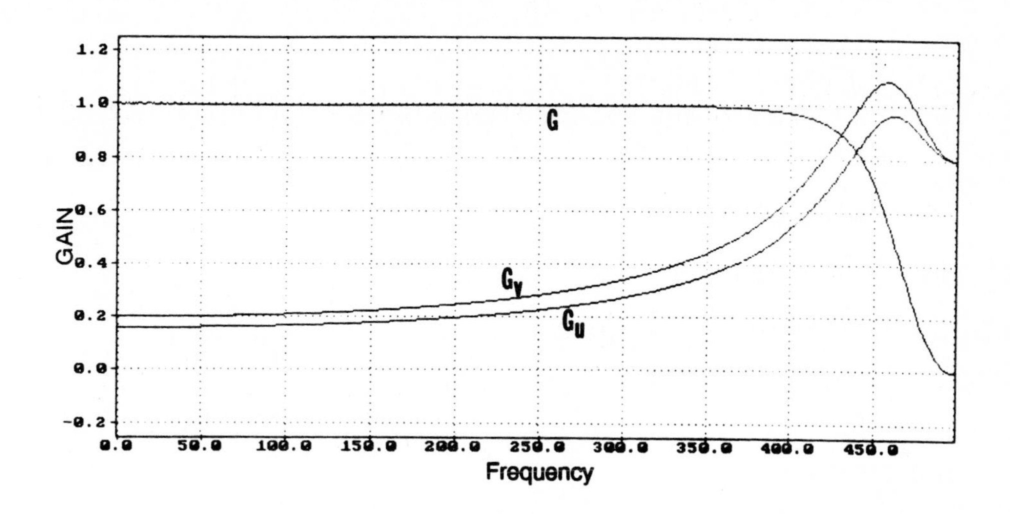

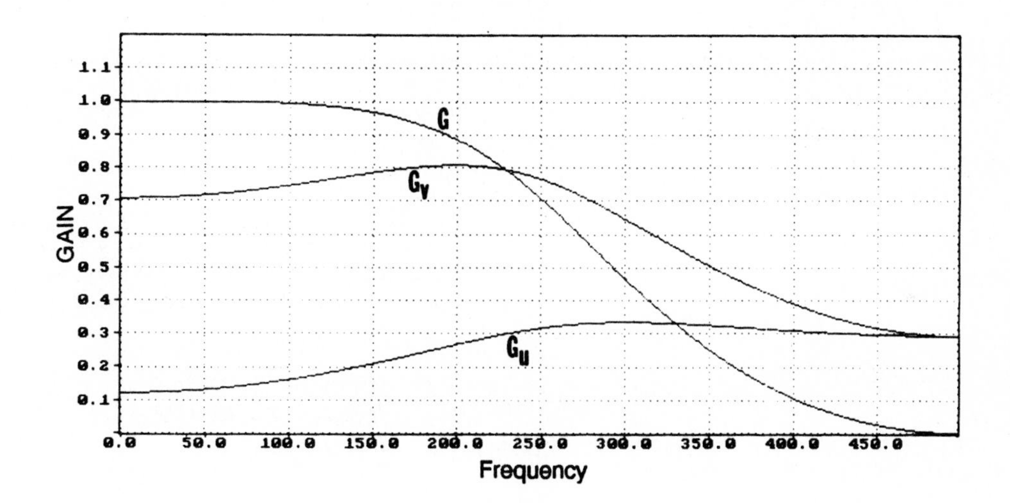

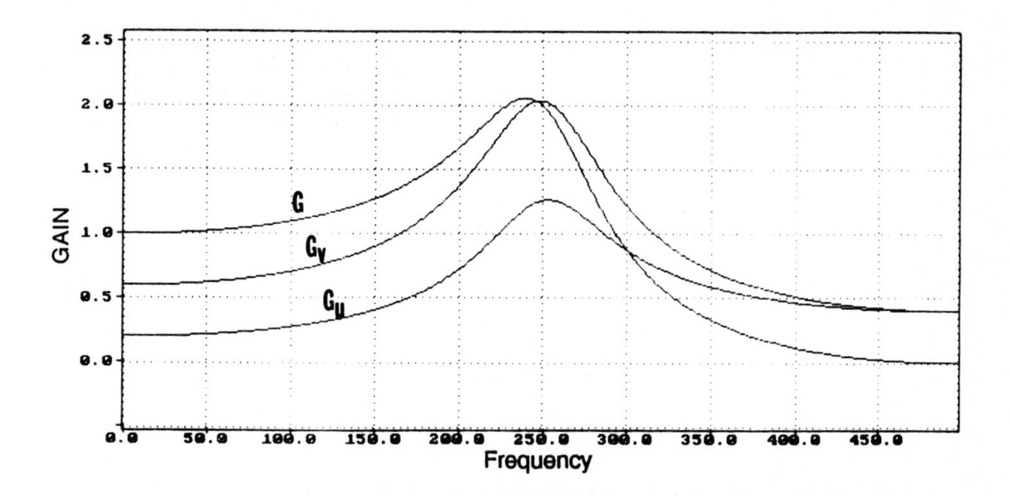

Figure 4.1-3. Total Gain and Gain at Internal Nodes of Lowpass Transpose Filter Network (of Figure 4.0-1) for $f_S = 1000 Hz$ *and (a)* $f_c = 450 Hz$ *and* $d = \sqrt{2}$*; (b)* $f_c = 250 Hz$ *and* $d = \sqrt{2}$*; (c)* $f_c = 250 Hz$ *and* $d = 1/2$.

SECOND ORDER DIRECT FORM I (TRANSPOSED) IMPLEMENTATION

Difference Equations:

$$y(n)=2\{\alpha\, x(n) + \tfrac{1}{2} v(n-1)\}$$

$$v(n)=2\{\alpha\mu\, x(n) + \gamma\, y(n) + \tfrac{1}{2} u(n-1)\}$$

$$u(n)=2\{\alpha\sigma\, x(n) - \beta\, y(n-1)\}$$

Data Structures:

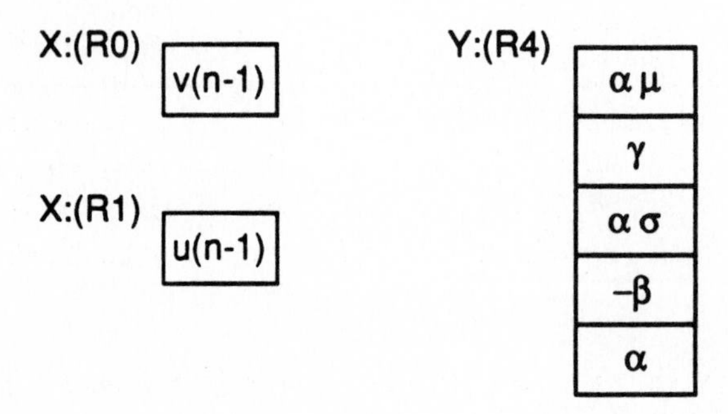

DSP56001 Code:

```
                                                ;Y1=x(n)  (Input)
                                                ;Y0=α . A=v(n-1)/2
MACR  Y0,Y1,A   X:(R1),B      Y:(R4)+,Y0        ;A=A+α x(n)
ASR    B        A,X0                            ;y(n)=X0=2A (scale mode on)
MAC   Y0,Y1,B                 Y:(R4)+,Y0        ;B=u(n-1)/2+α μ x(n)
MACR  X0,Y0,B                 Y:(R4)+,Y0        ;B=B+γ y(n)
MPY    Y0,Y1,B  B,X:(R0)      Y:(R4)+,Y0        ;v(n)=2B.  B=α σ x(n)
MACR   X0,Y0,B  X:(R0),A                        ;B=B-β y(n)
ASR    A        B,X:(R1)      Y:(R4)+,Y0        ;A=v(n)/2.  u(n)=2B
                                                ;Y0=α .  y(n)=X0 (Output)
```

Total Instruction Cycles:

7 Icyc @20 MHz=700ns

Figure 4.2-1. Single Second Order Section Transpose DSP56001 Code Implementation and Data Structures.

5.0 CASCADED DIRECT FORM

By placing any of the *direct form* second order filter networks from Figures 2.1-1, 2.2-1, 2.3-1, or 2.4-1 in series (i.e., connect the *y(n)* of one to the *x(n)* of the next), a cascaded filter is created. The resulting order *N* of the network is two times the number of second order sections. An odd order network can be made simply by adding one first order section in the chain. Note that in general, to achieve a particular response the filter parameters associated with each second order section are different. The reason for this is that in order to generate a predefined total response requires that each individual section have a different response. This will become more obvious in the next section where the special case of Butterworth lowpass filters will be discussed in detail.

5.1 BUTTERWORTH LOWPASS FILTER

The Butterworth filter response, is maximally flat in the passband at the expense of phase linearity and steepness of attenuation slope in the transition band. For lowpass or highpass cascaded second order sections, all sections have the same center frequency (this is not the case for bandpass filters). For this reason it is easy to design since all that remains to be determined are the damping factors d_k of each individual k^{th} section. The damping coefficients d_k are calculated from a simple formula for any order *N* of response.

The s-domain transfer function for the Nth order lowpass Butterworth filter is:

$$H(s) = \frac{1}{(s/\Omega_C)^2 + d_1(s/\Omega_C) + 1} \; \frac{1}{(s/\Omega_C)^2 + d_2(s/\Omega_C) + 1} \cdots$$

$$= \prod_{k=1}^{N/2} \frac{1}{(s/\Omega_C)^2 + d_k(s/\Omega_C) + 1} \qquad (5.1-1)$$

where d_k is the kth damping coefficient, $s = j\Omega$, and Ω_C is the common cutoff frequency [ ]. Only filter orders of even *N* will be considered in this discussion in order to minimize the complexity of mathematical results. The analysis can be extended to include odd values of *N* by inserting an additional term of $[(s/\Omega_c)+1]^{-1}$ in Equation (5.1-1). Equation (5.1-1) can be generalized if Ω_C of each second order section is an arbitrary value (corresponding to a different cutoff frequency Ω_k of each section). However, since we are limiting this discussion to Butterworth polynomials, Equation (5.1-1) will serve as the basis of all following derivations.

A filter of order *N* has *N/2* second order sections. The second order section of a Butterworth filter can be derived from the simple RCL network of Figure 1.1-1. However, the Butterworth damping factors are predetermined values which can be shown to yield a maximally flat passband response [ ]. The Butterworth damping coefficients are given by the following equation:

$$d_k = 2 \sin\frac{(2k-1)\pi}{2N} \qquad (5.1-2)$$

Equation (5.1-2) is the characteristic equation that determines a Butterworth filter response. Note that for a single ($k=1$) second order ($N=2$) section lowpass filter, $d_k = 2\sin(\pi/4) = \sqrt{2}$ as expected for a maximally flat response. For a fourth order filter with two second order sections, $d_1 = 2\sin(\pi/8)$ (where $k=1$ and $N=4$) and $d_2 = 2\sin(3\pi/8)$ (where $k=2$ and $N=4$).

Equation (5.1-1) represents an all pole response (the only zeros are at plus and minus infinity in the analog s-domain). The poles of a second order section are the roots of the quadratic denominator as given by:

$$p_{k1} = -d_k/2 - j(1 - d_k^2/4)^{1/2} \qquad (5.1-3a)$$

and

$$p_{k2} = -d_k/2 + j(1 - d_k^2/4)^{1/2} \qquad (5.1-3b)$$

Using (5.1-3) Equation (5.1-1) becomes:

$$H(s) = \prod_{k=1}^{N/2} \frac{1}{[(s/\Omega_C) - p_k][s/\Omega_C) - p_k^*]} \qquad (5.1-4)$$

where $p_k = p_{k1}$ and $p_k^* = p_{k2}$ (complex conjugate of p_k).

Equation (5.1-4) is useful in that the response of the system can be analyzed entirely by studying the poles of the polynomial. However, for purposed of transforming to the z-domain, Equation (5.1-1) can be used as previously done in Figure 2.1-1.

In order to examine the gain and phase response (physically measurable quantities) of the lowpass Butterworth filter, the transfer function $H(s)$ of Equation (5.1-1) will be converted into a polar representation. The magnitude of $H(s)$ is the gain $G(\Omega)$, while the angle between the real and imaginary components of $H(s)$, $\phi(\Omega)$, is the *arctangent* of the phase shift introduced by the filter:

$$G(\Omega) = \sqrt{H(s)H^*(s)}\Big|_{s=j\Omega}$$

$$= \prod_{k=1}^{N/2} \frac{1}{\sqrt{[(\Omega/\Omega_C)^2 - 1]^2 + (d_k\Omega/\Omega_C)^2}} \qquad (5.1-5)$$

$$\phi(\Omega) = \sum_{k=1}^{N/2} \tan^{-1} \frac{d_k\Omega/\Omega_C}{(\Omega/\Omega_C)^2 - 1} \qquad (5.1-6)$$

Equations (5.1-2), (5.1-5), and (5.1-6) completely describe the response characteristics of a Nth-order lowpass Butterworth filter in the continuous analog frequency domain. Since the quantity of interest is usually $20\log G(\Omega)$, Equation (5.1-5) can be transformed into a sum (over the second order sections) of $20\log(G_k)$, where G_k is the gain of the kth section. Similarly, the total phase is just the sum of the phase contribution by each section from Equation (5.1-6).

The *bilinear transformation* is used to convert the continuous frequency domain transfer function into the digital domain representation as done previously in *Section 2.1.* where θ the normalized digital domain frequency equal to $2\pi f/f_s$ can be thought of as the ratio of frequency to sampling frequency scaled by 2π. Substituting s from Equation (2.0-11) and Ω_C from Equation (2.0-12) into the kth section of Equation (5.1-1) yields the digital domain form of the Butterworth lowpass filter (for the kth second order section):

$$H_k(z) = \frac{\alpha_k(1 + 2z^{-1} + z^{-2})}{\frac{1}{2} - \gamma_k z^{-1} + \beta_k z^{-2}} \qquad (5.1-7)$$

where

$$\alpha_k = [\tan^2(\theta_c/2)]/A_k(\theta_c) \qquad (5.1-8a)$$

$$\beta_k = [1 - d_k \tan(\theta_c/2) + \tan^2(\theta_c/2)]/A_k(\theta_c) \qquad (5.1-8b)$$

$$\gamma_k = 2[1 - \tan^2(\theta_c/2)]/A_k(\theta_c) \qquad (5.1-8c)$$

and

$$A_k(\theta_c) = 2[1 + d_k \tan(\theta_c/2) + \tan^2(\theta_c/2)] \qquad (5.1-8d)$$

Equations (5.1-8a) thru (5.1-8d) provide a complete description of the digital lowpass N^{th} order Butterworth filter. Given θ_C and d_k, these formulas allow precise calculation of the digital coefficients α_k, β_k, and γ_k used to implement each kth second order section of the filter.

Following the game approach used in *Section 2.1* Equations (5.1-8a) thru (5.1-8d) can be further simplified into the following set of formulas:

$$\beta_k = \frac{1 - (d_k/2)\sin(\theta_C)}{2[1 + (d_k/2)\sin(\theta_C)]} \qquad (5.1-9a)$$

$$\gamma_k = (1/2 + \beta_k)\cos(\theta_C) \qquad (5.1-9b)$$

$$\alpha_k = (1/2 + \beta_k - \gamma_k)/4 \qquad (5.1-9c)$$

where d_k is given by Equation (5.1-2), and θ_C is the digital domain cutoff frequency (this is the actual operating cutoff frequency of the digital filter).

Figure 5.1-1 shows an example of a 6th order lowpass Butterworth filter (three second order sections) in both the analog domain the the digital domain. Note that the gain of the first (k=*1*) section is greater than unity near the cutoff frequency but that the overall composite response never exceeds unity. This fact allows for easy implementation of the Butterworth filter in cascaded *direct form* (i.e., scaling of sections is not needed as long as the sections are implemented in the order of decreasing k). Overflow at the output of any section is then guaranteed not to occur (the gain of the filter never exceeds unity). Observe from Figure 5.1-1 that the digital response is identical to the analog response but warped from the right along the frequency axis. One may imagine the zero at plus infinity in the analog response mapping into the zero at $f_S/2$ in the digital case. Also note that because of this mapping, the digital response falls off faster than the -12 dB/octave of the analog filter when the cutoff is near $f_S/2$.

In the case of the highpass filter, the above analysis is nearly identical except for coefficients (see Figure 3.0-1) have slightly different values. The bandstop is just the sum of a lowpass and highpass so it can be analyzed by these techniques. The bandpass however is more difficult and requires considerably more work [ ]. This is because the center frequency of each section is now different and the formula for calculating these frequencies is not as simple as the formulas for the previous filter types. In addition, to complicate matters further, scaling between sections becomes more of a problem since the offset of the center of each section reduces the final center response which must be compensated at some point in the filter network. It is for cases such as higher order Butterworth bandpass filter designs that commercially available filter design packages such as FDAS (Filter Design and Analysis System) available from Momentum Data Systems, are useful. The use of FDAS is discussed in *Sections 6* and *7*.

5.2 CASCADED DIRECT FORM NETWORK IMPLEMENTATION

Figure 5.2-1 shows the *cascaded direct form* network and data structures for the DSP56001 code implementation of Figure 5.2-2. Note that by cascading network diagrams of *Section 2* the set of delays at the output of one section can be combined with the set of delays at the input of the next section, thus reducing the total number of delays by almost a factor of two. For this reason, the *cascaded direct form* network becomes canonic as the filter order N increases. The DSP56001 code of Figure 5.2-2 shows an example of reading data from a user supplied memory-mapped analog to digital converter (ADC) and writing it to a memory-mapped digital to analog converter (DAC). The number of sections (*nsec*) in these examples is three, thus the filter order is six. The total instruction time for this filter structure is $600nsec+800$ [ns], including the data I/O moves (but excluding the interrupt overhead).

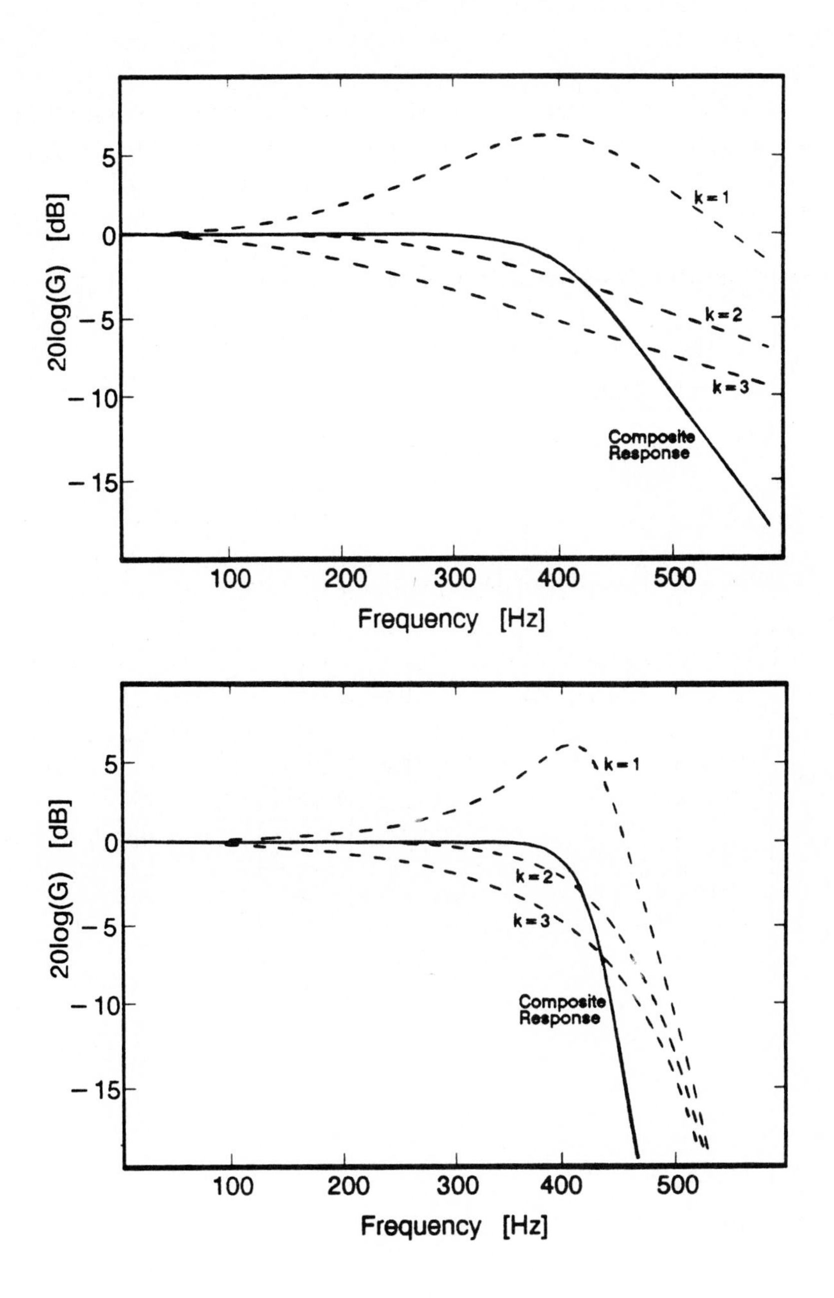

Figure 5.1-1. Composite Response of Cascaded Second Order Sections for Lowpass Butterworth Filter (k^{th} Section Damping Factor According to Equations [5.1-2]). (a) is Aanalog Case, (b) is Digital Filter with $f_S = 1200\ Hz$ and $f_C = 425\ Hz$.

SECOND ORDER CASCADED DIRECT FORM IMPLEMENTATION

Network Diagram:

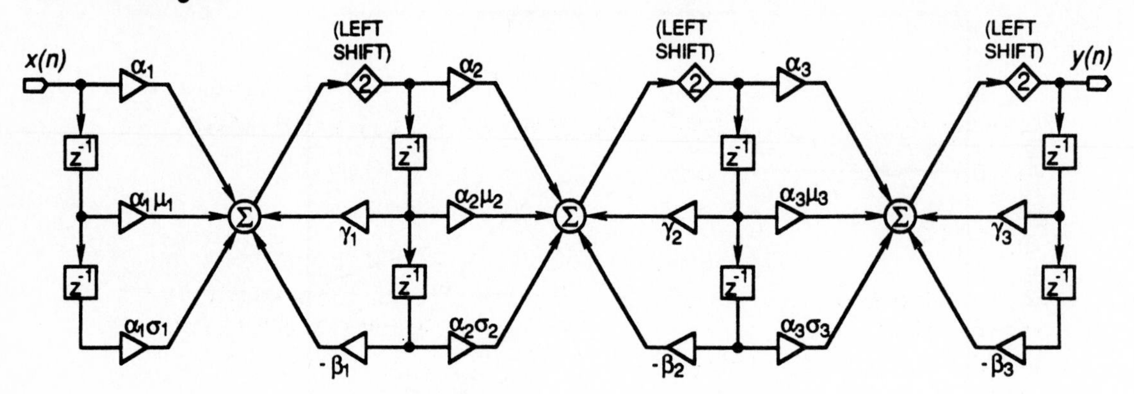

Difference Equations:

$$y_i(n)=2\{\alpha_i\,[x_i(n)+\mu_i x_i(n-1)+\sigma_i x_i(n-2)]+\gamma_i y_i(n-1)-\beta_i y_i(n-2)\}$$

$$x_{i+1}(n-k) = y_i(n-k)$$

Data Structures:

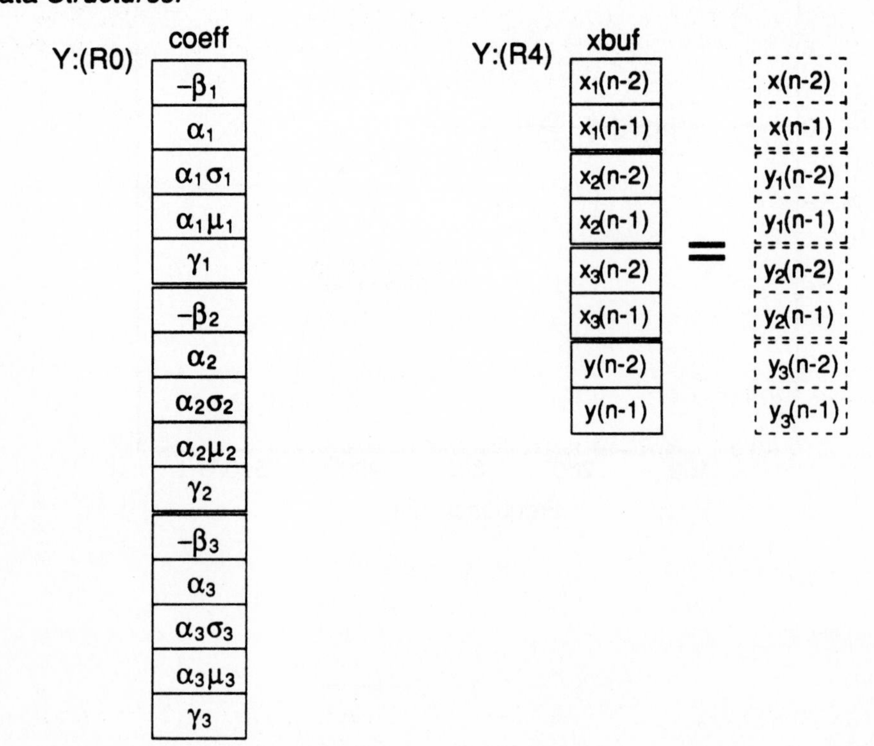

Figure 5.2-1. Network Diagram for Cascaded Direct Form Filter and Data Structutes used in Code Implementation (Three Section Example).

SECOND ORDER CASCADED DIRECT FORM IMPLEMENTATION

DSP56001 Code:

```
        MOVE   #coeff,R0                                    ;Pointer to coefficients
        MOVE   #1,M4                                        ;Modulo of length 2 for xbuf
        MOVE               X:(R0)+,X1                       ;β1
        MOVE               X:(R0)+,X0                       ;α1
        MOVE               (R4)+                            ;Point to next xbuf entry
                                                            ;Input: Y1=x(n)
        DO     X:nsec,Sectn                                 ;Loop on number of sections
        MPY    X0,Y1,A     X:(R0)+,X0   Y:(R4)+,Y0          ;A=αi xi (n)
        MAC    X0,Y0,A     X:(R0)+,X0   Y:(R4)+N4,Y0        ;A=A+αi σi xi (n-2)
        MAC    X0,Y0,A     X:(R0)+,X0   Y:(R4)+,Y0          ;A=A+αi μi xi (n-1)
        MAC    X0,Y0,A                  Y:(R4)-N4,Y0        ;A=A+γi yi (n-1)
        MAC    -X1,Y0,A    X:(R0)+,X1   Y1,Y:(R4)+N4        ;A=A-βi yi (n-2).  Save x(n)
        MOVE   A,Y1        X:(R0)+,X0                       ;yi (n)=2A, assumes scale
                                                            ;mode on.
Sectn                                                       ;X1=βi+1.  X0=αi+1

                                                            ;Output: y(n)=Y1
        MOVE               X:Buflen,M4                      ;Filter Order+1
        NOP
        MOVE                            Y1,Y:(R4)+N4        ;Save y(n)
```

Total Instruction Cycles:

(6*nsec + 10) Icyc @20 MHz=(0.6*nsec+1.0) ns

Figure 5.2-2. DSP56001 Code for Cascaded Direct Form Filter (nsec is the number of sections).

6.0 FILTER DESIGN AND ANALYSIS SYSTEM (FDAS)

The following sections will demonstrate the design of a cascaded filter in both the *direct form* and *canonic implementations* using a software package, *FDAS*, available from Momentum Data Systems, Inc. The filter example used in the next sections is a 6th order Butterworth lowpass filter with a cutoff frequency of approximately 225 Hz and a sample frequency of 1000 Hz. Figure 6.0-1 is the *Log Magnitude* (gain) plot from the system output (plotted on an HP Laserjet+ printer). Figure 6.0-2 is the Phase as a function of frequency in wrapped format ($-\pi$ wraps to $+\pi$). In addition, Figure 6.0-3 is a *Zero/Pole* plot and Figure 6.0-4 is the *Group Delay* which is the negative of the derivative of the phase with respect to frequency. *FDAS* will also generate an *Impulse Response, Step Response,* and a linear *Magnitude* plot. The results of the design are written to a file, *FDAS.OUT*, which contains much useful information. The coefficient data is written to *COEFF.FLT*. The DSP56001 code generator (*MGEN*) reads the *COEFF.FLT* file and generates a DSP56001 assembly source file, *COEFF.ASM*, which can be assembled by the DSP56001 assembler or linker software. Examples of all these files will be shown in the next two sections based on the 6th order Butterworth lowpass filter example design of this section.

6.1 CANONIC IMPLEMENTATION

Figure 6.1-1 is the output file associated with the *FDAS* design session. As can be seen, it contains information on the analog s-domain equivalent filter as well as the final digital coefficients (listed again in Figure 6.1-2) which have been properly scaled to prevent overflow at the internal nodes as well as the outputs of each cascaded section. This is all done automatically by the program in a matter of seconds. Executable code is generated by *MGEN* (also from Momentum Data Systems, Inc.) in Figure 6.1-3. The code internal to each cascaded section is five instructions long, so that 500 [ns] (assuming a DSP56001 clock of 20 MHz) is added to the execution time for each additional *second order* section.

6.2 DIRECT FORM I TRANSPOSE IMPLEMENTATION

Figure 6.2-1 is the output file from *FDAS* for a *transpose form* implementation. As before, it contains information on the analog s-domain equivalent filter as well as the final digital coefficients (listed again in Figure 6.2-2) which have been properly scaled to prevent overflow at the internal nodes as well as the outputs of each cascaded section. It should be noted that because of the stability of the internal node gain of the *transpose form* and because of the response of each individual second order Butterworth section, scaling was not done by the program because it was not needed. Executable code shown in Figure 6.2-3 is again generated by *MGEN*. The code internal to each cascaded section is seven instructions long, so that 700 [ns] (assuming a DSP56001 clock of 20 MHz) is added to the execution time for each additional *second order* section.

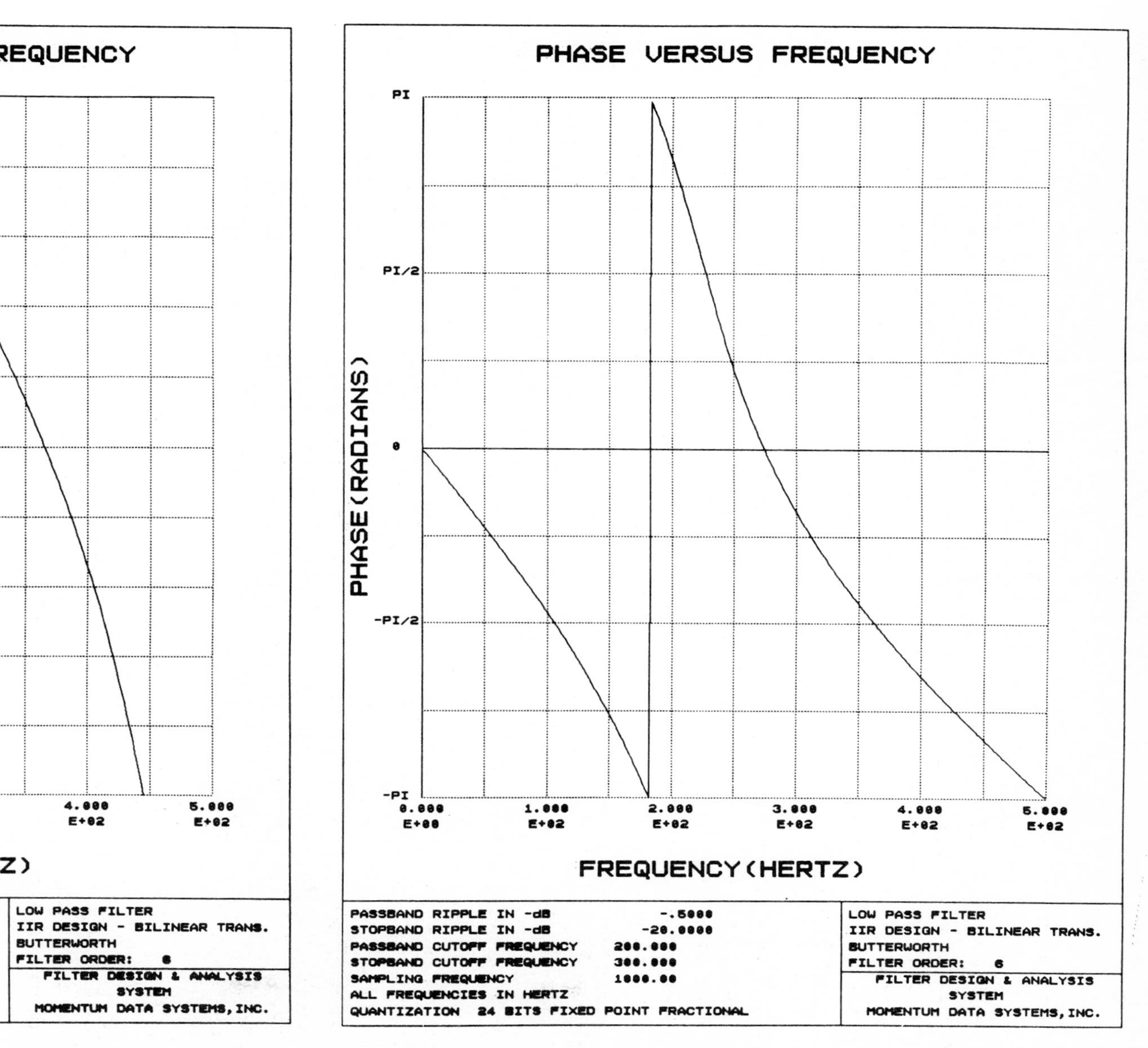

Figure 6.0-1. Log Magnitude Plot of Example Lowpass Butterworth Filter.

Figure 6.0-2. Phase versus Frequency Plot for Example Filter.

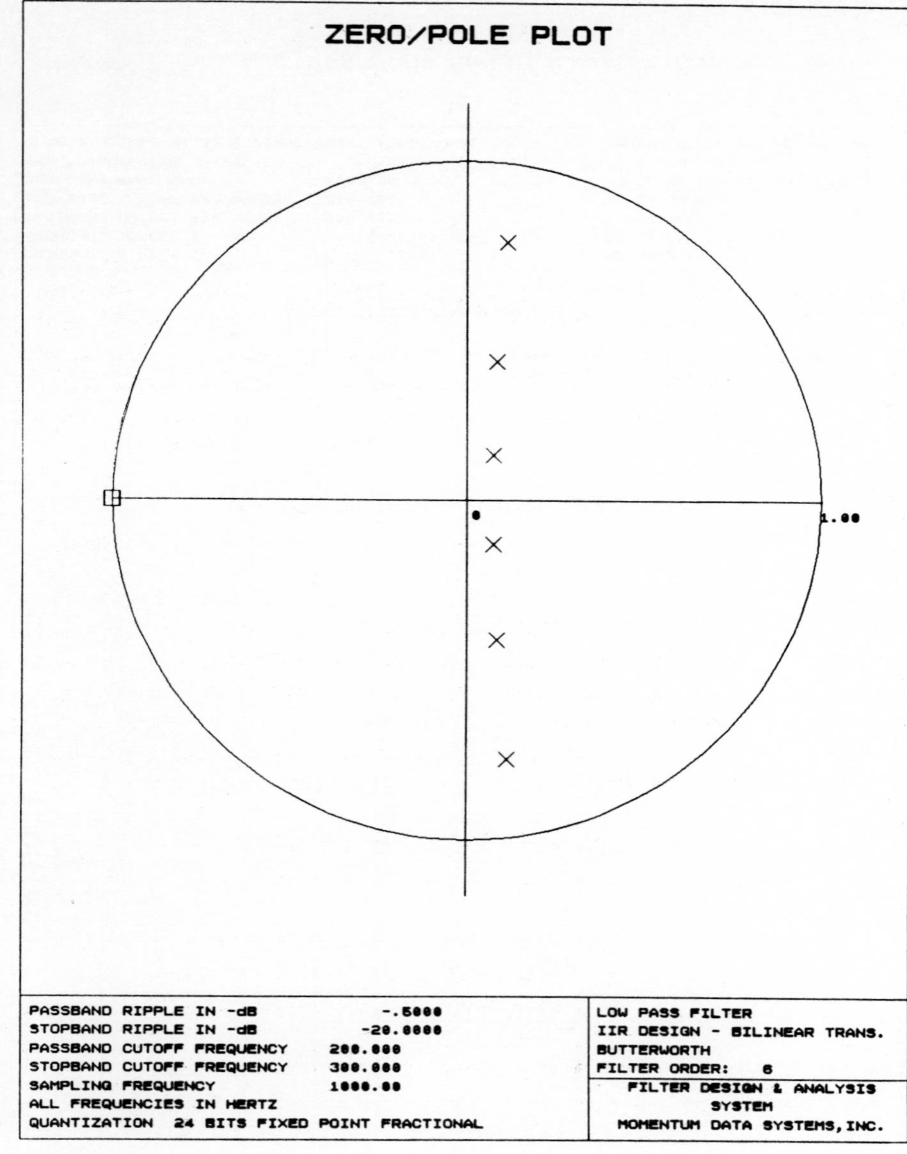

Figure 6.0-3. Zero/Pole Plot of 6th Order Lowpass Example Filter.

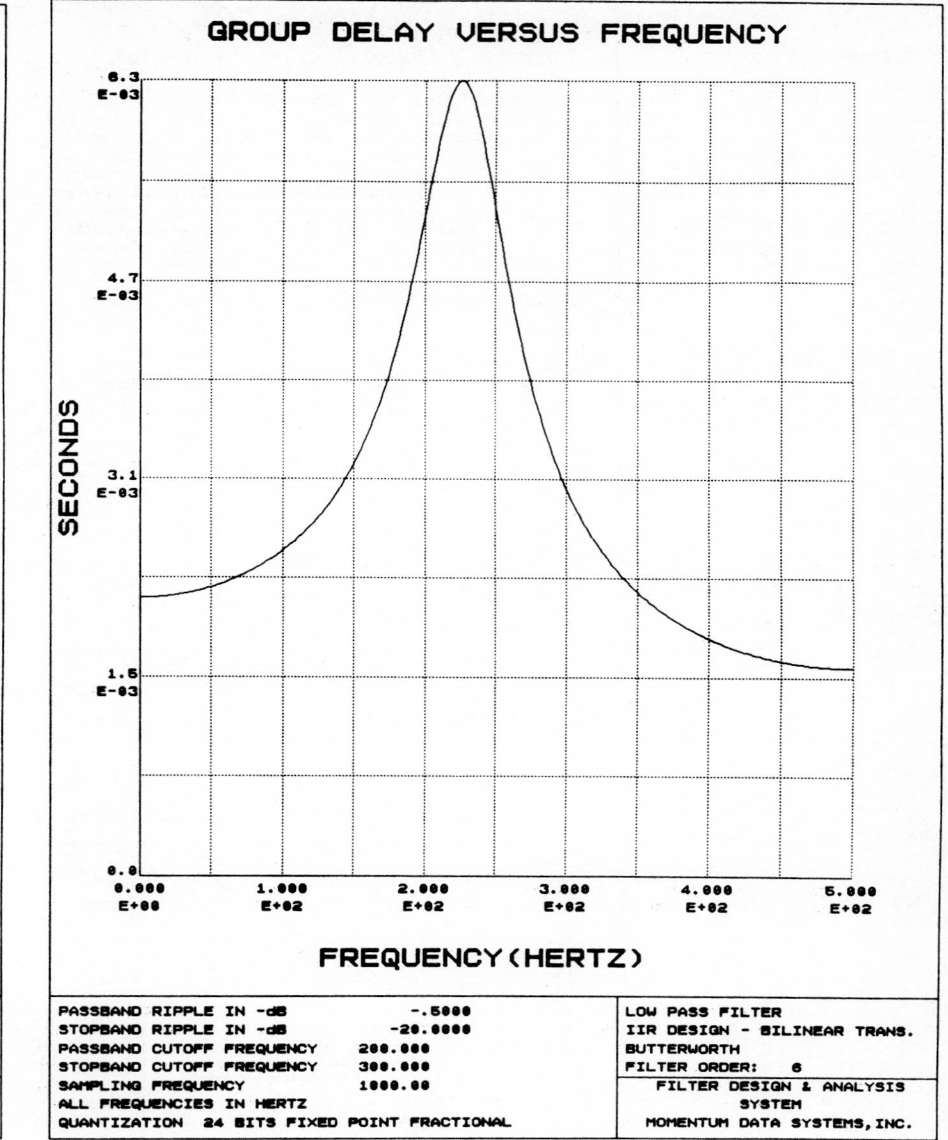

Figure 6.0-4. Group Delay versus Frequency for FDAS IIR Example.

```
FILTER TYPE                  LOW PASS
ANALOG FILTER TYPE           BUTTERWORTH
PASSBAND RIPPLE IN -dB          -.5000
STOPBAND RIPPLE IN -dB        -20.0000
PASSBAND CUTOFF FREQUENCY    200.000     HERTZ
STOPBAND CUTOFF FREQUENCY    300.000     HERTZ
SAMPLING FREQUENCY           1000.00     HERTZ
FILTER ORDER:      6
FILTER DESIGN METHOD: BILINEAR TRANSFORMATION

COEFFICIENTS OF Hd(Z) ARE QUANTIZED TO   24 BITS
QUANTIZATION TYPE: FIXED POINT FRACTIONAL
COEFFICIENTS SCALED FOR CASCADE FORM II


             NORMALIZED ANALOG TRANSFER FUNCTION T(s)

       NUMERATOR COEFFICIENTS                 DENOMINATOR COEFFICIENTS
************************************   ************************************
 S**2 TERM     S TERM      CONST TERM   S**2 TERM     S TERM      CONST TERM
.000000E+00  .000000E+00  .100000E+01  .100000E+01  .193185E+01  .100000E+01
.000000E+00  .000000E+00  .100000E+01  .100000E+01  .141421E+01  .100000E+01
.000000E+00  .000000E+00  .100000E+01  .100000E+01  .517638E+00  .100000E+01
      INITIAL GAIN  1.00000000


             UNNORMALIZED ANALOG TRANSFER FUNCTION T(s)

       NUMERATOR COEFFICIENTS                 DENOMINATOR COEFFICIENTS
************************************   ************************************
 S**2 TERM     S TERM      CONST TERM   S**2 TERM     S TERM      CONST TERM
.000000E+00  .000000E+00  .299809E+07  .100000E+01  .334500E+04  .299809E+07
.000000E+00  .000000E+00  .299809E+07  .100000E+01  .244871E+04  .299809E+07
.000000E+00  .000000E+00  .299809E+07  .100000E+01  .896290E+03  .299809E+07
      INITIAL GAIN  1.00000000


                    DIGITAL TRANSFER FUNCTION Hd(z)

          NUMERATOR COEFFICIENTS                    DENOMINATOR COEFFICIENTS
*************************************   *************************************
  Z**2 TERM      Z TERM      CONST TERM   Z**2 TERM     Z TERM       CONST TERM
.2520353794  .5040707588  .2520353794   1.00000  -.1463916302  .0225079060
.2364231348  .4728463888  .2364231348   1.00000  -.1684520245  .1765935421
.3606387377  .7212774754  .3606387377   1.00000  -.2279487848  .5921629667
      INITIAL GAIN  .876116210


                  ZEROES OF TRANSFER FUNCTION Hd(z)

  REAL PART      IMAGINARY PART     REAL PART      IMAGINARY PART      RADIUS
-.100000000E+01  .000000000E+00  -.100000000E+01  .000000000E+00  .100000000E+01
-.999290168E+00  .000000000E+00  -.100071034E+01  .000000000E+00  .999290168E+00
-.100000000E+01  .000000000E+00  -.100000000E+01  .000000000E+00  .100000000E+01


                  POLES OF TRANSFER FUNCTION Hd(z)

  REAL PART      IMAGINARY PART     REAL PART      IMAGINARY PART      RADIUS
.731958151E-01  .130959072E+00  .731958151E-01  -.130959072E+00  .150026351E+00
.842260122E-01  .411703195E+00  .842260122E-01  -.411703195E+00  .420230344E+00
.113974392E+00  .761034036E+00  .113974392E+00  -.761034036E+00  .769521258E+00


          NUMERATOR COEFFICIENTS - HIGHEST ORDER FIRST (in z)

.214893785E-01  .128936282E+00  .322340720E+00  .429787634E+00  .322340720E+00
.128936282E+00  .214893785E-01


          DENOMINATOR COEFFICIENTS - HIGHEST ORDER FIRST (in z)

.100000000E+01  -.542792439E+00  .887692610E+00  -.267088214E+00  .143235132E+00
-.184597152E-01  .235370025E-02


          IMPULSE RESPONSE  MIN = -.179722E+00, MAX =  .452734E+00
```

Figure 6.1-1. FDAS.OUT File of Example Filter for Cascaded Canonic Implementation.

```
FILTER COEFFICIENT FILE
IIR DESIGN
FILTER TYPE                    LOW PASS
ANALOG FILTER TYPE             BUTTERWORTH
PASSBAND RIPPLE IN -dB              -.5000
STOPBAND RIPPLE IN -dB            -20.0000
PASSBAND CUTOFF FREQUENCY     .200000E+03 HERTZ
STOPBAND CUTOFF FREQUENCY     .300000E+03 HERTZ
SAMPLING FREQUENCY            .100000E+04 HERTZ
FILTER DESIGN METHOD: BILINEAR TRANSFORMATION
FILTER ORDER             6 0006h
NUMBER OF SECTIONS       3 0003h
NO. OF QUANTIZED BITS  24 0018h
QUANTIZATION TYPE - FRACTIONAL FIXED POINT
COEFFICIENTS SCALED FOR CASCADE FORM II
        0 00000000     /* shift count for overall gain */
  7349395 00702493     /* overall gain               */
        0 00000000     /* shift count for section  1 values           */
  2114226 002042B2     /* section  1 coefficient B0                   */
  4228452 00408564     /* section  1 coefficient B1                   */
  2114226 002042B2     /* section  1 coefficient B2                   */
  1228022 0012BCF6     /* section  1 coefficient A1                   */
  -188810 FFFD1E76     /* section  1 coefficient A2                   */
        0 00000000     /* shift count for section  2 values           */
  1983261 001E431D     /* section  2 coefficient B0                   */
  3966523 003C863B     /* section  2 coefficient B1                   */
  1983261 001E431D     /* section  2 coefficient B2                   */
  1413078 00158FD6     /* section  2 coefficient A1                   */
 -1481374 FFE96562     /* section  2 coefficient A2                   */
        0 00000000     /* shift count for section  3 values           */
  3025257 002E2969     /* section  3 coefficient B0                   */
  6050514 005C52D2     /* section  3 coefficient B1                   */
  3025257 002E2969     /* section  3 coefficient B2                   */
  1912173 001D2D6D     /* section  3 coefficient A1                   */
 -4967423 FFB43401     /* section  3 coefficient A2                   */
 .2520353794097900D+00 3FD0215900000000  .25203538E+00 /* section  1 B0 */
 .5040707588195801D+00 3FE0215900000000  .50407076E+00 /* section  1 B1 */
 .2520353794097900D+00 3FD0215900000000  .25203538E+00 /* section  1 B2 */
 .1463916301727295D+00 3FC2BCF600000000  .14639171E+00 /* section  1 A1 */
-.2250790596008301D-01 BF970C5000000000 -.22507921E-01 /* section  1 A2 */
 .2364231348037720D+00 3FCE431D00000000  .23642325E+00 /* section  2 B0 */
 .4728463888168335D+00 3FDE431D80000000  .47284651E+00 /* section  2 B1 */
 .2364231348037720D+00 3FCE431D00000000  .23642325E+00 /* section  2 B2 */
 .1684520244598389D+00 3FC58FD600000000  .16845204E+00 /* section  2 A1 */
-.1765935420989990D+00 BFC69A9E00000000 -.17659356E+00 /* section  2 A2 */

 .3606387376785278D+00 3FD714B480000000  .36063876E+00 /* section  3 B0 */
 .7212774753570557D+00 3FE714B480000000  .72127753E+00 /* section  3 B1 */
 .3606387376785278D+00 3FD714B480000000  .36063876E+00 /* section  3 B2 */
 .2279487848281860D+00 3FCD2D6D00000000  .22794882E+00 /* section  3 A1 */
-.5921629667282104D+00 BFE2F2FFC0000000 -.59216306E+00 /* section  3 A2 */
```

Figure 6.1-2. COEFF.OUT File of Example Filter Design and Scaled for Cascaded Canonic Implementation.

Figure 6.1-3. COEFF.ASM File Generated by MGEN for Example Design and Cascaded Canonic Implementation.

```
1                        COEFF     ident   1,0
2                                  include 'head2.asm'
3
4                                  page    132,66,0,10
5                                  opt     cex,mex
6                        ;
7                        ; This program implements an IIR filter in cascaded canonic sections
8                        ; The coefficients of each section are scaled for cascaded canonic sections
9                        ;
10
11          00FFFF       datin     equ     $ffff                           ;location in Y memory of input file
12          00FFFF       datout    equ     $ffff                           ;location in Y memory of output file
13          00FFF0       m_scr     equ     $fff0                           ; sci control register
14          00FFF1       m_ssr     equ     $fff1                           ; sci status register
15          00FFF2       m_sccr    equ     $fff2                           ; sci clock control register
16          00FFE1       m_pcc     equ     $ffe1                           ; port c control register
17          00FFFF       m_ipr     equ     $ffff                           ; interrupt priority register
18          000140       xx        equ     @cvi(20480/(64*1.000))          ;timer interrupt value
19          FFD8F1       m_tim     equ     -9999                           ;board timer interrupt value
20          000003       nsec      equ     3                               ;number of second order sections
21                                 include 'cascade2.asm'
22                       ;
23                       ; This code segment implements cascaded biquad sections in canonic form
24                       ;
25
26                       cascade2  macro   nsec
27     m                 ;
28     m                 ;     assumes each section's coefficients are divided by 2
29     m                 ;
30     m                           mpy     y0,y1,a   x:(r0)+,x0  y:(r4)+,y0  ;x0=1st section w(n-2),y0=a12/2
31     m                           do      #nsec,_ends                     ;do each section
32     m                           mac     x0,y0,a   x:(r0)-,x1  y:(r4)+,y0  ;x1=w(n-1) ,y0=ai1/2
33     m                           macr    x1,y0,a   x1,x:(r0)+  y:(r4)+,y0  ;push w(n-1) to w(n-2),y0=bi2/2
34     m                           mac     x0,y0,a   a,x:(r0)    y:(r4)+,y0  ;push w(n) to w(n-1),y0=bi1/2
35     m                           mac     x1,y0,a   x:(r0)+,x0  y:(r4)+,y0  :get this iter w(n),y0=bi0/2
36     m                           mac     x0,y0,a   x:(r0)+,x0  y:(r4)+,y0  ;next iter:x0=w(n-2),y0=ai2/2
37     m                 _ends
38     m                           endm
39
40
41                       ;
42                       ; multiple shift left macro
43                       ;
44                       mshl      macro   scount
45     m                           if      scount
46     m                           rep     #scount
47     m                           asl     a
48     m                           endif
49     m                           endm
50          X:0000                 org     x:0
51          X:0000       states    ds      2*nsec
52          Y:0000                 org     y:0
```

```
53                      coef
54  d   Y:0000 FE8F3B              dc      $FE8F3B                         ;a(*,2)/2     =-.01125395    section number
55  d   Y:0001 095E7B              dc      $095E7B                         ;a(*,1)/2     = .07319582    section number
56  d   Y:0002 102159              dc      $102159                         ;b(*,2)/2     = .12601769    section number
57  d   Y:0003 2042B2              dc      $2042B2                         ;b(*,1)/2     = .25203538    section number
58  d   Y:0004 D02159              dc      $D02159                         ;b(*,0)/2-0.5=-.37398231     section number
59  d   Y:0005 F4B2B1              dc      $F4B2B1                         ;a(*,2)/2     =-.08829677    section number
60  d   Y:0006 0AC7EB              dc      $0AC7EB                         ;a(*,1)/2     = .08422601    section number
61  d   Y:0007 0F218E              dc      $0F218E                         ;b(*,2)/2     = .11821157    section number
62  d   Y:0008 1E431D              dc      $1E431D                         ;b(*,1)/2     = .23642319    section number
63  d   Y:0009 CF218E              dc      $CF218E                         ;b(*,0)/2-0.5=-.38178843     section number
64  d   Y:000A DA1A01              dc      $DA1A01                         ;a(*,2)/2     =-.29608148    section number
65  d   Y:000B 0E96B6              dc      $0E96B6                         ;a(*,1)/2     = .11397439    section number
66  d   Y:000C 1714B4              dc      $1714B4                         ;b(*,2)/2     = .18031937    section number
67  d   Y:000D 2E2969              dc      $2E2969                         ;b(*,1)/2     = .36063874    section number
68  d   Y:000E D714B4              dc      $D714B4                         ;b(*,0)/2-0.5=-.31968063     section number
69      Y:00C8                     org     y:200
70  d   Y:00C8 381249     igain    dc      3674697
71      000000            scount   equ     0                               ;final shift count
72                                 include 'body2.asm'
73
74      000040            start    equ     $40                             ;origin for user program
75
76      P:0000                     org     p:$0                            ;origin for reset vector
77      P:0000 0C0040              jmp     start                           ;jump to 'start' on system reset
78
79      P:001C                     org     p:$1c                           ;origin for timer interrupt vector
80      P:001C 0BF080              jsr     filter                          ;jump to 'filter' on timer interrupt
           000051
81
82      P:0040                     org     p:start                         ;origin for user program
83
84      P:0040 0003F8              ori     #3,mr                           ;disable all interrupts
85      P:0041 08F4B2              movep           #(xx-1),x:m_sccr        ;cd=xx-1 for divide by xx
           00013F
86      P:0043 08F4A1              movep           #$7,x:m_pcc             ;set cc(2;0) to turn on timer
           000007
87      P:0045 08F4B0              movep           #$2000,x:m_scr          ;enable timer interrupts
           002000
88      P:0047 08F4BF              movep           #$c000,x:m_ipr          ;set interrupt priority for sci
           00C000
89
90      P:0049 300000              move            #states,r0              ;initialize internal state storage
91      P:004A 200013              clr     a                               ;*   set memory to zero
92      P:004B 0606A0              rep     #nsec*2                         ;*
93      P:004C 565800              move            a,x:(r0)+               ;*
94
95      P:004D 4FF000              move                        y:igain,y1  ;y1=initial gain/2
           0000C8
96      P:004F 00FCB8              andi    #$fc,mr                         ;allow interrupts
97      P:0050 0C0050              jmp     *                               ;wait for interrupt
98
```

```
99                      filter
100      P:0051 0008F8          ori     #$08,mr                          ;set scaling mode
101      P:0052 300000          move              #states,r0             ;point to filter states
102      P:0053 340000          move              #coef,r4               ;point to filter coefficients
103      P:0054 09463F          movep             y:datin,y0             ;get sample
104
105                             cascade2 nsec                            ;do cascaded biquads
106   +                 ;
107   +                 ;      assumes each section's coefficients are divided by 2
108   +                 ;
109   +  P:0055 F098B0          mpy     y0,y1,a   x:(r0)+,x0  y:(r4)+,y0 ;x0=1st section w(n-2),y0=a12/2
110   +  P:0056 060380          do      #nsec,_ends                      ;do each section
                00005C
111   +  P:0058 F490D2          mac     x0,y0,a   x:(r0)-,x1  y:(r4)+,y0 ;x1=w(n-1) ,y0=ai1/2
112   +  P:0059 F418E3          macr    x1,y0,a   x1,x:(r0)+  y:(r4)+,y0 ;push w(n-1) to w(n-2),y0=bi2/2
113   +  P:005A F800D2          mac     x0,y0,a   a,x:(r0)    y:(r4)+,y0 ;push w(n) to w(n-1),y0=bi1/2
114   +  P:005B F098E2          mac     x1,y0,a   x:(r0)+,x0  y:(r4)+,y0 :get this iter w(n),y0=bi0/2
115   +  P:005C F098D2          mac     x0,y0,a   x:(r0)+,x0  y:(r4)+,y0 ;next iter:x0=w(n-2),y0=ai2/2
116   +                 _ends
117                             mshl    scount
118   +                         if      scount
121   +                         endif
122      P:005D 200011          rnd     a                                ;round result
123      P:005E 09CE3F          movep             a,y:datout             ;output sample
124      P:005F 000004          rti
125
126                             end
0     Errors
0     Warnings
```

```
FILTER TYPE                 LOW PASS
ANALOG FILTER TYPE          BUTTERWORTH
PASSBAND RIPPLE IN -dB           -.5000
STOPBAND RIPPLE IN -dB          -20.0000
PASSBAND CUTOFF FREQUENCY    200.000      HERTZ
STOPBAND CUTOFF FREQUENCY    300.000      HERTZ
SAMPLING FREQUENCY           1000.00      HERTZ
FILTER ORDER:      6
FILTER DESIGN METHOD: BILINEAR TRANSFORMATION

COEFFICIENTS OF Hd(Z) ARE QUANTIZED TO   24 BITS
QUANTIZATION TYPE: FIXED POINT FRACTIONAL
COEFFICIENTS SCALED FOR CASCADE FORM I (TRANSPOSE FORM)


          NORMALIZED ANALOG TRANSFER FUNCTION T(s)

     NUMERATOR COEFFICIENTS                DENOMINATOR COEFFICIENTS
**************************************  **************************************
 S**2 TERM     S TERM      CONST TERM   S**2 TERM     S TERM      CONST TERM
.000000E+00  .000000E+00  .100000E+01  .100000E+01  .193185E+01  .100000E+01
.000000E+00  .000000E+00  .100000E+01  .100000E+01  .141421E+01  .100000E+01
.000000E+00  .000000E+00  .100000E+01  .100000E+01  .517638E+00  .100000E+01
     INITIAL GAIN  1.00000000


          UNNORMALIZED ANALOG TRANSFER FUNCTION T(s)

     NUMERATOR COEFFICIENTS                DENOMINATOR COEFFICIENTS
**************************************  **************************************
 S**2 TERM     S TERM      CONST TERM   S**2 TERM     S TERM      CONST TERM
.000000E+00  .000000E+00  .299809E+07  .100000E+01  .334500E+04  .299809E+07
.000000E+00  .000000E+00  .299809E+07  .100000E+01  .244871E+04  .299809E+07
.000000E+00  .000000E+00  .299809E+07  .100000E+01  .896290E+03  .299809E+07
     INITIAL GAIN  1.00000000


                 DIGITAL TRANSFER FUNCTION Hd(z)

      NUMERATOR COEFFICIENTS                DENOMINATOR COEFFICIENTS
*************************************  **************************************
  Z**2 TERM     Z TERM      CONST TERM  Z**2 TERM     Z TERM        CONST TERM
.2190289497  .4380580187  .2190289497   1.00000  -.1463916302   .0225079060
.2520353794  .5040707588  .2520353794   1.00000  -.1684520245   .1765935421
.3410534859  .6821070910  .3410534859   1.00000  -.2279487848   .5921629667
     INITIAL GAIN  1.00000000


                 ZEROES OF TRANSFER FUNCTION Hd(z)

   REAL PART     IMAGINARY PART     REAL PART     IMAGINARY PART       RADIUS
-.999262530E+00  .000000000E+00  -.100073801E+01  .000000000E+00  .999262530E+00
-.100000000E+01  .000000000E+00  -.100000000E+01  .000000000E+00  .100000000E+01
-.999408962E+00  .000000000E+00  -.100059139E+01  .000000000E+00  .999408962E+00


                 POLES OF TRANSFER FUNCTION Hd(z)

   REAL PART     IMAGINARY PART     REAL PART     IMAGINARY PART       RADIUS
.731958151E-01  .130959072E+00  .731958151E-01  -.130959072E+00  .150026351E+00
.842260122E-01  .411703195E+00  .842260122E-01  -.411703195E+00  .420230344E+00
.113974392E+00  .761034036E+00  .113974392E+00  -.761034036E+00  .769521258E+00


          NUMERATOR COEFFICIENTS - HIGHEST ORDER FIRST (in z)

.188271907E-01  .112963161E+00  .282407928E+00  .376543916E+00  .282407928E+00
.112963161E+00  .188271907E-01


          DENOMINATOR COEFFICIENTS - HIGHEST ORDER FIRST (in z)

.100000000E+01 -.542792439E+00  .887692610E+00 -.267088214E+00  .143235132E+00
-.184597152E-01  .235370025E-02


          IMPULSE RESPONSE  MIN = -.179722E+00, MAX =  .452734E+00
```

Figure 6.2-1. FDAS.OUT File of Example Filter for Cascaded Transpose Form Implementation.

```
FILTER COEFFICIENT FILE
IIR DESIGN
FILTER TYPE                 LOW PASS
ANALOG FILTER TYPE          BUTTERWORTH
PASSBAND RIPPLE IN -dB           -.5000
STOPBAND RIPPLE IN -dB          -20.0000
PASSBAND CUTOFF FREQUENCY    .200000E+03 HERTZ
STOPBAND CUTOFF FREQUENCY    .300000E+03 HERTZ
SAMPLING FREQUENCY           .100000E+04 HERTZ
FILTER DESIGN METHOD: BILINEAR TRANSFORMATION
FILTER ORDER             6 0006h
NUMBER OF SECTIONS       3 0003h
NO. OF QUANTIZED BITS   24 0018h
QUANTIZATION TYPE - FRACTIONAL FIXED POINT
COEFFICIENTS SCALED FOR CASCADE FORM I
         1 00000001     /* shift count for overall gain */
   4194304 00400000     /* overall gain                 */
         0 00000000     /* shift count for section  1 values          */
   1837348 001C0924     /* section  1 coefficient B0                  */
   3674697 00381249     /* section  1 coefficient B1                  */
   1837348 001C0924     /* section  1 coefficient B2                  */
   1228022 0012BCF6     /* section  1 coefficient A1                  */
   -188810 FFFD1E76     /* section  1 coefficient A2                  */
         0 00000000     /* shift count for section  2 values          */
   2114226 002042B2     /* section  2 coefficient B0                  */
   4228452 00408564     /* section  2 coefficient B1                  */
   2114226 002042B2     /* section  2 coefficient B2                  */
   1413078 00158FD6     /* section  2 coefficient A1                  */
  -1481374 FFE96562     /* section  2 coefficient A2                  */
         0 00000000     /* shift count for section  3 values          */
   2860964 002BA7A4     /* section  3 coefficient B0                  */
   5721929 00574F49     /* section  3 coefficient B1                  */
   2860964 002BA7A4     /* section  3 coefficient B2                  */
   1912173 001D2D6D     /* section  3 coefficient A1                  */
  -4967423 FFB43401     /* section  3 coefficient A2                  */
  .2190289497375488D+00 3FCC092400000000  .21902905E+00 /* section  1 B0 */
  .4380580186843872D+00 3FDC092480000000  .43805810E+00 /* section  1 B1 */
  .2190289497375488D+00 3FCC092400000000  .21902905E+00 /* section  1 B2 */
  .1463916301727295D+00 3FC2BCF600000000  .14639171E+00 /* section  1 A1 */
 -.2250790596008301D-01 BF970C5000000000 -.22507921E-01 /* section  1 A2 */
  .2520353794097900D+00 3FD0215900000000  .25203538E+00 /* section  2 B0 */
  .5040707588195801D+00 3FE0215900000000  .50407076E+00 /* section  2 B1 */
  .2520353794097900D+00 3FD0215900000000  .25203538E+00 /* section  2 B2 */
  .1684520244598389D+00 3FC58FD600000000  .16845204E+00 /* section  2 A1 */
 -.1765935420989990D+00 BFC69A9E00000000 -.17659356E+00 /* section  2 A2 */
  .3410534858703613D+00 3FD5D3D200000000  .34105356E+00 /* section  3 B0 */
  .6821070909500122D+00 3FE5D3D240000000  .68210712E+00 /* section  3 B1 */
  .3410534858703613D+00 3FD5D3D200000000  .34105356E+00 /* section  3 B2 */
  .2279487848281860D+00 3FCD2D6D00000000  .22794882E+00 /* section  3 A1 */
 -.5921629667282104D+00 BFE2F2FFC0000000 -.59216306E+00 /* section  3 A2 */
```

Figure 6.2-2. COEFF.FLT File for Example Filter Design and Scaled for Cascaded Transpose Form.

Figure 6.2-3. COEFF.ASM File Generated by MGEN for Example Design and Cascaded Transpose Implementation.

```
1                      COEFF     ident   1,0
2                                include 'head1.asm'
3
4                                page    132,66,0,10
5                                opt     cex,mex
6                      ;
7                      ; This program implements an IIR filter in cascaded transpose sections
8                      ; The coefficients of each section are scaled for cascaded transpose sections
9                      ;
10
11       00FFFF        datin     equ     $ffff                          ;location in Y memory of input file
12       00FFFF        datout    equ     $ffff                          ;location in Y memory of output file
13       00FFF0        m_scr     equ     $fff0                          ; sci control register
14       00FFF1        m_ssr     equ     $fff1                          ; sci status register
15       00FFF2        m_sccr    equ     $fff2                          ; sci clock control register
16       00FFE1        m_pcc     equ     $ffe1                          ; port c control register
17       00FFFF        m_ipr     equ     $ffff                          ; interrupt priority register
18       000140        xx        equ     @cvi(20480/(64*1.000))         ;timer interrupt value
19       FFD8F1        m_tim     equ     -9999                          ;board timer interrupt value
20       000003        nsec      equ     3                              ;number of second order sections
21                               include 'cascade1.asm'
22                     ;
23                     ; This code segment implements cascaded biquad sections in transpose form
24                     ;
25
26                     cascade1  macro   nsec
27   m                 ;
28   m                 ;     assumes each section's coefficients are divided by 2
29   m                 ;
30   m                           do      #nsec,_ends                    ;do each section
31   m                           macr    y0,y1,a   x:(r1),b   y:(r4)+,y0  ;a=x(n)*bi0/2+wi1/2,b=wi2,y0=bi1/2
32   m                           asr     b         a,x0                  ;b=wi2/2,x0=y(n)
33   m                           mac     y0,y1,b   y:(r4)+,y0            ;b=x(n)*bi1/2+wi2/2,y0=ai1/2
34   m                           macr    x0,y0,b   y:(r4)+,y0            ;b=b+y(n)*ai1/2,y0=bi2/2
35   m                           mpy     y0,y1,b   b,x:(r0)+  y:(r4)+,y0  ;b=x(n)*bi2/2,save wi1,y0=ai2
36   m                           macr    x0,y0,b   x:(r0),a   a,y1        ;b=b+y(n)*ai2/2,a=next iter wi1,
37   m                                                                   ;y1=output of section i
38   m                           asr     a         b,x:(r1)+  y:(r4)+,y0  ;a=next iter wi1/2,save wi2,
39   m                                                                   ;y0=next iter bi0
40   m                 _ends
41   m                           endm
42
43
44       X:0000                  org     x:0
45       X:0000        state1    dsm     nsec
46       X:0004        state2    dsm     nsec
47       Y:0000                  org     y:0
48                     coef
49   d   Y:0000 0E0492           dc      $0E0492                        ;b(*,0)/2    = .10951447     section number
50   d   Y:0001 1C0924           dc      $1C0924                        ;b(*,1)/2    = .21902901     section number
51   d   Y:0002 095E7B           dc      $095E7B                        ;a(*,1)/2    = .07319582     section number
52   d   Y:0003 0E0492           dc      $0E0492                        ;b(*,2)/2    = .10951447     section number
```

```
53  d    Y:0004 FE8F3B              dc     $FE8F3B                               ;a(*,2)/2     =-.01125395     section number  1
54  d    Y:0005 102159              dc     $102159                               ;b(*,0)/2     = .12601769     section number  2
55  d    Y:0006 2042B2              dc     $2042B2                               ;b(*,1)/2     = .25203538     section number  2
56  d    Y:0007 0AC7EB              dc     $0AC7EB                               ;a(*,1)/2     = .08422601     section number  2
57  d    Y:0008 102159              dc     $102159                               ;b(*,2)/2     = .12601769     section number  2
58  d    Y:0009 F4B2B1              dc     $F4B2B1                               ;a(*,2)/2     =-.08829677     section number  2
59  d    Y:000A 15D3D2              dc     $15D3D2                               ;b(*,0)/2     = .17052674     section number  3
60  d    Y:000B 2BA7A4              dc     $2BA7A4                               ;b(*,1)/2     = .34105355     section number  3
61  d    Y:000C 0E96B6              dc     $0E96B6                               ;a(*,1)/2     = .11397439     section number  3
62  d    Y:000D 15D3D2              dc     $15D3D2                               ;b(*,2)/2     = .17052674     section number  3
63  d    Y:000E DA1A01              dc     $DA1A01                               ;a(*,2)/2     =-.29608148     section number  3
64                                  include 'body1.asm'
65
66       000040           start     equ    $40                                   ;origin for user program
67
68       P:0000                     org    p:$0                                  ;origin for reset vector
69       P:0000 0C0040              jmp    start                                 ;jump to 'start' on system reset
70
71       P:001C                     org    p:$1c                                 ;origin for timer interrupt vector
72       P:001C 0BF080              jsr    filter                                ;jump to 'filter' on timer interrupt
                000057
73
74       P:0040                     org    p:start                               ;origin for user program
75
76       P:0040 0003F8              ori    #3,mr                                 ;disable all interrupts
77       P:0041 08F4B2              movep          #(xx-1),x:m_sccr              ;cd=xx-1 for divide by xx
                00013F
78       P:0043 08F4A1              movep          #$7,x:m_pcc                   ;set cc(2;0) to turn on timer
                000007
79       P:0045 08F4B0              movep          #$2000,x:m_scr                ;enable timer interrupts
                002000
80       P:0047 08F4BF              movep          #$c000,x:m_ipr                ;set interrupt priority for sci
                00C000
81
82
83       P:0049 300000              move           #state1,r0                    ;point to filter state1
84       P:004A 310400              move           #state2,r1                    ;point to filter state2
85       P:004B 340000              move           #coef,r4                      ;point to filter coefficients
86       P:004C 0502A0              move           #nsec-1,m0                    ;addressing modulo nsec
87       P:004D 050EA4              move           #5*nsec-1,m4                  ;addressing modulo 5*nsec
88       P:004E 0502A1              move           #nsec-1,m1                    ;addressing modulo nsec
89       P:004F 200013              clr    a                                     ;initialize internal state storage
90       P:0050 0603A0              rep    #nsec                                 ;*    zero state1
91       P:0051 565800              move           a,x:(r0)+                     ;*
92       P:0052 0603A0              rep    #nsec                                 ;*    zero state2
93       P:0053 565900              move           a,x:(r1)+                     ;*
94
95       P:0054 F88000              move           x:(r0),a     y:(r4)+,y0       ;a=w1 (initially zero) ,y0=b10/2
96
97       P:0055 00FCB8              andi   #$fc,mr                               ;allow interrupts
98       P:0056 0C0056              jmp    *                                     ;wait for interrupt
99
```

```
100                       filter
101        P:0057 0008F8           ori      #$08,mr                                  ;set scaling mode
102        P:0058 09473F           movep              y:datin,y1                     ;get sample
103
104                                cascade1 nsec                                     ;do cascaded biquads
105    +                  ;
106    +                  ;     assumes each section's coefficients are divided by 2
107    +                  ;
108    +   P:0059 060380           do       #nsec,_ends                              ;do each section
              000061
109    +   P:005B FC81B3           macr     y0,y1,a   x:(r1),b    y:(r4)+,y0   ;a=x(n)*bi0/2+wi1/2,b=wi2,y0=bi1/2
110    +   P:005C 21C42A           asr      b         a,x0                     ;b=wi2/2,x0=y(n)
111    +   P:005D 4EDCBA           mac      y0,y1,b               y:(r4)+,y0   ;b=x(n)*bi1/2+wi2/2,y0=ai1/2
112    +   P:005E 4EDCDB           macr     x0,y0,b               y:(r4)+,y0   ;b=b+y(n)*ai1/2,y0=bi2/2
113    +   P:005F FC18B8           mpy      y0,y1,b   b,x:(r0)+   y:(r4)+,y0   ;b=x(n)*bi2/2,save wi1,y0=ai2
114    +   P:0060 19A0DB           macr     x0,y0,b   x:(r0),a    a,y1         ;b=b+y(n)*ai2/2,a=next iter wi1,
115    +                                                                       ;y1=output of section i
116    +   P:0061 FC1922           asr      a         b,x:(r1)+   y:(r4)+,y0   ;a=next iter wi1/2,save wi2,
117    +                                                                       ;y0=next iter bi0
118    +                  _ends
119
120        P:0062 09C73F           movep              y1,y:datout                    ;output sample
121                       _endp
122
123        P:0063 000004           rti
124                                end
0     Errors
0     Warnings
```

7.0 FIR FILTERS

As we discovered when discussing IIR filters, if phase distortion is of secondary importance to magnitude response the desired filter can generally be implemented with less memory, less computational complexity, and at the lowest cost using IIR structures. On the other hand, in applications requiring linear phase in the passband and a specific magnitude response, the specified filter is generally best implemented using FIR structures. Examples of applications requiring linear phase are:

(1) Communication systems such as modems or *Integrated Services Digital Networks* (ISDN) in which the data pulse shape and relative timing in the channel must be preserved.

(2) Ideal differentiators which provide a 90 degree phase shift at all frequencies in addition to a constant group delay.

(3) Hilbert transformers used to demodulate complex signals such as those used in high speed modems.

(4) Hi-fidelity audio systems where enthusiasts are always trying to minimize phase distortion of recorded music in order to reproduce the original sound with as much fidelity as possible.

(5) System synthesis where the system impulse response is known *a priori*. Also, as we shall see, FIR filters are important because they are all-zero filters (i.e., no feedback) and are therefore guaranteed to be stable.

7.1 LINEAR PHASE FIR FILTER STRUCTURE

The basic structure of a FIR filter is simply a *tapped delay line* in which the output from each tap is summed to generate the filter output. This is shown in Figure 7.1-1. This structure can be represented mathematically as:

$$y(n) = \sum_{i=0}^{N-1} h(i)\, x(n-i) \qquad (7.1-1)$$

where $x(n)$ is the most recent ($t=nT$) input signal sample, $x(n-i)$ are signal samples delayed by i sample periods (iT), $h(i)$ are the tap weights (or filter coefficients), and $y(n)$ is the filter output at time $t=nT$.

From this structure it is easy to see why the filter is termed *finite* - the impulse response of the filter will be identically zero after N sample periods because an impulse input (i.e., $x(n)=1$, $x(n-i)=0$ for $i \neq 0$) will have traversed the entire delay line at time $t=NT$. That is, the impulse response of FIR filters will only last for a finite period of time. This is in contrast to IIR filters which will "ring" in response to an impulse for an infinite period of time. The values of the coefficients represent the impulse response of the FIR filter as can be seen by evaluating Equation (7.1-1) over N sample periods for a single unit input pulse at time $t=0$.

There are no feedback terms in the structure (i.e., Equation (7.1-1) has no denominator) but rather only N zeroes. By taking the *Z-transform* of Equation (7.1-1) the *transfer function* of the filter becomes:

$$H(z) = \frac{Y(z)}{X(z)} = \sum_{i=0}^{N-1} h(i)\, z^{-i} \qquad (7.1-2)$$

Equation (7.1-2) is a polynomial in z of order N. The roots of this polynomial are the N zeroes of the filter.

The same procedures used to calculate the magnitude and phase response of IIR filters can be applied to FIR filters. Accordingly, the gain, $G(\theta)$, can be obtained by substituting $z = e^{j\theta}$ into Equation (7.1-2) (where $\theta = 2\pi f / f_s$ is the normalized digital frequency) then taking the absolute value so that:

$$G(\theta) = \left| \sum_{i=0}^{N-1} h(i)\, e^{-ji\theta} \right|$$

$$= \sum_{i=0}^{N-1} [h(i) \cos i\theta - j\, h(i) \sin i\theta] \qquad (7.1-3)$$

The phase response, $\phi(\theta)$, is found by taking the inverse tangent of the ratio of the *imaginary* to *real* components of $G(\theta)$ so that:

$$\phi(\theta) = \tan^{-1} \left[\frac{-\sum_{i=0}^{N-1} h(i) \sin i\theta}{\sum_{i=0}^{N-1} h(i) \cos i\theta} \right] \qquad (7.1-4)$$

where it has been implicitly assumed that the $h(i)$ are real.

Intuitively it can be reasoned that in order for any pulse to retain its general shape and relative timing (i.e., *pulse width* and *time delay* to its peak value) after passing through a filter, the delay of each frequency component making up the pulse must be the same so that each component recombines in phase to reconstruct the original shape. In terms of phase, this implies that the phase delay must be linearly related to frequency or:

$$\phi(\theta) = -\tau\theta \qquad (7.1-5)$$

This is illustrated in Figure 7.1-2 where a pulse having two components $\sin \omega t$ and $\sin 2\omega t$ passes through a linear phase filter having a delay of two cycles per *Hertz* and constant magnitude response equal to unity. That is, the fundamental is delayed by two cycles (4π) and the first harmonic by four cycles (8π). The group delay of a system, τ_g, is defined by taking the derivative of $\phi(\theta)$:

$$\tau_g \stackrel{\Delta}{=} -\frac{d\phi}{d\theta} \qquad (7.1-6)$$

Since θ is the normalized frequency, τ_g in Equation (7.1-6) is a dimensionless quantity and can be related to the group delay in seconds by dividing by the sample frequency f_s. For a linear phase system τ_g is independent of θ and is equal to τ. This can be seen by substituting Equation (7.1-5) into Equation (7.1-6).

In our example, τ_g is 2 cycles per Hz. Note that the pulse shape within the filter has been retained but is twice the width of the original pulse; that is, the change in width of the pulse is equal to the group delay which is the origin of its name. The negative sign indicates the phase is retarded or delayed, i.e., a *causal system.*

The impact of the requirement of linear phase on the design of a FIR filter can be seen by substituting Equation (7.1-4) into Equation (7.1-5) so that:

$$\tan\theta\tau_g = \left[\frac{\sum_{i=0}^{N-1} h(i)\sin i\theta}{\sum_{i=0}^{N-1} h(i)\cos i\theta}\right] \tag{7.1-7}$$

which can be rewritten as,

$$\sum_{i=0}^{N-1} h(i)(\cos i\theta\ \sin\theta\tau_g - \sin i\theta\ \cos\theta\tau_g) = 0$$

or

$$\sum_{i=0}^{N-1} h(i)\,[\sin\theta(\tau_g - i)] = 0 \tag{7.1-8}$$

The solution to Equation (7.1-8) is the constraint on τ_g for a FIR filter to be linear phase. The solution to Equation (7.1-8) can be found by expanding the *left hand side* (LHS) as follows:

$$LHS = h_o\sin\theta\tau_g + h_1\sin\theta(\tau_g - 1) + h_2\sin\theta(\tau_g - 2) + \ldots.$$

$$+h_{N-3}\sin\theta(\tau_g - N + 3) + h_{N-2}\sin\theta(\tau_g - N + 2) + h_{N-1}\sin\theta(\tau_g - N + 1)$$

and noting that if:

$$\tau_g = \frac{(N-1)}{2} \tag{7.1-9}$$

then for every positive argument there will be a corresponding negative argument. For example the argument for the h_1 term becomes:

$$\theta\left(\frac{N-1}{2} - 1\right) = \theta\left(\frac{N-3}{2}\right)$$

which is the negative of the argument for the h_{N-2} term, i.e.,

$$\theta\left(\frac{N-1}{2} - N + 2\right) = -\theta\left(\frac{N-3}{2}\right)$$

so that if:

$$h(i) = h(N-1-i)\ \ for\ \ 0 \le i \le N-1 \tag{7.1-10}$$

then Equations (7.1-9) and (7.1-10) represent the solution to (7.1-8). By substituting Equation (7.1-9) into (7.1-5) the phase of a linear phase FIR filter is given by:

$$\phi(\theta) = -\left(\frac{N-1}{2}\right)\theta \qquad (7.1-11)$$

Therefore, a nonrecursive filter (FIR), unlike a recursive filter (IIR), can have a constant time delay for all frequencies over the entire range (from *0* to $f_s/2$). It is only necessary that the coefficients (and therefore impulse response) be symmetrical about the midpoint between samples *(N-2)/2* and *N/2* for even *N* or about sample *(N-1)/2* for odd *N*, as described by Equation (7.1-10). When this symmetry exists, τ_g for the filter will be $\left(\frac{N-1}{2}\right)T$ seconds.

7.2 LINEAR PHASE FIR FILTER DESIGN USING THE FREQUENCY-SAMPLING (FFT) METHOD

Implementing a FIR filter in DSP hardware such as the DSP56001 is a relatively simple task. Determining a set of coefficients which describe a given impulse response of a filter can also be a straight forward procedure. However, deriving the optimal coefficients necessary to obtain a particular response in the frequency domain is not always so easy. The following section introduces a simple method to determine a set of coefficients based on a desired arbitrary frequency response (often referred to as the *Frequency Sampling Method*). This method has the distinct advantage that it can be done in real time. However, the most efficient determination of coefficients is best done by utilizing a software filter design system such as *FDAS (Filter Design and Analysis System* from Momentum Data Systems, Inc.) to perform numerical curve-fitting and optimization which is a procedure that necessitates the use of a computer. (Using *FDAS* is discussed in the next section.) For example, the *Inverse Fourier Transform Integral* is used to determine the FIR coefficients from a response specification in the frequency domain; this generally requires a numerical integration procedure. In this section, we will use the *Inverse Discrete Fourier Transform* (IDFT), which can be implemented using the *Inverse Fast Fourier Transform* (IFFT) algorithm. One reason for choosing this approach is to demonstrate a method that can be used to determine coefficients in real time since the FFT can be implemented in the same DSP hardware as the FIR filter (this is at least true for the DSP56001).

The question is "How must the filter be specified in the frequency domain so that Equations (7.1-10) and (7.1-11) are realized?". That is, starting with the definition of the filter in the frequency domain, a method for calculating the filter coefficients is required. Beginning with Equation (7.1-2), the *Z-transform* of the FIR filter, and evaluating this transfer function on the unit circle at *N* equally spaced normalized frequencies (i.e., $z = e^{j2\pi k/N}$), one can write:

$$H(k) = H(e^{j2\pi k/N}) = \sum_{i=0}^{N-1} h(i)\, e^{-j2\pi k i/N} \qquad (7.2-1)$$

where $0 \le k \le N-1$. *h(i)* can be solved in terms of the frequency response at the discrete frequencies, *H(k)*, by multiplying both sides of Equation (7.2-1) by $e^{j2\pi k m/N}$ and summing over *k*, as follows:

$$\sum_{k=0}^{N-1} H(k)\, e^{j2\pi km/N} = \sum_{k=0}^{N-1}\sum_{i=0}^{N-1} h(i)\, e^{-j2\pi ki/N}\, e^{j2\pi km/N}$$

$$= \sum_{i=0}^{N-1} h(i) \sum_{k=0}^{N-1} e^{-j2\pi ki/N}\, e^{j2\pi km/N}$$

$$= \sum_{i=0}^{N-1} h(i)\, N\, \delta_{im}$$

$$= N\, h(m) \qquad (7.2-2)$$

where δ_{im} is the *Kronecker delta* which is equal to one when $i=m$ but is zero otherwise. Equation (7.2-2) can be used to find $h(i)$ by simply setting $i=m$:

$$h(i) = \frac{1}{N} \sum_{k=0}^{N-1} H(k)\, e^{j2\pi ki/N} \qquad (7.2-3)$$

Equation (7.2-3) is the *Inverse Discrete Fourier Transform* (IDFT) of the filter response.

In general the discrete Fourier coefficients are complex therefore the $H(k)$ can be represented as:

$$H(k) = A(k)e^{j\phi(k)}$$

where

$$A(k) = |H(k)|$$

so that Equation (7.2-3) becomes:

$$h(i) = \frac{1}{N} \sum_{k=0}^{N-1} A(k)\, e^{j\phi(k)}\, e^{j2\pi ki/N} \qquad (7.2-4)$$

At this point the linear phase constraints, Equations (7.1-10) and (7.1-11) can be applied. The constraint that the coefficients be real and symmetrical (Equation (7.1-10)) implies that

$$h(i) = h(N-1-i) = h^*(N-1-i)$$

where * signifies the complex conjugate and therefore:

$$h^*(N-1-i) = \frac{1}{N} \sum_{k=0}^{N-1} A(k)\, e^{-j\phi(k)}\, e^{-j2\pi k(N-1-i)/N}$$

$$= \frac{1}{N} \sum_{k=0}^{N-1} A(k)\, e^{-j[\phi(k)+2\pi k(N-1)/N]}\, e^{j2\pi ki/N} \qquad (7.2-5)$$

Equation (7.2-5) will be identically equal to Equation (7.2-4) if,

$$\phi(k) = -\phi(k) - \frac{2\pi k(N-1)}{N} + 2\pi r \quad \text{for} \quad r = 0,1,2...$$

or

$$\phi(k) = \pi r - \pi k \frac{(N-1)}{N} \qquad (7.2-6)$$

Now, what are the constraints on r and $A(k)$ which will guarantee a purely real response for N even? By substituting Equation (7.2-6) into Equation (7.2-4):

$$h(i) = \frac{1}{N} \sum_{k=0}^{N-1} A(k) e^{j[\pi r - \pi k(N-1)/N + 2\pi k i/N]} \qquad (7.2-7)$$

Expanding Equation (7.2-7) for even N yields:

$$h(i) = \frac{1}{N}\{A(0)e^{j\pi r} + A(1)e^{j[\pi r - \pi(N-1)/N + 2\pi i/N]} + \ldots$$

$$+ A(N/2)e^{j[\pi r - \pi(N-1)/N + \pi i]} + \ldots + A(N-1)e^{j[\pi r - \pi(N-1)(N-1)/N + 2\pi(N-1)i/N]}\}$$

Consider the $A(k)$ and $A(N-k)$ terms; if $r=0$ for the $A(k)$ term and $r=1$ for the $A(N-k)$ term then the argument for the $A(N-k)$ term is the negative of the argument for the $A(k)$ term, given that N is even. That is,

$$e^{j[\pi - \pi(N-k)(N-1)/N + 2\pi(N-k)i/N]} = e^{j[\pi k(N-1)/N - 2\pi k i/N]} \; e^{j[\pi - \pi(N-1) + 2\pi i]}$$

$$= e^{-j[-\pi k(N-1)/N + 2\pi k i/N]} \; e^{j\pi[-(N-2) + 2i]}$$

$$= e^{-j[-\pi k(N-1)/N + 2\pi k i/N]} \, (1)$$

Therefore if $A(k) = (A(N-k)$ then all imaginary components in Equation (7.2-7) will cancel and the $h(i)$ will be purely real, which is the desired result. In general, for N even, the formulas for the *Frequency Sampling Method* can be reduced to:

$$h(i) = \frac{1}{N}\left\{A(0) + \sum_{k=1}^{N-1} A(k) e^{j[\pi r - \pi k(N-1)/N + 2\pi k i/N]}\right\} \qquad r = 0 \;\text{ for }\; 0 \le k < \frac{N}{2}$$

$$r = 1 \;\text{ for }\; \frac{N}{2} < k \le N-1$$

$$= \frac{1}{N}\left\{A(0) + \sum_{k=1}^{N/2-1} A(k)[e^{j[-\pi k(N-1)/N + 2\pi k i/N]} + e^{j[\pi - \pi(N-k)(N-1)/N + 2\pi(N-k)i/N]}]\right\}$$

$$= \frac{1}{N}\left\{A(0) + \sum_{k=1}^{N/2-1} A(k)[e^{j[-\pi k(N-1)/N + 2\pi k i/N]} + e^{-j[-\pi k(N-1)/N + 2\pi k i/N]}]\right\}$$

$$= \frac{1}{N}\left\{A(0) + \sum_{k=1}^{N/2-1} A(k)[e^{-j\pi k} e^{j[\pi k/N + 2\pi k i/N]} + e^{j\pi k} e^{-j[\pi k/N + 2\pi k i/N]}]\right\}$$

$$= \frac{1}{N}\left\{A(0) + 2\sum_{k=1}^{N/2-1} A(k)(-1)^k \cos\frac{\pi k}{N}(1+2i)\right\} \qquad (7.2-8)$$

given,

$$A(k) = A(N-k) \quad for \;\; 1 \le k \le \frac{N}{2} - 1$$

with, $$A(N/2) = 0$$

and,

$$\phi(k) = -\pi k \frac{(N-1)}{N} \qquad \text{for} \quad 0 \le k \le \frac{N}{2} - 1$$

$$\phi(k) = \pi - \pi k \frac{(N-1)}{N} \qquad \text{for} \quad \frac{N}{2} + 1 \le k \le N - 1$$

In summary it can be said that if a filter with an even number of symmetrical real coefficients is desired then the phase must be linear and the frequency response must be symmetrical about $N/2$ and zero at $N/2$. [1]

As an example consider the arbitrary filter specified in Figure 7.2-1. In this example $N=32$ and an arbitrary lowpass and bandpass combination is specified. Figures 7.2-2a and 7.2-2b show the result of transforming the polar coordinate filter specification into rectangular coordinates, yielding real and imaginary components of the transfer function. This transformation is accomplished simply by treating the magnitude of $H(k)$ as the length of a vector in polar coordinates and the phase as the angle. The x component (*real part*) is the product of the length (magnitude of $H(k)$) and the cosine of the phase. Likewise, the y component (*imaginary part*) is the product of the length and the sine of the phase angle. This transformation to rectangular coordinates is necessary to perform the *Inverse DFT* (or *Inverse FFT*) calculation:

$$h(n) = \frac{1}{N} \sum_{k=0}^{N-1} H(k)\, e^{jk\theta_n} \qquad (7.2-9)$$

where $H(k)$ is a complex number, and $\theta_n = 2\pi n/N$.

When the filter's transfer function, $H(k)$, is described as in Figures 7.2-1 and 7.2-2 (i.e., the real part symmetric about $N/2$ and the imaginary part antisymmetric about $N/2$), the $h(n)$ are *strictly real.* If the $h(n)$ were complex, the FIR filter would be much more difficult to implement, since twice as many terms would be present. Figure 7.2-3 shows the results of the *IDFT* applied to the example arbitrary filter specification. Note the symmetry of the coefficients in Figure 7.2-3 (they are symmetrical about $(N-1)/2$). This symmetry is to be expected for an even number of coefficients. Equation (7.2-10) has $N-1$ roots, since it is a polynomial of order $N-1$. These roots are plotted in Figure 7.2-4 and must be found from a numerical algorithm such as the *Newton-Raphson* root finding recursion relation or the *Mueller* method [2]. The $h(n)$ can now be used to specify the filter response in the continuous frequency domain by setting $\theta = 2\pi f/f_s$.

The transfer function has exactly the same form as the *DFT* of the $h(n)$, but now the frequency is continuous up to $f_s/2$:

$$H(\theta) = \sum_{n=0}^{N-1} h(n)\, e^{-jn\theta} \qquad (7.2-10)$$

where $\theta = 2\pi f/f_s$. The continuous frequency gain and phase response of our 32 coefficient example filter are plotted in Figures 7.2-5a and 7.2-5b, which are generated by:

$$G(\theta) = |H(\theta)| = [H(\theta)H^*(\theta)]^{1/2}$$

$$= \left\{ \left[\sum_{n=0}^{N-1} h(n)\cos n\theta \right]^2 + \left[\sum_{n=0}^{N-1} h(n)\sin n\theta \right]^2 \right\}^{1/2} \quad (7.2\text{-}11)$$

and

$$\phi(\theta) = -\tan^{-1}\left\{ \frac{\sum_{n=0}^{N-1} h(n)\sin n\theta}{\sum_{n=0}^{N-1} h(n)\cos n\theta} \right\} \quad (7.2\text{-}12)$$

where the $h(n)$ are the values obtained from Equation (7.2-9) or equivalently from Equation (7.2-8). Note that the gain plot (Figure 7.2-5a) intersects exactly the discrete frequency points originally specified in Figure 7.2-1. Clearly $G(\theta)$ and $\phi(\theta)$ are poorly behaved mathematically since they have many discontinuities. Due to the symmetry of linear phase FIR filters, analytic expressions can be found for $H(\omega)$. This is shown in section 7.3, Equations (7.3-4) and (7.3-5).

Figures 7.2-6 and 7.2-7 show, for comparison sake, the log magnitude response of the example filter with larger values of N. Note that the stopband attenuation is not improved as much as one may expect as the number of coefficients is increased, however the sharpness of the transition band is significantly enhanced. This is because the approach discussed thus far has implicitly assumed a rectangular window function. In order to improve this situation, a smoothing window, $w(n)$, can be used [11].

If the $h(n)$ are modified by a window function $w(n)$, $h_w(n) = h(n)w(n)$ for $0 \le n \le N-1$, the gain of Equation (7.2-11) (as shown in Figure 7.2-8a) can be greatly improved. The function $w(n)$ goes to zero at both ends ($n=0$ *and* $n=N$) and is unity at the center. $w(n)$ is symmetrical so as not to disrupt the linear phase characteristics of the filter. Figure 7.2-8b uses a window function described by $w(n) = \sin^2[\pi n/(N-1)]$ (also known as a *Hanning* window) to demonstrate the sensitivity of the gain to windowing. $w(n)$ is basically an envelope function used to tapper the ends of $h(n)$ smoothly. Note the rounding of the transition band edges in Figure 7.2-8b. This effect is the trade-off for windowing, however in most applications this trade-off is well worthwhile. The best passband transition slope performance is achieved with the rectangular window, as has been previously shown, but results in very poor stopband performance and often severe passband fluctuation (or ripple). All window functions have the effect of increasing the stopband attenuation and reducing the passband ripple at the expense of increasing the width of the transition region. It turns out that for most window functions, the passband ripple is relatively insensitive to N.

Windowing is described in virtually any digital signal processing text book. Many are listed in the reference section of this Applications Report. Also, windowing is discussed with practical examples in the Motorola Applications Note; *Implementation of Fast Fourier Transforms on Motorola's DSP56000/DSP56001 and DSP96002 Digital Signal Processors*; APR4/D.

7.3 FIR FILTER DESIGN USING FDAS

Figure 7.3-1 shows an example (log magnitude and impulse response) of a bandpass filter generated with the *Filter Design and Analysis System* (FDAS) software package from Momentum Data Systems using the Kaiser window.

A totally different approach to FIR filter design, the *Equiripple Method*, is based on finding an optimum approximation to the ideal or desired response, $D(\theta)$. The reason why an optimum approximation can be found is the inherent symmetry of the coefficients of linear phase filters.

Recall that the continuous frequency response of a FIR filter can be found by setting $z = e^{j\theta}$ in the z-transform so that:

$$H(\theta) = \sum_{n=0}^{N-1} h(n)e^{-j\theta n} \tag{7.3-1}$$

Now, if the linear phase factor $z = e^{-j\theta(N-1)/2}$ is factored out, Equation (7.3-1) can be written as:

$$\begin{aligned} H(\theta) &= e^{-j\theta(N-1)/2}\sum_{n=0}^{N-1} h(n)e^{j\theta[(N-1)/2-n]} \\ &= e^{-j\theta(N-1)/2}\{h(0)e^{j\theta(N-1)/2)} + h(1)e^{j\theta[(N-1)/2)-1]} + \ldots \\ & h(N-1)e^{-j\theta(N-1)/2)} + h(N-2)e^{-j\theta[(N-1)/2)-1]}\} \end{aligned} \tag{7.3-2}$$

Using *Euler's identity*, Equation (7.3-2) can be grouped into sine and cosine terms as follows:

$$\begin{aligned} H(\theta) = e^{-j\theta(N-1)/2}\{&[h(0)+h(N-1)]\cos\left[\theta\left(\frac{N-1}{2}\right)\right] + j[h(0)-h(N-1)]\sin\left[\theta\left(\frac{N-1}{2}\right)\right] \\ &+[h(1)+h(N-2)]\cos\left[\theta\left(\frac{N-1}{2}-1\right)\right] + j[h(1)-h(N-2)]\sin\left[\theta\left(\frac{N-1}{2}-1\right)\right] \\ &+\ldots\ldots\} \end{aligned} \tag{7.3-3}$$

If it is assumed that linear phase is achieved by even symmetry, i.e., $h(n) = h(N-1-n)$, and that N is even, Equation (7.3-3) reduces to:

$$H(\theta) = e^{-j\theta(N-1)/2}\sum_{n=0}^{N/2-1} 2h(n)\cos\left[\theta\left(\frac{N-2}{2}-n\right)\right]$$

which, with a change of variable, $k=N/2-n-1$, can be written as:

$$H(\theta) = e^{-j\theta(N-1)/2}\sum_{k=0}^{N/2-1} 2h\left(\frac{N}{2}-k-1\right)\cos\left[\theta\left(k+\frac{1}{2}\right)\right] \tag{7.3-4}$$

which is in the form:

$$H(\theta) = A(\theta)e^{jP(\theta)} \tag{7.3-5}$$

where $A(\theta)$ is a real valued amplitude function and $P(\theta)$ is a linear phase function. The linear functions $A(\theta)$ and $P(\theta)$ are to be contrasted with the inherently nonlinear *absolute value* and *arctangent* functions in Equations (7.2-11) and (7.2-12). It turns out [14] that for all four types of linear phase filters (N even or odd and symmetry even or odd), $A(\theta)$ can be expressed as a sum of *cosines.*

Since an ideal filter cannot be realized, an approximation must be used. If the desired frequency response, $D(\theta)$, can be specified in terms of a deviation, δ, from the ideal response, then the error function in Equation (7.3-6) can be minimized:

$$\| E(\theta) \| = \max_{\theta} | D(\theta) - A(\theta) | \tag{7.3-6}$$

by finding the best $A(\theta)$. Because $A(\theta)$ can be expressed as a finite sum of *cosines* as in Equation (7.3-4), then it can be shown [17] that the optimal $A(\theta)$ will be unique and will have at least *N/2+1 extremal frequencies* where *extremal frequencies* are points such that for:

$$\theta_1 < \theta_2 < \ldots < \theta_{N/2} < \theta_{N/2+1}$$

$$E(\theta_e) = -E(\theta_{e+1}) \qquad \text{for} \quad e = 1, 2, \ldots, \frac{N}{2} + 1$$

and

$$| E(\theta_e) | = \max_{\theta} [E(\theta)]$$

The foregoing states that the best approximation will exhibit an *equiripple* error function. The problem reduces to finding the extremal frequencies since once they are found, the coefficients, *2h(N/2-k-1)*, can be found by solving the set of linear equations:

$$D(\theta) \pm \delta = A(\theta_e) = \sum_{k=0}^{N/2-1} 2h\left(\frac{N}{2} - k - 1\right)\cos\left[\theta_e\left(k + \frac{1}{2}\right)\right] \tag{7.3-7}$$

The *Remez exchange* algorithm is used to systematically find the extremal frequencies [6]. Basically, a guess is made for the initial *N/2+1* extremal frequencies. (Usually, this guess consists of *N/2+1* equally spaced frequencies in the *Nyquist* range.) Using this guess, Equation (7.3-7) is solved for the coefficients and δ. Using these coefficients, $A(\theta)$ is calculated for all frequencies and the extrema and frequencies at which the extrema are attained are determined. If the extrema are all equal and equal to or less than that specified in the initial filter specification, the problem is solved. However, if this is not the case, the frequencies at which the extrema were attained are used as the next guess. Note that the final extrema frequencies do not have to be equally spaced. Clearly the *equiripple* design approach is calculation intensive. Don't be alarmed when filter design programs such as FDAS ask you to wait while it designs an equiripple FIR filter to match the specification input.

Other than looking neat, what is the benefit of *equiripple* designs over *window* designs? In general, *equiripple* designs will require fewer taps for straight forward requirements. When the specification requires a sharp cut-off and/or a large stopband attenuation or a narrow bandpass, the *equiripple* approach may fail to converge. In general when N is decreased, an *equiripple* design tends to maintain its transition band while sacrificing stopband attenuation. This is in contrast to *window* designs which tend to do the opposite. Of the window alternatives, the *Kaiser* window is to be preferred for designing filters. This is because the passband ripple and stopband attenuation can be varied relatively independently of the transition width [1]. For spectral analysis, the *Blackman-Harris* window is to be preferred [11].

Figure 7.3-2 shows an example of an *equiripple* design generated from the FDAS design software package. This example is the same one used for the Kaiser window example of Figure 7.2-1. Note that the number of coefficients in the equiripple design is far less (179 versus 291) than that generated by the Kaiser window method. Note however that the passband ripple is larger.

7.4 FIR IMPLEMENTATION ON THE DSP56001

The DSP56000 has several architectural features that make it ideally suitable for implementing FIR filters:

(1) *Dual Harvard Architecture* - two data memories with dedicated buses and address generation units allowing two addresses to be generated in a single cycle. If one address is pointing to data and another address is pointing to coefficients, a word of data and a coefficient can be fetched in a single cycle.

(2) *Modulo Addressing* - make the shifting of data unnecessary. If an address pointer is incremented (or decremented) with the modulo modifier in effect, data shifting can be accomplished by just "backing up" the address register by one to overwrite the dat that would normally be shifted out. This allows very efficient addressing of operands without wasting time shifting the data or reinitializing pointers.

(3) *Hardware DO Loops* - execute without any overhead once the loop is started. After a 3 cycle initialization of the DO loop, the body of the loop executes as if it was straight line code. Since the DO loop does not require any overhead cycles for each pass, the need for straight line code is eliminated.

For a four coefficient example of a linear phase FIR filter, the input-output difference equation can be found by expanding Equation (7.1-1):

$$y(n) = h_0 x(n) + h_1 x(n-1) + h_2 x(n-2) + h_3 x(n-3)$$

This difference equation can be realized with the discrete time filter structure shown in Figure 7.4-1, using a four tap example. The filter can be efficiently implemented on the DSP56001 by using modulo addressing to implement the shifting and parallel data moves to load the multiplier-accumulator. For the filter network shown in Figure 7.4-1, the memory map for the filter inputs and coefficients are shown in Figure 7.4-2. The DSP56001 code to implement the direct form FIR filter is simply:

```
CLR   A          X0,X:(R0)+    Y:(R4)+,Y0    ;Save input sample, fetch coef.
REP   #NTAPS-1                                ;Repeat next instruction.
MAC   X0,Y0,A    X:(R0)+,X0    Y:(R4)+,Y0    ;FIR Filters.
MACR  X0,Y0,A    (R0)-                        ;Round result and adjust R0.
```

A description of the code follows:

Register RO points to the input variable buffer, MO is set to 3 (modulo 4), R4 points to the coefficient buffer, and M4 is set to 3. The input sample is in XO.

The CLR instruction clears accumulator A and performs parallel data moves. The data move saves the most recent input value to the filter (assumed to be in XO) into the location occupied by the oldest data in the shift register and moves the first coefficient in the filter (h_0) into the data ALU.

The REP instruction repeats the next instruction NTAPS-1 times. Since there are four taps in this filter, the next instruction is repeated 3 times.

The MAC instruction multiplies the data in XO by the coefficient in YO and adds the result to accumulator A in a single cycle. The data move in this instruction loads the next input data variable into XO and the next coefficient into YO. Both address registers RO and R4 are incremented.

The MACR instruction calculates the final tap of the filter, rounds the result using convergent rounding and address register RO is decremented.

Note that address register R4 is incremented once before the REP instruction and three times due to the REP instruction for a total of four increments. Since the modulus for R4 is four, the value of R4 wraps around pointing back to the first coefficient.

The operation of RO is similar. The value input to the filter is saved and then RO is incremented pointing to the first state. The REP instruction increments RO three times. Since the modulo on RO is four and RO is incremented four times, the value in RO wraps around pointing to the new input sample. When the MACR instruction is executed, the value of RO is decremented pointing to the old $x(n\text{-}3)$. The next sample time overwrites the value of $x(n\text{-}3)$ with the new input sample, $x(n)$. Thus the shifting of the input data is accomplished by simply adjusting the address pointer and the modular addressing wraps the pointer around at the ends of the input data buffer. Instruction Cycle counts for this filter are NTAPS+3 cycles. Thus for a four tap filter, seven instruction are required.

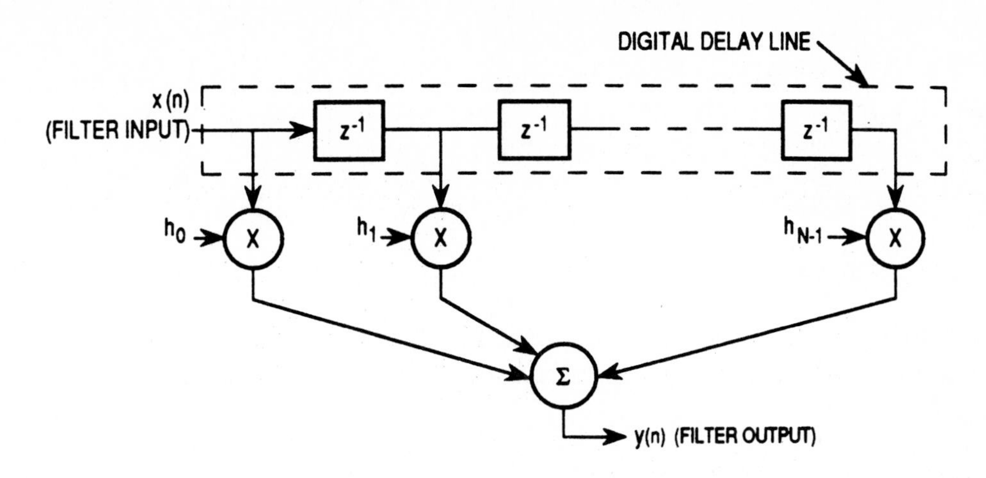

Figure 7.1-1. FIR Structure.

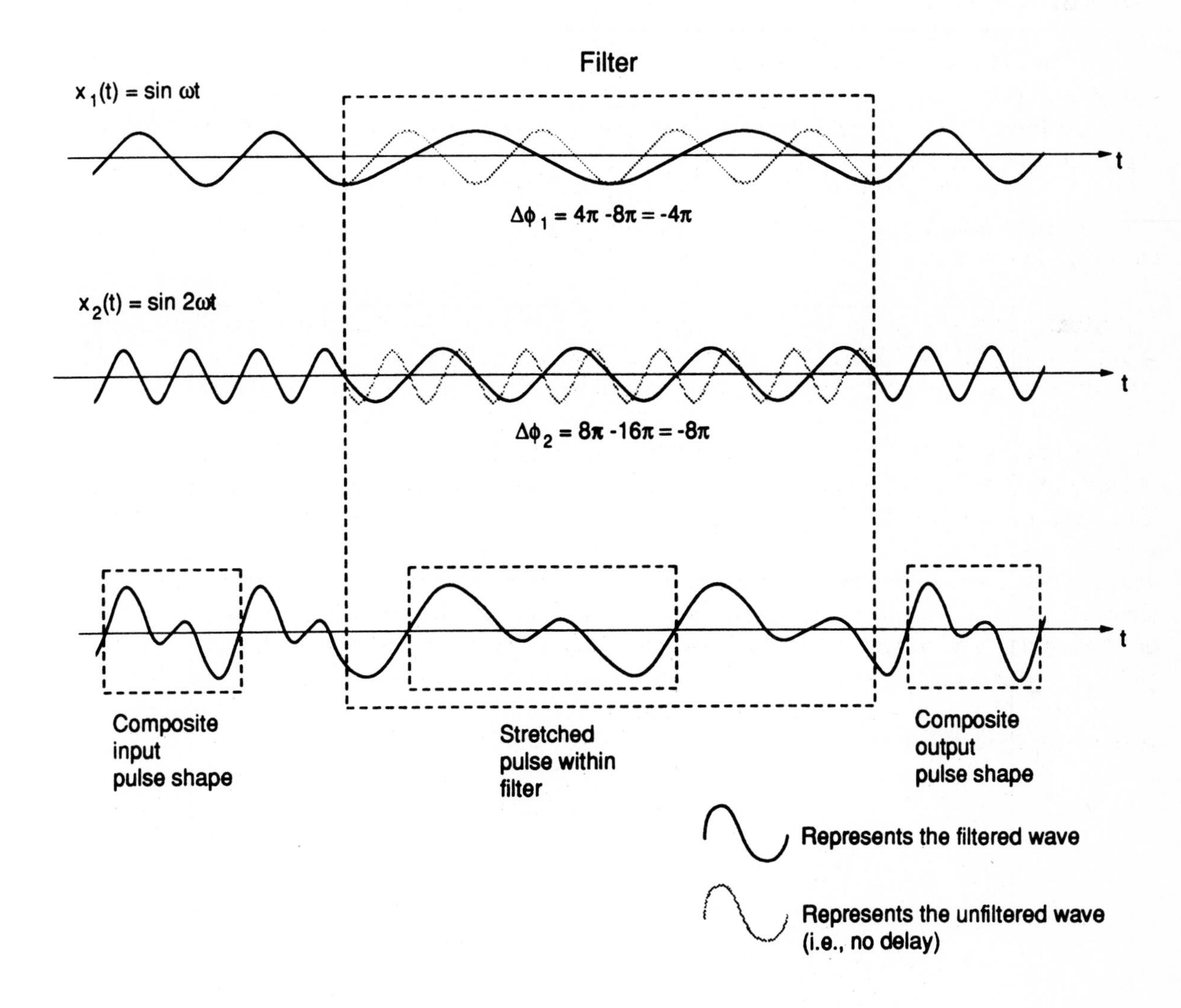

Figure 7.1-2. Signal Data Through a FIR Filter.

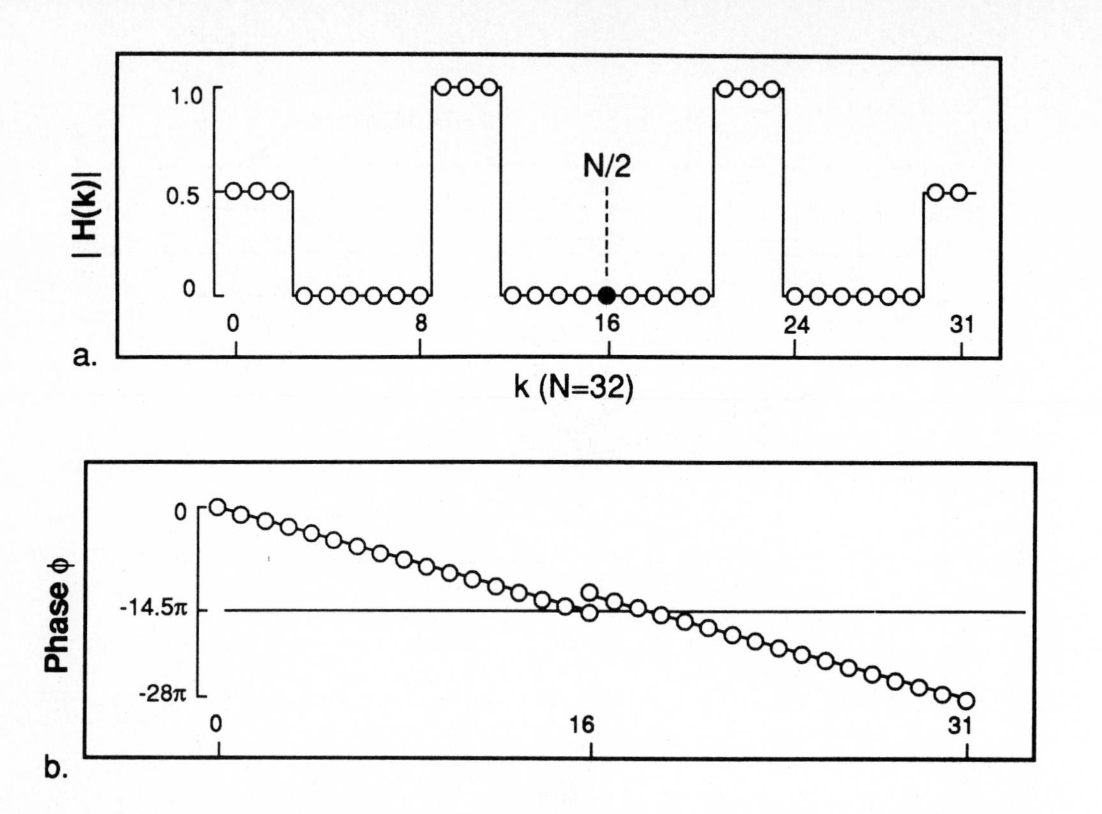

Figure 7.2-1a. Arbitrary Input Specifications for Frequency Response Magnitude.

Figure 7.2-1b. Necessary Phase Specification for Constant Time Delay.

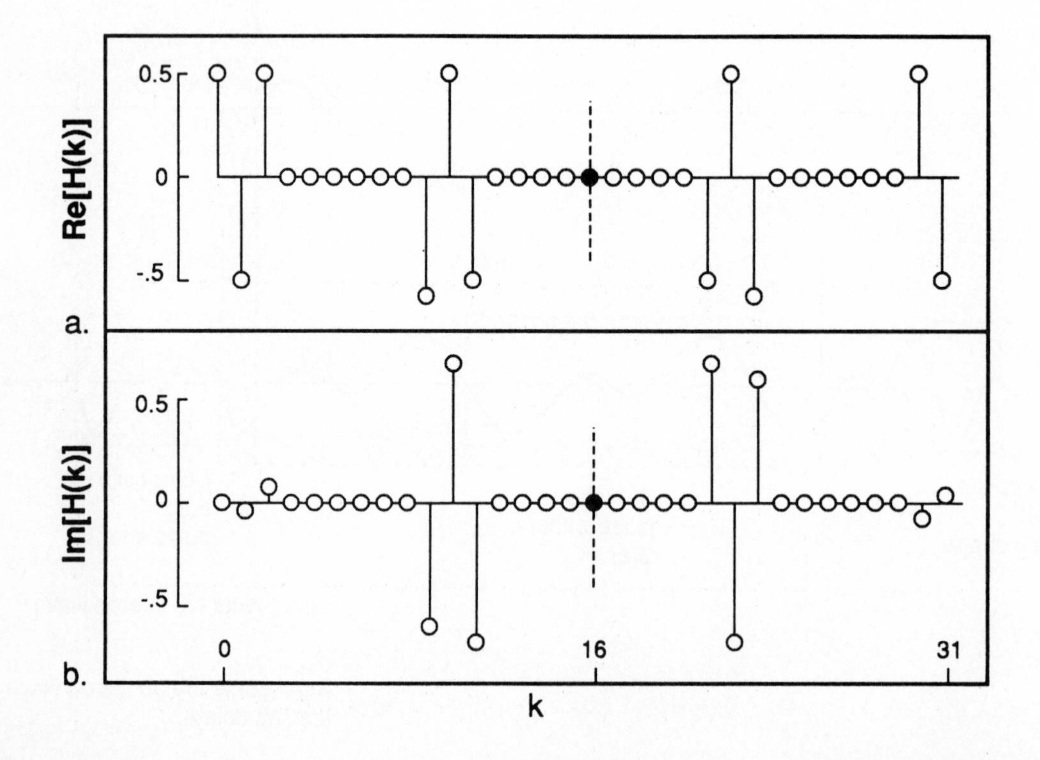

Figure 7.2-2a. Real Part of Response Transformed from Polar Coordinates.

Figure 7.2-2b. Imaginary Part of Response Transformed from Polar Coordinates.

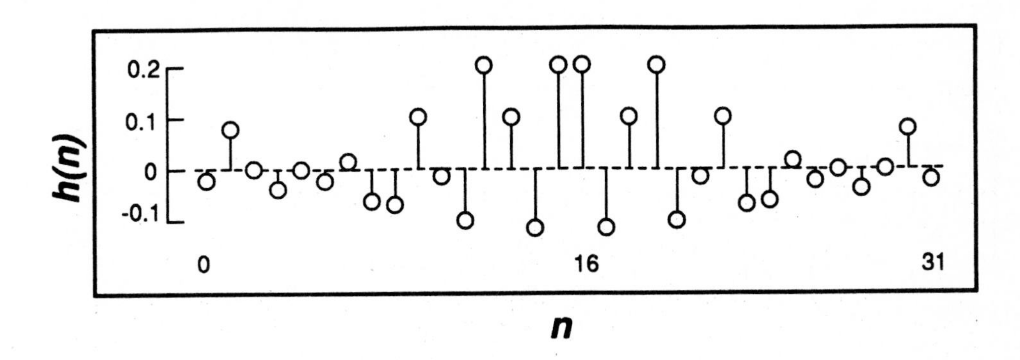

Figure 7.2-3. FIR Coefficients from Equation (7.1-2) for Example Filter.

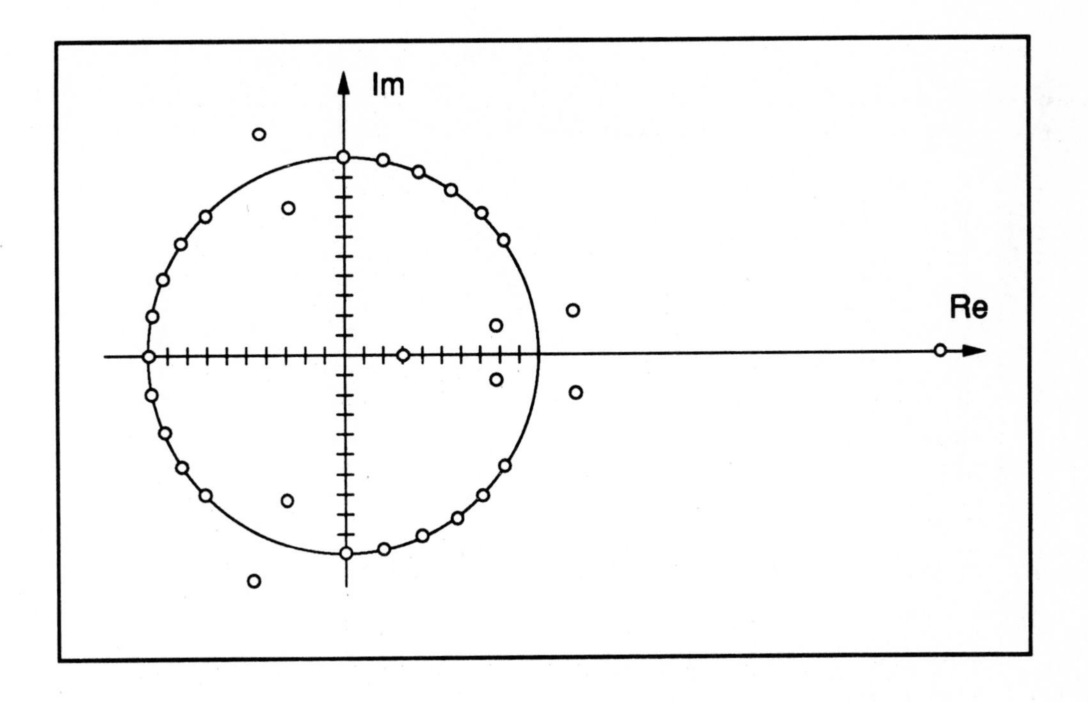

Figure 7.2-4. Roots (Zeros) of Equation (7.4-11) for Example Filter.

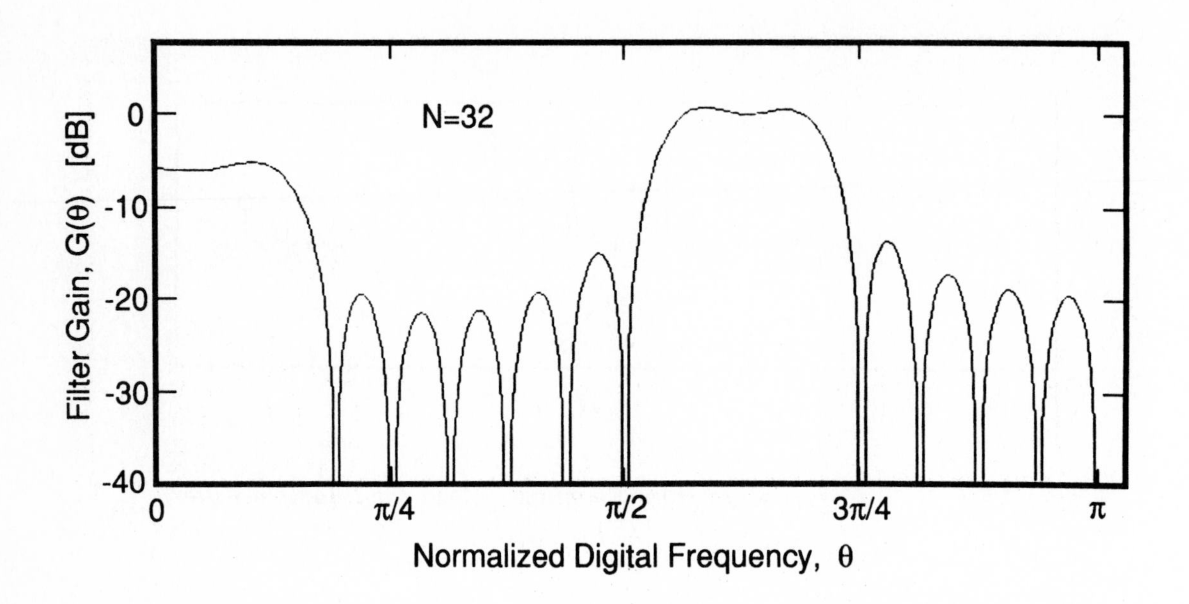

Figure 7.2-5a. Gain (Magnitude of H[z]) Plot of Example 32 Point FIR Filter.

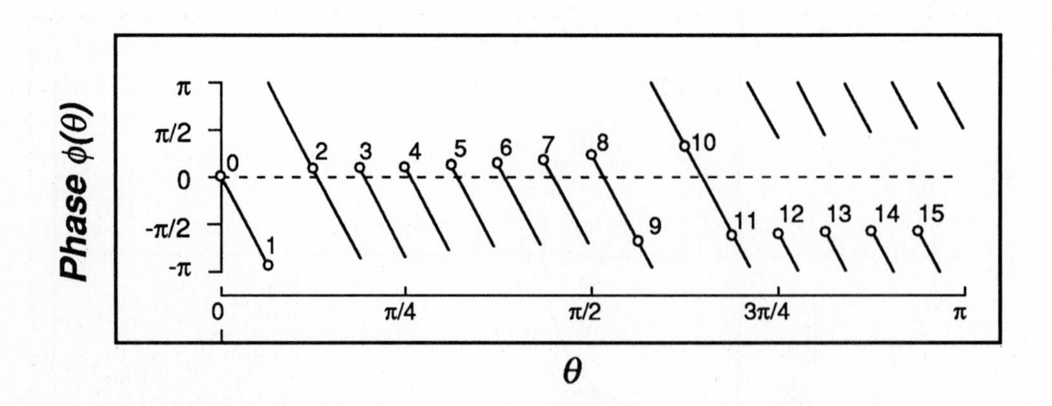

Figure 7.2-5b. Phase Response Corresponding to Gain Plot.

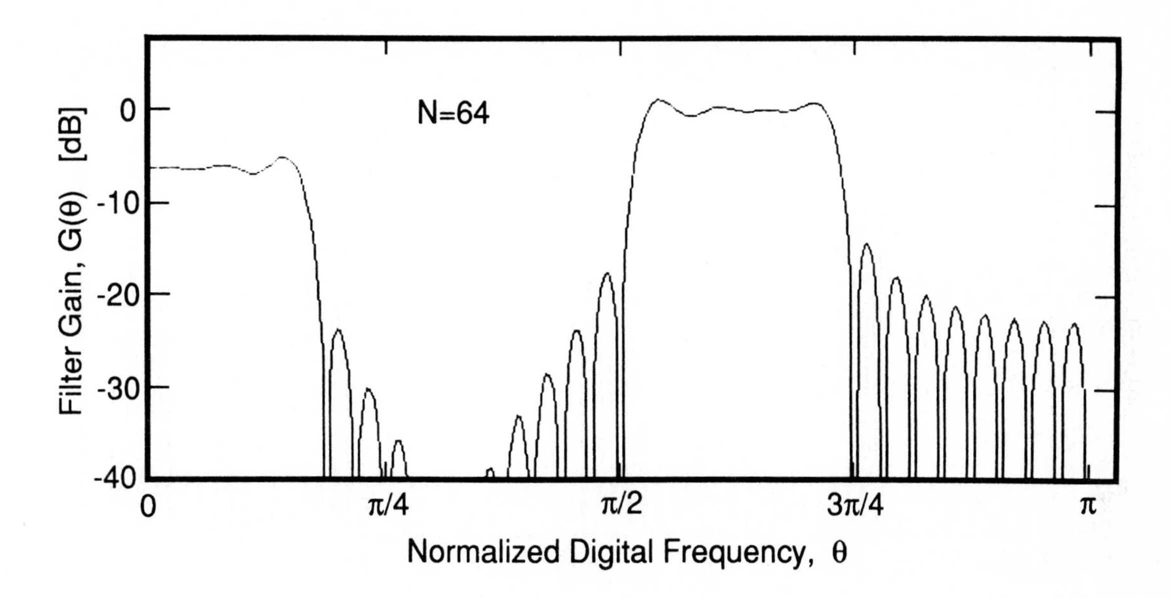

Figure 7.2-6. Same Example Filter Extended to 64 Points.

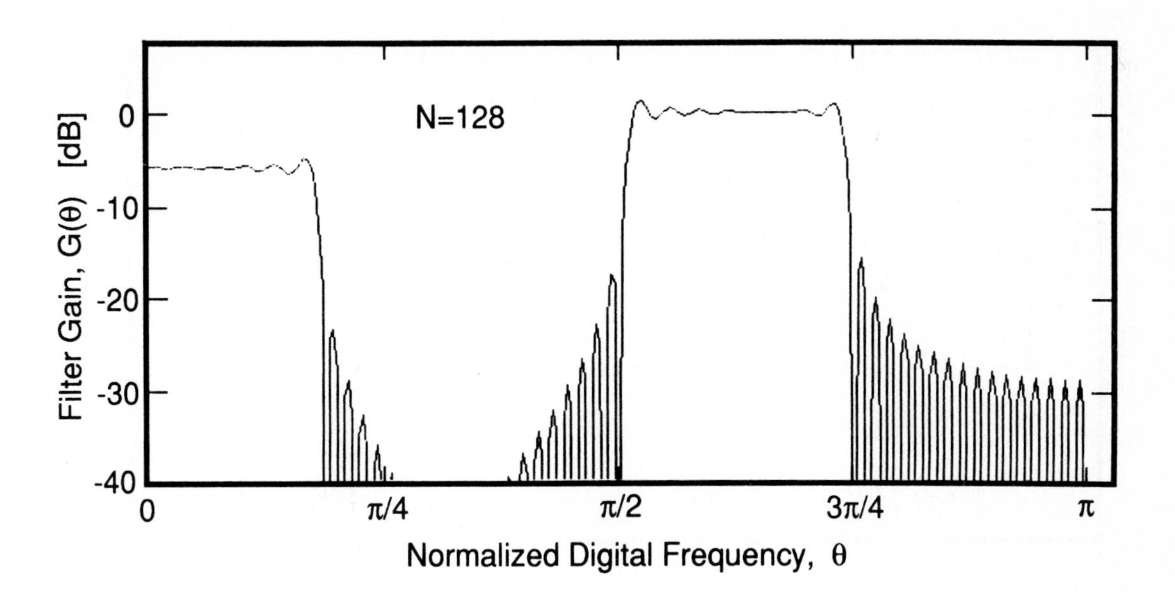

Figure 7.2-7. Same Example Filter Extended to 128 Points.

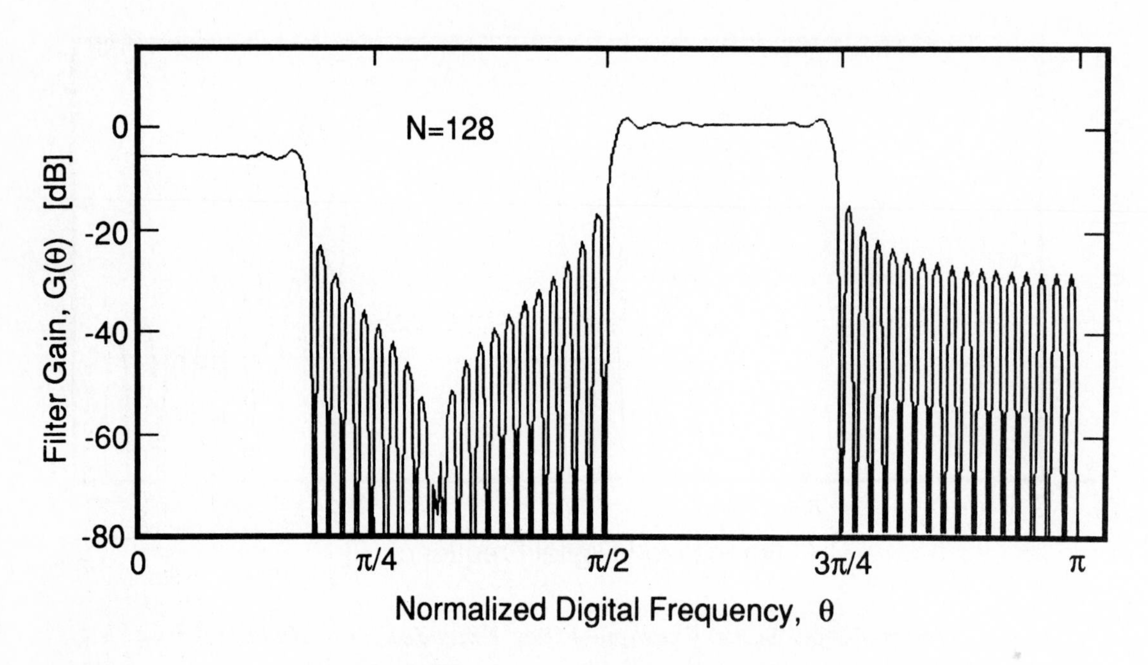

Figure 7.2-8a. Same Plot as Figure 7.2-7, but Different dB Scale. (Note that all previous gain plots use a rectangular window.)

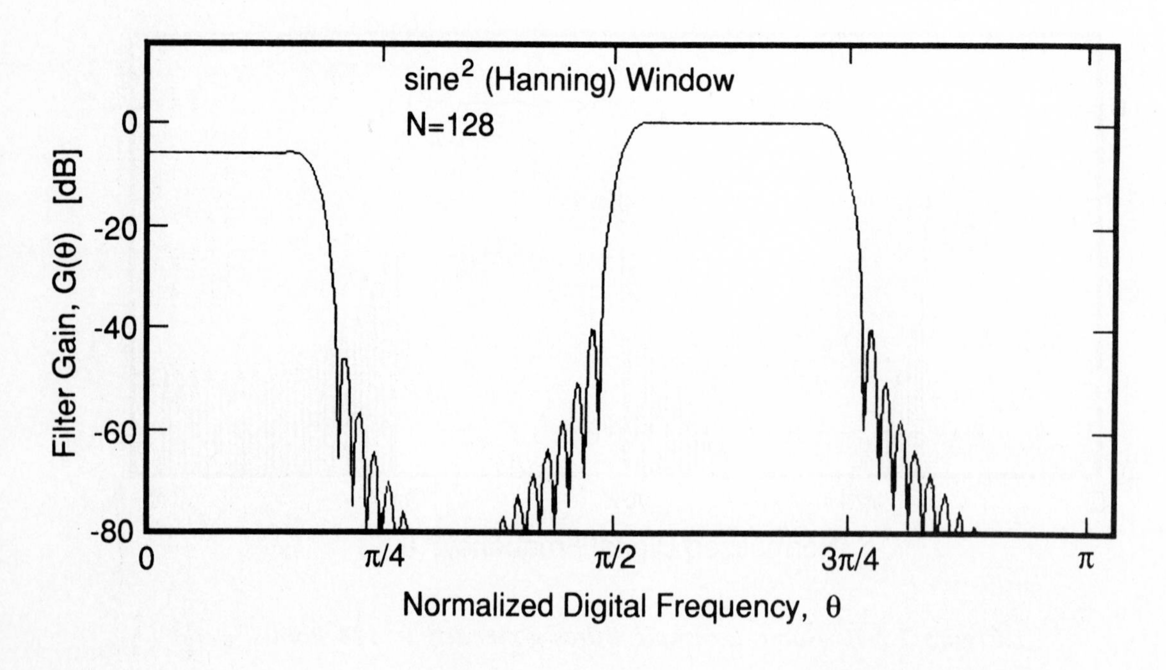

Figure 7.2-8b. Same as Above but with $\sin^2$ *(Hanning) Window.*

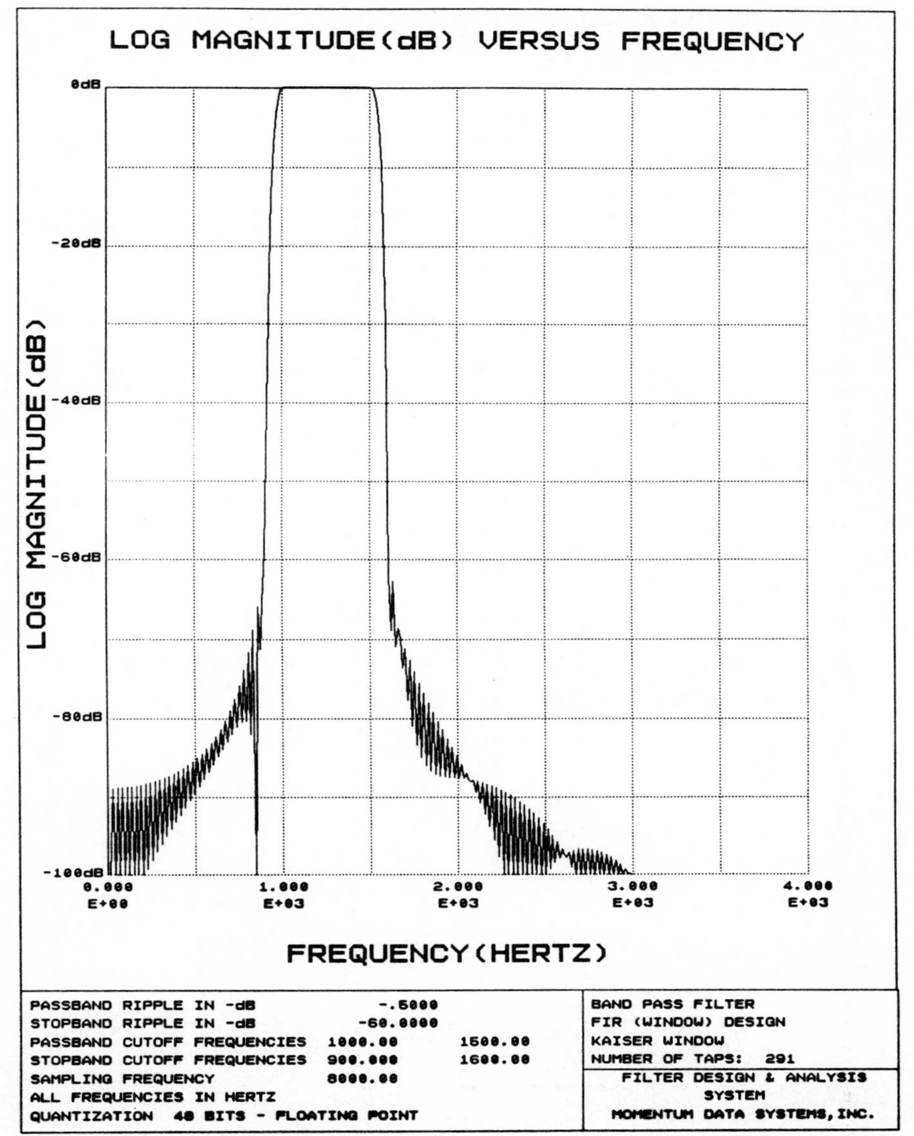

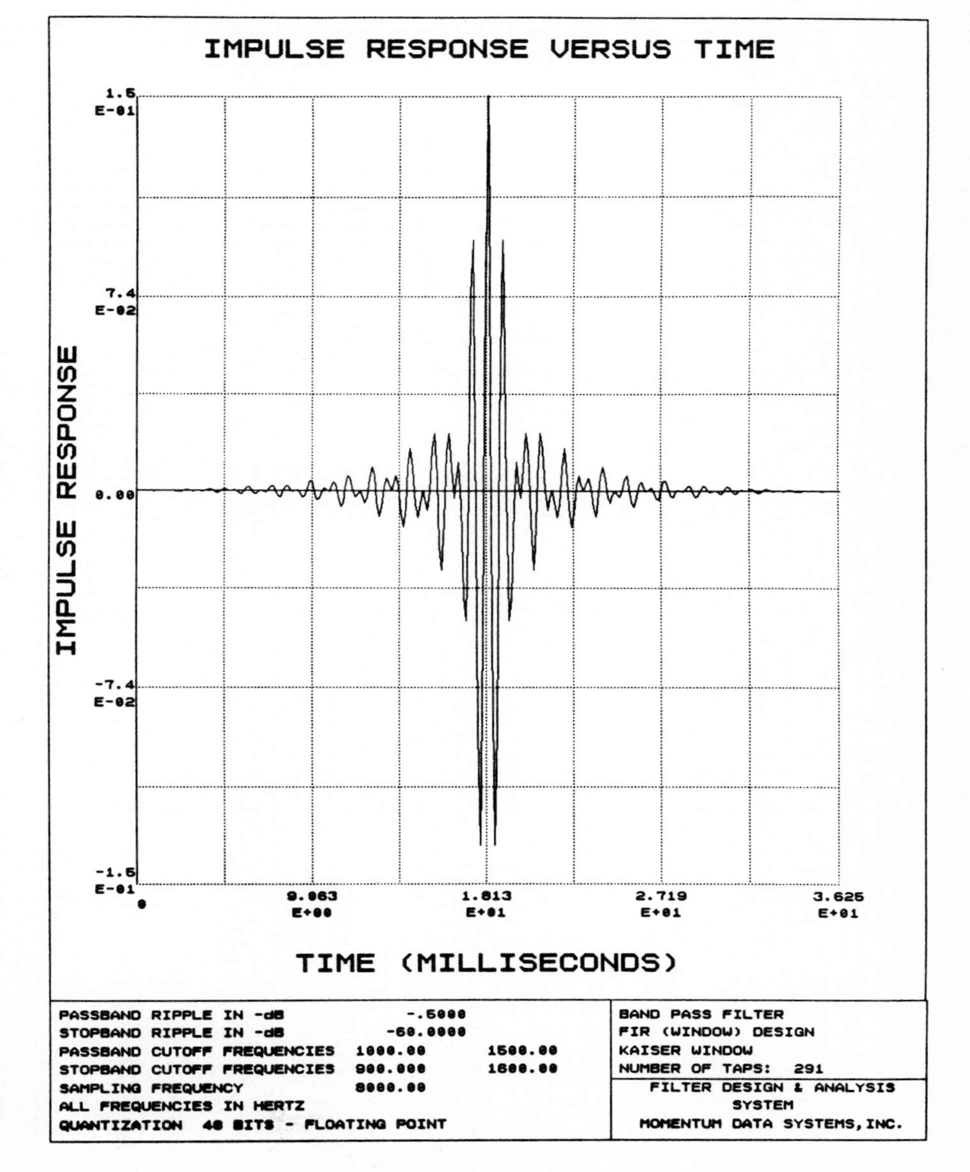

Figure 7.3-1. FDAS Output for FIR Bandpass Filter Example with a Kaiser Window.

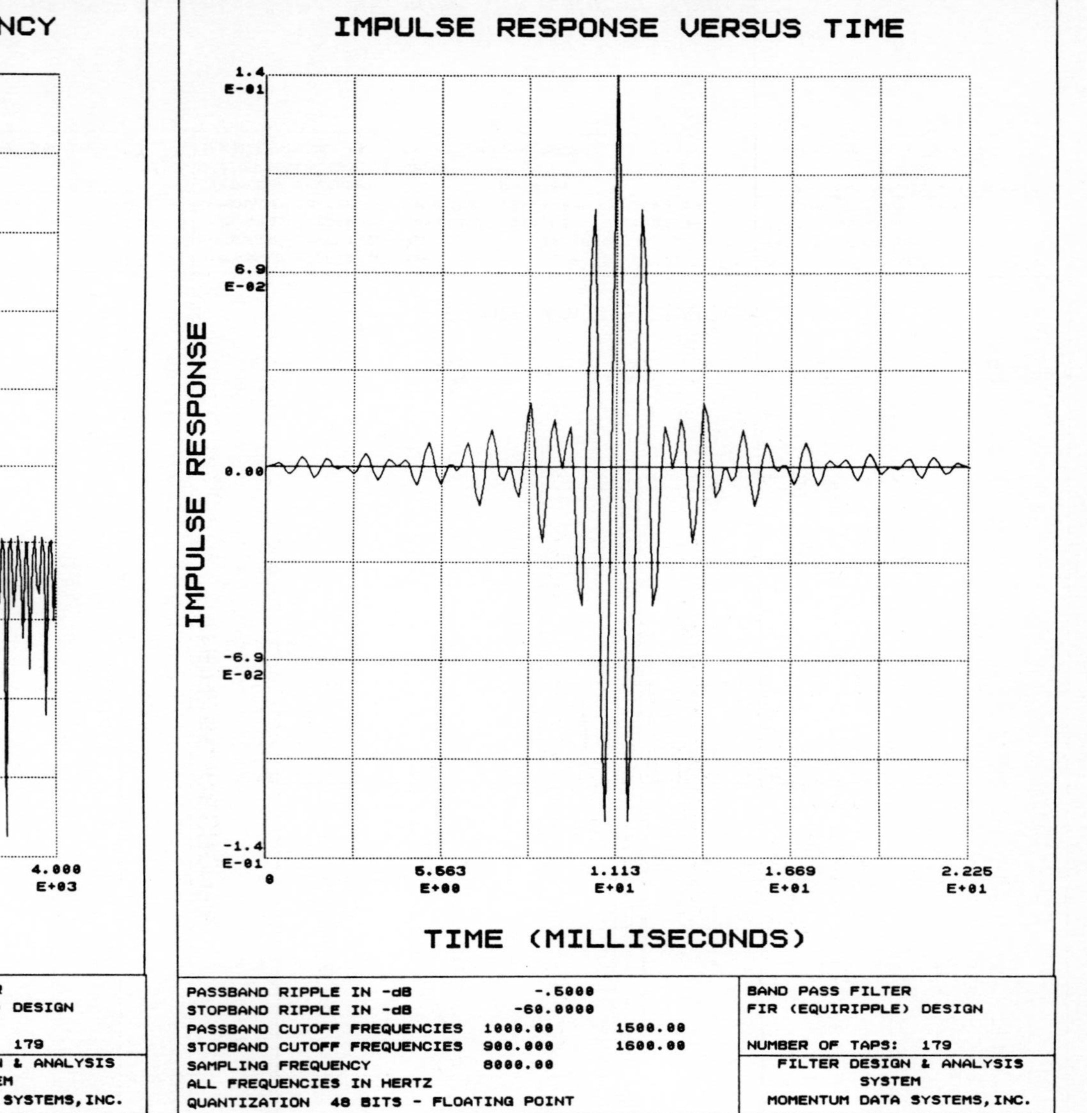

Figure 7.3-2. FDAS Output for FIR Bandpass Filter Example (same as Figure 7.3-1) but with Equiripple Design.

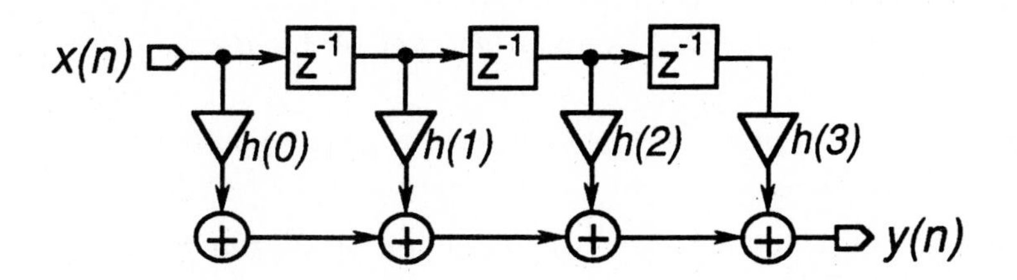

Figure 7.4-1a. Direct Form FIR Filter Network with Four Coefficients.

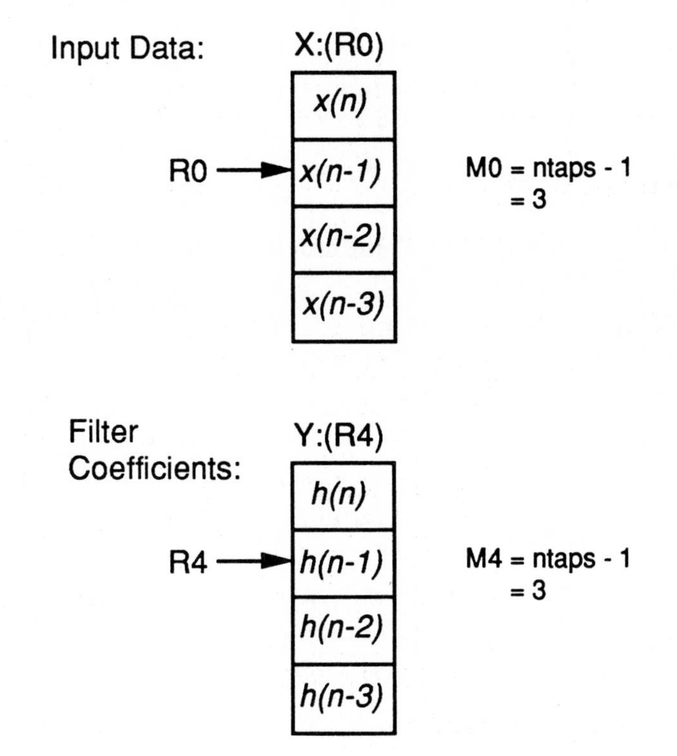

Figure 7.4-1b. Data Structures for Four Tap FIR Example.

REFERENCES

1. Antoniou, Andreas. *Digital Filters: Analysis and Design.* McGraw-Hill, 1974.
2. Beckett, Royce, et al. *Numerical Calculations and Algorithms.* McGraw-Hill Inc., 1967.
3. Berlin, Howard M. *Design of Active Filters with Experiments.* Indianapolis: Howard W. Sams & Co., Inc., 1977.
4. Bogner and Constantinides, editors. *Introduction to Digital Filtering.* Wiley, 1975.
5. Brophy, J.J. *Basic Electronics for Scientists.* New York: McGraw-Hill, 1966.
6. Cheney, E.W. *Introduction to Approximation Theory.* New York: McGraw-Hill, 1966.
7. Chrysafis, A. "Fractional and Interger Arithmetic Using the DSP56000 Family of General-Purpose Digital Signal Processors" (APR3/D). Motorola, Inc., 1988.
8. "Digital Stereo 10-Band Graphic Equalizer Using the DSP56001" (APR2/D). Motorola, Inc., 1988.
9. "Digital Filters on DSP56000/1." Motorola Technical Bulletin, 1988.
10. "Filter Design & Analysis System." Version 1.3, Momentum Data Systems, 1985.
11. Harris, Fredric J. "On The Use of Windows For Harmonic Analysis With The Discrete Fourier Transform." Proc. of IEEE, Vol. 66, No. 1, 51-83 (1978).
12. Jackson, Leland B. *Digital Filters and Signal Processing.* Boston: Kluwer Academic Publishers, 1986.
13. Lancaster, D. *Active Filter Cookbook.* Indianapolis: Howard W. Sams & Co., Inc., 1975.
14. McClellan, J.H. and T.W. Parks. "A Unified Approach To The Design Of Optimum FIR Linear-Phase Digital Filters." IEEE Trans. Circuits Systems CT-20, 697-701, (1973).
15. Moschytzm, G.S. and P. Horn. *Active Filter Design Handbook.* John Wiley & Sons, 1981.
16. Oppenheim, A.V. and R.W. Schafer. *Digital Signal Processing.* New Jersey: Prentice-Hall, 1975.
17. Parks, T.W. and J.H. McClellan. Chebyshev Approximation For Nonrecursive Digital Filters With Linear Phase." IEEE Trans. Circuit Theory CT-19, 189-194, (1972).
18. Proakis, John G. et al. *Introduction to Digital Signal Processing.* MacMillan, 1988.
19. Rabiner, L.R. and Gold, B. *Theory and Application of Digital Signal Processing.* New Jersey: Prentice-Hall, 1975.
20. Sohie, Guy R. L. "Implementation of Fast Fourier Transform on Motorola's DSP56000/DSP56001 and DSP96002 Digital Signal Processors" (APR4/D). Motorola, Inc., 1989.
21. Strawn, J., et al. *Digital Audio Signal Procesing - An Anthology.* William Kaufmann, 1985.
22. Williams, Arthur B. *Electronic Filter Design Handbook.* New York: McGraw Hill, 1981.

Index

(numbers refer to sections)

ORDER FORM FOR "DSP56000/1 TEXTBOOK SOFTWARE"

PLEASE PRINT:

Order Date: ______________ Your Phone #: ( ) ____________________________

Your Name: __________________________ Title: ____________________

University/Company name: ____________________________________

Bldg. Name: ______________ Room #: _______ Mail Drop #: ________________

Street Address: ____

City: ____________________ State: ______________ Country: ________________

Zip/Country Code: __________________

PROVIDE A CHECK, MONEY ORDER, OR PURCHASE ORDER FOR:

__________ copies times $40.00 equals $ __________ payment.

PLACE THIS ORDER FORM AND PAYMENT IN AN ENVELOPE AND MAIL TO:

Mohamed El-Sharkawy
Electrical Eng. Dept.
Bucknell University
Lewisburg, PA 17837

Cut along